P9-AFF-187

Industrial Control Handbook

There are thousands of laws legislators have spoken,
a few the Creator sent.
The former are being continually broken
the latter can't even be bent.

Anon

Industrial Control Handbook

Second Edition

E.A. Parr BSc, CEng, MIEE, MInstMC

Butterworth-Heinemann Ltd
Linacre House, Jordan Hill, Oxford OX2 8DP

A member of the Reed Elsevier plc group of companies

OXFORD LONDON BOSTON
MUNICH NEW DELHI SINGAPORE SYDNEY
TOKYO TORONTO WELLINGTON

First published in three volumes.
Volume 1 published by Collins 1986.
Volumes 2 and 3 published by Blackwell Scientific 1987, 1989
Second edition 1995

British Library Cataloguing in Publication Data

Parr, E. A.
Industrial Control Handbook. – 2Rev.ed
I. Title
670.427

ISBN 0 7506 2000 5

Library of Congress Cataloguing in Publication Data

Parr, E. A. (E. Andrew)
Industrial Control Handbook/E.A. Parr. – 2nd ed.
p. cm.
First published 3v. London: Collins, 1986–1989.
Includes index.
ISBN 0 7506 2000 5
1. Process control – Handbooks, manuals, etc.
I. Parr, E.A. (E. Andrew). Industrial control handbook. II. Title.
TS156.8.P375
670.42'7 – dc20 94–268571
CIP

Typeset by Vision Typesetting, Manchester
Printed in Great Britain by Hartnolls, Bodmin

Contents

Preface

I have two major faults in life, (Only two? says my wife): a dreadful memory and a magpie-like tendency to gather books and printed material. Over the years I have coped with the first by writing things I think are useful or interesting on sheets of A4 paper for later reference, and the family copes with the second by tolerating untidy piles of technical snippets in bookcases and filing cabinets all over our house. From my oddities, and my family's tolerance, has come this Industrial Control Handbook.

I have been involved with process control for longer than I care to remember, and this book is a collection of things I have found useful or interesting, plus many 'I wish I'd known that ...' facts. As readers may notice, I have a lot of sympathy for the poor maintenance engineer trying to solve a problem at 3 a.m. Coherent thought is difficult then and this book hopefully contains much material that will prevent 3 a.m. brainstorms.

As a result of my practical background, I have not written a mathematical book. I have been primarily concerned with things, facts and, most importantly, advice. Mathematical equations only appear where they are directly necessary for everyday work, are intriguing (as is the case with Fuzzy Logic) or the reasons for some obscure conclusion arising from the background maths.

Engineering, particularly process control, is above all a satisfying career. I enjoy engineering and the sense of achievement that results from seeing a complex plant come alive. I hope this book conveys some of the fascination of the world of industrial control and does a little to arrest the trend away from industry. The wealth of a nation arises from its industrial base, and this simple economic fact seems to have been forgotten. Whatever item you use each day, an industrial control system probably made its manufacture possible.

E. A. Parr
Minster on Sea
Kent

Acknowledgement

Some six years ago I suffered a major heart failure whilst working on an interesting control problem. That I survived it and fully recovered was the result of the prompt and skilful action from:

the medical and security staff of Sheerness Steel,
the Kent ambulance service,
the staff of the local Minster and regional Medway hospitals who ensured that I made a safe and rapid journey up the NHS casualty route,
the cardiac unit of St Thomas' Hospital in London who did some rather clever plumbing at very short notice.

To them all I offer my sincere thanks. Without their efforts there would have been one less engineer in the world, and this book would not have been written.

Chapter 1

Sensors and transducers

1.1. Instrumentation systems

Accurate measurement of strategic quantities such as flow, pressure or temperature is an essential part of the control or monitoring of the operation of any process. Figure 1.1 represents most industrial instrumentation systems. A physical quantity, called the *measurand* or *process variable* is converted by a measurement system into a measured value, usually an electrical or pneumatic signal, which can be used for display or control. The first ten chapters of this book are concerned with the conversion from process variable to measured value.

Figure 1.1 can be redrawn in more detail as Fig. 1.2. A sensing element, or *sensor*, is connected to the process and experiences a change which relates to the process variable being monitored. A platinum resistance thermometer, for example, experiences a change in resistance with temperature, or a flow-dependent differential pressure is developed across an orifice plate.

In many systems there may be a chain of sensing elements to measure just one variable. An orifice plate is used to measure flow, and produces a flow-dependent differential pressure. This pressure may be converted to a positional displacement by bellows, and then the displacement is converted to an electrical signal by a potentiometer. In such systems the first sensor is called the *primary sensor* (the orifice plate in our example).

The initial signal directly off the sensor is often small, so local signal processing or conditioning may be used. Typical signal conditioning is a bridge circuit to convert the change in resistance of a strain gauge to an electrical voltage, or a frequency-to-voltage converter to change the flow-dependent pulse train from a turbine flow meter to a flow-dependent voltage. Other simple processing examples are filtering and linear amplification. More complex processing could

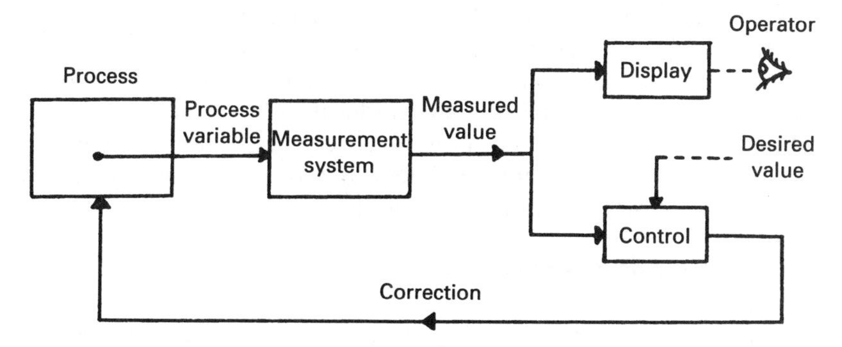

Fig. 1.1 *Industrial instrumentation and control system.*

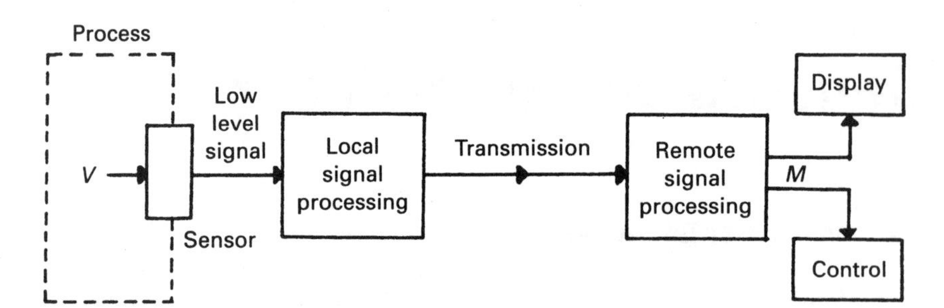

Fig. 1.2 *Elements of an instrumentation system.*

take place where a radio telemetry transmission path is used. The requirement here would probably be multiplexing, analog to digital conversion and some form of modulation.

The transmission path from process to display/control device is of prime importance, as it will inevitably introduce errors into the measurement system. Examples of transmission paths are electrical cables (for voltage and current), pneumatic pipes, fibre optic cables and radio links. Errors are introduced via interference (noise) and cable impedances into electrical systems, and by minute, but unavoidable, leaks in pneumatic systems. All transmission paths also inherently introduce a lag into the system; the measured value cannot react instantaneously to changes in the process variable. Capacitance and inductance cause lags in electrical systems, and finite pipe volumes lags in pneumatic systems.

In most systems further signal processing is performed local to the display or control device. In telemetry, obviously, a receiver and demodulator are required. Often linearisation is performed at this point (e.g. square-root conversion for differential pressure flow meters, described later in Chapter 5). Some form of computational correction may also be performed; applying cold junction compensation for thermocouples or temperature correction for mass flow measurements are typical examples.

The term *transducer* is often encountered. In strict terms a transducer is a device that converts one physical quantity into another, the second being an analog representation of the first. A thermocouple is a transducer that converts temperature to an electrical potential. It is more common, however, to use the term sensor for the actual measurement device (i.e. the primary sensor) and transducer for the entire measuring system local to the plant (including local signal processing). There are, however, no strict rules, and in many cases the terms sensor and transducer are used interchangeably. The word *transmitter* is also often used to mean transducer or sensor.

1.2. Signals and standards

Primary sensors produce a wide variety of signals: strain gauges give a very small resistance change, resistance thermometers a larger resistance change, thermocouples a voltage of a few millivolts, position measuring potentiometers several volts, and so on. Commercial transducers (with the word 'transducer' used to indicate signal conditioning as described in the previous section) are designed to give standard output signals for transmission to the control and display devices.

There are obvious maintenance and design advantages to standardised signals. If all the instrumentation, say, is designed around one standard, there can be commonality of spares and no need for specialised fault-finding aids.

The commonest electrical standard is the 4 to 20 mA current loop. As its name implies, this uses a variable current signal with 4 mA representing one end of the signal range and 20 mA the other. The current loop is totally floating from earth (and may not work correctly if an earth is applied to the signal lines). This gives excellent noise immunity as common mode noise has no effect and errors caused by different earth potentials around the plant are avoided. Because current, rather than voltage, is used, line resistance has no effect.

Several display/control devices can be connected *in series* (as in Fig. 1.3b), providing the total resistance does not rise above some value specified for the transducer (usually around 1 kohm).

Transducers using 4–20 mA can be current sourcing (with their own power supply, as in Fig. 1.4a) or designed for two-wire operation where the signal wires also act as the power supply connections, as in Fig. 1.4b. In the latter (and obviously cheaper to install) case, a separate power supply local to the display provides, say, 24 V DC. The transducer senses the current being drawn, and adjusts a shunt regulator to give the correct current for the process variable while maintaining sufficient voltage for its own internal electronics. The loop current then returns through the display device to the negative side of the power supply. Many commercial controllers incorporate a suitable power supply to allow self-powered or two-wire transducers to be used. Alternatively, one power supply can feed several transducers with some loss of isolation. Voltage-to-current and

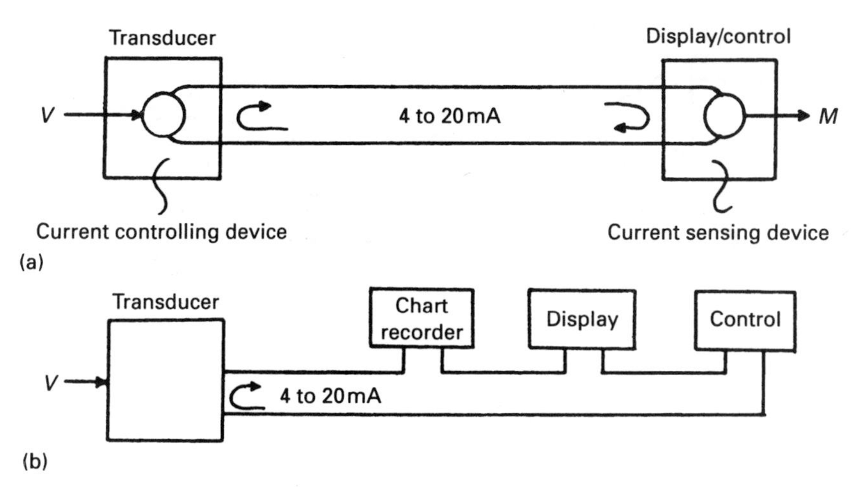

Fig. 1.3 *The 4 to 20 mA current loop transmission system. (a) Principle of 4 to 20 mA current loop. (b) Series connection of display/control devices.*

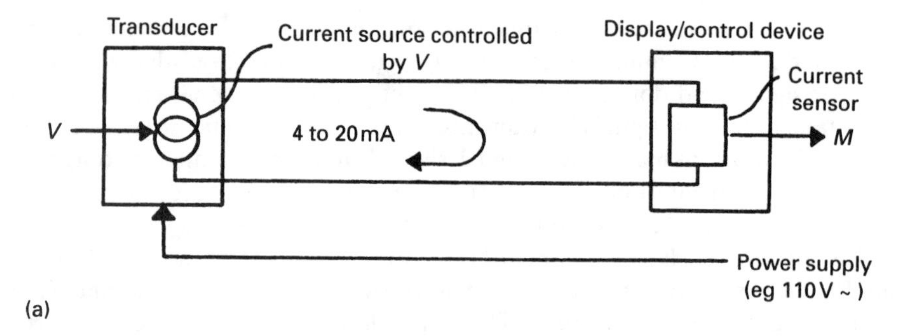

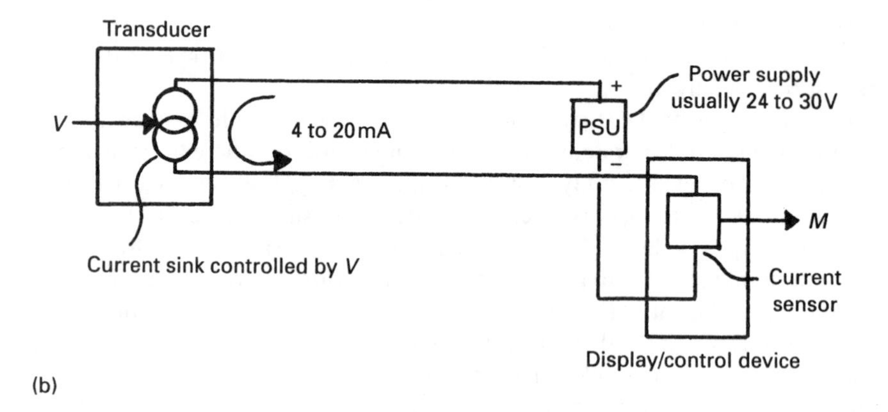

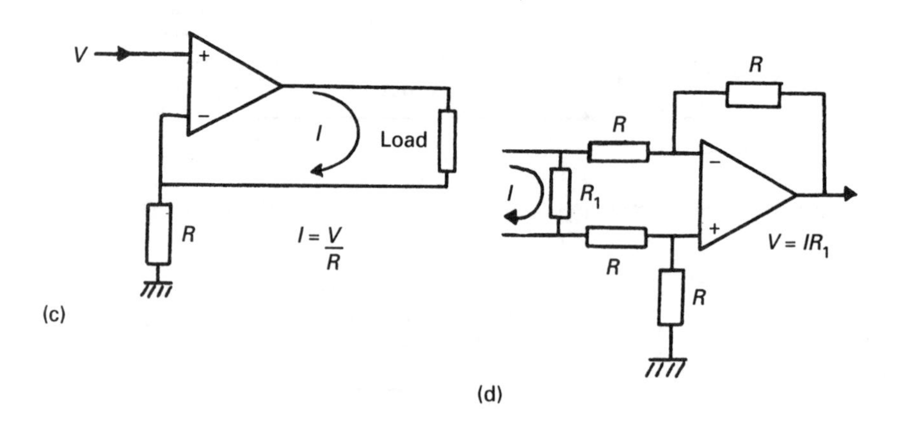

Fig. 1.4 *Current loop circuits. (a) Self-powered transducer. (b) Two-wire operation. (c) Voltage-to-current conversion. (d) Current-to-voltage conversion.*

current-to-voltage circuits are often based on the DC amplifier circuits of Fig. 1.4c, d. These circuits are discussed further in Chapter 11.

The use of an offset zero (4 mA) has several advantages (not least of which is the provision of sufficient current for a two-wire transducer to continue working at 'zero' output). If a zero voltage or current for the bottom of the range was chosen, an open-circuit or short-circuit line would look like a bottom-range signal. Any line fault on a 4–20 mA line will cause a substantial 'negative' signal which is easily detected at the controller or display device. In addition, the signal is decidedly unipolar, giving no ambiguities around zero, and obviating the need for a negative power supply which would be needed to give a zero voltage or current output.

Although 4–20 mA is by far the commonest electrical standard, others may be encountered. Among these are 10–50 mA and 1–5 V, again using an offset zero. Often 4–20 mA signals are converted to 1–5 V at the display device or controller by a series 250 ohm resistor.

Pneumatic signals also use an offset zero, the commonest standards being 3–15 p.s.i. or its metric equivalent, 0.2–1 bar (20–100 kPa). An offset zero improves the speed of response of a pneumatic system as well as bringing similar advantages to an electrical offset zero. A pneumatic line loses pressure by venting to atmosphere, and follows an exponential decay curve with respect to time. The use of an offset zero reduces the time taken to go from full scale to zero.

1.3. Definitions and terms

1.3.1. Introduction

The process variable, V, of Fig. 1.5 is measured by some instrumentation system to give a measured value, M, which is an analog representation of the value of V. There will, inevitably, be errors in this representation. Perfect representation can never be achieved, but there will always be a maximum allowable error. A temperature measurement in a chemical process, for example, may need to be accurate to $\pm 2\,°C$, or the level measurement in a vat can tolerate a ± 10 mm error. The process control engineer needs to be able to predict the errors in a measurement system in order to ensure that adequate accuracy is obtained but money is not wasted on a system that is unnecessarily accurate.

Transducer specifications use every-day terms like accuracy, error and repeatability in a precise way. The rest of this section defines the terms used on transducer data sheets.

1.3.2. Range and span

The *range* of a value is specified by its maximum and minimum values. On Fig. 1.5 the input range is V_{min} to V_{max}, and the output range M_{min} to M_{max}. A pressure transducer, for example, could have an input range of 0–100 kPa, and an output range of 4–20 mA. A single sensor can be similarly defined; a type K thermocouple could have an input range of 200–500 °C and an output range of 8–20 mV.

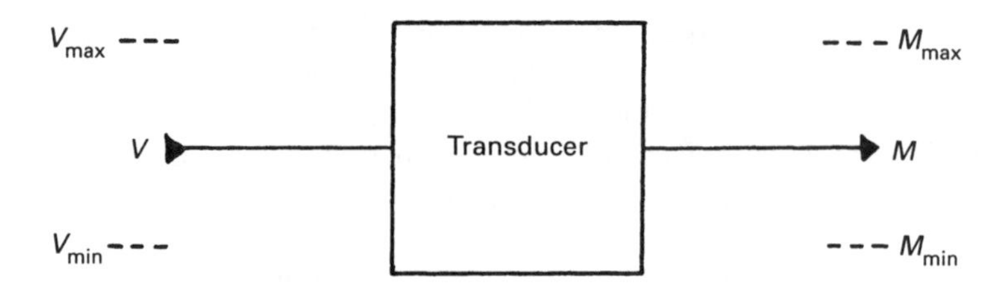

Fig. 1.5 *Range and span of a transducer.*

The *span* is the difference between maximum and minimum values: the input span is ($V_{max} - V_{min}$) and the output span ($M_{max} - M_{min}$). For example, the type K thermocouple just mentioned has an input span of 300 °C and an output span of 12 mV.

1.3.3. Linear and non-linear devices

If the relationship between M and V is plotted on a graph, a result similar to Fig. 1.6a will probably be obtained. The ideal relationship will be a straight line, Fig. 1.6b, which has the form:

$$M = KV + Z \tag{1.1}$$

where K is the *sensitivity* or *scale* factor given by:

$$K = \frac{M_{max} - M_{min}}{V_{max} - V_{min}} \tag{1.2}$$

K will have the units of M/V (e.g. mA/kPa for a pressure transducer with current output). Z is the zero offset given by:

$$Z = M_{min} - K \,.\, V_{min} \tag{1.3}$$

Z can be positive or negative.

A device that can be represented with tolerable error by equation 1.1 is said to be a linear device. If the relationship between V and M cannot be represented by equation 1.1, it is said to be non-linear. Devices with known non-linearities can be made linear by suitable signal conditioning. An orifice plate, for example, has a square law response between flow V and differential pressure M, i.e.:

$$M = A \,.\, V^2 \tag{1.4}$$

where A is a constant. If the orifice plate signal is passed through a circuit that performs a square-root operation, a linear relationship is obtained. Many devices can be linearised by considering them to be of the form:

$$M = A + BV + CV^2 + \ldots \tag{1.5}$$

where A, B, C, etc. are constants (which can be positive or negative), and designing a suitable compensating circuit.

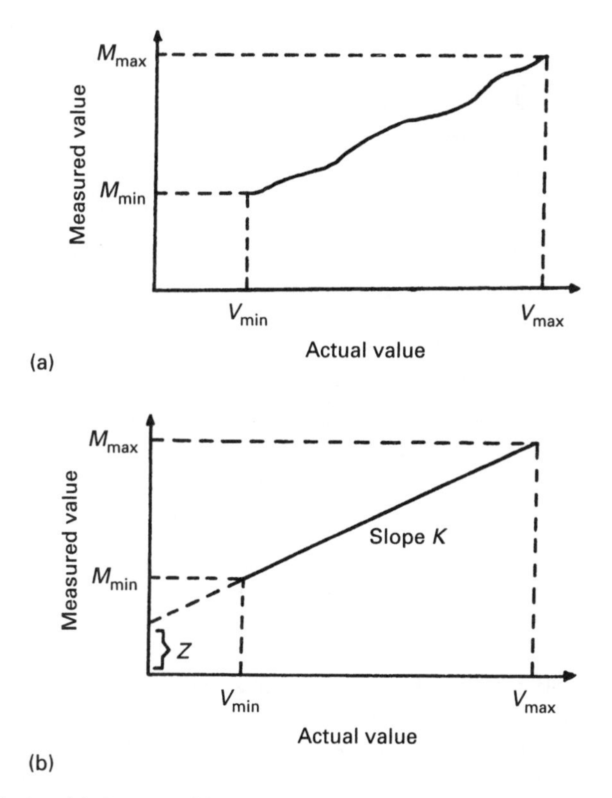

Fig. 1.6 *Relationship between* M *and* V. *(a) Typical transducer response. (b) Idealised transducer response.*

1.3.4. Accuracy and error

The *accuracy* of an instrument is a measure of how close the measured value is to the process variable. Accuracy is a rather loose term, and the precise term *error* is more generally used. This is defined as the maximum difference which may occur between the process variable and the measured value.

Error can be expressed in many ways. The commonest are absolute value (e.g. the maximum error on a temperature measurement may be defined as $\pm 2\,^{\circ}$C regardless of value), as a percentage of actual value of the process variable, or as a percentage of full scale of the measuring device (usually called FSD for full-scale deflection). The device in Fig. 1.7, for example, is a pressure-measuring transducer with an exaggerated non-linearity. It has an absolute maximum error of 0.5 p.s.i., which can be expressed as 6.25% of actual value or 3.33% of FSD.

1.3.5. Resolution

Many devices have an inherent 'coarseness' in their measuring capabilities. A wirewound potentiometer, used for position measurement as in Fig. 1.8a, has an inherent step size determined by the gauge of wire used. This gives a response

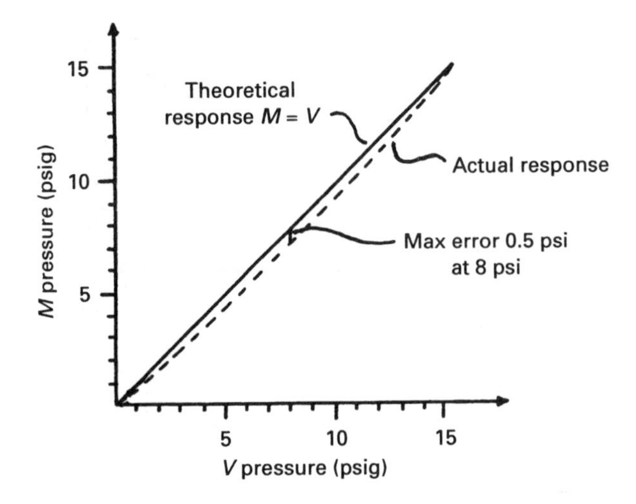

Fig. 1.7 *Transducer error as difference between ideal and actual response. Note that* M *is expressed in the same units and range as* V, *although in practice* M *would be an electrical signal of, say, 4 to 20 mA.*

similar to Fig. 1.8b. Even a cermet type pot has an ultimate coarseness caused by the grain size of the material used.

Totally digital systems (e.g. shaft encoders or equipment using analog-to-digital and digital-to-analog converters, see Chapter 13) also have an easily definable step-like response. Less obviously, almost all supposedly analog systems come up against some limit which prevents a more precise reading being made.

The term *resolution* is used to define the 'steps' in which a reading can sensibly be made. The resolution can be larger or smaller than the inherent error. A well-constructed steel ruler kept in a temperature-controlled environment may

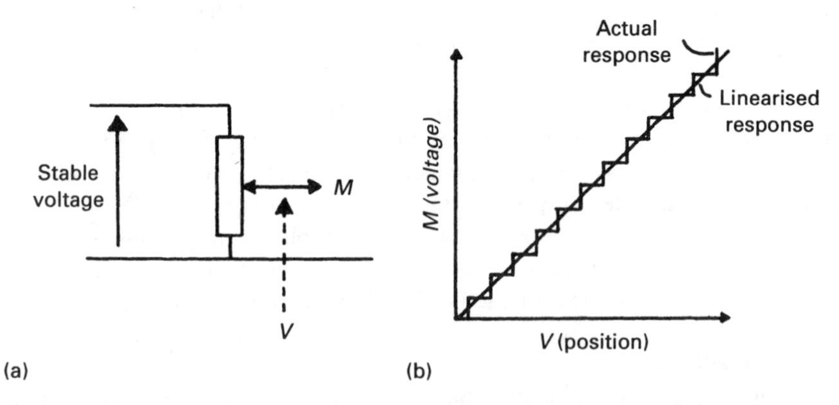

Fig. 1.8 *Transducer with finite resolution. (a) Potentiometer circuit:* P *is movement of slider,* M *is output voltage. In theory* V = KM *where* K *is a constant. (b) Actual response showing finite resolution.*

have an absolute error of less than $\pm$ 0.1 mm. A cheap plastic ruler may have an absolute error of more than $\pm$ 2 mm. If read by eye, the resolution of both would, however, be the same at about $\pm$ 0.5 mm, or half a scale division. Resolution and error are additive in terms of overall *system* error. The measuring accuracy of the plastic ruler is $\pm$ 2.5 mm.

Resolution becomes important when comparisons are to be made. A low-accuracy device with good resolution can be used to compare two values or to indicate if a value is increasing or decreasing. The cheap plastic ruler above, for example, could indicate the longer of two items with a resolution of 0.5 mm.

1.3.6. Repeatability and hysteresis

In many applications, the accuracy of a measurement is of less importance than its consistency. Where material is cut to some length, for example, the consistency of cutting accuracy may be of more importance than the absolute accuracy. Similarly a process may be required to keep at the *same* temperature for some period, but the actual temperature need not be known to any great degree of accuracy. The consistency of a measurement is defined by the terms repeatability and hysteresis.

Repeatability is defined as the difference in readings, obtained when the same measuring point is approached several times from the *same* direction.

Hysteresis occurs when the measured value depends on whether the process variable approached its current value by increasing or decreasing its previous value. The commonest example, backlash in gears or mechanical linkages, is illustrated in Fig. 1.9, giving a hysteresis error, h, between increasing and decreasing readings. Stiction (where a certain minimum force is needed to move an object) is also a common cause of hysteresis.

The effect of hysteresis can be reduced, or eliminated, by careful system design. The use of sprung gears (effectively two gear wheels on the same shaft tensioned by a spring) and similar pre-tensioning of couplings can remove mechanical hysteresis. An alternative approach, common in position control systems, is to use a unidirectional approach as in Fig. 1.9c. At time T the position is required to go from C to A which is a reversal, so the system introduces a deliberate overshoot to allow position A to be approached in the same direction as the other previous movements.

1.3.7. Environmental and ageing effects

The specifications for a transducer normally detail the possible errors outlined above for a new calibrated instrument under fixed operating conditions (e.g. constant ambient temperature, fixed electrical supply voltage or instrument air supply pressure). The accuracy of the transducer will be adversely affected by changes in its environment, and will progressively degrade with age. Both these effects will manifest themselves as a zero shift (or zero error), as in Fig. 1.10a, or a sensitivity change (or span error), as in Fig. 1.10b. Both may change with time, an effect known as *drift*.

Environmental effects are usually defined as the percentage error for some

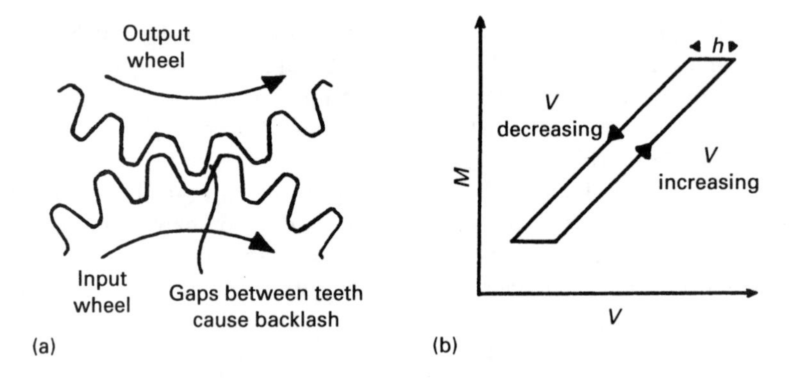

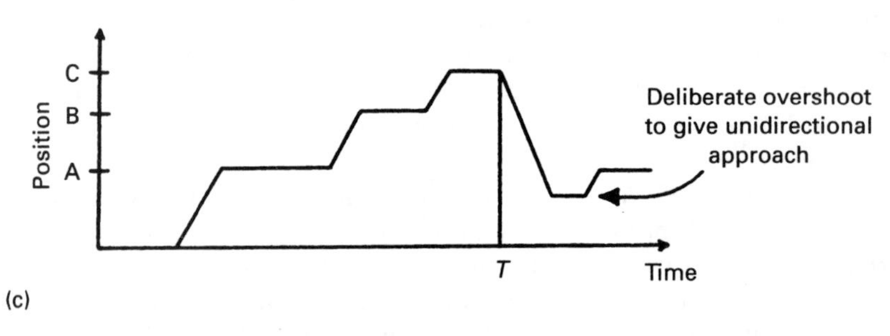

Fig. 1.9 *Causes and effects of hysteresis. (a) Gearwheels with exaggerated play. (b) Effect of backlash. (c) Eliminating the effect of hysteresis in a position control system by means of unidirectional approach.*

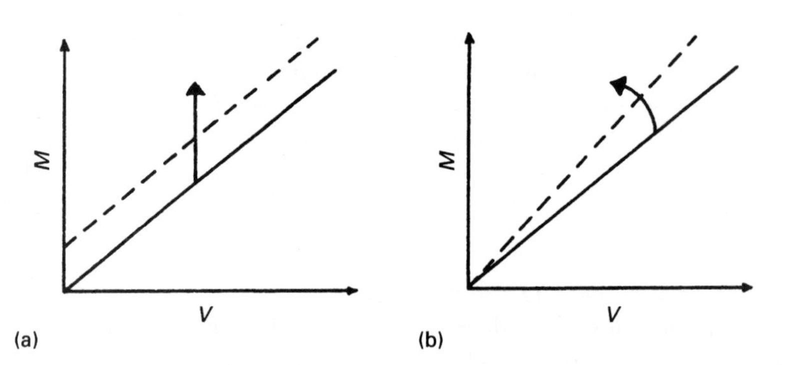

Fig. 1.10 *The effects of environmental changes and ageing. (a) Zero shift. (b) Span shift.*

environmental change. A differential pressure transmitter, for example, may be affected by the static pressure. A transducer with a 4–20 mA current output may be slightly load dependent. Almost all instruments are ambient temperature sensitive to some extent. Specifications list the possible error sources for a transducer, allowing the process control engineer to assess their importance in a particular application. Temperature effects may be crucially important for plant-mounted equipment, which may experience ambient changes from $-15\,^{\circ}C$ to $+50\,^{\circ}C$, but are of little relevance for equipment in an air-conditioned control room.

Known environmental effects can be eliminated by the inclusion of suitable compensation. The commonest example of this is probably the use of cold junction compensation with thermocouples: see Section 2.5.5.

Age effects are less important with modern solid-state electronics-based instruments than with earlier vacuum-tube or pneumatic devices. Age effects can, however, be eliminated by planned maintenance and recalibration at regular intervals. Device specifications will, again, allow the engineer to devise suitable schedules to keep age effects within acceptable limits.

1.3.8. Error band

In many instruments, all the effects of Sections 1.3.3–1.3.7 will be individually small and possibly difficult to measure. It is therefore becoming increasingly common for manufacturers to specify a *total error*, or *error band*, figure which includes *all* the above effects. This can be specified in the same way as error was defined earlier; as absolute or percentage of FSD (Fig. 1.11a) or as a percentage of value (Fig. 1.11b). In all cases the measured value lies within the specified limits.

1.4. Dynamic effects

1.4.1. Introduction

The definitions in Section 1.3 refer to the static characteristics of transducers, i.e. the measurements of signals which are considered not to vary. In many

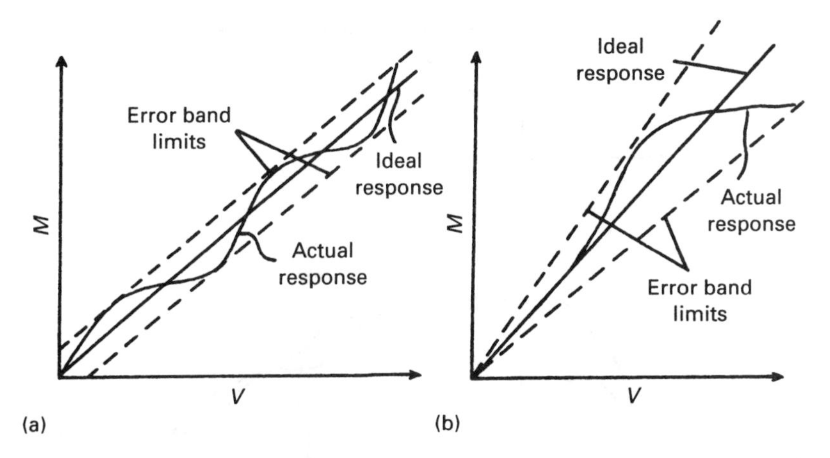

Fig. 1.11 *Definitions of total error and error band specifications. (a) Absolute value or per cent FSD. (b) Percentage of value.*

applications the process variable will be changing rapidly and it will be necessary for the transducer to follow these variations. A pressure transducer, for example, may be needed to follow the pressure variations inside the cylinders of an internal combustion engine where pressure spikes of less than 1 ms must be recorded.

1.4.2. First-order systems

If a temperature sensor at an ambient temperature of 20 °C is suddenly plunged into water at 80 °C, it will experience a step change of process variable as shown in Fig. 1.12a. The indicated temperature (ignoring errors) will, however, follow a curve as shown in Fig. 1.12b, the delay being caused by the time taken for the sensor to heat up to the temperature of the water.

Curves similar to Fig. 1.12b arise when the rate of change of the output is proportional to the difference between the current value, V, and the final value, V_f. This obviously applies to many thermal transducers. We can write:

$$\frac{dV}{dt} = K(V_f - V) \tag{1.6}$$

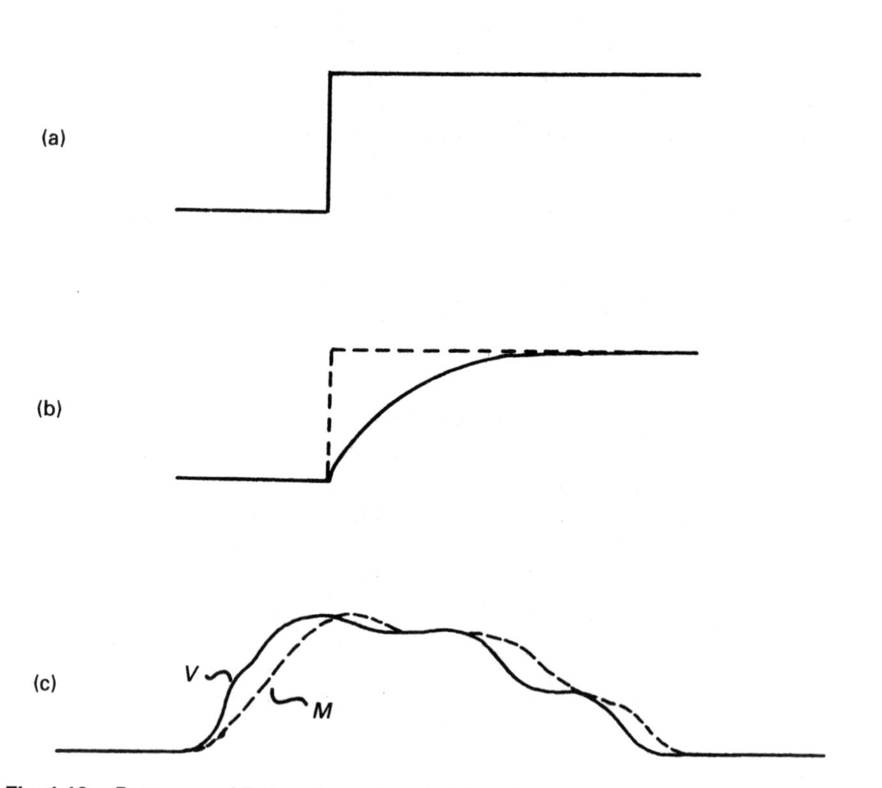

Fig. 1.12 *Response of first-order system. (a) Step change in V. (b) Observed change in M. (c) Response to slow-changing V.*

where K is a constant.

Solving equation 1.6 gives:

$$V = V_f - (V_f - V)\exp(-Kt) \tag{1.7}$$

This can be more conventionally expressed for practical purposes by replacing the constant K by $1/T$ where T is called the *time constant*. Equation 1.7 then becomes:

$$V = V_f - (V_f - V)\exp(-t/T) \tag{1.8}$$

The term $\exp(-t/T)$ approaches zero as t becomes large, so V approaches V_f as shown in Fig. 1.12b. Equation 1.8 is the response of a *first-order linear system*.

The time constant T determines the delay, V reaching 63% of V_f in a time T. The table that follows shows the percentage difference reduced for various multiples of T after a step change of input.

Time	*Percentage difference reduced*
T	63
$2T$	86
$3T$	95
$4T$	98
$5T$	99

It follows that for a step change in input, significant errors in measured value will occur. If the temperature transducer in Fig. 1.12 has a time constant of 4 s, the indicated temperature will be:

Time(s)	*Reading (°C)*
4	57.8
8	71.6
12	77.0
16	78.8
20	79.4

A step change is a worst-case input, but even a more natural input (Fig. 1.12c) will give a significant delay. It is good practice, however, to design a system such that a step-change signal gives an acceptable delay in the knowledge that all real inputs will comfortably meet the design criteria.

1.4.3. Second-order systems

A second-order response occurs when the transducer contains an element which is analogous to a mechanical spring/viscous damper or an electrical tuned circuit. The response of such a system to a step input of height a is given by the second-order equation:

$$\frac{d^2x}{dt^2} + 2b\omega_n\frac{dx}{dt} + \omega_n{}^2x = a \tag{1.9}$$

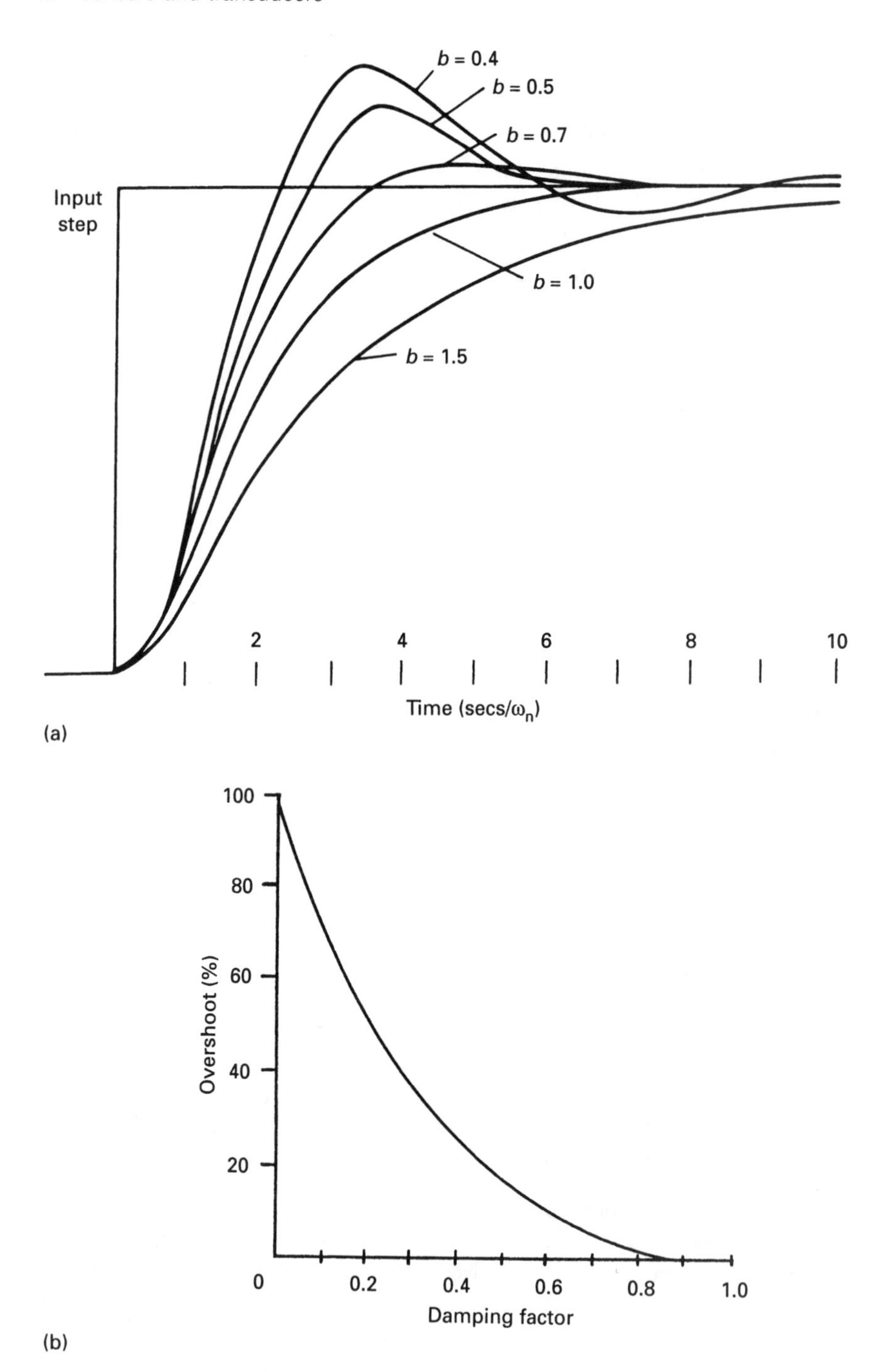

Fig. 1.13 *Effect of different damping factors. (a) Step response for various damping factors. (b) Overshoot related to damping factor.*

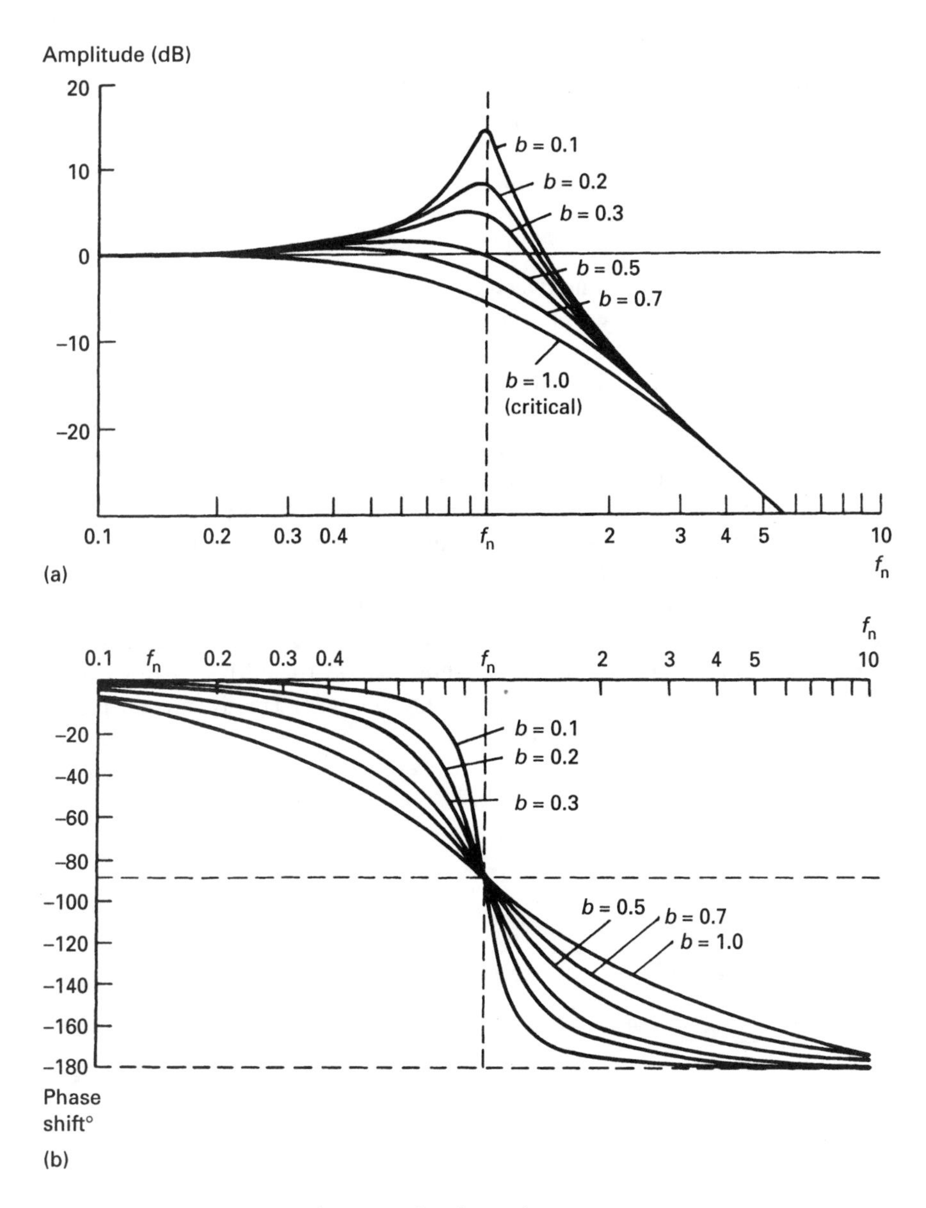

Fig. 1.14 *Bode diagrams for second-order system.*

where b is called the *damping factor* and ω_n is called the *natural frequency*. The final value of x is given by:

$$x = a/\omega_n^2 \tag{1.10}$$

The step response depends on both b and ω_n, the former determining the overshoot and the latter the speed of response as shown on Fig. 1.13a. It can be seen that for values of $b < 1$ damped oscillations occur causing a first overshoot as on Fig. 1.13b. The time to the first overshoot is given by:

$$T = \frac{1}{2f_n\sqrt{(1 - b^2)}} \tag{1.11}$$

where $f_n = \omega_n/2\pi$. For $b > 1$ no overshoot occurs. The special case $b = 1$ is called critical damping and represents the fastest response without overshoot for a given ω_n.

If an overshoot can be tolerated, other values of b may be advantageous. Figure 1.14 shows the Bode diagrams for a second-order system (normalised for ω/ω_n). In most applications, the fastest possible response will be required from a transducer, and this implies as wide a frequency response as possible.

An ideal response will remain within a given deviation from a normalised unity gain to as high a frequency as possible. Examination of Fig. 1.14a shows that this occurs for $b = 0.7$ (and not $b = 1$ as might first be thought). Similarly, Fig. 1.14b shows that the phase shift for $b = 0.7$ is lower than for $b = 1$ for frequencies below ω_n (and the relationship between phase shift and frequency is more linear). Figure 1.13b shows that $b = 0.7$ gives an overshoot of less than 10%. If this can be tolerated, a selection of $b = 0.7$ gives the best response, and many transducers are specified with this value of damping.

Chapter 2

Temperature sensors

2.1. Introduction

The measurement and control of temperature are possibly the most common operations in process control. To measure temperature qualitatively we need to define a temperature scale. This is done by choosing two temperatures at which some readily identifiable physical effect occurs, and assigning numerical values to these temperatures. Other temperatures can then be found by interpolation.

The Fahrenheit and Celsius (or Centigrade) scales use the freezing and boiling points of water as the two reference points.

	Fahrenheit	*Celsius*
Freezing point	32	0
Boiling point	212	100

It follows that

$$F = \tfrac{9}{5}C + 32 \quad \text{and} \quad C = \tfrac{5}{9}(F - 32).$$

The SI unit of temperature is the Kelvin (it should be noted that, unlike °F and °C, the degree symbol (°) is not used on the Kelvin scale). The lower of the two defining points on the Kelvin scale is absolute zero. This is the lowest theoretical temperature and is defined as 0 K. For comparison, 0 K = − 273.16 °C. The second defining point is the triple point of water (this being the unique temperature at which gaseous, liquid and solid phases can be in equilibrium). This is defined as 273.16 K, which corresponds to 0.01 °C. The definition is chosen such that a change in temperature of 1 K corresponds to a change of 1 °C and

$$K = {}^{\circ}C + 273.15$$

In industrial applications, the Celsius scale is most widely used, but conversion to the Kelvin scale is often needed when gaseous volumes or pressures are converted to some standard pressure.

2.2. Principles of temperature measurement

There are, in general, four types of temperature sensor based on the following physical properties, which are temperature dependent:

(1) Expansion of a substance with temperature, which produces a change in length, volume or pressure. In its simplest form this is the common mercury-in-glass or alcohol-in-glass thermometer.

(2) Changes in electrical resistance with temperature, used in resistance thermometers and thermistors.
(3) Changes in contact potential between dissimilar metals with temperature; thermocouples.
(4) Changes in radiated energy with temperature; optical and radiation pyrometers.

Transducers based on these principles will be described in following sections.

2.3. Expansion thermometers

2.3.1. Coefficients of expansion

In Fig. 2.1a we have a rod of length L_0 at some temperature T_0. If this is heated to some higher temperature, T_1, the rod will increase to a new length L_1 given by:

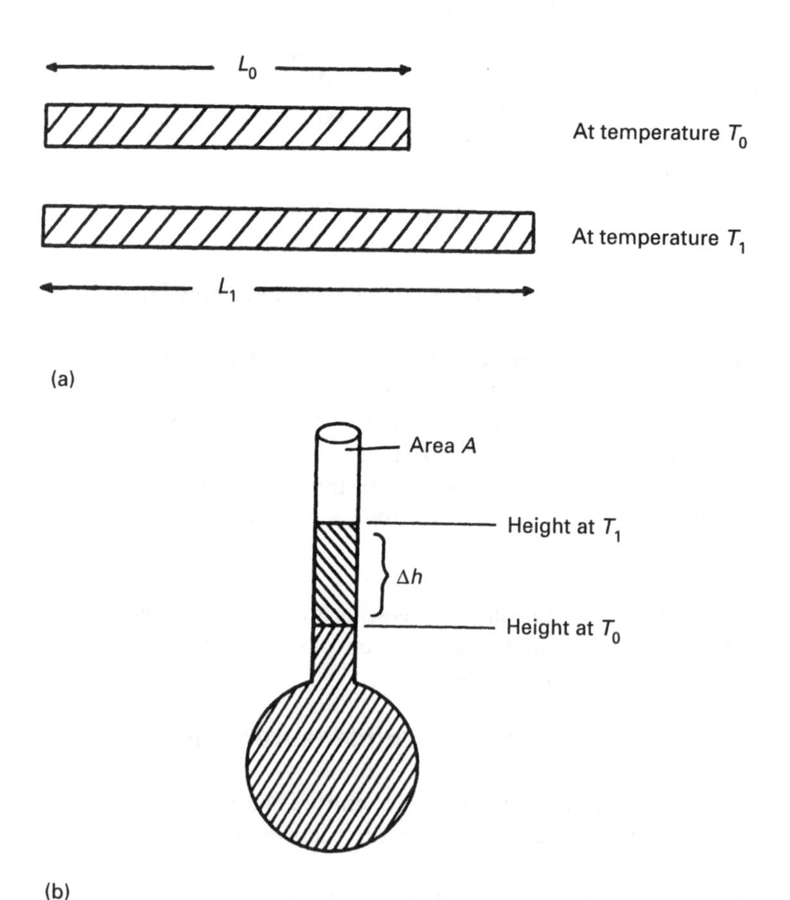

Fig. 2.1 *Expansion of liquids and solids with temperature. (a) Expansion of a solid rod with temperature. (b) Expansion of liquid with temperature.*

$$L_1 = L_0(1 + \gamma(T_1 - T_0)) \tag{2.1}$$

where γ is defined as the coefficient of linear thermal expansion. (The relationship does, in reality, include terms in $(T_1 - T_0)^2$ and higher terms, but the above equation is accurate for all practical purposes).
Typical values of γ are, per degree Celsius:

Steel	6.7×10^{-6}
Copper	16.6×10^{-6}
Aluminium	25×10^{-6}

In Fig. 2.1b, a volume V_0 of liquid is at temperature T_0. If this is heated to temperature T_1, its volume will increase to V_1, again given by:

$$V_1 = V_0(1 + \alpha(T_1 - T_0)) \tag{2.2}$$

where α is the coefficient of cubical thermal expansion. Again, the full equation has higher-order terms which may be neglected. Typical values of α are, per degree Celsius:

Mercury	0.56×10^{-4}
Alcohol	0.35×10^{-3}

In Fig. 2.1b, the increase in volume ΔV appears as a change Δh in the length of the liquid column in a capillary tube. If A is the cross-sectional area of the tube:

$$\Delta h = \frac{\Delta V}{A} \tag{2.3}$$

If A is made small, significant changes in the column length can be obtained for small changes in temperature. This is the basis of mercury-in-glass and alcohol-in-glass thermometers. In practice, a small correction needs to be made for the expansion of the container holding the liquid.

2.3.2. Bimetallic thermometers

In Fig. 2.2a, two dissimilar metals, A and B, have been bonded together. Metal A has a high coefficient of expansion and metal B a low coefficient of expansion. (The alloy Invar is often used for metal B.) As the temperature rises, the greater expansion of metal A causes the bar to bend, producing a deflection d which is a function of temperature. The change, d, is small in Fig. 2.2a, but can be increased and made more linear by the use of a coiled bimetallic spring as shown in Fig. 2.2b.

Bimetallic thermometers are cheap but of quite low accuracy. They are not widely used in industry, mainly because they cannot provide remote indication. Temperature sensing switches (thermostats) are often based on Fig. 2.2b with switch contacts replacing the pointer.

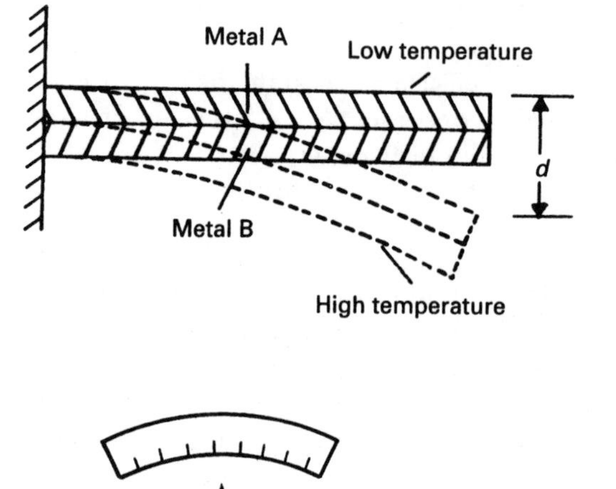

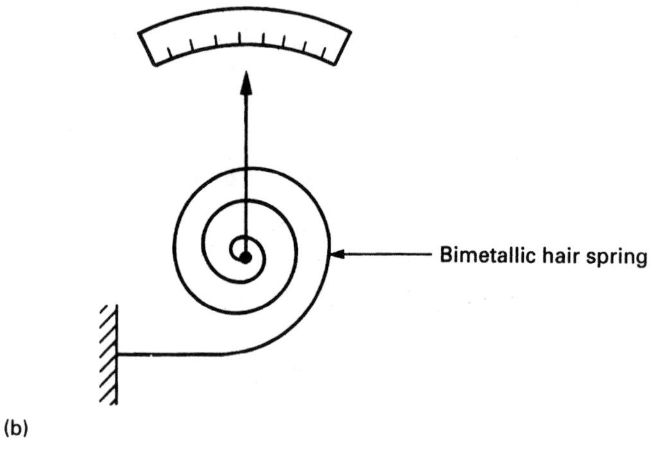

Fig. 2.2 *Bimetallic thermometers. (a) Bimetallic strip. (b) Indicating bimetallic thermometer.*

2.3.3. Gas-pressure thermometers

If a gas is contained in a vessel then

$$\frac{PV}{T} = \text{constant} \tag{2.4}$$

where P is the gas pressure, V is the gas volume, and T is the gas absolute temperature. If V is constant (i.e. a sealed container), then

$$P = \alpha T \tag{2.5}$$

where α is a constant, dependent on the gas and the initial pressure in the container. The equation implies that the pressure is linear with absolute temperature.

Gas-pressure thermometers can be used with remote indication provided the volume of the capillary tube connecting the bulb to the indicator is small by comparison with the bulb itself. A typical instrument, using nitrogen, can achieve an accuracy of 2% while giving remote indication up to 25 m from the bulb.

Because no electrical power is used, they are particularly suited for use in hazardous areas.

2.3.4. Vapour-pressure thermometers

Some of the restrictions of the gas-pressure thermometers can be overcome by using the vapour pressure of a volatile liquid. Methyl chloride, for example, has the vapour-pressure curve of Fig. 2.3a. As can be seen, there is a 1 : 10 pressure variation over a 100 °C temperature variation.

The construction of a typical instrument is shown in Fig. 2.3b. The volatile liquid is entrapped in the bulb via a non-volatile liquid which also fills the capillary tube to the indicator. The vapour pressure is thus relayed directly to the indicator, and the temperature or volume of the capillary tube has no effect.

The major constraint on the distance between bulb and indicator is speed of response. A practical limit is about 100 m, at which the instrument will exhibit a time constant of about 20 s. Like the gas-pressure instrument, the vapour-pressure thermometer is well suited to hazardous environments.

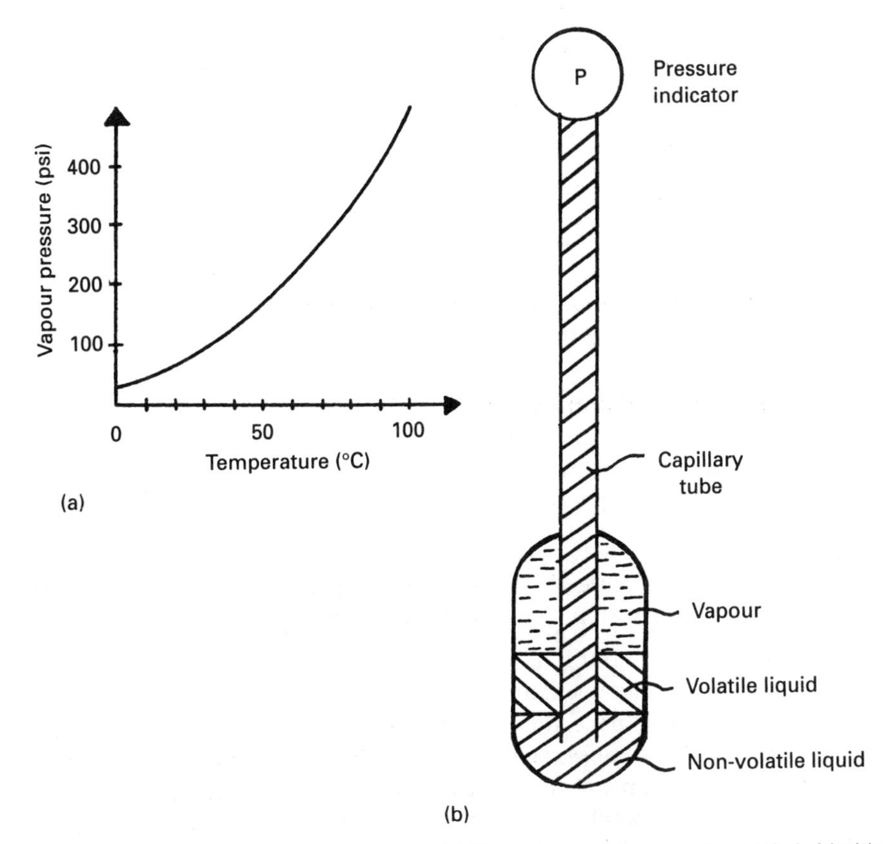

Fig. 2.3 *Vapour-pressure thermometers. (a) Vapour-pressure curve for methyl chloride. (b) Practical thermometer.*

2.4. Resistance thermometers

2.4.1. Basic theory

The electrical resistance of most metals increases approximately linearly with temperature. If a metal wire has a temperature R_0 at 0 °C, then the resistance at T°C will be given by:

$$R_T = R_0(1 + \alpha T + \beta T^2 + \ldots) \tag{2.6}$$

In almost all industrial applications, terms higher than the square can be ignored, and for most the relationship

$$R_T = R_0(1 + \alpha T) \tag{2.7}$$

will suffice if $0 < T < 150$ °C.

The constant α is called the temperature coefficient of resistance. Typical values are:

Metal	α
Platinum	0.0039
Copper	0.0043
Nickel	0.0068

Figure 2.4 shows the variations in resistance with temperature for various metals. In each case, a resistance of 100 ohm at 0 °C is used as a reference. As can be seen, a nickel-based thermometer is most sensitive, but is the most non-linear (due to the βT^2 and higher terms). Platinum is the least sensitive, but the most linear. In practice, the choice of material for a specific application will be determined by the accuracy required, the ability to resist contamination, and the cost. Platinum-based resistance thermometers are probably the most widely used (although they are the most expensive).

2.4.2. Resistance temperature detectors

A temperature transducer based on the above principles is called a resistance temperature detector (RTD), and is specified in terms of its resistance at 0 °C, and the change in resistance from 0 °C to 100 °C. This is known as the *fundamental interval.*

Platinum RTDs are constructed with a resistance of 100 ohm at 0 °C (and are often referred to as PT100 sensors). This gives a resistance of 138.5 ohm at 100 °C, and hence a fundamental interval of 38.5 ohm. In the UK, the relevant standard for RTDs is BS 1904 which specifies calibration methods and tolerances for the sensor. PT100 sensors can be used over a temperature range of − 200 °C to 800 °C with an accuracy of ± 0.5% between 0 °C and 100 °C and ± 3% at the extremes of the temperature range.

RTDs are available in many shapes and sizes; a typical sensor is shown in Fig. 2.5. The construction is a trade-off between protection against the atmosphere or

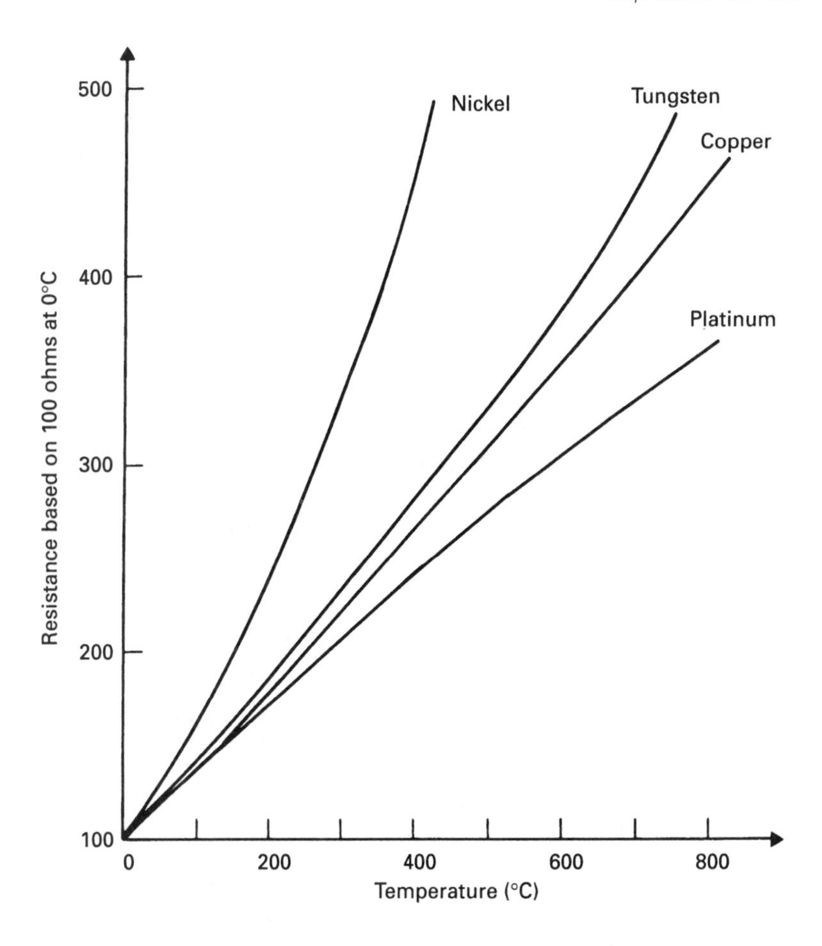

Fig. 2.4 *Change of resistance with temperature for various materials.*

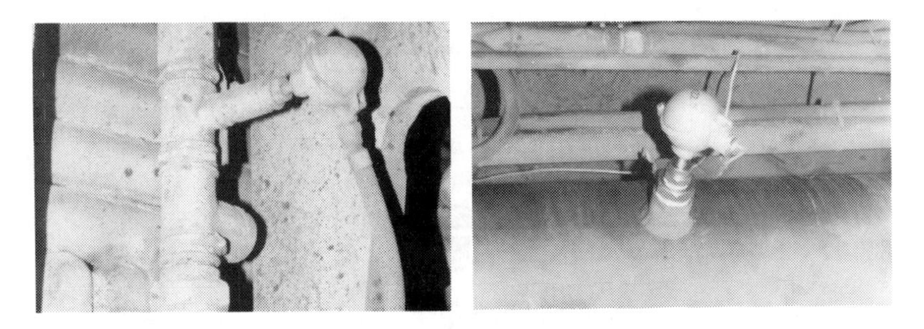

Fig. 2.5 *Examples of RTD sensors.*

fluid whose temperature is to be measured, and physical size (which determines the time constant of the RTD's response to a temperature change).

Figure 2.6 shows various constructions of RTDs. These are designed to protect the wire from mechanical shock while not applying any stress on the wire (which would cause resistance changes in a similar way to a strain gauge). Constructions with the wire in direct contact with the fluid give a fast response but little protection against corrosion. The sensors in Fig. 2.6 are totally enclosed, but the increased mass gives longer time constants.

2.4.3. Practical circuits

An RTD exhibits a change in resistance with temperature. Before it can be used for measurement or control, this change in resistance must be converted to a change in voltage or current. The electrical power dissipated in the RTD for this conversion must be strictly limited to avoid errors due to I^2R heating of the sensor. Typically 10 mW dissipation will cause a temperature rise of 0.3 °C,

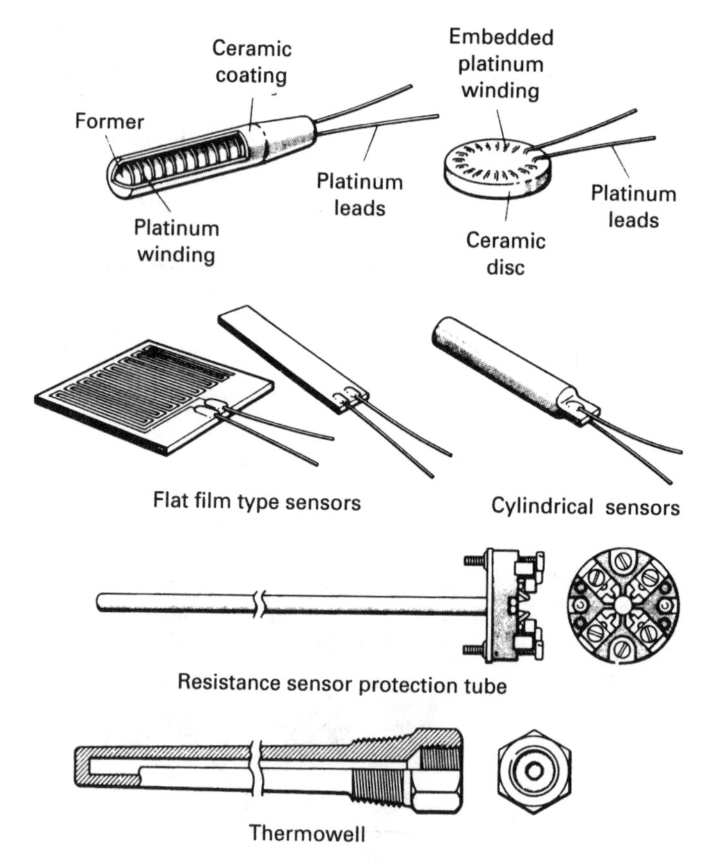

Fig. 2.6 *Platinum resistance thermometer construction (courtesy TC Ltd).*

which implies low values of current (less than 10 mA) and voltage (below 1 V).

The simplest circuit, shown in Fig. 2.7, uses a constant current source to convert the resistance change to a voltage change, V_T, where:

$$V_T = IR_0(1 + \alpha T) \tag{2.8}$$

V_T feeds a unity gain differential amplifier, with V_r corresponding to IR_0. The output voltage is then proportional to T °C.

The commonest circuits, however, are based on the Wheatstone bridge of Fig. 2.8. If the measuring circuit has a high impedance (so that it does not load the bridge), simple circuit analysis shows that:

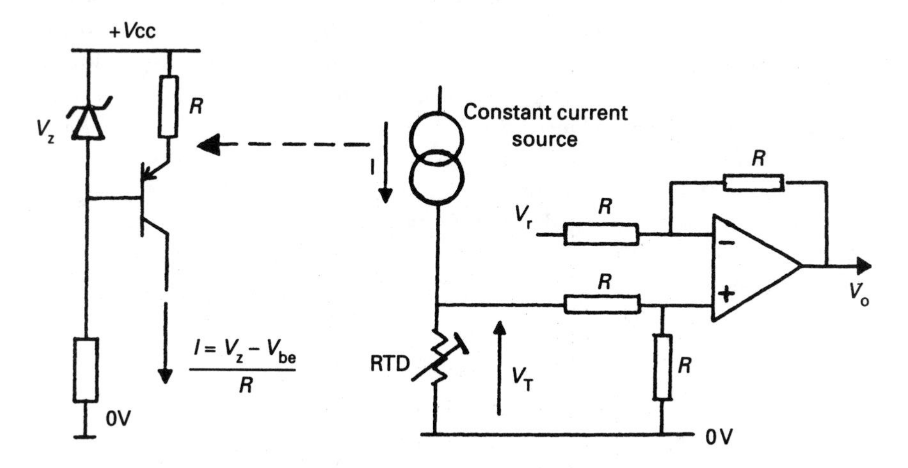

Fig. 2.7 *Resistance thermometer circuit.*

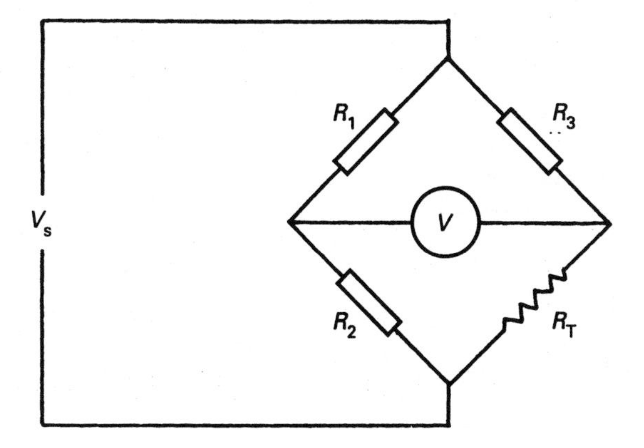

Fig. 2.8 *Wheatstone bridge.*

$$V = V_s\left(\frac{R_T}{R_T + R_3} - \frac{R_2}{R_1 + R_2}\right) \tag{2.9}$$

Unfortunately, because R_T appears in both the numerator and denominator of the left-hand term, V does not change linearly with changes in R_T. There are three common ways of overcoming this non-linearity, shown in Fig. 2.9.

The non-linearity can be reduced to acceptable levels by making $R_3 \gg R_T$ and $R_1 \gg R_2$ (typically by a factor of 100). This has the side-effect of reducing the bridge voltage by a factor of 100 as well, but this can easily be re-established by means of a DC amplifier.

Non-linearity does not, in itself, imply inaccuracy. Orifice plates, for example, have a very non-linear response, but are used to measure flow with minimal errors. In Fig. 2.9b, the non-linear output from the bridge is processed by a suitable linearising circuit to give an output voltage which is linearly related to temperature. The linearising can be performed by an Op Amp circuit (see chapter 11) or by a microprocessor 'intelligent' instrument.

Figure 2.9c uses an electronic circuit to balance the bridge. This is a modern version of the original Wheatstone bridge which was nulled by adjusting the resistance in one of the arms; the desired parameter being inferred by the new resistor value. The null circuit measures the out-of-balance voltage V_b, and increases, or decreases, the current injected into the one arm of the bridge to bring

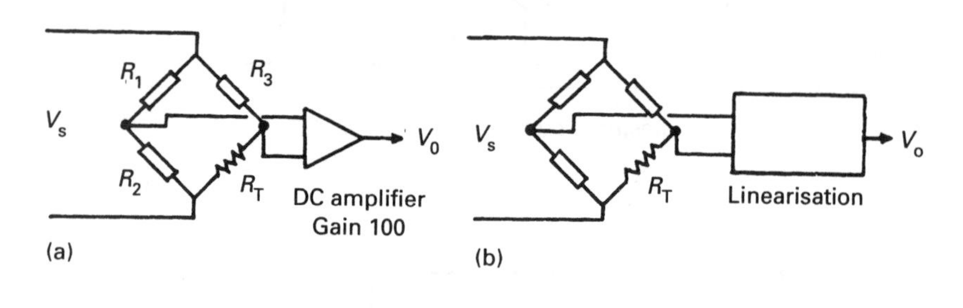

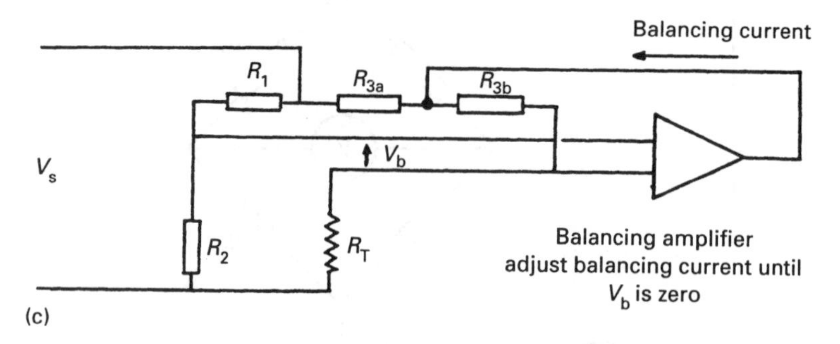

Fig. 2.9 *Methods of overcoming non-linearity of bridge output signal. (a) Making* $R_3 \gg R_T$ *and* $R_1 \gg R_2$. *(b) Linearisation of bridge output. (c) Null balancing technique.*

V_b to zero. With the bridge balanced, the change in R_T is simply obtained from the new value of I.

In most industrial applications, the RTD will be situated remote from its measurement electronics. If the connecting leads are more than a few metres in length, they will introduce an unknown resistance, r, into each lead, as shown in Fig. 2.10a. This unknown resistance is itself subject to change caused by temperature and strain effects, and is a possible source of error. This can be overcome by using a four-wire connection to the RTD, as shown on Fig. 2.10b. Each lead will experience the same conditions, so the changes introduced into the RTD leads will be matched by the changes in the dummy leads. The bridge output will only be dependent on changes in temperature of the RTD, not the leads.

In many applications, the three-wire connections of Fig. 2.11 give adequate compensation with a small cost saving. Figure 2.11a is the commonest industrial circuit for RTDs.

2.4.4. Thermistors

RTDs utilise the small, but essentially linear, increase in resistance of metals with increasing temperature. Semiconductor materials, however, exhibit a large, but very non-linear, decrease in resistance with increasing temperature. Temperature sensors based on semiconductors are called thermistors.

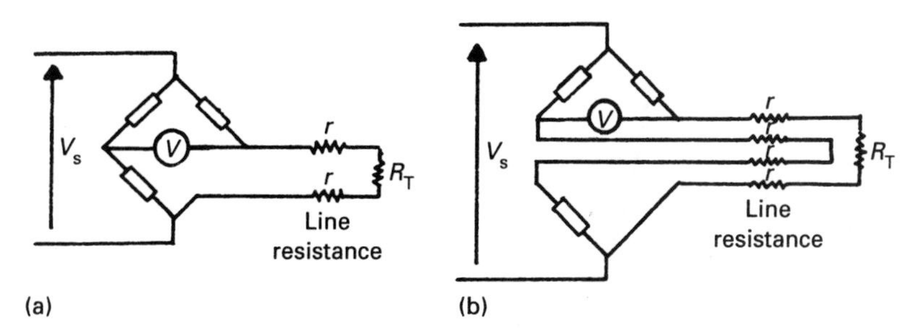

Fig. 2.10 *The effects of line resistance on RTDs. (a) Two-wire circuit. (b) Four-wire circuit.*

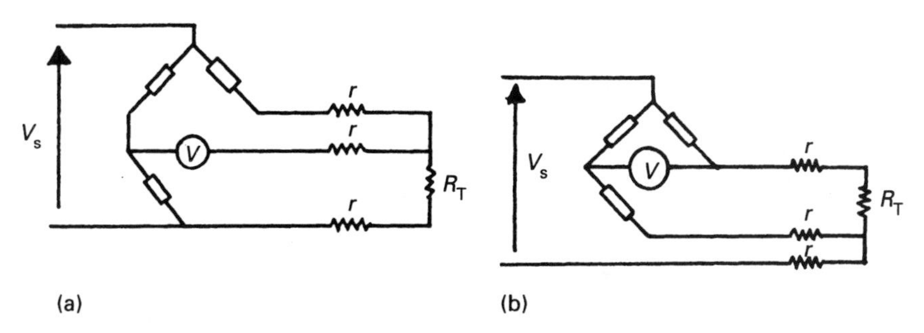

Fig. 2.11 *Three-wire connection commonly used in industry. (a) First arrangement. (b) Second arrangement.*

The variation of resistance for a typical thermistor is shown in Fig. 2.12. The resistance of this device decreases from over 10 k ohm at 0 °C to less than 200 ohm at 100 °C, a change of 50 : 1. The figure also shows the non-linearity inherent in thermistors.

The resistance of a thermistor is defined by:

$$R = A \exp\left(\frac{B}{T}\right) \tag{2.10}$$

where A and B are constants for the particular thermistor, and T is the

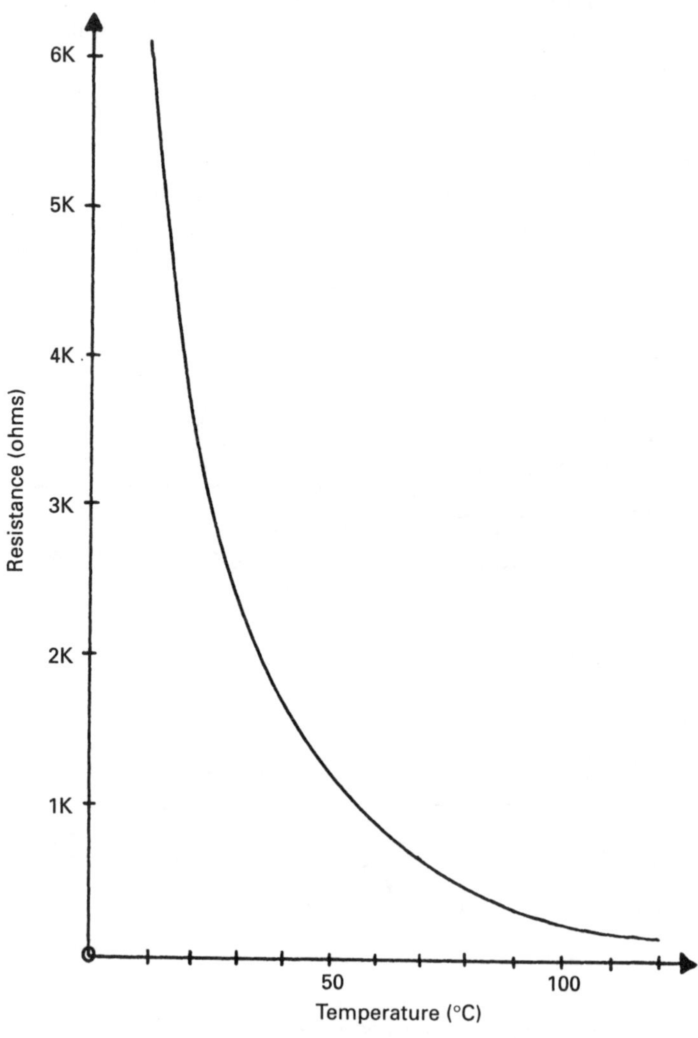

Fig. 2.12 *Resistance/temperature curve for a NTC thermistor.*

temperature in degrees Kelvin. The constant B, which has the dimensions of temperature, is called the characteristic temperature and is typically between 3000 K and 5000 K.

Equation 2.10 is not particularly useful, as it is not easy to see what value of resistance will be obtained in the more practical temperature range of − 100 °C to + 200 °C. Data sheets usually define the resistance R_0 at some temperature T_0 (often 0 °C, 273 K). The resistance at any other temperature is then:

$$R = R_0 \exp B\left(\frac{1}{T} - \frac{1}{T_0}\right) \tag{2.11}$$

where B is the characteristic temperature. Note that T and T_0 are both in degrees Kelvin. The device of Fig. 2.12, for example, has a resistance at 25 °C of 3 k ohm and a value for B of 4020 K.

Thermistors come in a wide variety of shapes, sizes and enclosures. Most are considerably smaller than RTDs, and consequently have a faster response. Although the response of thermistors is non-linear, they can be used for temperature measurement over a limited range (say 0 °C to 100 °C). A typical circuit, shown in Fig. 2.13, produces a voltage at point A given by:

$$V_A = \frac{R_1}{R_1 + R_T} V_{cc} \tag{2.12}$$

This gives a non-linear response to changes in R_T which roughly compensates for the non-linearity of the thermistor. RV_1 sets a voltage at its slider equivalent to V_A at 0 °C. RV_2 sets the full-scale current through the meter.

The sensitivity of thermistors makes them ideally suited for temperature alarm circuits. In these applications, where all that is required is a signal that a temperature has gone above (or below) some preset, the non-linearity is of little importance.

The thermistors described so far are known as NTC thermistors for negative temperature coefficient. It is also possible to manufacture PTC (positive temperature coefficient) thermistors which exhibit a non-linear increase in

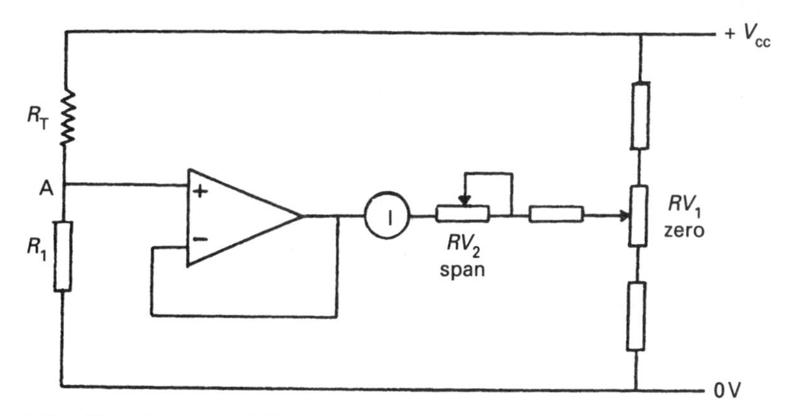

Fig. 2.13 *Thermistor-based thermometer.*

resistance with temperature. These have the response of Fig. 2.14, which shows that the resistance increases suddenly at some temperature (called the reference temperature). The response of PTC thermistors makes them unsuitable for temperature measurement, but they are widely used as temperature alarm devices for, say, motor windings.

2.5. Thermocouples

2.5.1. The thermoelectric effect

In Fig. 2.15, two dissimilar metals are joined at two points as shown. If one end is heated to a temperature T_1, and the other end kept at a lower temperature T_2, a current will flow around the circuit. The current depends on the metals and the temperatures T_1, T_2. This phenomenon, discovered by the Victorian scientist Seebeck, is called the thermoelectric (or Seebeck) effect, and can be used as an accurate measurement of temperature. Devices using this effect are called thermocouples.

The effect arises because an electrical potential arises across the junction of two dissimilar metals. This potential depends on the temperature of the junction, and occurs because of different electrical and thermal properties of the metals. Somewhat simplified, electrons at a higher temperature T_1 have more thermal

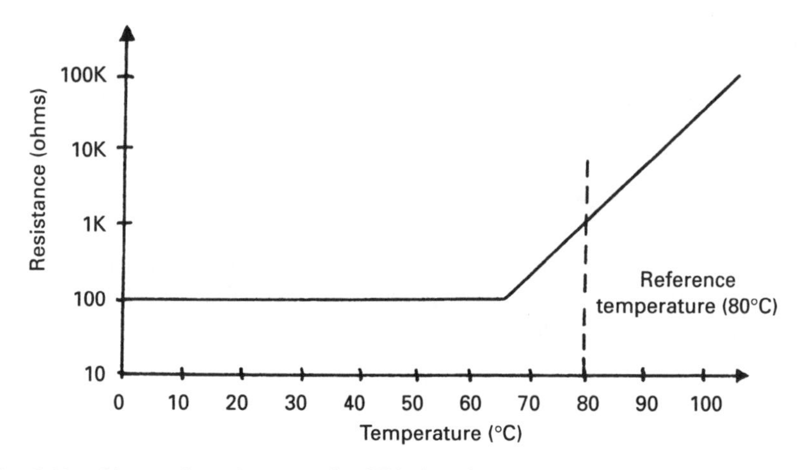

Fig. 2.14 *Change in resistance of a PTC thermistor.*

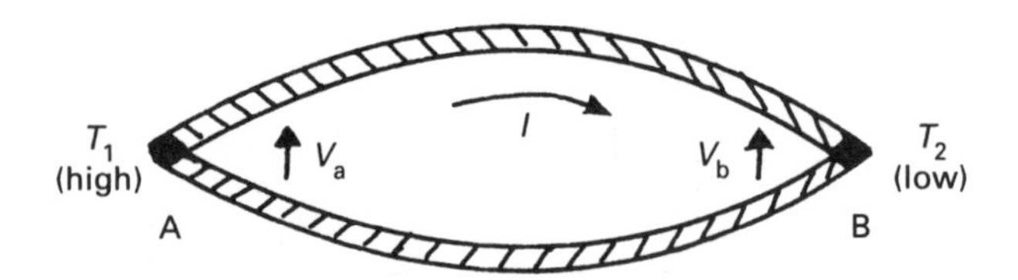

Fig. 2.15 *Simple thermocouple.*

energy than those at the cooler end in each metal, and there is a drift of electrons towards T_2. The difference in this drift between the two metals produces the voltage. The potential is small, typically a few tens of millivolts.

At junction A, there is a voltage V_a which is a function of T_1. Similarly, at junction B there is a voltage V_b which is a function of T_2. The current arises because V_a differs from V_b. Obviously if $T_1 = T_2$ no current will flow. Inherently, therefore, a thermocouple is a *differential* temperature-measuring device.

2.5.2. Thermocouple laws

The current flowing in the circuit, as in Fig. 2.15, is not a convenient indication of temperature, as it depends on the size and length of the wire. The potentials across the junctions are, however, independent of the wire size or length, being solely determined by the metals and the temperature. Introducing a voltmeter into the circuit can introduce errors, however, as new dissimilar metal junctions will be formed in the circuit.

The effects of introducing measuring instruments into the circuit (and other important considerations) are described by five thermocouple laws, illustrated in Fig. 2.16.

Law 1 states that the thermoelectric effect depends only on the temperatures of the junctions, and is unaffected by intermediate temperatures along the wires. In Fig. 2.16a, the thermocouple wires pass through an area at temperature T_3. The thermocouple effect in this circuit, however, still only depends on T_1 and T_2. This is extremely important in practical locations where the temperature of connecting leads is not known.

Law 2 allows additional metals to be introduced in the circuit without affecting the potentials *provided* junctions of each metal are at the same temperature. In Fig. 2.16b new metals are added with junctions CD and EF. These will not affect the circuit provided $T_c = T_d$ and $T_e = T_f$. At each junction, a contact potential will exist, but will be equal and opposite (and hence cancel) if the junction temperatures are the same. Thermocouple cables can be run through connectors, terminal strips and such devices without error, provided temperature differences do not occur across the device.

The third law is an extension of law 2, and states that a third metal can be introduced at either junction, as in Fig. 2.16c, without effect, provided that both junctions of the third metal (T_c, T_d) are the same. This has obvious practical implications, as it allows mechanically strong junctions to be made by using brazed, welded or soldered joints. Figure 2.16c also represents the commonest measuring technique, with the third wire being the millivoltmeter.

Law 4, illustrated in Fig. 2.16d, is called the law of intermediate metals, and can be used to determine the voltage of, say, a thermocouple based on iron/copper given tables for constantan/copper and iron/constantan.

The final law, called the law of intermediate temperature, is of particular importance when interpolating thermocouple tables. The contact potentials across the junctions are dependent on *absolute* (degrees Kelvin) temperature, and have the form:

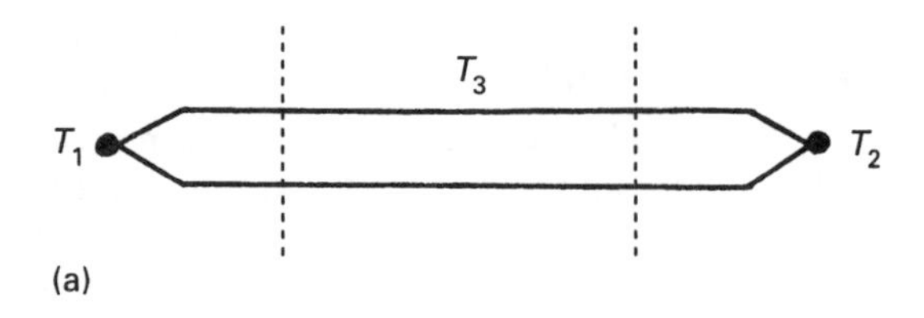

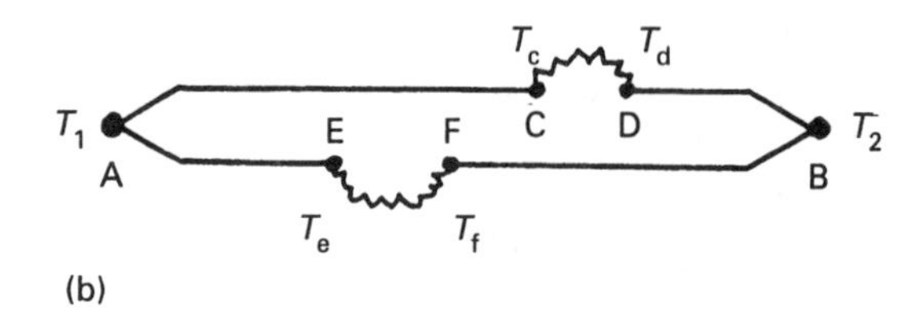

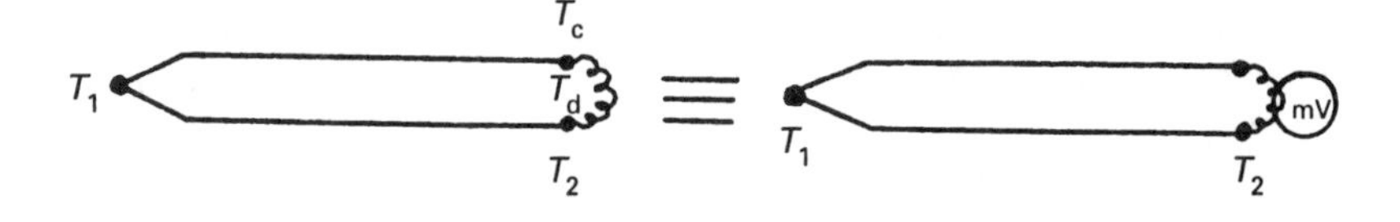

X X Z

T_1 Y T_2 + T_1 Z T_2 = T_1 Y T_2

(d)

X X X

T_1 Y T_3 + T_3 Y T_2 = T_1 Y T_2

(e)

Fig. 2.16 *The five thermocouple laws. (a) First thermocouple law. (b) Second thermocouple law. (c) Third thermocouple law. (d) Fourth thermocouple law (intermediate metals). (e) Fifth thermocouple law (intermediate temperatures).*

$$E_T = AT + BT^2 + CT^3 + \dots \tag{2.13}$$

where A, B, C, etc. are constants, and T is the temperature. For most applications, terms above the square are ignored.

The thermocouple is a temperature differential measuring device. If T_1, T_2 are the temperatures of the junctions, the resultant voltage is given by:

$$E_{T_1T_2} = E_{T_1} - E_{T_2}$$
$$= A(T_1 - T_2) + B(T_1^{\,2} - T_2^{\,2}) \tag{2.14}$$

This is a non-linear response, so the same temperature difference does not result in the same voltage. For example, if a type R thermocouple is used, and T_1 is 1250 °C, T_2 is 50 °C, the resulting voltage is 13.626 mV. If T_1 is reduced to 1200 °C and T_2 to 0 °C (maintaining the differential of 1200 °C), the voltage changes to 13.224 mV, a difference of 3% (and a temperature error of about 25 °C).

Tables of thermocouple voltages are published for various temperatures, measured as Fig. 2.16c, with T_2 at some reference voltage (usually 0 °C or 20 °C). The law of intermediate temperatures allows these tables to be used to deduce the voltage for any other values of T_1, T_2. Suppose we have $T_1 = 1100$ °C and $T_2 = 40$ °C with a type R thermocouple. The tables (referenced to 0 °C) give a voltage of 11.846 mV for 1100 °C and 0.232 mV for 40 °C. The resultant voltage is 11.614 mV. Note that it is *incorrect* to say 1100 − 40 °C = 1060 °C, and read the value for 1060 °C (11.304 mV) from the table. Section 2.5.7 will deal with using tables.

2.5.3. Thermocouple types

Although almost any pair of dissimilar metals can be used to make a thermocouple, over the years various standards have evolved. These have well documented voltage/temperature relationships, and their use gives interchangeability between different manufacturers. These are shown in Table 2.1. Figure 2.17 compares thermocouple outputs and shows their useful ranges.

All that is required to manufacture a thermocouple is to joint the requisite materials. In practice, of course, the thermocouple junction needs protection and

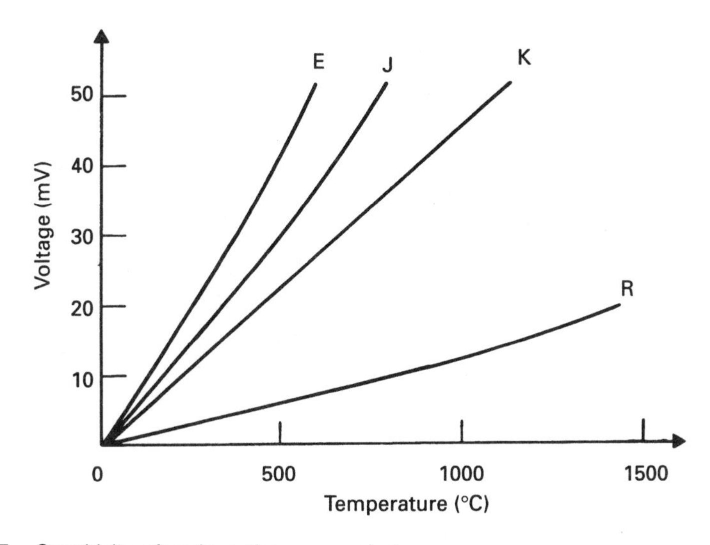

Fig. 2.17 *Sensitivity of various thermocouple types.*

Table 2.1 *Thermocouple types*

Type	*Material+*	*Material−*	$\Delta V/°C$ *at* 100 °C (μV)	*Usable range (°C)*	*Comments*
E	Chromel (90% nickel, 10% chromium)	Constantan (57% copper, 43% nickel)	68	0 to 800	Highest-output thermocouple
T	Copper	Constantan	46	−185 to +300	Used for cryogenics and mildly oxidising or reducing atmospheres (e.g. boiler flues)
K	Chromel	Alumel (94% nickel, 3% manganese, 2% aluminium, 1% silicon)	42	0 to 1100	General purpose, widely used
J	Iron	Constantan	46	20 to 700	Used with reducing atmospheres. Iron tends to rust and oxidise; can be improved with chrome/nickel/titanium steel
R	Platinum/13% rhodium	Platinum	8	0 to 1600	High temperatures (e.g. steel making). Used in UK in preference to type S
S	Platinum/10% rhodium	Platinum	8	0 to 1600	As type R, but used outside UK
V	Copper	Copper/nickel	–	–	Compensating cable for type K to 80°C can also be used for type T
U	Copper	Copper/nickel	–	–	Compensating cable for types R and S to 50°C

rigidity. These are provided by a sheath, usually constructed from magnesium oxide to give a fast thermal response. Such assemblies are called mineral insulated metal sheathed thermocouples. The actual junction can be arranged in one of the options of Fig. 2.18. An insulated junction gives an electrically isolated output (with an impedance to ground in excess of 100 Mohm) and total protection from the measured atmosphere. A grounded junction gives a fast thermal response. The fastest response is obtained with an exposed junction, but this can only be used where the atmosphere does not attack the thermocouple wires.

Thermocouples are available in a variety of sizes and shapes, some of which are shown in Fig. 2.19. They can be protected by an enclosure called a thermowell, but this obviously introduces a large lag.

2.5.4. Extension and compensating cables

Equation 2.13 emphasises that a thermocouple is a differential measuring device, and that to be useful one temperature (usually the lower) must be known. This can cause measurement problems if due care is not taken with the installation.

Figure 2.20 shows a typical thermocouple installation where a probe is used to measure the temperature in an oven. In Fig. 2.20a, the probe terminals are connected back to the control room by ordinary copper wires. By law 3, and Fig. 2.16c, the voltage seen at the control room will be determined by the oven temperature, and the *terminal* temperature of the thermocouple probe. The latter is unknown, but will probably be high due to heat conducted out of the oven, and will vary from day to day.

An improved connection, Fig. 2.20b, uses matching leads of the same material as the thermocouple to connect the probe to the control room. The voltage now

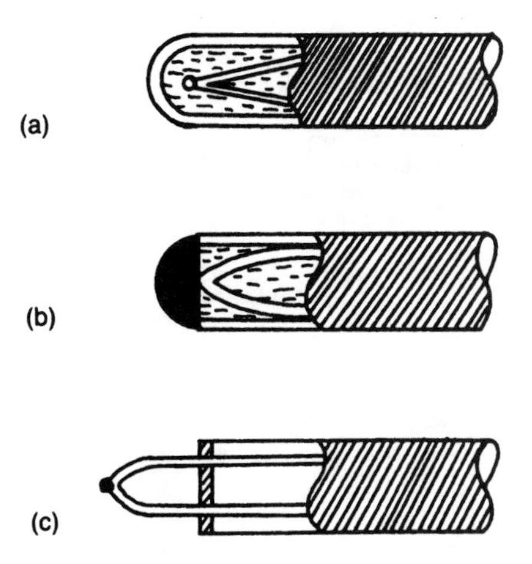

Fig. 2.18 *Types of measuring junction configuration. (a) Insulated junction. (b) Grounded Junction. (c) Exposed junction.*

Type 3. General purpose probe

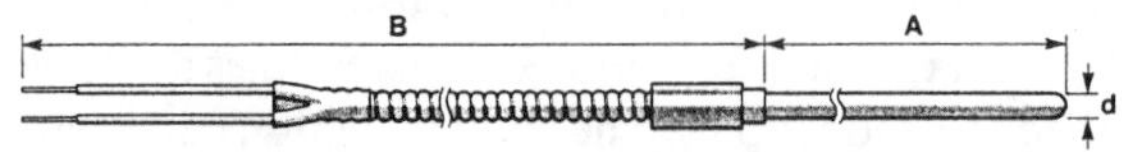

Suitable for general purpose applications up to 400°C these assemblies are available in thermocouple conductor calibration codes: K, T, J and E as simplex and duplex units.
Type 3 assemblies are supplied as standard with seamless welded closed end sheaths in AISI type 316 stainless steel. Other sheath materials are available from our range of stainless steels, Inconel*, Incoloy* and other alloys.
If the limiting operating temperature of 400°C at the tip of these assemblies is too low then our type 12 assemblies may be more suitable. SEE PAGES 24 & 25.
Junctions are supplied grounded to the sheath as standard with insulated junctions available as an option.

Type No:	Lead arrangement
3AX	PVC leads
3AY	Teflon leads
3AZ	Fibreglass leads
3AS	Stainless steel braid over fibreglass leads
3AF	Galvanised steel conduit over fibreglass leads
3AG	Stainless steel conduit over fibreglass leads
3H11	Thermocouple plug termination on sheath
3M11	Miniature thermocouple plug termination on sheath
3TH	Screw top weatherproof head (3P10) on sheath end

Dimension A (sheath length): to suit application
Dimension B (lead length): to suit application
Dimension d (sheath diameter): inches: 3/16, 1/8, 1/4
mm: 1.5, 3.0, 3.0, 5.0 or 6.0

Optional extras.
Insulated junction. Pin or spade termination.
Reduced tip. 120 degree ground tip.
90 degree ground tip.

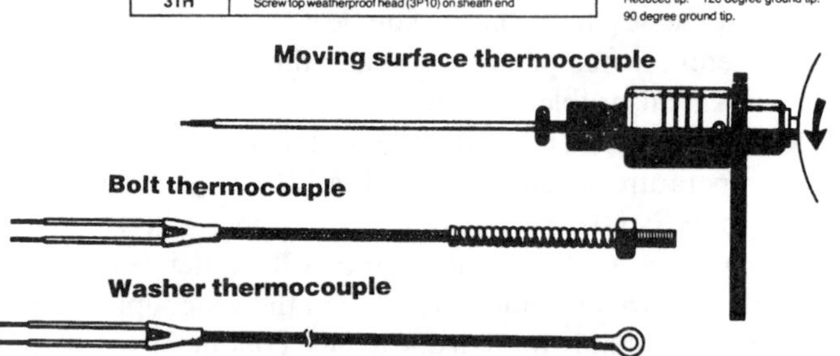

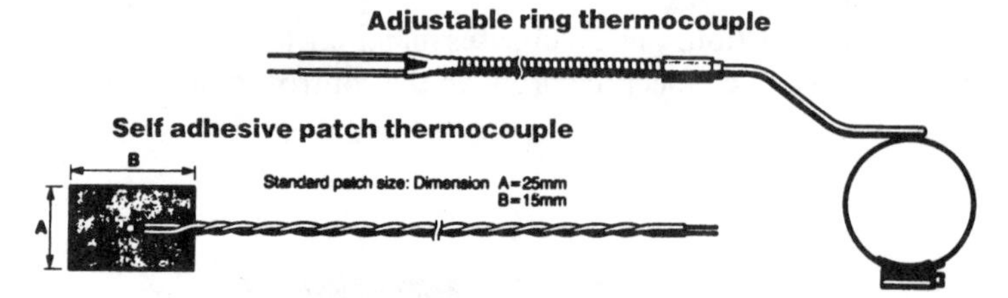

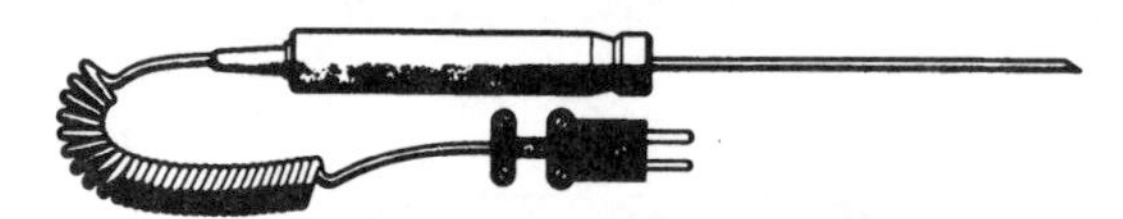

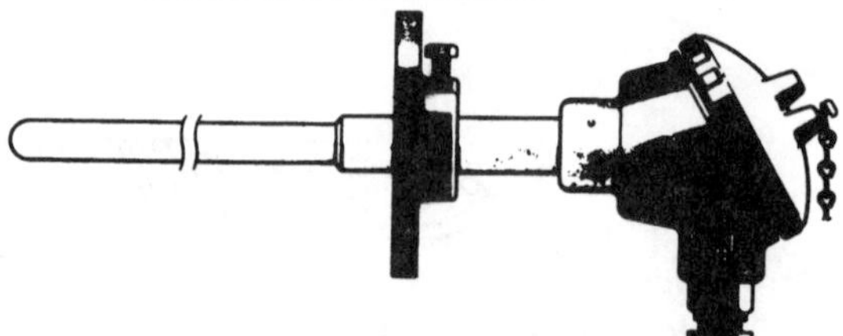

Fig. 2.19 *A selection of thermocouple types (courtesy TC Ltd).*

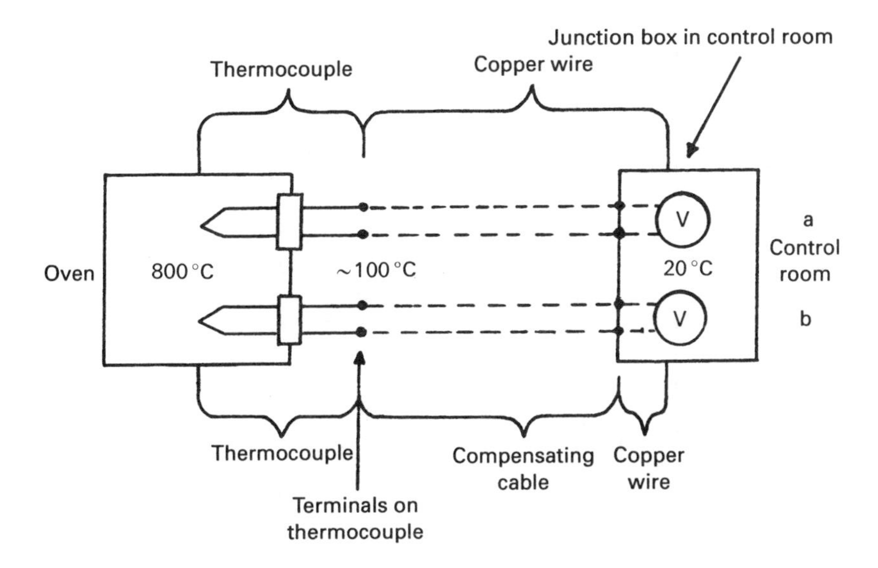

Fig. 2.20 *The use of compensating cable.*

seen is a function of the oven temperature and the control-room temperature. The latter will be around 20 °C, and will be known and relatively stable. Meaningful measurements can now be made.

The connection leads do not have to be made to as tight a standard as the thermocouples, and are known as extension cables. With type R and S thermocouples, however, the cost of extension cables would be prohibitively high and an alternative approach is used. In most applications, the temperature difference between ends of the connection cable is small compared with the temperature being measured (in Fig. 2.20, for example, it is about 80 °C). The cost of connecting cables can be reduced considerably by using metals which match the thermocouple over a *limited* temperature range. Such cables are called compensating cables. Type U on Table 2.1, for example, is a copper/nickel cable manufactured to match types R and S thermocouples up to 50 °C. If the thermocouple terminals are likely to exceed this, an alternative compensating cable must be used, or extension cables taken to an area below 50 °C.

Compensation/extension cables are colour coded according to type. There are different standards in different countries, as shown in Table 2.2, but fortunately there are no ambiguities. A red sheath with brown/blue leads, for example, can only be a type K extension cable to UK standard BS 1843.

It is essential, of course, to maintain the correct polarity from thermocouple, through the compensating/extension cable to the measuring device.

2.5.5. Cold junctions and cold junction compensation

Figure 2.20b allows measurement of the thermocouple temperature, but for accurate results the control-room temperature must be known and some simple

Table 2.2 *Colour coding of thermocouple cables*

Type	*British BS* 1843			*American ANSI*			*German DIN* 43714			*French NFC* 42–323		
	Sheath	+	–	*Sheath*	+	–	*Sheath*	+	–	*Sheath*	+	–
E	Brown	Brown	Blue	Violet	Violet	Red	Black	Red	Black	–	–	–
T	Blue	White	Blue	Blue	Blue	Red	Brown	Red	Brown	Blue	Yellow	Blue
K	Red	Brown	Blue	Yellow	Yellow	Red	Green	Red	Green	Yellow	Yellow	Violet
J	Black	Yellow	Blue	Black	White	Red	Blue	Red	Blue	Black	Yellow	Black
R	Green	White	Blue	Green	Black	Red	–	–	–	–	–	–
S	Green	White	Blue	Green	Black	Red	White	Red	White	Green	Yellow	Green
V*	Red	White	Blue	–	–	–	–	–	–	Red	Yellow	Brown
U*	Green	White	Blue	Green	Black	Red	White	Red	White	Green	Yellow	Green

*Compensating cable

arithmetic must be performed on thermocouple tables. The operation is simplified if the junction of the extension cable/measuring instrument (called the cold junction) is kept at a fixed temperature.

A possible approach (called an ice cell) is to use an ice/water mix to keep the junction at 0 °C. This allows 0 °C referenced tables to be used directly. An ice cell is, however, inconvenient for industrial use.

A more convenient solution, called cold junction compensation (shown in Fig. 2.21), is to measure the temperature of the cold junction by an RTD or thermistor and correct the thermocouple indicated temperature. This can be done by adding a correction temperature to the instrument, as in Fig. 2.21a, or by modifying the thermocouple signal directly, as in Fig. 2.21b.

Where thermocouple signals are to be transmitted over long distances, a combined cold junction compensator/amplifier may be used. A typical device is shown in Fig. 2.22. This unit accepts a millivolt signal from a type R thermocouple and outputs a 4–20 mA current signal corresponding to 0–1400 °C.

2.5.6. Thermocouple temperature indicators

The main problem encountered with thermocouples is the low signal level. Electrical noise on the signal lines will almost certainly be several orders of magnitude higher than the temperature signal itself. The measuring device must therefore have a high input impedance and very high common-mode noise

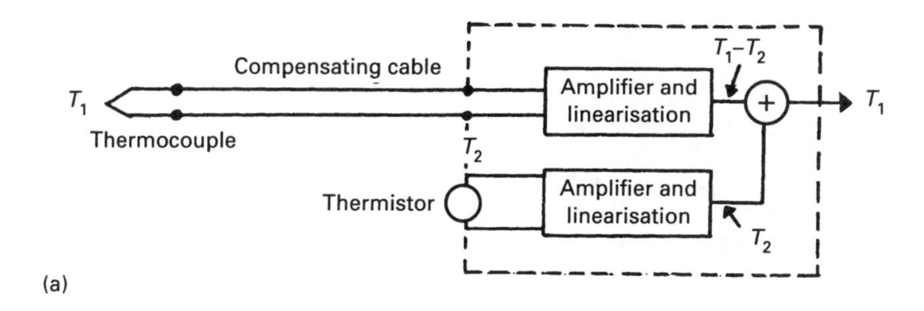

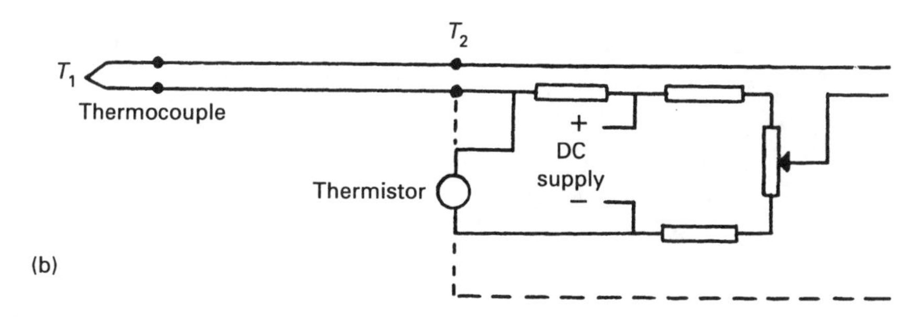

Fig. 2.21 *Cold junction compensation. (a) Addition of cold junction temperature. (b) In-line compensation.*

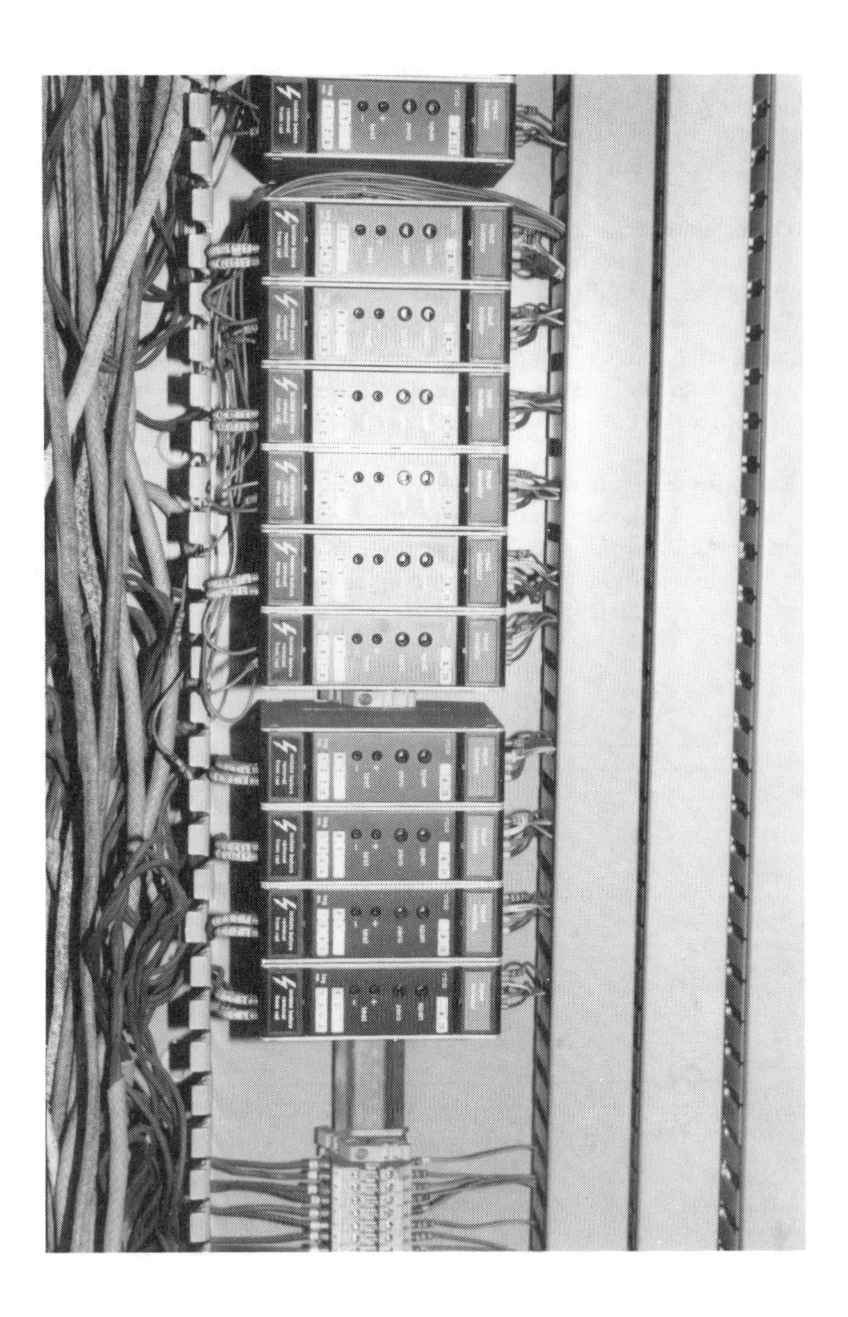

Fig. 2.22 *Combined cold junction compensation and isolation amplifiers. Manufactured by TCS Ltd.*

rejection. Particular care needs to be taken with the cable installation with regard to screening and the avoidance of ground loops.

Equation 2.13 shows that linearisation will be needed in all but the least critical applications. This must be performed in the instrument, either via Op Amp diode break circuits (see Chapter 12) in analog instruments, or via look-up tables and direct computation in digital instruments. In most instruments, the cold junction compensation will be included in the instrument itself.

2.5.7. Using thermocouple tables

Thermocouple tables are provided by the manufacturers. These detail the voltages obtained at various temperatures. Table 2.3 shows an extract from the table for a type R thermocouple. These tables may be used for maintenance and calibration in two circumstances: for checking the voltages from a thermocouple with a millivoltmeter (Fig. 2.23a) or for injecting a test voltage from a millivolt source into a temperature indicator (Fig. 2.23b). In each case, the ambient temperature must be known to apply the law of intermediate temperatures. In the examples below, Type R tables are used.

For Fig. 2.23a, there are four steps:

(1) Measure the ambient temperature (a mercury-in-glass thermometer will suffice). Read the corresponding voltage from the tables. For an ambient of 22 °C, say, the tables give 0.123 mV.
(2) Measure the thermocouple voltage. Let us assume this is 9.64 mV.
(3) Add the ambient voltage: $9.64 + 0.123 = 9.763$ mV.
(4) Find the temperature on the table corresponding to this.

Table 2.3 *Extract from thermocouple tables for type R (platinum–platinum/13% rhodium) (Referenced to cold junction at 0 °C. Voltage in μV)*

	Temperature (°C)									
	0	1	2	3	4	5	6	7	8	9
0:	0	5	11	16	21	27	32	38	43	49
10:	54	60	65	71	77	82	88	94	100	105
20:	111	117	123	129	135	141	147	152	158	165
800:	7949	7961	7973	7986	7998	8010	8023	8035	8047	8060
810:	8072	8085	8097	8109	8122	8134	8146	8159	8171	8184
820:	8196	8208	8221	8233	8246	8258	8271	8283	8295	8308
830:	8320	8333	8345	8358	8370	8383	8395	8408	8420	8433
840:	8445	8458	8470	8483	8495	8508	8520	8533	8545	8558
850:	8570	8583	9595	8608	8621	8633	8646	8658	8671	8683
930:	9589	9602	9614	9627	9640	9653	9666	9679	9692	9705
940:	9718	9731	9744	9757	9770	9783	9796	9809	9822	9835
950:	9848	9861	9874	9887	9900	9913	9926	9939	9952	9965
960:	9978	9991	10004	10017	10030	10043	10056	10069	10082	10095

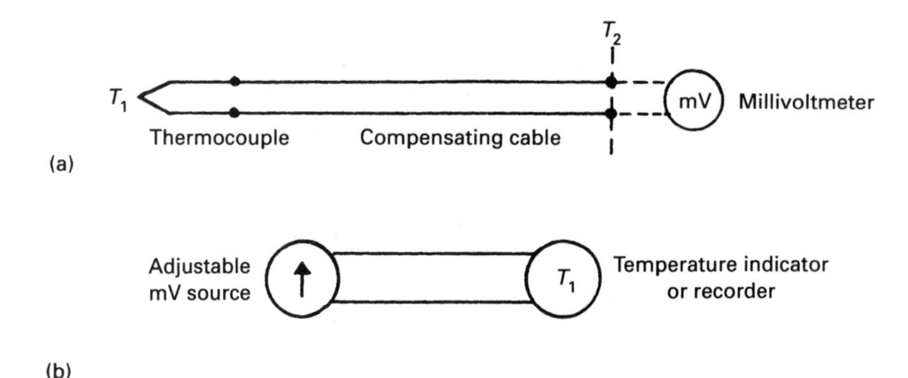

Fig. 2.23 *Using thermocouple tables. (a) Checking thermocouple output. (b) Checking temperature measuring device.*

The tables show 9.757 mV at 943 °C, and 9.770 mV at 944 °C, so our temperature is about 943.5 °C. The precise temperature could be found by interpolation, but 1 °C is sufficiently accurate for most purposes.

Note that reading the temperature for 9.64 mV (934 °C) and adding 22 °C (to give 956 °C) gives an error of 13 °C. The non-linearity of Equation 2.13 requires the above procedure to be followed.

With Fig. 2.23b, the injection voltage required to give a particular temperature indication is to be found.

(1) Measure the ambient temperature and find the corresponding table voltage; say 18 °C, which gives 0.1 mV.
(2) Find the table voltage for the test temperature; say 850 °C, which gives 8.57 mV.
(3) Subtract the ambient voltage from the test temperature to give the required injection voltage (8.57 − 0.1 = 8.47 mV).

Note again that an injection voltage corresponding to 850 °C − 18 °C = 832 °C (8.345 mV) will give an incorrect result. In this case the instrument would display 849 °C, an error of 10 °C.

2.6. Radiation pyrometry

2.6.1. Introduction

When an object is heated, it radiates electromagnetic energy. At low temperatures this radiation can be felt; as the temperature rises it starts to emit visible radiation (i.e. light), passing from red heat, through yellow to white heat. Intuitively, this radiation can be used to measure temperature; qualitatively we can say an object glowing yellow is hotter than an object glowing dull red. Pyrometers use the same radiation to measure temperature.

Pyrometers allow non-contact measurement of temperature, which is essential where the temperature of a moving object is to be measured, or where an environment exists that would destroy a more conventional sensor.

Light, radio waves, X-rays, infrared and ultraviolet are all electromagnetic waves and part of the electromagnetic spectrum (Fig. 2.24). The difference between them is simply their frequency, which ranges from below 10^4 Hz for radio waves, through 10^{14} Hz for visible light, to over 10^{20} Hz for gamma and X-rays. Electromagnetic radiation can also be described by its wavelength, λ, which is related to the frequency, f, by:

$$C = f\lambda \tag{2.15}$$

where C is the speed of light (3×10^{10} cms^{-1}).

Although we are only physically aware of radiated energy when an object achieves a temperature of about 200 °C, *all* objects at a temperature above absolute zero (0 K) emit radiation whose power and frequency distribution are temperature dependent.

2.6.2. Black-body radiation

Any object is constantly receiving energy from its surroundings. Figure 2.25

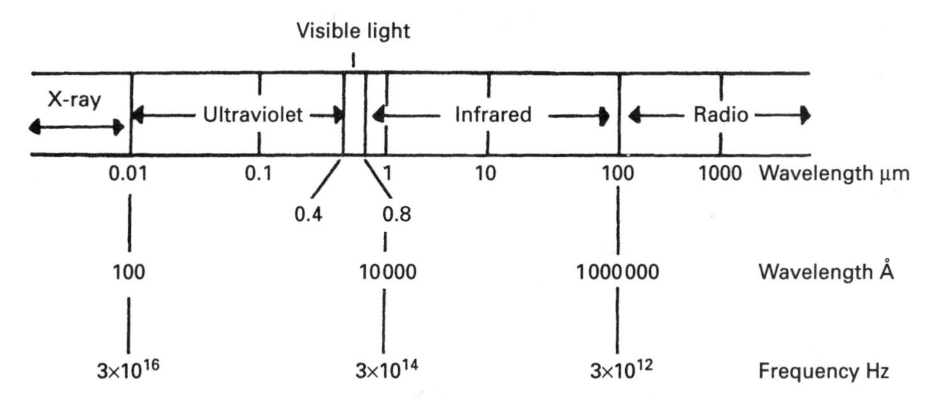

Fig. 2.24 *The electromagnetic spectrum.*

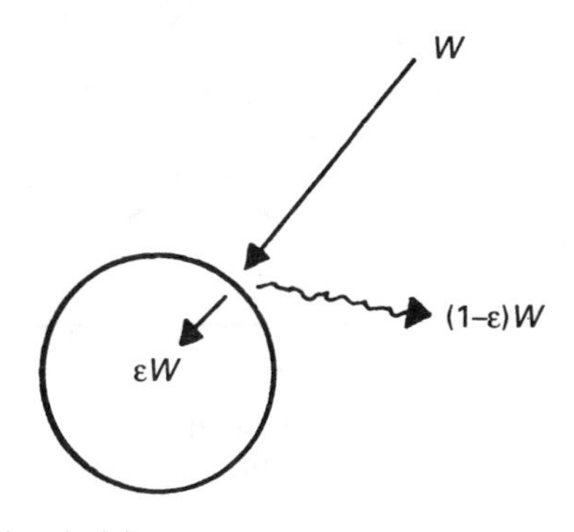

Fig. 2.25 *Definition of emissivity.*

shows an object receiving energy W; of this a proportion εW is absorbed (and a proportion $(1 - \varepsilon)W$ reflected), where ε is defined as the emissivity of the object. If the object absorbs all the incident energy (i.e. $\varepsilon = 1$), it is called a black body.

The term is actually a misleading one, as it implies the object is 'black' in colour. A black body, like all objects, is also an emitter of radiation and when sufficiently raised in temperature will emit visible light. The sun, for example, is an almost perfect black body with a temperature of 6000 K; a hole in the side of an enclosed oven is a good approximation to a black body.

Any object at a steady temperature and not subject to convection or conduction losses must be radiating the same amount of energy as it receives (otherwise there would be a net gain or loss of energy and the temperature would rise or fall). The energy radiated is not at one specific frequency, but is spread over a range of frequencies. The relative radiated power density of a black body is plotted against wavelength for various temperatures in Fig. 2.26. Note that logarithmic scales are used for both power density and wavelength axis. Pyrometry is mainly used in the range 0.3 to 100 μm, which approximately corresponds to the visible light and infrared regions.

Figure 2.26 shows that the total radiated power (which is the area under any curve) increases dramatically with temperature. The total power at any temperature is given by Stefan's law, which states that:

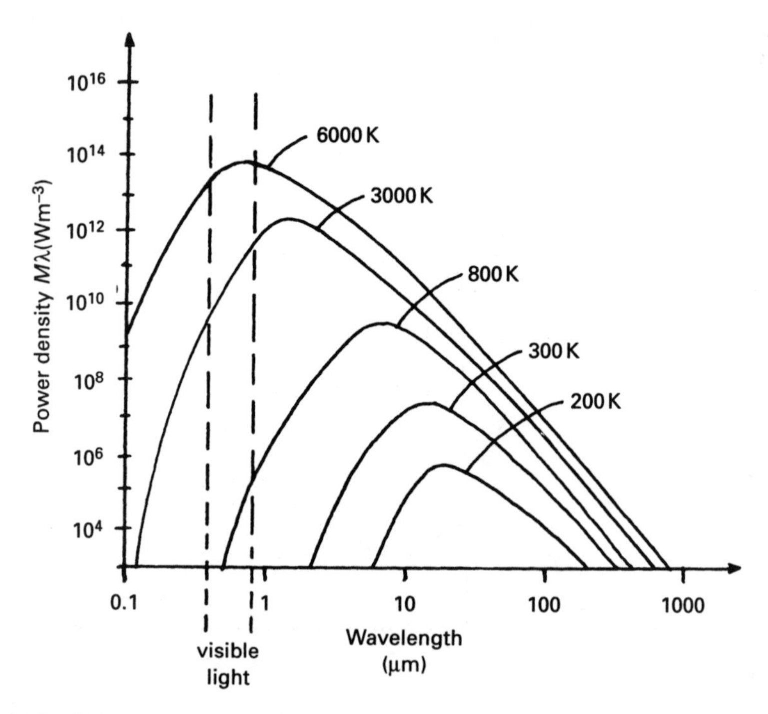

Fig. 2.26 *Relative radiated power density for a black body at different temperatures.*

$$M = \sigma T^4 \text{ W m}^{-2} \tag{2.16}$$

where σ is Stefan's constant (5.67×10^{-8} W m^{-2} K^{-4}) and T is measured in Kelvin.

As well as increasing in magnitude, the curves peak at shorter wavelengths (higher frequencies) for increasing temperature. At around 800 K, there is a significant amount of power radiated in the visible region of the spectrum, and we see the object start to glow.

Very few real-life objects are black bodies: most reflect radiation to some degree. If a real-life object absorbs less radiation than a black body, it must also radiate less (or there will be a net loss of energy and the object's temperature will fall). This is expressed as Kirchhoff's law:

> If a body at a certain temperature T absorbs a certain fraction ε of the radiation of wavelength λ inclined upon it, then it will emit the same fraction ε of the radiation at that wavelength that a black body of the same temperature T would emit.

This can be expressed more succinctly by:

$$\varepsilon = \frac{\text{radiation emitted by an object at temperature } T}{\text{radiation emitted by black body at temperature } T}$$

at any wavelength λ. The value of ε varies with the wavelength of the radiation. Refractory brick, for example, has an emissivity which varies as shown in Fig. 2.27. Obviously the emissivity introduces a large potential error into pyrometry.

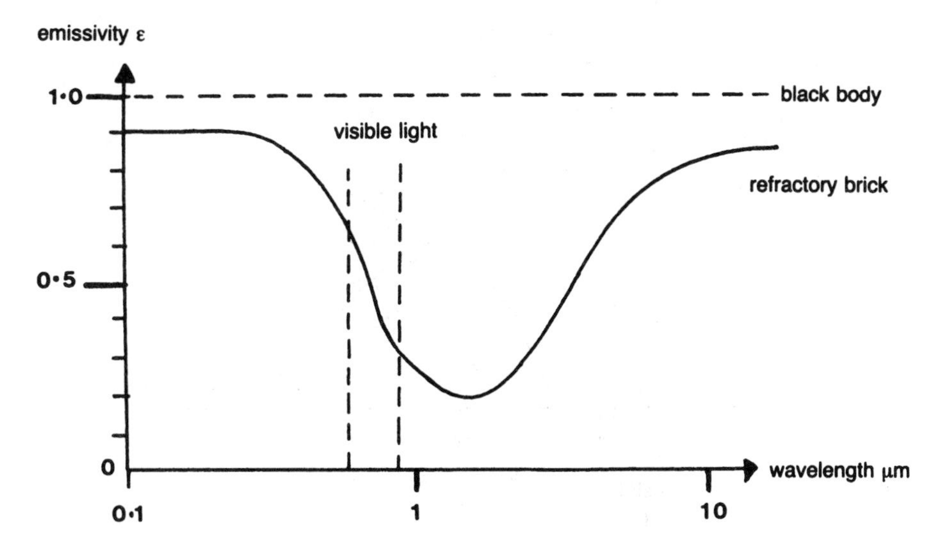

Fig. 2.27 *Emissivity of refractory brick.*

2.6.3. Principles of pyrometry

A pyrometer is, in theory, a very simple device, as shown on Fig. 2.28a. The object whose temperature is to be measured is 'viewed' through a fixed aperture by a temperature measuring device. Part of the radiation emitted by the object falls on the temperature sensor, and causes its temperature to rise. The object's temperature is deduced from the sensor temperature. The temperature sensor itself must be of small thermal mass to give reasonable sensitivity. Usually a circular ring of thermocouples in series (called a thermopile) is arranged as shown in Fig. 2.28b, with a diameter of about 2–10 mm. Alternatively a small resistance thermometer (called a bolometer) may be used.

One of the advantages of pyrometers is that the measurement is independent of the distance from the object (provided the object fills the field of view). Any point on the surface of the object radiates in all directions. The amount of radiated energy received by the sensor will be proportional to the solid angle subtended by the sensor, as shown in Fig. 2.29a. Less energy will be received, from a given point, at position B than at position A. The angle varies inversely as the square of the distance from the object to the sensor.

There is another factor at work, however. As the sensor moves further away, the surface area scanned increases, as shown on Fig. 2.29b, so the total radiation being scanned increases with distance. Simple geometry shows that the area scanned increases as the square of the distance.

These two effects cancel, and the radiation received by the sensor remains constant with distance. In practice, the object must fill the field of view, and absorption of radiation by air causes a slight fall-off with distance.

The design of a practical instrument is shown in Fig. 2.30. This is a hand-held

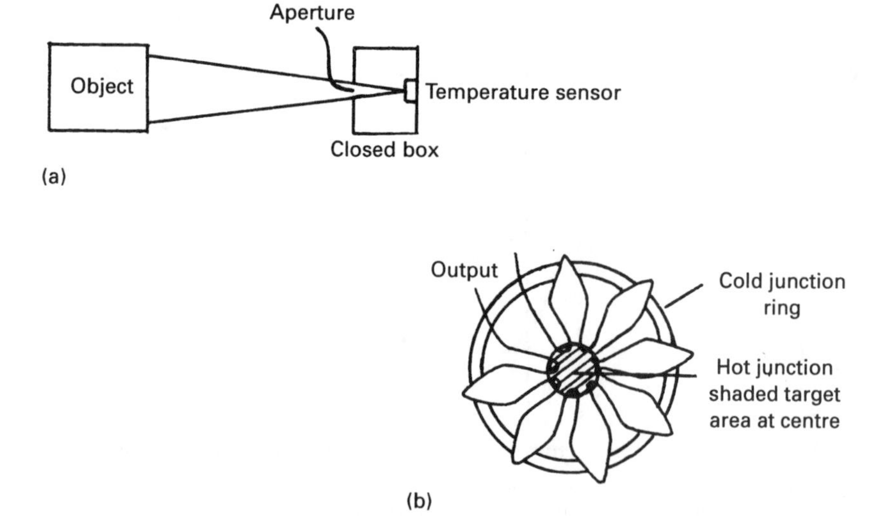

Fig. 2.28 *Optical pyrometer. (a) Principle of optical pyrometry. (b) Thermopile.*

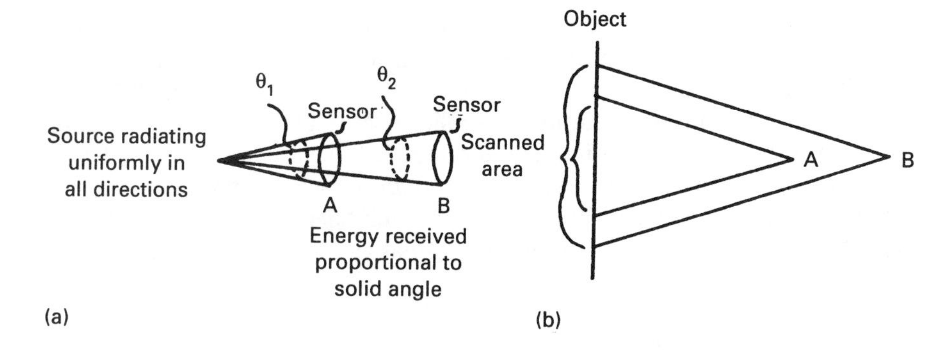

Fig. 2.29 *Pyrometry and distance from sensor to object. (a) Energy received from given point decreases with distance. (b) Scanned area increases with distance.*

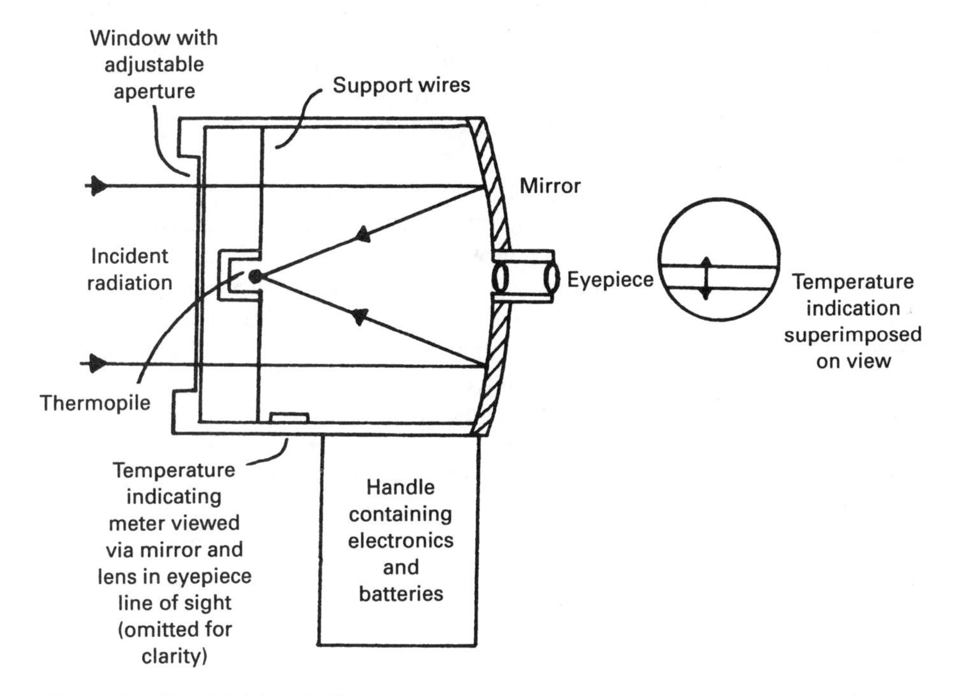

Fig. 2.30 *Hand-held optical pyrometer.*

device, and is aimed at the target object by means of an eyepiece at the rear of the instrument. A concave mirror focuses the radiation on to the thermopile, which is supported on the centreline of the device. The mirror focusing reduces the overall length of the instrument, and increases the sensitivity. The thermopile output is converted to a temperature indication by electronics in the handle. The meter is viewed through the eyepiece, and is superimposed on the field of view by suitable optics.

The output from the thermopile is obviously small and very non-linear. It is also prone to errors due to drifts in the electronic amplifier and linearisation circuit. The latter error can be overcome by the arrangement of Fig. 2.31. The detector is housed in a temperature-controlled oven, and the incoming radiation is chopped by a rotating disc. The detector has a very small thermal time constant, so its temperature rises and falls and produces an alternating voltage as shown. This can be amplified by a drift-free AC amplifier. The resulting high-level signal can then be rectified to give a DC signal and linearised for display.

2.6.4. Types of pyrometer

Although all pyrometers operate on radiated energy, there are various ways in which the temperature is deduced from the radiation absorbed by the sensor. The simplest method is to measure the total power of the radiation, effectively the total area under any curve on Fig. 2.26. The unit of Fig. 2.30 is of this type.

The curves of Fig. 2.26 also show that the power radiated at any frequency also increases with temperature. Monochromatic radiation pyrometers use an optical filter to restrict the measurement to a narrow frequency band. Although this reduces the sensitivity, it allows the choice of a frequency at which the emissivity approaches unity.

Lack of knowledge of emissivity is the major source of error in pyrometry. This can be overcome by a two-colour pyrometer which compares radiation at two frequencies. Examination of Fig. 2.26 will show that the ratio of the power at wavelengths of, say, 3 and 4 μm is temperature dependent. Since both frequencies have similar emissivities, the ratio will be independent of the actual emissivity value. This can be achieved by the arrangement of Fig. 2.32. The incoming radiation is split equally to pass through two filters to two detectors. The outputs of the two detectors are ratioed electronically.

An optical pyrometer, shown in Fig. 2.33a, does not use a temperature detector as such. The target object is viewed through a telescope optical system, and a hot wire filament is superimposed on the field of view. The current through the filament, and hence its temperature, can be controlled by the operator. The

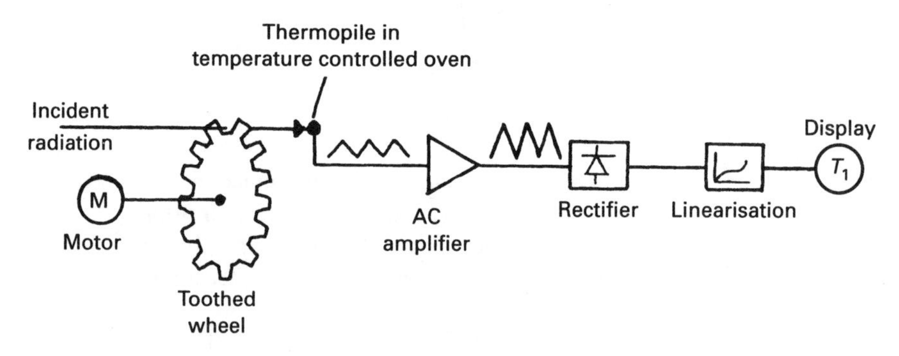

Fig. 2.31 *Chopper pyrometer.*

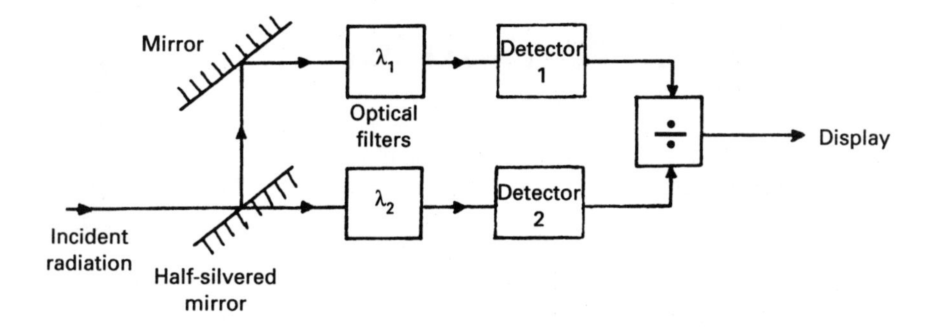

Fig. 2.32 *Overcoming unknown emissivity by measurements at different wavelengths.*

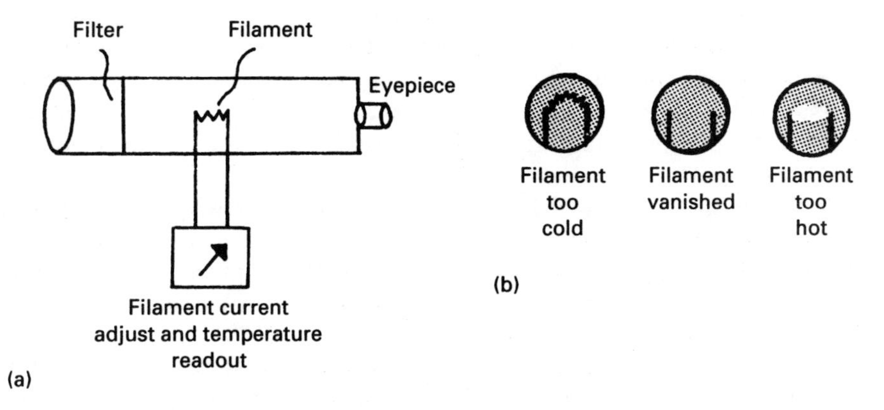

Fig. 2.33 *Disappearing filament pyrometer. (a) Principle of instrument. (b) View through eyepiece.*

current is adjusted until the filament merges into the background, as shown in Fig. 2.33b. At this point the brightness of the filament is the same as the target object, and the temperature can be deduced from the filament current.

Chapter 3

Pressure transducers

3.1. Introduction

3.1.1. Definition of terms

Pressure is a very loose and general term. It is scientifically defined as the force per unit area that a fluid exerts on the walls of the container that holds it. For measurement purposes, however, it is necessary to define the conditions under which the measurement is made.

All pure pressure transducers are concerned with the measurement of *static pressure*, i.e. the pressure of a fluid at rest. If the fluid is in motion, its pressure will depend on its flow velocity and is termed the *dynamic pressure*. Section 5.2 describes flowmeters which use this principle. Although an orifice/differential pressure transmitter may be used to measure flow from the fluid's dynamic pressure, the pressure transmitter itself is measuring the static pressure difference between the two pipes from the orifice tapping.

Differential pressure (Fig. 3.1.a) is the difference between two pressures applied to the transducer. Differential pressure transducers (commonly called Delta P transmitters) are widely used in flow measurement. Conceptually, the differential pressure transducer can be considered the 'basic' pressure measurement device, other types being variations.

If, for example, one pressure input to a differential pressure transducer is left open, the device indicates *gauge pressure* (Fig. 3.1b), which is pressure related to atmospheric pressure. Atmospheric pressure can vary, however, so if a more accurate measurement is required, the pressure measurement can be made with respect to a fixed pressure in a sealed chamber. Such a device is called a *sealed gauge pressure transducer*. In UK engineering terms, the suffix 'g' is added to indicate gauge pressure, e.g. 37 p.s.i.g. *Absolute pressure* (Fig. 3.1c) is measured with respect to a vacuum. This is effectively a differential pressure measurement with one pressure at zero. *Head pressure* (fig. 3.1d) is a term used in liquid level

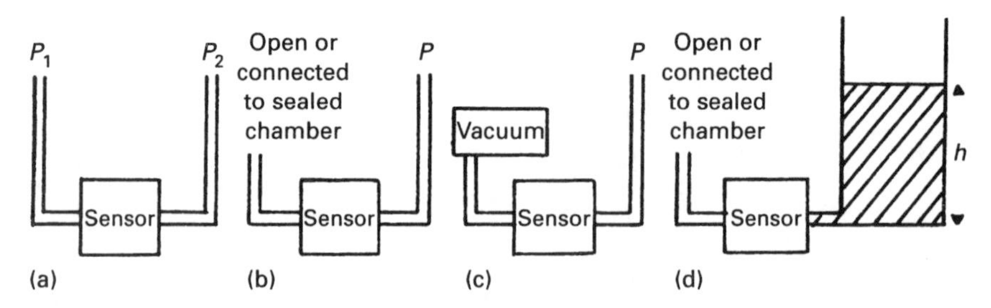

Fig. 3.1 *Definitions of 'pressure'. (a) Differential. (b) Gauge. (c) Absolute. (d) Head.*

measurement, and refers to pressure in terms of the height of the liquid (e.g. inches water gauge, millimetres mercury). It is effectively a gauge pressure measurement, but changes in atmospheric pressure affect both the liquid and the pressure-measuring devices equally, provided the liquid is held in a vented vessel.

Vacuum measurement is a form of pressure measurement, but one requiring special devices and techniques. Vacuum transducers are therefore described separately in Section 3.8.

3.1.2. Units

There is a wide and confusing range of units for pressure measurement. Scientifically, pressure is force per unit area. This is described in the SI system as newtons per square metre ($N\,m^{-2}$), which has been named the pascal (Pa). This is not a widely used unit, the kilopascal (kPa) or the 'non-standard' $N\,cm^{-2}$ being more convenient. In the UK, pounds per square inch (p.s.i.) is widely used. In many applications, the *atmosphere* (14.7 p.s.i.) is used, or the *bar* (100 kPa), which approximates to an atmosphere. Measurements in inches water gauge (WG) or millimetres mercury are also common. Table 3.1. gives conversions between common units.

Table 3.1 *Common pressure conversions.*

1 p.s.i (lbf/in^2)	=6.895 kPa
	=27.7 inches WG
1 lbf/ft^2	=47.88 Pa
1 kgf/cm^2	=98.07 kPa
1 inch WG	=249.0 Pa
	=5.2 lbf/ft^2
	=0.036 p.s.i.
1 foot WG	=2.989 kPa
	=62.43 lbf/ft^2
	=0.433 p.s.i.
1 torr (mmHg)	=133.3 Pa
1 bar	=100 kPa
	=14.5 p.s.i.
	=750 mmHg
	=401.8 inches WG
	=1.0197 kgf/cm^2
1 atmosphere	=1.013 bar
	=14.7 p.s.i.
1 kilopascal	=0.145 p.s.i.
	=20.89 lbf/ft^2
	$=1.0197 \times 10^{-3}$ kgf/cm^2
	=4.141 inches WG
	=7.502 torr (mmHg)
	=0.01 bar
	$=9.872 \times 10^{-3}$ atm

3.2. U-tube manometers

Although manometers are not widely used nowadays in industry, they give a useful insight into the principle of pressure measurement. They are also used as a standard against which other devices may be calibrated.

If a U-tube is filled with a liquid (commonly water, alcohol or mercury, depending on the pressure to be measured), the liquids will naturally adopt the same height in both tubes, as shown in Fig. 3.2a. If a pressure is applied to each leg, as shown in Fig. 3.2b, the liquid level will fall on the high-pressure side and rise on the low-pressure side to give a height difference, h.

The head pressure of a column of liquid is given by:

$$p = \rho g h \tag{3.1}$$

where p is the pressure (Pa), ρ is the density, g is the acceleration due to gravity (9.8 m s^{-2}), and h is the column height in metres.

In the UK, weight density is used, so equation 3.1 becomes

$$p = \rho h \tag{3.2}$$

where p is in pounds per square foot, ρ is in pounds per cubic foot and h is in feet.

In Fig. 3.2b, the pressures in both tubes must balance, so

$$P_1 = P_2 + \rho g h \tag{3.3}$$

or

$$h = (P_1 - P_2)/\rho g \tag{3.4}$$

i.e. the difference in column heights is proportional to the differential pressure.

Manometers are usually scaled with a datum at zero, and 'half scaled' divisions above and below the datum (e.g. a device measuring inches water gauge would have divisions spaced 0.5 in apart, but labelled 1 in, 2 in, etc.). This allows the pressure to be read off one tube.

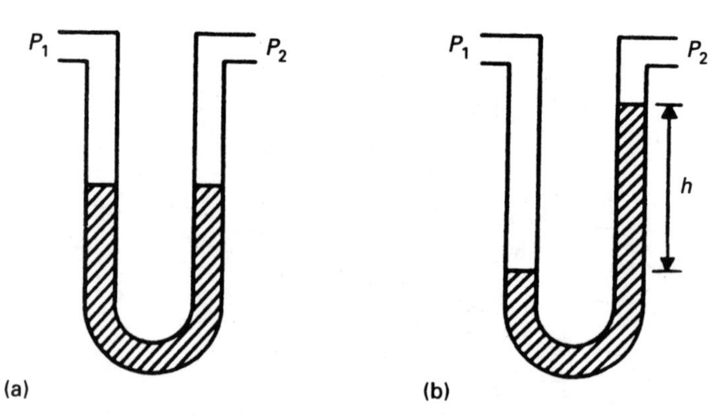

Fig. 3.2 *The U-tube manometer. (a) $P_1 = P_2$. (b) $P_1 > P_2$.*

Scaled manometers can be constructed by having unequal areas in the two tubes, as in Fig. 3.3. The differential height H still obeys equation 3.4, but the movements from the zero (d and h as shown) are not equal. If A_1, A_2 are the areas of the tubes, the volume displacements in each tube will be equal, so

$$A_1 h = A_2 d \tag{3.5}$$

or

$$h = \frac{A_2}{A_1} d \tag{3.6}$$

Since $h + d = H$, substituting in equation 3.4 gives:

$$d = \frac{P_1 - P_2}{\rho g(1 + A_2/A_1)} \tag{3.7}$$

The scaling of the manometer can be adjusted by varying A_2/A_1. Sometimes 'range tubes' are provided with manometers which are screw-in tubes of different sizes to allow one area to be altered.

An inclined tube manometer (Fig. 3.4) gives increased sensitivity for low-pressure

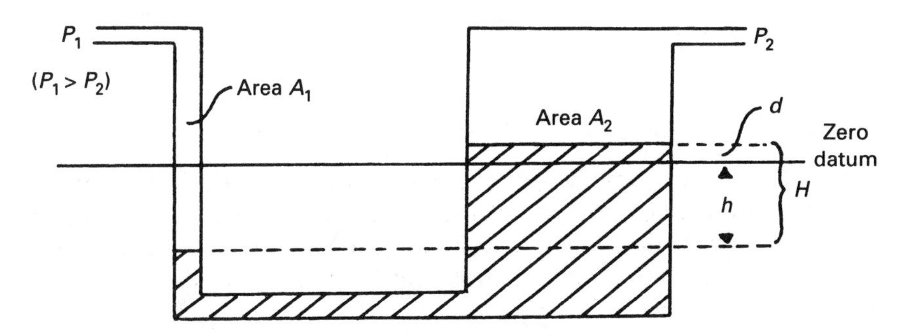

Fig. 3.3 *Scaled (unequal area) manometer.*

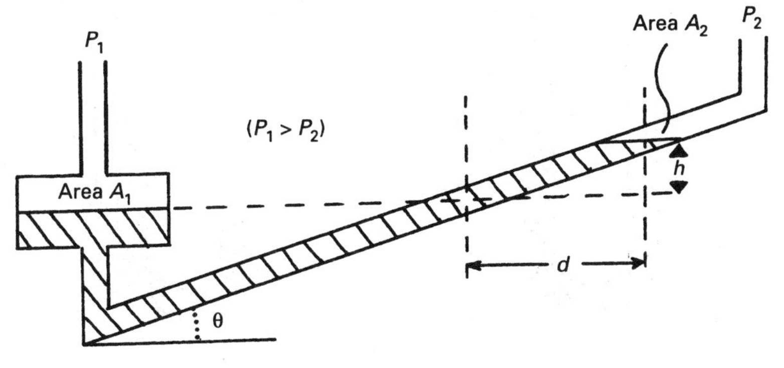

Fig. 3.4 *Inclined tube manometer.*

measurements. The pressure is read off an inclined measurement scale from the distance d. If the area of the tube A_2 is considerably less than the area of the reservoir, h is given by equation 3.4 and d by simple trigonometry:

$$d = \frac{(P_1 - P_2)}{\rho g \tan \theta} \tag{3.8}$$

3.3. Elastic sensing elements

3.3.1. The Bourdon tube

The Bourdon tube, dating from the mid nineteenth century, is still the commonest pressure measuring device where remote indication over a long distance is required or where very high or low pressures are to be measured. The tube itself is manufactured by flattening a circular-cross-section tube to the section shown in Fig. 3.5, and bending to a C shape. One end is fixed, left open and connected to the pressure to be measured. The other is closed and left free.

If a pressure is now applied to the inside of the tube, it will tend to straighten, causing the free end of the tube to move up and to the right. This movement is converted to a circular pointer movement by a quadrant-and-pinion mechanical linkage. The movement depends on the pressure difference between the inside

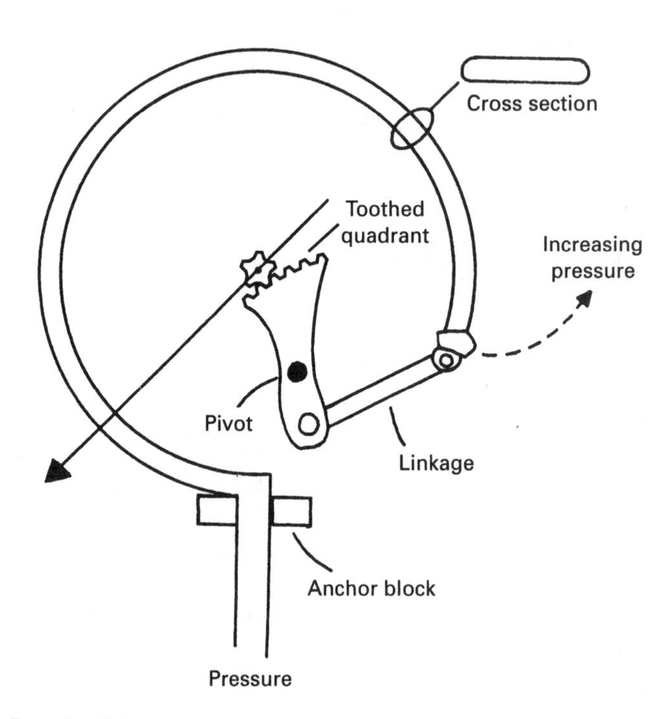

Fig. 3.5 *Bourdon tube.*

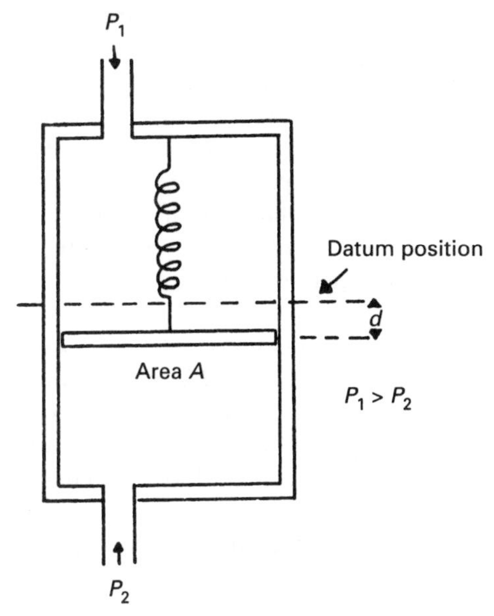

Fig. 3.6 *Differential pressure transducer.*

and outside of the tube, so the Bourdon tube inherently measures gauge pressure.

Bourdon tubes are usable from 0–5 p.s.i. (0–30 kPa) to about 0–10 000 p.s.i. (0–50 MPa approx.). The lower ranges employ a spiral tube to increase the sensitivity.

Where an electrical output is required, the Bourdon tube can be coupled to a potentiometer giving an electrical output that varies with applied pressure between 0 and 100% of the applied voltage.

3.3.2. Various sensing elements

The Bourdon tube is only one example of an elastic sensing element. Pressure transducers using this principle convert a differential pressure to a displacement which can be measured by any of the methods in Sections 3.3.3. and 3.3.4.

Figure 3.6 is a schematic representation of a differential pressure transducer. Input pressures P_1, P_2 ($P_1 > P_2$) are applied to a cylinder containing a movable plate of area A. A force is produced on the piston of $A.(P_1 - P_2)$. The plate is restrained by a spring of stiffness K, so the plate moves until the force exerted by the spring equals the force caused by the pressure differential, i.e.

$$Kd = A.(P_1 - P_2) \tag{3.9}$$

or

$$d = \frac{A.(P_1 - P_2)}{K} \tag{3.10}$$

The displacement is proportional to the pressure differential.

The arrangement of Fig. 3.6 is obviously impractical (the plate edges would be impossible to seal without excessive friction, for example). Figure 3.7 shows common practical elements; all of these convert a pressure differential to a linear displacement. The designs provide a seal between the low- and high-pressure sides, and are constructed to minimise frictional effects and give a linear pressure-to-displacement relationship.

The linear movement can be amplified mechanically to drive a pointer (the common domestic aneroid barometer, for example, uses the capsule of Fig. 3.7d) or converted to an electrical signal as described below.

3.3.3. Electrical displacement sensors

The displacement of all the devices in Fig. 3.7 is small: at most a few millimetres. For remote indication this displacement must be converted to an electrical signal (pneumatic force balance transducers are described in Section 3.5.2). The devices described below are all position measuring devices, and are dealt with in more detail in Section 4.5.

Strain gauges are often used with the simple diaphragm of Fig. 3.7a where the

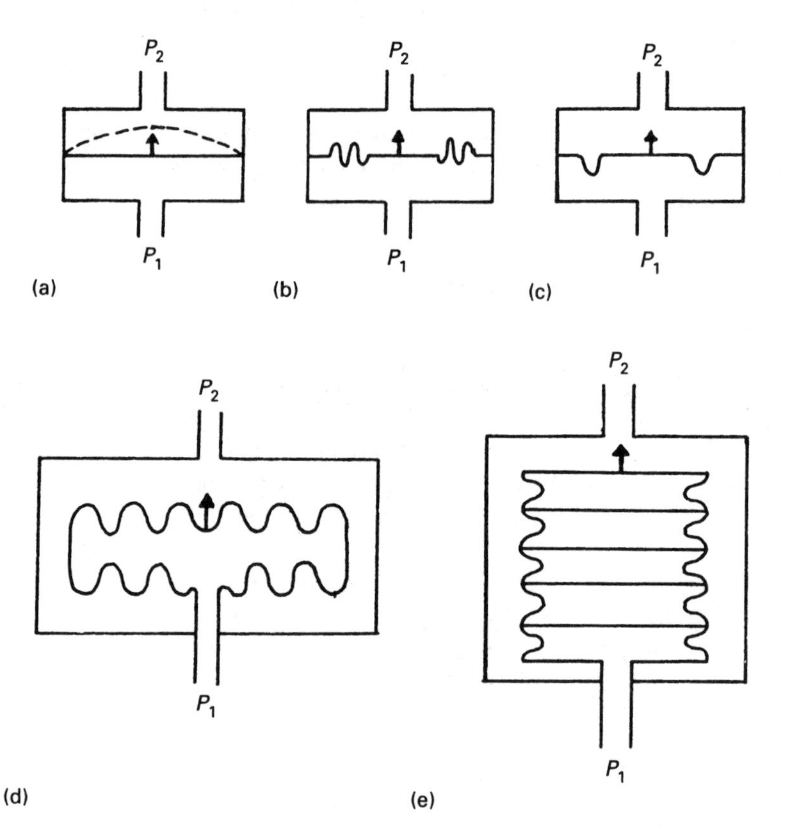

Fig. 3.7 *Elastic pressure sensing elements. (a) Diaphragm. (b) Corrugated diaphragm. (c) Catenary diaphragm. (d) Capsule. (e) Bellows.*

displacement is usually small. Care must be taken to provide correct temperature compensation. Often the strain gauge(s) is (are) simply fixed direct to the surface of the diaphragm itself.

Figure 3.8 shows various arrangements of *capacitive sensors*. In Fig. 3.8a the diaphragm acts as one plate of a capacitor. As the diaphragm moves with respect to the fixed plate, the change in plate separation causes a change in capacitance. Figure 3.8b uses the same principle, but the movable plate moves between two fixed plates, causing one capacitance to increase and one to decrease. This increases the sensitivity, and is convenient for use with an AC bridge circuit. Figure 3.8c uses the displacement to move one plate at a constant distance from the fixed plate. The change in area causes a change in capacitance.

The *LVDT (linear variable differential transformer)* of Fig. 3.9 uses the displacement from the elastic sensing element to move a ferromagnetic core in a differential transformer. The displacement alters the coupling between the primary (fed typically at a few kilohertz) and the two secondaries, causing the voltages E_1, E_2 to differ. The output voltage magnitude $(E_1 - E_2)$ and phase therefore depend on the core movement. A phase-sensitive rectifier gives a DC output signal proportional to the core displacement from the centre (balanced) position.

The elastic sensing displacement in Fig. 3.10 moves a ferromagnetic plate in front on a coil wound on an E core. Movement of the plate causes a change in the inductance of the coil, which can be measured by an AC bridge. This arrangement is called a *variable reluctance* or *variable inductance transducer*.

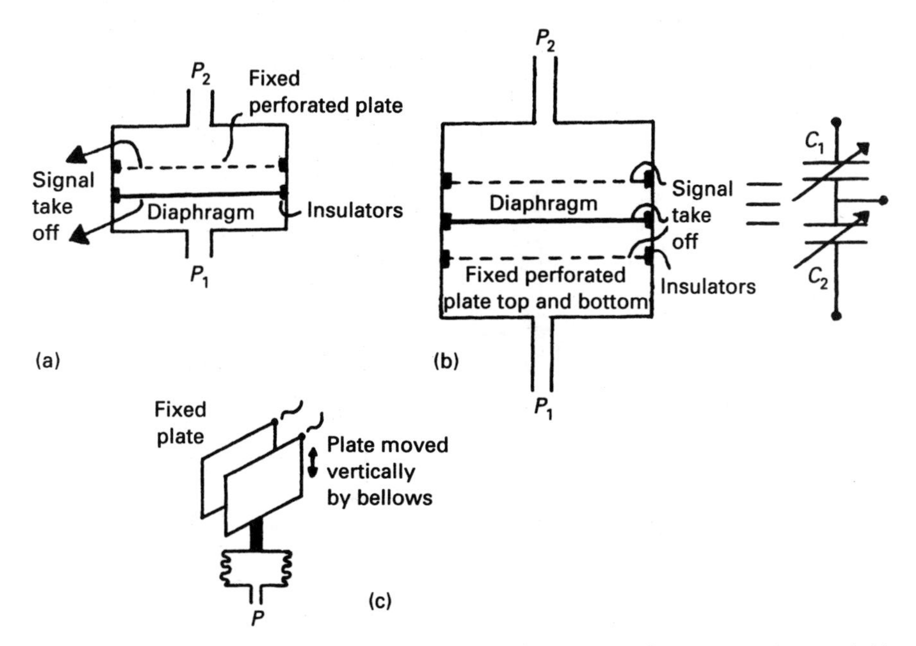

Fig. 3.8 *Variable capacitor pressure sensing elements. (a) Single capacitor, variable spacing. (b) Double capacitor, variable spacing. (c) Variable area.*

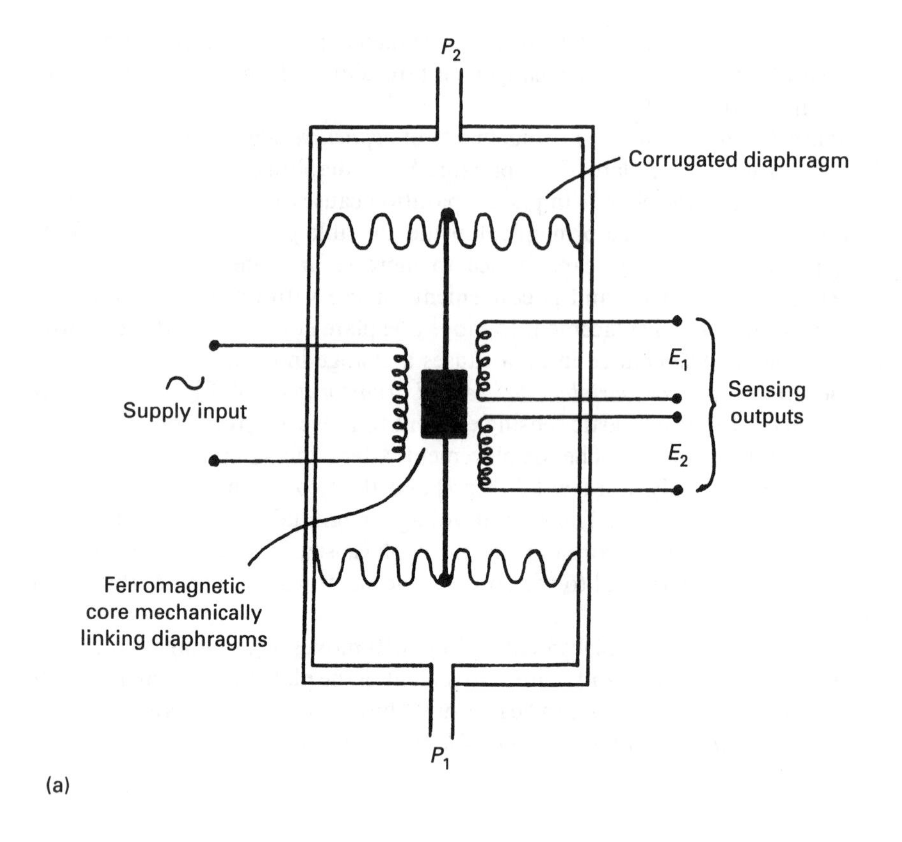

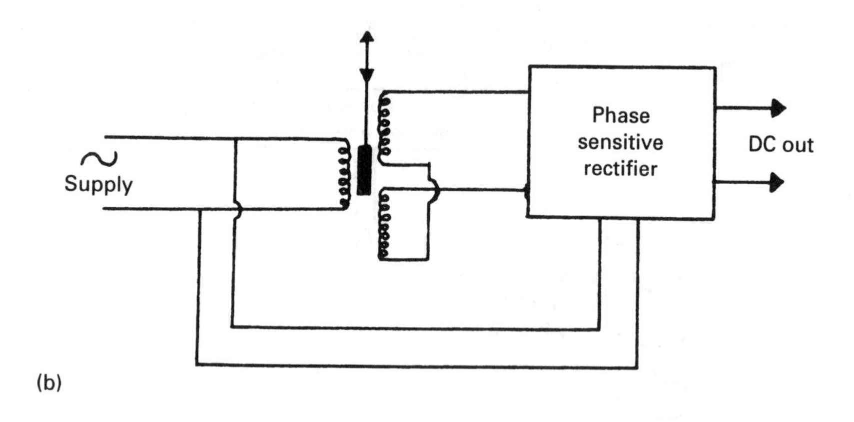

Fig. 3.9 *LVDT pressure transducer. (a) Mechanical arrangement. (b) Electrical circuit.*

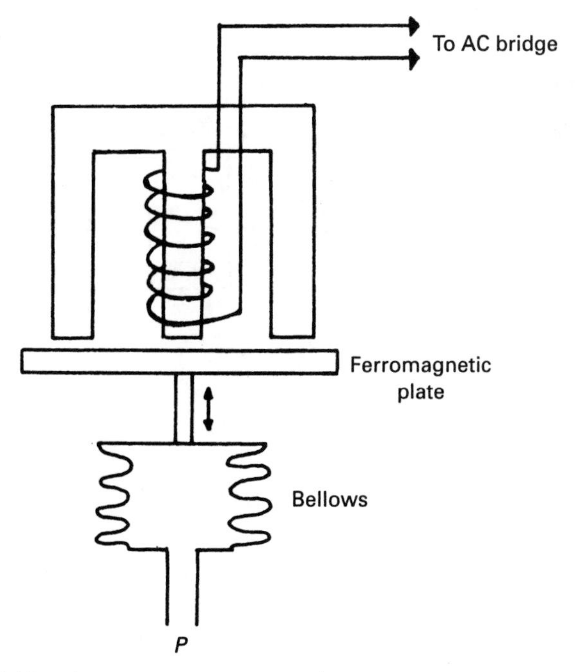

Fig. 3.10 *Variable reluctance pressure transducer.*

3.4. Piezo elements

The piezo-electric effect occurs in quartz crystals. When a suitably prepared crystal has a force applied to it (Fig. 3.11a), electrical charges of opposite polarity appear on the faces. The charge is proportional to the applied force.

To be of use, an output voltage must be produced from the charge. This is provided by the circuit of Fig 3.11b, which is called a charge amplifier. C_s is the stray capacitance of the connecting cable and the crystal itself. Pressure changes across the diaphragm cause a charge q across the crystal:

$$I_1 = \frac{dq}{dt} \tag{3.11}$$

$$I_2 = -C\frac{dV}{dt} \tag{3.12}$$

These are equal, so:

$$\frac{dq}{dt} = -C\frac{dV}{dt} \tag{3.13}$$

or

$$q = -CV + K \tag{3.14}$$

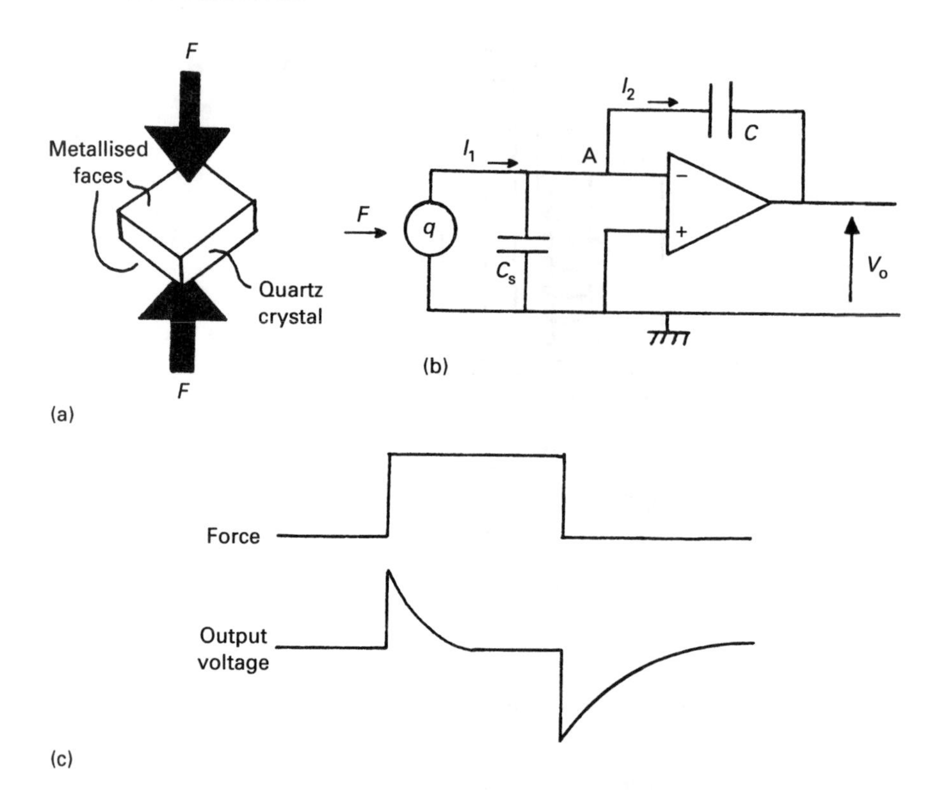

Fig. 3.11 *Piezo-electric force transducer. (a) Piezo-electric effect. (b) Charge amplifier. (c) Low-frequency response of charge amplifier.*

Where K is a constant of integration which can be nulled out, giving:

$$V = -\frac{q}{C} \tag{3.15}$$

Since q is proportional to the applied force, V is proportional to the pressure difference across the diaphragm.

In practice, the charge leaks away due to the non-infinite input impedance of the amplifier, giving responses similar to Fig. 3.11c even with a FET amplifier. Piezo-electric transducers are therefore unsuitable for measuring *static* pressures. They do, however, have a fast response and are widely used for measuring fast *dynamic* pressure changes. A piezo-electric transducer, for example, can easily follow the pressure variations inside a car-engine cylinder head.

Suitably prepared crystals also exhibit a change in resistance with applied force: the piezo-resistive effect. Using this effect it is possible to construct a full Wheatstone bridge on a single chip, as Fig. 3.12. Using a self-nulling balance circuit, an output current proportional to the applied force (and hence the differential pressure) is produced. Unlike the piezo-electric sensor, the piezo-resistive device can measure static pressure. A typical device is shown in Fig. 3.13.

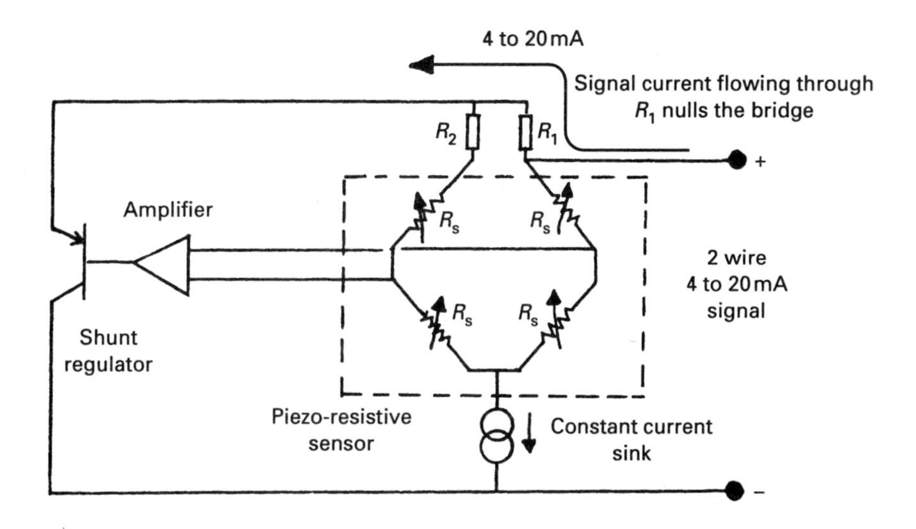

Fig. 3.12 *Two-wire piezo-resistive transducer.*

Fig. 3.13 *Piezo-resistive differential pressure transducer. Compare the mechanical simplicity with Fig. 3.15.*

Both the piezo-electric and piezo-resistive devices are essentially *force* measuring sensors and need minimal displacement of the elastic sensing elements. This obviates the need for complex linkages and gives a more linear response as the output does not depend on the movement or elastic sensitivity of the mechanical sensing element.

3.5. Force balancing systems

3.5.1. Electrical systems

Friction and non-linear spring constants can cause errors in the elastic sensor/displacement transducers of Section 3.3. The force balance principle provides a force to return the elastic sensor to a 'datum' position. The measurement of the force then indicates the pressure; providing a signal which is independent of the elastic characteristics of the sensor itself.

An electrical force balance transducer is shown in Fig. 3.14. A differential pressure is applied to a conventional diaphragm, causing a movement of arm X which in turn moves arm Y. A displacement transducer (capacitive, inductive or LVDT) senses this movement. A servo amplifier applies a current to the solenoid at the other end of arm Y. The current is automatically adjusted until arm Y (and hence arm X and the diaphragm) are returned to the datum position.

The force exerted by the solenoid now balances the differential pressure force

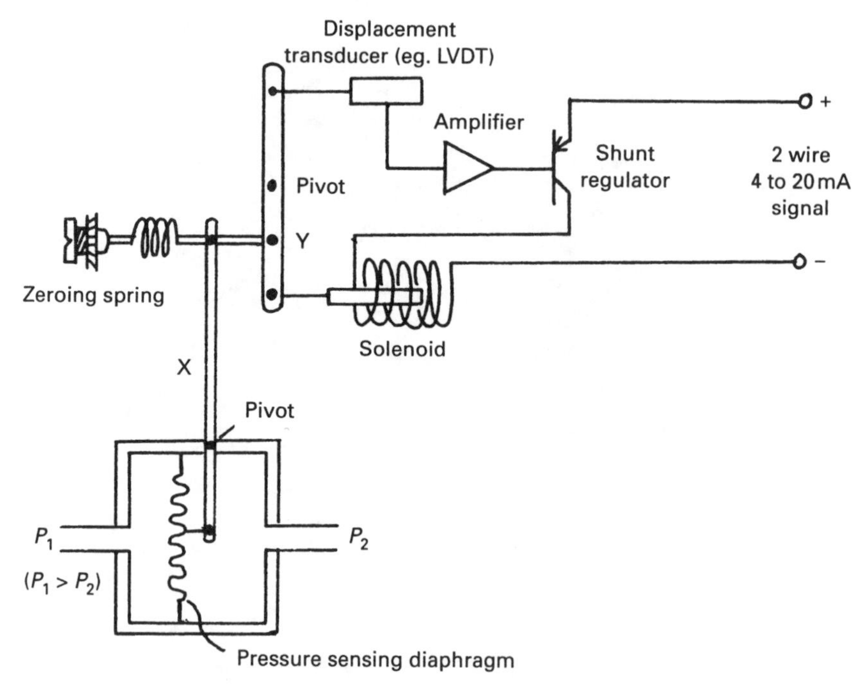

Fig. 3.14 *Force balance pressure transducer.*

on the diaphragm. The solenoid force is directly proportional to the coil current, so an ammeter in the coil can be used to indicate differential pressure. A transducer using this principle is shown in Fig. 3.15.

3.5.2. Pneumatic systems

A pneumatic force balance system is based on the ubiquitous flapper/nozzle (described in Chapter 15) and is shown schematically in Fig. 3.16. As before, the pressure differential initially moves the diaphragm, which in turn moves arms X and Y. As the flapper moves, say, towards the nozzle, the air loss through the nozzle decreases, causing the pressure in the bellows (and the output pressure) to increase.

The increased pressure on the bellows causes arm Y, and hence arm X and the diaphragm, to move back to the datum position. The output pressure is thus proportional to the pressure differential across the diaphragm. Like the electrical arrangement of Section 3.5.1, the output pressure does not depend on the elastic constant of the diaphragm.

3.6. Pressure transducer specifications

The specification sheet for a pressure transducer will contain the usual details (hysteresis, error, linearity, etc.) plus other terminology that is unique to pressure measurement. A pressure transducer may be required to measure a differential pressure superimposed on a large pressure: flow measurement with an orifice plate in a high-pressure line, for example. The pressure that any transducer connection can stand with *respect to atmosphere* is called the *proof pressure* or the static pressure. A typical differential pressure transmitter may have a range of 0–0.1 bar and a proof pressure of 100 bar: 1000 times the range.

Overpressure is referenced to the actual measurement range, and is usually

Fig. 3.15 *Force balance differential pressure transducer.*

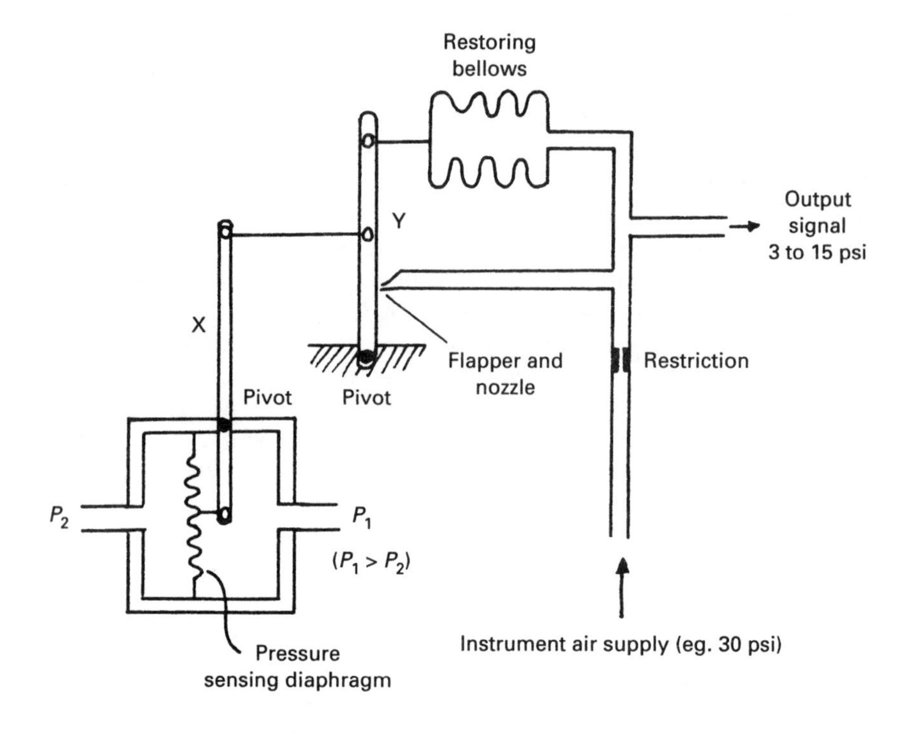

Fig. 3.16 *Pneumatic force balance transducer.*

defined in three ways. First, and lowest, there is the overpressure which does not cause a zero or span shift. Next comes the overpressure which causes a zero or span shift. Finally, the overpressure which causes permanent damage, usually considerably higher. Devices can usually withstand a short pressure pulse, so often a dynamic overpressure is also specified, which is a high-pressure pulse (greater than the static permanent damage pressure) specified for some maximum time. Figure 3.17 shows a common overload protection arrangement where a movable plug seals a bleed hole between HP and LP diaphragms.

Most pressure transducers employ elastic sensors, which inherently have a natural frequency and a tendency to give damped oscillations. The specifications for a device will give the natural frequency (typically 0.1–3 Hz) and the damping factor (usually adjustable). The factors determine the response to fast pressure changes (although the user should also remember the slugging effect of long pipe runs from plant to transducer).

3.7. Installation notes

If a long pipe run is used between plant and a pressure transducer, a significant volume of fluid will be moved in the pipes as the pressure changes. Time constants of several seconds can result, so piping runs should be kept short.

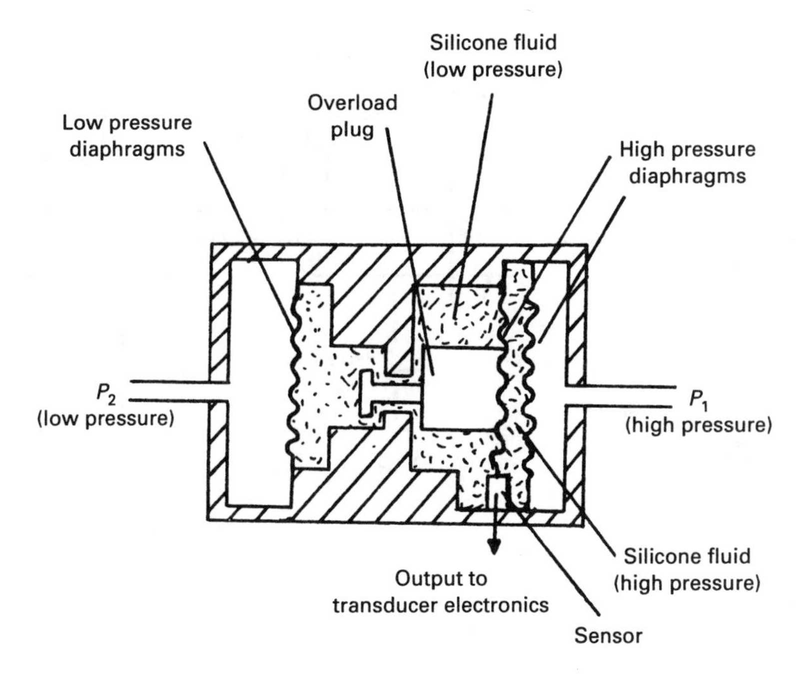

Fig. 3.17 *Overload protection on differential pressure transducer.*

Air or gas entrapped in piping can cause significant errors on liquid pressure measurement applications. Similarly, condensed moisture can form in piping on gas pressure systems. Ideally piping should rise from the pipe to the transducer on gas systems, and fall directly from the pipe to the transducer on liquid systems. Where this is not feasible, vent and drain cocks should be fitted, as in Fig. 3.18. Under no circumstances should the piping form traps where air bubbles or liquid sumps can form (as in Fig. 3.18c).

When a pressure transducer is situated below a pipe and liquid pressure measurement is being made, an error will be introduced from the static head of liquid in the signal pipe. The pressure error is ρgh, as shown in Fig. 3.19. If this amounts to less than 10% of the pressure range being measured, it can usually be removed by zero adjustment of the transducer.

Pressure transducers will inevitably need to be changed on line, so isolation valves should be fitted. Figure 5.14 shows a typical arrangement for a differential transmitter. Where a low pressure range device is used with a high static pressure, care must be taken not to damage the device by overpressure caused by pipe pressure getting 'locked in' to one side. The bypass valve should always to opened first and shut last.

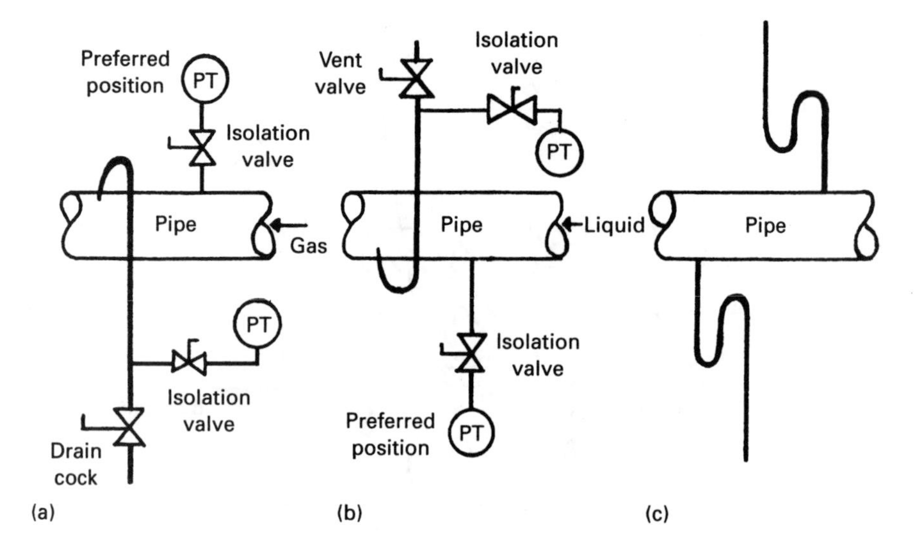

Fig. 3.18 *Installation of pressure transmitters. (a) Gas pressure measurement. (b) Liquid pressure measurement. (c) Faulty piping with potential traps.*

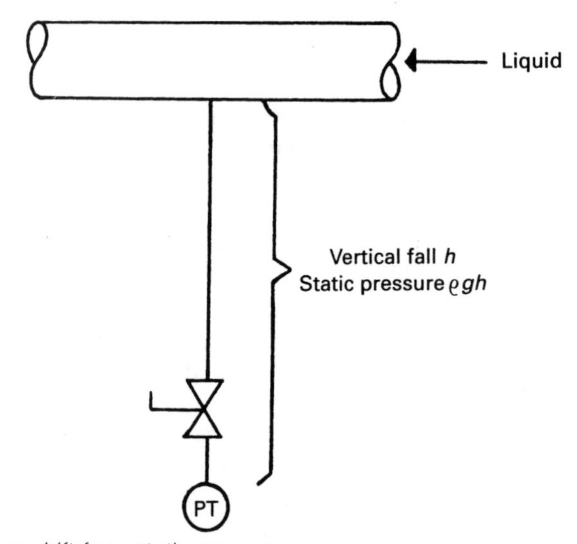

Fig. 3.19 *Zero shift from static pressure.*

3.8. Vacuum measurement

3.8.1. Introduction

Vacuum measurement is normally made in terms of the height of a column of mercury supported by the vacuum (mmHg). Atmospheric pressure therefore corresponds to about 760 mmHg, and 'absolute' vacuum to 0 mmHg. The term 'torr' is generally given to 1 mmHg. .

Conventional absolute pressure transducers are usable without difficulty down to 20 torr, and to 1 torr with special diaphragms. At lower pressures, the scales should be considered logarithmic, i.e. the range 1–0.1 torr equates in range to 0.1–0.01 torr. Measurement down to below 10^{-7} torr is sometimes required.

3.8.2. Pirani gauge

Heat is lost from a hot body by conduction, convection and radiation. If a hot wire is surrounded by a gas, the first two losses are pressure dependent because the heat transfer depends on the number of gas molecules per unit volume.

In the Pirani gauge, constant energy is supplied to a wire in the vacuum to be measured. As the level of vacuum varies, the differing heat losses cause the temperature of the wire to change. In Fig. 3.20a, the temperature of the wire is sensed by a thermocouple; the voltage indicating the pressure. Alternatively the temperature change can be sensed by the change in resistance of the wire with temperature, as in Fig. 3.20b, or by using a self-balancing bridge.

The range of the Pirani gauge can easily be changed by altering the energy supplied to the wire. A typical instrument would cover the range $5–10^{-5}$ torr in three ranges. Care must be taken not to overheat the wire, which causes a change in characteristics due to surface oxidation or even wire breakage.

3.8.3. Ionisation gauge

The ionisation gauge (shown in Fig. 3.21) is superficially similar to an electronic triode. Electrons emitted from the heated filament cause ionisation current to flow which is proportional (within limits) to the absolute pressure. The current can be measured with high-sensitivity microammeters. Ionisation gauges cover the range 10^{-3} to 10^{-12} torr.

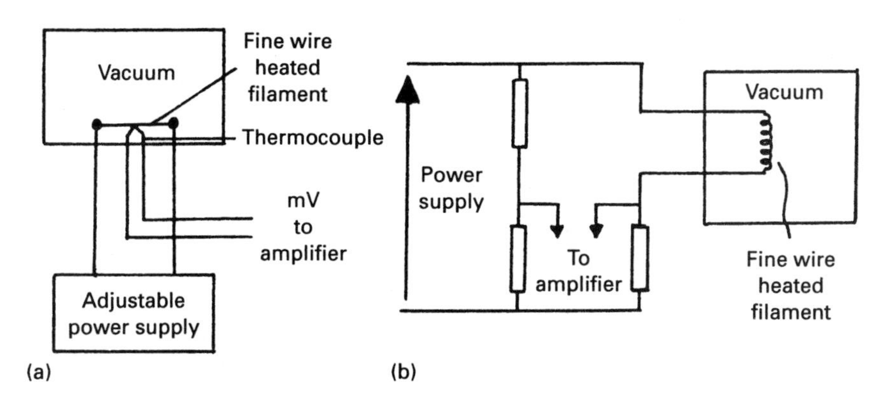

Fig. 3.20 *The Pirani vacuum gauge. (a) Thermocouple sensor. (b) Combined filament and resistance sensor.*

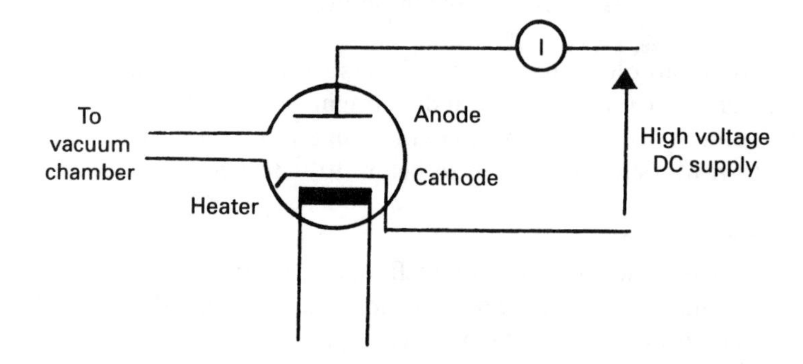

Fig. 3.21 *The ionisation vacuum gauge.*

Chapter 4

Position transducers

4.1. Introduction

The measurement of position is of fundamental importance in many systems: steel rolling mills, radio astronomy and numerically controlled machine tools all require accurate control of the position of some equipment. In addition, transducers for pressure, weight, temperature, level and other physical variables often convert the measurement to a displacement which is then converted to an electrical signal by a position or displacement sensor.

Position transducers can measure linear displacement or angular displacement, although the two can be easily interchanged by screws, rack-and-pinions, and similar mechanisms. The linear position of the tool in a lathe, for example, can be measured by an angular position transducer connected to the lead screw.

The measurement of position can be *absolute* or *incremental*. An absolute transducer measures the position at all times with respect to some fixed datum. An incremental transducer gives a signal corresponding to distance moved, and as such does not correctly indicate position after a power failure. The simplest, and commonest, incremental encoder is a pulse counter. Incremental measurement is simple and cheap, but a datum must be established after each power-up and at regular intervals to avoid ambiguity. Care must also be taken to prevent cumulation of errors each time a reversal of direction occurs.

4.2. The potentiometer

The simplest displacement transducer is the humble potentiometer. The wiper of the potentiometer is mechanically linked to the object whose displacement is to be measured, and an electrical output voltage directly proportional to wiper position is produced. Potentiometers, or 'pots' as they are more commonly known, can directly measure linear or angular displacement as shown in Figs. 4.1 and 4.2.

The circuit of a pot-based displacement transducer is shown in Fig. 4.3. The ends of the potentiometer are connected to a stable voltage, V_i. If L is the maximum displacement of the pot, and d the slider position, the unloaded output voltage will be:

$$V_o = \frac{d}{L} . V_i \qquad (4.1)$$

It is often convenient to consider the fractional displacement x, given by

$$x = \frac{d}{L} \qquad (4.2)$$

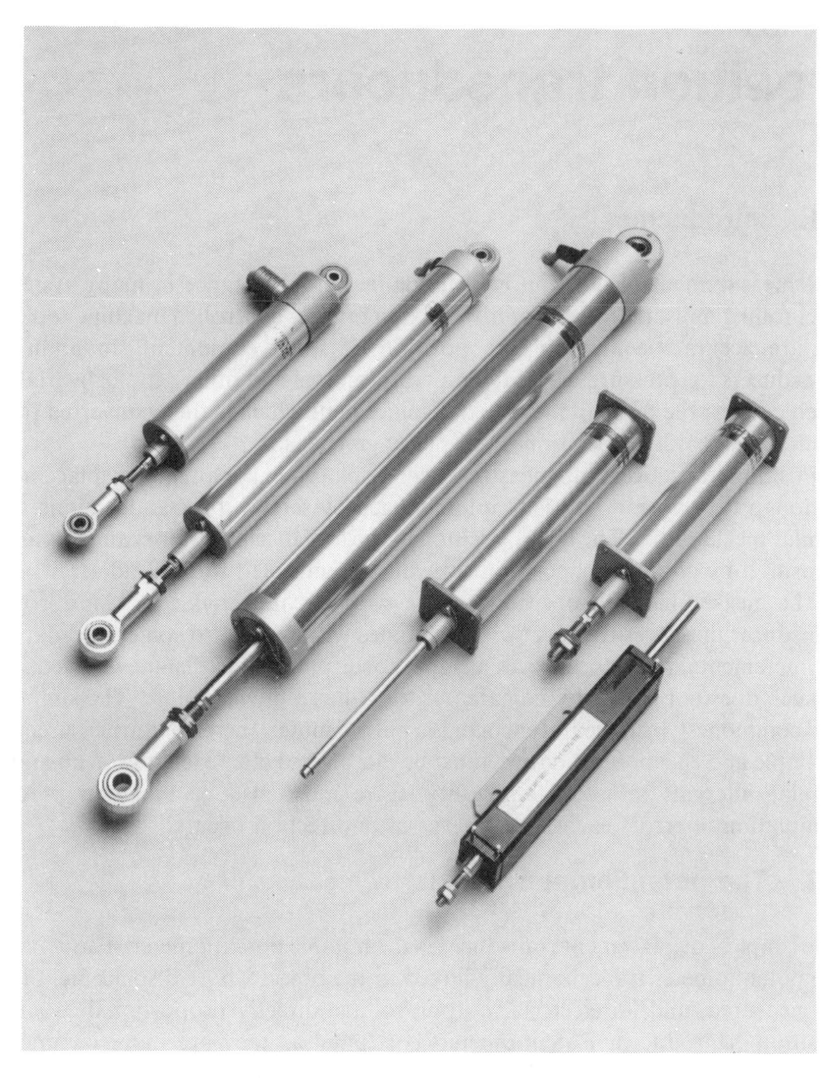

Fig. 4.1 *Industrial linear potentiometers. Note universal joints at ends. (Photo courtesy of Penny & Giles.)*

where $0 < x < 1$. Equation 4.1 can be rewritten.

$$V_o = x \cdot V_i \tag{4.3}$$

The above equations assume no current is drawn from the slider. If a finite load is connected, the relationship between position and output voltage becomes non-linear, as shown in Fig. 4.4a. For a load resistor R_L and potentiometer resistance R, the output voltage for a fractional displacement x is:

$$V_o = \frac{xV_i}{1 + x(1-x)(R/R_L)} \tag{4.4}$$

Fig. 4.2 *Industrial rotary potentiometers. (Photo courtesy of Penny & Giles.)*

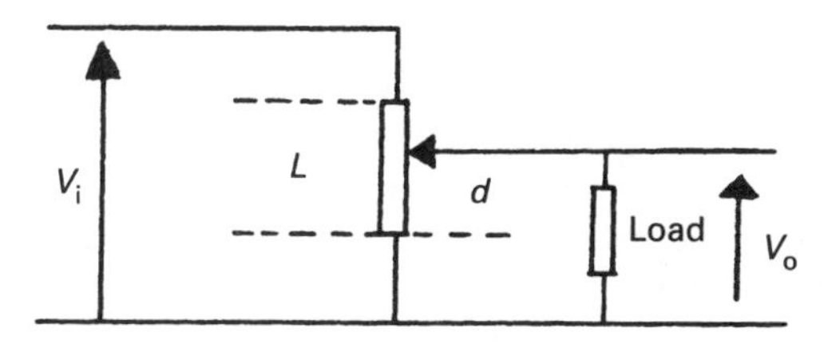

Fig. 4.3 *Simple potentiometer circuit.*

The maximum error occurs at $x = 2/3$, and if R_L is significantly larger than R, the error is about $25R/R_L\%$, as a percentage of true value or $15R/R_L\%$ as a percentage of full scale. A 10 k ohm pot with a 100 k ohm load, for example, will have an error of 1.5% FSD due to loading.

Loading errors are reduced by having a high impedance load or a low-value

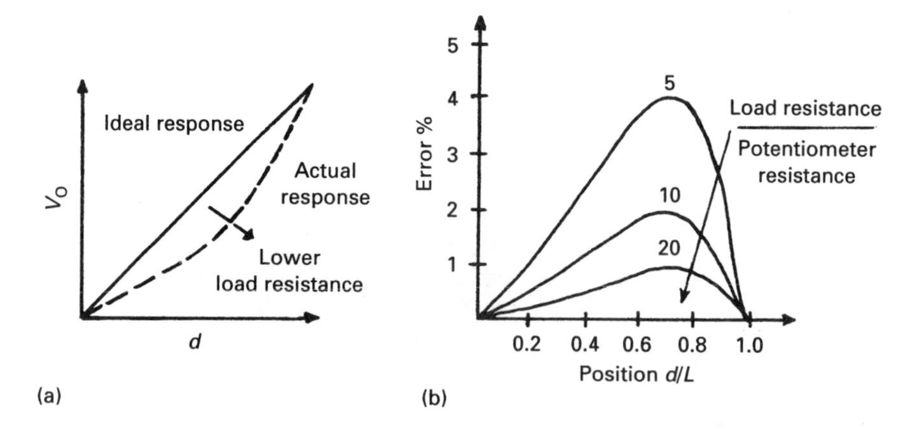

Fig. 4.4 *Effect of non-infinite load resistance on output signal. (a) Position/output voltage relationship. (b) Error induced for various values of load.*

potentiometer, but the latter is limited by the maximum allowable dissipation in the potentiometer itself. The dissipated power is given by:

$$P = V_i^2/R \tag{4.5}$$

A typical value of maximum power for a potentiometer is 1 W, so for a 10-V supply the minimum potentiometer value would be 100 ohm. The maximum dissipated power is usually quoted for some temperature (often 50 °C) and the potentiometer has to be derated considerably for higher ambient temperatures.

Loading errors are further augmented by linearity and resolution errors from the potentiometer itself, and system errors due to backlash and friction. Pots inherently have finite resolution; from the wire size for wire-wound pots and the grain size for metal film and cermet pots. The common carbon potentiometer (used for adjustment purposes) tends towards a dirty track with much movement and should be avoided. With care (and some expenditure) resolutions and linearities of 0.01% can be obtained.

There are safety implications in using potentiometers for position measurement in closed loop systems. Examination of Fig. 4.3 will show that a dirt spot on the track giving an open circuit wiper, or a track break above the wiper will cause the output to read full-scale low, whereas a track break below the wiper will cause the output to read full-scale high. The symptom of a track break or dirty track is often a high-speed dither of a closed loop system about the failure point of the pot.

The pot is essentially a mechanical device and suffers from friction effects. As a result it puts loading on the measuring device which induces errors and increases effects such as backlash. Careful mechanical design of linkages and gearboxes is needed to reduce mechanical errors to an acceptable level.

Being a mechanical moving device, the life of a pot is limited, failure usually being caused by wear of the wiper or the track. The life is reduced further if the potentiometer operates permanently over a small range of its travel.

Potentiometers with a specified non-linear response can be manufactured by varying the track resistance along its length. Logarithmic, square root and trigonometrical (sine, cosine) responses are readily available.

4.3. Synchros and resolvers

4.3.1. Basic theory

Figure 4.5a represents a simple transformer, with an AC input voltage V_i. The output voltage V_o will be KV_i where K is a constant dependent on the turns ratio and the losses in the transformer. In Fig. 4.5b the secondary has been rotated through 90°. The flux from the primary induces equal and opposite voltages in the secondary, giving a net output voltage of zero. In Fig. 4.5c the secondary has been rotated further to 180° from the initial position. The output voltage will be the same as Fig. 4.5a, but the output voltage is *antiphase* to the input, i.e. $V_o = -KV_i$.

Figures 4.5a–c are special cases of Fig. 4.5d where the secondary is rotated to an angle θ with respect to the primary. The output voltage will be given by:

$$V_o = KV_i \cos\theta \tag{4.6}$$

This relationship is shown in Fig. 4.6. It is important to appreciate the significance of equation 4.6. The output voltage can only be in phase with the

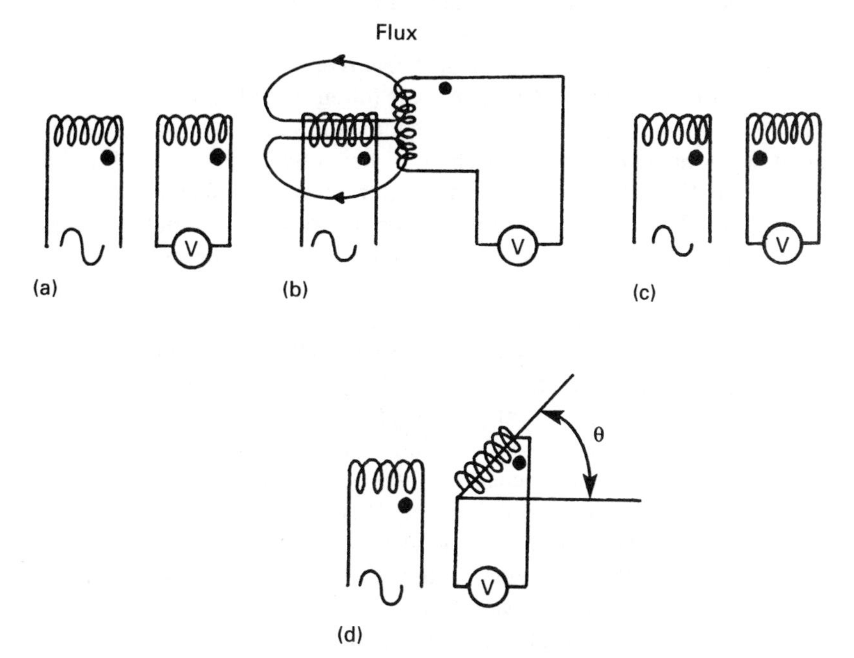

Fig. 4.5 *Basic theory of resolvers and synchros. (a) Coils aligned. (b) Coils at 90°. (c) Coils aligned but reversed. (d) Generalised case.*

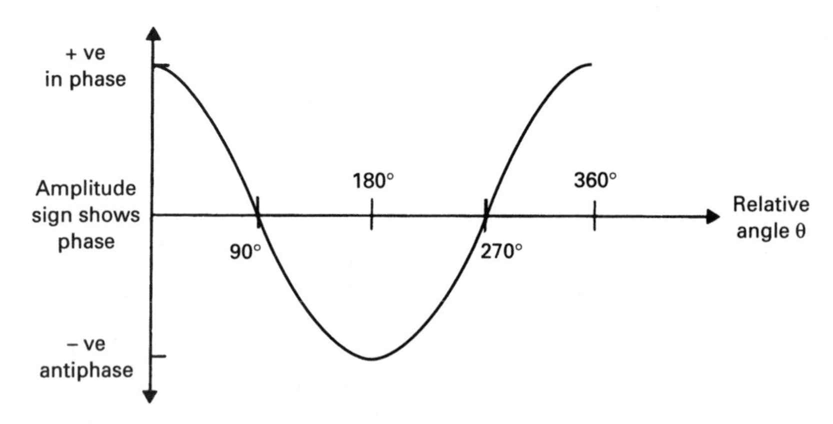

Fig. 4.6 *Relationships between relative angle and output voltage, amplitude and phase.*

input (cos θ positive; $\theta > 270°$ or $\theta < 90°$) or antiphase with the input (cos θ negative; $90° < \theta < 270°$). At the points $\theta = 90°$ and $\theta = 270°$, the output voltage is theoretically zero (although in practice leakage flux and manufacturing tolerances will cause a very small voltage to be present). Output waveforms for various angles are shown in Fig. 4.7.

Figures 4.5–7 form the basis of two types of position transducer; synchros and resolvers. These are both angular measuring devices, but can be used to measure linear displacement by suitable mechanical linkages.

4.3.2. Synchro torque transmitter and receiver

The synchro's transmitter and receiver pair (often described by the trade name 'Selsyn' link) is widely used where remote indication of some angular reading is required. Remote indication of a mechanical weigh scale as shown in Fig. 4.8 is a typical application. The synchro transmitter receiver replaces a mechanical linkage as there is a 1 : 1 relationship between the transmitter input shaft and the receiver output shaft. The devices are called *torque* units because the receiver exerts a positioning torque directly on its shaft. *Control* units, described in the next section, are used as part of a position control system.

A torque transmitter looks superficially like a small electric motor. Internally it has two major components, a stator with three separate windings at 120° star connected, and a rotor which can be considered to be one winding. The rotor connections are brought out via slip rings. The transmitter can therefore be represented as Fig. 4.9a. The schematic of Fig. 4.9b is sometimes used.

In use, an AC supply (usually 400 Hz or 50 Hz, 110 V) is applied to the rotor connections R_1, R_2. Voltages will be induced in the stator coils whose phase and magnitude uniquely define the angle of the rotor. To use these voltages a torque receiver is connected to the transmitter.

The construction of a torque receiver is almost identical to that of a torque transmitter. Electrically they are identical but the torque receiver has low friction bearings and the addition of a damping disc to the rotor to prevent oscillations

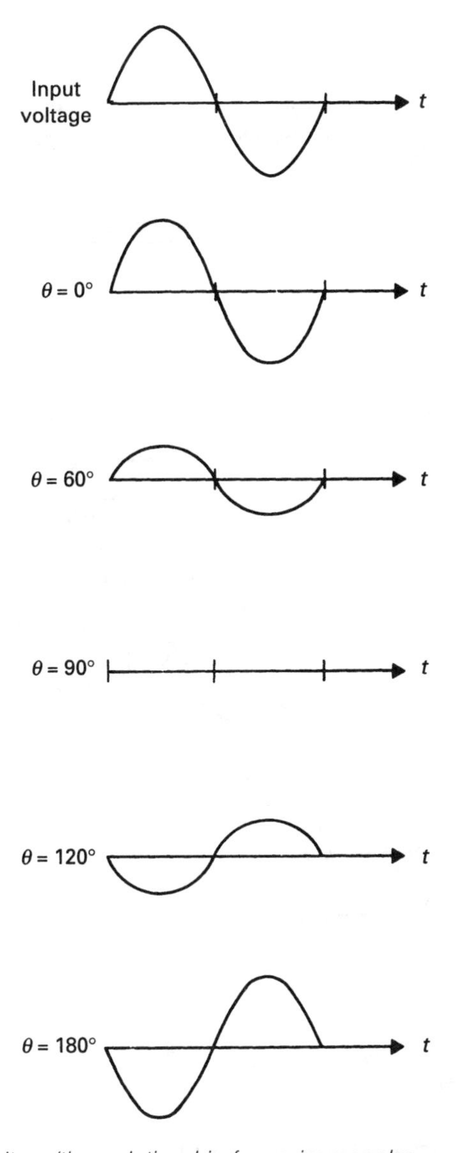

Fig. 4.7 *Output voltage/time relationship for various angles.*

and limit acceleration. The receiver terminals are connected to the corresponding transmitter terminals as shown in Fig. 4.10.

The AC voltage applied to the transmitter rotor induces voltages in the stator windings as explained above, which causes current to flow through the stator windings of the receiver. These currents produce a magnetic field at the receiver which is aligned at the same angle as the transmitter rotor.

The AC voltage applied to the receiver rotor also produces a magnetic field. If

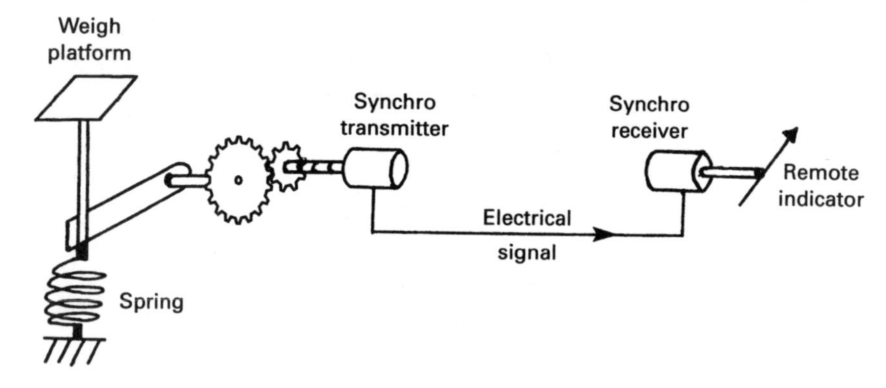

Fig 4.8 *Synchro transmitter/receiver.*

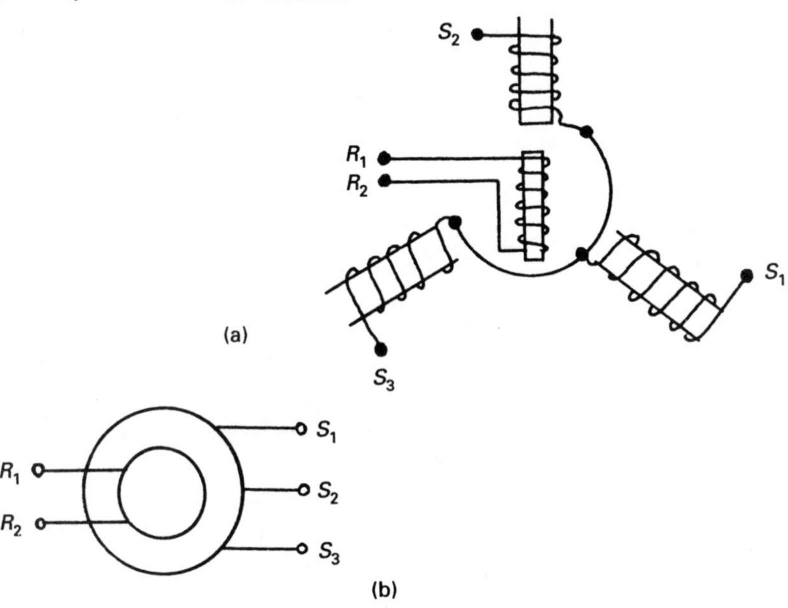

Fig. 4.9 *The synchro torque transmitter. (a) Construction of a torque transmitter. (b) Schematic representation.*

this field does not correspond in direction with the stator-produced field, a torque (caused by magnetic repulsion/attraction) will be experienced by the receiver rotor. This torque will cause the rotor to rotate until the rotor and stator fields are aligned; i.e. the receiver rotor is at the same angle as the transmitter rotor.

If the transmitter rotor is rotated to some new angle, the magnetic field from the receiver stator will also move causing the receiver rotor to rotate until it is again at the same angle as the transmitter rotor. The receiver shaft thus follows the movements of the transmitter shaft.

If the receiver is purely driving an indicator (which requires little or no torque), positioning accuracy is about 0.25°. If the receiver is required to produce torque

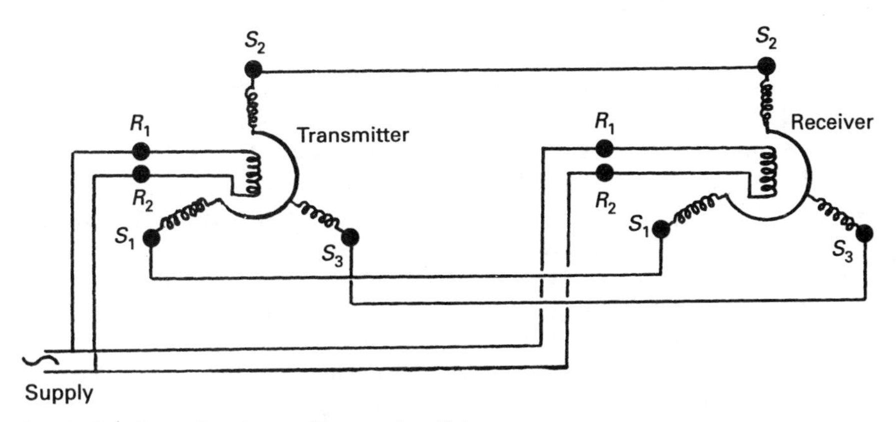

Fig. 4.10 *A synchro transmitter-receiver link.*

(to move a pilot valve, say), the torque is obtained by a standing error between the receiver rotor position and the receiver stator field. If more than a nominal torque is required, a closed loop system using a control transformer is needed, as described in section 4.3.4.

A more accurate indicator, called a Magslip, is shown in Fig. 4.11. The indicator is connected to a balanced L-shaped soft iron rotor. The stator is constructed of three windings identical to a synchro receiver. A fixed polarising coil induces the field into the rotor, which aligns itself with the stator field. The polarising coil also exerts a pulsating pull on the rotor, which causes the rotor to 'dither' slightly on its bearing. This, along with the absence of slip rings, reduces the effect of stiction and friction and gives increased accuracy.

4.3.3. Zeroing of synchros and lead transpositions

The zero position of a synchro is defined as the rotor R_1 end being aligned with stator winding 2 as shown in Fig. 4.10. In this position a minimum voltage exists between S_1 and S_3 (although note that another, identical, minimum occurs at 180° from zero). In practice, however, actual knowledge of synchro zero is not required: all that is needed is to align the receiver shaft with the transmitter shaft.

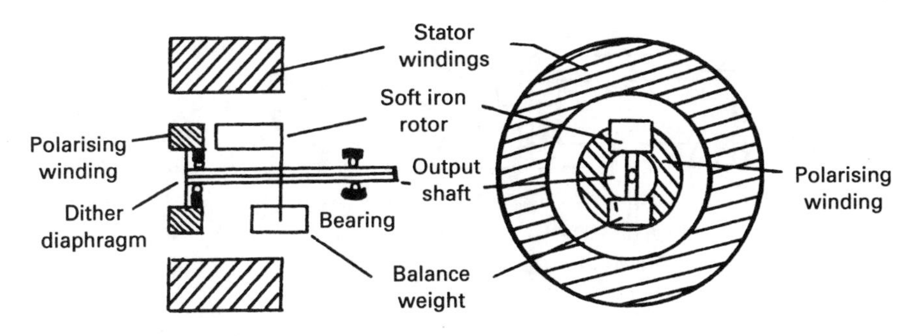

Fig. 4.11 *The Magslip indicator.*

This can be achieved by rotating the whole body of the transmitter or receiver.

Synchros are manufactured with a small lip, as shown in Fig. 4.12, and are mounted into an accurately machined hole. Small mounting brackets locate on to the lip and allow the synchro body to be turned by hand for zeroing. For more accurate zeroing the synchro can be mounted in a frame which can be rotated by a gear or screw-thread mechanism. Electrical zeroing can also be performed by the differential transmitter, described in Section 4.3.5.

The zero and direction of receiver shaft rotation can be altered (deliberately or accidentally) by interchanging of leads. The results of a few of the many possible interchanges are shown in Fig. 4.13. The most important of these is the interchanging of the rotor leads on one device (which shifts the zero by 180°) and the interchanging of two stator leads (which reverses the direction of rotation of the receiver). Where reversed direction is required, the preferred exchange is S_1–S_3 which leaves the nominal synchro zero unchanged.

The power factor of a synchro link is poor because of the magnetising current of the stator winding, which can be in excess of 60% of the stator current. This leads to a loss of accuracy, but can be overcome by the addition of three suitable matched capacitors across the stator lines.

4.3.4. Control transformers and closed loop control

A torque receiver only produces a small torque, and can only drive small loads such as indicators or pilot valves. To drive larger loads such as gun turrets or rolls in a steel mill, a closed loop system is required, with a synchro-based error measuring device to indicate the error between the set point and the actual position.

The basis of such a closed loop system is the *control transformer*, which is used as shown in Fig. 4.14a. The set point is set mechanically on the shaft of a transmitter, which gives an electrical set point on S_1–S_3. The actual position is set on the shaft of the control transformer; the error is an AC signal of the same frequency as the driving supply, with the magnitude indicating the size of the

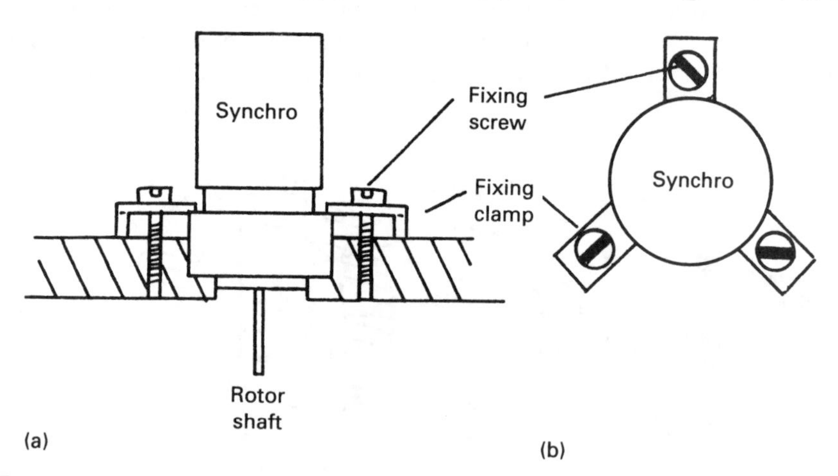

Fig. 4.12 *Mounting and zero adjustment of synchros. (a) Side view. (b) Top view.*

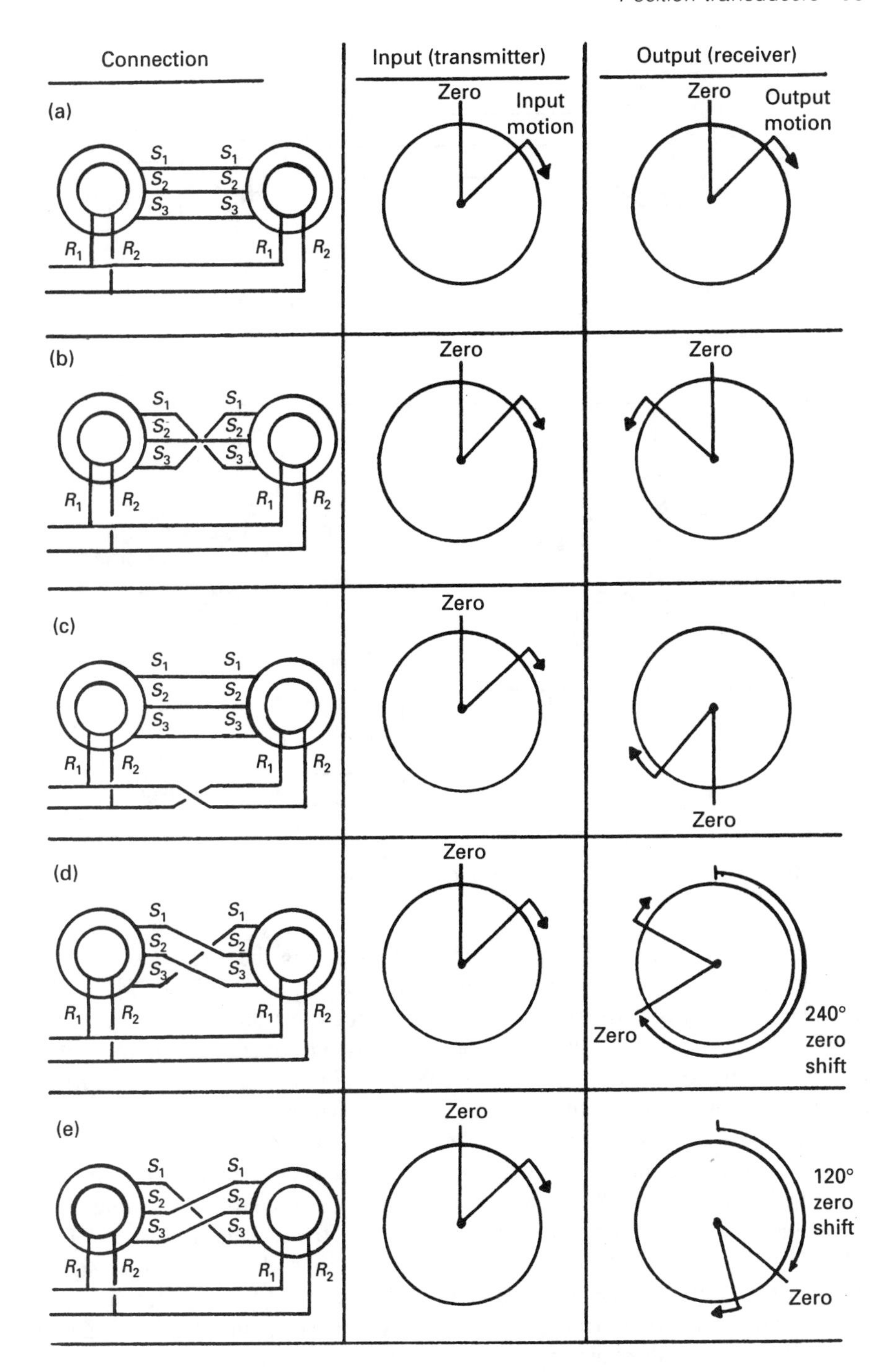

Fig. 4.13 *Various synchro connections. (a) Standard. (b) Crossed stator. (c) Crossed rotor. (d) Cyclic shift. (e) Cyclic shift.*

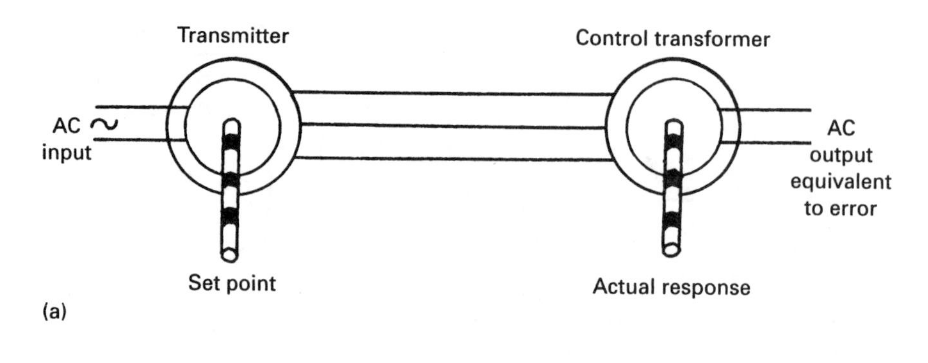

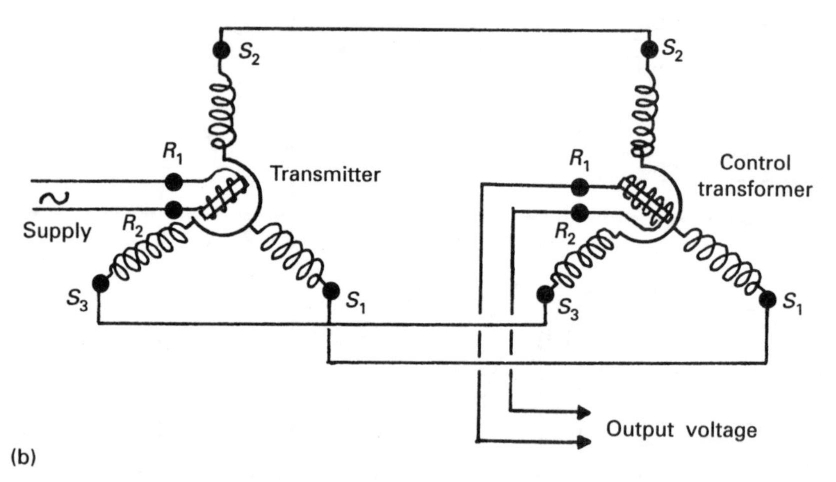

Fig. 4.14 *The control transformer. (a) Control transformer operation. (b) Control transformer connection.*

error, and the phase (in phase or antiphase) indicating the direction.

Internally, a control transformer is similar to a transmitter, so Fig. 4.14a can be redrawn as Fig. 4.14b. As explained in the previous section, at any transmitter rotor angle, a magnetic field will be produced at the same angle by the stator windings in the control transformer. This field will indicate a voltage in the rotor windings which will be a minimum when the rotor windings are at right angles to the field. The aligned position for the control transformer is therefore 90° displaced from the aligned position for a receiver. (There will, of course, be a second aligned position 180° displaced. Care must be taken to avoid ambiguity.)

The basis of a closed loop system using a control transformer is shown in Fig. 4.15. The set-point angle is set on the transmitter, and the AC error signal between the actual and set point is given by the control transformer. The AC error signal is converted to a DC error signal by a phase-sensitive rectifier (giving a positive signal for in-phase input, negative signal for antiphase). The DC error signal becomes the speed reference for a thyristor drive, controlling the speed of the positioning motor.

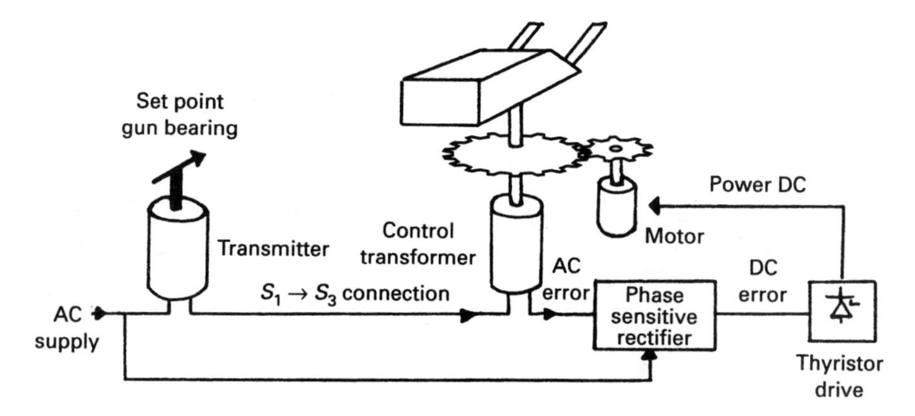

Fig. 4.15 *Position control based on control transformer.*

There are various phase-sensitive rectifier circuits, and the commonest are shown in Fig. 4.16. The circuit of Fig. 4.16a, known as a Cowan circuit, is a half-wave rectifier. When V_{ref} is negative, the diodes are back biased and V_o follows V_i from the control transformer. When V_{ref} is positive, the diodes conduct, shorting out the signal. The output therefore consists of positive half cycles if the input is in phase with V_{ref}, and negative half cycles when it is antiphase. Resistors R_1 and R_2 limit the currents through the diodes when they are conducting.

The Op Amp circuit of Fig. 4.16b gives a full wave rectified output. This is a modification of the inverting/non-inverting amplifier (described in Chapter 11), with the FET switch being operated on alternate cycles by V_{ref}. If V_i is in phase with V_{ref}, a positive full wave output is given; if it is antiphase a negative output.

Both of the circuits shown in Figs. 4.16a and b need filtering to give a DC signal. This filtering will degrade the frequency response of the position control system, and should be designed for as high a cut-off frequency as possible. For this reason, 400 Hz AC supplies are used in preference to 50 or 60 Hz supplies when a fast response is required (400 Hz synchros are also smaller than their 50 Hz counterparts).

Figures 4.14 and 4.15 imply that a control transformer is used with the same transmitter as a torque receiver. In practice, low stator currents are required for maximum accuracy. Control transformers are therefore wound with fine wire to give a higher impedance and are used with *control transmitters* which are similar to torque transmitters but again wound with high impedance wire.

The error in a closed loop synchro system can be reduced by gearing up at both ends of the link, as shown in Fig. 4.17a. This reduces the error in proportion to the gear ratio; but introduces the probability of false positions (with an 18 : 1 ratio, for example, the control transformer could home every 20° of the output position).

To overcome this ambiguity while maintaining the accuracy of the geared system, a coarse/fine synchro system as in Fig. 4.17b is commonly used. Assuming an 18 : 1 gear ratio, the system drives on the coarse 1 : 1 chain until the changeover circuit detects a coarse error of less than 20°. At this point it changes

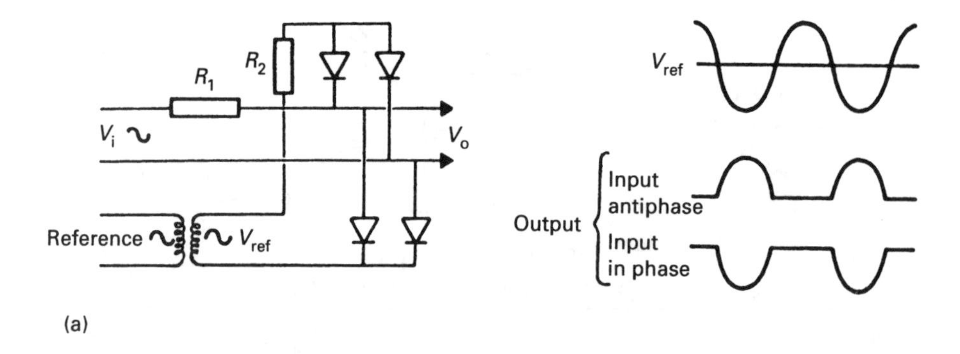

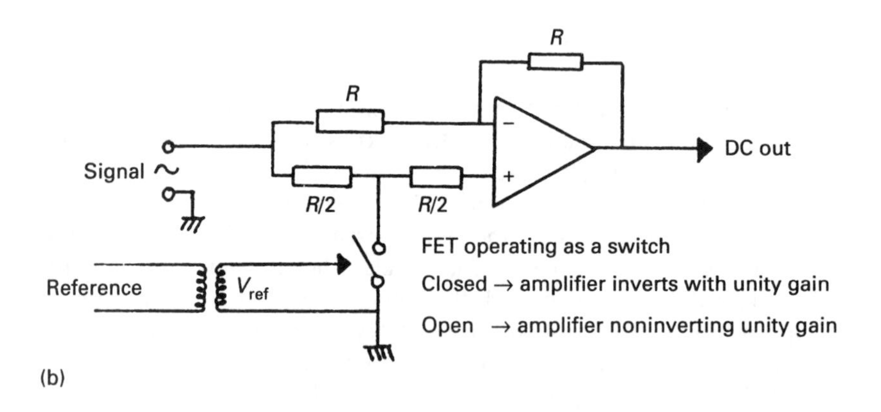

Fig. 4.16 *Phase-sensitive rectifiers. (a) The Cowan half-wave phase-sensitive rectifier. (b) Full-wave rectifier based on operational amplifier.*

over and drives home with the increased accuracy of the fine chain. As the final positioning is done on the geared fine chain, increased accuracy is obtained without the possibility of false home positions.

4.3.5. Differential synchros

Differential synchros are constructed with three rotor windings, as in Fig. 4.18. There are two types of differential synchro: differential transmitters and differential receivers. Both are electrically similar, but the differential receiver incorporates flywheel damping (as outlined earlier for torque receivers).

The differential transmitter is connected between a transmitter and a receiver (or control transformer), as in Fig. 4.19a. The signal it transmits is the difference between the signal on its stator lines and its own rotor. On Fig. 4.19a the stator lines are indicating 90° and the rotor is at 30°, so the new transmitted signal is $90° - 30° = 60°$. Differential transmitters are often used for zeroing purposes in place of the less elegant zeroing of Fig. 4.12. By interchanging S_1, S_3 and R_1, R_3, the differential transmitter adds angles as in Fig. 4.19b.

The differential receiver takes two electrical signals as shown in Fig. 4.20, and

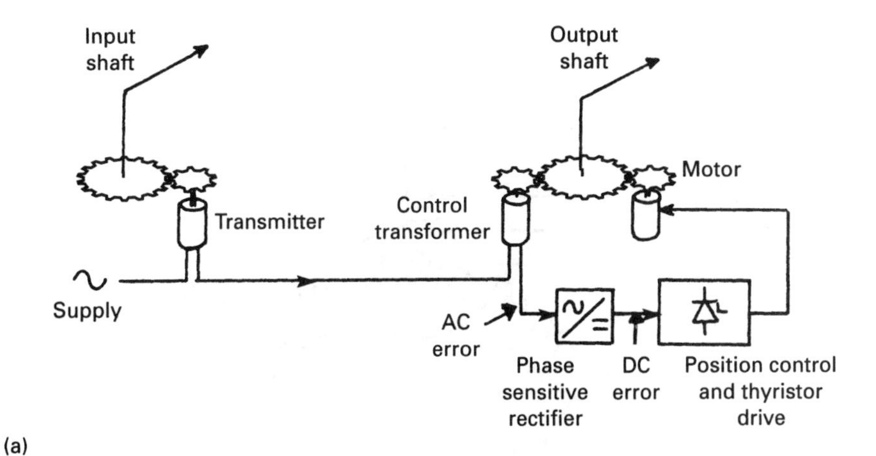

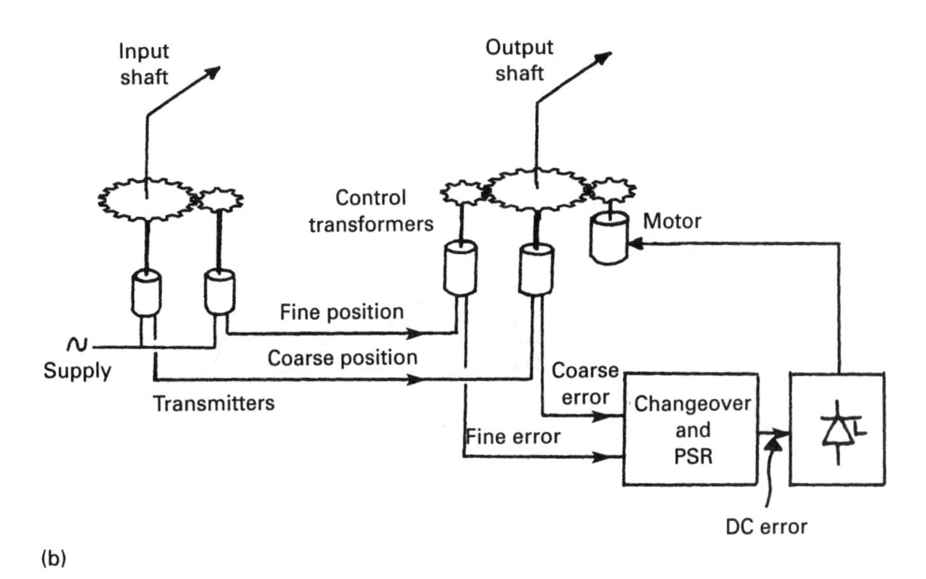

Fig. 4.17 *Methods of increasing accuracy on synchro links. (a) Increasing accuracy with gearing. (b) Coarse/fine changeover system.*

gives a rotor displacement equal to the difference between the two electrical signals. It is not as widely used as the preceding synchro devices.

Note the difference between the differential transmitter and receiver. The transmitter has an electrical and an angular input and an electrical output. The receiver has two electrical inputs and an angular output.

4.3.6. Synchro identification

There are five common sizes of synchro: 23, 18, 15, 11, 09. This figure denotes the outside diameter in tenths of an inch. There are two standards codes for synchros, MIL spec and Muirhead.

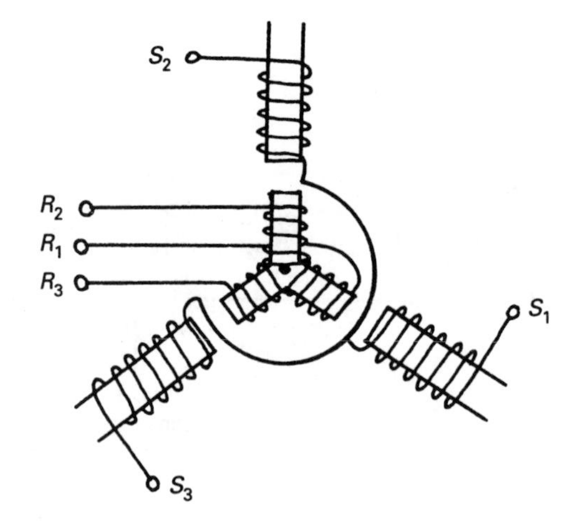

Fig. 4.18 *The differential synchro.*

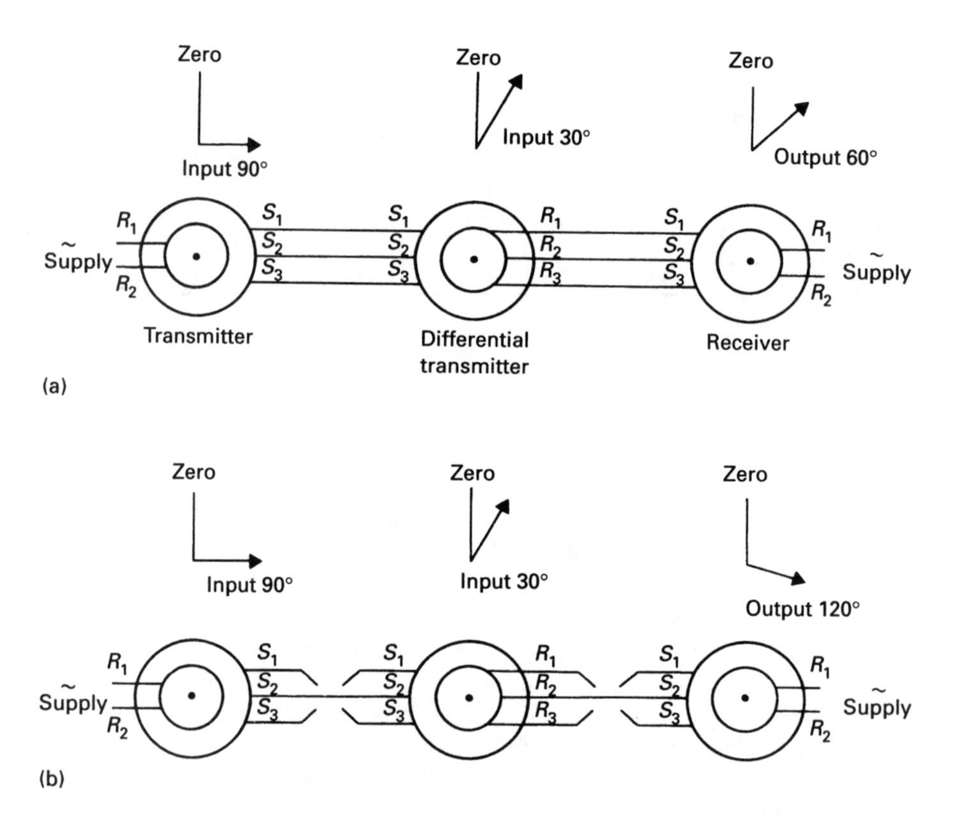

Fig. 4.19 *Differential transmitter applications. (a) Differential transmitter subtracting angles. (b) Differential transmitter adding angles.*

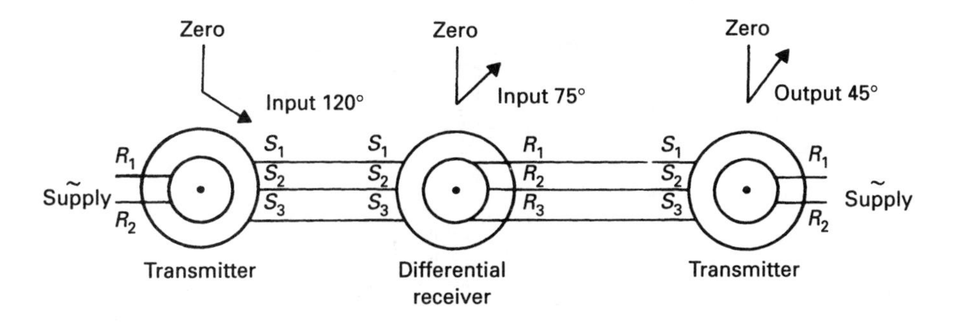

Fig. 4.20 *The differential receiver.*

A MIL spec code has the form 11CT4b. The first numbers specify the size, followed by a code indicating the type as below:

CX control transmitter
CT control transformer
CDX control differential transmitter
TX torque transmitter
TR torque receiver
TDX torque differential transmitter
TDR torque differential receiver

The next digit indicates the frequency, 4 for 400 Hz, 6 for 60 Hz. The final letter is a modification code.

The Muirhead code has the form 18M2B3. The first two digits specify the size, as before, followed by a letter indicating the accuracy:

M Military (standard)
C Commercial (reduced accuracy)
R Reference (high accuracy)
The next digit specifies the synchro type:
1 control transmitter
2 control transformer
3 control differential transmitter
4 torque receiver
5 torque differential transmitter
8 torque differential receiver
9 torque transmitter

The final letters/digits are a modification code.

4.3.7. Resolvers

Resolvers (coded R in the MIL code) have two stator coils at right angles and a rotatable rotor coil, as in Fig. 4.21a (often two rotor coils at right angles are

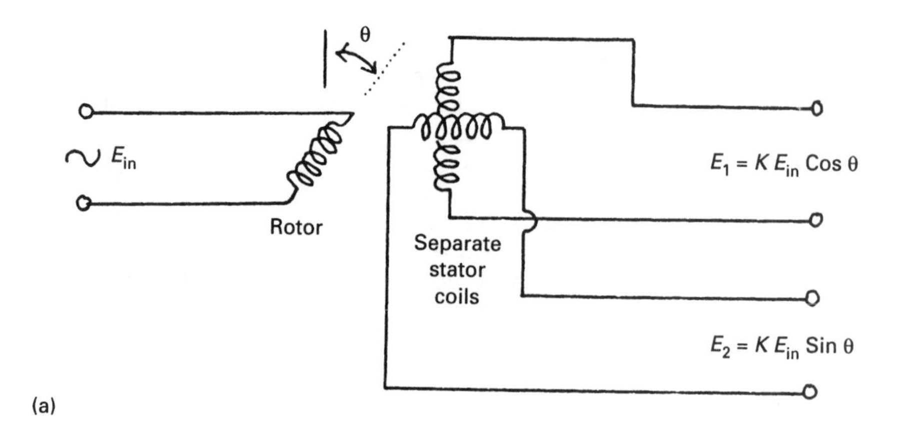

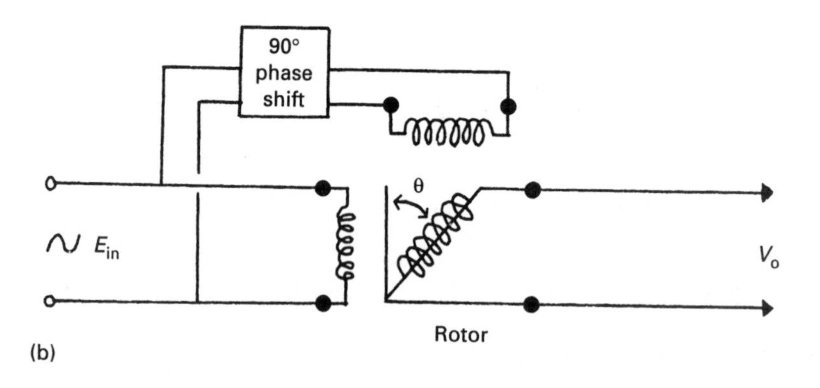

Fig. 4.21 *The resolver. (a) Drive signal applied to rotor, position read from stators. (b) Drive signal applied to stators, position read from rotor.*

provided, but only one is used for position control and indication). The voltages induced into the two stator coils are simply:

$$E_1 = K\,E_{in} \cos\theta$$

$$E_2 = K\,E_{in} \sin\theta \qquad (4.7)$$

Resolvers are used for coordinate conversion and conversion from rectangular to polar coordinates. They are also widely used with solid-state digital converters. One advantage of resolvers is that $E_1{}^2 + E_2{}^2$ is a constant, which makes it easy to check for an open circuit signal.

Resolvers can also be used with AC applied to the stator windings and the position read off the rotor winding. In this operating mode the stators are fed with AC signals with a 90° phase shift, as in Fig. 4.21b. If the frequency of the drive signals is $\omega/2\pi$, they can be represented as $\sin(\omega t)$ and $\cos(\omega t)$. The output rotor signal is then:

$$V_o = K\,V_i(\sin \omega t \cos \theta + \cos \omega t \sin \theta) \tag{4.8}$$

where V_i is the input signal amplitude (both stators assumed the same), θ is the rotor angular position, and K is a coupling constant.

By trigonometrical manipulation:

$$V_o = K\,V_i \sin(\omega t + \theta) \tag{4.9}$$

which is a constant-amplitude signal with frequency identical to the drive signal but a phase shift equal to the rotor position. This can be converted to a DC signal by a phase-sensitive rectifier.

A close relative of the resolver is the inductive potentiometer of Fig. 4.22. This gives an output of $K \cos \theta$ or, with special construction, $K\theta$. The output signal is only unique over the range $0 < \theta < 180°$, so the inductive potentiometer has a restricted angular range.

4.3.8. Solid-state converters

Hybrid encapsulated circuits are available to convert the stator signals from synchros or resolvers direct to a digital number in binary, BCD, or angle in BCD form. A typical device, shown in Fig. 4.23, can resolve to one part in 4096 (12 binary bits). Coarse/fine changeover circuits are also available.

These devices, known as SDCs (for synchro-to-digital converter), are particularly attractive for computer-based position measurement/control systems. Resolver SDCs usually drive as in Fig. 4.21b and incorporate the quadrature oscillator.

4.4. Shaft encoders

4.4.1. Absolute encoders

An absolute-position shaft encoder can be represented by Fig. 4.24a. An angular displacement is applied to the shaft input, and a set of parallel digital output lines give a unique and unambiguous indication of the shaft position. The digital outputs can be binary (as in Fig. 4.24b), BCD or angular-coded BCD (0–360, say, for one shaft revolution). Resolution of one part in 4000 (12-bit binary) is easily obtainable.

Most absolute shaft encoders use optical methods similar to Fig. 4.25. For

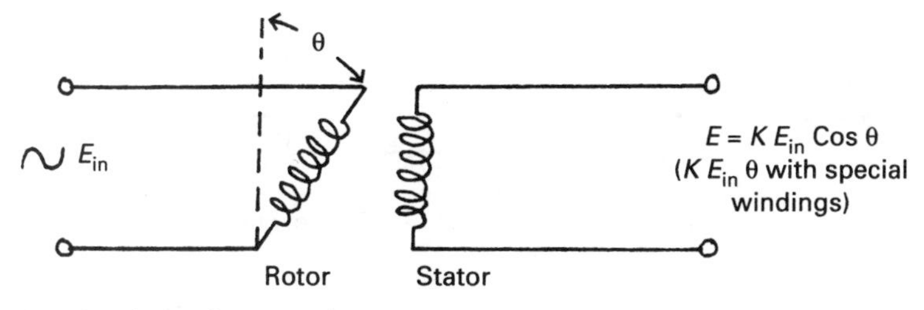

Fig. 4.22 *The inductive potentiometer.*

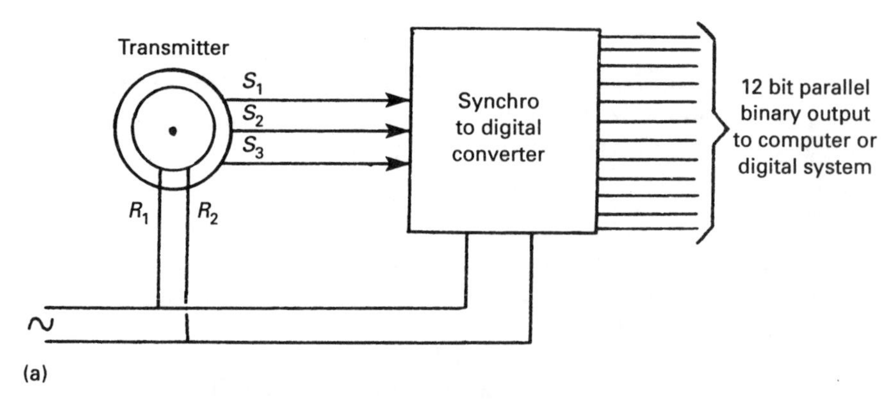

Fig. 4.23 *Solid-state synchro converters. (a) SDC circuit arrangement. (b) Photo of separate SDC module. Analog Devices 12-bit SDC/RDC 1767 and 14-bit SDC/RDC 1768 are synchro- and resolver-to-digital converters. The hybrid devices combine high tracking rates with a velocity output and internal transformer isolation. (Photo courtesy of Analog Devices.)*

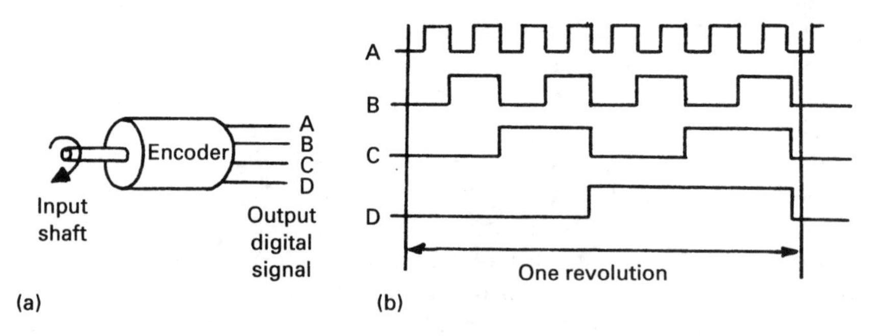

Fig. 4.24 *The shaft encoder. (a) Schematic. (b) 4-bit output signal.*

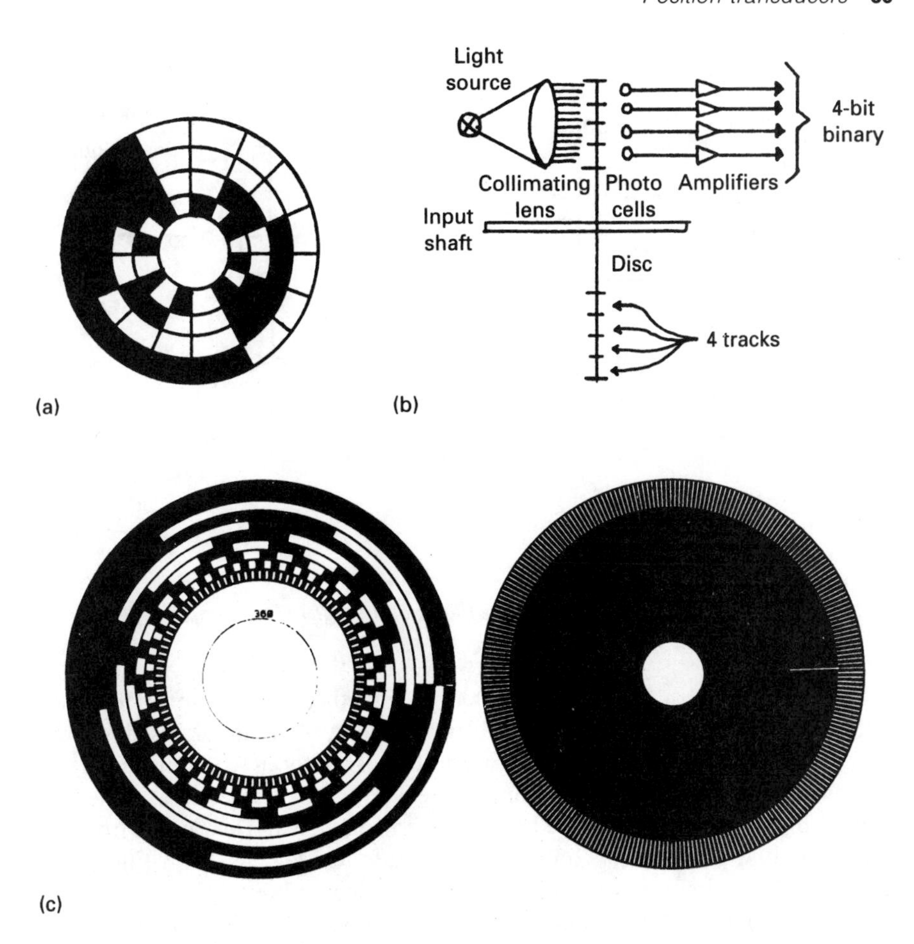

Fig. 4.25 *Construction of shaft encoders. (a) 4-Bit coded disc (inner track Least Significant Bit). (b) Side view. (c) Commercial absolute and incremental encoder discs.*

simplicity of explanation, Fig. 4.25 has only a resolution of one part in 16, but the techniques are identical for higher resolutions. A transparent disc (Fig. 4.25a) has coded tracks according to some required pattern. In Fig. 4.25 a binary pattern is used, but in practice a unit distance code, described below, is usually employed. The disc is illuminated on one side, and photocells sense the track pattern on the other side. The track coding corresponding to the current angular position can be read directly off the photocell amplifiers. Figure 4.25a shows a four-track disc. Commercial units employ up to twelve tracks; a typical example is shown in Fig. 4.25c. An additional unmasked photocell is often included to act as a lamp failure detector and to give a reference level for the track photocells.

A simple binary-coded shaft encoder can give false readings as the outputs change state. Suppose a four-track encoder is going from 0111 to 1000. It is unlikely that all PECs will change together, so the outputs could go

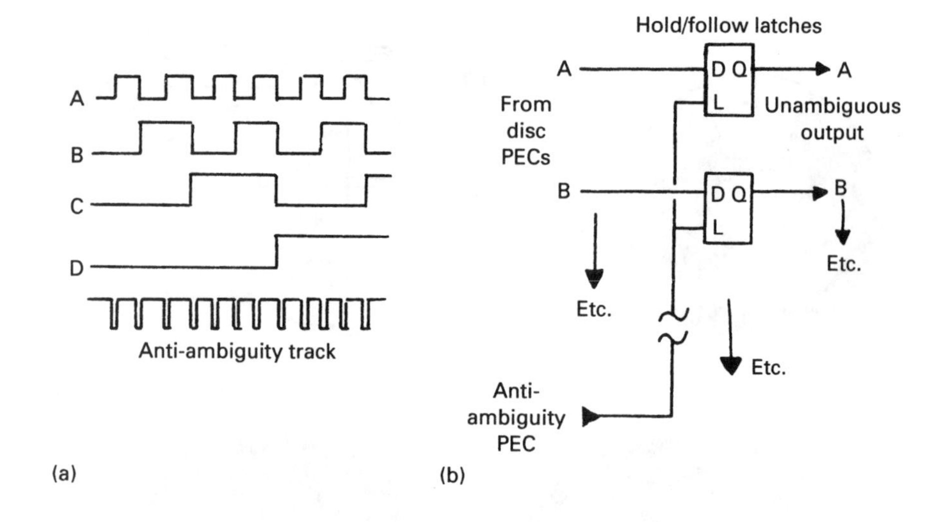

Fig. 4.26 *Use of anti-ambiguity track. (a) Disc coding. (b) Decoding circuit.*

0111→0000→1000 or 0111→1111→1000, or any other combination of four bits.

There are two solutions to this problem. The first is commonly used where a BCD output is required, and utilises an additional photocell to inhibit changes in the output around transition points. The additional track, often called an anti-ambiguity track, is arranged as shown in Fig. 4.26a. The output from the corresponding photocell is connected to hold/follow latches, as in Fig. 4.26b. Usually all the circuit is contained within the encoder body.

Where a binary output is needed, a Gray code disc is used. In this coding, only one bit changes at a time so there is no ambiguity. Because only one bit changes, the term *unit distance code* is often used. A four-bit Gray code, for example, goes:

Decimal	*Gray code*
0	0000
1	0001
2	0011
3	0010
4	0110
5	0111
6	0101
7	0100
8	1100
9	1101
10	1111
11	1110
12	1010

13	1011
14	1001
15	1000

It will be noted that the code is symmetrical about the 7/8 transition, and for this reason such codes are often called reflected codes. Construction of unit distance codes and conversions to and from binary are described in Chapter 13.

Most shaft encoders today use optical techniques, but other designs are available. A cheaper but less robust encoder can be made with a copper-covered disc etched to the required pattern. The track data is then read off by small copper or carbon brushes. These brush shaft encoders have no internal circuits and can work with a wide range of signal levels. The inevitable brush friction, however, gives a relatively high shaft torque and a limited life due to brush or track wear.

4.4.2. Incremental encoders

Incremental encoders are the simplest position-measuring devices but do not give an unambiguous indication of position as do potentiometers, synchros or absolute shaft encoders. Incremental encoders give, instead, a pulse output with each pulse corresponding to some predetermined distance. These pulses are counted by some external counter which then indicates the distance moved.

In many applications it is possible to construct incremental encoders as part of the plant. Incremental encoders, though cheap and simple, do have some disadvantages. The first is that position measurement is normally lost after a power failure (unlike previous devices which correctly, and unambiguously, give the position on resumption of power). Systems using incremental encoders must therefore incorporate some way of establishing a datum position from which counting can recommence.

Simple devices can also gain or lose a count on each reversal of direction if the counter sense is solely selected by the *supposed* direction of movement. Such devices are particularly prone to cumulation of errors during overshoots and oscillations. More complex encoders, described below, give direct indication of count direction.

Commercial encoders are based on the principle of light source, toothed wheel and single photocell. One major problem, though, is that for fine resolution the teeth, and hence the photocell head, become very small. The toothed wheel, therefore, is usually replaced by a Moiré fringe assembly which is best understood by reference to Fig. 4.27. Moiré fringes give excellent resolution without the need for very small photocells. A linear displacement device is shown first for ease of explanation.

The unit consists of two transparent plates which have grating patterns as shown in Figs. 4.27a and b. These are arranged as in Fig. 4.27c with a lamp and photocell. As the moving grating moves with respect to the fixed grating, diagonal dark/light areas move *vertically* past the photocell giving the required pulse output train. The light patterns are known as Moiré fringes, and the pulse/distance relationship depends on the number of gratings on the fixed and

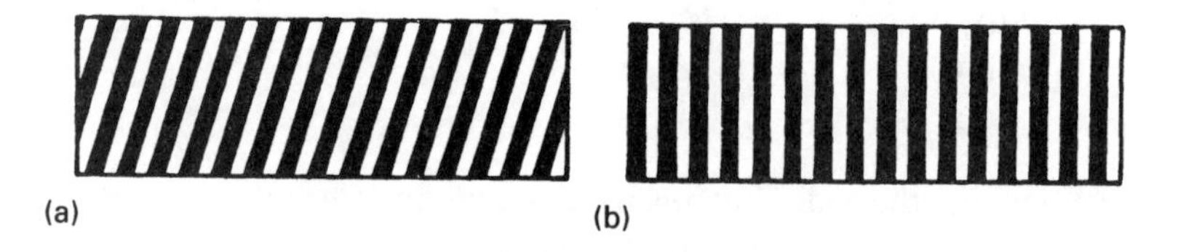

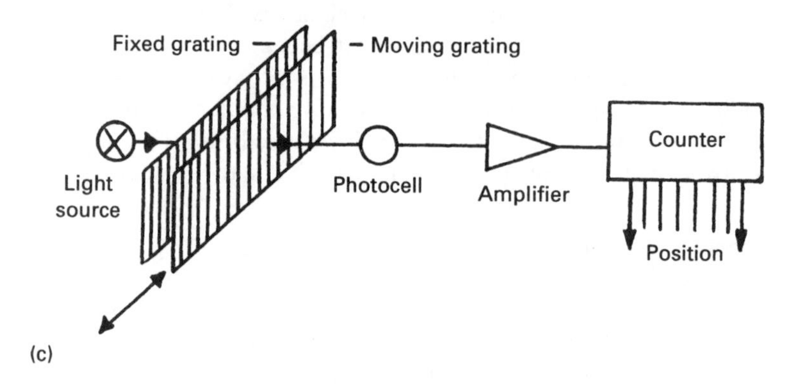

Fig. 4.27 *Linear Moiré fringe position transducer. (a) Fixed grating. (b) Moving grating. (c) Arrangement.*

moving plates. Devices similar to Fig. 4.27c are widely used on numerically controlled machine tools, giving a resolution of 0.01 mm with a total travel of several metres. Where a long travel is needed, a reflective grating is often attached to the body of the machine, and the photocell moves with the carriage.

There are several ways of producing Moiré fringes with angular rotation, but all use a rotating disc moving in front of a fixed disc, as in Fig. 4.28. The angular equivalent of Fig. 4.27 uses one disc with skewed lines. A technique called a vernier Moiré fringe device uses N segments on one disc and $N + 1$ on the other.

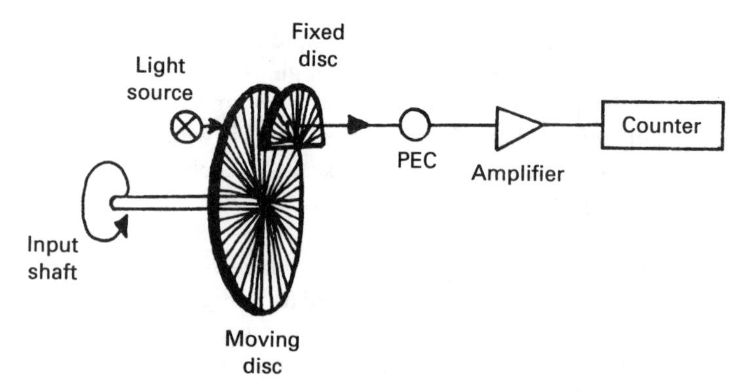

Fig. 4.28 *Angular incremental encoder. Note that only a small segment of the fixed disc is needed.*

As the moving disc rotates, m light/dark transitions are produced where m is the number of segments on the moving disk. Other techniques use two identical discs, either on the same axis or with the fixed disc slightly offset. The aim of all of these is to produce a known number of light/dark transitions per revolution.

With both the linear device of Fig. 4.27 and the angular devices of Fig. 4.28, the actual photocell output signal will not be a square wave, but will probably be somewhat sinusoidal with slow rising and falling edges. The photocell signal is therefore usually passed through a Schmitt trigger circuit to give a clean pulse output with sharp edges.

A single pulse output train carries no information as to direction. In simple low-accuracy applications, the direction can be obtained indirectly by, say, having the count direction selected by the state of the reversing contactor feeding the AC motor driving the shaft. There is still the possibility of a missed or gained pulse, however, so regular zeroing at some datum position is needed.

An incremental encoder can provide directional information with the addition of a second photocell arranged so its output is shifted by 90° in position, as in Fig. 4.29. Let us call these two signals A and B. For clockwise rotation, let us say output A leads B by 90°, as in Fig. 4.29a. If the direction is reversed, output B will lead A by 90°. Clockwise rotation is indicated by '(positive edge on B) and A high'. Similarly, anticlockwise rotation is indicated by '(positive edge on A) and B high'. These signals can be generated by RC differentiation and used to set or reset a flipflop, as in Fig. 4.30. The output of this flipflop becomes the direction signal for the counter indicating position (digital circuits are described in Chapter 13). The counter will follow reversals in direction without cumulative error (but zeroing at a datum position will still be required after a power failure). Obviously

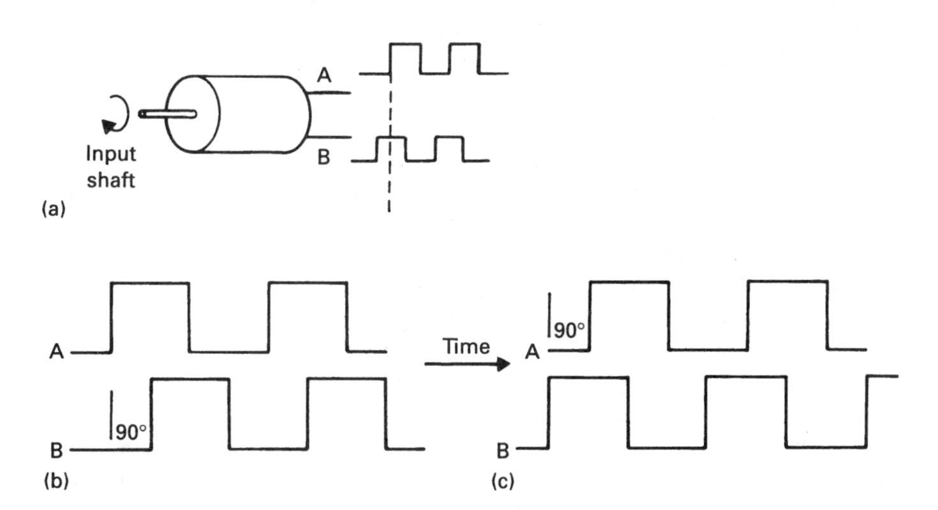

Fig. 4.29 *Incremental encoder with two outputs for directional information. (a) Directional incremental encoder. (b) Clockwise rotation. (c) Anticlockwise rotation.*

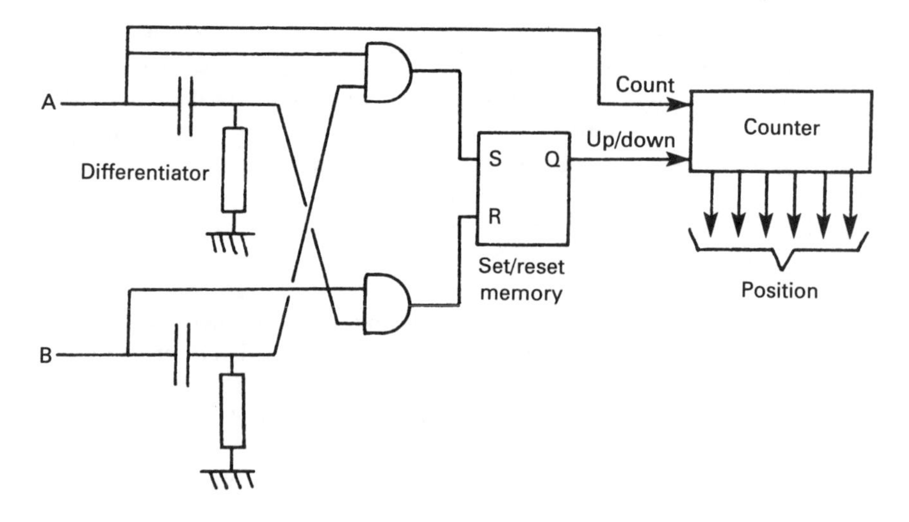

Fig. 4.30 *Using a two-output incremental encoder.*

a similar technique can be applied to the linear incremental encoder of Fig. 4.27.

4.5. Small displacement transducers

4.5.1 Introduction

Previous transducers can be used to measure distances and displacements of any required magnitude by suitable mechanical coupling. This section deals with transducers that measure small displacements. These can be position transducers in their own right (measuring the position of a pilot valve spool, for example) or as an integral part of a transducer measuring some other process variable. The commonest example of the latter are pressure transducers which invariably convert the pressure to movement of a diaphragm.

4.5.2. Linear variable differential transformer (LVDT)

The commonest small displacement transducer is the LVDT, shown schematically in Fig. 4.31a. The unit consists of a transformer with two secondary windings and a movable core. With the core in the centre position, the voltage in the two secondaries will be equal, and the output voltage will be zero.

If the core moves in a positive direction, voltage V_1 will increase and V_2 decrease, giving an output voltage which is dependent on the displacement, and in phase with the primary voltage. If the core moves in a negative direction, voltage V_1 will decrease and V_2 increase, giving an antiphase output voltage dependent on the displacement. The output signal thus has an amplitude indicating the magnitude of the displacement, and phase indicating the direction, as in Fig. 4.31b. This AC signal can be converted to a DC signal by means of a phase-sensitive rectifier similar to those described earlier in Section 4.3.4 (Fig. 4.16).

A related device for small angular measurements is the E transformer of Fig.

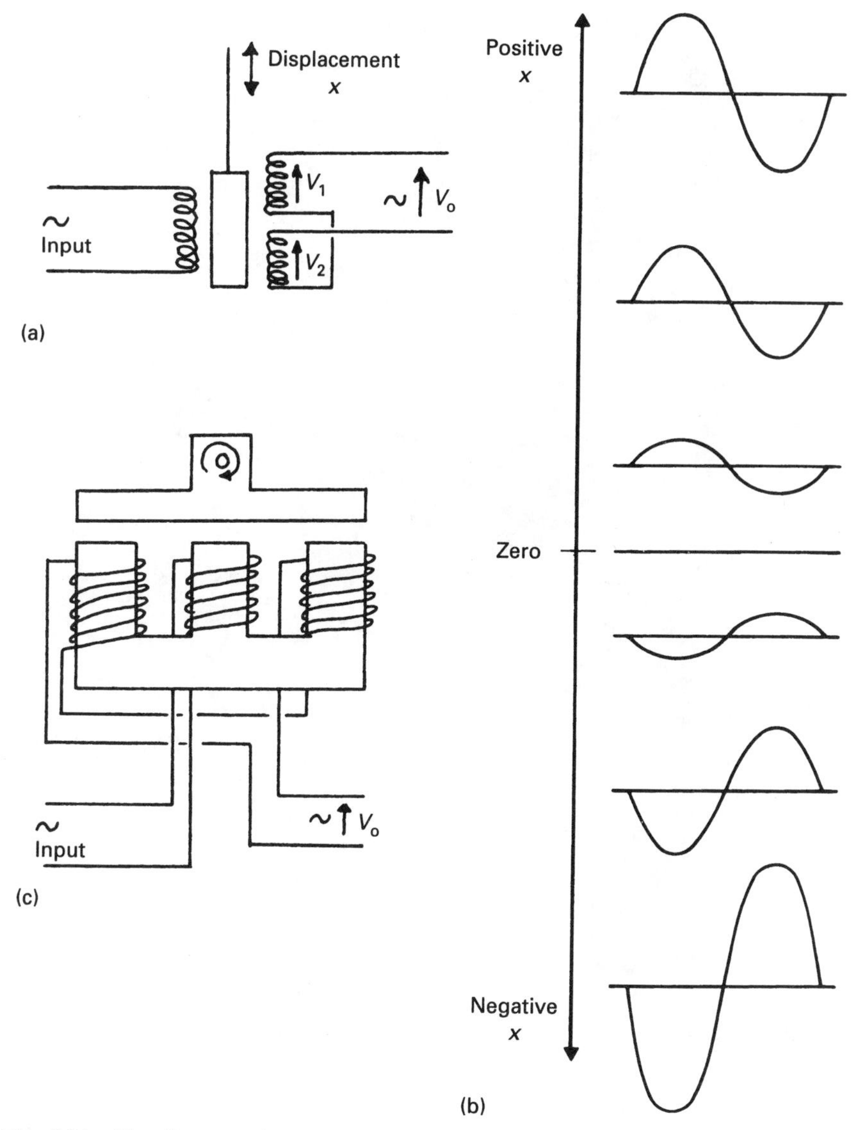

Fig. 4.31 *The linear variable differential transformer (LVDT). (a) LVDT. (b) Output signals. (c) Angular LVDT.*

4.31c. This uses rotation of the pole piece to alter coupling to the two secondaries, giving an output identical to Fig. 4.31b.

A commercial device, such as Fig. 4.32 , is a simpler device to use than Fig. 4.31 might imply. It is powered by a DC supply and incorporates its own high-frequency oscillator (typically 10 kHz) and phase-sensitive rectifier. Giving a DC output, and needing a DC supply, it behaves like a potentiometer.

There are many advantages to LVDTs. They have total electrical isolation and

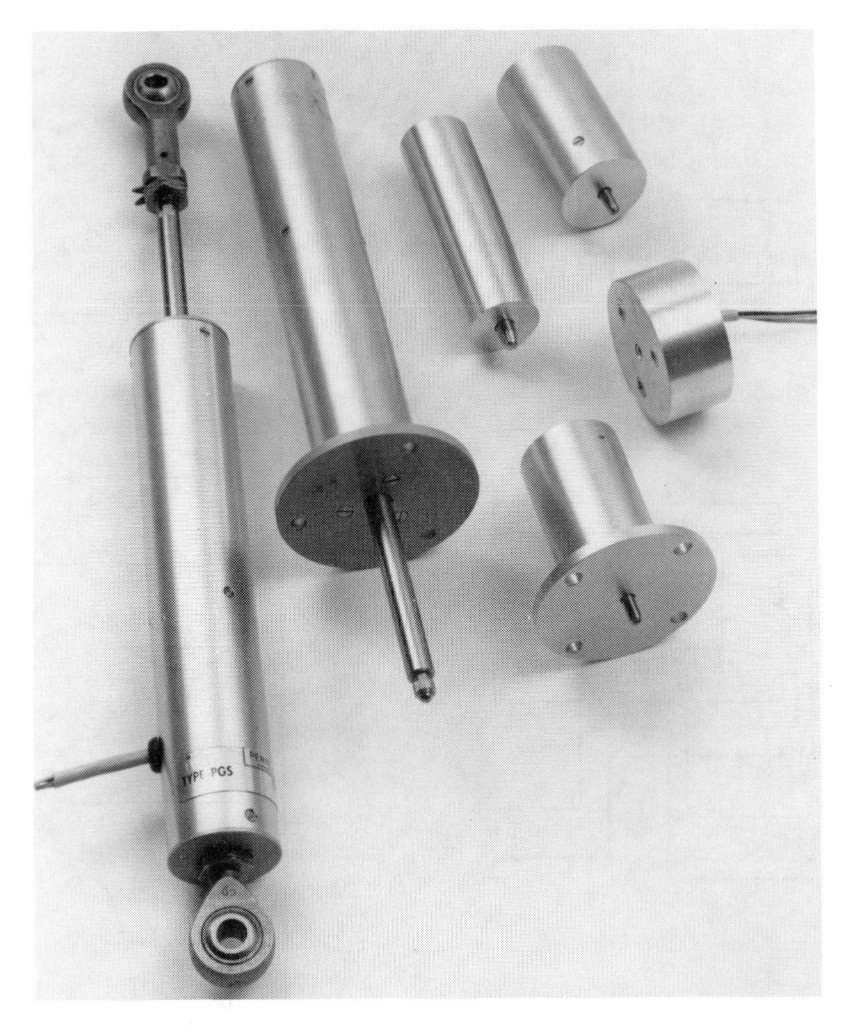

Fig. 4.32 *A selection of typical LVDTs. (Photo courtesy of Penny & Giles.)*

are contactless devices, so friction effects and wear are minimal. Unlike potentiometers they have a virtually infinite life and do not go intermittent with age. Less obviously, the coils can be sealed against liquid ingress with the movable core left open. This allows LVDTs to be used in corrosive atmospheres without the need for sealing collars around the spindle. Submersible transducers can operate at depths of over 150 m.

LVDTs are available for a wide range of displacements from 0.1 mm to about 500 mm. They are very stable and have virtually infinite resolution (although the display/control devices connected to them will define some level of resolution themselves). They are, however, more expensive than simple potentiometers.

4.5.3. Variable inductance transducers

The inductance of a coil can be varied either by a movable core, as in Fig. 4.33a, or by a movable pole piece on an E core, as in Fig. 4.33b. The inductance in both cases is related to the input displacement, and can be measured by an AC bridge or by the current produced by the application of a known AC voltage.

The variable inductance transducer (also known as the variable reluctance transducer) has a unidirectional output and is far less common than the LVDT. Figure 4.33b is, however, the basis of many proximity detectors (see Section 4.6).

4.5.4. Variable capacitance transducers

The capacitance of a parallel plate capacitor such as Fig. 4.34a is given by:

$$C = \frac{\varepsilon A}{d} \text{ farads} \quad (4.10)$$

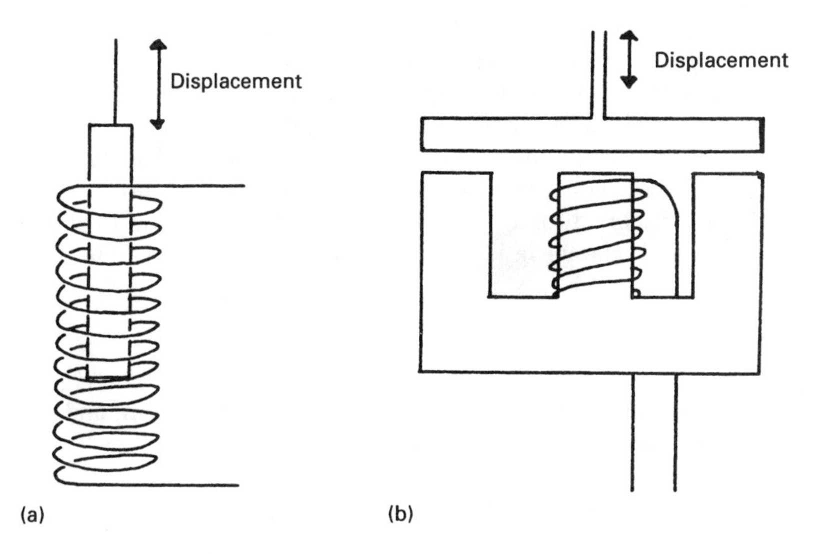

Fig. 4.33 *Variable inductance displacement transducers. (a) Movable core. (b) E core.*

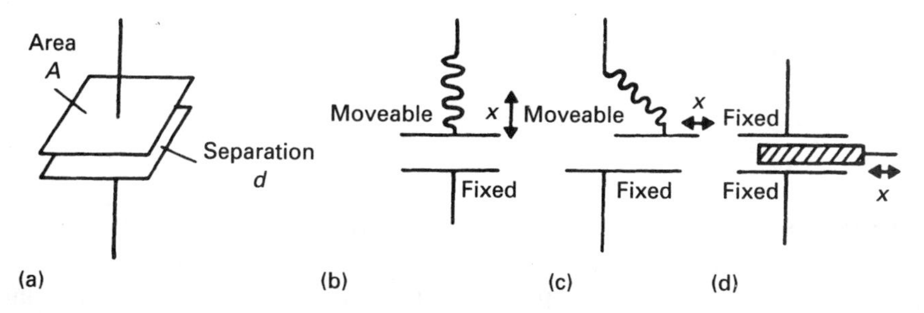

Fig. 4.34 *Fundamentals of variable capacitance displacement transducers. (a) Capacitor basics. (b) Variable separation. (c) Variable area. (d) Variable dielectric.*

where ε is the permittivity of the material between the plates (8.854 pF m^{-1} for free space), A is the cross-sectional area, and d is the separation.

Each of the terms in equation 4.10 can be varied to provide a displacement transducer. In Fig. 4.34b the separation is varied, and in Fig. 4.34c the area. An angular version of Fig. 4.34c is used as a tuning control on many radio receivers, varying the C of an LC tuned circuit. In Fig. 4.34d a dielectric material is moved between the plates altering ε. Examination of equation 4.10 shows that Figs 4.34c and d give a linear relationship between capacitance and displacement.

Unfortunately, for practical devices the capacitance and the change given by a displacement is small. A 30 mm square capacitor with a plate separation of 0.5 mm has a capacitance of 16 pF, and this changes by 4 pF for 0.1 mm displacement (d decreasing; the change is 2.7 pF for d increasing because of the non-linearity of equation 4.10).

Such changes are difficult to measure directly. Possible methods are via the usual voltage/current relationship for a capacitor, or varying C in an LC tuned oscillator to give a variable frequency. More commonly, however, a differential circuit is used. In Fig. 4.35a, the movable centre plate is displaced, causing the capacitance of one side to increase and the other to decrease. Similarly the variable area tubular capacitor of Fig. 4.35b causes opposite changes in the two fixed plates. These devices can be connected directly into an AC bridge, as in Fig. 4.35c. One particular advantage of this arrangement is that a linear relationship between V_o and displacement is obtained even for Fig. 4.35c, as shown below:

$$C_1 = C_0 \frac{d}{d + x}, \qquad C_2 = C_0 \frac{d}{d - x} \tag{4.11}$$

where C_0 is the zero-position capacitance, x is the displacement, and d is the zero-position separation.

$$V_2 = \frac{(d + x)V_i}{(d - x) + (d + x)} = \frac{(d + x)V_i}{2d} \tag{4.12}$$

The voltage at the junction of the resistors is $V_i/2$, so:

$$V_o = V_i\left(\frac{d + x}{2d} - \frac{1}{2}\right) \tag{4.13}$$

$$= \frac{V_i x}{2d} \tag{4.14}$$

The variable area arrangement gives a similar result.

There are few pure displacement transducers based on the principles above, but many transducers for other process variables use a small displacement as an intermediate stage. Pressure transducers, for example, are often based on Fig. 4.35a with the pressure transducer's diaphragm being used as the movable plate.

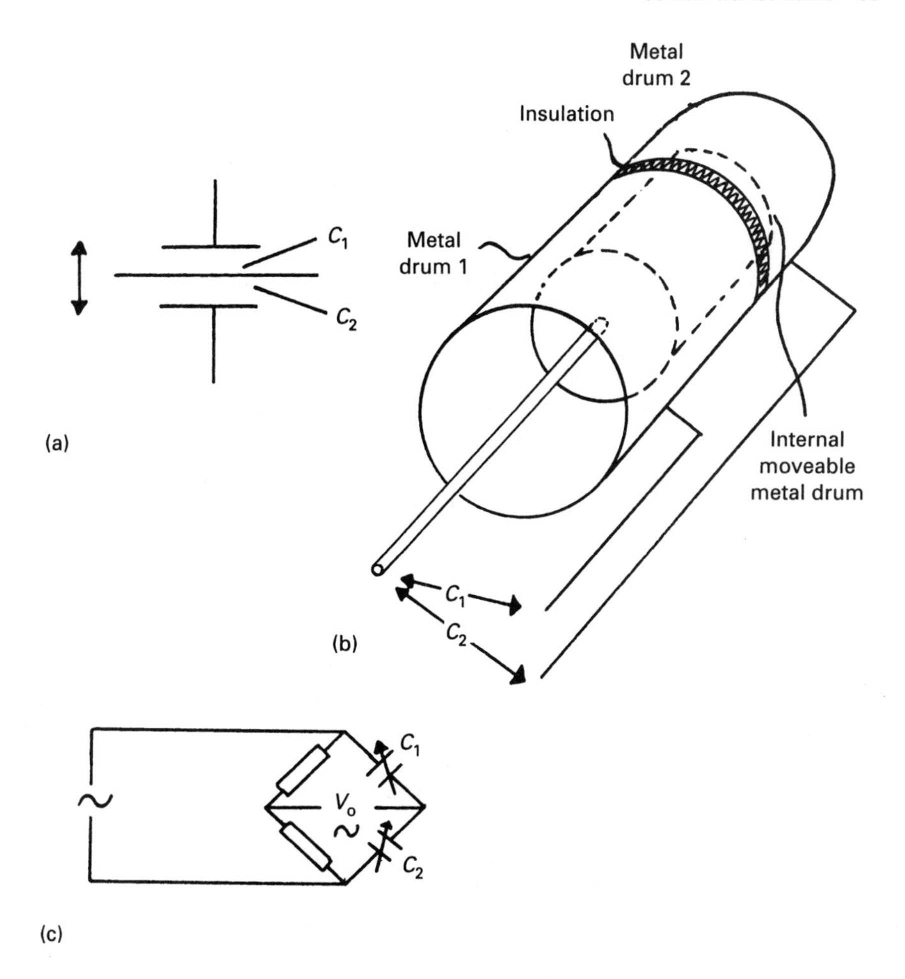

Fig. 4.35 *Linearising the response from capacitive displacement transducers. (a) Two-capacitor sensor. (b) Practical transducer. (c) Measuring circuit.*

4.6. Proximity detectors

Sequencing applications often require indication of the state of items such as valves, flow dampers, etc. In many cases, all that is needed is an on/off signal at the limits of travel (e.g. valve fully open/fully shut). Traditionally these applications use mechanical limit switches. These devices, however, are bulky and have a limited life (particularly where the environment is hostile because of dust, moisture, temperature or vibration).

A proximity detector can be considered as a solid-state limit switch, and is usually based on the inductive transducer of Fig. 4.33b. The transducer contains a coil whose inductance changes as it comes close to a metal surface. This inductance change is detected by a circuit within the transducer, causing the output to switch to some preset level. In a 'limit switch' application, all that is required by way of a 'striker' is a small metal plate that moves in front of the detector at the correct position.

AC-powered devices use two wires and behave just like a switch (apart from a small leakage current in the off state, typically 1 mA). DC-powered devices use three wires: two for the supply and an output which switches between the supplies. DC-powered circuits use a high-frequency internal oscillator which gives a fast response. Detection at 2 kHz is possible with devices such as Fig. 4.36.

Sensing distances up to 20 mm are feasible, but 5–10 mm is more common. Maximum sensitivity is obtained with steel or iron, the sensitivity being reduced by over 50% for materials such as brass, copper or aluminium.

Inductive detection can obviously only be used with metallic targets. Capacitive proximity detectors work on the change of capacitance caused by the target, and accordingly work with paper, glass, wood and other materials. Their one major disadvantage is that they need adjustable sensitivity to cope with different applications (unlike inductive versions which are fit-and-forget). Sensing distances up to 50 mm are feasible.

Retro-reflective photocells (see Section 8.7) are another alternative to inductive detection. Devices superficially similar to Fig. 4.36 contain an infrared transmitter and photocell in the head, and detect the target by reflected light. Sensitivity obviously depends on the target surface, but is typically 100 mm for non-rusty mild steel.

4.7. Integration of velocity and acceleration

Velocity is the time integral of acceleration, and displacement the time integral of velocity. Given a transducer for velocity or acceleration, displacement can be obtained by one, or two, integrations. Care must be taken to avoid integration of offset signals. The technique is not widely used in industry, but is the basis of maritime and astronomical inertial navigation systems.

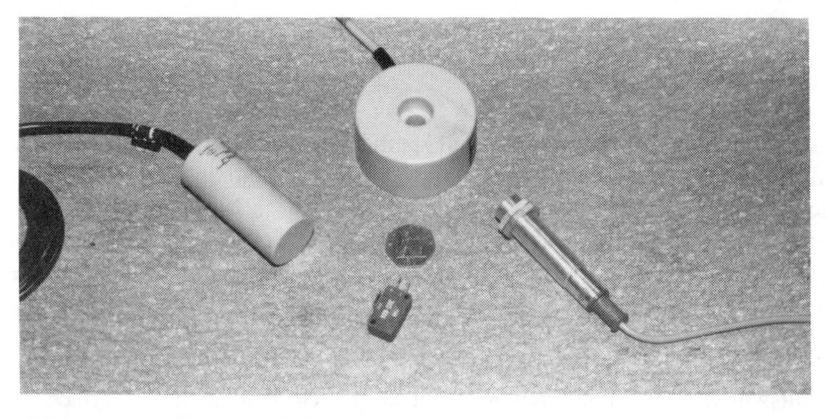

Fig. 4.36 *A selection of proximity detectors.*

Chapter 5

Flow transducers

5.1. Introduction

The measurement of flow is an essential part of almost every industrial process, and many techniques have evolved. The word 'flow' is a general, and not very precise, term that is used to describe distinctly different quantities.

Volumetric flow is the commonest, and is used to measure the volume of fluid past a given point per unit time (e.g. $m^3 s^{-1}$). It may be indicated at the temperature and pressure of the fluid, or *normalised* to some standard temperature and pressure by the relationship:

$$V_n = \frac{P_m V_m T_n}{P_n T_m} \tag{5.1}$$

where V_n is the normalised volumetric flow at pressure P_n and absolute temperature T_n, and V_m is the measured flow at pressure P_m and absolute temperature T_m.

Mass flow is the mass of fluid past a given point per unit time (e.g. $kg\ s^{-1}$). In the case of gases, there is an obvious relationship to normalised volumetric flow as density is dependent on temperature and pressure:

$$M = \rho_n V_n \tag{5.2}$$

where M is mass flow, V_n is normalised volumetric flow and ρ_n is the density at the normalised conditions.

Velocity of flow is the velocity with which the fluid is moving past a given point. Care is needed in the measurement because the velocity may not be equal across the pipe or duct. Point velocity of flow measurement is used, for example, in checking flows over car bodies in wind tunnels.

Flow measurement is often a measurement of some other property. Fiscal measurement for domestic gas, for example, is concerned with charging for the supply of energy. This is achieved by measuring normalised volumetric flow or mass flow and applying an 'energy content' conversion factor measured separately by the gas-supplying authority.

There are several classes into which flow measurement can be divided. An important one is liquids, gases or slurries. Flow measurement in liquids is the simplest because in most cases the liquid can be considered incompressible (thereby simplifying the analysis). With gas flow measurements it is nearly always necessary to make correction for temperature and pressure and to make allowance for the compressibility of the gas. Slurries are liquids with suspended solids and can vary from mud-like substances to relatively clear liquids carrying large pieces of solid matter. The indeterminate nature of the fluid causes measuring difficulties. Another division is closed pipes or open ducts. Most industrial measurement is done in closed pipes, but water and sewage authorities perform measurements in open ducts or conduits. This requires different techniques.

5.2. Differential pressure flowmeters

5.2.1. Basic theory

If a constriction is placed in a pipe through which a fluid is flowing as in Fig. 5.1, a differential pressure will be developed across the constriction which is dependent on the volumetric flow. This simple principle is the basis of the commonest flow measuring devices, namely orifice plates and venturi tubes. To be used as a flow transducer, however, it is necessary to be able to calculate the volumetric flow from the differential pressure. This is done by considering the energy in the fluid.

The energy in a unit mass of fluid has three components:

(a) Kinetic energy due to its motion, given by $v^2/2$ where v is the velocity.
(b) Potential energy gh where g is the acceleration due to gravity and h is the height above some datum.
(c) Energy due to the pressure of the fluid (often called, rather confusingly, flow energy). This is similar to potential energy, and is given by P/ρ where P is the pressure and ρ the density at the fluid's temperature and pressure.

The energy of a unit mass of liquid at any point is the sum of these three components, and because there is no net gain or loss of energy across the constriction we can say:

$$\frac{v_1^2}{2} + gh_1 + \frac{P_1}{\rho_1} = \frac{v_2^2}{2} + gh_2 + \frac{P_2}{\rho_2} \tag{5.3}$$

This is the fundamental equation for differential pressure flowmeters.

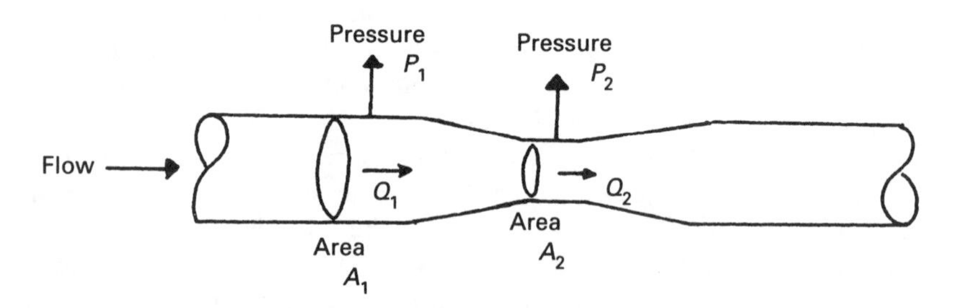

Fig. 5.1 *The basis of differential pressure flowmeters; because $Q_1 = Q_2$, the flow velocity must increase in the region of A_2 and cause a pressure difference between P_1 and P_2.*

5.2.2. *Turbulent flow and Reynolds number*

If equation 5.3 is to be used as the basis for flow measurement, there is an underlying assumption that the velocity of the fluid is the same at all points across the pipe. At low flows this may not be so because of frictional effects at the pipe wall. This leads to a velocity profile similar to Fig. 5.2a which is called streamline, or laminar, flow. If the flow velocity increases, turbulence starts and at a sufficiently high flow the velocity profile becomes equal across the pipe, as shown in Fig. 5.2b. In general, differential pressure flowmeters can only be used where the flow is turbulent.

For a given fluid, the flow condition is indicated by its Reynolds number, R_e, given by:

$$R_e = \frac{vD\rho}{\eta} \tag{5.4}$$

where v is the fluid velocity, D is the pipe diameter, ρ is the fluid density, and η is the fluid viscosity. (The kinematic viscosity, η/ρ, is sometimes given in the literature rather than individual values of ρ and η.) The Reynolds number is a ratio, and hence dimensionless. The larger the value of R_e, the more turbulent the flow. In general, if $R_e < 2000$ the flow is laminar. If $R_e > 10^5$ the flow is fully turbulent.

5.2.3. *Incompressible fluids*

If we consider an incompressible fluid in Fig. 5.1 and equation 5.3, $\rho_1 = \rho_2 = \rho$. For simplification, we will also assume that the pipe is horizontal so $gh_1 = gh_2$. Equation 5.3 simplifies to:

$$\frac{v_2^2 - v_1^2}{2} = \frac{P_1 - P_2}{\rho} \tag{5.5}$$

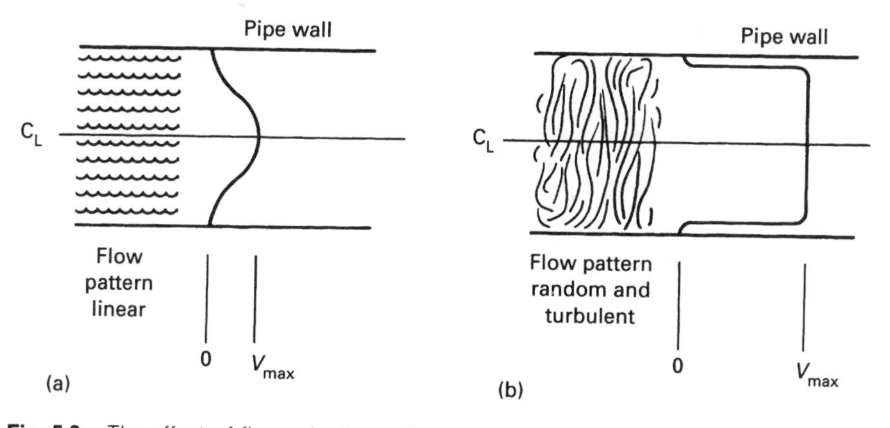

Fig. 5.2 *The effect of flow velocity on flow pattern and velocity profile. (a) Streamline (laminar) flow. Velocity profile is not constant if flow stagnates at walls. (b) Turbulent flow. Velocity profile is uniform across pipe.*

The ingoing flow, Q_1, is $A_1 v_1$, and the flow at the constriction Q_2 is $A_2 v_2$. Because flow is conserved, $Q_1 = Q_2 = Q$ and $v_1 = Q/A_1$, $v_2 = Q/A_2$. Substituting into equation 5.5 gives:

$$Q = \frac{A_2}{\sqrt{1 - (A_2/A_1)^2}} \sqrt{\frac{2(P_1 - P_2)}{\rho}} \tag{5.6}$$

In practice, this equation needs modification because A_1, A_2 do not correspond exactly to the pipe area and the area of the constriction. In the case of an orifice plate, Fig. 5.3, the minimum area (called the plane of vena contracta) occurs downstream of the constriction.

A more practical equation is:

$$Q = C_D \frac{A_2}{\sqrt{1 - (A_2/A_1)^2}} \sqrt{\frac{2(P_1 - P_2)}{\rho}} \tag{5.7}$$

where C_D is called the discharge coefficient, and is effectively a 'frig factor'. Typical values are 0.97 for a venturi tube and 0.6 for orifice plates.

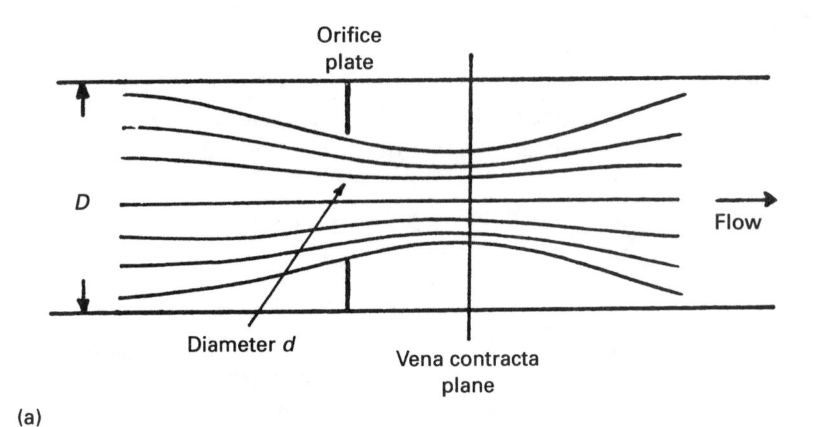

Fig. 5.3 *Pressure curve for an orifice plate. (a) Orifice plate. (b) Pressure curve.*

In practical calculations it is easier to work with diameters of the pipe (D) and the restriction (d). If we define m as the area ratio A_2/A_1 ($m = d^2/D^2$) and E as the velocity of approach factor $1/\sqrt{(1 - m^2)}$, equation 5.7 can be written in a more convenient form.

$$Q = C_D E \frac{\pi d^2}{4} \sqrt{\frac{2(P_1 - P_2)}{\rho}} \tag{5.8}$$

This equation is used for the design of orifice plates. The user can simplify it further to:

$$Q = K\sqrt{\Delta p} \tag{5.9}$$

where Δp is the observed differential pressure and K is a constant for a specific application. Note the non-linear square-root relationship.

The difficulty in using equation 5.8 is establishing a value for C_D. This can be established by experiment with independent flow measurement. Alternatively British Standards Institute BS 1042, Part 1: 1964, Methods for the Measurement of Fluid Flow in Pipes, Orifice Plates, Nozzles and Venturi Tubes, provides graphs from which C_D can be calculated. A mathematical approach (the Stolz equation) is also given in ISO 5167 : 1980.

5.2.4. *Compressible fluids*

With gases, the fluid density does not remain constant through the constriction, so $\rho_1 \neq \rho_2$ in equation 5.3. It is also necessary to consider mass equality rather than volumetric flow equality through the constriction. These changes complicate the calculations for gas flowmeters.

The equation 5.6 is modified to:

$$Q_m = C_D \varepsilon \frac{A_2}{\sqrt{1 - (A_2/A_1)^2}} \sqrt{2\rho_1(P_1 - P_2)} \tag{5.10}$$

where ρ_1 is the density at pressure P_1 (upstream) and ε is defined as the expansion ratio or expansibility factor. C_D is the discharge coefficient factor defined earlier. Note that Q_m is *mass* flow rate.

Calculation of ε is complex as it depends on γ (the ratio of specific heats C_p/C_v), P_2, P_1, A_2, A_1. These all interact so it is not simply a matter of plugging values into an equation. BS 1042 uses a nomograph to calculate ε (unfortunately in imperial units). ISO 5167 gives an equation which is used repetitively for successive approximations of ε until a value is obtained to the required accuracy. This approach, called regression, is well suited to computer analysis but rather laborious for hand calculation.

A practical version of equation 5.10 (BS 1042) is:

$$Q_m = \varepsilon E C_D A_2 \sqrt{2\rho_1(P_1 - P_2)} \tag{5.11}$$

where E is the velocity of approach factor defined earlier.

Note that the upstream pressure is effectively included in equations 5.10 and

5.11 via ρ_1. A gas differential pressure flowmeter is calibrated for one particular upstream pressure.

5.2.5. Orifice plates

An orifice plate is used to make an abrupt change in the pipe area, and simply consists of a circular plate usually inserted between pipe flanges as shown in Fig. 5.4. This produces a pressure differential which is usually measured at D upstream and $D/2$ downstream where D is the pipe diameter.

Figures 5.3 and 5.4 also show that the final downstream pressure is lower than the upstream pressure; the orifice plate has caused a permanent loss of pressure called the head loss. This can be as high as 50% of the upstream pressure. In applications where this cannot be tolerated, a venturi tube (described in Section 5.2.6) is used.

There is more to manufacturing an orifice plate than drilling a hole in a circular plate. There will be a considerable force on the plate, which must be sufficiently rigid to resist distortion. BS 1042 recommends a *maximum* thickness of $0.1D$. The upstream edge of the plate must have a sharp edge as shown in Fig. 5.4 (which implies that an orifice plate has an upstream and downstream face which must be identified). The edge will suffer abrasive effects from the fluid flow, and should be constructed of materials such as stainless steel to avoid excessive wear.

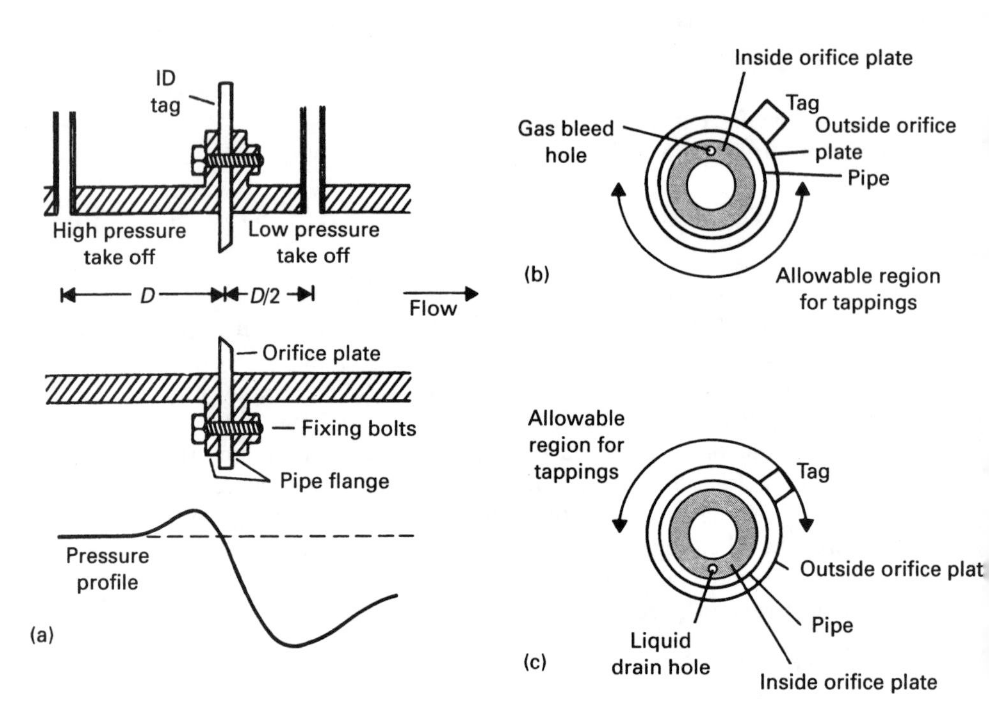

Fig. 5.4 *Flow measurement with orifice plates. (a) Orifice plate arrangement. (b) Liquids. (c) Gases.*

A small hole must be drilled in the plate as shown in Figs. 5.4b and c. For liquids, the hole should be vertically above the opening to allow passage of trapped air or gas. For gases or vapours the hole should be below the opening, flush with the pipe wall, to allow condensate to pass. To avoid errors, the pressure tappings must be displaced by at least 90° from the hole as shown. The identification tag is essential for future users of the system. This should show its identification in the system (e.g. FE207) and the internal diameter.

Care must be taken when installing orifice plates in pipe runs. Close proximity of bends and control valves can cause local pressure variations, so clear pipe runs of at least $10D$ are required both upstream and downstream of the orifice plate. This can be relaxed if straightening vanes similar to Fig. 5.5 are used.

When a gas flow measuring orifice plate is used in conjunction with a flow control valve, as in Fig. 5.6, the orifice plate must be situated upstream of the valve. Downstream of the valve the fluid pressure will vary considerably according to the setting of the control valve, causing significant errors via changes of the density, ρ, in equation 5.11.

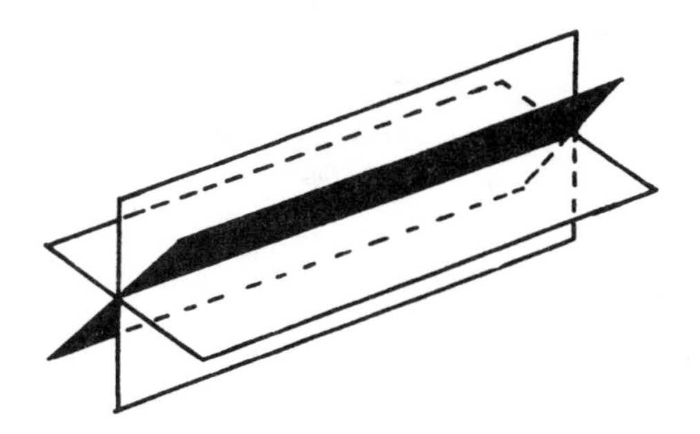

Fig. 5.5 *Flow-straightening vanes.*

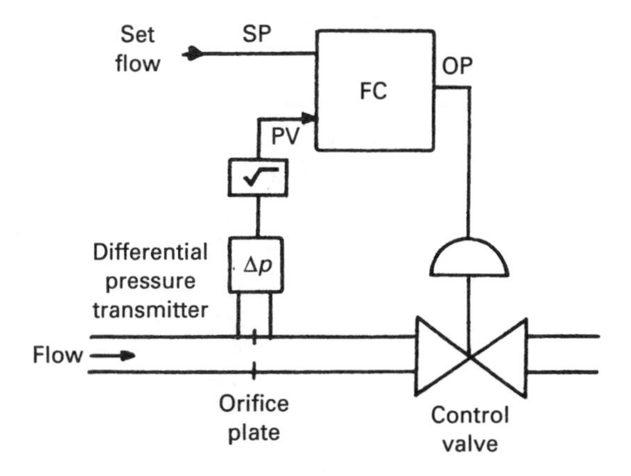

Fig. 5.6 *The use of an orifice plate in a flow control loop.*

There are various arrangements of pressure tappings used with orifice plates. The commonest of these are shown on Fig. 5.7. The $D - D/2$ arrangement (sometimes called radius or throat taps) approximate to the theoretical conditions of Fig. 5.4. Plate taps and carrier rings are complete assemblies and are used where it is not feasible to drill the pipe or the flange assembly. The nozzle arrangement of Fig. 5.7f has a more predictable discharge coefficient and lower head loss but is obviously more expensive to manufacture.

Orifice plates are used in many applications and are probably the commonest flow measuring device. Typical installations are shown in Fig. 5.8.

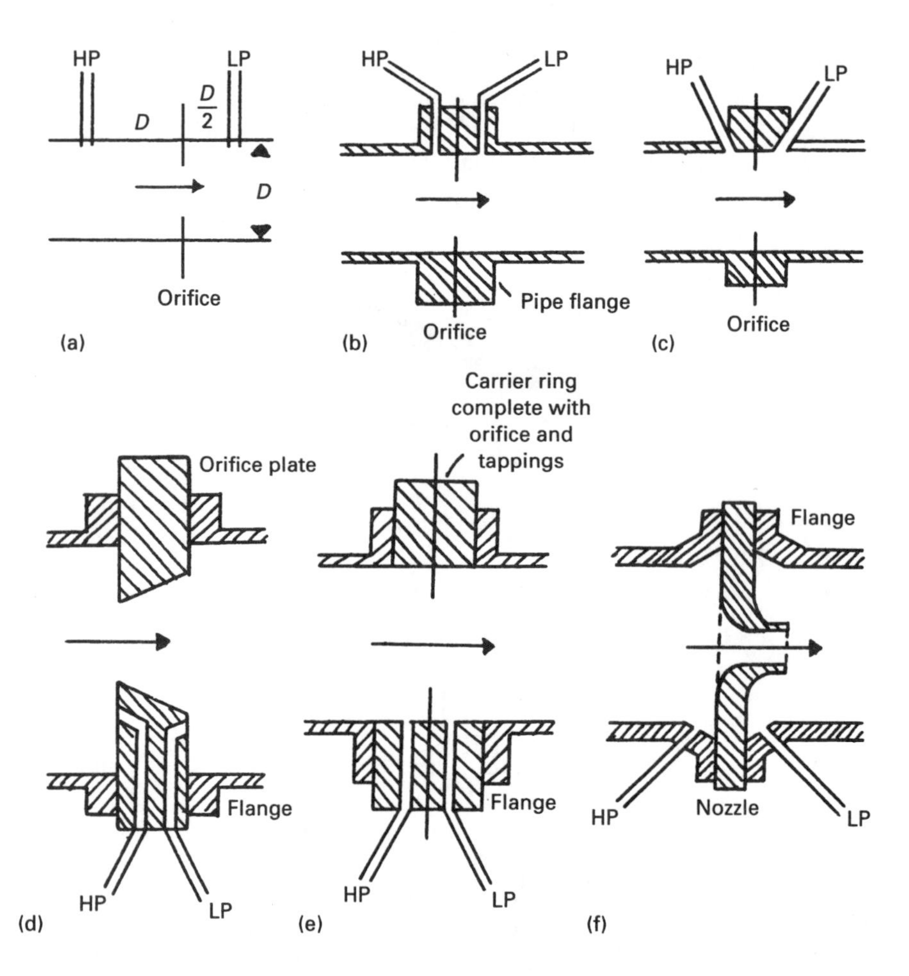

Fig. 5.7 *Various installation arrangements for orifice plates. (a) D–D/2 tapping (very common). (b) Flange taps, normally used on large pipes with substantial flanges. Tappings typically ±25 mm from orifice. (c) Corner taps, drilled obliquely through flange. (d) Plate taps (horizontal scale exaggerated). (e) Orifice carrier (can be factory made and needs no site drilling). (f) Nozzle (gives lower head loss).*

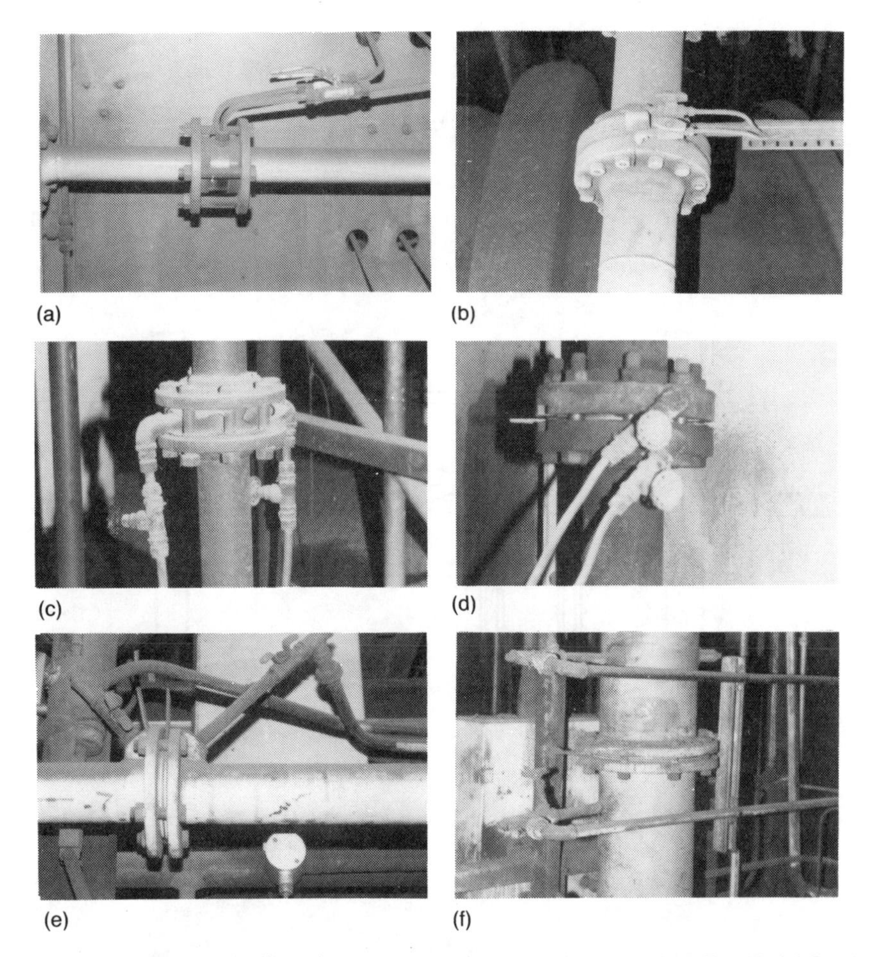

Fig. 5.8 *A selection of orifice plates; note tappings and details such as Tag ID. (a) Carrier Assembly. (b) Flange Tapping. (c) Carrier Assembly. (d) Flange Tapping. (e) Corner Tapping. (f) D–D/2.*

5.2.6. Dall tubes and venturi tubes

The orifice plate produces a large head loss. If this is unacceptable, a smoother obstruction must be used. The two commonest devices are the venturi tube (Fig. 5.9a), which is a manufactured assembly, and the Dall tube (Fig. 5.9b), which is effectively an insert in a pipe section.

The smoother transition lowers the head loss to between 5 and 10% (compared with around 50% for an orifice plate) but the differential pressure is reduced. The output from a venturi tube is about one-third that of an orifice plate operating under similar conditions. The output from a Dall tube is midway between a venturi and an orifice plate.

The venturi and Dall tubes are superior to orifice plates from a theoretical instrumentation viewpoint, but have practical drawbacks. They are obviously

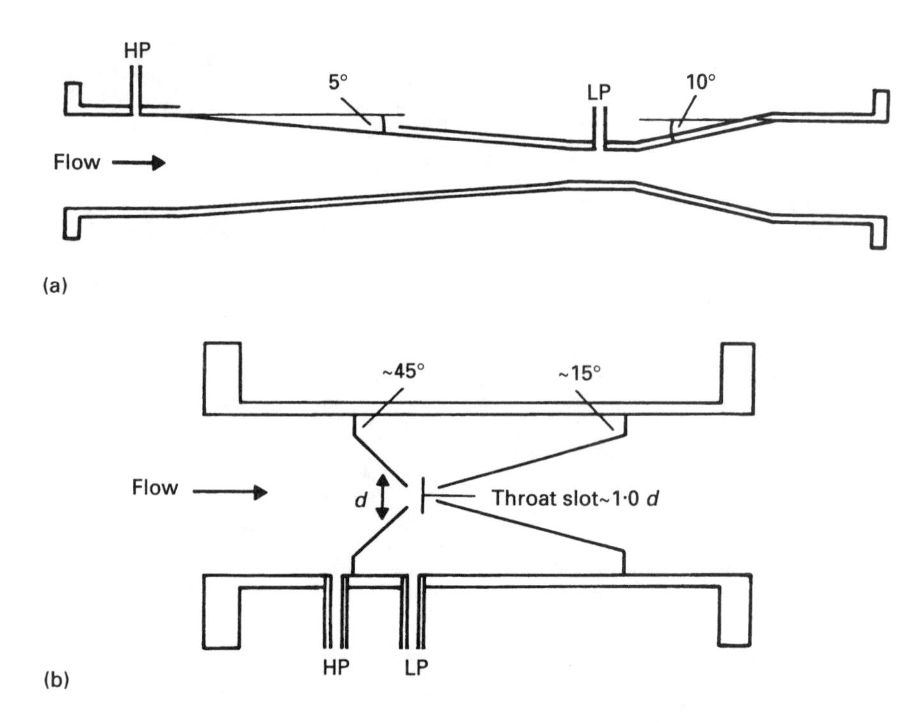

(c)

Fig. 5.9 *Low head loss primary sensors. (a) Venturi tube (note the length!). (b) Dall tube. (c) Photograph of Dall tube (Photo courtesy of Kent Industrial Measurements.)*

expensive to manufacture, and difficult to fit. A venturi for a 25 cm pipe with 15 cm throat would be over 2 m in length, and would need at least 1 m free of obstruction or bends up- and downstream of the device.

The flow equations derived for orifice plates apply equally to venturi and Dall tubes. Typical values for C_D are in the range 0.94 to 0.97.

5.2.7. The Pitot tube

Previous differential pressure devices have measured volumetric or mass flow through a pipe. The Pitot tube, shown in its simplest form in Fig. 5.10, measures flow velocity at one point. It is used for velocity measurements in, say, car and aircraft body testing in wind tunnels, and, slightly modified, as an air speed indicator.

At the tip of the impact probe, the fluid is brought to rest. This leads to a rise in pressure at P_1 with respect to P_2, which can be evaluated by considering the flow and kinetic energy at the impact and static probe.

Assuming an incompressible fluid, and no difference in height between the static and impact probe, the energy per unit mass is calculated:

(a) at the impact probe P_1/ρ since there is no kinetic energy:
(b) at the static probe $P_2/\rho + V^2/2$.

These will be equal so:

$$\frac{P_1}{\rho} = \frac{P_2}{\rho} + \frac{V^2}{2}$$

or

$$V = \sqrt{\frac{2(P_1 - P_2)}{\rho}} \tag{5.12}$$

The relationship for a compressible fluid is, naturally, more complex, but rather surprisingly the effect is small. Equation 5.12 can be used for gases at velocities up to 150 m s^{-1} with an error of about 2%.

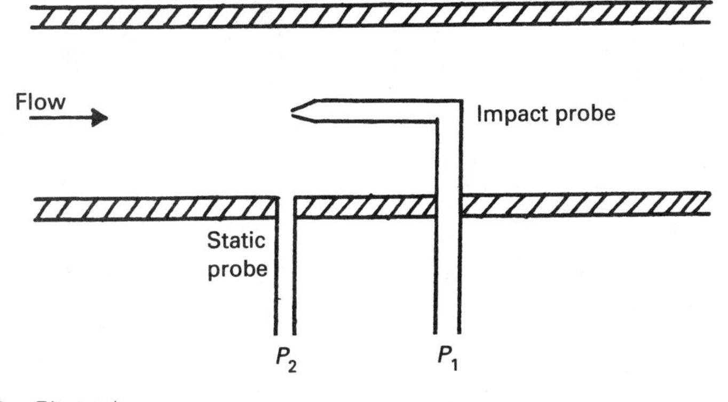

Fig. 5.10 *Pitot tube.*

A practical Pitot tube, shown in Fig. 5.11, consists of two concentric tubes. The inner, open at one end, acts as the impact probe and the outer as the static probe via holes around its circumference. Because of the small radial separation of the probes, the measurement is unaffected by the type of flow (turbulent or laminar) in the pipe. Alignment is important; a 20° offset gives about 2% error.

Pitot tubes can be used to infer volumetric flow if the flow is turbulent, or an averaging tube can be employed (such as the Annubar of Fig. 5.12). (The name

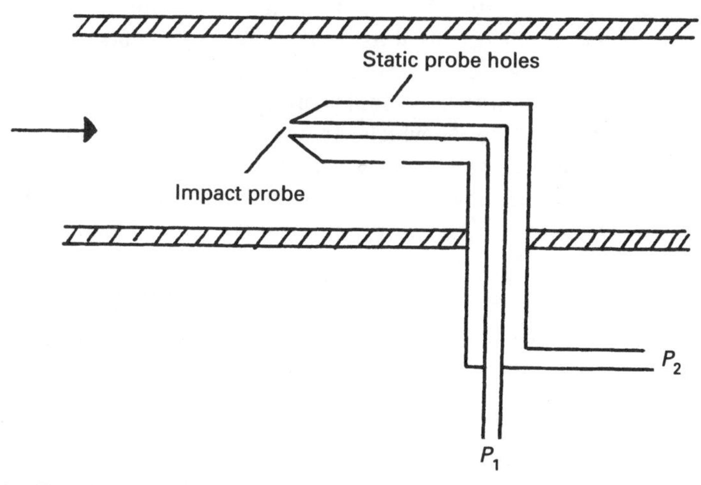

Fig. 5.11 *Practical insertion Pitot tube.*

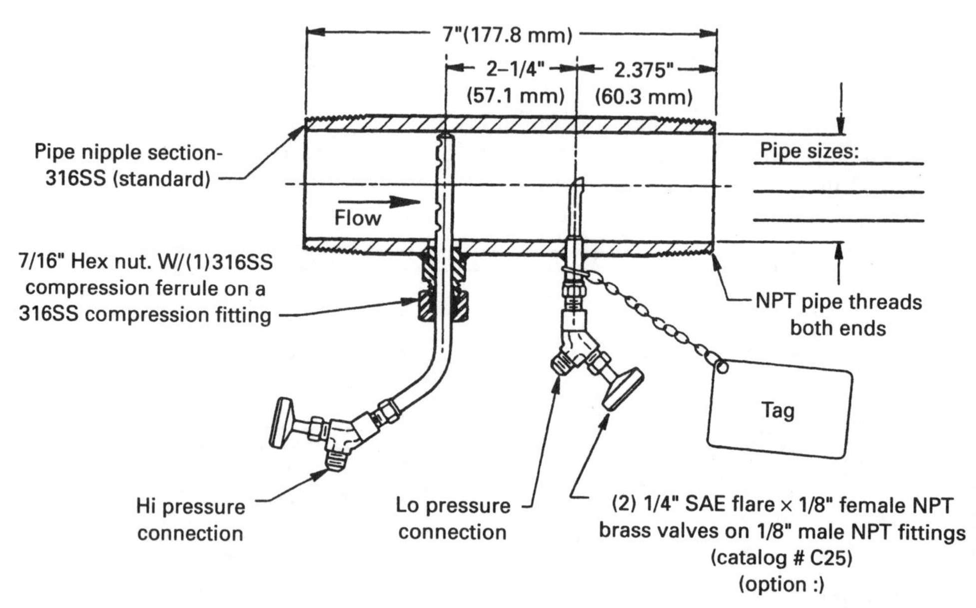

Fig. 5.12 *Annubar pipe insert. (Annubar is a trade name of the Dietrich Corporation; illustration courtesy of their UK representatives, Auxitrol Ltd.)*

Annubar is a registered trade mark of the Dietrich Standard Corporation.) Figure 5.13 shows a typical Annubar installation. An alternative approach uses a single Pitot tube which is traversed across the pipe at regular intervals by an electric motor to enable an average flow to be computed.

The lack of complications such as discharge coefficients and expansibility factors and a negligible head loss makes the Pitot tube an attractive sensor. Its main disadvantage is a low differential pressure which is often at the lower limit of availability of industrial pressure transducers. The small diameter of the impact tube also makes it vulnerable to blockage from fluid-borne particles.

The design of Pitot tubes is covered in the BSI publication BS 1042, Part 2: 1943.

5.2.8. Measurement of differential pressure

Conversion of the differential pressure to an electrical signal requires a differential pressure (Δp) transducer. The construction of these devices is described in Chapter 3; this section describes practical details. Orifice plate connections are shown; piping for venturi or Pitot tubes is similar.

The Δp transmitter is connected via a manifold block as shown in Fig. 5.14. This allows the transmitter to be isolated for removal or maintenance by means of the valves B and C. Valve A is an equalising valve and is used during zero adjustments when B and C are closed and A is open. During normal operation A is closed. When these valves are being operated, care must be taken to ensure that full pipe pressure does not get 'locked in' one leg (which could damage the diaphragm in the transmitter). The order should always be: open A, close B, C; or open B, C, close A.

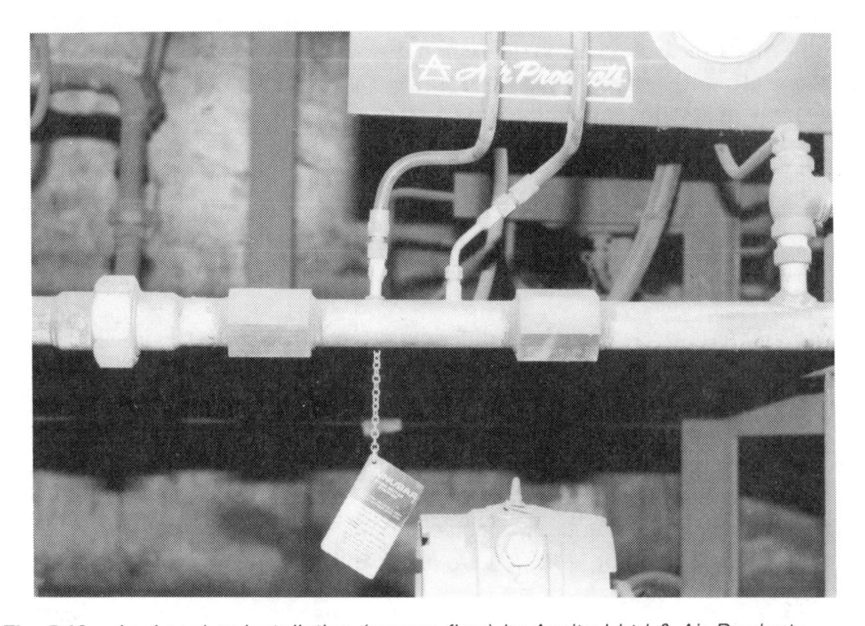

Fig. 5.13 *An Annubar installation (oxygen flow) by Auxitrol Ltd & Air Products.*

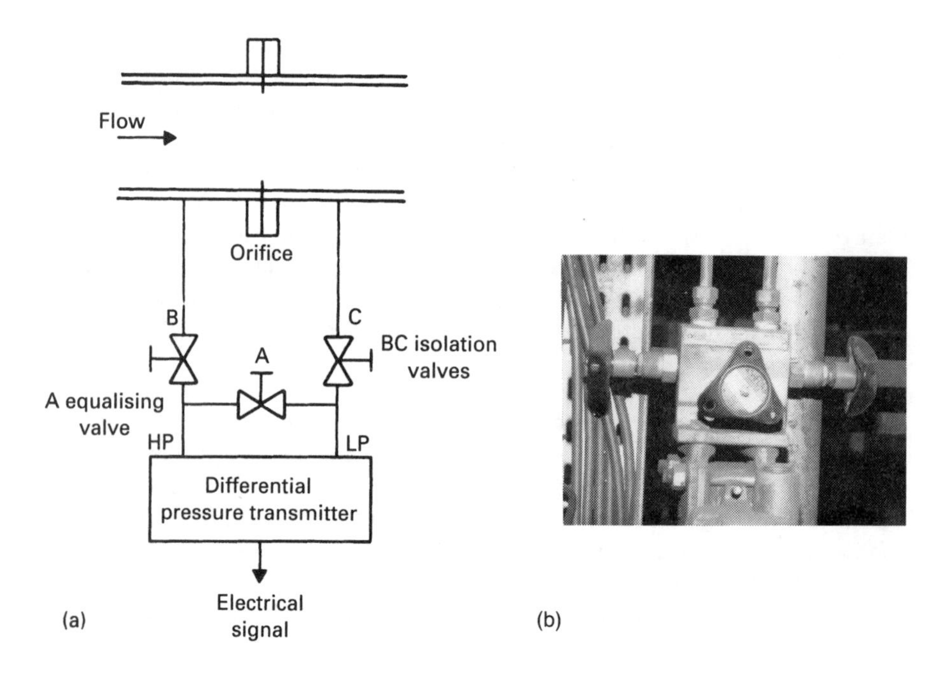

Fig. 5.14 *Connection of differential pressure flow sensor to Delta P transmitter. (a) Valve arrangement. (b) Manifold for differential pressure transmitter.*

Care must be taken with the piping installation to avoid problems from condensed liquids (with gas measurement) or air pockets (with liquid measurement). Ideally a gas measurement system should have the Δp transmitter above the orifice plate with straight pipe drops (so the pipes drain). Similarly liquid measurement systems should have the Δp transmitter below the orifice plate with straight pipe rises (so air bubbles rise back into the pipe).

If these ideal arrangements are not possible, the piping should be arranged as in Fig. 5.15 with the vent/drain cocks. In the liquid arrangement of Fig. 5.15a the line pressure must be such that the Δp pipes fill with the vent cocks open. If both Δp pipes are not full, there will be errors due to the different liquid heads in the two pipes.

The square-law relationship of differential pressure flowmeters necessitates a square-root extraction somewhere in the installation. Usually this is performed electronically on the Δp transmitter output; either via a multibreak point Op Amp circuit (see Chapter 11) or digitally via look-up tables.

The square-root extraction determines the lowest measurable flow, as a small zero offset makes for a large error at low flow. The effect of the square root can be seen in the accompanying table which shows (as a percentage of full scale) true flow, true Δp, Δp with a $+0.5\%$ zero error and indicated flow.

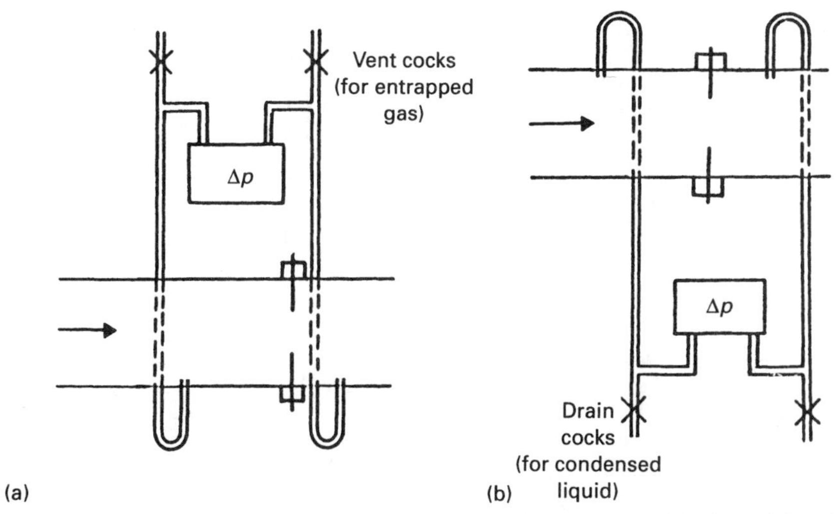

Fig. 5.15 *Preventing build up of liquid sumps and gas pockets if straight pipe connections are not feasible. (a) Liquid flow. (b) Gas flow.*

Flow	Δp	Δp + *offset*	*Indicated flow*
100	100	100.50	100.25
75	56.25	56.75	75.33
50	25.0	25.5	50.5
25	6.25	6.75	25.98
10	1.0	1.5	12.25
5	0.25	0.75	8.66
2	0.04	0.54	7.35
1	0.01	0.51	7.14

With a −0.5% zero error, the indicated flow will be zero at a true flow of about 7%. On a 4–20 mA signal, 0.5% corresponds to a zero error of less than 0.1 mA. Zero offset is not just a setting-up problem; zero drift also occurs as a result of hysteresis, temperature effects and ageing.

This low flow error limits differential pressure flow measurement to a turndown ratio(high flow/low flow) of at most 6 : 1. If a greater range is required, two or more Δp transmitters are used with range switching between them.

5.3. Turbine flowmeters

As its name suggests, a turbine flowmeter consists of a small turbine (usually four-bladed) placed in the flow as shown in Fig. 5.16. Within a specified flow range (usually about 10 : 1 turndown) the speed of rotation is directly proportional to flow velocity.

The turbine blades are constructed of ferromagnetic material and pass beneath

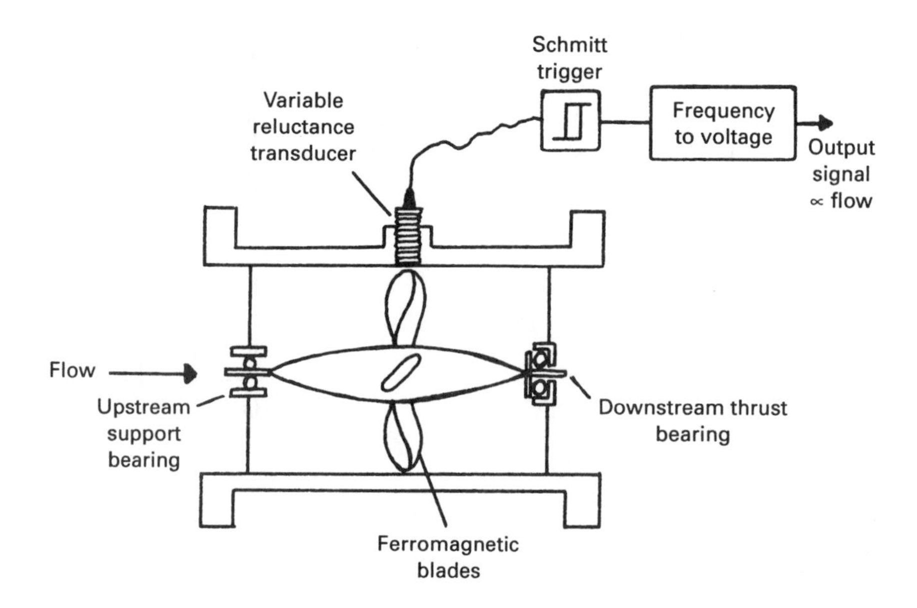

Fig. 5.16 *Turbine flowmeter.*

a magnetic detector operating as a variable reluctance transducer, which produces an output voltage approximating to a sine wave of the form:

$$E = A\omega \sin N\omega t \tag{5.13}$$

where A is a constant, ω is the angular velocity (proportional to flow velocity) and N is the number of blades. Both the output amplitude and frequency are proportional to flow velocity, although frequency-dependent circuits are normally employed to give a flow-dependent current or voltage output.

The lower flow limit is determined by frictional effects on the rotor or an unacceptably low amplitude sine wave from the magnetic detector. Other non-linearities occur from magnetic drag from the pick-up and viscous drag from the fluid itself. A typical accuracy is about $\pm 0.5\%$ with a $10:1$ turndown.

The flowmeter can be affected by swirling of the fluid itself. This can be overcome by the use of straightening vanes (Fig. 5.5) in the pipe upstream of the flowmeter.

Turbine flowmeters are relatively expensive, and less robust than differential pressure devices. Having mechanical rotating parts, they are prone to damage from suspended solids. Their main advantages are a linear output, superior turndown and a pulse signal that can be used directly for flow totalisation. Figure 5.17 shows a typical turbine flowmeter.

5.4. Variable area flowmeter

A variable area flowmeter consists of a tapered float in a vertical tapered glass tube as shown in Fig. 5.18. Fluid flows vertically past the float, which rises in the

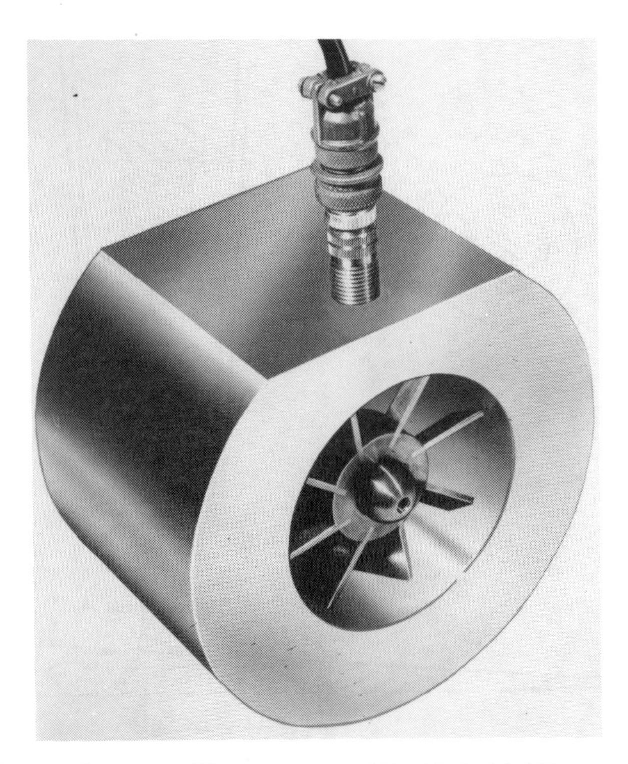

Fig. 5.17 *Turbine flowmeter. (Photo courtesy of Kent Industrial Measurements.)*

tube to a position which is dependent on the volumetric flow. The float forms an obstruction in the pipe, with cross-section as in Fig. 5.18a. There is therefore a pressure drop across the float Δp which will produce an upward force $\Delta p \,.\, A_1$. There will also be a downward force due to the displacement of the float (Archimedes' principle). If the float is in equilibrium, these forces will balance, i.e.

$$\Delta p \,.\, A_1 + V\rho_2 g = V\rho_1 g \tag{5.14}$$

Where V is the volume of the float, ρ_2 is the density of the fluid, and ρ_1 is the density of the float. This can be rewritten

$$\Delta p = h(\rho_1 - \rho_2)g \tag{5.15}$$

where h is the float thickness.

From equation 5.10 we have:

$$Q = KA_2\sqrt{\Delta p} \tag{5.16}$$

where A_2 is the area of the annulus and K is a constant.

From equation 5.15, the float will move to maintain a constant differential pressure Δp. From equation 5.16, with constant Δp, the area of the annulus, A_2, is proportional to the volumetric flow.

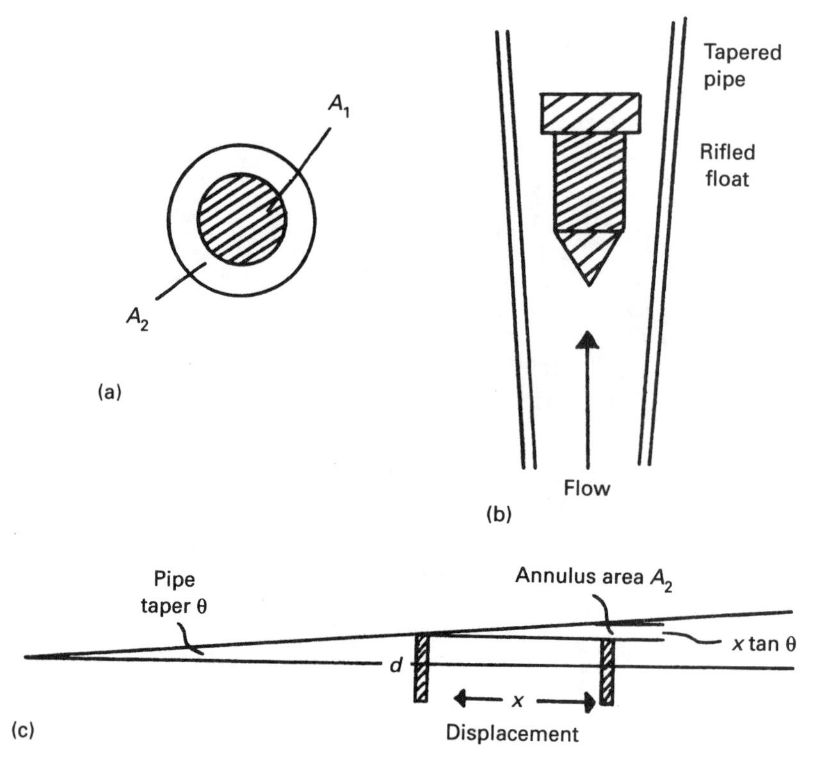

Fig. 5.18 *Variable area flowmeter. (a) Cross-section of float and pipe. (b) Side view. (c) Trigonometrical relationships.*

The measured variable is the float position, x. This is related to A_2 by Fig. 5.18c, where θ is the angle of the tube. θ is very small, typically a few degrees, so we can write:

$$A_2 = 2\pi \, \mathrm{d}x \tan\theta \tag{5.17}$$

Substituting back into 5.16 gives:

$$Q = 2\pi K \, \mathrm{d}x \tan\theta \sqrt{\Delta p} \tag{5.18}$$

or

$$Q = Bx \tag{5.19}$$

where B is a constant. The displacement, x, is directly proportional to flow.

A practical device often has a slight rifle around its perimeter to induce a stabilising spin. The variable area flowmeter is simple, linear and cheap, and has a good turndown ratio of about 10 : 1. It cannot be easily adapted for remote indication (although devices using the float as the core of an LVDT are available). Obviously, the fluid must be a clean and clear gas or liquid. If the float is made of a ferromagnetic material, a magnetic detector placed alongside the tube can be used to give a low (or high) flow alarm.

5.5. Vortex shedding flowmeter

If a bluff (non-streamlined) body is placed in a pipe as in Fig. 5.19, the flow cannot follow the surface of the object, and vortices detach themselves from the downstream side. The frequency of this vortex shedding is directly proportional to the volumetric flow rate, i.e.:

$$Q = Kf \tag{5.20}$$

where K is a constant, Q is the volumetric flow and f is the observed frequency. K is effectively constant for flows with Reynolds numbers in excess of 10^3. Surprisingly, K is largely independent of fluid density and viscosity, and is determined by the pipe diameter D, the obstruction diameter d, and the obstruction shape (e.g. rectangular, circular). A rectangular cross-section with $d/D = 0.25$ and $d/l = 1.5$ is usually used where l is the thickness of the obstruction as shown in Fig. 5.19b. The vortex shedding frequency is typically a few hundred hertz.

The vortex shedding flowmeter is an attractive device (Fig. 20). It has no moving parts, low head loss (compared with an orifice plate), an ability to work to relatively low Reynolds numbers, a linear response, and a turndown ratio of

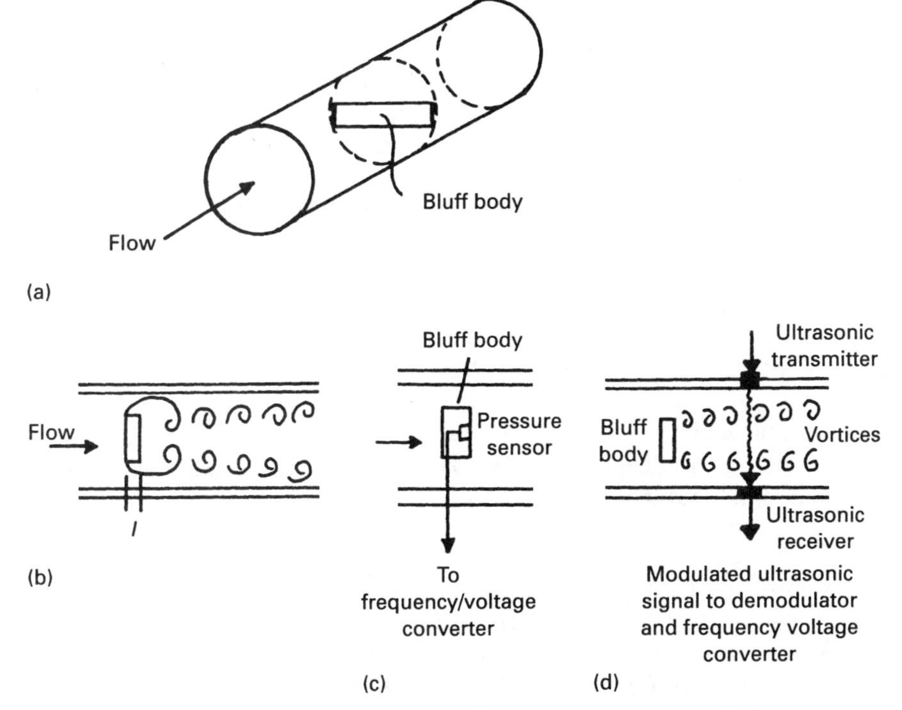

Fig. 5.19 *Vortex shedding flowmeters. (a) Construction. (b) Vortex formation. (c) Pressure sensor on downstream face of bluff body. (d) Ultrasonic method.*

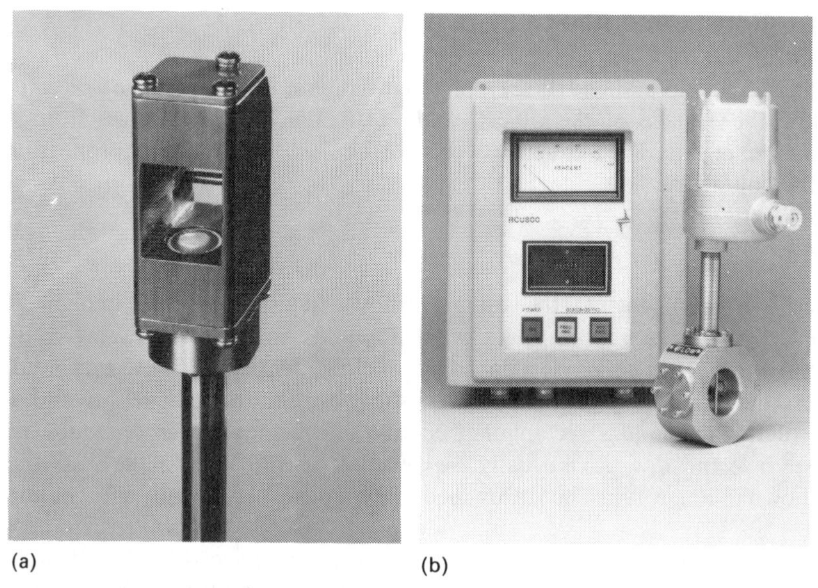

Fig. 5.20 *Practical vortex shedding flowmeters. (a) Vortex shedding head. (b) Complete unit. (Photos courtesy of Scheme Engineering.)*

about 15 : 1. Its one drawback is that secondary sensors are needed to detect the vortices.

The vortices manifest themselves as sinusoidal pressure variations. These can be detected by a sensitive diaphragm on the trailing edge of the bluff body. The movement of the diaphragm can be converted to a sinusoidal electrical signal by capacitance variations or via strain gauges. Alternatively, an ultrasonic beam across the pipe downstream of the bluff body can be modulated by the vortices. Whatever the detection method, the resulting sinusoidal output is converted to a constant-amplitude square wave prior to conversion to a linear voltage or 4–20 mA current signal by a frequency-to-voltage converter.

5.6. Electromagnetic flowmeter

In Fig. 5.21a, a conductor length l is moving with velocity v perpendicular to a magnetic field of flux density B. By Faraday's law of electromagnetic induction, a voltage E is induced where:

$$E = Blv \tag{5.21}$$

The electromagnetic flowmeter is based on the above equation, with the fluid forming the conductor. In Fig. 5.21b, a conducting fluid is passing down a pipe with mean velocity $\bar{v}$ (the use of mean velocity covers both laminar and turbulent flow conditions). An insulated section is inserted in the pipe as shown, and a magnetic field B is applied perpendicular to the flow. Two electrodes are inserted through the insulated section into the fluid, to form, with the fluid, a moving

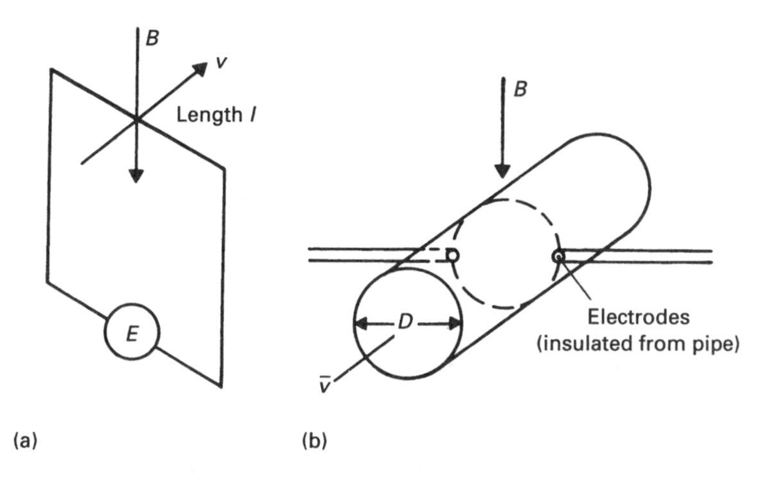

Fig. 5.21 *Electromagnetic flowmeter. (a) Electromagnetic principles. (b) Schematic. (c) Electromagnetic flowmeters. Note insulated PTFE lining. (Magflo unit. photo courtesy of Danfoss Ltd.). (d) Electrodes being installed after lining of sensor. (Photo courtesy of Danfoss Ltd.)*

conductor of length D where D is the pipe diameter. A voltage E will occur across the electrodes given by:

$$E = BD\bar{v} \tag{5.22}$$

The electromagnetic flowmeter therefore measures average flow velocity.

Figure 5.21 implies a steady field and a DC voltage. In practice, an AC field (and hence an AC-induced voltage) is used to prevent electrolysis effects at the electrodes, and also to minimise errors from DC voltages arising from thermoelectric and electrochemical effects which are of the same order of magnitude as the induced voltage. There is also the possibility that a DC current might actually distort the flow (magnetohydrodynamic effects).

The electromagnetic flowmeter is a flow velocity measuring device, but in general volumetric flow will be proportional to average velocity providing the flow characteristic (laminar/turbulent) does not change.

Although the electromagnetic flowmeter is linear, and has a good turndown ratio of about 15 : 1 and effectively zero head loss, it has several disadvantages. It is bulky and expensive to install (both the instrument itself and the physical installation/cabling) and can only be used on fluids with a conductivity in excess of 1 mS m^{-1}. This excludes all gases and many liquids. The main use of the device is in the flow measurement of 'difficult' liquids such as slurries with a large solids content.

5.7. Ultrasonic flowmeters

5.7.1. Doppler flowmeter

The Doppler effect occurs when there is a relative motion between a sound transmitter and receiver as shown in Fig. 5.22a. If the transmitted frequency is f_t Hz, v_s the velocity of sound, and v_r the relative velocity, then the observed frequency f_r is given by:

$$f_r = f_t \frac{(v_r + v_s)}{v_r} \tag{5.23}$$

i.e. the pitch of the note rises. If the transmitter and receiver are moving apart with relative velocity v_r,

$$f_r = f_t \frac{(v_r - v_s)}{v_r} \tag{5.24}$$

i.e. the pitch falls. The effect is commonly heard as a change in the pitch of a car engine as it passes. Equations 5.23 and 5.24 apply for both a stationary transmitter/moving receiver and a moving transmitter/stationary receiver.

Figure 5.22a assumes sound waves, although Doppler shift also occurs in electromagnetic radiation. Doppler shift of microwaves is used for speed measurement in radar 'speed traps', and Doppler shift of visible light (red shift) is used in astronomy to measure the velocity of recession of other galaxies.

A Doppler flowmeter injects an ultrasonic (high frequency, typically a few hundred kilohertz) sound wave into a moving fluid as shown in Fig. 5.22c. A small part of this sound wave is reflected off solid matter, vapour and air bubbles or eddies/vortices back to a receiver mounted alongside the transmitter. As it passes through the fluid, the frequency is subject to two changes (one travelling upstream against the flow, and one downstream with the flow).

The received frequency is therefore:

$$f_r = f_t \frac{(v_s + v \cos \theta)}{(v_s - v \cos \theta)} \tag{5.25}$$

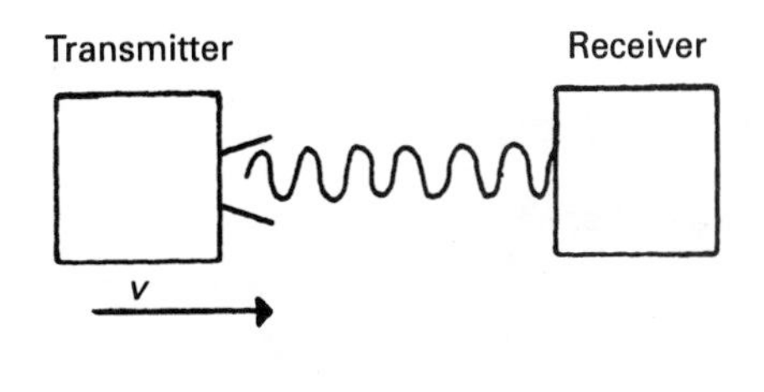

(a)

Oscillator
TX
Target
f_t
f_r
v
Envelope frequency ∝ v
RC
Mixer
Demod-ulator
Frequency to voltage

(b)

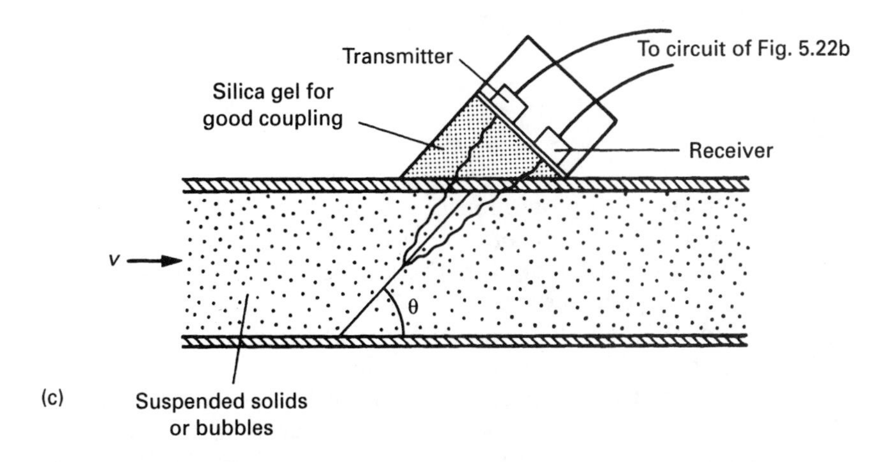

(c)

Fig. 5.22 *Ultrasonic flowmeters. (a) Doppler shift principle. (b) Block diagram of Doppler velocity measurement. (c) Practical clip-on ultrasonic flowmeter. (d) Ultrasonic flowmeters. (Photo courtesy of Danfoss Ltd.)* (continued overleaf)

(d)

Fig. 5.22 *contd.*

where v is the average fluid velocity and θ the angle of the ultrasonic beam to the fluid flow. If Δf is the change in frequency $(f_r - f_t)$ and $v \ll v_s$, equation 5.25 can be rewritten

$$\Delta f = \frac{2f_t}{v_s} v \cos \theta \tag{5.26}$$

The change in frequency is proportional to the average fluid velocity.

Figure 5.22b is a block diagram of a complete flowmeter. The transmitted and received frequencies are added together. This produces an amplitude-modulated signal with an envelope frequency of $\Delta f/2$. This is rectified and smoothed to produce a sinusoidal signal of frequency $\Delta f/2$, which is converted to a DC voltage or current proportional to the average flow velocity.

The Doppler flowmeter is linear, and can be installed without the need to break into the pipe (it is the only real 'clip-on' flowmeter). It can be used with all fluids, and is well suited for use with difficult fluids such as corrosive liquids or heavy slurries. The flowmeter works with laminar or turbulent flow (and even with transitional cases). It is rather expensive and over complex for straightforward flow measurement applications, however.

5.7.2. Cross-correlation flowmeter

There are many difficulties in measuring the flow of slurries: density changes, abrasion, blockage of sensors, etc. The cross-correlation flowmeter is designed to use the properties of suspended solids to measure the flow.

In Fig. 5.23, a slurry is flowing down a pipe from left to right with average velocity V. Two ultrasonic beams, a distance X apart, are passed across the pipe. Random variations in the phase and amplitude of the received signals will occur due to the random distribution of the solids in the fluid. If the distance X is not too large, output 2 will be related to output 1 but delayed by a time T_d which is given by:

$$T_d = X/V \tag{5.27}$$

Obviously the two outputs will not be identical, but with the correct choice of X (usually one pipe diameter) a correlation circuit can determine the value of the delay T_d which gives the maximum on the cross-correlation function as shown on Fig. 5.23c. With T_d known, the flow velocity can be deducted from equation 5.27.

Although ultrasonics are most commonly used with cross-correlation flowmeters, other techniques are available. Random variations of conductivity, pressure and even temperature have all been the basis of successful cross-correlation flowmeters. Cross-correlation techniques have also been successfully used for linear measurement of speed with two retro-reflected infrared beams. A typical application is the measurement of the velocity of hot steel strip in a high-speed rolling mill.

5.8. Hot-wire anemometer

If a fluid passes over a hot object, heat is removed. It can be shown that the power loss is given by:

$$P = A + B\sqrt{v} \tag{5.28}$$

where v is the flow velocity and A, B are constants. A is related to radiation and B to conduction power loss.

Figure 5.24 shows a flowmeter based on equation 5.28. A hot wire is inserted in the fluid, and maintained at a constant temperature by the self-balancing bridge. If the flow increases, say, more heat is removed, the temperature of the wire falls and its resistance falls. This unbalances the bridge, which is detected and the bridge voltage is increased until the temperature of the wire is restored.

With a constant wire temperature, the heat dissipated by the wire is equal to the power loss, i.e.

$$I^2R = A + B\sqrt{v} \tag{5.29}$$

or

$$v = K(I^2R - A)^2 \tag{5.30}$$

where K is a constant.

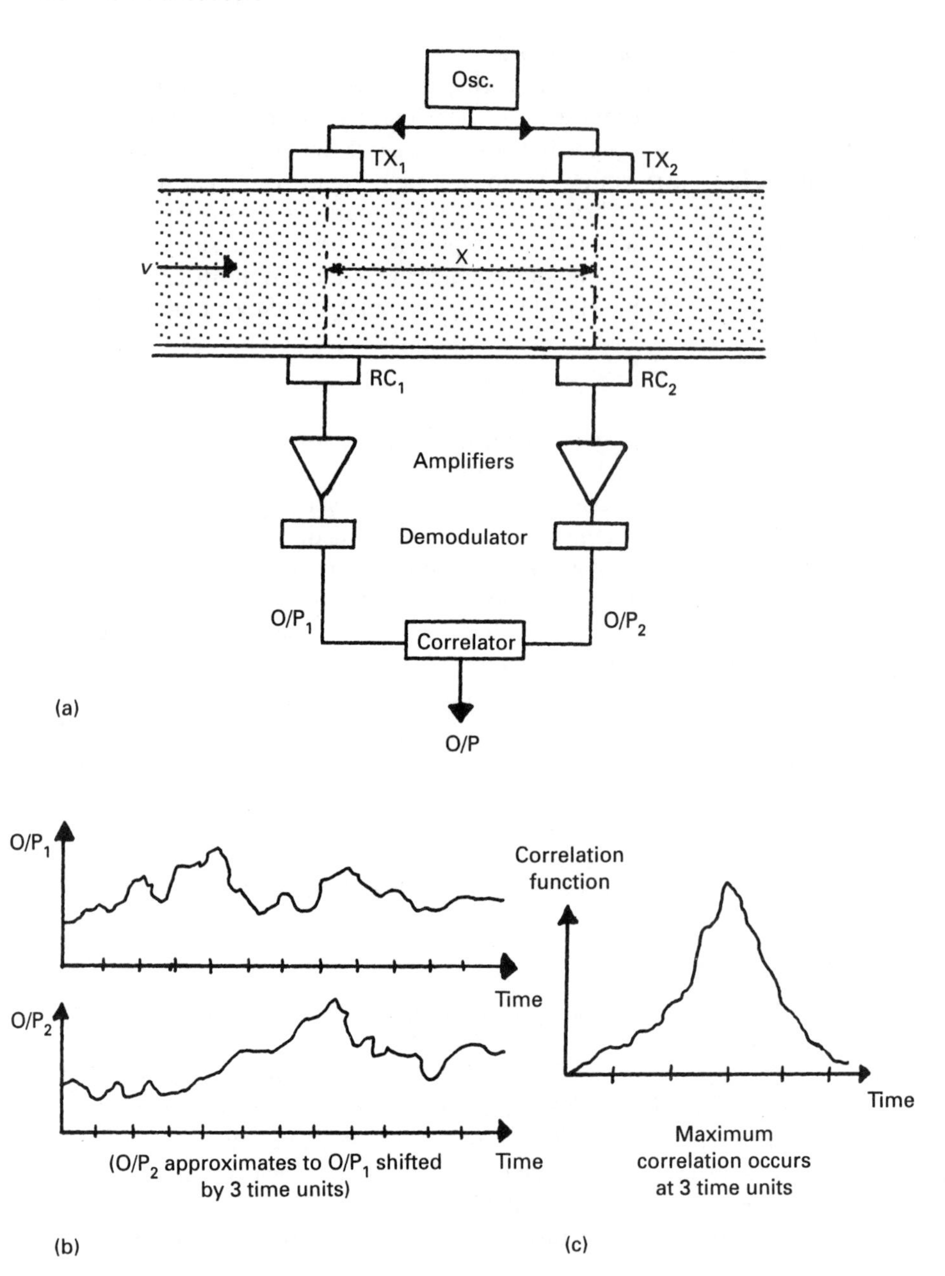

Fig. 5.23 *Cross-correlation flowmeter. (a) Transducer schematic. (b) Output signals. (c) Correlation function. (d) Cross-correlation flowmeter and electronics rack. (Photo courtesy of Kent Industrial Measurements.)*

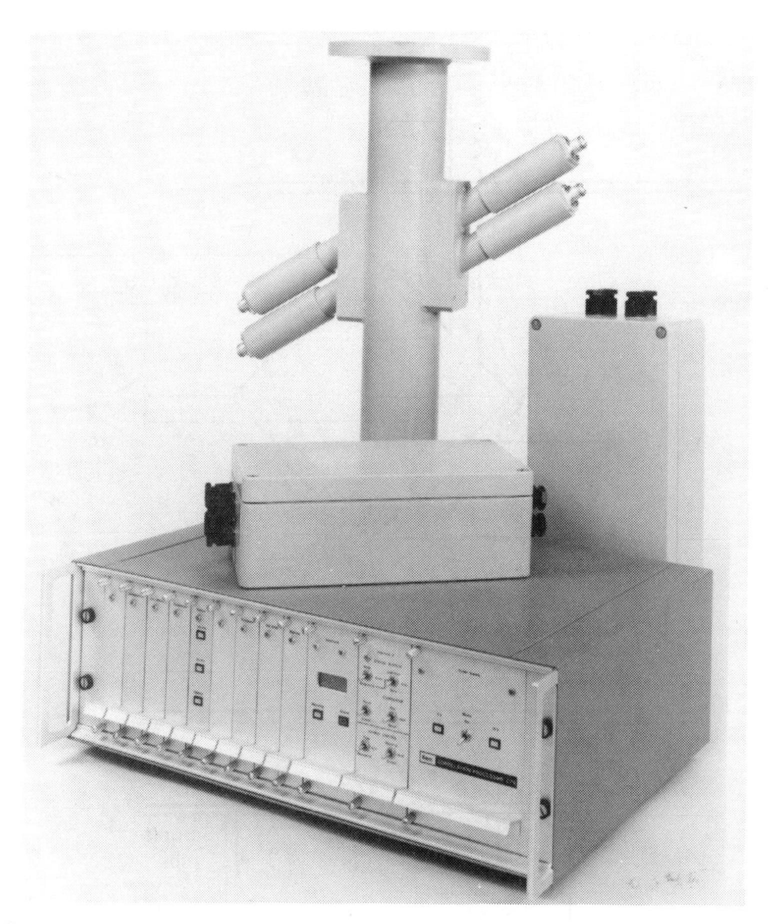

(d)

Fig. 5.23 *contd.*

The current I is measured (or converted to a voltage by the lower resistor in the bridge) from which the flow velocity can be deduced. Obviously the relationship is non-linear, and compensation for changes in fluid temperature is required.

5.9. Injection flow measurement

In some applications, an *ad hoc* flow measurement is needed where there is no real knowledge of the required range of flow. Figure 5.25 shows a method which can be readily adapted to measure gaseous flow over a wide flow range without a previous detailed study of the system.

An impurity gas is bled into the pipe at a known rate. Typical injection gases are CO_2 for town gas flow measurement, and oxygen for air (the oxygen content of air being fixed, it is easy to detect an increase). A gas analyser is used to sample

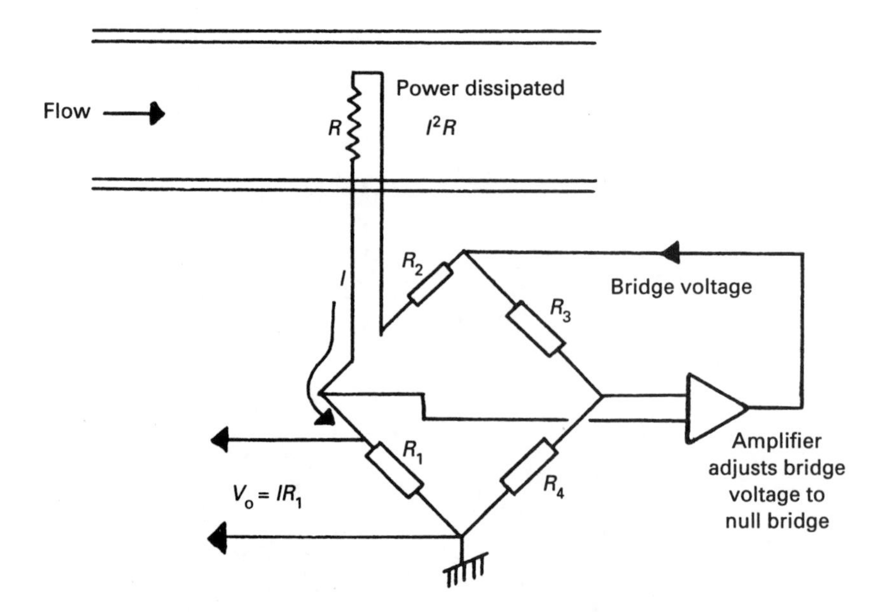

Fig. 5.24 *The hot-wire anemometer.*

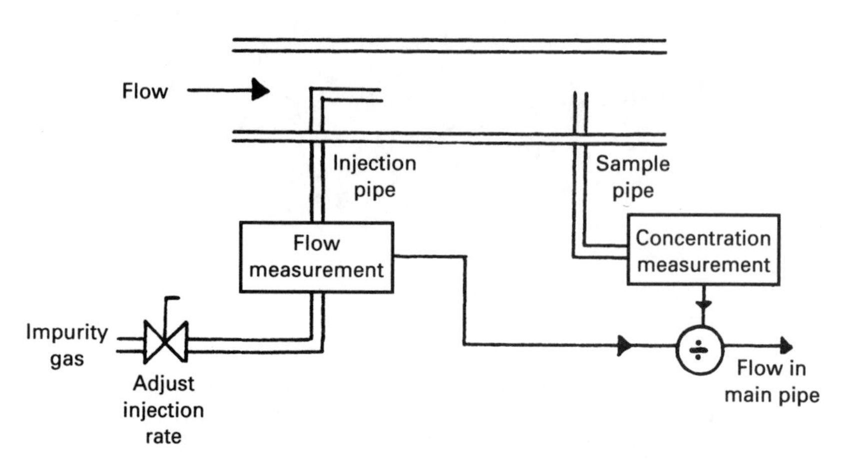

Fig. 5.25 *Injection-flow measurement.*

the percentage of the impurity gas downstream of the injection point (at a point where mixing is complete). As the injection flow rate, and the resulting percentage impurity are known, it is a simple calculation to find the main gas mass flow rate.

5.10. Flow in open channels

Measurement of flow in open channels is required in municipal water and sewage operations. This is most commonly performed by the use of weirs (see Fig. 5.26)

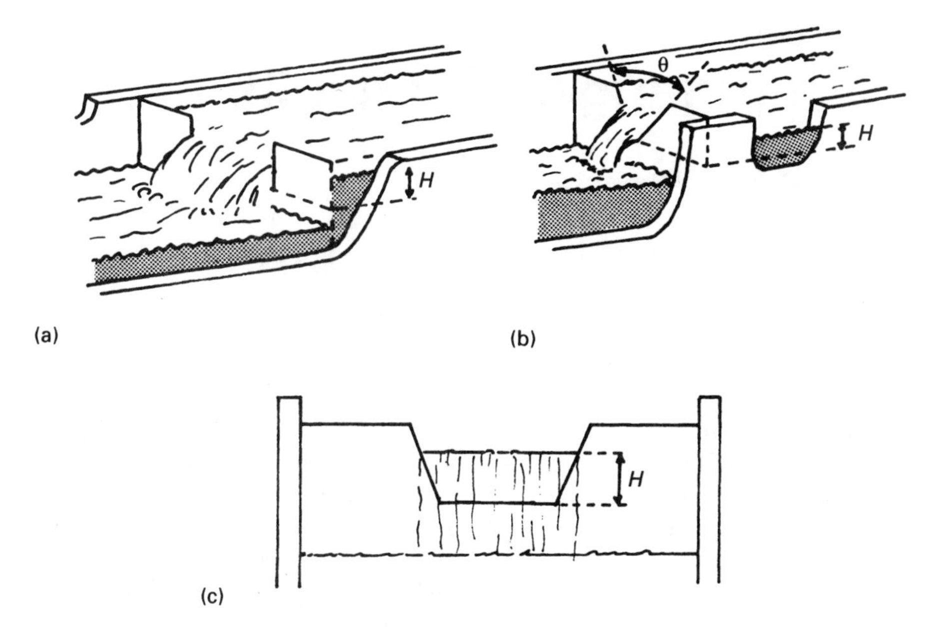

Fig. 5.26 *Flow in open channels. (a) Rectangular weir. (b) V-notch or Thompson Weir. (c) Trapezoid or Cipoletti weir.*

to produce a fluid head which is dependent on the volumetric flow. The fluid head can be measured by a float system or by measuring the pressure head via a pressure pipe.

The relationships are non-linear. For the rectangular and Cipoletti weir the relationship is:

$$Q = K H^{3/2} \tag{5.31}$$

where K is a constant dependent on the weir dimensions. For the V-notch (or Thomson) weir, the relationship is:

$$Q = K H^{5/2} \tag{5.32}$$

In all cases, the velocity of the fluid is kept low, or corrections must be made for the kinetic energy of the flow. Design of open-channel flowmeters is covered in BS 3680.

Chapter 6

Strain gauges, loadcells and weighing

6.1. First principles

6.1.1. Introduction

Measurement of weight is an essential part of most industrial processes, particularly where batch manufacturing is used. Weight is also the method of determining the value of most goods, so there is often an economic and legal requirement for accurate weight measurement. This chapter is concerned with techniques for measuring weight, and the closely related topic of strain measurement.

There are essentially two techniques of weight measurement, shown diagrammatically in Figs. 6.1 and 6.2. In a null balance weigher the applied weight causes a deflection of some sort in the structure of the weigher. An opposing force is then applied to bring the structure back to its unloaded, datum, position. In the steady state, the applied force matches the opposing force and the latter can be measured by a secondary transducer.

The earliest, and simplest, weighers are based on the arm balance (or steelyard)

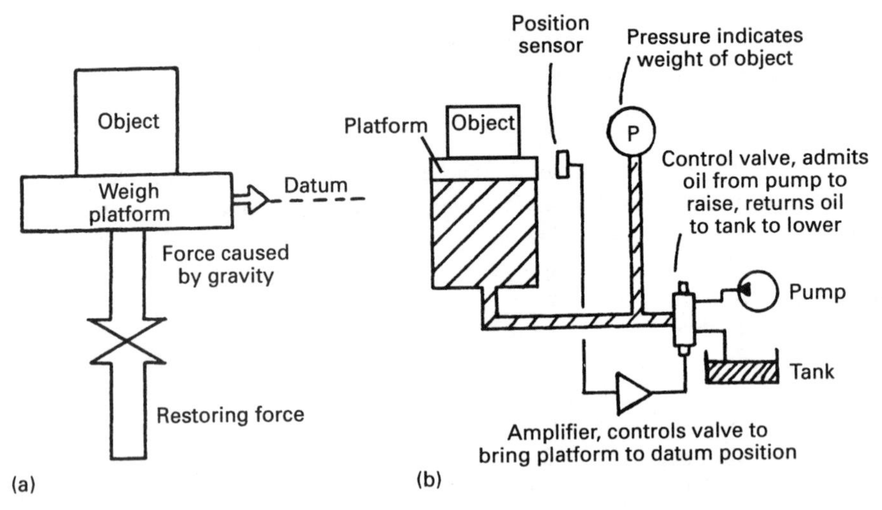

Fig. 6.1 *Force balance (null balance) weigher. (a) Principle. (b) Hydraulic implementation.*

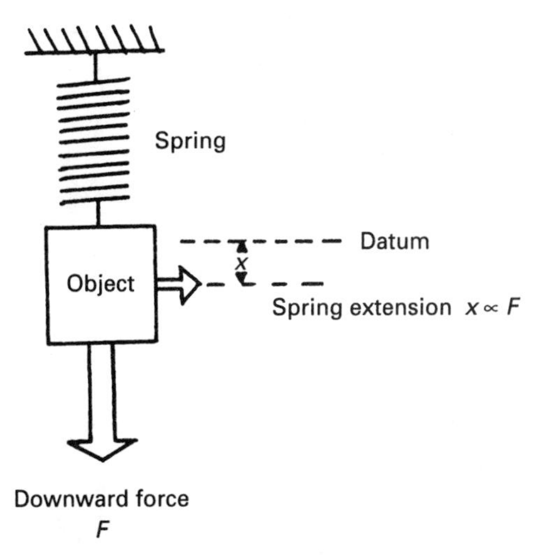

Fig. 6.2 *Strain weigher (elastic weigher).*

and these use the null balance principle. The unknown weight is balanced by a known weight, either by using a set of standard weights or by moving a known weight along the arm until a balance position is found. A more modern equivalent is shown in Fig. 6.1b. The load is balanced on a hydraulic ram, whose position is monitored by a LVDT (see Section 4.5.2). The position measurement is used to admit, or release, fluid until the null position is reached. The hydraulic pressure then indicates the load as the upwards force is (pressure $\times$ ram area). The hydraulic pressure can be measured by any of the transducers described in Chapter 3.

The second technique is the strain weigher, shown in Fig. 6.2. The applied load again causes a deflection of the weigher structure. This deflection is measured by a secondary transducer, and the load is deduced from a knowledge of the mechanical properties of the structure.

The simplest weigher is the spring balance. A spring is defined by its 'spring constant' which relates its extension to the load (e.g. 2 cm kg^{-1}). An applied load can easily be calculated by observing the spring extension and applying the spring constant. The simple spring balance can be made to give an electrical output by the use of a position measuring device, but in most industrial applications the need to attach peripheral equipment (such as piping) precludes large deflections. Most industrial weighers, however, are based on the strain weigher principle, but the secondary 'position' transducers are chosen to give minimal deflection.

6.1.2. Stress and strain

When any object is subjected to a force, deformation will occur. To be of any use, the relationship between the force and the deformation must be quantified. In Fig. 6.3a a tensile force is applied to a rod of cross-sectional area A and length l. This produces a deformation in length Δl. The effect of the force is assumed

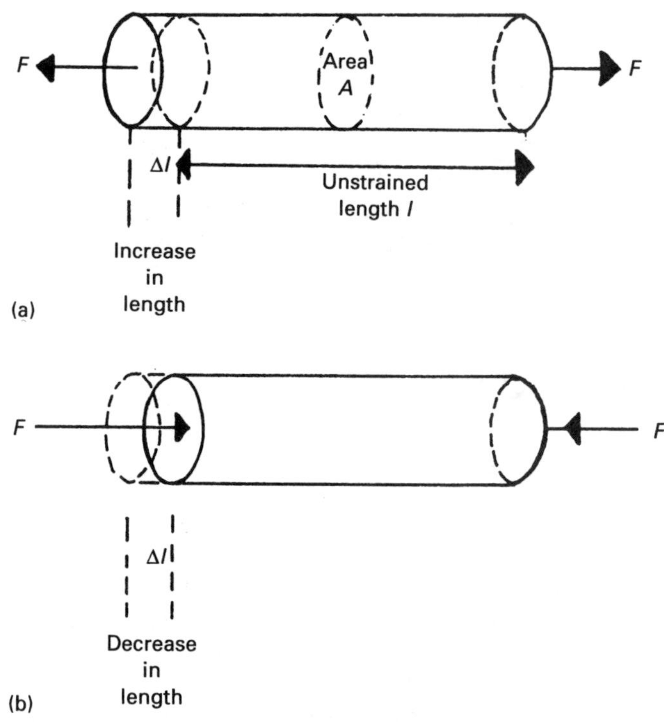

Fig. 6.3 *Tensile and compressive strain. (a) Tensile strain. (b) Compressive strain.*

uniform over the cross-section of the rod, and intuitively the larger the area the force acts over, the less deformation will occur. The effect of the force is called the *stress*, and is defined as force per unit area. i.e.

$$\text{Stress} = F/A \tag{6.1}$$

Stress has the units of N/m^2 (i.e. it has the same units as pressure).

Intuitively, again, it would be expected that the change in length, Δl, will be proportional to the unstrained length l as the stress is equal at all points of the rod. The deformation is called the *strain* and is defined as the fractional change in length:

$$\text{Strain} = \Delta l/l \tag{6.2}$$

As Δl and l both have the dimensions of length, strain is a dimensionless quality. Often, however, the dimensionless unit 'microstrain' is used. A 10 m rod, for example, exhibiting a stress-induced length change of 4.5×10^{-3} m, is exhibiting 45 μstrain. Changes in dimensions are usually small, and the μstrain is a convenient unit.

In Fig. 6.3b a compressive force is applied to a rod, and this produces a reduction in length (assuming that the rod stays straight, and bowing does not occur). Stress and strain are defined as above, with a change of sign for F and Δl to indicate the change of direction.

If an object is subjected to an increasing stress, the strain will obviously increase, resulting in a relationship as indicated in Fig. 6.4. In the region AB, the object behaves as a spring; the relationship is linear and there is no hysteresis (i.e. the object returns to its original size when the stress is removed; the deformation is not permanent). Beyond B, the object suffers permanent deformation. (In a bar or rod, necking starts to occur causing the stress to increase in a length of the object: see Fig. 6.4b. The strain increases non-linearly with stress until it breaks at point C.)

The linear portion AB is called the elastic region and point B the elastic limit. Typically AB will correspond to 10 000 μstrain. Obviously all mechanical and civil structures must be kept within the elastic region.

The inverse slope of the line AB is called the elastic modulus or modulus of elasticity, and is constant for a specific material. For simple tensile or

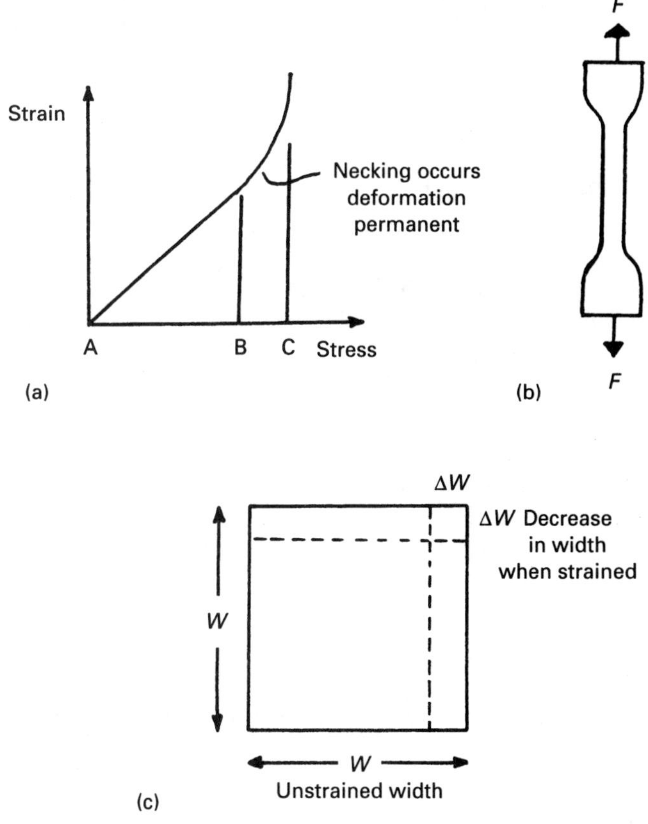

Fig. 6.4 *The effect of stress on an object. (a) Relationship between stress and strain. (b) Necking in a rod under tensile stress. (c) Poisson's ratio; change in cross-sectional area with stress.*

compressive stress the term Young's modulus is also used. The slope is defined by:

$$\text{Young's modulus} = \frac{\text{Stress}}{\text{Strain}} = \frac{F/A}{\Delta l/l} \tag{6.3}$$

As strain is dimensionless, Young's modulus has the dimensions of N m^{-2}, i.e. pressure, and is commonly given in pascals.
Typical values are:

Steel	210 GPa
Copper	120 GPa
Aluminium	70 GPa
Plastics	30 GPa

When an object suffers tensile strain, not only does it experience a length change, but it also experiences a decrease in cross-sectional area. Similarly an object undergoing compressive strain will exhibit an increase in cross-sectional area. This effect is quantified by Poisson's ratio, usually denoted by the Greek letter v. If an object has a length L and width W in its unstrained state, and experiences a change ΔL and ΔW when strained, the Poisson's ratio is defined as:

$$v = \frac{\Delta W/W}{\Delta L/L} \tag{6.4}$$

Typically v is between 0.2 and 0.4.

Poisson's ratio can be used to calculate the change in cross-sectional area. In Fig. 6.4, the unstrained area is $W.W$, and the change in area $2.W.\Delta W$ (neglecting the term ΔW^2). The fractional change is $2.\Delta W/W$, where ΔW can be calculated, knowing the tensile (or compressive) stress, via Young's modulus and Poisson's ratio.

6.1.3. Shear strain

A third type of strain is shown in Fig. 6.5. A force is applied to the top of a cube, causing it to distort. This is known as shear strain. To quantify shear strain we must first define shear stress.

Stress has the dimensions of force per unit area, so the shear stress on Fig. 6.5 is defined as:

$$\text{Shear stress} = F/A \tag{6.5}$$

where A is the cross-sectional area of the face.

Shear strain is the ratio between the displacement, BC, and the height of the cube AB, i.e.

$$\text{Shear strain} = \frac{\text{BC}}{\text{AB}} = \tan\theta \tag{6.6}$$

As strains are small, shear strain approximates closely to θ radians.

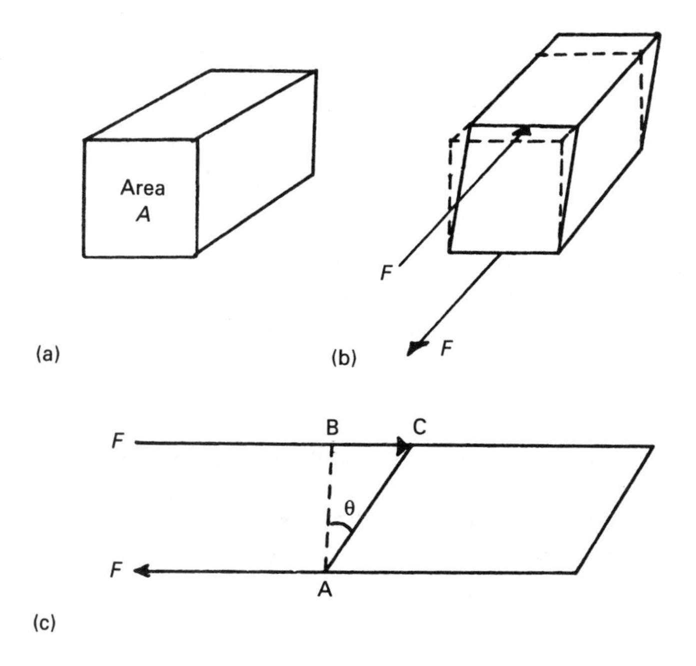

Fig. 6.5 *Shear strain. (a) Unstrained object. (b) Object with shear strain. (c) Side view and definition of shear strain.*

We can again define a modulus of rigidity, which has the same form as equation 6.3 except shear stress and shear strain are used. Typical values are:

Steel	75 GPa
Copper	45 GPa
Aluminium	25 GPa

6.2. Strain gauges

6.2.1. Introduction

The electrical resistance of a conductor such as that of Fig. 6.6 is proportional to its length, and inversely proportional to the cross-sectional area, i.e.

$$R = \rho \frac{L}{H.W} = \rho \frac{L}{A} \tag{6.7}$$

where ρ is a constant called the resistivity of the material.

When the conductor suffers tensile stress, L will increase (as determined by equation 6.3) and A will decrease (ΔW and ΔH being determined by equation 6.4). The resistivity also increases slightly as the material is deformed. The cumulative effect of these changes is to increase the resistance

$$R + \Delta R = (\rho + \Delta\rho)\frac{(L + \Delta L)}{(A - \Delta A)} \tag{6.8}$$

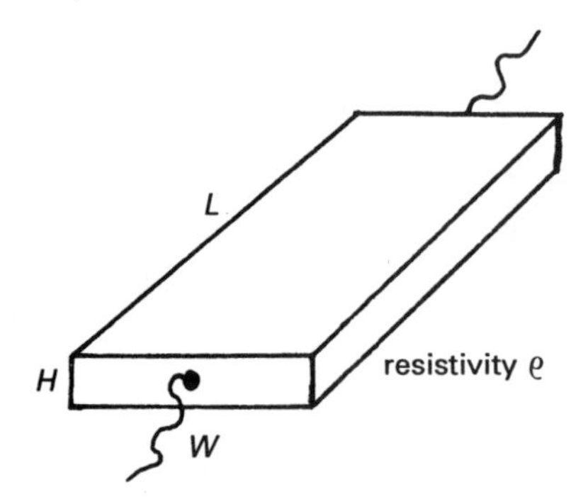

Fig. 6.6 *Electrical resistance of a slab conductor.*

Similar arguments show that a compressive stress will cause the resistance to decrease.

Ignoring second-order effects, the fractional change in resistance $\Delta R/R$ is proportional to $\Delta L/L$, i.e.

$$\frac{\Delta R}{R} = G\frac{\Delta L}{L} \tag{6.9}$$

where G is a constant called the gauge factor.

Strain, however, is given by $\Delta L/L$, so equation 6.9 can be rewritten:

$$\Delta R = G \, . \, R \, . \, e \tag{6.10}$$

where e is the strain exhibited by the conductor. This is the fundamental strain gauge equation.

Equation 6.10 allows us to relate an observed change in resistance to a strain (and hence to an applied force). A transducer using this principle is called a strain gauge. In practice, the force is not applied directly to the gauge. Usually the gauge is attached to some stressed member with epoxy resin. The strain experienced by the gauge will then be identical to the strain induced in the member.

6.2.2. Foil gauges

Practical strain gauges are not a slab of material as implied by Fig. 6.6. They must, of necessity, be flimsy devices so that their strain matches exactly that of the object to which they are attached.

Strain gauges must also ignore strains in unwanted directions. An active axis is defined for any gauge, and this is the direction in which the sensitivity is highest. Similarly, a passive axis is defined along which the gauge is least sensitive (usually at 90° to the active axis). These are related by the cross sensitivity:

$$\text{Cross sensitivity} = \frac{\text{Sensitivity along passive axis}}{\text{Sensitivity along active axis}} \tag{6.11}$$

In a well-designed strain gauge, cross sensitivities as low as 0.002 can be obtained.

Early gauges were constructed from thin wire, but modern gauges are photo etched from a metal film deposited on a thin polyester or plastic backing. This allows a device to be manufactured which has predictable properties. The backing material assists handling and attachment, and serves to insulate the gauge from the (probably metallic) object under test. These devices are called foil strain gauges.

A typical device is shown enlarged in Fig. 6.7. This has a length of 8 mm, an unstrained resistance of 120 ohm and a gauge factor of 2. Standard gauge resistance values are 120 and 350 ohm; most have a gauge factor of about 2. Leads are attached to pads on the foil.

Gauges are available in lengths from about 0.25 mm up to about 50 mm, although 5–10 mm gauges are commonest. The gauge length should be chosen such that the strain is even along the gauge, although there are obvious handling problems with small devices. Small gauges also tend to have degraded performance.

Standard gauges can be used up to about 10 000 μstrain, although permanent resistance changes can occur with cyclical strains above about 2000 μstrain. Strains up to 10% (100 000 μstrain) can occur in plastic and can be measured by wire gauges constructed from annealed constantan.

Foil gauges are effectively glued to the object under test, but the operation requires far more care than this simple statement might imply. It is imperative that the strain experienced by the gauge matches exactly the strain of the object. Many glues and epoxy resins have a 'creeping' property that allows strain relief in the gauge and causes responses as shown in Fig. 6.8. This property is common in adhesives designed to give a flexible joint. Resins designed specifically for strain gauges should be used.

The surface of the object must be clean and grease-free, and solvent degreasing is required just before attaching the gauge. It is essential to avoid touching the area with fingers and dragging in dirt from surrounding areas. Cleanliness is essential: any contaminated area can lead to a loss of strain coupling. The area next needs abrading with silicon carbide paper to give a good key for the resin.

Guide lines are now usually marked with a 4H pencil to assist with aligning the gauge. This is necessary as placing the gauge is a once-and-only operation; it must be right first time and cannot be skewed or slid into position. A conditioner is now applied to prepare the surface and remove any excess material from the guide lines.

Finally, the resin is applied in a thin layer with no entrapped air bubbles. The gauge must not be contaminated by finger contact and should be peeled down in the correct position first time, immediately after the application of the resin. Gauge handling is simplified by attaching a short length of adhesive tape to the top (non-contact) face of the gauge. The tape should not be removed until the resin has set. If weatherproofing is required, a coating of silicon rubber compound can be applied after lead connections have been made.

The change in resistance in a strain gauge is very small. Strain values in metals are normally less than 1000 μstrain (failure can occur at 10 000 μstrain). For a typical gauge of 120 ohm resistance and gauge factor 2, from equation 6.10 this

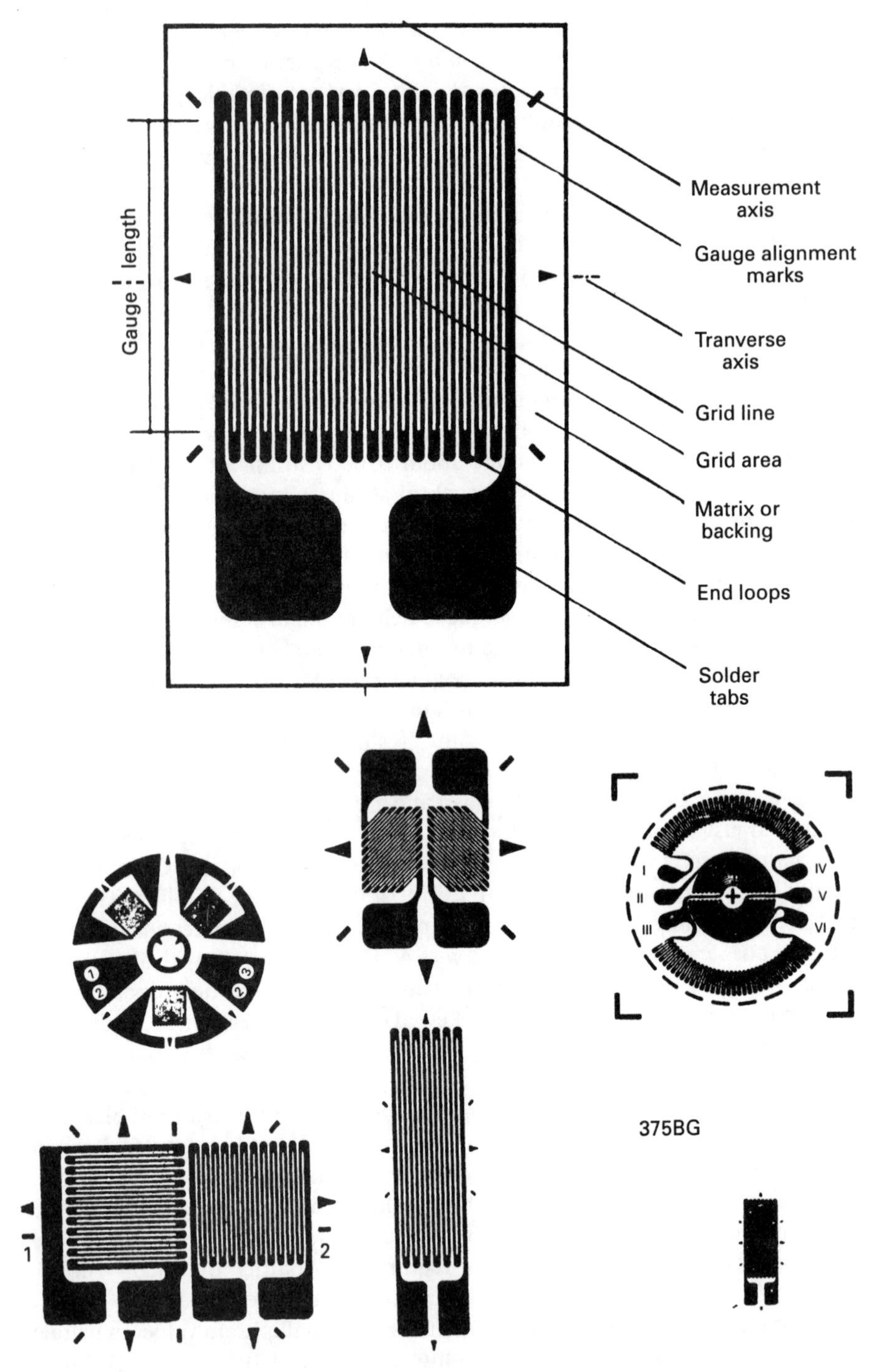

Fig. 6.7 *Typical strain gauges and terminology. Insert 375BG is shown full size. (Courtesy of Welwyn Strain Measurement Ltd. UK representatives of Measurement Group, Vishay, USA.)*

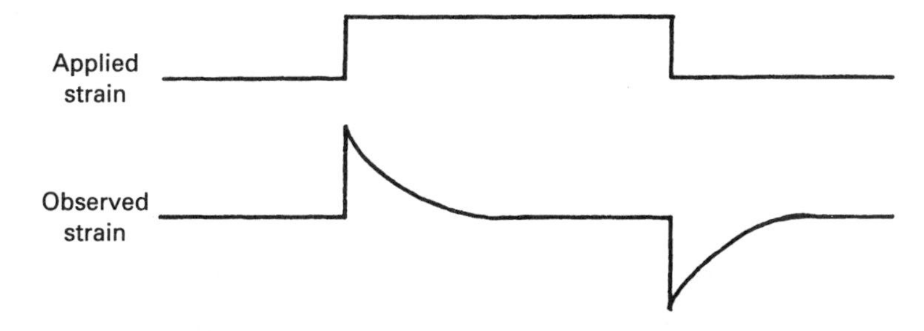

Fig. 6.8 *The effect of 'creep' in the mounting adhesive.*

corresponds to a resistance change of 0.24 ohm. In many applications resistance changes as low as 0.01 ohm must be measured.

Resistance also changes with temperature (see Section 2.4). These temperature effects are similar to, or larger than, the strain-induced effects. Bridge circuits, described in Section 6.3, provide temperature compensation. Temperature changes will also cause dimensional changes in the object under test, which are indistinguishable from strain-induced changes. Temperature-compensated gauges are available with coefficients of linear expansion matched to those of common materials such as mild steel.

6.2.3. Semiconductor strain gauges

The change in resistance with stress (as opposed to the change in resistance with strain discussed so far) is called the piezo-resistance effect. The effect is small in metals, but is significant in suitably doped semiconductor materials. Devices based on this principle are called semiconductor strain gauges (the external strain inducing a stress in the crystal).

Their main advantages are large gauge factors, typically 250, and small size (less than 0.5 mm), which allow them to be used, for example, in medical transducers. They are, however, very temperature sensitive. Both the gauge factor and the actual gauge resistance vary with temperature. The gauge factor also varies with the strain in the crystals. Semiconductor gauges are therefore usually restricted to applications requiring measurement of small, localised, dynamic strains.

6.3. Bridge circuits

6.3.1. Wheatstone bridge

The small change in resistance in a strain gauge is superimposed on a much larger unstrained resistance. Typically the change will represent one part in 1000 to 5000, which makes it difficult to produce an output voltage (or current) by direct means.

The classical laboratory method of measuring unknown resistance is the Wheatstone bridge shown in Fig. 6.9. In the normal laboratory method, R_a and

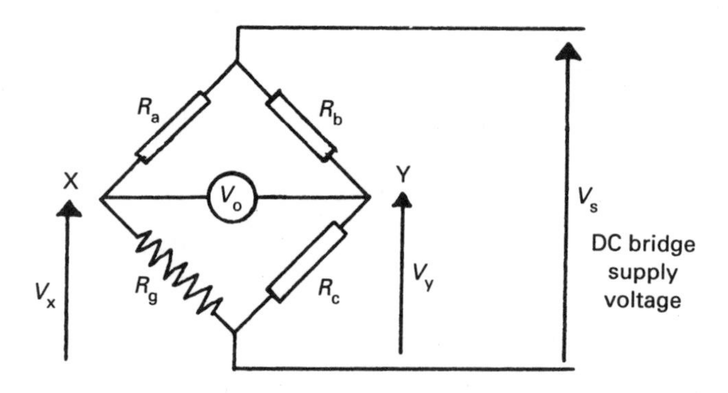

Fig. 6.9 *Measurement of strain with a Wheatstone bridge.*

R_b are equal, and the calibrated resistance box R_c is adjusted until V_o is zero (as measured by a sensitive galvanometer). Voltages V_x and V_y must be equal, so:

$$\frac{R_g}{R_g + R} V_s = \frac{R_c}{R_c + R} V_s \tag{6.12}$$

or

$$R_g = R_c \tag{6.13}$$

the value of the unknown resistor can be read from the calibrated resistance box R_c.

In strain-gauge work, however, we are not so much interested in the actual *value* as in its *change* from some norm. Suppose R_b and R_c are made fixed, and equal, resistances, such that V_y is half V_s. If R_a is made equal to the unstrained resistance of the gauge, V_x will also be 0.5 V_s, and V_o will be zero. If the gauge is strained, R_g will increase to $R_g + \Delta R$ causing a rise in V_x, and a non-zero reading of V_o. We must relate this change in V_o to the strain, e, causing ΔR. Let us assume that the unstrained gauge resistance is R, and to simplify the calculation we will represent the fractional change in resistance ($\Delta R/R$) by x. Assuming the device measuring V_o draws minimal current:

$$V_y = 0.5V_s \tag{6.14}$$

$$V_x = \frac{R_g}{R_g + R_a} V_s \tag{6.15}$$

$$= \frac{R(1 + x)}{R(1 + x) + R} V_s \tag{6.16}$$

$$= \frac{(1 + x)}{(2 + x)} V_s \tag{6.17}$$

Now $V_o = V_x - V_y$ (6.18)

$$= \frac{(1 + x)}{(2 + x)} V_s - \frac{V_s}{2} \tag{6.19}$$

$$= \frac{V_s\ x}{2(2 + x)} \tag{6.20}$$

Typically, x will be negligible compared with 2 (for the very large strain of 1000 μstrain calculated in Section 6.2.2, x was 0.002). Equation 6.20 thus approximates to:

$$V_o = \frac{V_s\ x}{4} = \frac{\Delta R}{4R} x \tag{6.21}$$

But $\Delta R = eGR$, so:

$$V_o = \frac{eGR_x}{4R} V_s \tag{6.22}$$

$$= \frac{eG}{4} V_s \tag{6.23}$$

Equation 6.23 implies that the change in output voltage is linear and proportional to the strain, and the sensitivity is dependent on the supply voltage V_s and the gauge factor G. It is independent of the unstrained gauge resistance. This result is, however, conditional on the assumption that x is small (which is usually a more than reasonable assumption).

It is instructive to put some values into equation 6.23. For the gauge of Fig. 6.9, G has the value 2. Choice of bridge voltage will be discussed later, but let us assume $V_s = 24$ V. For 1000 μstrain, we get:

$$V_o = \frac{1000 \times 10^{-6} \times 2 \times 24}{4}$$

$$= 12 \text{ mV}$$

i.e. very small. The bridge output voltage requires considerable amplification before it can be used for indication or control.

6.3.2. Temperature compensation

The circuit of Fig. 6.9 cannot distinguish between resistance changes caused by strain and those caused by temperature effects. In the arrangement of Fig. 6.10a, gauge 1 is aligned with the strain, and gauge 2 is aligned across the strain. Gauge 1 will exhibit a change in resistance as predicted by equation 6.10 (plus temperature effects) whereas gauge 2 will only have a small change due to strain (as predicted by equation 6.11) plus a change due to temperature. As both gauges are in close proximity it is reasonable to assume that both are at the same temperature.

Let x be the fractional change in resistance due to strain (occurring in gauge 1 only) and t the fractional change in resistance due to temperature. As before, $V_x = 0.5\ V_s$:

$$V_x = \frac{R_1}{R_1 + R_2} V_s \tag{6.24}$$

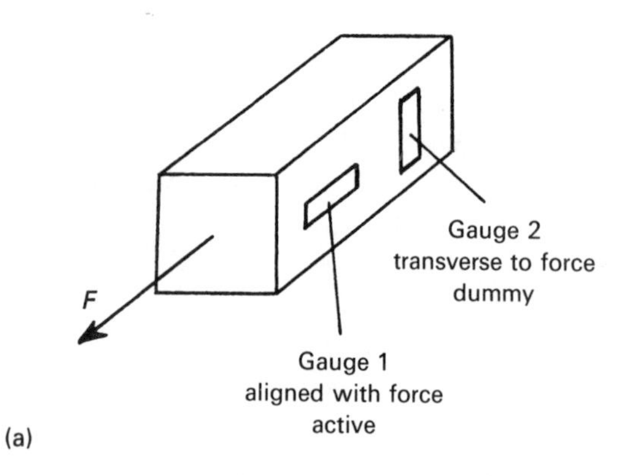

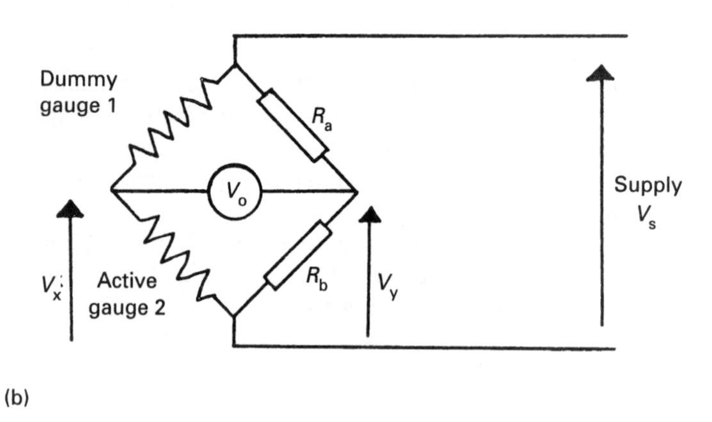

Fig. 6.10 *The use of a dummy gauge for temperature compensation (a) Mounting arrangement. (b) Wheatstone bridge circuit.*

$$= \frac{R(1 + x)(1 + t)}{R(1 + x)(1 + t) + R(1 + t)} V_s \qquad (6.25)$$

Terms in $(1 + t)$ cancel, leaving:

$$V_x = \frac{R(1 + x)}{R(1 + x) + R} V_s \qquad (6.26)$$

$$= \frac{(1 + x)}{(2 + x)} V_s \qquad (6.27)$$

which is the same result as equation 6.20, leading, therefore, to the same result as equation 6.23. The inclusion of the dummy gauge has compensated completely for resistance changes caused by variations in temperature.

6.3.3. Multigauge bridges

Many applications utilise gauges in all four bridge arms. In Fig. 6.11a, R_1 and R_3 are aligned with the strain, and R_2 and R_4 transversely. Connected as shown in Fig. 6.11b, V_x will increase and V_y decrease with strain, giving twice the sensitivity of Fig. 6.10. R_3 and R_4 act as dummy gauges. As all four gauges are affected equally by temperature changes, the bridge is temperature compensated.

The arrangement of Fig. 6.11c is used to detect bending strain and ignore axial strain. As the object bends, material above the neutral axis will suffer positive strain and that below negative strain. If the gauges are again connected as in Fig. 6.11b, bending strain as shown will cause V_x to increase and V_y to decrease.

Axial strain affects all four gauges equally, and consequently will not affect V_o. Similarly, temperature changes will cause equal changes in all gauges and will be ignored.

6.3.4. Bridge balancing

Gauges are manufactured to a resistance tolerance of about 0.5%; about ± 0.6 ohm for a 120 ohm gauge. As this is greater than the change caused by strain, some form of zeroing will be required so that V_o is zero in the unstrained state. It is possible to incorporate this in the amplifier stages after the bridge, but it is common to include bridge balancing in the bridge circuit itself.

There are three common ways of achieving this, shown in Fig. 6.12. In all three, the potentiometer is adjusted to zero V_o in the unstrained state. Arrangements 6.12b and 6.12c are preferred if the potentiometer is remote from the bridge, as errors induced by the resistance of the interconnecting cables (and changes in the cable resistance caused by temperature) will not cause a span or zero shift in V_o.

6.3.5. Bridge connections

Equation 6.23 predicts that the bridge sensitivity is proportional to the bridge supply voltage. Increasing the supply voltage, however, increases the bridge dissipation (which rises as the square of the voltage). In the extreme this could permanently damage the gauges, but errors caused by gauge heating will occur at much lower powers.

Gauge manufacturers specify a maximum dissipation, gauge current or bridge voltage. In general, higher resistance gauges can be used at higher voltages. Typical bridge voltages and currents are 15–30 V and 20–100 mA.

Equation 6.23 also indicates that V_s must be stabilised. If the bridge is remote from the power supply unit (PSU), bridge-balancing potentiometer and amplifier electronics, error can be introduced from the resistance of the connecting cables.

There are many possible ways of compensating for lead resistance, and one solution is shown in Fig. 6.13. The bridge supply is provided on lines 1 and 2 from a stabilised power supply. The actual bridge voltage is sensed on lines 3 and 4, which provide feedback so the PSU gives the correct voltage at the bridge (rather than at the PSU terminals). Leads 5 and 6 are part of the apex balancing circuit, but as both leads have the same resistance, no error is introduced. The other apex of the bridge is brought back on lead 7. Leads 3, 4 and 7 are sensing leads carrying

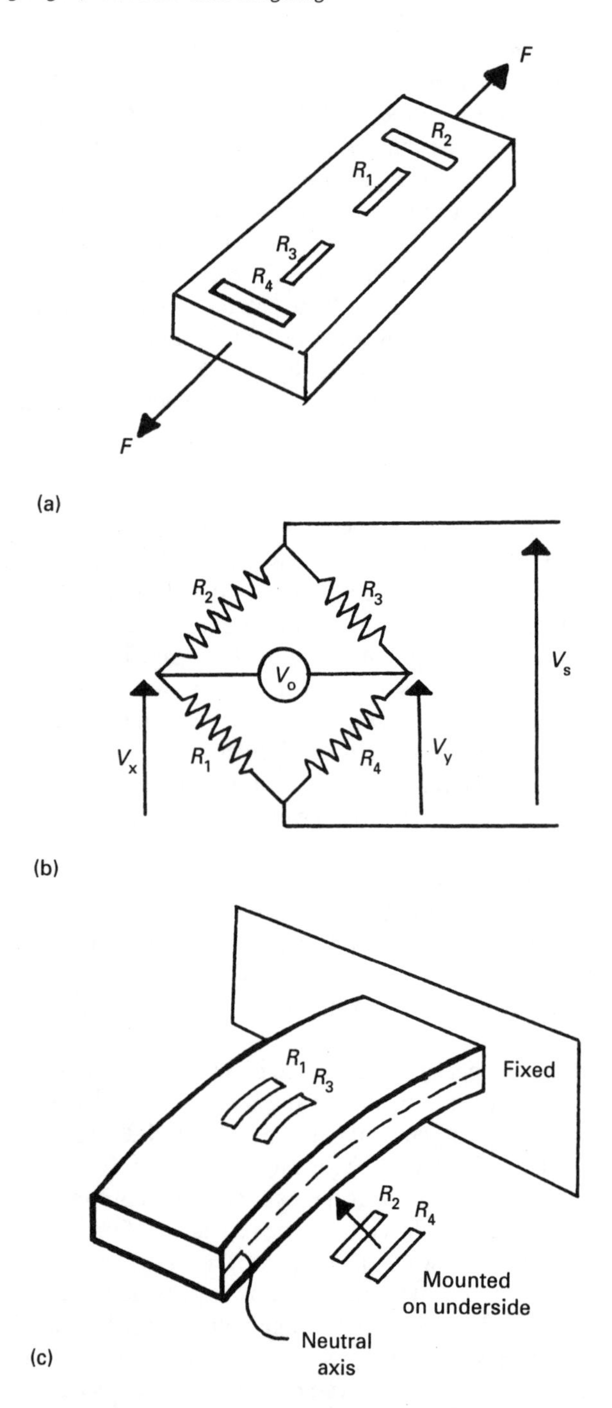

Fig. 6.11 *Four-gauge applications. (a) Four-gauge mounting. (b) Four-gauge bridge giving increased sensitivity. (c) Measuring bending strain in presence of axial strain.*

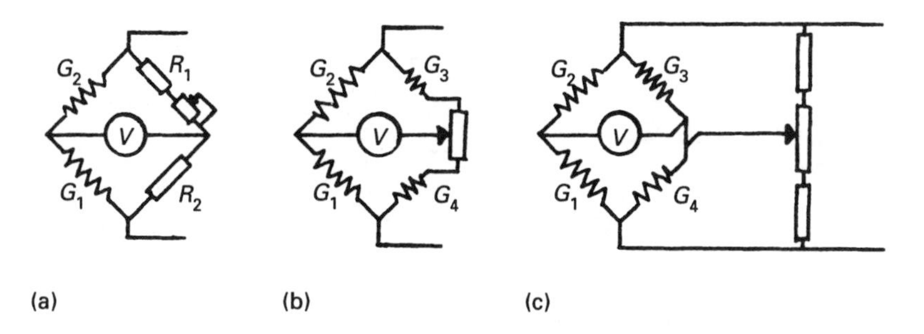

Fig. 6.12 *Methods of bridge balancing. (a) One leg balancing ($R_1 < R_2$). (b) Apex balancing. (c) Parallel balancing.*

minimal current, so the errors caused by the lead resistance are negligible.

6.3.6. Bridge amplifiers

The bridge output voltage of a few millivolts needs to be amplified before it can be used. Care must be taken to avoid common mode induced noise, and it is usual to employ a differential amplifier similar to Fig. 6.14. This circuit is discussed in detail in Chapter 11, but if $R_a = R_b = R_1$, and $R_c = R_d = R_2$, the output voltage is:

$$V_o = \frac{R_2}{R_1}(V_2 - V_1) \tag{6.28}$$

$$= -\frac{R_2}{R_1}V_b \tag{6.29}$$

where V_b is the bridge output voltage. The circuit amplifies the difference between its input terminals and ignores common mode voltage. The gain is simply $-R_2/R_1$.

Any real-life voltage source can be represented by a perfect voltage source in

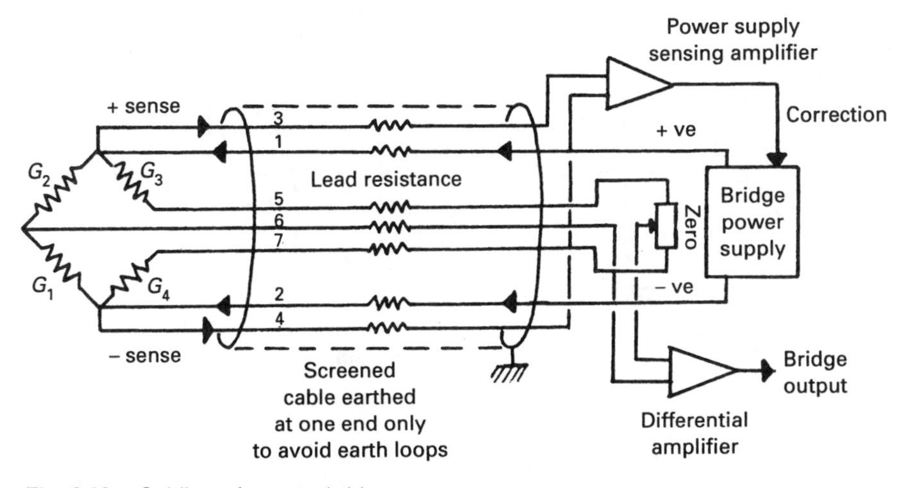

Fig. 6.13 *Cabling of remote bridge.*

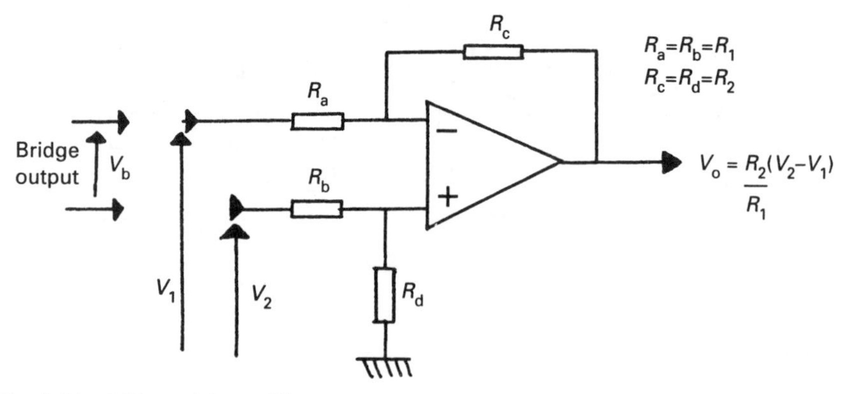

Fig. 6.14 *Differential amplifier.*

series with a resistor (Thévenin's theorem). The voltage from each apex of a bridge such as Fig. 6.15a can be considered as an open circuit voltage V_x in series with a resistor. The value of this resistor is simply $R_g/2$ where R_g is the nominal strain-gauge resistance.

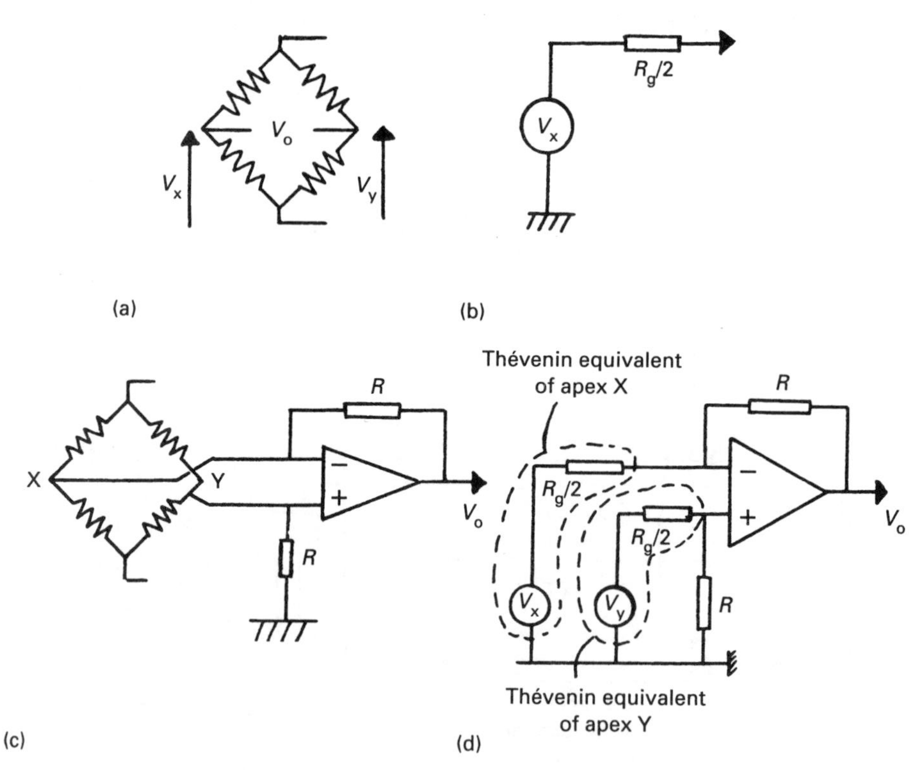

Fig. 6.15 *Thévenin representation of Wheatstone bridge. (a) Bridge circuit. (b) Thévenin equivalent circuit of an apex. (c) Bridge connected to differential amplifier. (d) Equivalent circuit.*

In Fig. 6.15c, the bridge is connected direct to the amplifier without resistors R_a and R_b of Fig. 6.14. Thévenin's theorem allows us to draw this circuit as Fig. 6.15d with the equivalent circuit for each bridge apex voltage. The output voltage from the amplifier is:

$$V_o = -\frac{2R}{R_g}(V_y - V_x) \tag{6.30}$$

6.3.7. Torque measurement

Torque is a twisting force, usually encountered on shafts, bars, pulleys and similar rotational devices. It is defined as the product of the force and the radius over which it acts. The shaft in Fig. 6.16a, for example, is experiencing a torque of 4.2 Nm.

Torque produces distortion of a shaft which can be visualised as a 'wind-up' along the shaft. This wind-up is a form of shear strain, and in Fig. 6.16a the shaft has been distorted by an angle θ. Torque-induced shear strain is defined by θ if θ is expressed in radians.

A shear modulus G can be defined for the shaft. T and θ are then related by the expression:

$$T = \frac{\pi G r^3}{2}\theta \tag{6.31}$$

where r is the shaft radius. Note that the shear strain varies inversely as the third power of the shaft radius.

Shear strain induced by torque can be measured by a strain gauge mounted at 45° to the axis as in Fig. 6.16. In the unstressed state, the gauge will lie along the line AB. When torque is experienced, the gauge lies along the line AC. The line AC is longer than AB, so the gauge experiences a strain:

$$e = \frac{\mathrm{AC} - \mathrm{AB}}{\mathrm{AB}} \tag{6.32}$$

$$= \theta/2 \tag{6.33}$$

where θ is small (and expressed in radians).

In practice, two or more gauges are used to provide increased sensitivity and temperature compensation. In Fig. 6.16c, for example, gauge 1 is experiencing tensile strain and gauge 2 compressive strain.

There are practical problems in connecting measuring equipment to gauges on a rotating shaft. This is usually achieved by slip rings or by using an AC bridge supply and coupling both bridge supply and output signal via a transformer. If the centrifugal forces are not large and an out-of-balance load can be tolerated on the shaft, an entire battery-based power supply, amplifier and low-powered radio telemetry transmitter can be strapped to the shaft.

Torque measuring devices are used for determining the power of engines, motors, and other rotating devices. Such instruments are commonly called dynamometers.

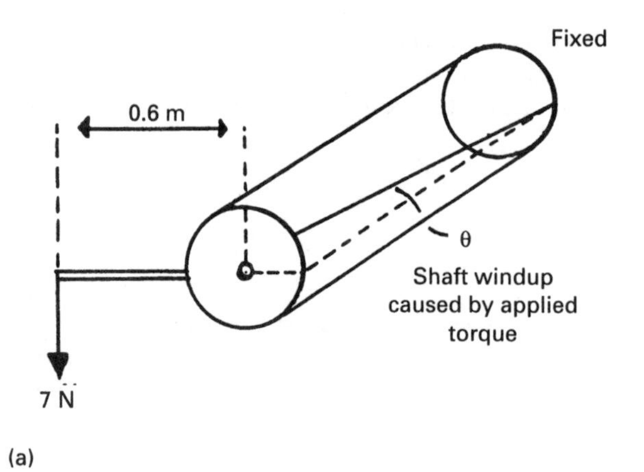

(a)

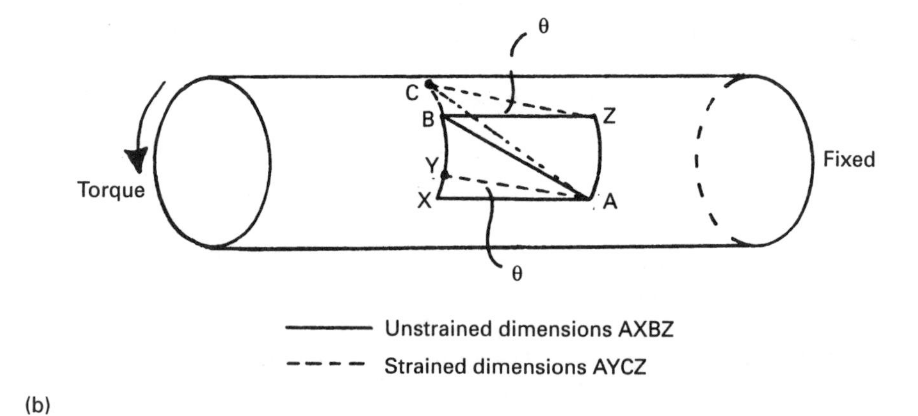

(b)

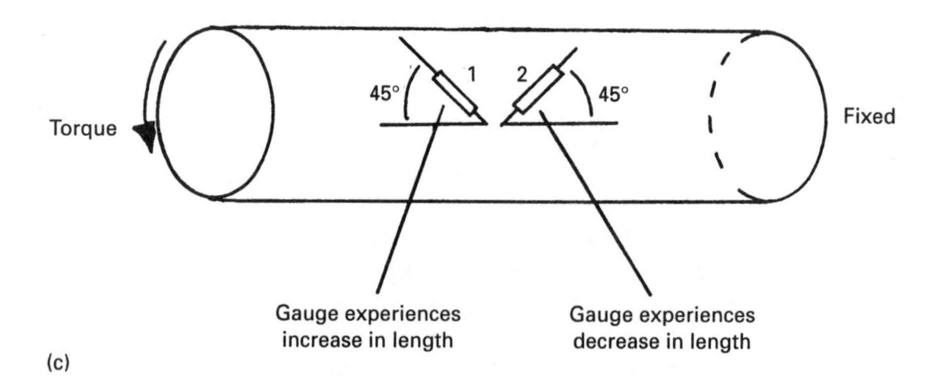

(c)

Fig. 6.16 *Measurement of torque. (a) Definition of torque and torque-induced strain. (b) Dimensional effects of torque-induced strain. (c) Mounting strain gauges to measure torque-induced strain.*

6.4. Magnetoelastic devices

The strain gauge is not the only force measuring device. An important class of devices uses a change in magnetic permeability with applied force as a load sensing mechanism. The principle, known as the magnetoelastic effect, is shown in Fig. 6.17.

In Fig. 6.17a, a magnetic field, H, is applied at 45° to a slab of steel. This induces equal flux densities B_v and B_h, giving a net vector **B** in the same direction as H. Permeability, however, is decreased by an applied force. In Fig. 6.17b the applied force causes a reduction in B_v (B_h being unchanged). The net flux vector **B** is no longer aligned with H, and the deviation is related to the force F.

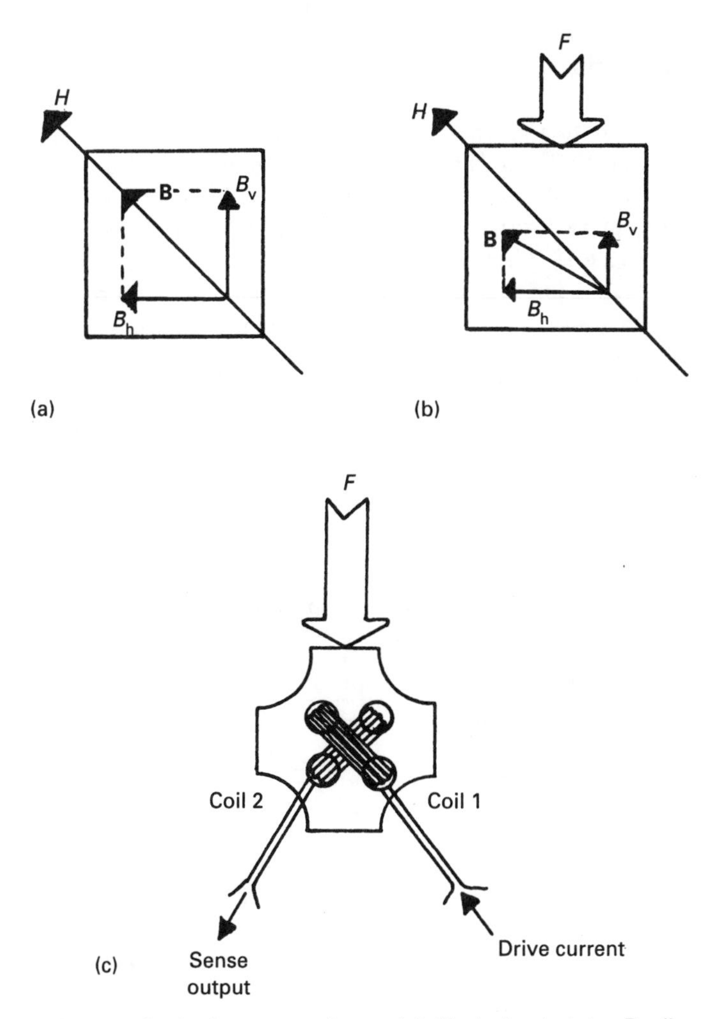

Fig. 6.17 *Magnetoelastic force transducer. (a) Unstrained state, B aligns with H. (b) Strained state, B no longer aligns with H. (c) Practical device.*

This principle can be applied as shown in Fig. 6.17c. Two coils are wound through holes in a transformer steel core. The coils are precisely aligned at 90° so in the unstressed state there is no coupling between them. Coil 1 is driven by an AC source, but no voltage is induced in coil 2.

When a force is applied, the shift of the **B** vector causes a voltage to be induced in coil 2; the magnitude of the voltage being dependent on the applied force.

Magnetoelastic transducers are primarily marketed by the Swedish company ASEA under the trade name Pressductor. These devices are physically much more robust than strain gauge devices and have an output signal of several hundred millivolts (compared with the few millivolts of a strain gauge bridge) and as such are less affected by noise. The drive electronics are, however, more complex, and the device's non-linear response requires more complex linearisation circuits.

6.5. Loadcells

Strain gauges and (to a lesser extent) magnetoelastic devices are strain measuring devices. A weighing device is required to measure weight (or more pedantically force). A force measuring device is called a loadcell, and usually works by converting an applied force to a measurable strain. Common sensing arrangements are shown in Fig. 6.18. These would all be enclosed in a protective case. All utilise four strain gauges for maximum sensitivity and temperature compensation (magnetoelastic transducers all use a sensor similar to that shown in Fig. 6.17c).

The column type is only used for large loads as the strain is small. The proof ring, proving frame and cylinder give reasonable outputs at low loads. All rely on two strain gauges in compression and two in tension. The gauges are arranged in a circuit, as shown in Fig. 6.19. The span resistors are adjusted on test to give a

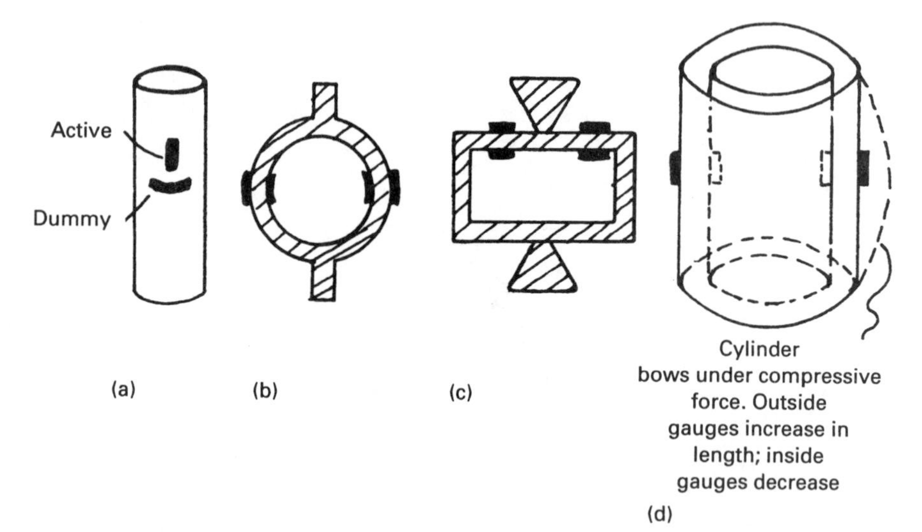

Fig. 6.18 *Various load sensors with gauge mounting positions. (a) Simple column. (b) Proof ring. (c) Proving frame. (d) Cylinder.*

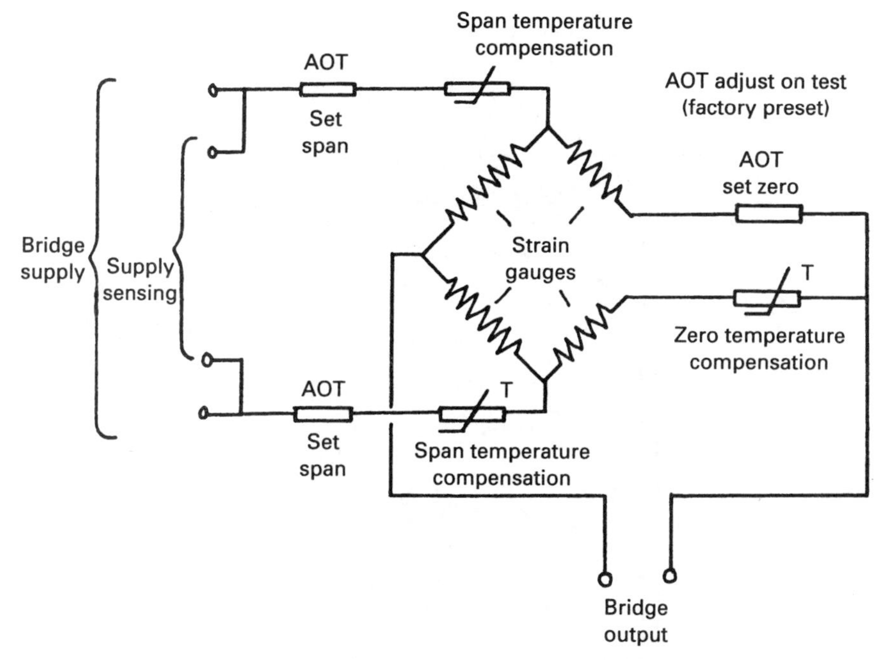

Fig. 6.19 *Schematic diagram of industrial weighing loadcell. Note that a six-core screened cable is required.*

predetermined output for the rated load. The temperature compensation resistors compensate for changes in Young's modulus of the cell with temperature (not for bridge changes which are, of course inherently ignored by the bridge).

Coupling of the load to the cell requires care. A typical arrangement is shown in Fig. 6.20. A pressure plate, effectively a metal/rubber/metal sandwich, applies the load to a knuckle on the top of the loadcell and avoids errors caused by slight misalignment. A flexible diaphragm seals the cell against dust and the weather. Practical devices are shown in Fig. 6.21. The cell in Fig. 6.21a is a compressive cell and that in Fig. 6.21b is a tensile cell.

Only rarely is a single loadcell used as it is difficult to keep a load vertical on a single point support. Guide rods and tie bars can be used, but care must be taken to avoid errors from friction. Single cells can be used in suspension weighers (Fig. 6.22a). Multicell systems are commoner, the best arrangement being the three-cell system of Fig. 6.22b. This inherently spreads the load roughly equally among the three cells, whose outputs are summed electronically. A four-cell system such as in Fig. 6.22c can be used, but care must be taken to ensure that the load is shared at all times among the cells and that no cell ever loses contact with the load platform.

The choice of a loadcell is obviously determined by the maximum expected load, but there are other factors to consider. An overload capacity should be included to cover mechanical failures and shock loads from, say, material falling into a hopper resting on loadcells. Typical overload allowance is 100–500%. The

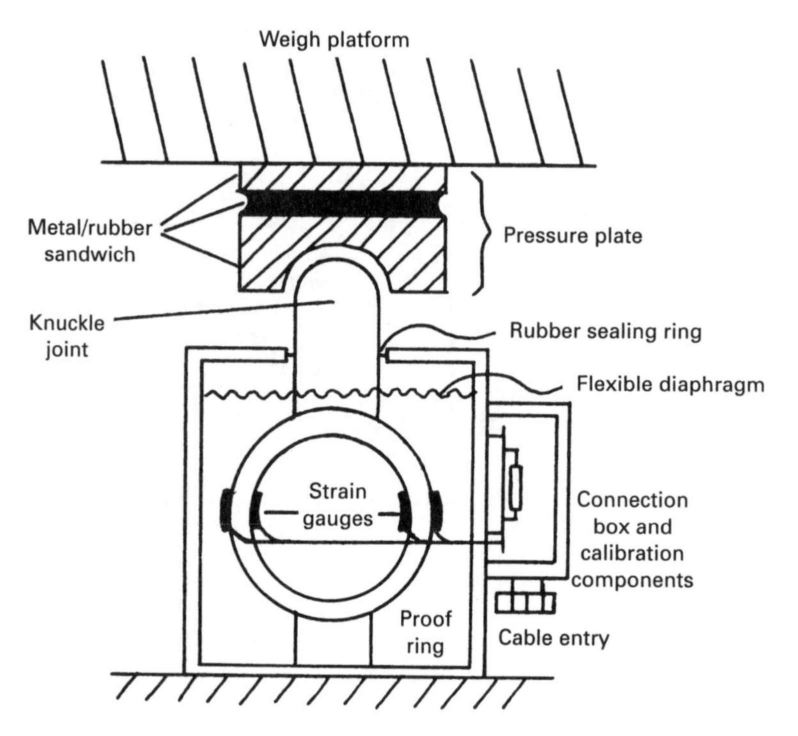

Fig. 6.20 *Construction of typical loadcell.*

(a) (b)

Fig. 6.21 *(a) Practical load cell. Note knuckle joint and pressure plate. (b) S type load cell.*

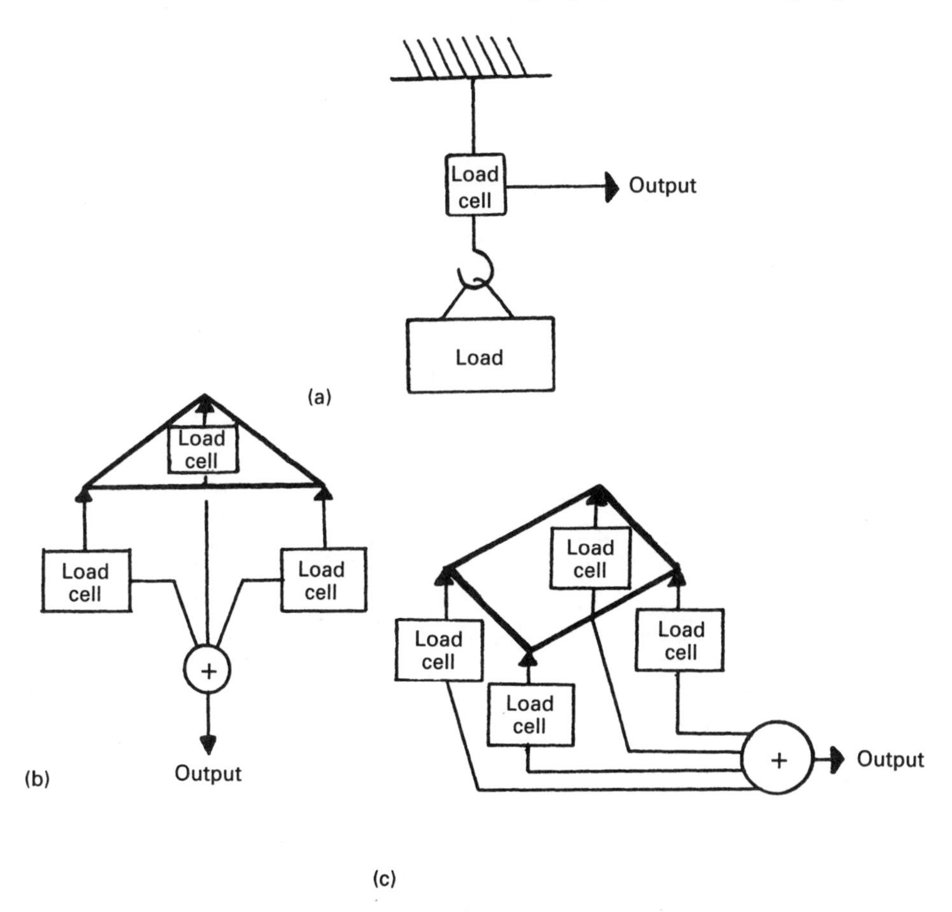

Fig. 6.22 *Mechanical coupling of loadcells. (a) Single-point support. (b) Three-point support (preferred). (c) Four-point support. Care must be take to ensure cells are always loaded.*

system should also be examined for possible side loads, which can cause damage to the sealing diaphragms and the sensing element itself. Acceleration of a moving weigh car can place severe side loads on loadcells supporting a heavy weight of material.

Usually the loadcells are the only electrical route to ground from the weigh platform. It is advisable to incorporate a flexible earth strap to the platform, not only for electrical safety but also to provide a route to ground for welding currents should mechanical repairs be necessary on the weigh platform. Welding current passing through a loadcell body will cause instant failure.

6.6. Weight controllers

A weighing system is usually more than a collection of loadcells and a display system. Figure 6.23 illustrates two common weighing systems. Figure 6.23a is a

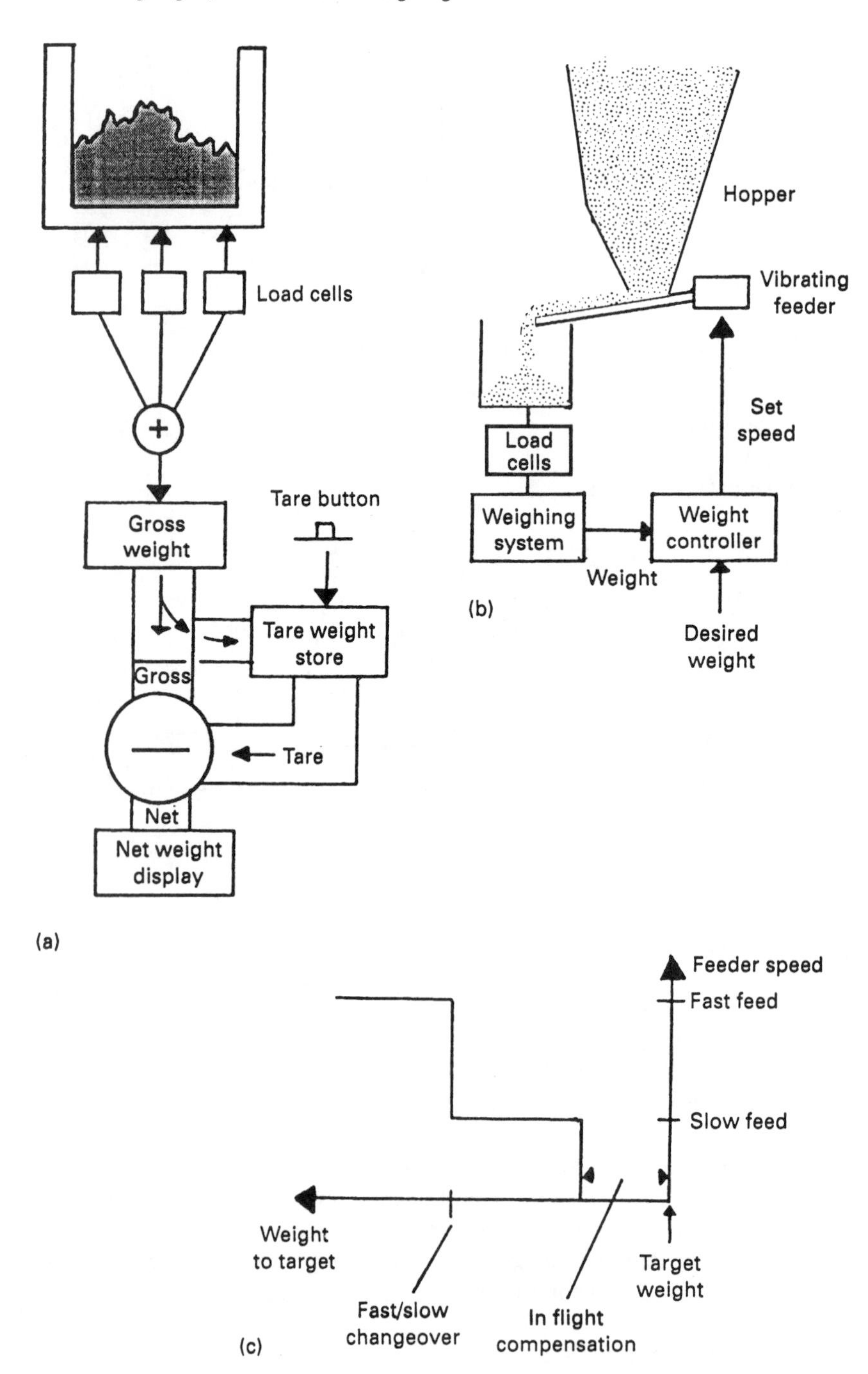

Fig. 6.23 *Weighing control systems. (a) Tare weigher. (b) Batch feeder system. (c) Feeder speed set-up. (d) Photograph of batch feeder.*

(d)

Fig. 6.23 *contd.*

Taring system, used where a 'recipe' of several materials is to be combined in a hopper. The gross display shows the total weight indicated by the cells, including the weight of the hopper itself.

When the tare button is pressed, the current gross weight is stored and subtracted from the gross to give a net weight display. This can be used to obtain a weight for each new material in a batch, the tare button being pressed before every addition.

Figure 6.23b is a batch feeder system, where material is fed to a hopper by a variable speed vibrating feeder. The speed is controlled by the weigh system along the lines of Fig. 6.23c. A two-speed system is commonly used, with a changeover from fast to dribble at some fixed point before the target weight. The feeder turns off at a second fixed point just before the target weight to allow for material in flight between the feeder and the hopper. This is known variously as anticipation, preact or in-flight compensation. Setting up of the four parameters of Fig. 6.23c is an obvious compromise between speed and accuracy.

Other features found in weight control schemes are indication of stable weight (sometimes called motion detection), automatic taring and storage of batch menus for automatic mixing.

6.7. Belt weighers

Solid material is often carried by conveyor belts, and an indication is usually required of feed rate in weight per unit time (e.g. tons per minute). This can be provided by the scheme of Fig. 6.24.

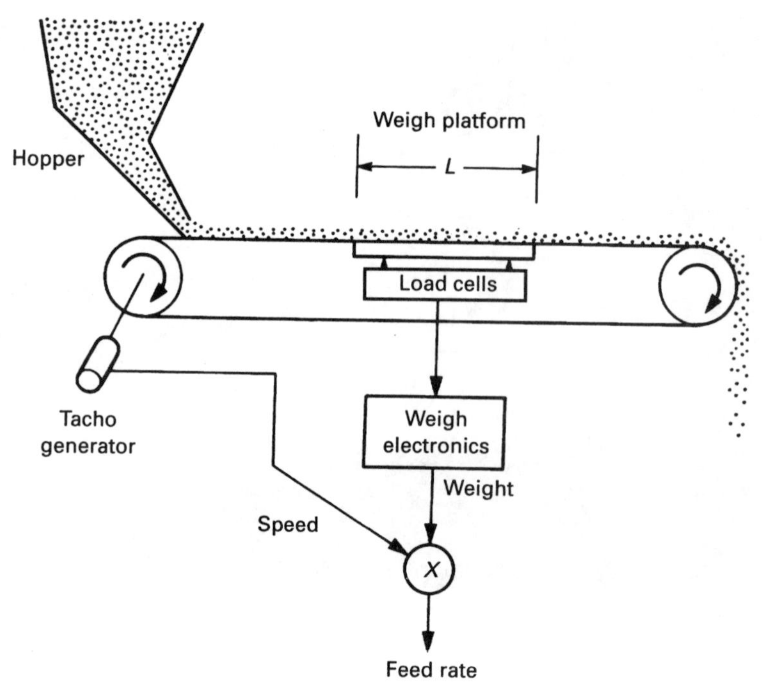

Fig. 6.24 *Belt weigher schematic.*

The conveyor passes over a weigh platform of known length L metres, and the conveyor speed is measured by a suitable device (e.g. a tachogenerator). If the weight controller indicates a load of M kg, and the belt is moving at V m s^{-1}, the feed rate is simply $M \, . \, V/L$ kg s^{-1}.

The feed rate can obviously be used to control either the feeder or the belt speed, but the inherent transit delay can make it a difficult system to control.

Chapter 7

Level measurement

7.1. Introduction

Wherever liquids, or bulk solids, are used, stored or conveyed, some type of level indication will be needed. There are many techniques for the measurement of level, and the choice of sensor is more difficult than for other process variables.

Level is often used to infer volume. This is straightforward with a container of uniform cross-section such as the tank of Fig. 7.1a where the volume is given by:

$$V = hA \tag{7.1}$$

The relationship for the pressure vessel of Fig. 7.1b is, however distinctly non-linear, as shown in Fig. 7.1c.

Level can also be used to infer mass by calculating the volume and multiplying by the density. This is subject to the same problems of container geometry as volume measurement, plus the additional complication that the density may be affected by temperature or the chemical composition of the liquid/bulk solid concerned.

In many applications where volume or mass measurement is really needed, the simplest solution is often simply to weigh the tank, container or silo by suspending it from loadcells or mounting it on a weigh scale. Care must be taken to avoid friction effects from inflow or outflow pipe connections, but flexible couplings are available. Weighing gives volume or mass measurement which is unaffected by the shape of the container.

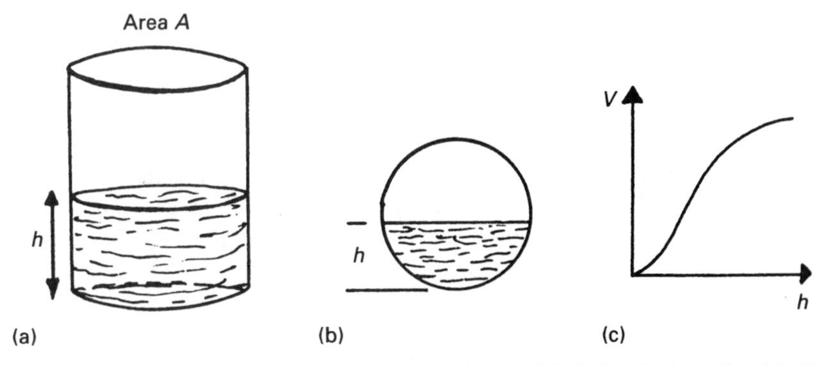

Fig. 7.1 *Relationship between level and volume. (a) Cylindrical tank with linear relationship between depth and volume. (b) Spherical pressure tank. (c) Relationship between depth and volume for pressure tank.*

The conditions inside the tank must also be considered. Temperature, pressure and corrosive atmospheres tend to affect level sensors more than, say, resistance thermometers or thermocouples. The liquid (or solid) characteristics must also be considered. A fluid that wets or clings to a sensor will tend to give a possibly dangerous high indication. Level measurement will also be adversely affected by turbulence or frothing on the surface (a domestic example of which is self-perpetuating 'hammer' in hot-water header tanks where ripples affect the float-operated shut-off valve). It may be necessary to provide a protected 'harbour' or still well for the level sensor.

A tank containing liquid can also resonate. A tank of diameter L metres will oscillate with period:

$$T = 2\pi \sqrt{\frac{L}{2g}} \quad \text{(secs)} \tag{7.2}$$

Note that this is independent of density. Rectangular vessels can exhibit two differing resonant periods. Hydraulic resonance is often observed by parents of young children at bath time! Inserting the length of a bath (1.5 m) into equation 7.2 gives a period of about 1.7 s.

There are, in general, four interfaces that need to be considered for level control:

(a) liquid/gas;
(b) solid/gas;
(c) liquid 1/liquid 2 (non-miscible, e.g. water/oil);
(d) solid/liquid (rare, e.g. identifying sludge depth in oil tank).

Most applications cover the first two types. Level sensors, in general, work by identifying the position of the interface or responding in some way to the bulk of one of the materials.

7.2. Float-based systems

Float-based measurement systems are the simplest level transducers. The basic principle was discovered many centuries ago by Archimedes, and is illustrated in Fig. 7.2. A floating body experiences two forces: a downward force of gravity and an opposing force due to its buoyancy.

$$\text{Downward force} = g \times \text{mass of float} \tag{7.3}$$

$$\text{Upward force} = g \times \text{mass of liquid displaced} \tag{7.4}$$

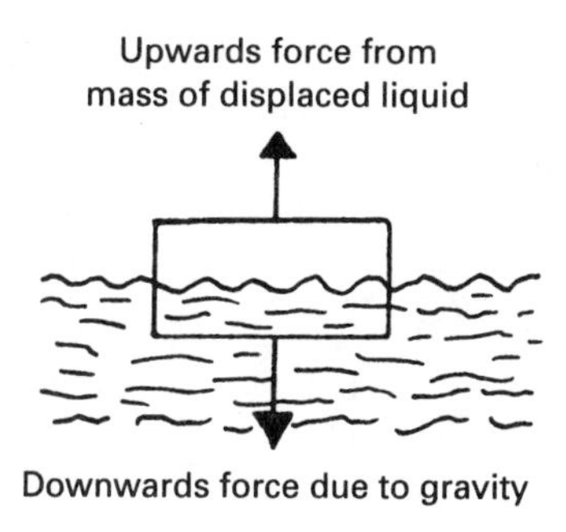

Fig. 7.2 *Archimedes' principle.*

In equilibrium these balance, so

$$g \times \text{mass of float} = g \times \text{mass of liquid displaced} \tag{7.5}$$

$$\text{Mass of liquid displaced} = \rho \times A \times d \tag{7.6}$$

where ρ is the liquid density, A the float area and d the submerged depth of the float. The float is therefore submerged at a constant depth:

$$d = \frac{M}{\rho A} \tag{7.7}$$

where M is the mass of the float.

Simple float systems are based on a rigid arm as shown in Fig. 7.3a,b, and convert the liquid level to an angle which can be measured by an angular position transducer (commonly a potentiometer, see Section 4.2). While these arrangements are simple, the output is non-linear with the liquid depth being given by:

$$h = H - L \sin \theta \tag{7.8}$$

The arrangements of Fig. 7.3b,c use a vertically moving float and accordingly give a linear output. Figure 7.3b uses a counterweight to couple the float to the position transducer, and Fig. 7.3c a torque-controlled electric motor (e.g. a stalled DC motor operating in current limit). Other possibilities include a spring-powered take-up drum. In each case the output is a signal proportional to level.

Equation 7.7 shows that the float submersion depth is sensitive to changes in M and ρ (A presumably remains constant). At low levels the masses of the floats in Figs. 7.3b and c increase because of the length of the suspension wire. To minimise this error (and errors due to changes in ρ), A should be large implying a 'cheesebox'-shape float as in Fig. 7.3d. Errors can also occur if the suspension wire is not vertical due to, say, liquid flow in the tank. Restraining guide wires or a float running on a guide tube should be used in these circumstances.

Figure 7.3 requires either the position measuring device to be mounted inside

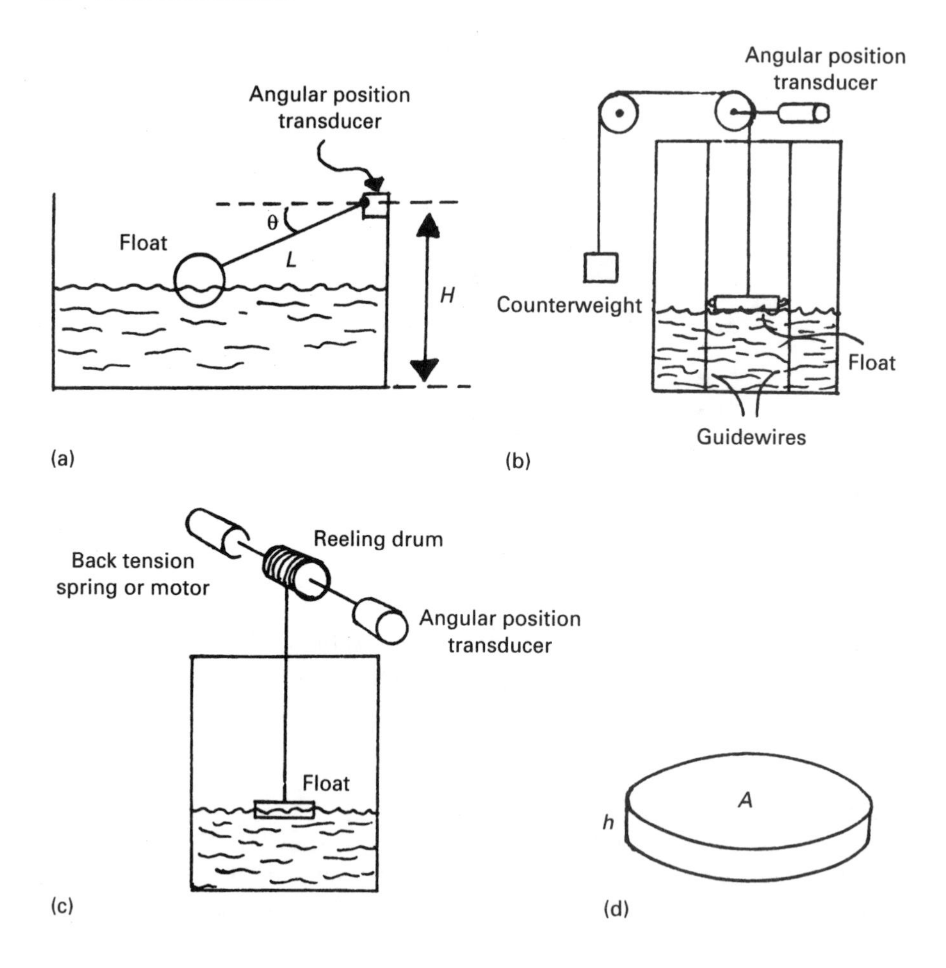

Fig. 7.3 *Float-based level measurement. (a) Lever arm float-level sensor. (b) Counterweight float. (c) Reeling drum float. (d) Ideal float shape: large surface area, low height.*

the tank (with attendant maintenance problems, and totally impossible in applications where the tank atmosphere is corrosive or explosive) or the float displacement conveyed outside the tank via flexible bellows. Both cases present difficulties if the tank is substantially above atmospheric pressure.

Figure 7.4 shows a float-based system that can operate at very high pressures, albeit with a limited measurement range. A ferromagnetic float is contained in a glass tube similar to a sight glass (which is itself the simplest level gauge). The float moves within the coils of an LVDT position transducer (see Section 4.5.2) to give an electrical output dependent on level.

Figure 7.5 is an alternative float transducer. In this application the 'float' is a fixed sealed tube connected to the top of the tank via a force transducer (e.g. a loadcell, see Section 6.5). By equation 7.4 the float will experience an upward

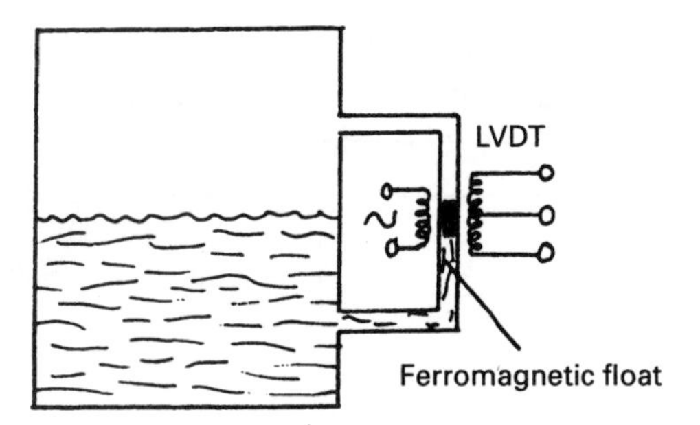

Fig. 7.4 *Float level measurement usable at high pressures.*

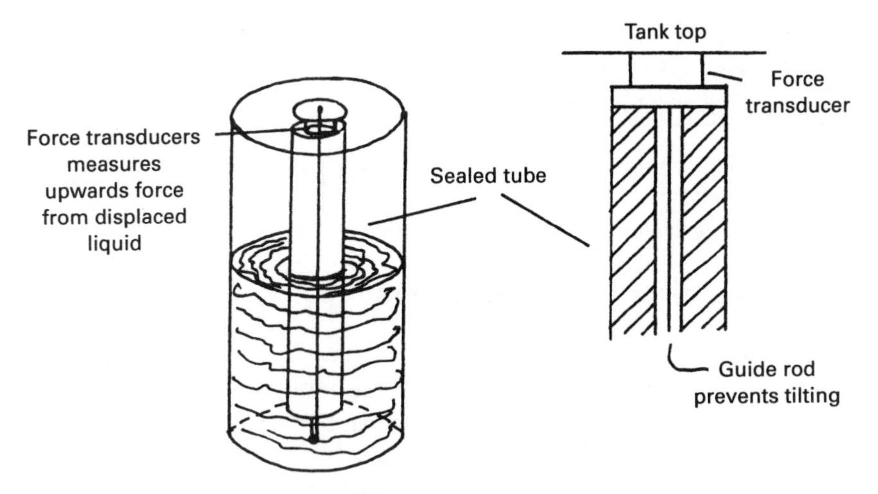

Fig. 7.5 *Level measurement with force transducer.*

force dependent on the length of tube submerged, i.e. the liquid level. Having no moving parts, this arrangement is attractive in dirty locations. Care must again be taken to keep the tube vertical. The same technique can be used to twist a torque tube whose shear strain can be measured.

Surprisingly, floats can also be used to measure bulk solid level via the principle of Fig. 7.6 (which also shows that 'level' is an imprecise term for bulk solids!). The technique does not provide a continuous indication of level, but a sampled level at regular intervals. The 'float' is wound to the top of the silo, then lowered until the slack wire limit gives a signal to the sequencing logic. The length of supporting cable paid out then gives the distance between the solid surface and the top of the silo. The depth of solid can be found by subtraction from the silo height.

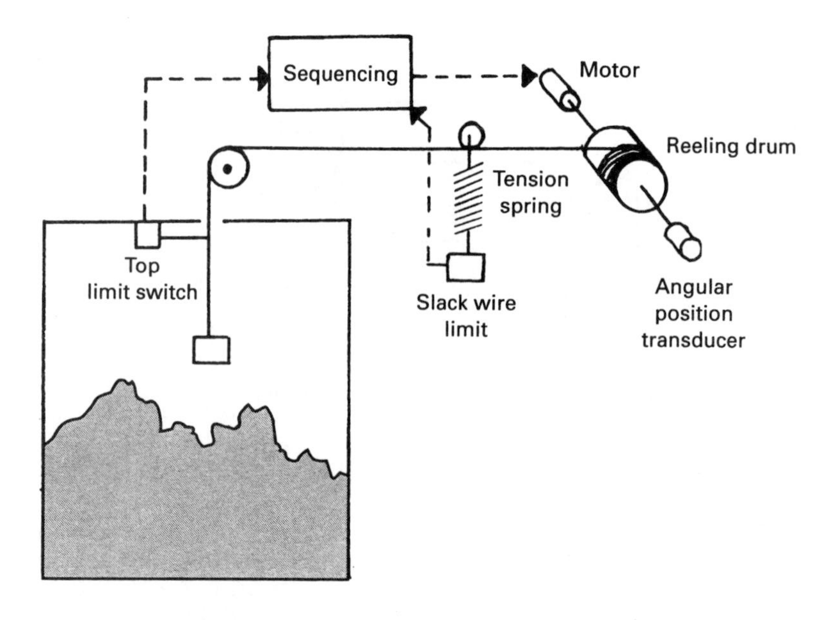

Fig. 7.6 *Measurement of bulk solid level.*

7.3. Pressure-operated transducers

7.3.1. Direct measurement

A very common way of measuring level is the transformation of level to a differential pressure which can be converted to an electrical signal by any of the transducers in Chapter 3. The basic principle is shown in Fig. 7.7a. The absolute pressure at the bottom of the tank has two components: atmospheric pressure, and pressure due to the head of liquid. The absolute pressure is therefore given by:

$$P = \rho g h + \text{atmospheric pressure} \tag{7.9}$$

where ρ is the density of the liquid.

In practice, a gauge pressure transducer is used which measures pressure with respect to atmosphere. The gauge pressure is given by:

$$P = \rho g h \tag{7.10}$$

which is linearly related to liquid level (and is independent of the tank shape and construction). One problem, however, is that the level, h, is measured with respect to the level of the pressure transducer itself and not the tank bottom. The system of Fig. 7.7b will therefore indicate a level far in excess of its true value. Zero suppression can, of course, be used to give a true reading if the difference in height between the tank bottom and the pressure transducer is known.

Simple gauge pressure transducers cannot be used where the tank is at some

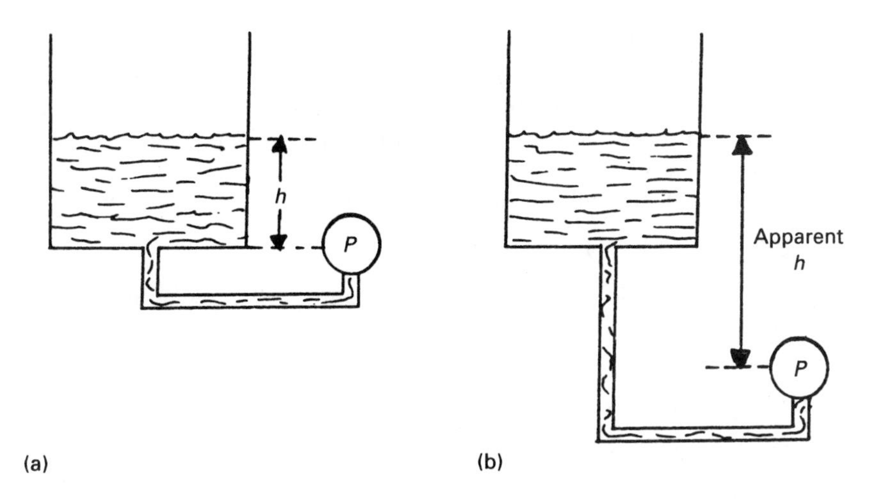

Fig. 7.7 *Level measurement from hydrostatic pressure. (a) Gauge pressure proportional to level. (b) Head error.*

pressure other than atmosphere. In these circumstances a differential transducer must be used with one leg connected to tank bottom (below the liquid surface) and one to the tank top as in Fig. 7.8a. The pressure seen by the transducer is then linearly related to tank level because the static pressure on both legs cancels.

For Fig. 7.8a to work, the LP leg must be kept totally free of liquid. This is often difficult to ensure, particularly in boiler applications where the space above the liquid is a condensable vapour. Figure 7.8a also retains the problem that a height

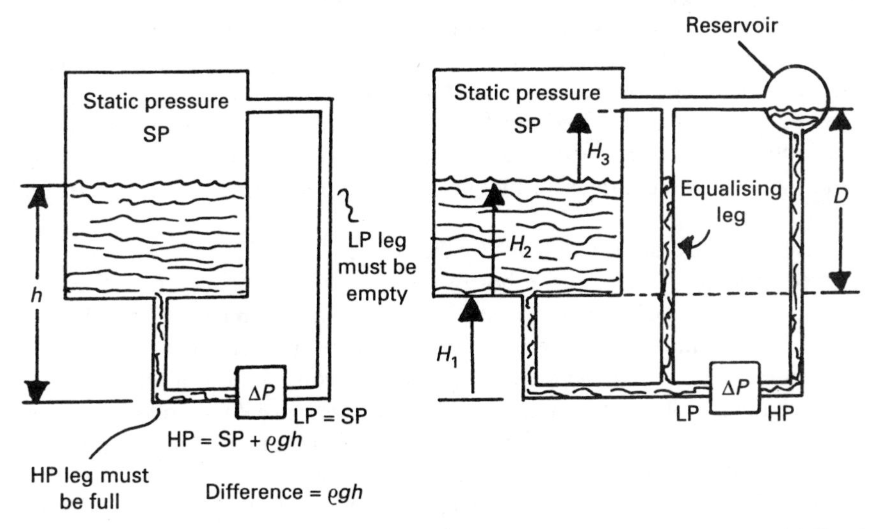

Fig. 7.8 *Differential pressure-based level measurement in pressurised tanks. (a) Simple method. (b) Level measurement with condensable vapour.*

differential between tank bottom and the transducer will give an offset in the output signal. Figure 7.8b overcomes both these problems by deliberately filling the leg to the tank top with liquid. In boiler level applications this can be arranged to occur naturally by keeping the measuring pipes unlagged and providing a small reservoir at the top of the tank top leg.

The pressure at the tank bottom leg is given by:

$$LP = \rho_1 g(H_1 + H_2) + \rho_2 g H_3 + SP \tag{7.11}$$

where ρ_1 is the liquid density, ρ_2 is the vapour density (often negligible), and SP is the static tank pressure.

The pressure in the tank top is given by:

$$HP = \rho_1 g(H_1 + H_2 + H_3) + SP \tag{7.12}$$

The differential pressure is therefore:

$$\Delta P = H_3(\rho_1 - \rho_2)g \tag{7.13}$$

If D is the height difference between the two tank tappings, this can be rearranged to:

$$\Delta P = (D - H_2).(\rho_1 - \rho_2)g \tag{7.14}$$

which is independent of the transducer position H_1. In most applications the terms in ρ_2 can be neglected.

Note that the LP leg of the transducer goes to the tank bottom and the transducer output is reversed, giving full scale for zero level and zero signal for maximum level. This can, of course, be reversed by suitable signal conditioning.

7.3.2. Gas reaction methods

The pressure-based methods in the previous section put some constraints on the user. In particular the pressure transducer must be at, or below, the lowest level in the tank. This is not always feasible, particularly with totally pneumatic systems. There can also be problems with applications where the liquid is corrosive or at a high temperature, as the fluid must come into direct contact with the pressure transmitter diaphragm. Level measurement in the food industries must also avoid stagnant liquid in measuring lines as this can be an ideal breeding ground for bacteria.

These problems can be overcome by the techniques of Fig. 7.9a. An inert gas is bubbled at a slow rate into the bottom of the tank. The pressure required to maintain flow is identical to the static pressure at the pipe entry and hence proportional to tank level. The pressure is measured by a conventional pressure transmitter, but because the measuring line only contains the purge gas, errors due to changes in level are negligible. The purge gas, usually argon or nitrogen, also keeps the measuring lines free of liquid, preventing corrosion or potential bacterial contamination.

Figure 7.9a is shown for a tank open to atmosphere with a gauge pressure transmitter. With a pressurised tank, a differential pressure transmitter is used, as in Fig. 7.9b. Obviously the purge gas supply pressure must exceed the tank

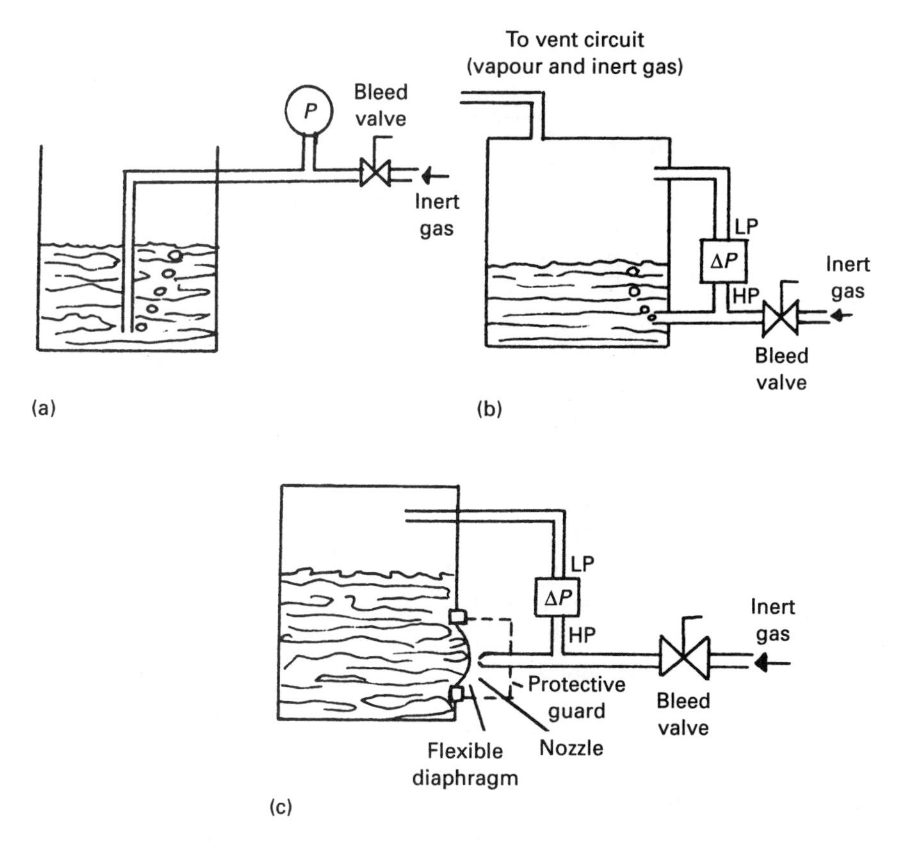

Fig. 7.9 *Gas reaction methods. (a) Gas bubble in open tank. (b) Sealed tank. (c) Sealed tank with flexible diaphragm.*

pressure, and some vent mechanism must be used to avoid raising the tank pressure with the purge gas.

A common variation on the principle is shown in Fig. 7.9c. A measuring head with a flexible diaphragm is placed in the tank side, and controls the pipe pressure via a flapper/nozzle link. The purge gas pressure at the flapper/nozzle is then dependent on the diaphragm pressure and hence tank level. The gas vents to atmosphere and does not enter the tank. This arrangement can be used as shown for open tanks or modified for pressurised tanks by taking the LP leg of a differential pressure transmitter to the tank top. In practice, the differential pressure transmitters would be positioned to avoid the LP leg filling with condensed vapour (e.g. by mounting on the tank top).

7.3.3. Collapsing resistive tube

Figure 7.10 shows an interesting pressure-based level transducer that gives a direct electrical output. A flexible sealed tube is lined on opposite inside faces with resistive material. The tube is maintained at a pressure slightly above tank

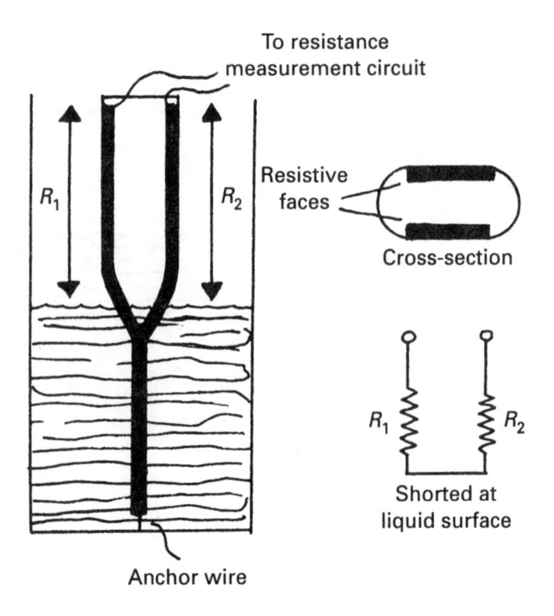

Fig. 7.10 *Collapsing tube method.*

pressure. The tube inflates, keeping the resistive faces apart. When the tube is submerged, it collapses below the surface, bringing a length of the resistive faces together. The electrical resistance of the tube therefore varies with level. The device can, with care, also be used for bulk solid level measurement.

7.4. Direct electrical probes

7.4.1. Capacitance sensing

The basis of a capacitance sensing level probe is shown in Fig. 7.11a. An insulated rod is inserted into the tank. In practice, a metal rod coated with PVC or PTFE is used to prevent corrosion. The capacitance between the rod and the tank walls has two components: C_1 above the liquid (or solid) surface and C_2 below. These capacitances depend on the geometry of the installation (e.g. rod diameter and the distance to the wall) and the dielectric constants of the liquid and the vapour above the surface.

As the liquid level rises, C_1 will decrease and C_2 will increase in value. The two capacitors are effectively in parallel, and as liquids and solids have a higher dielectric constant than vapours, the net result is an increase in capacitance with level.

The change in capacitance is, however, small, so the probe is usually used in conjunction with a measuring bridge/amplifier circuit, as in Fig. 7.11b. Normally a drive frequency of around 100 kHz is used. The detection circuit usually needs

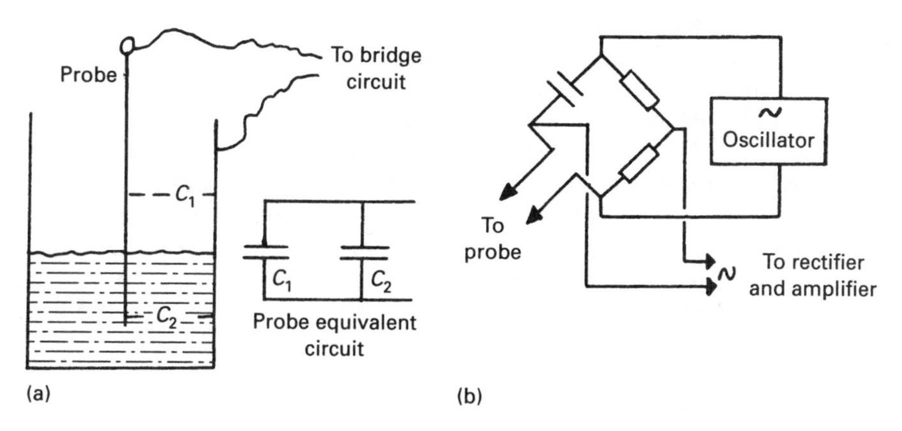

Fig. 7.11 *Capacitive probe method. (a) Physical arrangement. (b) AC bridge.*

to be fairly close to the probe (typically 50 m) to prevent the signal change in capacitance being swamped by the capacitance of the connecting cable.

The physical construction of the probe and tank is critical. Obvious errors will occur if the tank sides are not parallel to the probe, and the simple arrangement of Fig. 7.11a will obviously not work with non-conducting tanks (e.g. glass fibre or plastic). In these applications the capacitors may be formed from two parallel rods, or constructed as a rod and concentric tube.

The capacitance is not only dependent on level, but is also affected by change in the value of the dielectric constant of the liquid and/or vapour. If an uncoated probe is used, changes in the resistance of the liquid will also be perceived as changes in level. The transducer will only work satisfactorily with liquids or solids with relatively high dielectric constant. Liquids that are frothing, foaming or boiling have a low dielectric constant (typically less than 2), which is also probably variable.

Capacitance probes work with a wide range of liquids and solids, and are probably the most versatile of all level sensors. They do, unfortunately, have a rather bad reputation from early units which were prone to drift and suffered from difficult installation requirements (e.g. measuring circuits within a metre of the probe). The more serious objection still remains, however, in that the probe has to be designed specifically for each and every application with regard to tank geometry, dielectric constant, and so forth. A capacitance level probe is not a universal transducer in the way that a thermocouple or pressure transducer is.

Capacitance probes are probably the best solution to the measurement of bulk solid level (with the possible exception of direct weighing). Care must be taken, though, to avoid errors due to changes in dielectric constant from moisture or material compaction (the density of a bulk powder tends to increase towards the bottom of a solid). An unexpected hazard which can damage the measuring circuits is the build-up and discharge of high-voltage static electricity inside silos.

7.4.2. Resistive probes

If the resistance of the liquid is reasonably consistent, level can be inferred by observing the resistance between two metal rods inserted into the liquid. This requires uncoated probes, usually of stainless steel to overcome corrosion problems. Although a DC supply can, in theory, be used, in practice an AC supply is used to prevent electrolysis and plating effects. A bridge measuring circuit (similar to Fig. 7.11b but with resistance rather than capacitance) is usually employed local to the probe. The circuit is, in reality, measuring a combined resistance/capacitance impedance for the probe. Resistive probes are less versatile than capacitance probes, and tend to be found mainly in level sensing (on/off) applications.

7.5. Ultrasonic methods

Ultrasonic methods are based on high-frequency sound waves produced by the application of a suitable AC signal to a piezo-electric crystal. A typical device is shown in Fig. 7.12. Operation at frequencies up to 1 MHz is feasible, but 50 kHz is more typical in industrial applications (the high frequencies are used in, for example, medical ultrasonic scanners). The operating frequency is chosen to correspond with the resonant frequency of the transmitter and receiver.

The principle of ultrasonic level measurement is shown in Fig. 7.13. An ultrasonic transmitter and receiver are placed at the top of the tank, and an ultrasonic beam is directed at the surface of the liquid (or solid) contents. The level in the tank is inferred from the reflected signals.

The reflected signal will be a delayed version of the transmitted signal, with a delay of $2d/v$ seconds, where d is the distance of the surface from the tank top, and

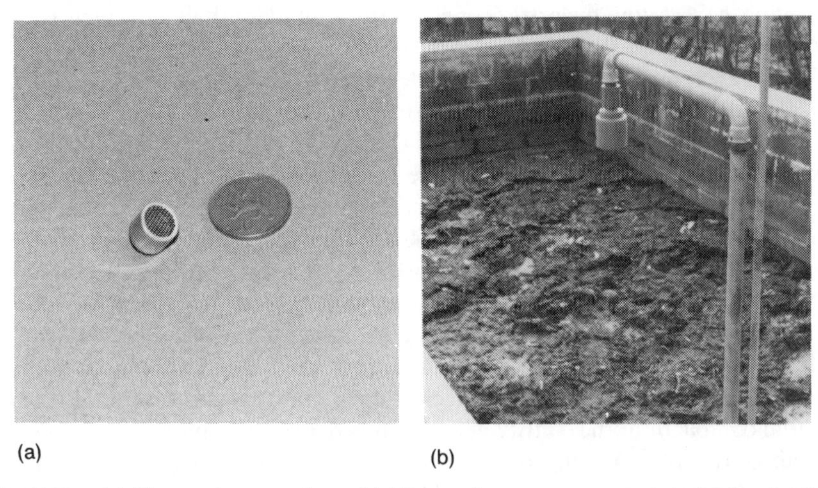

(a) (b)

Fig. 7.12 *(a) Ultrasonic transmitter. (b) Ultrasonic measurement of sludge level. (Photo courtesy of Hycontrol Ltd.)*

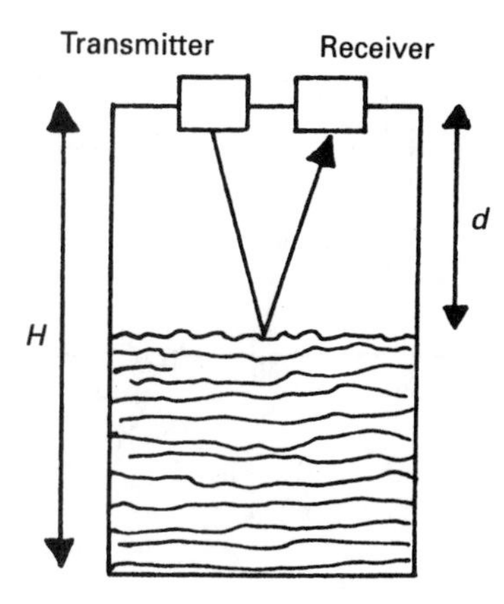

Fig. 7.13 *Principle of ultrasonic level measurement.*

v is the velocity of sound in the medium above the surface (which need not be air; the principle could be applied to non-miscible liquids, or to determine the position of a liquid/sludge interface. Maritime variations are, of course, Asdic and depth sounders).

The velocity of sound in air, for example, is about 300 m s^{-1}. For a level which can vary from 0 to 10 m, the delay will be 0 to 70 ms. There are two ways of measuring the delay time, the choice being determined by the range of measurement compared with the wavelength of the sound. If d is large (as it usually is in level measurement), a narrow sound pulse is transmitted as either a sound burst or a narrow spike. The receiver will see two pulses: one direct from the transmitter and one reflected off the surface. The time delay can be measured as shown in Fig. 7.14a, and the position of the surface computed directly.

The accuracy of the pulsed method decreases as the delay approaches the pulse width. The arrangement of Fig. 7.14b is then used. A broadband transmitter and receiver are used, with the transmitter frequency swept by a voltage-controlled oscillator. As before, the receiver receives two signals, one direct and one reflected. At one particular frequency where the difference in path length is an exact multiple of the wavelength, the two signals will combine to give a peak. By noting the wavelength at which this peak occurs, the reflected path distance can be computed from the formula:

$$2d = v/f \tag{7.15}$$

where d is the distance to the surface, v the velocity of propagation, and f the frequency at which the peak occurs. Equation 7.15 gives distances of a few millimetres, so this technique is usually found in thickness gauges and medical instruments. Note also that the result is ambiguous, as peaks will be seen at multiples of the wavelength.

Both methods of measurement rely on a knowledge of the velocity of propagation. This varies according to the material (1440 m s^{-1} for water,

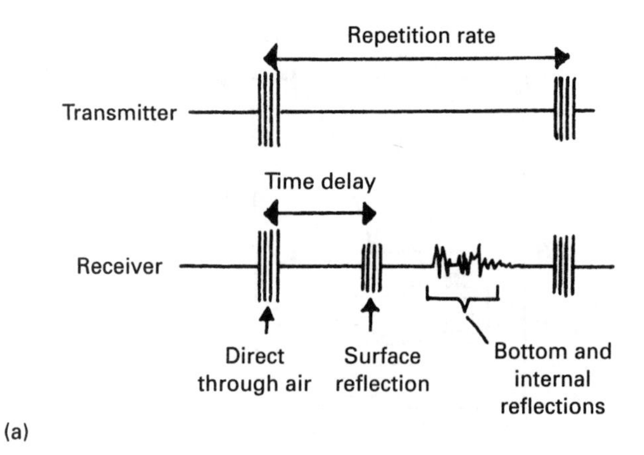

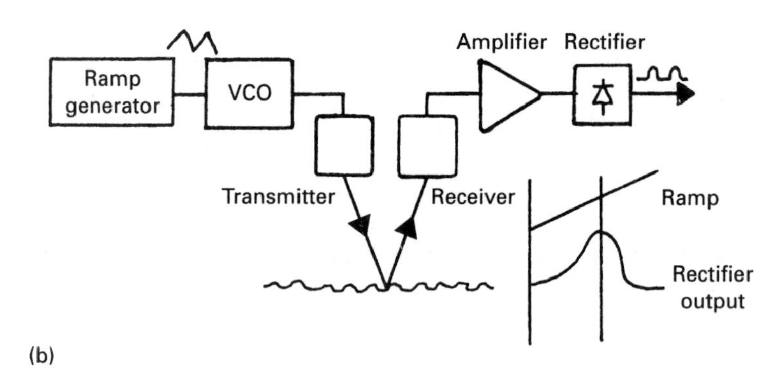

Fig. 7.14 *Ultrasonic techniques. (a) Pulsed transmitter. (b) Swept oscillator method for small distances.*

5000 m s^{-1} for steel) but also, for gases, with pressure and temperature. The velocity in air, for example, varies by about 1% for a 30 °C change in temperature. If environmental changes are likely to cause errors, these external effects can be measured separately and suitable compensation applied. Alternatively, a second calibration receiver can be located a known distance from the transmitter to measure v directly.

A less obvious source of error can arise from unwanted multiple reflections from walls. If the position of the transducer is correctly situated, these will occur *after* the surface reflection. Multiple secondary reflections determine the maximum pulse rate for systems based on Fig. 7.14a.

7.6. Nucleonic methods

7.6.1. Principles

Radioactive isotopes (such as cobalt 60) spontaneously emit gamma or beta radiation. As this radiation passes through material, it is attentuated according to the relationship:

$$I = I_0 \exp(-\mu\rho d) \tag{7.16}$$

where I_0 is the initial intensity; μ is a constant for the material, called the mass attenuation coefficient; ρ is the density of the material; and d is the thickness of material between the source and the measuring point.

Equation 7.16 allows a conceptually simple level transducer to be constructed, as in Fig. 7.15a. A strip radioactive source is placed on one side of the tank, and a point detector is placed on the other side. Gamma radiation above the liquid (or solid) surface will be slightly attentuated by the air and tank walls. Radiation passing through the liquid will be greatly attenuated. The intensity of the radiation seen by the detector will therefore be related to the liquid level in the tank (being a maximum when the tank is empty).

Figures 7.15b and c show variations on the same idea. Figure 7.15b is a point source and linear detector and Fig. 7.15c is a linear source and detector. Figure 7.16 is a typical installation, used to measure liquid steel level on a steel casting plant.

The strength of a radioactive source decays exponentially with time, and is described by its 'half-life', which is the time for the source strength to fall to half its initial value. Nucleonic level sensors therefore have a 'built-in' span drift, and consequently require periodic adjustment. A common source is cobalt 60. This has a half-life of 5.3 years and will exhibit a 1% change in about a month. Sources are normally usable (with readjustment of the sensitivity) for about one half-life. Other common isotopes are caesium 137 (half-life 37 years) and americium 241 (half-life 458 years).

Nucleonic level sensors have many advantages, and in some applications they are the only solution to a level measurement problem. Because no contact with the container is needed, nucleonic sensors can be used with liquids or solids regardless of temperature, pressure, or corrosion effects.

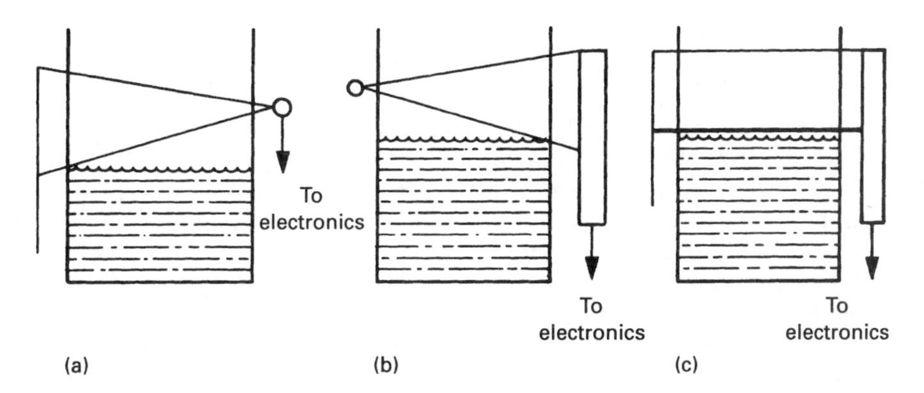

Fig. 7.15 *Radioactive source level measurement. (a) Line source, point detector. (b) point source, line detector. (c) Collimated line source, line detector.*

Fig. 7.16 *Radioactive source level measurement in steel casting. The source is in the large housing to the left of the mould, and the GM tube sensor on the right.*

There are some disadvantages, of course. Nucleonic systems have to be designed individually, and the sources and detectors chosen to suit the size of tank and the attenuation coefficient of the material being measured. Safety and legislation aspects are discussed in Section 7.6.3.

7.6.2. Radiation detectors

Level measurement systems using radioactive sources require some form of detector to convert the attenuated radiation to an electrical signal. Two detectors are commonly used, the Geiger-Müller (GM) tube and the scintillation counter.

The GM tube is constructed with a sealed cylinder containing an axial wire anode, as shown in Fig. 7.17a. The cylinder is earthed and the wire kept at a high positive potential (typically 300–500 V). The tube is filled with a halogen gas. As a radioactive particle passes through the tube, the gas is ionised forming electrons and positive ions which are attracted, respectively, to the wire and the tube wall. Collisions between the electrons and the gas release secondary electrons, and an avalanche effect occurs where a discharge occurs along the whole length of the central wire. This produces a negative voltage pulse at the anode. The pulse is of constant amplitude. Because a pulse occurs for each particle traversing the tube with sufficient energy to start the avalanche, the pulse rate is dependent on the strength of the radiation.

The pulse rate is also dependent on the voltage of the anode, as shown in Fig. 7.17b. There is, however, a region called the 'plateau', where the tube is relatively insensitive to changes in anode volts. GM tubes are normally operated at a voltage in the centre of the plateau.

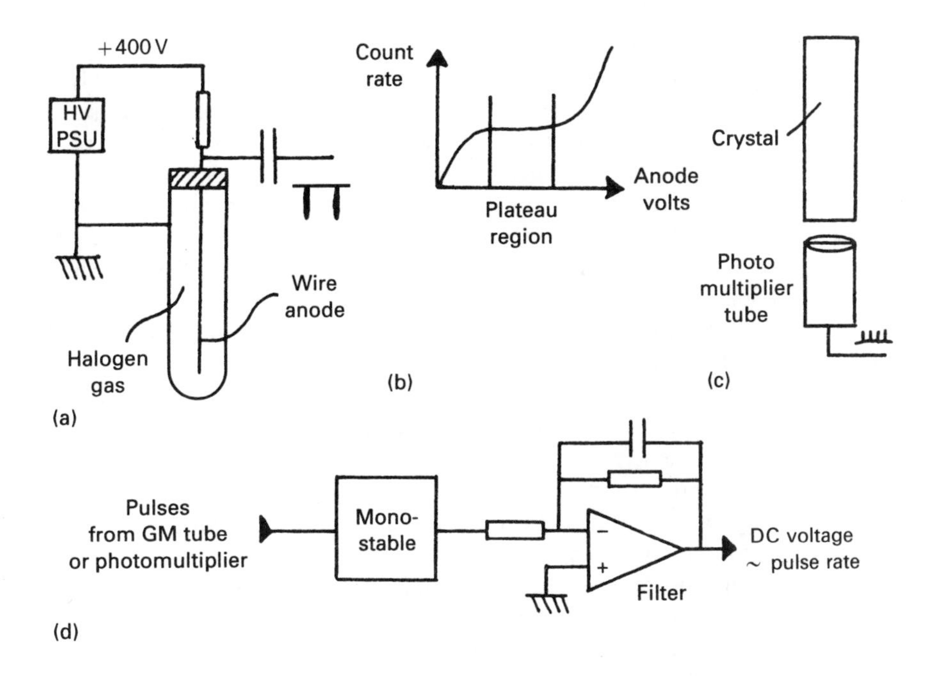

Fig. 7.17 *Radiation detectors. (a) Geiger-Müller (GM) tube. (b) Operating voltage for GM tube. (c) Scintillation counter. (d) Detection circuit.*

Scintillation counters detect ionising particles by their passage through a crystal. When an ionising particle passes through a material, atoms are raised to an excited state, subsequently emitting a short pulse of light as they return to their ground state. Most materials reabsorb this light pulse again. Scintillation crystals, however, are transparent to their own excited radiation, so the passage of ionising radiation can be observed as short pulses of light. Common crystals are NaJ(T1) and ZnS(Ag) phosphors.

The light pulses are very weak, so the construction illustrated in Fig. 7.17c is used to detect them. The scintillation crystal is 'viewed' by a photomultiplier tube (see Section 8.4.5) which gives random pulse outputs whose average frequency depends on the received radiation. Like the GM tube, the photomultiplier needs a stabilised HT supply.

Both the GM tube and the scintillation counter give a semi-random pulse train whose average repetition rate is radiation dependent. This must be converted to a DC voltage before it can be used to indicate radiation strength and hence liquid, or solid, level. One common way of achieving this is shown on Fig. 7.17d. The narrow pulses from the primary sensor are used to fire a monostable to give broader, fixed width, pulses. The output from the monostable is smoothed by an Op Amp filter circuit to give a DC signal which is related to the radiation intensity.

7.6.3. Safety and legislation

The most serious objection to the use of radiation sources is undoubtedly the inherent health hazard, which brings as a side-effect a plethora of legislation covering their use, storage, transportation and disposal. The user must accept that radioactive sources are dangerous, and treat them with respect. In the UK, sites using sources have to be registered, and records kept for many statutory bodies including the Factory Inspectorate, Health and Safety Executive, Department of the Environment, Radiochemical Inspectorate, Department of Health and Social Security and the National Radiological Protection Board. To this army of legislative might is added the inevitable suspicions of personnel in the area of the application. The use of radioactive sources brings more clerical and public-relations problems than technical difficulties.

Almost all sources used in industry are sealed, i.e. the source is contained permanently in an enclosure which can be totally sealed when not in use (unsealed sources are used in medical and laboratory applications and are subject to different legislation). When in use, a shutter is opened allowing a collimated beam to be emitted (typically $40^\circ \times 7^\circ$).

Source strength is determined by the number of disintegrations per second (d.p.s.) at the source. Two standards are in common use; the curie, Ci (3.7×10^{10} d.p.s.) and the bequerel, Bq (1 d.p.s.). It follows that 1 mCi = 37 MBq. A typical industrial source would be 500 mCi (18.5 GBq) caesium 137, although sources up to 5 Ci (185 GBq) are used occasionally.

The biological effects of radiation are complex, and there is no real 'safe dose'. The legislation is, in general, based on the concept of 'as low as is reasonably achievable' (ALARA) and 'acceptable levels' of biological effects. It is significant that the genetic and carcinogenic effects of radiation exhibit no lower threshold and all exposure can be considered potentially harmful. Legislation has become noticeably more restrictive over the years.

The absorbed dose (AD) is the energy density absorbed. Two units are, again, used; the millirad (Mrad) and the milligray (mGy):

$$1 \text{ mrad} = 6.25 \times 10^4 \text{ MeV g}^{-1}$$
$$1 \text{ mGy} = 6.25 \times 10^6 \text{ MeV g}^{-1}$$

It follows that 1 mGy = 100 mrad.

The absorbed dose is not, however, directly related to biological damage as it ignores the differing effects of α, b and γ radiation. A quality factor, Q, is defined (20 for α, 1 for β, γ) which allows a dose equivalent, DE, to be calculated from:

$$DE = Q \times AD \tag{7.17}$$

There are, yet again, two units in common use: millirem (mrem) and millisievert (mSv). These are defined by:

$$\text{mrem} = \text{mrad} \times Q \tag{7.18}$$

$$\text{mSv} = \text{mGy} \times Q \tag{7.19}$$

It again follows that 1 mSv = 100 mrem.

The AD and DE represent energy density absorbed, so the strength of radiation, the dose rate, is time related and expressed in mrad h^{-1}, mGy h^{-1}, mrem h^{-1} or μSv h^{-1}. Although it is possible to calculate dose rate from source strength, it is usually more convenient to refer to manufacturers' data sheets. A cobalt 60 source, for example, gives 3.39 mGy h^{-1} per MBq at 1 cm. Dose rates fall off as the inverse square of distance: doubling the distance gives one-quarter the dose rate.

Medical effects are related to both the dose rate and the total lifetime dose (DE). It follows that there are two aspects to protection from radiactive sources: the instantaneous dose rate and the dose equivalent (the cumulative effect) must both be controlled.

For industrial users, the legislation recognises three classes of person:

(i) Classified workers who are trained personnel wearing dose monitoring devices (usually film badges). These workers are allowed dose rates up to 2.5 mrem h^{-1} and an annual dose up to 5 rem. Medical records and dose history must be kept.
(ii) Supervised workers who operate under the supervision of classified workers, and are permitted dose rates of 0.75 mrem h^{-1} and an annual dose of 1.5 rem.
(iii) Unclassified personnel: this refers to other workers and the general public, for whom permitted dose rates are 0.25 mrem h^{-1} and the annual dose is 0.5 rem.

The legislation emphasises that the above are *not* design criteria, and the ALARA principle should apply. Dose rates are dependent on source strength, shielding and distance, and total dose on dose rate and time of exposure.

The legislation covering radioactive sources is lengthy and complex, so this discussion should only be regarded as a very brief summary. Although the underlying theory of nucleonic level and thickness transducers is straightforward, the legal aspects are a minefield for the unwary. Professional advice should always be sought at the design stage.

7.7. Level switches

Many level measurement and control applications involve 'surge' tanks where the level varies to cope with a sudden rise, or fall, of supply or demand. In these situations, tight control is not needed and would even negate the purpose of the tank by directly coupling inflow and outflow. All that is usually needed is an indication that the level has reached, or fallen to, some predetermined level.

Any of the previous transducers can be used as a level switch, but simpler devices are available. The commonest are simple float-operated switches which will work with most liquid/tank combinations. Solids can be detected by horizontally mounted capacitance probes. Other alternatives are rotating paddles or vibrating reeds which 'seize' when immersed and operate a microswitch.

The above sensors have moving parts which are prone to jamming when contaminated. Two non-moving alternatives are shown on Fig. 7.18. The heated probe contains a heater and a temperature sensor. The heat loss will be greater (and hence the temperature lower) when the probe is immersed. The light-reflective probe uses total internal reflection (see Section 8.3.2) to detect immersion of the plate.

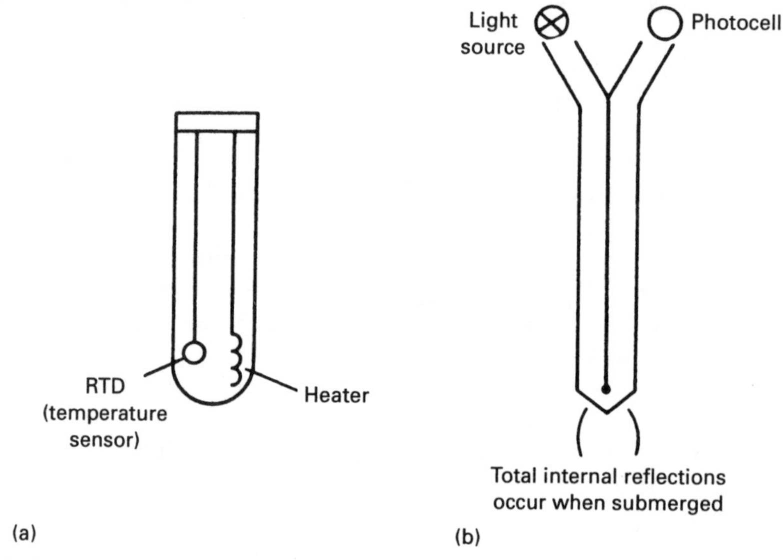

Fig. 7.18 *Unusual level switches. (a) Level switch working on heat loss from submerged object. (b) Optical level switch.*

Chapter 8

Optoelectronics

8.1. Introduction

Optical devices are becoming increasingly common in instrumentation and control. Such devices are covered by the multidiscipline term of optoelectronics. Optical sensors range from simple measurement of light intensity (photometry) to non-contact transducers for the measurement of position or temperature. Light emitters (such as LED or LCD devices) are used to display data to operators. Data transmission via fibre optic cable gives a high-speed alternative to electrical and pneumatic signals with almost perfect noise immunity and no explosion risk in hazardous atmospheres.

Optoelectronics is therefore a very broad subject, spanning across lasers, lenses, instrumentation, electronic circuits and the physics of electromagnetic radiation. This chapter covers the principles behind the optoelectronic devices that may be encountered by a process control engineer.

8.2. Electromagnetic radiation

The term optoelectronics implies the use of visible light. In reality, light is energy in the form of electromagnetic (EM) radiation, identical in nature to radio waves, gamma rays, infrared radiation and X-rays. A full description of EM radiation would require a fairly detailed knowledge of theoretical physics, and as such is beyond the scope of this book. Users of optoelectronics, however, do not need a deep understanding of the underlying theory.

Electromagnetic radiation propagates through space in a manner somewhat analogous to waves propagating through a liquid. Any EM radiation can therefore be described by its *frequency* (i.e. the number of oscillations per second passing a fixed point), its *wavelength* (the distance between successive maxima) and the *velocity of propagation*. These are related by:

$$c = \lambda f \tag{8.1}$$

where c is the velocity in metres per second, λ is the wavelength in metres and f is the frequency in hertz.

The velocity of EM radiation is constant for all frequencies, but is dependent on the transmission medium. The velocity is highest in a vacuum, for which c is 2.998×10^8 m s^{-1}. The velocity in glass is about 1.9×10^8 m s^{-1}.

The ratio of the EM velocity in vacuum to the velocity in a medium is defined as the *index of refraction* of the medium, i.e.

$$\mu = \frac{c}{v} \tag{8.2}$$

where μ is the index of refraction, c is the velocity of EM radiation in vacuum, and v is the velocity of EM radiation in the medium. The index of refraction for glass is about 1.57.

Electromagnetic radiation is described by its frequency or wavelength. The range is large, as shown by the EM spectrum of Fig. 8.1. which also demonstrates that the only difference between radio waves and visible light is the frequency of the radiation. Although EM radiation can be described by frequency or wavelength, it is usual to describe radiation used in microwave or broadcasting by its frequency, and radiation in optoelectronics by its wavelength. VHF radio, for example, is described as having a frequency between 30 and 300 MHz, whereas visible light is described as the small range with a wavelength between 0.38 μm and 0.78 μm.

Optoelectronics is concerned with EM radiation with wavelengths in the range 0.3μm to 30 μm, i.e. from just inside the ultraviolet range, through the visible light region to well into the infrared. In the rest of this chapter we will use the loose term 'light' for convenience, although the actual range of EM radiation is larger.

The wavelength of EM radiation is often expressed in Ångstrom units (Å) which is defined as 10^{-10} m. Red light therefore has a wavelength of 7800 Å and violet a wavelength of 4000 Å.

Visible light is perceived as having colour. Figure 8.1 shows that the colour is determined by the wavelength. Red light has the longest wavelength at 0.78 μm, with violet at 0.4 μm. The range from 0.8 μm to 100 μm is called the infrared, and that from 0.01 μm to 0.4 μm the ultraviolet.

8.3. Optics

8.3.1. Reflection and mirrors

Most optoelectronics devices incorporate mirrors and lenses to gather and focus light. In this section a simple description of the devices is given.

For most practical purposes, light can be considered to travel in straight lines. This can be demonstrated by the experiment in Fig. 8.2, where the hole in an

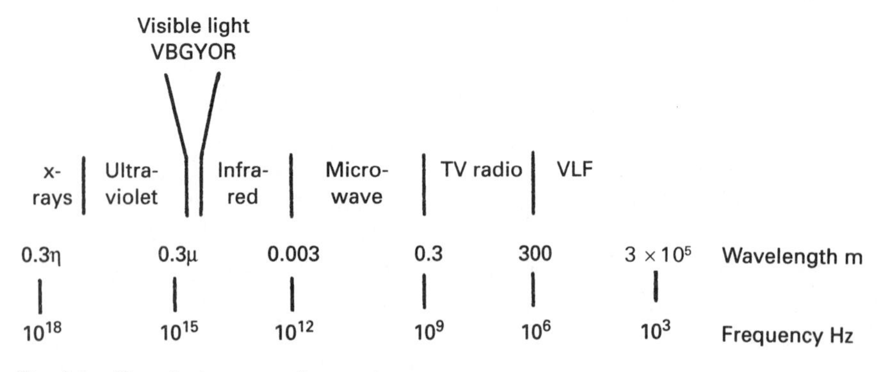

Fig. 8.1 *The electromagnetic spectrum.*

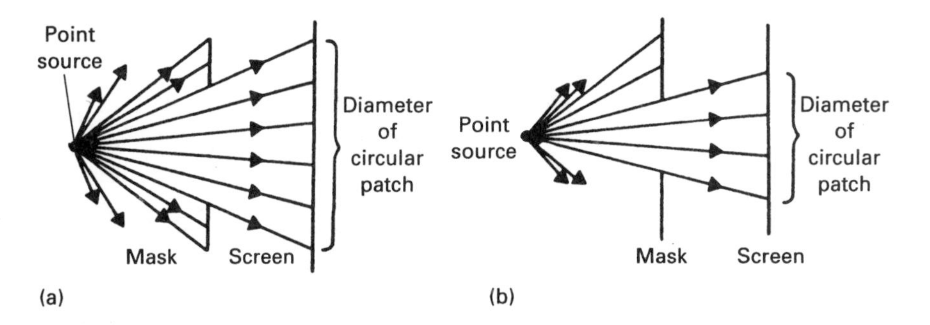

Fig. 8.2 *Demonstration of the straight-line nature of light. (a) Large hole. (b) Small hole.*

opaque screen is progressively reduced, resulting in a corresponding reduction in size of the circle of light on the screen. (If the hole size is reduced below 0.01 mm, other effects start to occur and a diffuse patch of light is seen. Provided we are not concerned with the passage of light through very small apertures, these effects need not concern us.) If light travels in straight lines, the analysis of optical devices is simplified considerably by considering light to be a collection of light rays and looking at what happens to a few of them.

In Fig. 8.3a, for example, A is a point object in front of a mirror. The object will emit light in all directions, so there will be a myriad of light rays. To analyse the effect of the mirror we need only consider any two, as shown in Fig. 8.3a. These strike the mirror at angles α and β, and are reflected with the same angles as shown. To an observer, the reflected rays appear to come from a point A^1, the same distance behind the surface as A is in front.

Figure 8.3b illustrates the general case of an object of non-point size. By considering two rays from each end of the object, it can be seen that an image A^1, B^1 is formed apparently behind the mirror.

Figure 8.4a shows a curved concave mirror with centre of curvature at point C and radius R. To examine its action we consider two rays of light, one parallel to the centre line of the mirror and one to the centre of curvature of the mirror. The ray parallel to the centre line is reflected through point F where FD is $R/2$ providing d is small compared with R. The ray through the centre of curvature of the mirror will be reflected back along its own path as shown. A small object A will therefore form an image at point B.

The distance from the mirror to the focus, FD, is called the focal length. Any ray parallel to the centre line of the mirror will pass through point F, which is correspondingly known as the focus of the mirror. Obviously, any light originating at point F will emerge as a parallel beam, a property used, for example, in searchlights and car headlamps.

The behaviour of light from a finite-sized object can be analysed by considering two rays from each end as shown in Fig. 8.4b. An inverted image of the object is produced at B. If a small screen is placed at this point an image will be seen. It also follows that an object placed at B will produce an image at A.

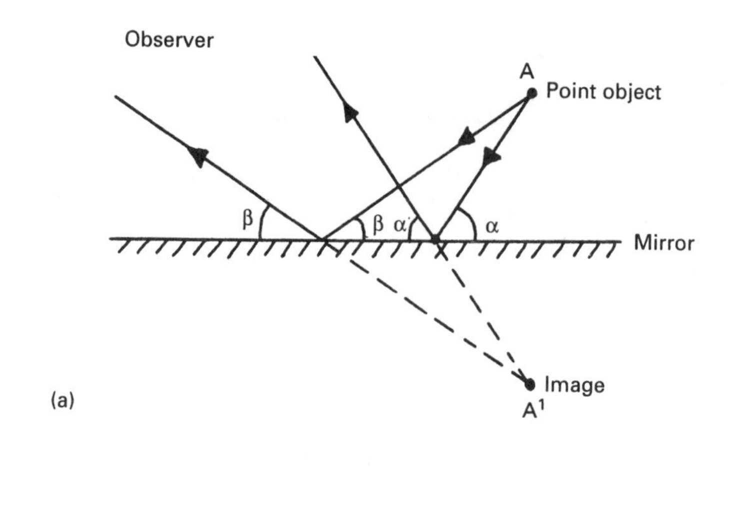

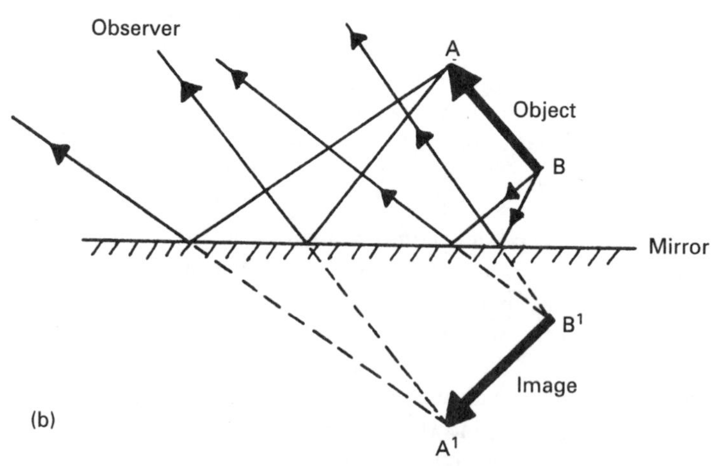

Fig. 8.3 *Reflections from a mirror. (a) Point object. (b) Finite object.*

Figure 8.4c shows the analogous case for a convex mirror with, again, centre of curvature C and radius R. As before, by considering the action of a ray parallel to the centre line and one to the centre of curvature, it is found that an image is formed at point B *behind* the mirror. Point F is, again, the focus of the mirror. Figure 8.4d illustrates the formation of an image from a finite-sized object. The image is, again, formed behind the screen.

There is a fundamental difference in the images of Figs. 8.4b and 8.4d. In Fig. 8.4b the image is formed by convergence of light rays at a point. Such an image can be focused on to a screen and is called a *real* image. In Fig. 8.4d, the rays *appear* to diverge from a point and the image cannot be focused on a screen. Such an image is said to be a *virtual* image.

The analysis of Fig. 8.4 is based on mirrors with a circular section and a diameter that is small by comparison with the radius. If these conditions are not

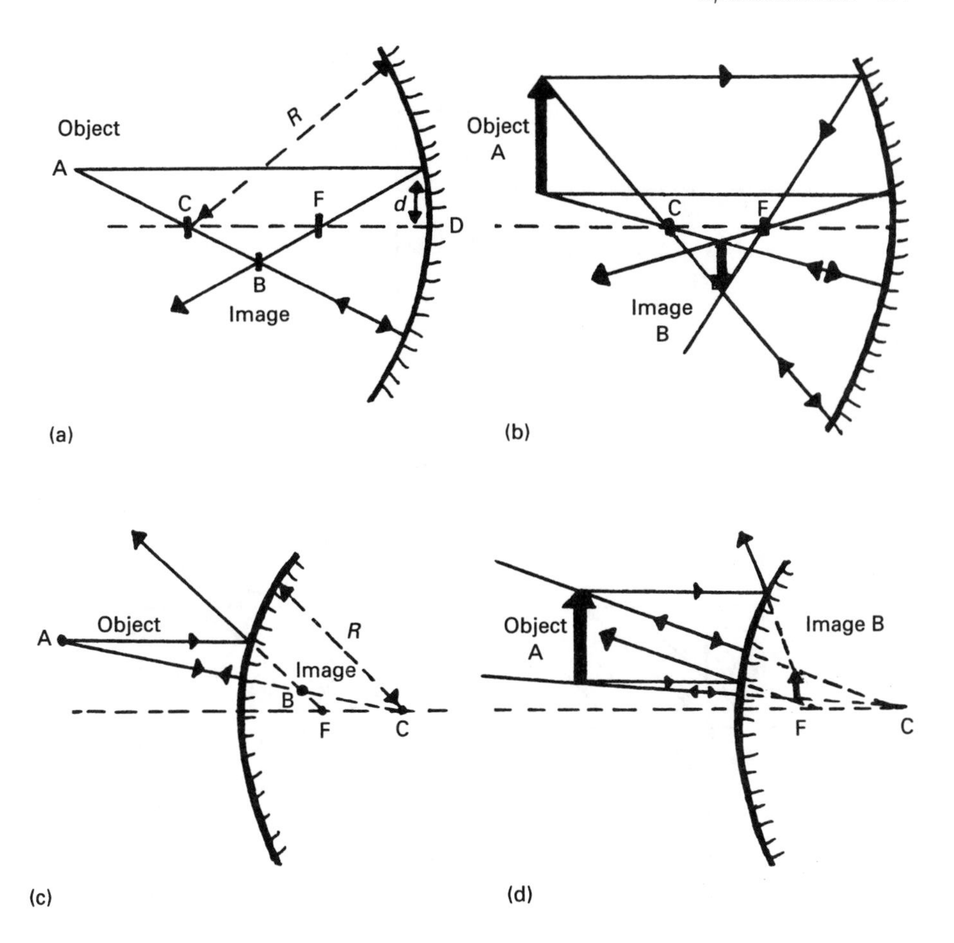

Fig. 8.4 *Reflections from curved mirrors. (a) Point object, concave mirror. (b) Finite object, concave mirror. (c) Point object, convex mirror. (d) Finite object, convex mirror.*

met, the image will be distorted. A parabolic mirror, however, will behave as described regardless of the diameter. The focus of the mirror is at the focus of the parabola, as shown in Fig. 8.5. A parabolic mirror is, of course, more expensive to manufacture.

8.3.2. Refraction and lenses

The speed of light is dependent on the material through which it travels as described by the index of refraction defined by equation 8.2. One consequence of this speed change is that the direction of a ray of light changes as it passes from one medium to another (e.g. from air to glass). If the light passes from an optically less dense medium (e.g. air to glass), the ray bends towards the normal. If the light passes from a more dense medium (e.g. glass to air), the ray bends away from the normal. Both these cases are illustrated in Fig. 8.6a.

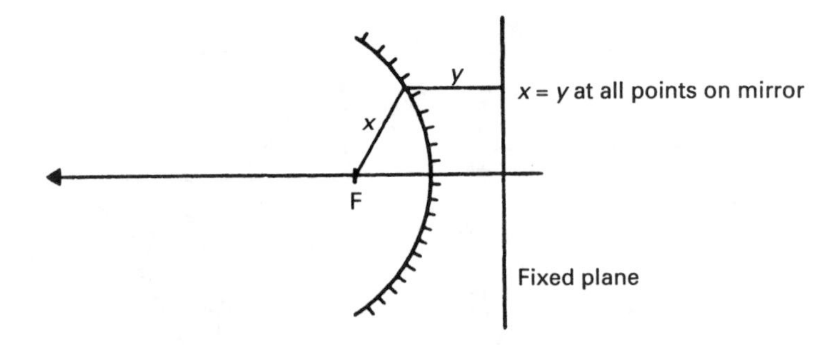

Fig. 8.5 *Construction of a parabolic mirror.*

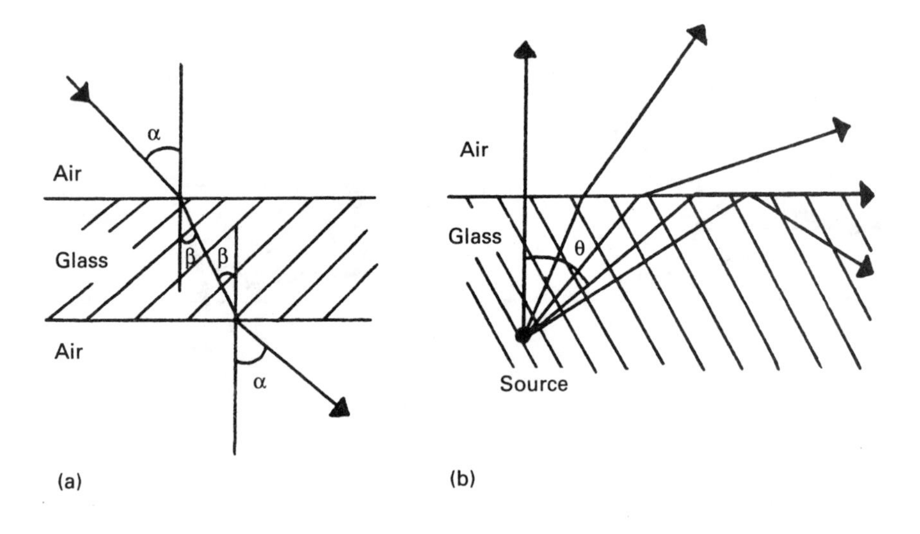

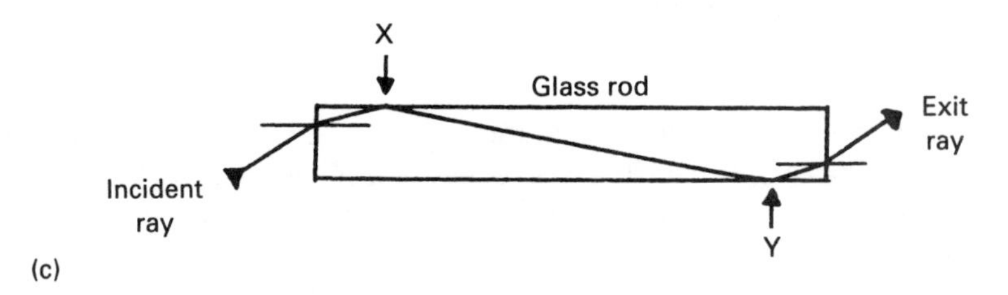

Fig. 8.6 *The refraction of light. (a) Principles of refraction. (b) Total internal reflection. (c) Conveying a light beam down a glass rod. Total internal reflection occurs at points X and Y.*

The behaviour of the ray is described quantitatively by Snell's law which states that:

$$\frac{\sin \alpha}{\sin \beta} = \frac{\text{Velocity of light in medium 1}}{\text{Velocity of light in medium 2}} = \frac{\mu_2}{\mu_1} \tag{8.3}$$

If medium 1 is air, the value of μ_1 is unity for all practical purposes so:

$$\frac{\sin \alpha}{\sin \beta} = \mu \tag{8.4}$$

where μ is the refractive index of the denser medium. Angle α is called the angle of incidence, and angle β the angle of refraction.

Figure 8.6b shows light passing from glass to air at increasing angles. Eventually at an angle θ the refracted ray just skirts the surface of the glass. At angles greater than θ, a ray of light will be internally reflected back into the glass. The angle θ is called the critical angle. The value of θ is given by:

$$\sin \theta = 1/\mu \tag{8.5}$$

For glass, with a μ of 1.57, the value of θ for a glass/air interface is about 40°.

Internal refraction allows a light beam to be conveyed along a glass tube as shown in Fig. 8.6c, which is the basis of fibre optic light guides (described further in Section 8.6).

Lenses are based on the phenomenon of refraction. Figure 8.7a shows a convex lens. A light ray entering on one side will exit in a different direction. The lens

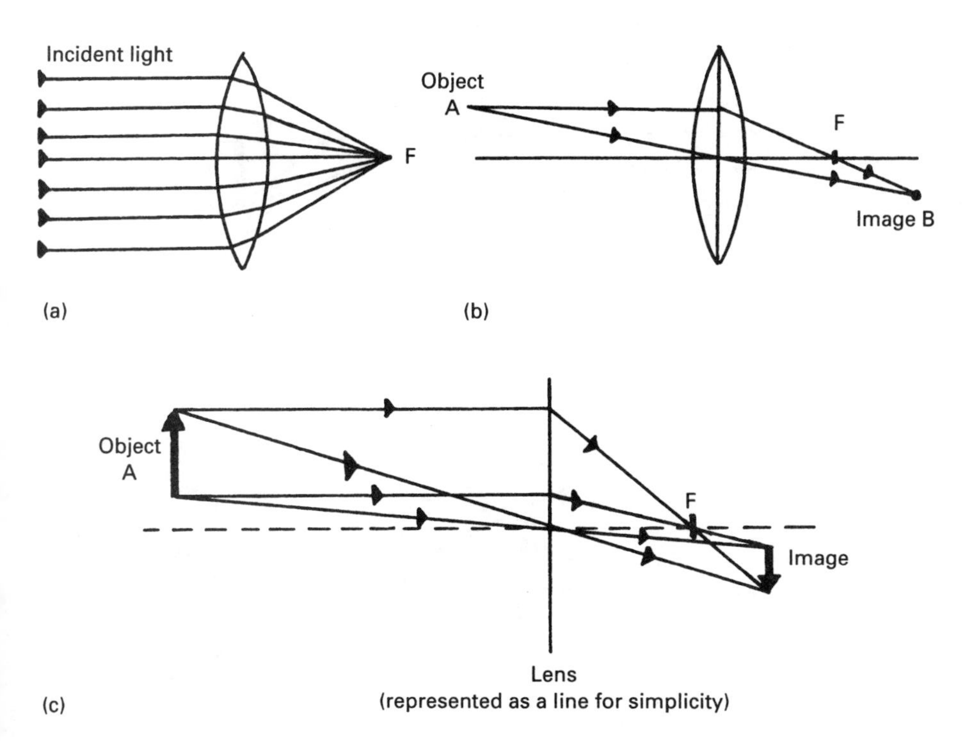

Fig. 8.7 *The convex lens. (a) The action of a lens. (b) Formation of an image of a point object. (c) Formation of an image of a finite object.*

surfaces are shaped such that any ray parallel to the axis of the lens passes through the same point called, not surprisingly, the focus of the lens. The distance from the lens to the focus is called the focal length of the lens.

The action of a lens can be analysed by considering two rays from a point in a similar way to that demonstrated for mirrors (Fig. 8.7b). A ray parallel to the axis of the lens will pass through the focus, and a ray through the lens centre will not be diverted (because entry and exit angles are equal). An object at point A will produce an image at point B. The image of a finite-sized object can be found by considering rays from each end (Fig. 8.7c). Under the conditions shown a real image is formed.

The position and size of the image can be found by simple geometry. Referring to Fig. 8.8, it follows that:

$$d = DF/(D - F) \tag{8.6}$$

and

$$h = dH/D \tag{8.7}$$

Note that if $D = 2F$, $h = H$. If D lies between $2F$ and F, h is larger than H (a magnified image is produced). If D is less than F, the rays after the lens diverge as shown in Fig. 8.8b. Under these circumstances a virtual image is formed on the

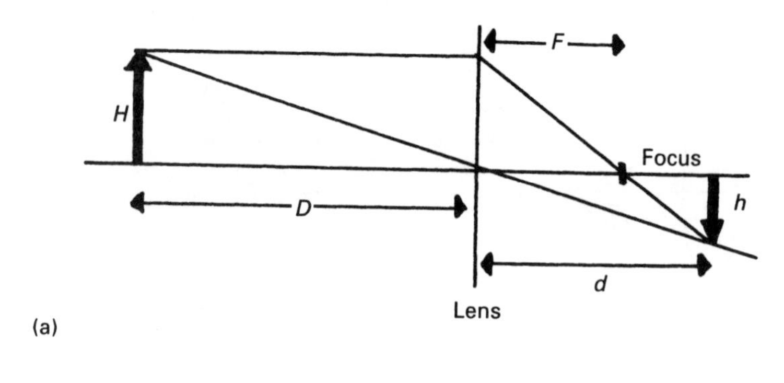

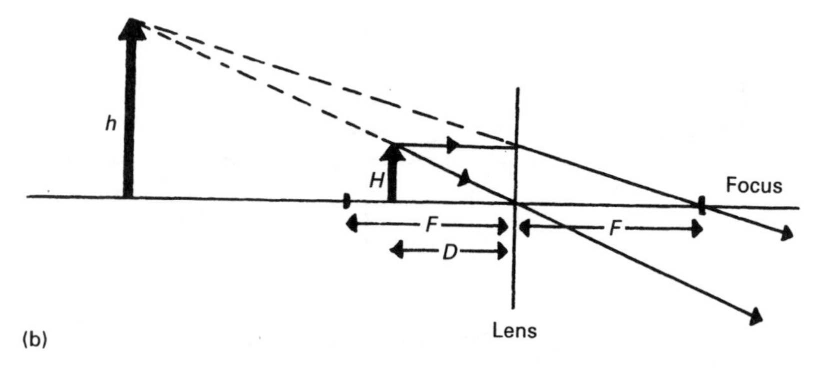

Fig. 8.8 *Analysis of the operation of a convex lens. (a) Mathematical analysis of image formation. (b) Magnifying operation.*

same side of the lens as the object (this is the basis of a hand-held magnifying glass).

Light rays diverge from a concave lens, but the action can be analysed in a similar manner. Figure 8.9a analyses two rays from a point source and Fig. 8.9b a finite object. A concave lens always produces a reduced virtual image.

8.3.3. Prisms and chromatic aberration

The index of refraction, μ, of glass was given in Section 8.2 as 1.57. In practice μ is found to vary slightly with the wavelength of light. This means that violet light is bent more on crossing an air/glass interface.

White light is a combination of all the colours of the spectrum, so a ray of white light will be split into its constituent colours on passing from air to glass (Fig. 8.10a). The effect can be increased by a prism (Fig. 8.10b). The splitting of light into its constituent colours is called spectroscopy, and can be used for chemical analysis, a topic covered further in Sections 8.9.1 and 10.4.

The variation of the refractive index with wavelength causes the focal length of

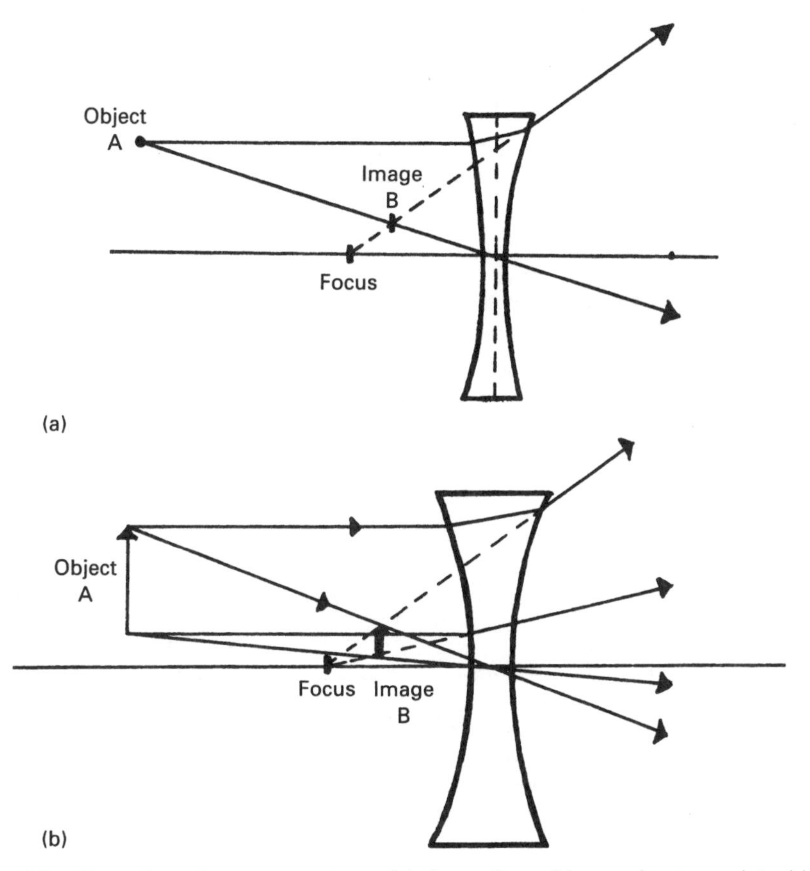

Fig. 8.9 *Operation of a concave lens. (a) Formation of image from a point object. (b) Formation of image from a finite object.*

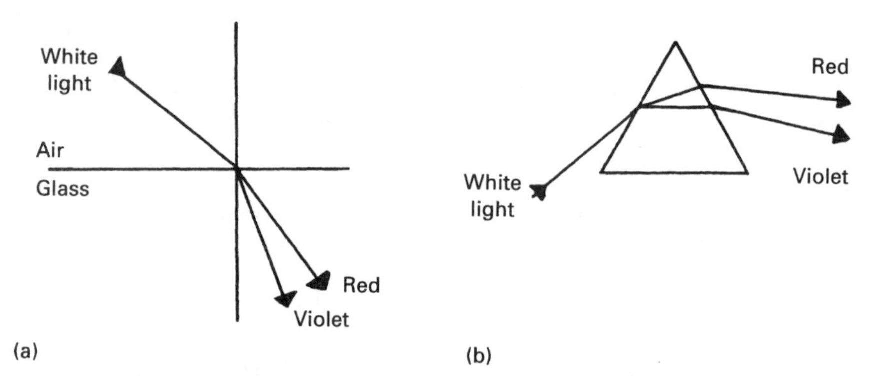

Fig. 8.10 *The variation of refractive index with wavelength. (a) The effect of passing white light from air into glass. (b) A prism.*

a lens to be colour dependent, as shown in Fig. 8.11a. This is known as chromatic aberration and appears in cheap optical instruments as a coloured haze around objects.

Chromatic aberration can be almost eliminated by combining a convex and concave lens (Fig. 8.11b). The two lenses are made of glass with different refractive index. By careful design, the chromatic aberration of the two lenses can be made to cancel, giving a focal point independent of wavelength.

8.4. Sensors

8.4.1. Photoresistors

An optoelectronic sensor is a device which converts a light signal into an electrical signal. The cheapest, and easiest to apply, sensor uses light to change the resistance of a semiconductor and is accordingly called a photoresistor.

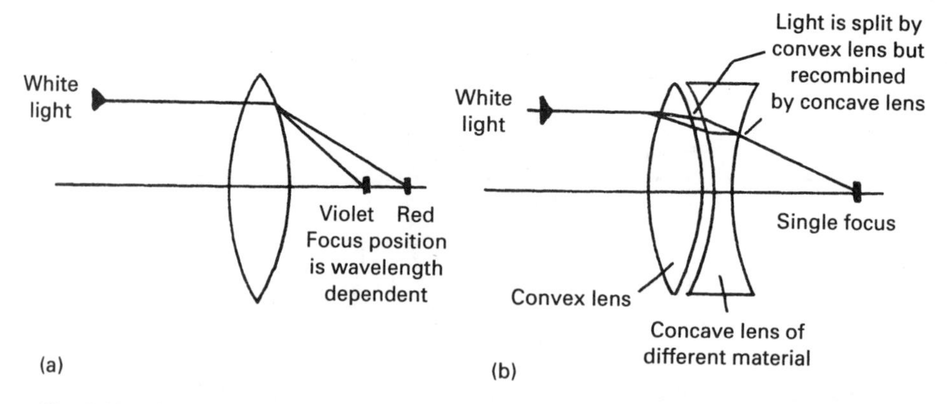

Fig. 8.11 *Chromatic aberration. (a) The effect of chromatic aberration. (b) The cure.*

The resistance of a semiconductor is determined largely by the number of electrons in the conduction energy band. Normally a few electrons are excited across the energy gap from the valence band to the conduction band, but the resistance of a semiconductor at room temperature is quite high.

If light of sufficient energy is absorbed by the semiconductor, more electrons are excited and can pass from the valence band to the conduction band, creating hole (in the valence band) and electron (in the conduction band) pairs. The absorbed light creates pairs in proportion to the intensity, causing a decrease in the semiconductor resistance.

The wavelength of the light is rather critical. If the wavelength is too long, the energy is too low to create the electron/hole pairs. If the wavelength is too small, it will all be absorbed at the surface and cause no net change in resistance. The spectral response of a typical photoresistor is shown in Fig. 8.12.

Useful photoresistive materials are:

Material	*Energy gap (eV)*	*Range (Å)*	*Peak sensitivity (Å)*
Cadmium sulphide	2.54	4 000–8 000	5 200
Cadmium selenide	1.74	6 800–7 500	7 000
Lead sulphide	0.4	5 000–30 000	20 000
Lead selenide	0.3	7 000–58 000	40 000
Indium antimonide	0.16	6 000–70 000	55 000

Visible light, it will be remembered, covers the range 4000 Å to 8000 Å. Photoresistors cover the range from visible light to well into the infrared. The commonest of the above materials is cadmium sulphide (CdS).

A typical device is the CdS-based ORP12 which is arranged in a thin layer to give a large surface area and minimum thickness. The change in resistance is large; an ORP12 will go from around 2M in the dark to around 100R in normal room lighting. This large change allows simple circuits to be used.

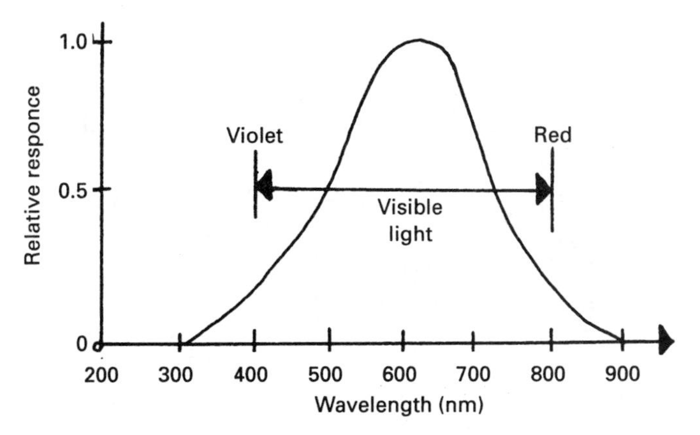

Fig. 8.12 *Response of cadmium sulphide photoresistor.*

The resistance change is, however, very non-linear, and the actual resistance value is dependent on temperature as well as light intensity. It is therefore important to minimise dissipation in the device. Photoresistors also have a significantly large time constant: typical values range from about 0.1 ms for a lead sulphide cell to over 100 ms for a cadmium sulphide cell.

8.4.2. Photodiodes

A photodiode consists of a back-biased *p-n* junction which, under dark conditions, behaves as a normal back-biased diode. With these conditions the only current flowing through the diode of Fig. 8.13a will be the usual leakage current (typically 1 μA).

Absorbed light will generate electron/hole pairs as described above, and the current through the diode will increase to a typical value of 100 μA. A photodiode has the response similar to Fig. 8.13b which shows that it can be considered as a constant current device with the current determined by the light intensity.

The current/intensity relationship is quite linear, and the response is fast; typically 0.2 μs but devices as fast as 1 ns are available. In general, photodiodes are the smallest optical sensor which, in conjunction with their high speed, makes them well suited for fibre optic data transmission and similar applications. Typical operating wavelengths are 8000 Å to 11 000 Å (silicon) and 13 000 Å to 20 000 Å (germanium).

The relatively low level current can easily be converted to a high-level voltage using a DC amplifier, as in Fig. 8.13c. The light-dependent diode current flows through R to give an output voltage IR which is directly related to light intensity.

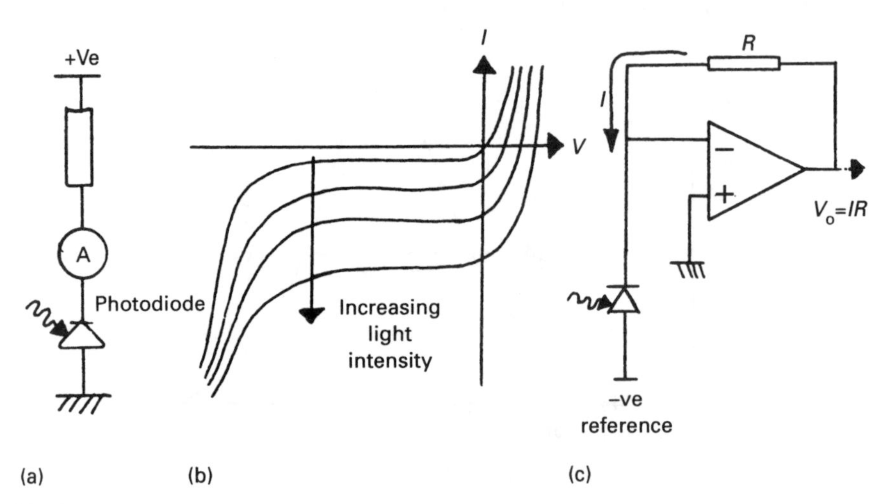

Fig. 8.13 *The photodiode. (a) Basic circuit. (b) V/I/light relationship. (c) Practical circuit.*

8.4.3. Phototransistors

A phototransistor is a variant of the photodiode, and is shown schematically in Fig. 8.14a. A photodiode is connected between collector and base so that the light-dependent diode current becomes the transistor base current. The transistor multiplies this by its current gain, β, to give a far larger collector/emitter current change.

Unfortunately the dark current is also multiplied by β. A typical device has the response of Fig. 8.14b, with a change from dark to light of 0.1 mA to 10 mA.

Using similar ideas to Fig. 8.14a, it is also possible to construct photothyristors, photoFETs, and photoDarlingtons, but these are relatively rare.

8.4.4. Photovoltaic cells

A photovoltaic cell is a *p-n* junction which is designed to generate a voltage when light is absorbed. The mechanism by which this occurs is complex, and a full description is beyond the scope of this book. The voltage generated is small; typically 0.4 V in sunlight for a silicon cell, with an available current between 30 mA and 100 mA. The power can be increased by series/parallel combinations, often called solar cells.

A photovoltaic cell can be used in two modes, open circuit voltage and short circuit current. In the open circuit voltage mode the voltage is logarithmically related to light intensity, which is useful in photometry applications where a large light range is to be covered. The short circuit current is linearly related to intensity and is used where increased accuracy is required over a small light range.

Photovoltaic cells are expensive and slow (with time constants similar to photoresistors). Their main application is in battery-less photometry and solar-powered circuits (a small solar cell can easily supply CMOS circuits).

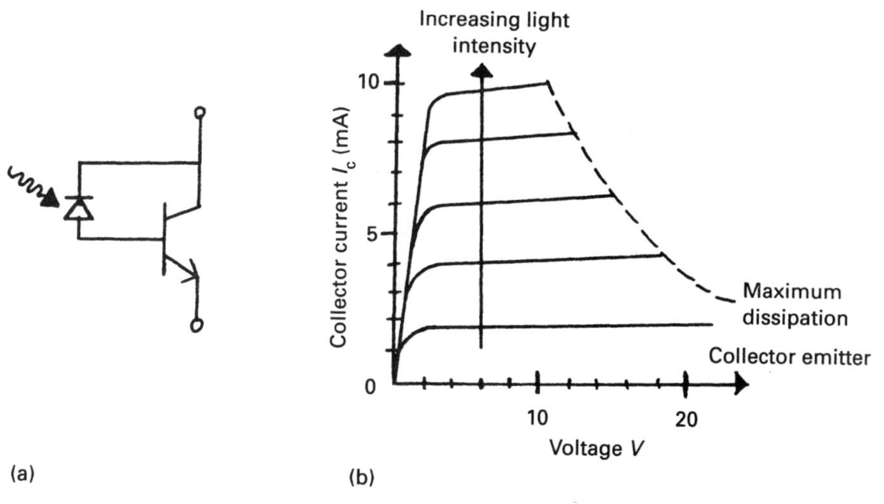

Fig. 8.14 *The phototransistor. (a) Construction. (b) V/I/Light relationship.*

8.4.5. Photomultipliers

The most sensitive optical sensor is the photomultiplier tube shown schematically in Fig. 8.15a. At the input end a cathode coated with photoemissive material is held at a large negative voltage (typically 1 kV). At the output end of the tube, the anode is held near 0 V via the load resistor R.

Between the cathode and anode there are several intermediate electrodes called *dynodes*. There are four dynodes on Fig. 8.15a for simplicity, but in practice many more are used. The dynode voltages are equally spaced between the anode and cathode, usually by a resistor chain as in Fig. 8.15b.

When light strikes the photoemissive cathode, electrons are released in quantities dependent on the light intensity. They are attracted to the more positive dynode, D_1. The dynodes are coated with a material from which electrons are easily detached. When the electrons from the cathode strike D_1,

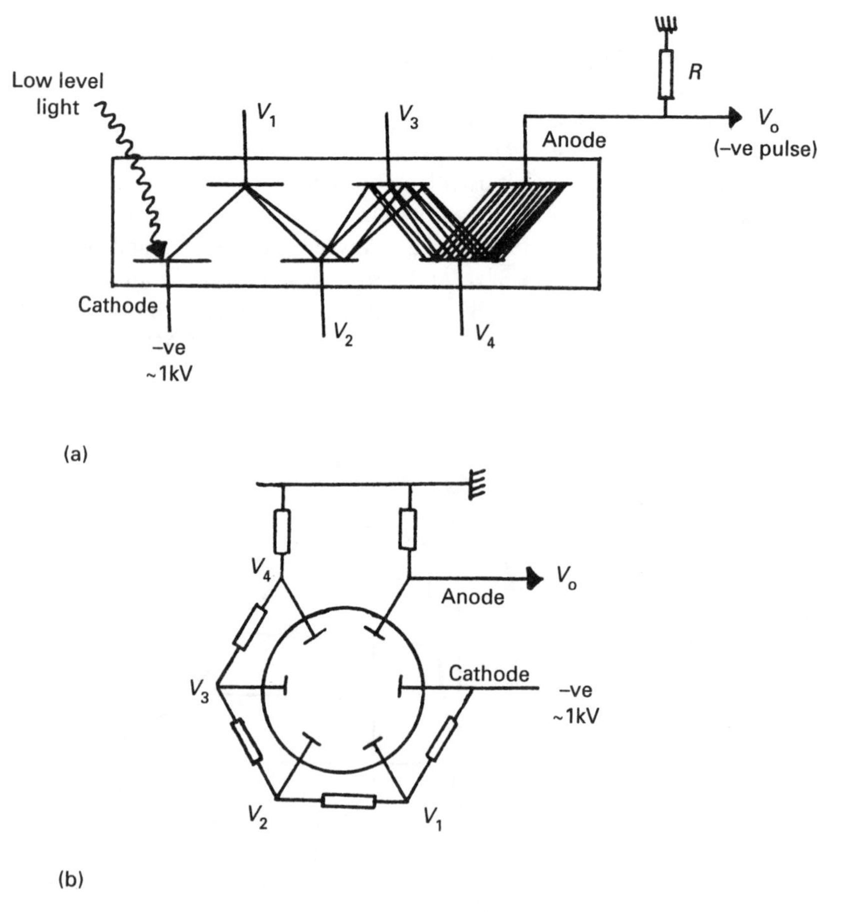

Fig. 8.15 *The photomultiplier. (a) Construction and principle of operation. (b) Derivation of dynode voltages.*

more electrons are released, which are now attracted to D_2. The process is repeated along the tube, so the number of electrons arriving at the anode is a large multiple (typically 10^6) of those released by the light absorbed at the cathode.

The electrons arriving at the anode cause a current to flow through the load resistor R, giving a negative output voltage proportional to the (low-level) input light intensity.

Photomultipliers are used where very low light levels are to be measured. A typical application is in spectroscopy, a topic covered further in Section 10.4.

8.4.6. Integrated circuit devices

All the basic light sensors require additional components to give a useful signal. It is increasingly common for semiconductor manufacturers to construct a complete photocell circuit (sensor and amplifier) in a single IC. Usually the IC incorporates a lens to focus the light on to the photosensor part of the circuit. Most of these, however, are designed to give an on/off digital indication (light above some threshold value/light below threshold value) rather than a photometry device with output proportional to light level.

8.5. Light emitters

8.5.1. Light-emitting diodes

Semiconductor light emitters are *p–n* junctions that emit light when forward biased. As a *p–n* junction forms a diode, the device conducts in one direction (emitting light) and blocks current in the reverse direction. Semiconductor light emitters are therefore usually called light-emitting diodes (LEDs).

The mechanism by which the light is produced is the recombination of electrons and holes between the valence and conduction bands, with the released energy appearing in the form of light. Fortunately it is not necessary to appreciate the underlying physics to use LEDs. The light emitted by an LED lies within a narrow band of wavelengths dependent on the materials used. Common devices are:

Material	*Wavelength range (nm)*
Gallium arsenide	890–980 (near infrared)
Gallium phosphide	530–580 (green to yellow)
Gallium phosphide/zinc oxide doping	620–700 (red)
Gallium arsenide/gallium phosphide	600–650 (orange)

The efficiencies of diodes based on these materials vary, but the actual perceived light intensity is more dependent on the eye's sensitivity to different colours. The maximum sensitivity is in the green/yellow/orange part of the spectrum, as can be seen in Fig. 8.16a. This is fortunate as gallium phosphide LEDs give a lower light intensity than other materials.

An LED is a current operated device, and operates with a relatively constant voltage of around 1.5–2.0 V. The operating current for most LEDs is around 20 mA. When operated from a voltage source, a current limiting resistor is needed (Fig. 8.16b) with value

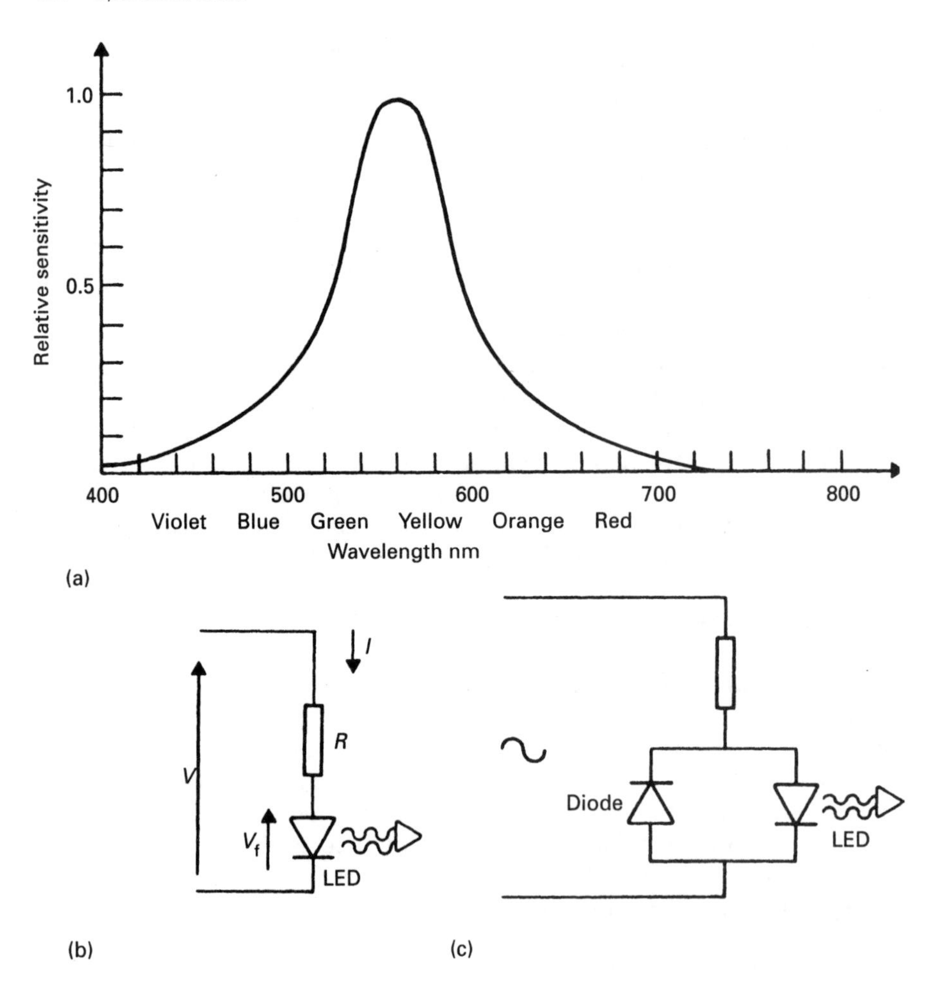

Fig. 8.16 *Light-emitting diodes (LEDs). (a) Spectral response of the human eye. (b) Basic circuit. (c) AC circuit.*

$$R = \frac{V - V_f}{I} \tag{8.8}$$

The light intensity is approximately linearly related to the current, but this is not as apparent as might be thought. The eye's perception of intensity is logarithmic, so once an LED has attained a reasonable intensity, the apparent brilliance appears fairly independent of current.

Although an LED behaves as a diode, its reverse voltage is low (typically 5 V). If an LED is to be driven from an AC source, a protection diode should be used (Fig. 8.16c).

LEDs are available in a variety of sizes and configurations. Although not as bright as incandescent bulbs, they operate at far lower currents. Seven-segment LED displays are widely used for numerical indication, a topic discussed further in Chapter 17.

8.5.2. Liquid crystal displays

Liquid crystal displays (LCDs) are based on materials which can exhibit crystal-like structures in a liquid state. Such materials are normally transparent, but if an electric field is applied, complex interactions between the internal molecules and free ions cause the molecules to align and the liquid to become opaque. The advantage of LCDs is that they respond to an electric field and need no current. Operating power levels are correspondingly very low (typically about 0.1 mW per cm^2), which makes them very attractive for battery-powered devices.

The construction of an LCD cell is very simple, as shown in Fig. 8.17a, consisting of two metallised glass plates separated by spacers. The gap between the plates is filled with the liquid. When an electric potential is applied to the plate, the liquid turns opaque.

An LCD requires an external light source. This can be provided by an integral light emitter (called the transmissive mode, Fig. 8.17b) or by ambient light via a mirror (called the reflective mode, Fig. 8.17c).

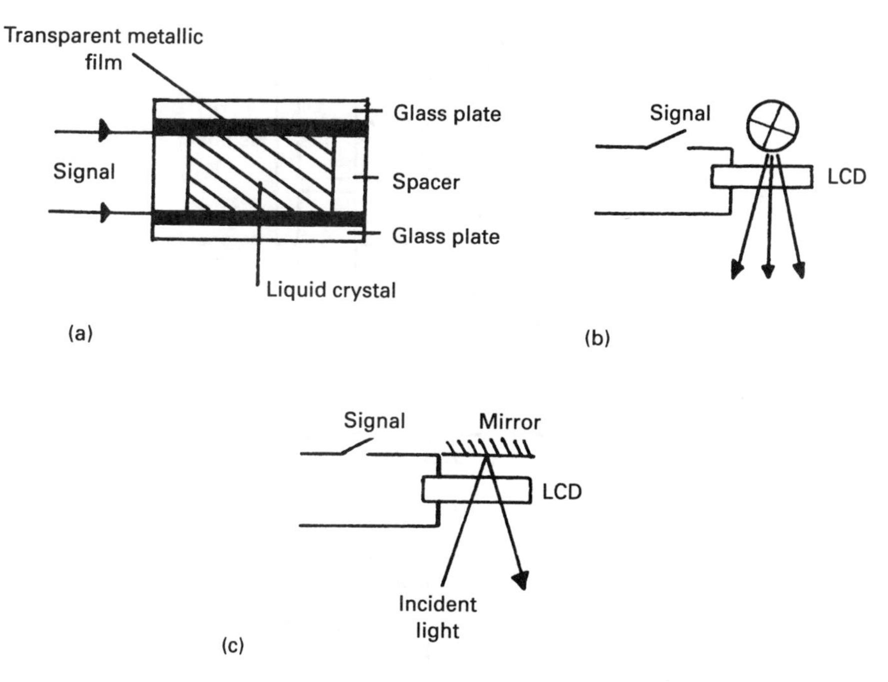

Fig. 8.17 *The liquid crystal display. (a) Construction. (b) Transmissive mode. (c) Reflective mode.*

An LCD will operate on DC, but in practice polarisation and electrolysis effects give a greatly reduced life. It is usual for LCDs to be run on pseudo AC, which gives greatly improved life expectancy. A common way of providing AC drive is shown on Fig. 8.18a. A simple square wave oscillator is used to drive the cell back plane and segment via an exclusive OR gate. Operating waveforms are shown in Fig. 8.18b. Using this arrangement a DC component of less than 100 mV is achievable.

8.5.3. Incandescent emitters

The commonest light source is the common light bulb used in domestic lighting, car headlamps, torches, etc. When a material is heated, EM radiation is emitted (see section 2.6), and if the temperature is sufficiently high, part of the radiation will be in the visible part of the spectrum. The radiation, however, covers a large wavelength band far into the infrared, so only a small part of the input energy appears as visible light.

Incandescent emitters (or bulbs!) are inefficient, generate heat and have a limited life (which can be extended considerably by under-running, or by using lamp-warming current through the filament in the off state to prevent thermal shock). They also have the undesirable characteristic of taking a large inrush current when first turned on. Their advantages are simplicity and much higher intensity than any other emitter.

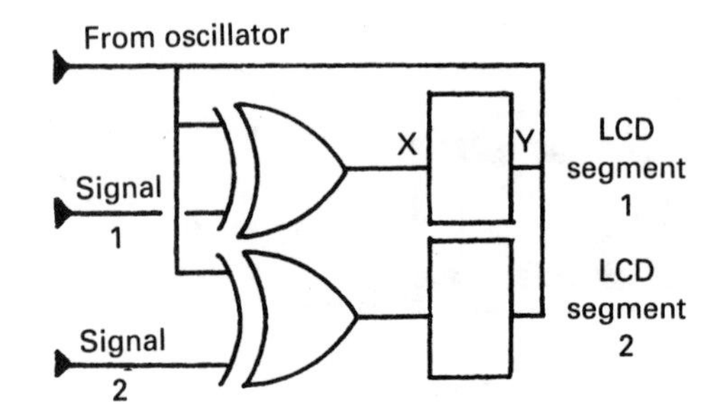

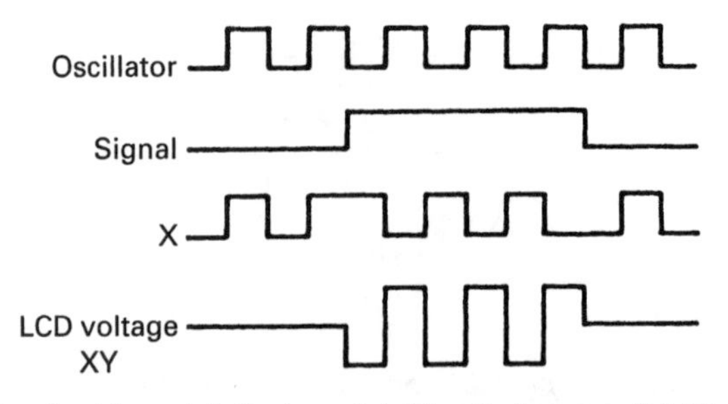

Fig. 8.18 *Driving liquid crystal displays. (a) Circuit diagram. (b) Waveforms.*

8.5.4. Atomic sources

Figure 8.19 shows an atom which can be considered as consisting of a nucleus surrounded by shells of orbiting electrons. The electron shells are at fixed distances from the nucleus. If external energy is supplied, an electron can be excited from an orbit to a higher energy orbit (from, say, X to Y). The electron will only stay in this state for a short time before resuming its lower energy orbit. In falling back it loses energy which is emitted as EM radiation. In atomic sources of EM radiation, therefore, energy is input to change the orbits of electrons and the energy re-emerges as EM radiation. The wavelength of the radiation depends on the energy difference between electron orbits, so for any given material there are several possible transitions, each with its own characteristic wavelength (a topic discussed further in Section 10.4).

In a neon lamp, for example, the energy is provided by the electric current flowing through ionised gas. For neon, the major transition produces light in the orange part of the spectrum giving the characteristic reddish/orange neon light.

In domestic fluorescent tubes the major transition produces ultraviolet radiation which is absorbed by the tube coating. Electrons in the coating are excited by ultraviolet radiation, but visible light is emitted when they resume their low-energy state.

Phosphorescence occurs in some materials and is used for luminous paint. The absorption of EM radiation in the form of light can excite electrons in some materials. Normally the time taken for an electron to resume its low energy state is very short (less than 10^{-8} s). In some materials, however, electrons can stay in a high-energy state for several minutes, slowly emitting light as they return to their normal orbits.

Many industrial displays are based on atomic light sources. Gas discharge displays, Nixies (a registered trade mark of the Burroughs Corporation) and phosphorescent displays are all variants of the above principles.

8.6. Fibre optics

Internal reflection of light was introduced in Section 8.3.2. In Fig. 8.20 a light ray

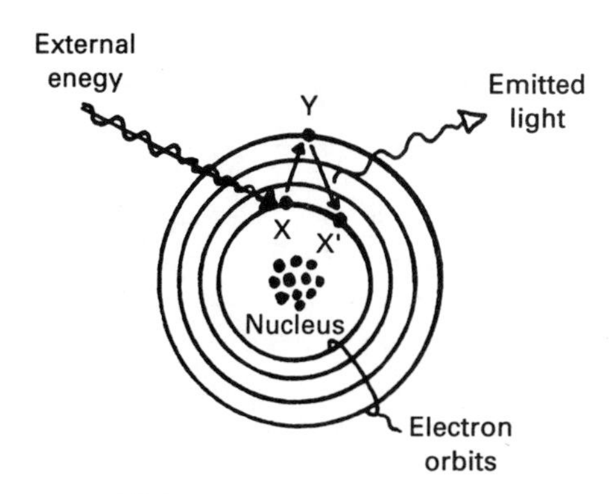

Fig. 8.19 *Atomic sources of light.*

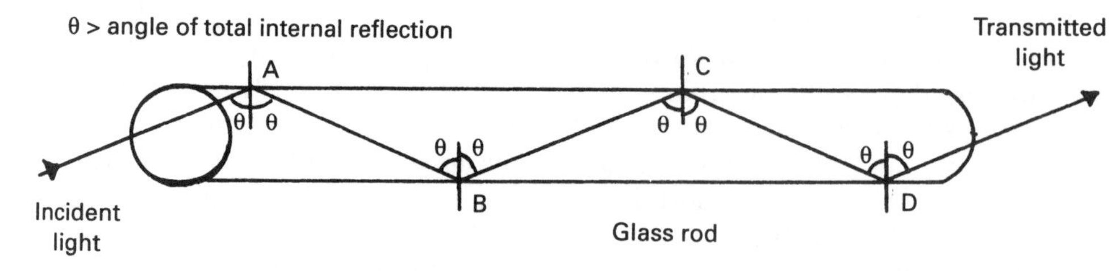

Fig. 8.20 *Principle of fibre-optic transmission.*

enters a glass rod at a shallow angle. At point A the ray strikes the side of the rod, but because the angle θ is greater than the critical angle, total internal reflection occurs and the ray is contained within the rod. Similar reflections occur at points B, C and D until the ray emerges, unattenuated, from the end of the rod.

The principle of Fig. 8.20 allows signals to be carried optically through glass or plastic rather than electrically down a wire. Figure 8.20 implies a rigid rod; in practice the use of many small diameter fibres allows the construction of an optical 'cable' which is as flexible and strong as its electrical counterpart. The use of such 'cables' is known as fibre optics.

The use of fibre optics has several advantages. Noise in instrumentation is an electromagnetic phenomenon, so fibre optic transmission is totally free from noise interference, crosstalk or ground-loop problems and gives total electrical isolation between transmitter and receiver. Fibre optic cables can also be run with total safety through flammable and explosive atmospheres because sparking cannot result from cable breakage or damage.

Fibre optic cables also have wider bandwidth and lower transmission losses than coaxial cables when high frequencies are used. In electrical transmission the available bandwidth varies inversely as the square of the transmission distance; with fibre optics it varies inversely as the distance.

There are two types of fibre-optic construction. *Step index* fibres (Fig. 8.21a)

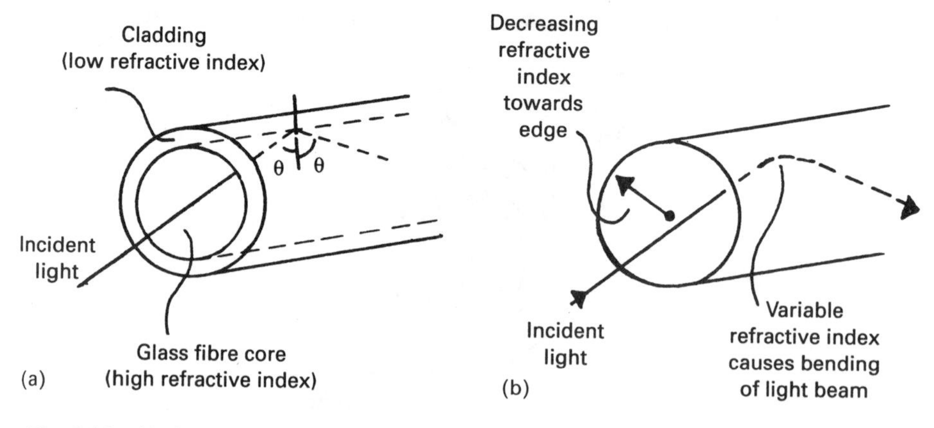

Fig. 8.21 *Various constructions of fibre-optic cable. (a) Step index. (b) Graded index.*

are identical in principle to Fig. 8.20, consisting of a central fibre of high refractive index surrounded by a thin cladding of low refractive index. Internal reflection takes place at junction between the fibre and the cladding.

Graded index fibre has a refractive index which varies gradually from a high value at the centre to a low value around the periphery. This causes the light rays to follow the gentler curves of Fig. 8.21b. Graded index fibres have lower transmission losses and other advantages described below, but are obviously more expensive to manufacture.

The optical signal is attenuated as it passes down a fibre optic cable. This attenuation occurs from four main causes. The first two of these are directly related to the cable length and are quoted together by manufacturers as an attenuation in dB km^{-1}.

Material absorption occurs because impurities and manufacturing defects cause the optical signal to be gradually absorbed along the cable. Losses can be minimised by choosing an operating wavelength at which the losses are least. Fibre optic manufacturers quote a recommended operating region for their cable.

Scattering losses occur because of slight irregularities at the fibre/cladding interface, which means that a small proportion of the light rays strike the interface at an angle less than the critical angle and are absorbed by the cladding.

Losses also occur when the fibre follows a tight curve as this decreases the angle of incidence on the outer side of the cable as shown in Fig. 8.22a. The curvature of the cable allows light which was previously outside the critical angle to be absorbed.

The final loss occurs at couplings between cable sections and between the cable and the transmitter and receiver. In Fig. 8.22b the overall attenuation is given by:

$$\text{Attenuation} = 10 \log\left(\frac{P_T}{P_R}\right) \text{dB} \tag{8.9}$$

The attenuation consists of $(\alpha_T + \alpha_L + N\alpha_C + \alpha_R)$ where α_T and α_R are the transmitter and receiver to fibre losses (typically 3 dB for a plastic polymer system or 1 dB for a glass system); α_L is the cable attenuation (typically 150 dB km^{-1} for polymer and 5 dB km^{-1} for glass) and $N\alpha_C$ is the attenuation for N in line connectors (each connector loss being similar to the transmitter/receiver loss).

In Fig. 8.23, a light beam is just being conveyed down the fibre and being reflected at just the critical angle each time. To do this, it must enter at angle θ_a which is known as the angle of acceptance. If n_a, n_f, and n_c are the index of refraction of air, the fibre and the cladding, analysis will show that

$$\sin \theta_a = \frac{n_f}{n_a}\sqrt{1 - \left(\frac{n_c}{n_f}\right)^2} \tag{8.10}$$

For all practical purposes, $n_a = 1$, so

$$\sin \theta_a = \sqrt{n_f^2 - n_c^2} \qquad (8.11$$

The term 'numerical aperture' is often given to $\sin \theta_a$. Light rays entering at an

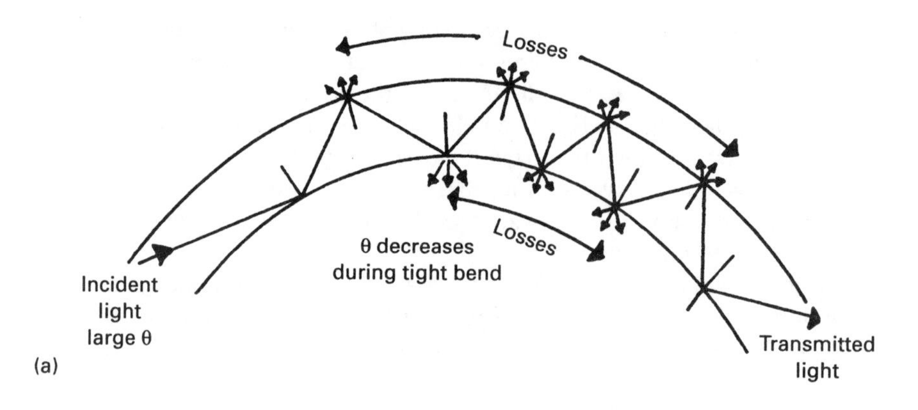

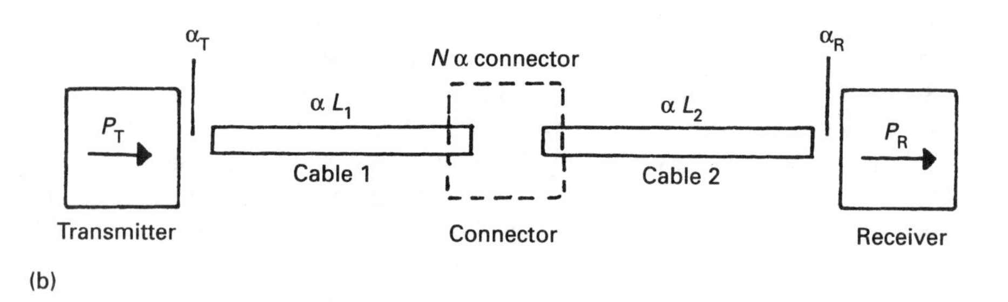

Fig. 8.22 *Losses in fibre-optic transmission systems. (a) Losses caused by tight bend. (b) Transmission and coupling losses.*

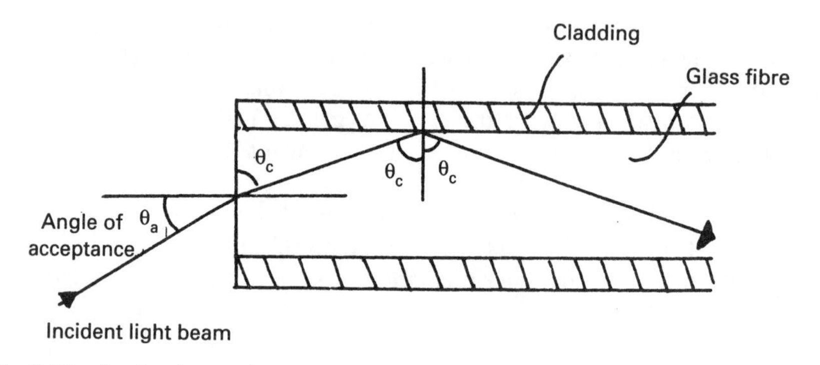

Fig. 8.23 *Angle of acceptance.*

angle greater than θ_a will not be conveyed down the cable. Typical values for the numerical aperture are 0.25 to 0.5, which gives angles of acceptance of between 15° and 30°.

The elements of a fibre optic system are therefore as shown on Fig. 8.24. Data (in digital or some modulated form, e.g. AM, FM, PWM) is passed via a drive

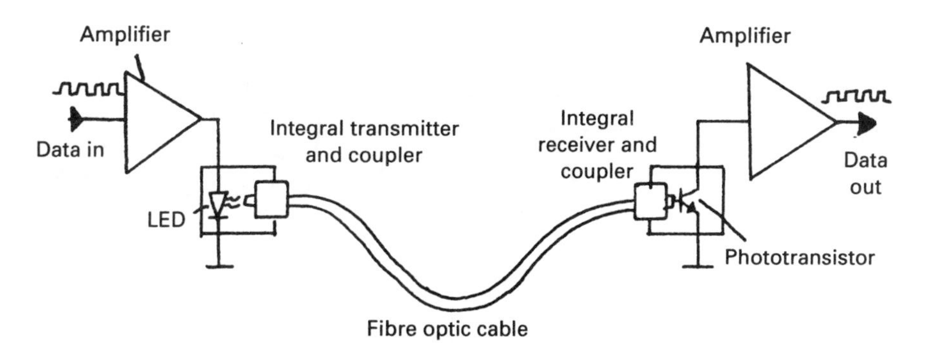

Fig. 8.24 *Fibre-optic data transmission system.*

amplifier to an LED operating at a wavelength suitable for the cable (typically 600–850 nm). The light travels down the fibre optic cable and is received by a photodiode. The resulting diode current changes are amplified to give the original signal. A data transmission system similar to Fig. 8.24 can have a bandwidth in excess of 100 MHz.

8.7. Photocells

Photocells are devices used to detect the presence or absence of objects. They can be used, for example, to count objects passing down a conveyor belt or to replace a limit switch in some circumstances. In essence the principle is very simple and is illustrated in Fig. 8.25a.

A light source is placed at the focus of a lens to produce a parallel beam of light. This is received by another lens and focused on to a light sensor whose output drives a relay via a suitable amplifier. As objects break the light beam, the output relay de-energises.

The circuit of Fig. 8.25a is, however, sensitive to all light, and not just that emerging from the photocell transmitter. It would, for example, probably respond to changes in ambient lighting. The circuit of Fig. 8.25b is more commonly used and employs a modulated light source.

The source is driven from an oscillator at a high frequency (typically several tens of kilohertz) to produce a chopped light beam. In some photocells this chopping is done mechanically with a motor-spun mirror. The chopped light beam is received by a sensor whose output is an AC signal. This is amplified by an AC bandpass amplifier which will reject all frequencies other than the one used by the transmitter. The amplifier output is rectified and used to drive the output relay. Changes in ambient light produce a low frequency or DC change in the sensor output, and are correspondingly rejected by the bandpass amplifier. Temperature effects on the sensor are similarly ignored.

Figure 8.26 uses separate transmitters and receivers, both of which need a power supply. If this is inconvenient, a retroreflective unit can be used, one example of which is shown in Fig. 8.27a. This incorporates the transmitter and

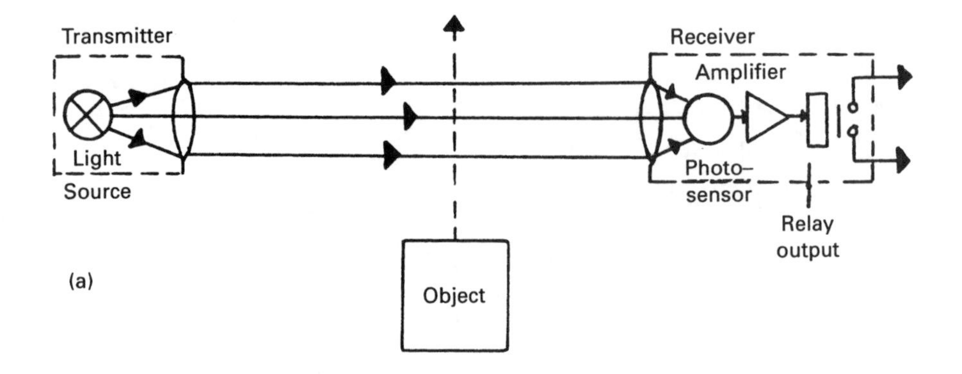

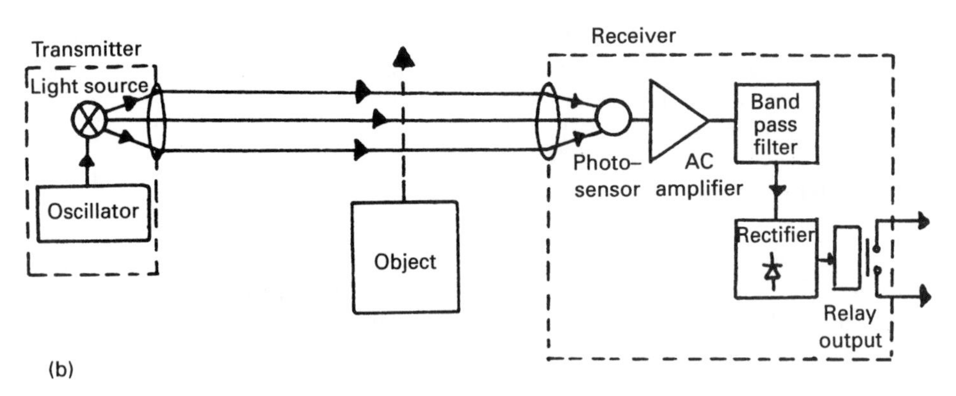

Fig. 8.25 *Photocell systems. (a) Simple transmitter and receiver. (b) Modulated photocell system.*

Fig. 8.26 *Practical photocell transmitter and receiver.*

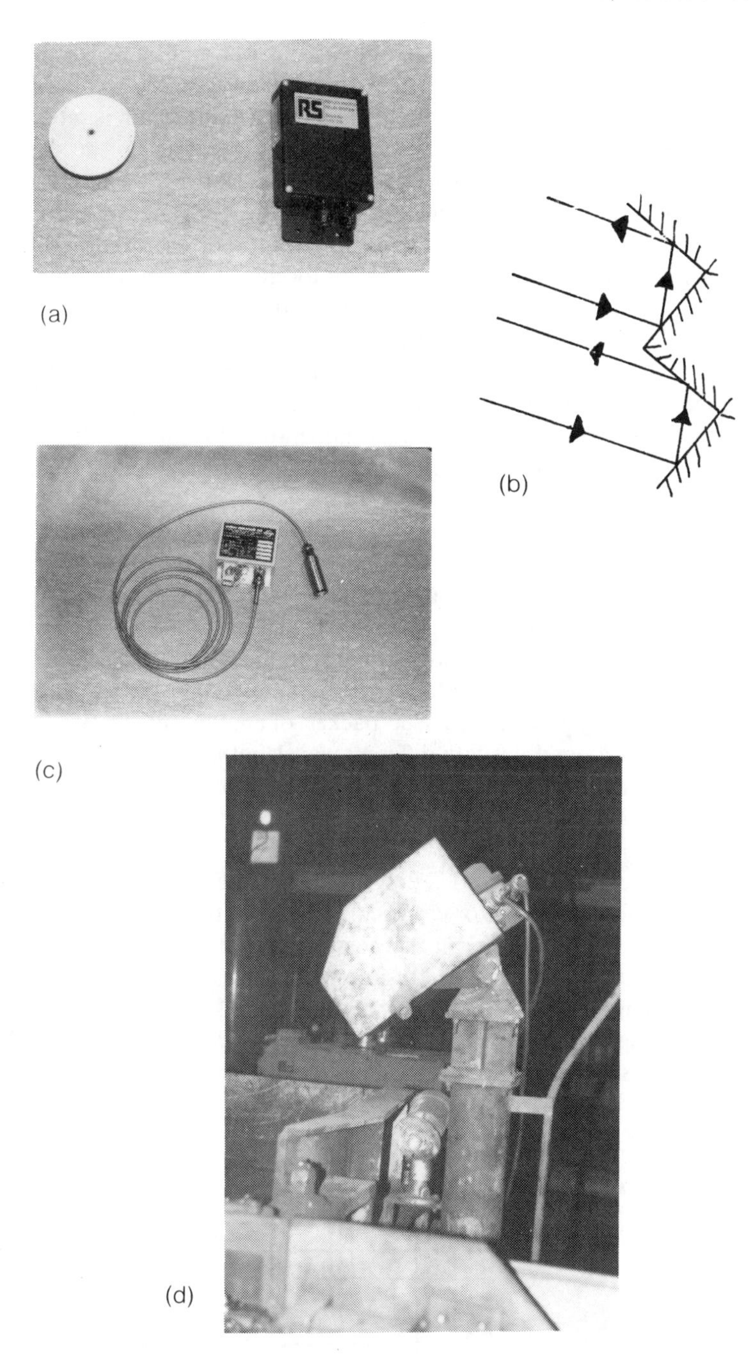

Fig. 8.27 *Various photocells. (a) Retroreflective unit and mirror. (b) Prismatic mirror used on retroreflective photocell. (c) Unit with fibre optic coupled head. (d) Infrared hot metal detector.*

receiver in one unit, and utilises a mirror to give an out-and-back light beam. The mirror is constructed of inverted prisms as shown in Fig. 8.27b (similar to bicycle reflectors) which have the characteristic that any arriving beam is returned back along its path. Perfect alignment of the mirror is therefore not needed.

Retroreflective photocells can also be used to detect the presence of objects by using the object surface itself to backscatter the light. The range of such a detector is small, typically 100 to 300 mm depending on the object's reflectivity.

Fibre optics can be combined with photocells to advantage, allowing the 'delicate' electronics to be removed to less hazardous environments. Figure 8.27c shows a photocell that has a separate lens and detector head which focuses light on to the end of a fibre optic cable. The electronics transmitter and receiver circuits are mounted at the other end of the fibre optic cable. Apart from allowing the head to be mounted in, say, a high-temperature location while keeping the electronics cool, the fibre optic head assembly is much smaller than any integral unit.

8.8. Lasers

Light emission caused by the change of energy level of electrons in atoms was briefly described in Section 8.5.4. Possible transitions for hydrogen atoms are shown in Fig. 8.28. This shows that an energy input of at least 10 eV is needed to lift an atom to its first state; on falling back it emits light of frequency 1216 Å (ultraviolet). The light is emitted as a 'packet' of energy called a photon. Other transitions are also possible, with different wavelengths as shown. The larger the energy change, the higher the frequency. The relationship is given by:

$$E_2 - E_1 = hf \qquad (8.12)$$

where h is Planck's constant. It follows that for a particular material emitting light via excitation from heat or electrical influences, there will be many specific wavelengths corresponding to all the possible transitions of Fig. 8.28.

Laser light, however, is unique. It is monochromatic, i.e. it consists of light of just one wavelength. It is also coherent: all the photons emitted from the laser are exactly in phase. The difference between coherent and non-coherent light is shown in Fig. 8.29. As a result of the coherence, monochromaticity and the

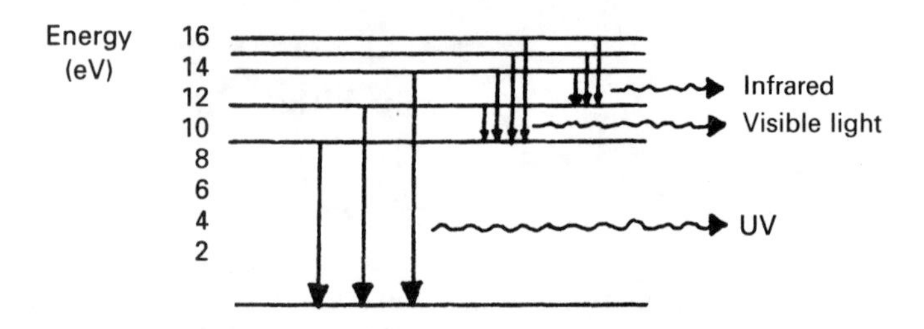

Fig. 8.28 *Possible energy states for hydrogen atoms.*

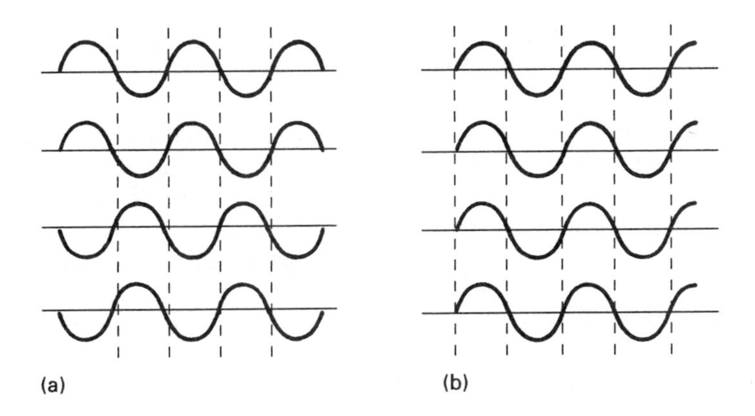

Fig. 8.29 *Coherent and non-coherent light. (a) Single-frequency non-coherent light. Components out of phase, partial cancellation results. (b) Coherent light. All components are in phase and consequently reinforce each other.*

mechanism by which it is produced, laser light is very intense and emerges as an exceptionally parallel beam. A typical divergence is less than 0.001 radian.

The construction of a typical ruby laser is shown in Fig. 8.30a, and consists of a ruby rod with the ends machined parallel; one end is fully silvered to give an almost perfect mirror. The other is partially silvered. The ruby rod is surrounded by a flash tube.

Ruby has the energy states of Fig. 8.30b. There is an energy state 1 at about 1.8 eV above the base state, and a large number (many hundreds) of energy states above state 1 which collectively form an energy 'band'.

To start laser action, it is first necessary to get more atoms into the excited state than remain in the ground state. This is called a population inversion, and is achieved by 'pumping' energy into the rod by firing the flash tube for a short

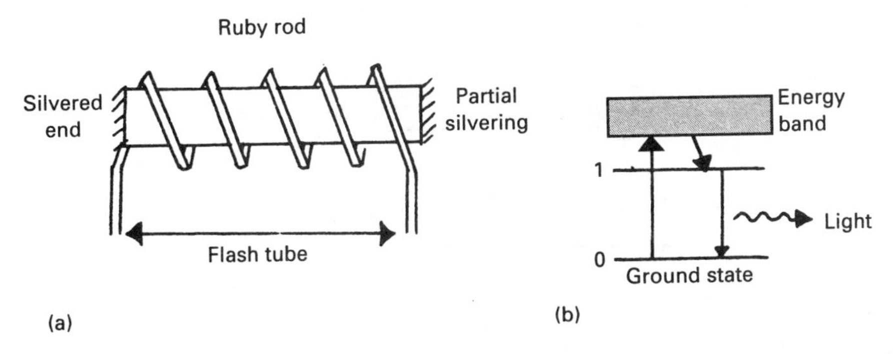

Fig. 8.30 *A ruby pulsed laser. (a) Schematic construction. (b) Energy states and transitions of a ruby laser.*

period. This takes the atoms in the ruby rod into the higher energy band of Fig. 8.30b, from which they all spontaneously fall back to state 1 where they are transiently stable.

Eventually an atom will fall back to the base state, emitting a photon of light as it does so. If this photon strikes another excited atom (as is probable in a population inversion), the latter falls to the base state, emitting another photon which is exactly in coherence with the impinging photon. A positive feedback effect now takes place with a rapid build-up of photons, all of identical frequency and coherent. The silvered ends make the resultant rapidly intensifying light pulse traverse the rod several times before it emerges from the half-silvered end.

The transition from state 1 to the ground state is 1.8 eV, which by equation 8.12 corresponds to a wavelength of 6943 Å. This is in the red part of the spectrum, so an intense coherent red pulse of light lasting about a millisecond is produced.

The ruby laser of Fig. 8.30 is called a pulse laser, for obvious reasons. For most industrial applications, a continuous beam is needed. This is not feasible with the ruby laser because the rod would not be able to absorb the energy necessary to maintain a permanent population inversion. To achieve this, a four-level laser system is used as shown in Fig. 8.31.

Atoms are pumped from the ground state to a high energy band from where they fall back to energy state 2. Laser action takes place between states 2 and 1, with atoms returning spontaneously to the ground state. To get laser action it is only necessary to maintain a population inversion between states 1 and 2, which is achievable with relatively low pumping energies. Figure 8.32 shows a typical industrial laser.

8.9. Miscellaneous topics

8.9.1. Spectroscopy

Light is emitted when excited atoms fall to a lower energy state, with the frequency being given by equation 8.12. The energy levels differ from element to element, so it is possible to identify the presence of elements, and their relative

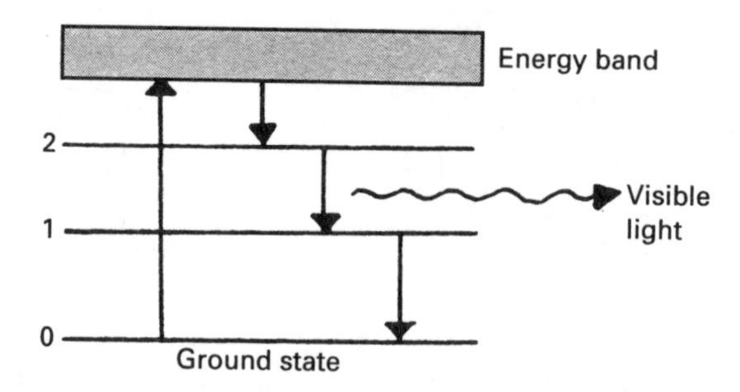

Fig. 8.31 *Energy states for a continuous laser.*

Fig. 8.32 *Ferranti MFK–1200 Watt CO_2 laser cutting mild steel plate.*

proportions, by observing the light emitted from a substance. Light from silicon, for example, contains a distinct spectral line at 2124 Å, and light from sodium contains two lines in the orange part of the spectrum (hence the colour of sodium street lights). These spectral lines can be used for chemical analysis (particularly in metallurgy). This analytical technique is called spectroscopy which is discussed later in Section 10.4.

8.9.2. Flame failure devices

Flame failure detection is required for gas and oil burners to prevent the explosion hazard from an accumulation of unburnt fuel. For small burners a simple thermal detector will suffice, but in large boilers the thermal time constant is too long to shut the fuel off quickly.

Gas and oil flames emit ultraviolet light which 'flickers' at a predictable rate. This phenomenon allows an ultraviolet flame failure detector to be constructed as in Fig. 8.33. Ultraviolet light from the flame is detected by a suitable detector whose output is amplified by an AC amplifier which passes the flicker but rejects the background UV. The amplifier output is rectified and used to energise an output relay.

Failure of the flame (or the detector head itself) will cause the output of the AC amplifier and rectifier to fall to zero and the relay to de-energise.

8.9.3. Photometry

Photometry is concerned with the measurement of light intensity, and as such is encountered in topics as diverse as photography and internal building illumination.

It is obvious that there is a difference between 'brightness' and the capability of

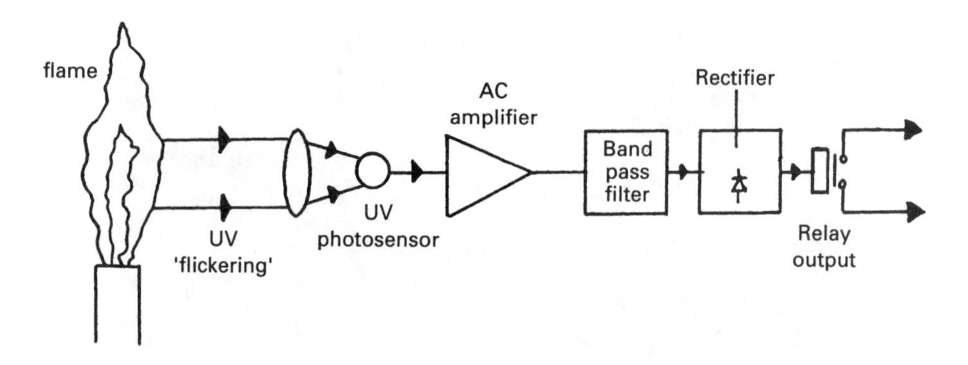

Fig. 8.33 *Flame failure detector. Principle of operation.*

a light source to illuminate an area. A 100-W tungsten filament bulb and a 20-W fluorescent tube have similar lighting capability, but the former is obviously brighter to the eye.

Lighting capability is defined as the luminous intensity and is measured in candelas (cd) approximating to the light from a standard candle. It being difficult to make a standard candle, the SI standard is based on the light emitted from a square centimetre of platinum at its melting point of 1773 °C, which has a luminous intensity of 60 cd.

Brightness is defined as the luminance, and is the luminous intensity divided by the area of the source. The SI source above, for example, has a luminance of 60 cd/cm^2. A fluorescent tube has a luminance of approximately 1 cd/cm^2 compared with 650 cd/cm^2 for a tungsten filament bulb. The much larger area of the fluorescent tube compared with the incandescent filament, however, gives a similar luminous intensity.

It is also possible to define the luminous flux, which relates the lighting capability of a light source to the area of the surface to be illuminated and the distance of the source. The light flux, ϕ, is measured in lumens, 1m, and is defined as the light flux emitted in one steradian from a uniform candela source. (This can be envisaged as the light flux through an area of 1 m^2 at a distance of 1 m from a candela source: see Fig. 8.34a.) By simple geometry, a given source of I candelas has a total luminous flux of 4 πI lumens.

From the definition of light flux, it is possible to define the level of illumination at a given surface area (e.g. a desk top or work table). The level of illumination, in lux, is simply defined as the flux per square metre of surface, 1m m^{-2} (obviously the area in Fig. 8.34a has a level of illumination of 1 lx).

Suppose we have a 50-candela source (approximately a 60-W bulb). This has a total luminous flux of 628 lm. Assuming this is radiated equally in all directions, at a distance of 2 m, say, the flux will spread over an area of about 50 m^2 ($4\pi r^2$) giving a level of illumination of about 12.5 lx (628/50).

Because the area over which the luminous flux is dispersed increases as the square of the distance, it follows that luminous intensity falls off as the square of

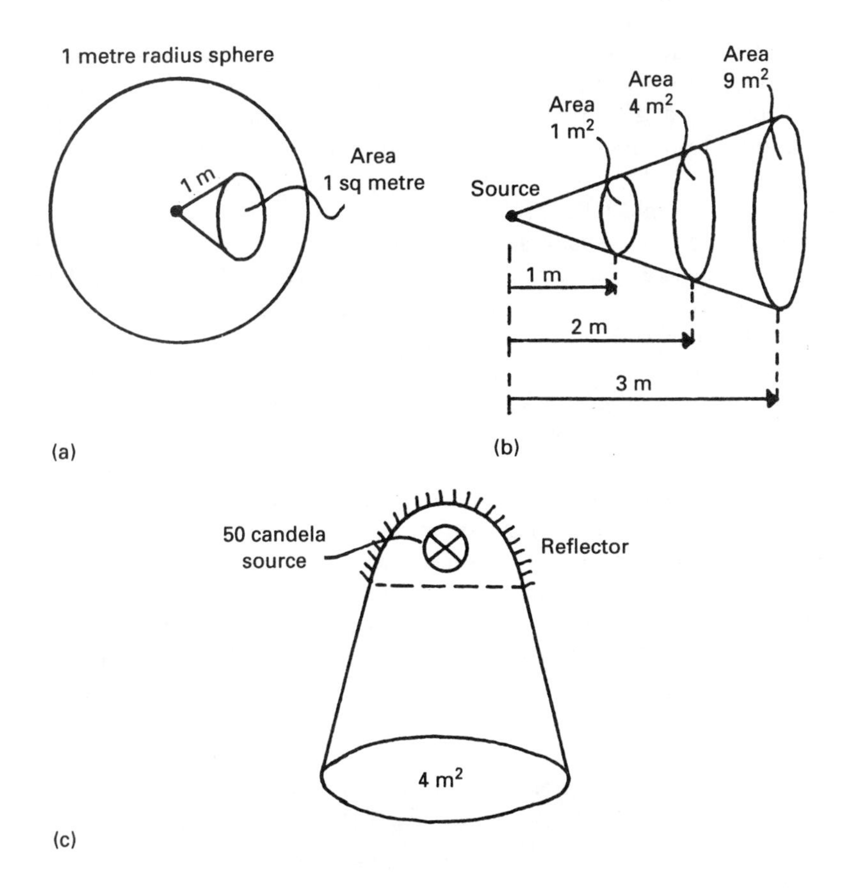

Fig. 8.34 *Photometry principles. (a) Definition of level of illumination. (b) The inverse square law of illumination. (c) The effect of a reflector.*

the distance, as illustrated by Fig. 8.34b. A tripling of the distance gives one-ninth of the illumination.

Adding a reflector to a lamp can increase the level of illumination dramatically. In Fig. 8.34c a reflector is used to focus the light from our 50-candela source on to a 4 m^2 area. The whole 628 lm are now spread over an area of 4 m^2, giving a level of illumination of 157 lx.

Light levels inside buildings are measured in lux. Recommended levels go from 100 lx in warehouses and general factory areas to about 750 lx for drawing offices. Light levels are generally measured by a light meter using a photovoltaic cell.

8.10. Barcodes

All articles bought in supermarkets are labelled with a barcode which identifies the country of origin, manufacturer and the item itself. These barcodes are read at the check out by a scanner, and the price found automatically from the store's computer and added to the customer's bill. This is a typical application of bar code technology which is being widely introduced in industry and commerce as a way of tracking and keeping inventory control of items with minimal human intervention. Even the ISBN numbers used to identify books are now read by barcode readers in libraries and used to keep track of library loans for the UK Public Lending Right (PLR) scheme.

The structure of a bar code is shown on Fig. 8.35. It consists of a series of bars and spaces used to encode alphanumeric symbols in a machine readable format. At either end is a quiet zone to allow the reading system to sense the start and finish of the bar code. A quiet zone is needed because not all bar code representations are of a consistent length. The bearer bars are added when there is a danger that a misaligned read scan may not catch all of the bar code data. The bearer bars will then give a broad pulse which the reader will detect and cause the read to be rejected.

There is not a single universal bar code, and many different codings are in use throughout the world. Some of the commoner ones are shown on Fig. 8.36. Code 39 uses nine bar elements per digit, three of which are wide (hence the name). The nine elements are always made up of four spaces and five bars.

EAN stands for European Article Number and was derived from the earlier American UPC or Universal Product Code. Both use characters constructed from two bars and two spaces occupying seven positions. Only numeric data can

Fig. 8.35 *Bar code structure.*

Fig. 8.36 *Common bar codes.*

be represented. EAN and UPC have long data streams which can be subdivided into subsets by longer twin bars. Supermarkets in the UK use EAN product coding, with three groups denoting country of manufacture (UK is 5, France is 3 and so on), the manufacturer (012427 is the Scottish soup maker Baxter, for example) and the product itself (020108 is Royal Game Soup.) EAN is also used for ISBN book markings.

Interleaved 2 of 5 is again a numeric-only coding and it represents characters in pairs, one by the bars and one by the spaces. Each character uses five positions, two of which are wide.

Considerable self-checking is built into these codings. The structure itself has a definite machine readable format which is easily checked by a machine (e.g. the 3/9 relationship in Code 39). In addition the last digit in the code is usually a check digit which is formed using ideas similar to the CRC method described in Section 19.2.7.

Industrial systems, unlike supermarkets, will normally use automatic bar code marking systems and unattended readers. A bar code is read by scanning a light beam across the code and detecting the reflection. Visible or infrared light can be used with LEDs or low powered lasers as the source. Infrared is attractive for industry because reads can be made through oil and grease coatings. Usually the scanning is performed by a vibrating mirror.

The reflection can be specular (as occurs wth a mirror) or diffuse (as occurs from a sheet of paper). Bar code readers rely solely on diffuse reflection. A bar code reading system can thus be represented by Fig. 8.37. The light beam should not strike the bar code at 90° as might be first thought, as the resultant specular reflection could dazzle the receiver. Usually angles between 60° and 80° are used.

A bar code is scanned continuously, not just once, and a good read is declared

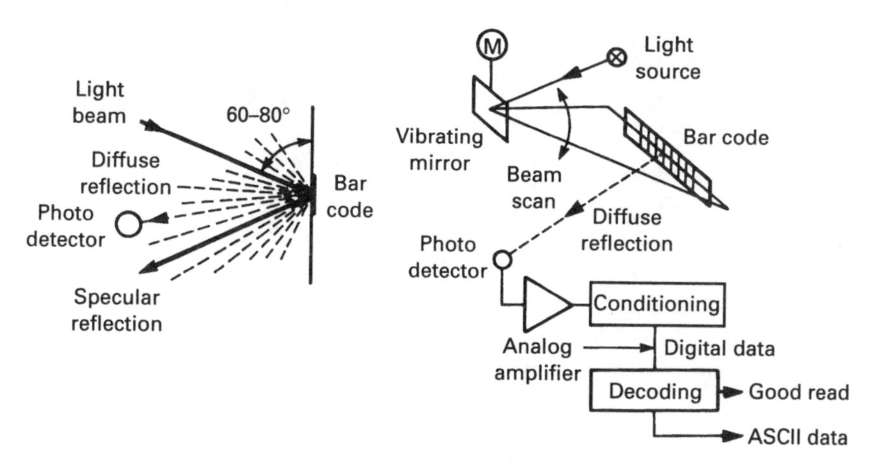

Fig. 8.37 *Bar code reader schematic.*

when the same information has been received several times. The number of identical reads needed is set by the designer, but five is a typical number.

The physical relationship between the scan and the bar code determines how many reads can be attempted. A bar code can be arranged vertically (called ladder orientation) or horizontally (called picket fence).

A reader at a given distance from a target bar code will have a fixed scan length. In Fig. 8.38, a reader with a scan length of 20 cm and a scan time of 5 m s (200 scans per second) is being used to read a bar code sized 16 cm by 8 cm. The bar code is moving transversely at 65 cm s^{-1}. In the ladder orientation it will remain in view for $8/65 = 0.123$ s which will allow 24 reads (each taking 5 m s). With the picket fence arrangement it can only travel 4 cm whilst remaining fully in view, which, by a similar calculation, will only allow 12 reads. The chance of getting five (say) identical reads increases with the number of reads, so for this application the ladder arrangement is obviously the correct choice.

The reliability of a bar code system is measured by the First Read Rate which is how many times the first scan gives the correct data. The FRR determines how many identical readings are needed before a good read is declared. Typical FRRs are 90% for which three to five identical reads would be needed. Systems can be made to operate at much lower FRRs by increasing the number of identical reads required.

This section is based on material provided by Allen Bradley, one of the leaders in the industrial application of bar code systems. The Allen Bradley PLC family includes bar code readers which can be directly connected into a PLC rack.

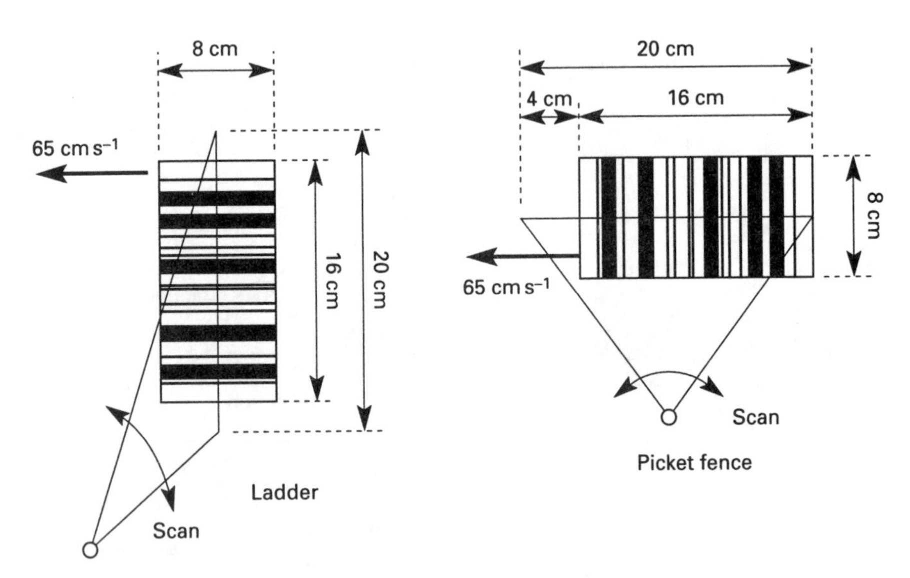

Fig. 8.38 *Bar code read timings.*

Chapter 9

Velocity, vibration and acceleration

9.1. Relationships

Velocity is the rate of change of distance, and acceleration is the rate of change of velocity. Velocity therefore has the units of m s^{-1} and acceleration m s^{-2}. Given any one quantity the others can be derived by integration or differentiation. DC amplifier circuits are described in Chapter 11. Practical integrator circuits, however, suffer from long-term drift, and differentiators, by their nature, are very noise sensitive, so position, velocity and acceleration transducers are usually employed directly. Examples of non-direct position measurement are inertial navigation systems in spacecraft, where position in three axes is obtained from double integration of acceleration signals.

It is instructive to consider an object moving sinusoidally with respect to a fixed point (Fig. 9.1.). The position will be given by:

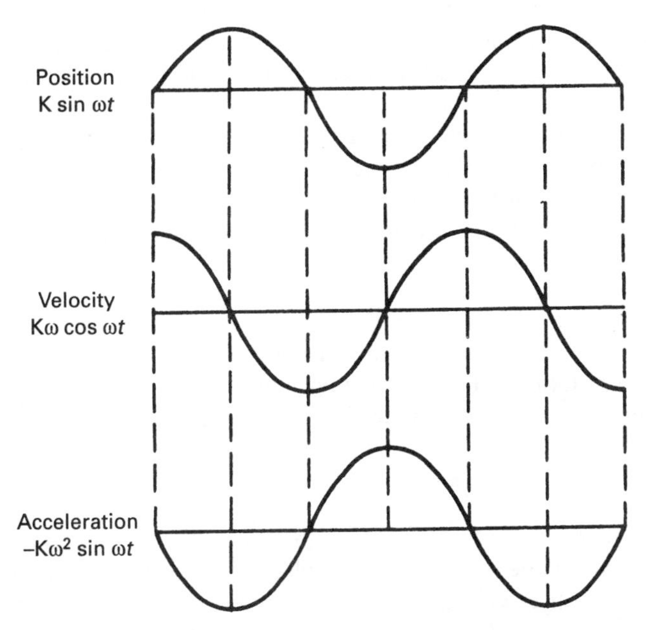

Fig. 9.1 *The relationship between position, velocity and acceleration.*

$$x(t) = K \sin \omega t \quad \text{m} \tag{9.1}$$

where K is the peak displacement and ω the angular frequency in rad s^{-1} (frequency in hertz is, of course, $\omega/2\pi$). The velocity is given by differentiation:

$$v(t) = K\,\omega \cos \omega t \quad \text{m s}^{-1} \tag{9.2}$$

and the acceleration by another differentiation

$$a(t) = -K\,\omega^2 \sin \omega t \quad \text{m s}^{-1} \tag{9.3}$$

Velocity is at a maximum at the centre point of the motion, and acceleration at the extremes. Velocity and acceleration are both proportional to K, the amplitude. Note, however, that velocity is proportional to the frequency, and acceleration proportional to the square of the frequency.

It follows that very large accelerations can occur with small displacements. It is, in fact, the acceleration required by the needle that limits the frequency response of a hi-fi unit. Similarly, small-amplitude high-frequency vibration of mechanical plant can cause damaging accelerations and forces. The measurement of vibration is essential in large rotating plant such as fans, compressors, etc.

Acceleration is often measured with respect to the acceleration due to gravity (9.8 m s^{-2}). An acceleration of 29.4 m s^{-2} can, for example, be represented as 3 g. This notation is generally used for large transient acceleration.

9.2. Velocity measurement

9.2.1. Tachogenerator

Many applications require measurement of angular velocity (e.g. shaft speeds of motor-driven plant) and in many others it is convenient to convert a linear movement to an angular movement (e.g. by a rack-and-pinion) for measurement purposes.

The commonest, and simplest, angular velocity transducer is the tachogenerator. This is effectively a simple DC generator. The principle is discussed further in Chapter 12, but is shown here in Fig. 9.2. A wire, length l, is moving with velocity v perpendicular to a magnetic field of flux density B. A voltage e is induced in the wire where:

$$e = Blv \tag{9.4}$$

In Fig. 9.2, a coil of n turns is rotating in a field B. Each side of each turn parallel to the axis of rotation will have a voltage induced as above. The turns are not, however, moving perpendicular to B. If the coil is at an angle θ, the induced voltage is:

$$e = Blv \sin \theta t \tag{9.5}$$

But $v = r\omega$ where r is the coil radius, and ω the angular velocity (in rad s^{-1}). The coil angle, θ, also varies with time as ωt where t is in seconds. Both halves of the coil have an identical voltage induced, so the net coil output is:

$$e = 2Blr\omega n \sin \omega t \tag{9.6}$$

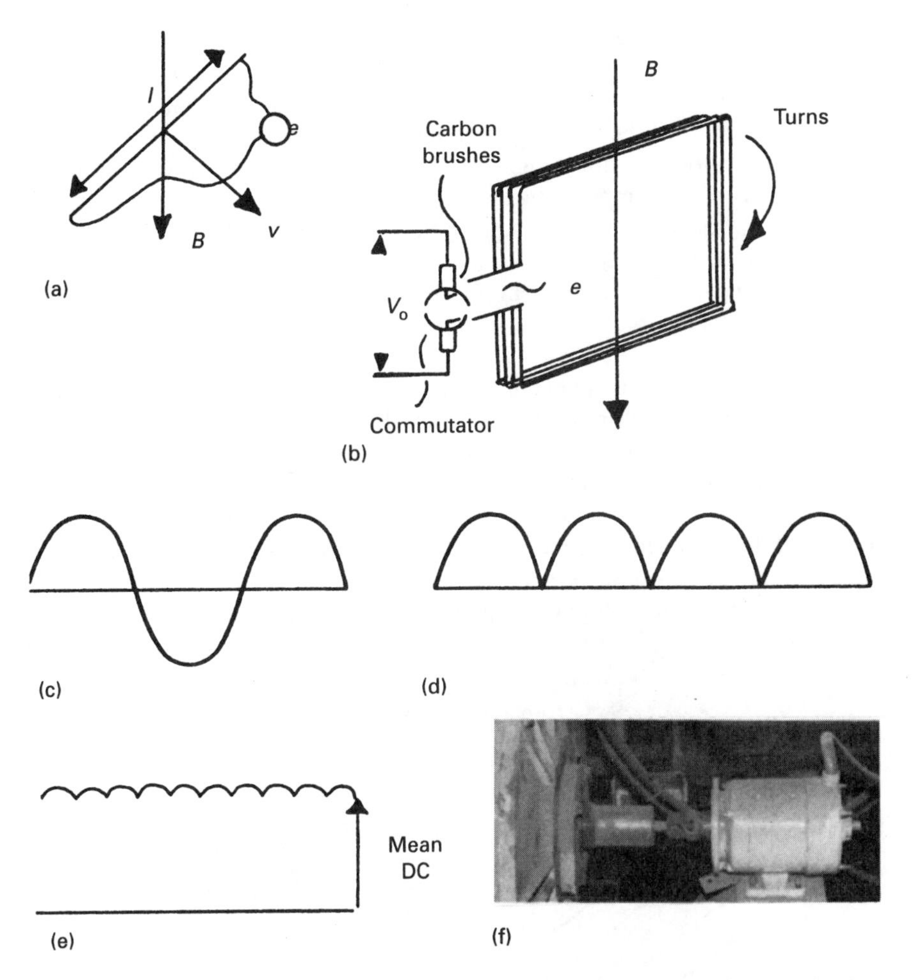

Fig. 9.2 *The tachogenerator. (a) Wire moving in a magnetic field. (b) Coil rotating in a magnetic field. (c) Induced coil voltage, e. (d) Output voltage; converted to DC by commutator. (e) Output from multiple-commutator segment generator. (f) Photograph of tachogenerator, note flexible coupling.*

This is a sinusoid whose frequency and peak amplitude are proportional to the angular frequency, ω, as shown in Fig. 9.2c.

If an AC output tachogenerator is used, the coil voltage can be measured via slip rings. It is usually more convenient, though, to have a DC output. The commutator segments on Fig. 9.2b reverse the coil connections as the coil voltage passes through zero to give a rectified DC output voltage as in Fig. 9.2d. The DC level is proportional to angular velocity, but the ripple content is unacceptably large and, worse, the ripple frequency is speed dependent.

The ripple can be reduced to acceptable levels by using more coils and commutator segments to give an output voltage similar to Fig. 9.2e. Ten to twenty commutator segments are common; the larger the number the less the ripple, but the more complex and expensive the device. Ripple will be about 1–2% of the output voltage in a well-designed tachogenerator, and can be reduced by subsequent filtering if the inherent filtering lag can be tolerated. Ripple frequency will be 2.N.shaft frequency where N is the number of coils.

Tachogenerators are not designed to supply a large current, and should be connected to a high impedance load (typically 100 kohm). They are usually calibrated in terms of revolutions per minute; a common standard is 10 V per 1000 rpm. Speeds up to 10 000 rpm can be measured directly, the limiting factor being centrifugal force on the commutator segments.

9.2.2. Drag cup

The drag cup converts angular velocity to angular displacement, and is commonly found in motor-car speedometers. If remote indication is required, the angular displacement can be converted to an electrical signal by any of the angular displacement transducers in Chapter 4.

The principle of the drag cup is shown in Fig. 9.3. A concentric cylindrical magnet and keeper cup are rotated by the input shaft. The magnet has four poles as shown in the cross-sectional insert. The drag cup itself is a conducting, non-ferromagnetic (e.g. copper) cup connected to the output shaft and fitted, again concentrically, between the magnet and the keeper.

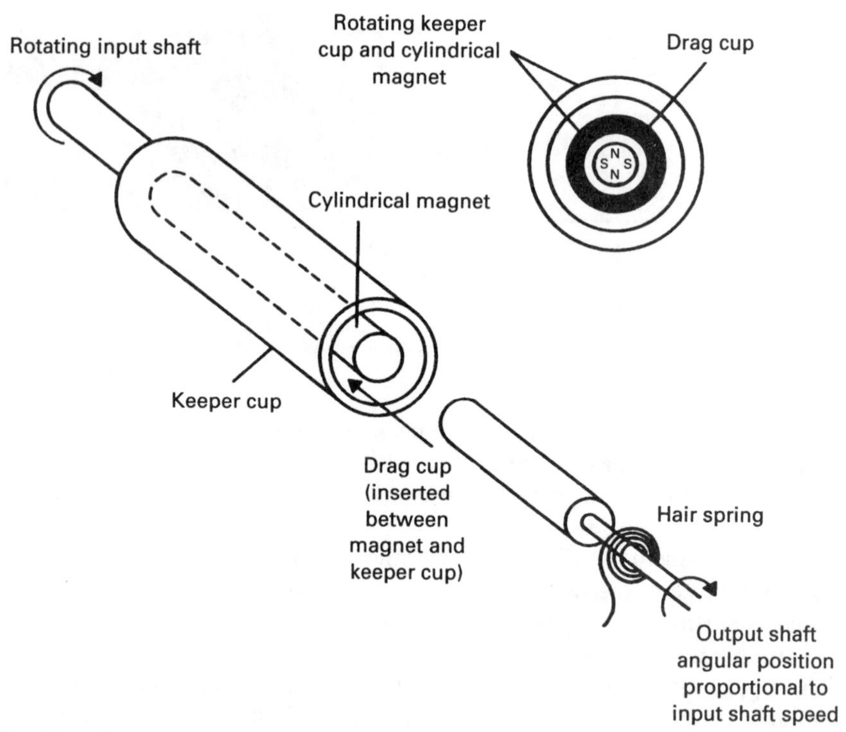

Fig. 9.3 *The drag cup angular velocity transducer.*

As the input shaft rotates, the flux lines between the magnet and keeper move with respect to the drag cup. This induces voltages as in equation 9.4, and induced currents flow in the drag cup. These currents, reacting with the magnetic field, produce a torque on the cup which, in the absence of any restraining torque, would rotate in the same direction as the input shaft. (The action is very similar in principle to the AC induction motor described in Chapter 12.)

The output shaft is restrained by a torsional spring which provides a torque proportional to shaft rotation. The shaft assumes an angle at which the torque from the spring balances the torque induced by the rotating magnet. The shaft angle is then proportional to input shaft angular speed.

9.2.3. Pulse tachometers

A pulse tacho is a device which produces a constant amplitude pulse train output whose frequency is related to input shaft speed. They are usually specified by the number of pulses per revolution of the input shaft. A device, for example, with 180 pulses per revolution would have an output frequency of 9 kHz at 3000 rpm.

There are two common ways of converting the pulse train to a signal for control or display. Figure 9.4a uses a monostable with constant width output pulse which is edge triggered by the tacho pulses. The monostable output is passed through a low-pass filter to give a DC output proportional to speed. The maximum speed is determined by the monostable pulse width.

An alternative technique, shown in Fig. 9.4b, directly counts the tacho pulses over a fixed period of time. (As such it is, of course, a sampling system whose sampling rate must be chosen with due consideration for the dynamics of the rest of the equipment being controlled or monitored.) The counter output can be transferred directly to a display (as in Fig. 9.4b) or converted to an electrical signal via a digital-to-analog converter (DAC). Digital counters and DACs are discussed in Chapter 13.

One interesting application of pulse tachos is in the area of DC motor speed control. Phase locked loop (PLL) ICs generate an output voltage proportional to the frequency difference between two pulse trains. In Fig. 9.5 the set speed is converted to a pulse train by a voltage-controlled oscillator (VCO) and compared with the output of a pulse tacho by a PLL. The output from the PLL is used as feedback for a P + I controller and power amplifier. P + I controllers and closed loop control are discussed in Chapter 18. Using the technique of Fig. 9.5, excellent speed accuracy (better than 0.01%) can be achieved.

Most pulse tachos use optical techniques and are identical to the optical incremental position encoders described in Section 4.4.2. In many numerically controlled machine tools, in fact, the same device serves as both position and velocity transducer.

Ad hoc optical pulse tachos can also be made by attaching reflective tape to a shaft and detecting reflections as in Fig. 9.6. Commercially available devices based on Fig. 9.6 allow non-contacting speed measurement with hand-held instruments.

Magnetic variable reluctance transducers can also be used with a toothed

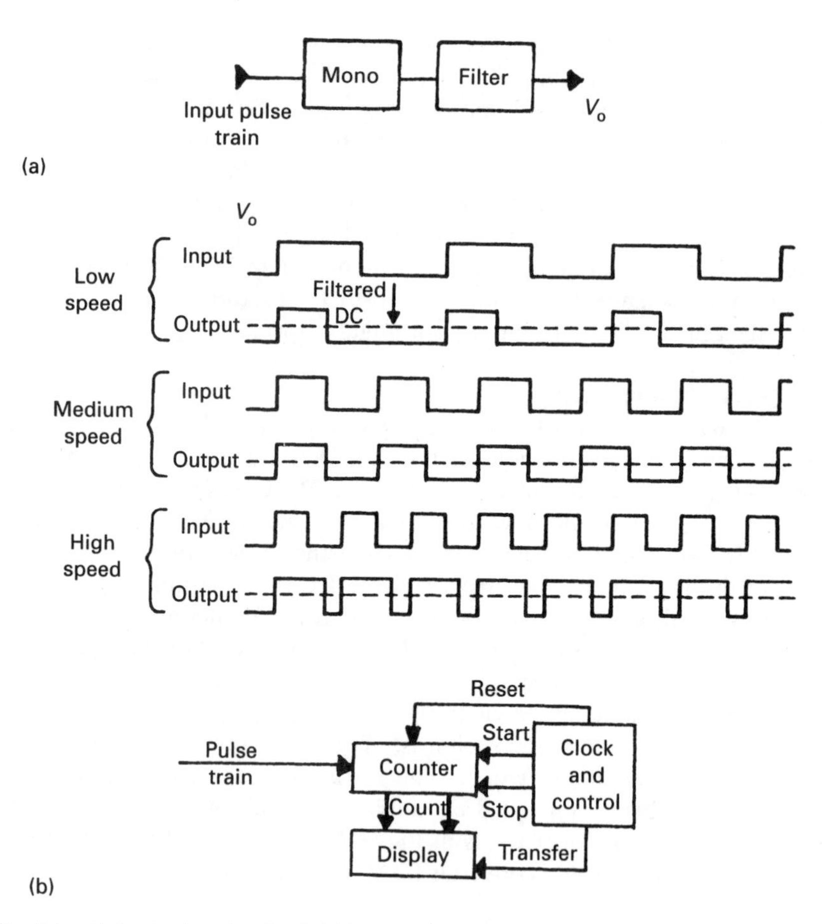

Fig. 9.4 *Pulse tacho circuits. (a) Monostable/filter circuit. (b) Timed counter circuit.*

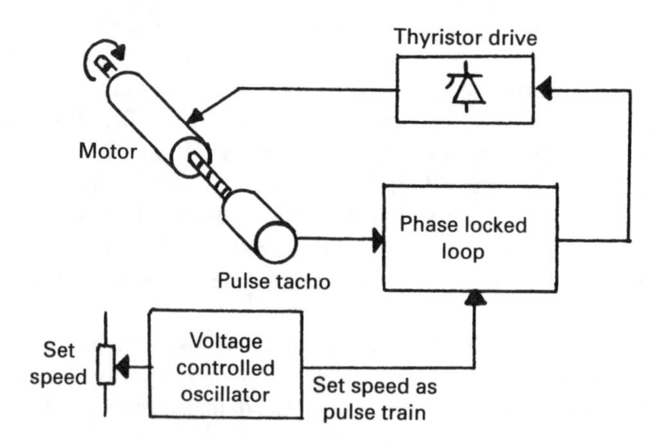

Fig. 9.5 *Pulse tacho-based speed control.*

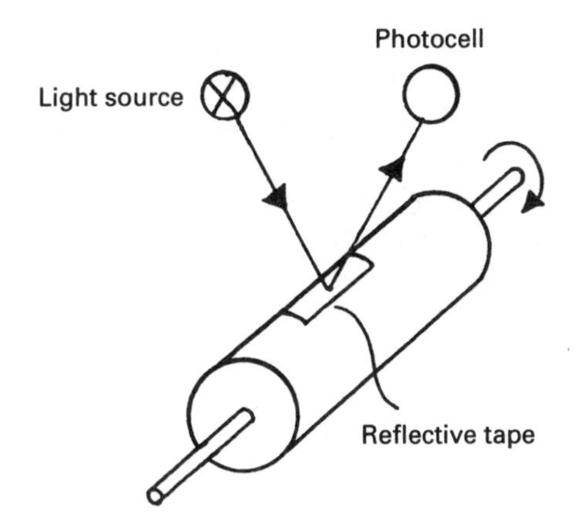

Fig. 9.6 Ad hoc *pulse tacho using photoreflective method.*

wheel, as shown in Fig. 9.7. As the teeth pass the magnet, the flux through the coil changes, causing an output voltage as shown in Fig. 9.7b. One disadvantage with the variable reluctance tacho is the fact that its output voltage amplitude is also speed dependent. Additional electronics (e.g. Schmitt triggers) are required to give a constant voltage pulse train. The variable level of output voltage also sets a minimum speed.

9.2.4. Tachometer mounting

The coupling of a tachometer to a shaft is very critical. Vertical or horizontal misalignment can produce side forces on the tachometer bearings and early

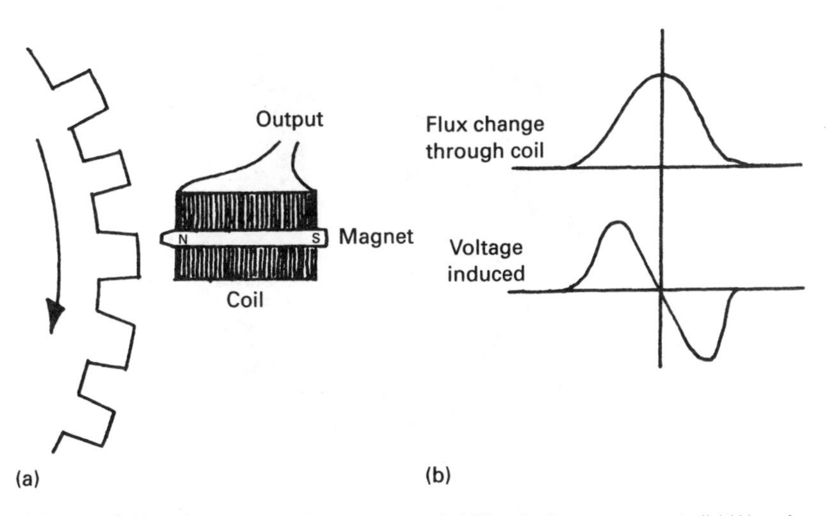

Fig. 9.7 *Variable reluctance tachogenerator. (a) Physical arrangement. (b) Waveforms.*

failure. Angular misalignment can cause a cyclic error between the input and the tachometer shaft. Above all there must be no backlash in the coupling, particularly in speed control systems. Backlash in a closed loop speed control appears as vicious speed hunting which damages motors, couplings and gearboxes.

Tachos are usually coupled by hysteresis-less universal joints similar to Fig. 9.8 or (sideways displaced) by toothed belts. Tension adjustment on belts is critical to avoid bearing wear. Axial mounted tachos require vertical or horizontal adjustment; often this is achieved with shims.

9.2.5. Doppler systems

Doppler shift occurs when there is relative motion between a source of sound, or electromagnetic radiation (radio, radar or light). It is commonly observed as a change in pitch of a car horn or train whistle as the vehicle passes.

In Fig. 9.9a, the source S is emitting waves (sound or electromagnetic radiation) with a frequency f. If the velocity of propagation is c, and the wavelength λ, these are related by:

$$f = c/\lambda \qquad (9.7)$$

The stationary observer will see f wavefronts pass per unit time, i.e. the observed frequency is the same as the transmitted frequency.

If the observer is now moving towards the source with velocity v, as in Fig. 9.9b, each wavefront will be observed earlier, as the apparent velocity of

Fig. 9.8 *Flexible coupling at centre of photograph linking motor and tachogenerator. The tacho is a pulse tacho and the unit to the left of the coupling is a mechanical overspeed switch.*

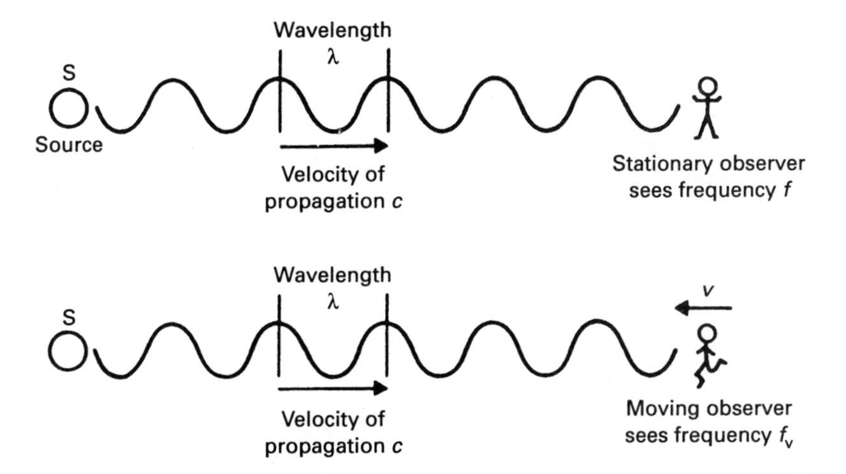

Fig. 9.9 *The Doppler effect. (a) Stationary observer. (b) Moving observer.*

propagation is $c + v$. The observed frequency is:

$$f_v = (c + v)/\lambda \tag{9.8}$$

i.e. a higher frequency. The frequency shift is:

$$f_v - f = \frac{c + v}{\lambda} - \frac{c}{\lambda} \tag{9.9}$$

$$= v/\lambda \tag{9.10}$$

$$= fv/c \tag{9.11}$$

Equations 9.10 and 9.11 show that the frequency shift is proportional to the relative velocity and the original frequency. A similar result is obtained for a receding observer, except that the frequency is lowered. Identical results are obtained for a fixed observer and a moving source since the Doppler effect arises from *relative* motion.

The Doppler shift allows remote measurement of velocity (it has, for example, been used to measure the rotational velocity of planets by observing the shift in frequency of radar pulses reflected from opposite sides). The principle is shown in Fig. 9.10. A transmitter (ultrasonic, radar or light) emits a frequency f_t at velocity c. The object, moving at velocity v, reflects part of this which is Doppler-shifted to f_r. Two Doppler shifts occur as the object is acting both as a moving observer and as a moving source, so

$$f_r - f_t = \frac{2f_t}{c} v \tag{9.12}$$

The transmitted and received frequencies are mixed to give a beat frequency f_b equal to the shift $f_r - f_t$. The beat frequency is proportional to v, and can be converted to a display or signal by a counter or frequency-sensitive rectifier.

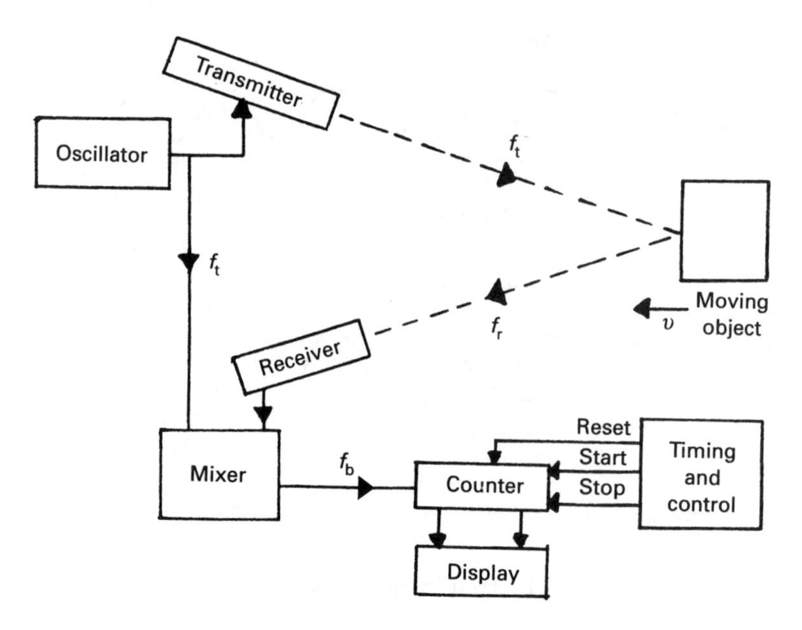

Fig. 9.10 *Doppler shift velocity measurement.*

It is instructive to observe the magnitude of the effect, which is dependent on the ratio v/c. For a radar signal of 10 GHz (10^{10} Hz) travelling at the speed of light 3×10^8 m s^{-1} and an object moving at 20 m s^{-1}, the shift is 1.333 kHz (i.e. an audio frequency).

9.3. Accelerometers

9.3.1. Seismic mass accelerometers

If a body of mass M kg experiences a force F newtons, it will accelerate at a m s^{-2}, these three quantities being related by Newton's second law of motion:

$$F = Ma \tag{9.13}$$

It also follows that a body which is accelerating will experience a force; we feel a force that pushes us into our seats when a car accelerates, and a force forwards into the seat belt when it brakes. This force is proportional to the acceleration and can be used to measure the acceleration.

The principle is shown in Fig. 9.11. A mass (called a seismic mass) is restrained by a spring of stiffness K. In the rest position, the mass is at position A. If the body accelerates at a constant rate a, the mass experiences a force according to equation 9.13. This causes the mass to move to the left to position B, until the force caused by acceleration balances the restoring force of the spring. In balance:

$$Ma = dK \tag{9.14}$$

i.e. the displacement is proportional to the acceleration.

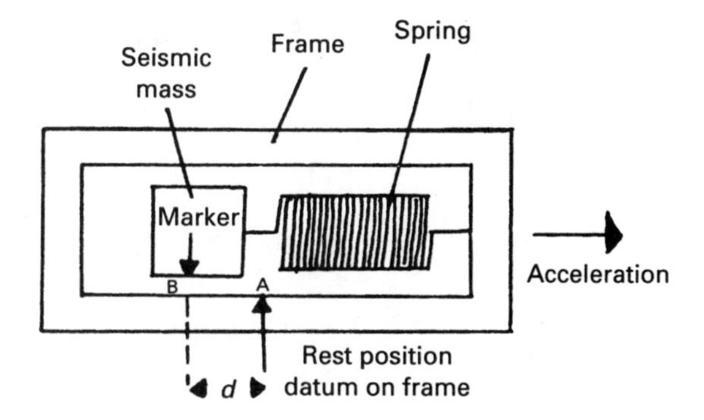

Fig. 9.11 *Seismic mass accelerometer.*

The displacement can be measured by any displacement transducer: strain gauges, potentiometers and LVDTs are common as we shall see in Section 9.3.3.

9.3.2. Second-order systems

Figure 9.11 is essentially similar to the suspended mass of Fig. 9.12 with acceleration and gravity fulfilling the role of a displacing force. If the mass in Fig. 9.12 is pulled down and released, we would, by experience, expect it to oscillate with decreasing amplitude until it returns to its original position. It is reasonable to expect similar oscillations from the seismic mass in an accelerometer.

In Fig. 9.12b, a seismic mass accelerometer is used to measure a step change in

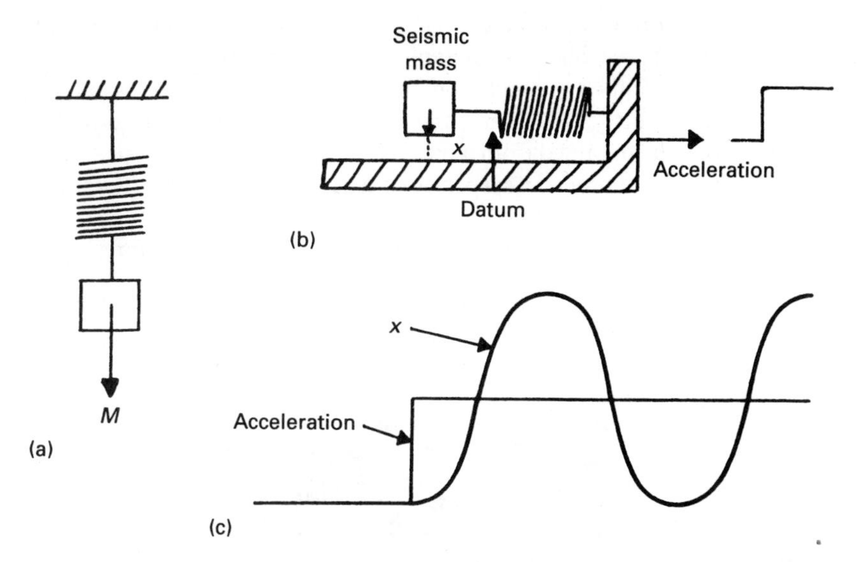

Fig. 9.12 *Step response of undamped accelerometer. (a) Oscillating spring/mass. (b) Seismic mass accelerometer with step change of acceleration. (c) Response of accelerometer.*

acceleration, a. The mass will lag behind the frame, giving a net mass acceleration of $[a - d^2x/dt^2]$.

Equation 9.14 can be rewritten:

$$M\left(a - \frac{d^2x}{dt^2}\right) = xK \tag{9.15}$$

or

$$\frac{d^2x}{dt^2} + \frac{Kx}{M} = a \tag{9.16}$$

If we denote K/M by ω_n^2 (for reasons which will become apparent), equation 9.16 becomes:

$$\frac{d^2x}{dt^2} + \omega_n^2 x = a \tag{9.17}$$

the solution of this second-order equation is a sinusoid of the form:

$$x = \frac{Ma}{K}(1 - \cos \omega_n t) \tag{9.18}$$

which is shown in Fig. 9.12c. The theory predicts a constant, non-decaying sinusoid of frequency $\omega_n/2\pi$ (called the natural frequency). Note that ω_n is $\sqrt{(K/M)}$, and the mean value is given by the steady state displacement Ma/K.

In practice, of course, friction will cause the oscillations to decay, but obviously the seismic mass/spring accelerometer cannot be used in the simple form of Fig. 9.11. Some form of controlled, predictable damping is needed. Simple constant frictional force will damp the oscillation but give an offset error. A viscous force, proportional to velocity, will damp the oscillation and cause no error. This characteristic can be obtained from a liquid dashpot (and is commonly found as shock absorbers on motor-car suspensions where they are used to damp out oscillations from the car springs).

The revised arrangement is shown in Fig. 9.13. The opposing force now is

$$Kx + C\frac{dx}{dt}$$

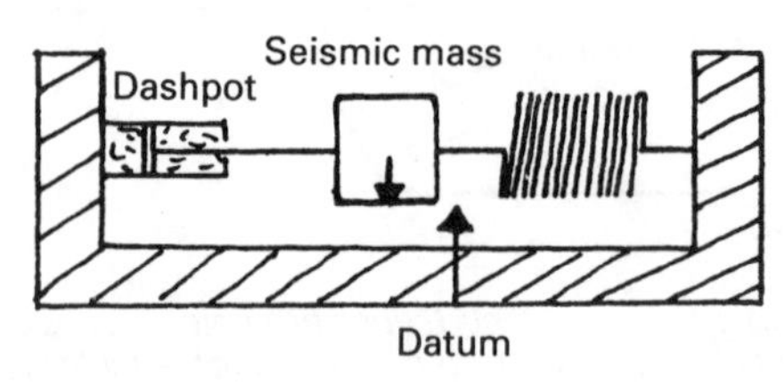

Fig. 9.13 *Accelerometer with damping.*

so equation 9.15 becomes:

$$M\left(a - \frac{\mathrm{d}^2x}{\mathrm{d}t^2}\right) = Kx + C\frac{\mathrm{d}x}{\mathrm{d}t} \tag{9.19}$$

which can be rearranged:

$$\frac{\mathrm{d}^2x}{\mathrm{d}t^2} + \frac{C}{M}\frac{\mathrm{d}x}{\mathrm{d}t} + \frac{K}{M}x = a \tag{9.20}$$

The solution of this equation has two parts, a transient part plus a steady state part (as t tends towards infinity). The steady state part is simply $x = Ma/K$ as before. To examine the transient part we define a damping factor b, where:

$$b = \frac{C}{2\sqrt{mK}} \tag{9.21}$$

Defining ω_n as before, equation 9.20 becomes:

$$\frac{\mathrm{d}^2x}{\mathrm{d}t^2} + 2b\omega_n\frac{\mathrm{d}x}{\mathrm{d}t} + \omega_n{}^2x = a \tag{9.22}$$

The mathematical analysis of this equation is somewhat lengthy and beyond the scope of this chapter, but the predicted result depends on the value of b.

For $b < 1$ the system will exhibit oscillations that decay exponentially as shown in Fig. 9.14a. The system is said to be underdamped. The case $b = 0$ corresponds to no damping and to equation 9.17 and Fig. 9.12c.

For $b < 1$ the system does not oscillate, and rises as in Fig. 9.14c. The system is said to be overdamped. The case of $b = 1$ marks the transition from overdamping to underdamping, as in Fig. 9.14b, and is called critical damping.

Figure 9.15 shows superimposed step responses for various damping factors. Note that the curves are normalised with respect to ω_n. Figure 9.15 also shows that the actual oscillation period increases from ω_n with increasing damping.

It might be thought that the best response occurs for a damping factor $b = 1$, but in practice slight underdamping is preferred. Examination of Fig. 9.15 shows that the shortest time to settle within any specified error band occurs for $b = 0.7$ (corresponding to a first overshoot of about 8%).

Figure 9.16 shows the Bode diagram for various values of damping factor.

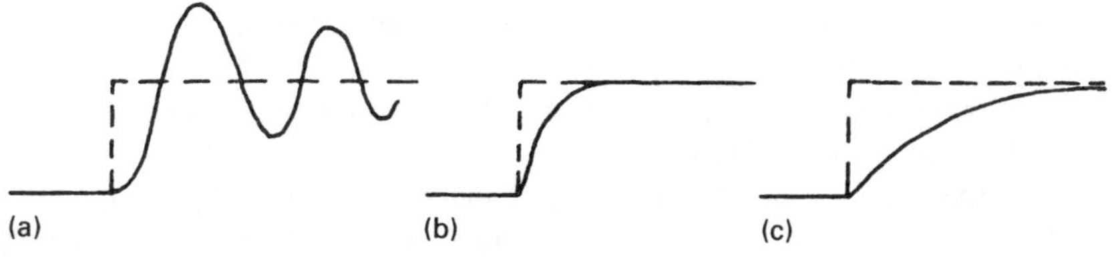

Fig. 9.14 *The effect of damping. (a) Underdamped ($b < 1$). (b) Critically damped ($b = 1$). (c) Overdamped ($b > 1$).*

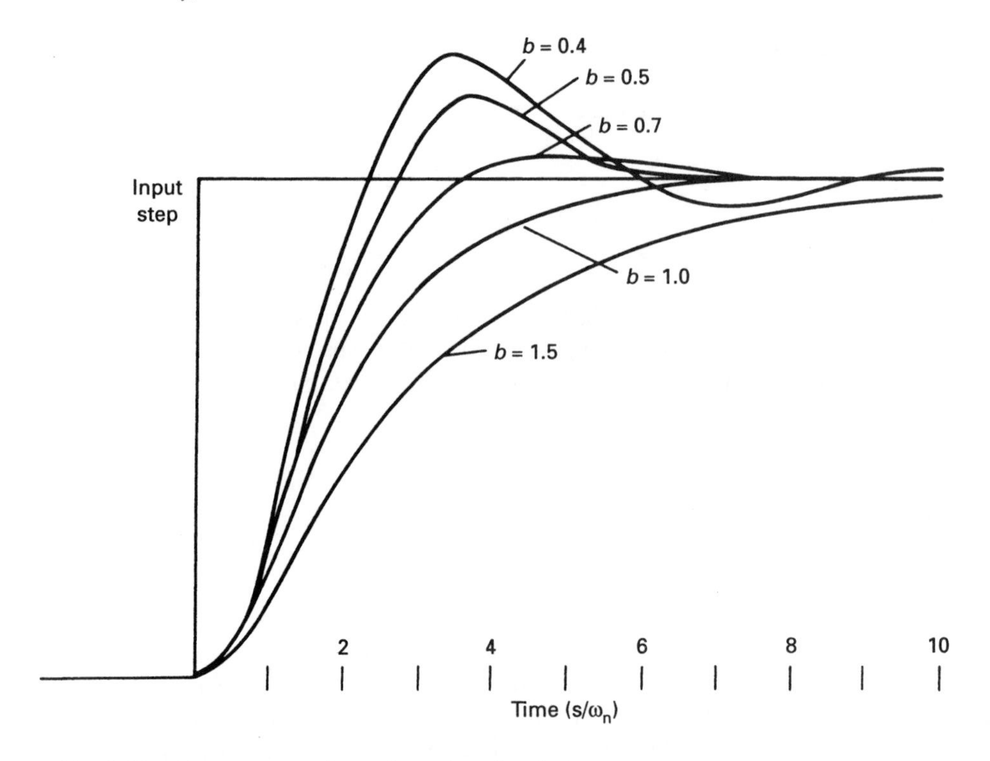

Fig. 9.15 *Step response for various damping factors.*

Again a damping factor of 0.7 gives the largest usable bandwidth. The Bode diagram also demonstrates the importance of knowing the frequency range of the acceleration being investigated. In general a second-order transducer is usable up to one-fifth of its resonant frequency.

9.3.3. Practical accelerometers

The transducer in Fig. 9.17 uses a seismic mass suspended from thin flexure plates. The mass is restrained by thin wires which also act as strain gauges for deflection measurement. As the device accelerates to the right, for example, gauges A and B will decrease in length, and C and D increase in length. These can be connected into a bridge circuit as described in Section 6.3, to give a voltage output that represents acceleration. The use of four gauges gives temperature compensation and an increased output signal. The spring is provided by the flexure plates and the strain gauges themselves. Viscous damping is obtained by filling the transducer with oil; the motion of the flexure plates and the mass itself giving the required damping force.

An alternative approach is the swinging mass of Fig. 9.18. For simplicity of drawing, the support frames and bearing have been omitted. The sensing axis runs perpendicular to the line joining the two shafts. As the device accelerates, the two masses rotate, the restoring force being provided by springs. The acceleration

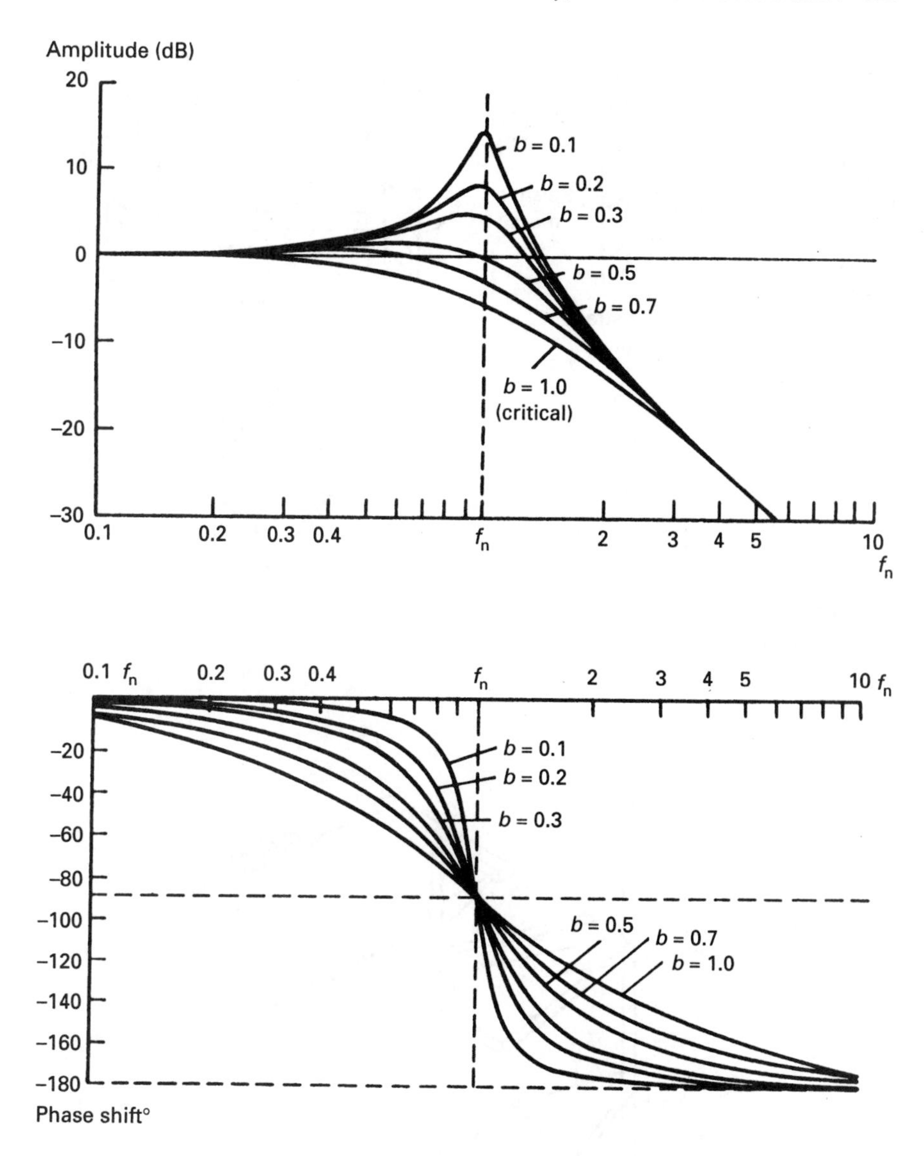

Fig. 9.16 *Bode diagram for second-order system*

is measured by an angular position transducer on one shaft, and damping provided by a dashpot on the other.

Accelerometers must only respond to acceleration along one axis, and ignore acceleration in other directions. For complete measurement of acceleration in three dimensions, three transducers are needed. The ability of an accelerometer to ignore non-axial acceleration is defined by its cross-axis sensitivity (also called the transverse sensitivity). It is usually defined as the ratio of the output of the

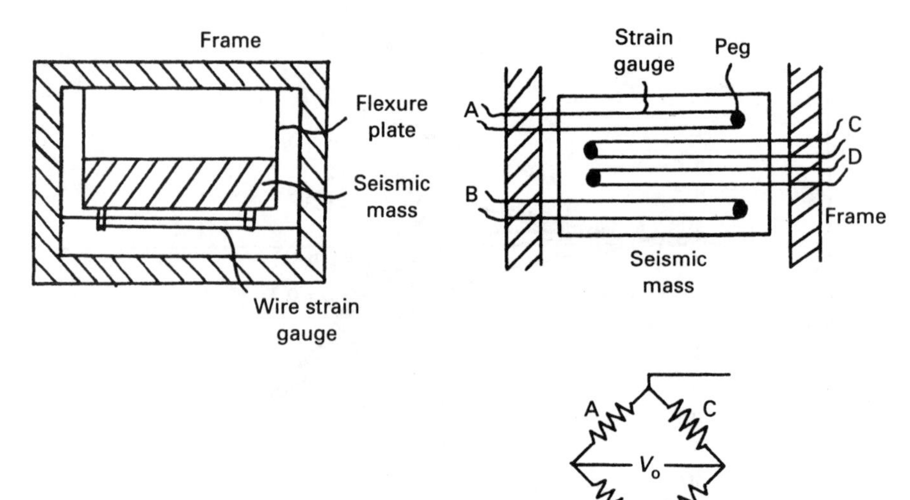

Fig. 9.17 *Strain gauge accelerometer. (a) Side view. (b) Top view. (c) Gauge connections in bridge.*

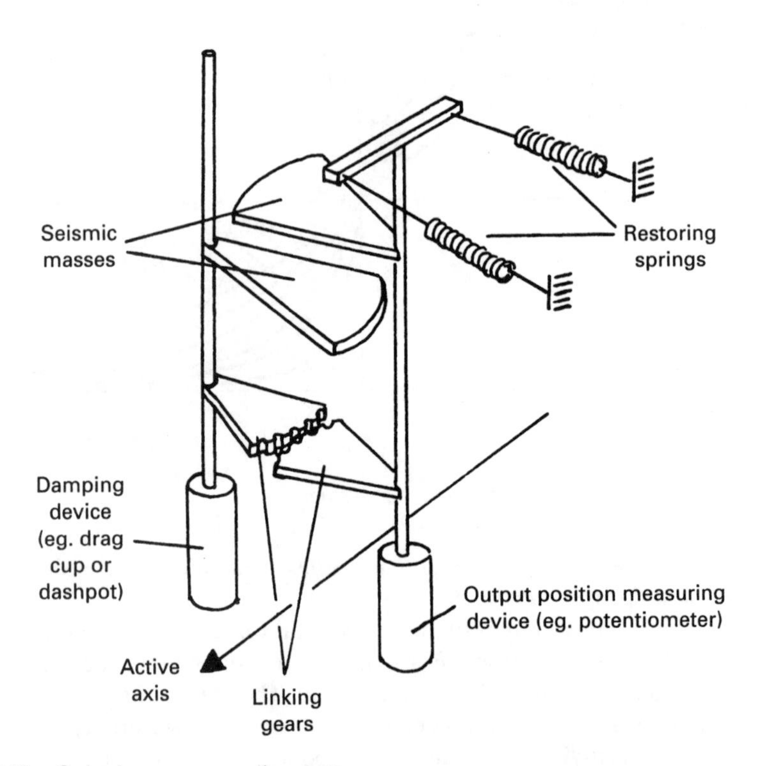

Fig. 9.18 *Swinging mass accelerometer.*

device for acceleration perpendicular to, and along, the sensing axis. Typical values are around 2%. In Fig. 9.18 cross-axis sensitivity is reduced by the use of two seismic masses, and in Fig. 9.17 the flexure plates only permit movement along the sensing axis.

Figure 9.19 uses a force balance technique, matching the Ma force with an electromagnetic force. As the transducer accelerates to the right, the mass lags and moves with respect to the housing. The deflection is measured by an LVDT, and an amplifier increases the current in the restoring coil to bring the mass back to its rest position. The coil current is then a measure of the acceleration.

The piezo-electric effect, described in Section 3.4, can also be used to measure acceleration. A piezo-electric crystal is essentially a force-measuring device, and as such can be used to measure the Ma force directly. The principle is shown in Fig. 9.20. The seismic mass is in direct contact with the crystal, the latter being kept in permanent compression by the pre-tensioning screw. Acceleration along the sensing axis produces an Ma force on the crystal, giving an output as described in Section 3.4.

The piezo-electric transducer requires a charge amplifier (also described in Section 3.4) and has poor low-frequency response. It is best suited for measurement of high-frequency, high-g acceleration, as found in impact testing. Piezo-electric transducers have the advantage of robustness, simplicity and small size.

9.4. Vibration transducers

Consider the seismic mass transducer of Fig. 9.21a, which is being driven by a sinusoidal displacement of constant amplitude but variable frequency. The

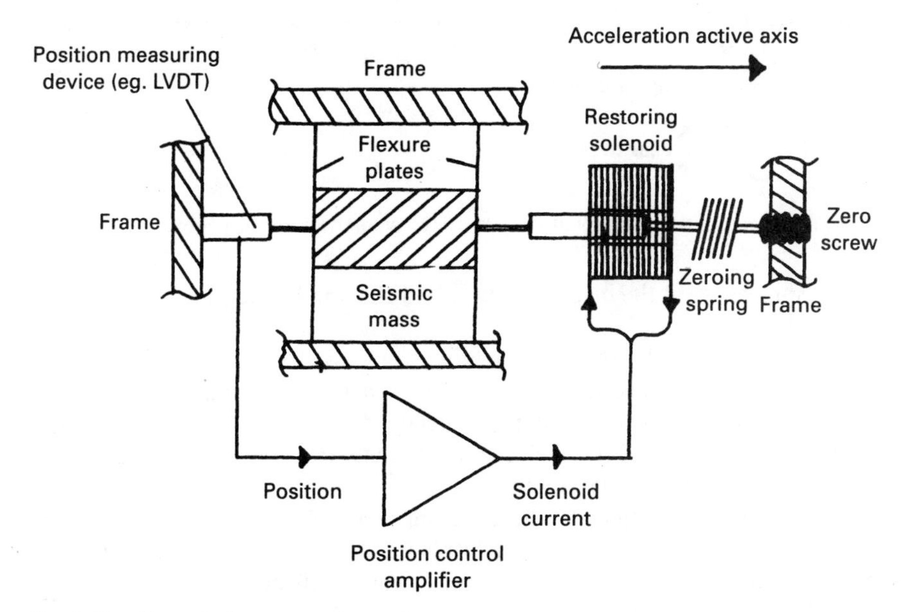

Fig. 9.19 *Force balance accelerometer.*

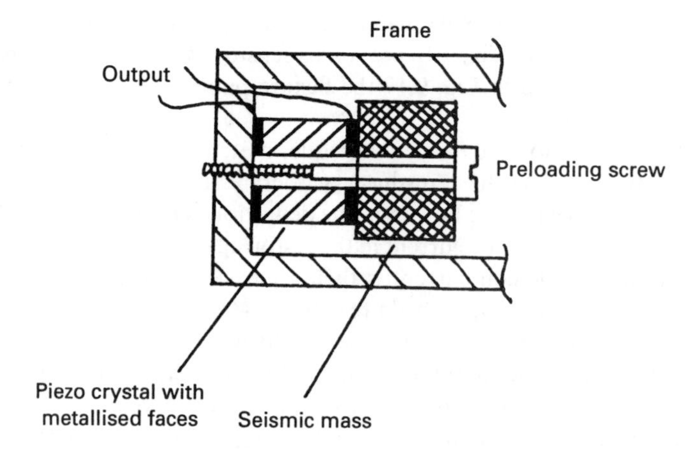

Fig. 9.20 *Piezo-electric accelerometer.*

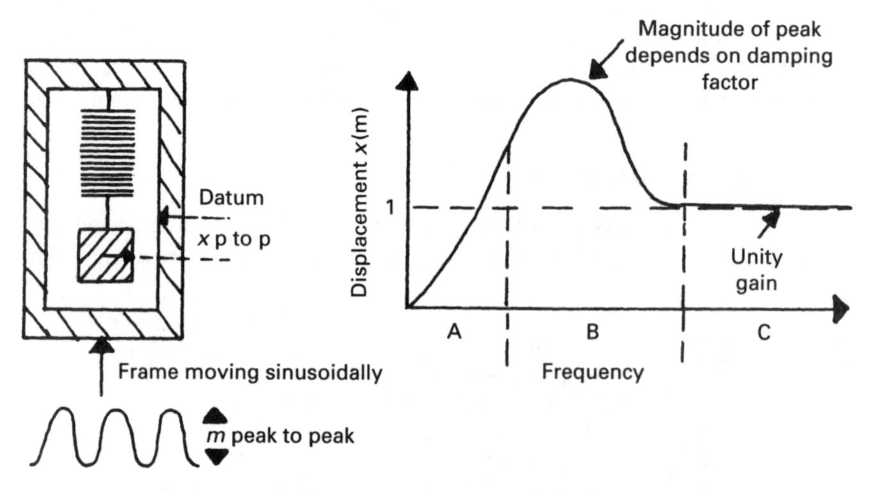

Fig. 9.21 *Principle of vibration transducers. (a) Schematic diagram (b) Frequency response.*

relationship between the displacement x and the applied frequency is shown in Fig. 9.21b.

In the region A, the device is acting as an accelerometer, the amplitude increasing because the acceleration increases with frequency. Region A extends to up to about one-fifth of the resonant frequency. As the frequency increases, there is a resonant region B after which the mass ceases to move in space, remaining fixed while the frame moves about it.

This corresponds to region C where the amplitude of the apparent displacement x follows the frame movement (but displaced by 180°). This region starts at about five times the resonant frequency, and extends, in theory, to infinite frequency.

A vibration transducer operating in region C becomes a displacement transducer, and as such can be used to measure high-frequency vibration if a position transducer is attached between the seismic mass and the frame. If a velocity transducer is used, vibration velocity can be measured.

Vibration transducers are widely used for fault monitoring and protection of large rotating plant. They must inherently be small and of low mass to avoid loading the device to which they are attached. Like accelerometers they are second-order devices, and are usually designed for a damping factor of 0.7.

Chapter 10

Analytical instrumentation

10.1 Introduction

Knowledge of the chemical composition of a gas or liquid is an important part of the control in many industrial processes. Examination of the oxygen or CO content in flue gas, for example, can be used to control the operation of burners and lead to lower fuel consumption. In this chapter we examine various common methods of analytical analysis.

10.2 Conductivity

10.2.1 Conductivity probes

Contrary to popular thought, pure de-ionised water does not conduct electricity easily and has a very high resistance. The ability of a liquid to conduct electricity is determined by the number of charged ions in solution. Most acid solutions, for example, have a large number of ions and hence conduct electricity easily.

Conductance is a measure of how easily electricity can flow through a material, and hence is the inverse of resistance:

$$G = 1/R = i/v \tag{10.1}$$

where R is the resistance and v and i are the voltage and current in the circuit.

The unit of conductance is the siemen (S). A resistance of 10 ohms has a conductance of 0.01 siemens. The older term mho (which is identical in definition to the siemens) may also be encountered.

The conductivity of a substance (liquid or solid) is defined as the conductance of length 1 cm and cross-sectional area 1 cm^2 obtained by measuring the resistance R as shown on Fig. 10.1. The conductance of a sample will increase with increasing electrode area and decrease with increasing separation, i.e.

$$\text{conductivity} = d\,.\,G/A \tag{10.2}$$

where d is the separation, A the area and G the measured conductance given by $G = 1/R$. Because d has the units of length, and A the units of length squared, the units of conductivity are siemens/cm.

Although pure water has a very low conductivity, normal tap water conducts electricity because of the various compounds in solution in the water. It follows that a measurement of the conductivity of a liquid can be used to indicate the presence and concentration of materials of interest. Figure 10.2 shows the conductivity of salt (sodium chloride) and sulphuric acid at various concentrations.

Unfortunately the conductivity is affected not only by the substance of interest, but also by temperature and contaminants. The conductivity of a salt solution,

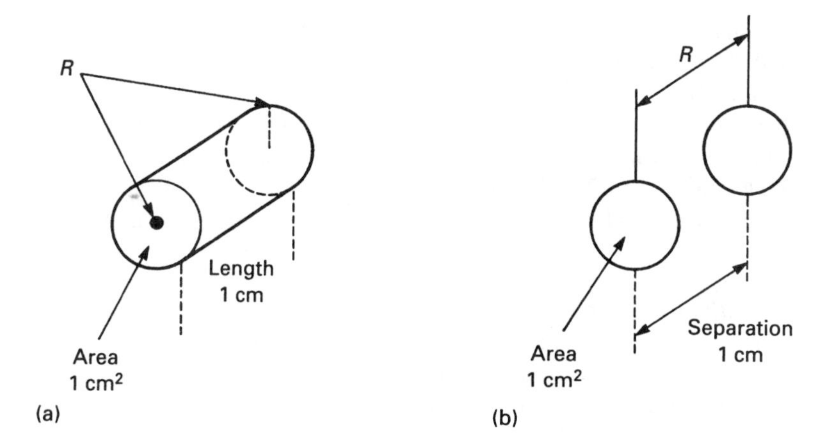

Fig. 10.1 *Definition of conductivity. (a) Solid. (b) Liquid.*

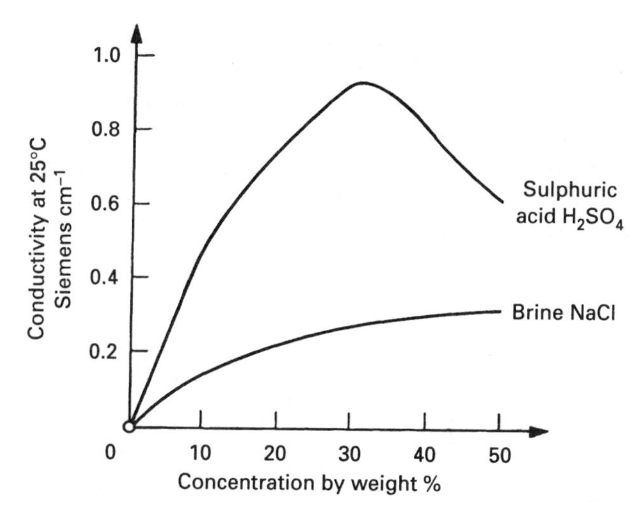

Fig. 10.2 *Typical conductivity/concentration curves.*

for example nearly doubles for a change in temperature from 0 °C to 20 °C. To minimise this effect the temperature of the solution must be controlled, or the temperature measured and a correction applied to the reading.

A conductivity cell simply consists of two electrodes which are placed in the solution. The measured conductance will depend on the area of electrodes and their separation. A cell has a 'cell constant', this is the ratio of the electrodes' separation and area, i.e.

$$\text{cell constant} = d/A \tag{10.3}$$

The cell constant has the units of inverse length. Cell constants normally range from 0.01 cm^{-1} to 100 cm^{-1}. Low cell constants are used for solutions with low conductivity, a cell constant of 0.01 cm^{-1} would be used with fluids of conductivity around 0.05 μS/cm, and a higher range cell for conductivities up to 100 000 μS/cm.

Dependent on the range of interest, the conductivity is either measured by applying a stable known voltage and measuring the current, or by use of a balancing bridge circuit. In both cases DC cannot be used or electrolysis will occur which may produce unwanted gassing and electrode erosion. A sinusoidal or square wave AC waveform must be used.

10.2.2. Inductive probes

The simple conductivity probe described in the previous section has direct connection with the solution of interest. It can therefore be attacked if the solution is at all aggressive. Simple probes therefore need careful maintenance procedures.

Inductive probes are an alternative which allow conductivity measurements to be made without direct contact with the sample. The principle is shown in Fig. 10.3.

The probe contains two totally separate insulated coils in a housing which will resist the aggressive properties of the solution. A stable AC voltage is applied to the primary coil, and this induces current flow in the liquid which will depend on the liquid's conductivity. The induced current also passes through the second pickup coil and induces a voltage which is proportional to the current flowing in the liquid. This current is directly proportional to the conductivity of the fluid.

Like all analytical techniques, the reading will be affected by temperature.

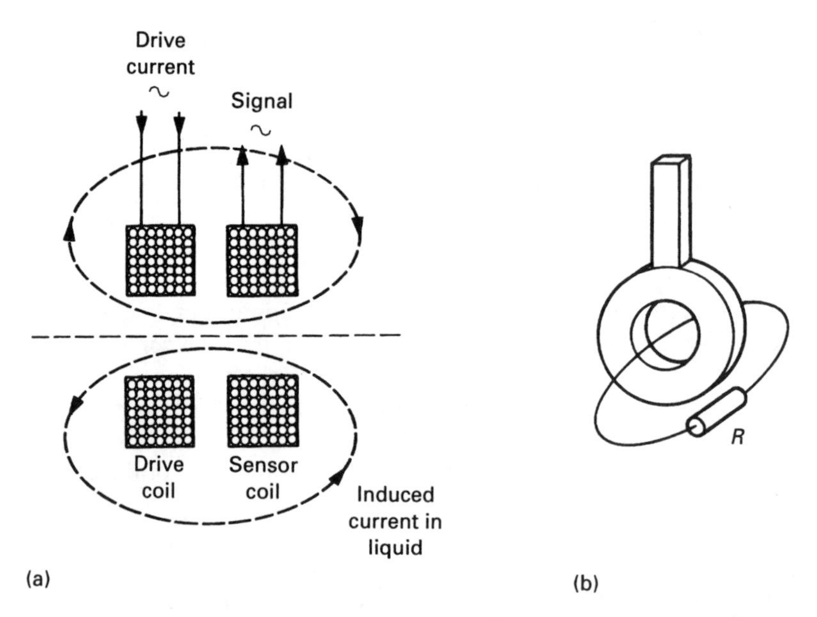

Fig. 10.3 *The inductive conductivity probe. (a) Principle of operation. (b) Calibration.*

Inductive probes therefore normally include a temperature sensor in the head which can be used to apply a correction to the reading.

The inductive probe has many maintenance advantages. It does not suffer from erosion and is easy to calibrate. All that is required is a loop of wire of known resistance connected as Fig. 10.3b. With simple conductivity probes, calibration must be done with standard liquid solutions, a messy, and potentially hazardous procedure.

10.3. Combustion measurements

10.3.1. Introduction

The rise of energy costs over the past decade plus increasing concern with the environment has meant that combustion process have come under increasing pressure to be more efficient and less polluting.

Air contains 20.8% oxygen, the remaining 79% being primarily nitrogen. When a hydrocarbon fuel (gas, oil or coal) is burnt, the oxygen combines with the carbon and hydrogen in the fuel to produce water and, with complete combustion, carbon dioxide.

With efficient burning, almost all the carbon in the fuel will be converted to carbon dioxide, giving a flue gas with around 12% carbon dioxide concentration. If the burning process is not efficient, (caused by poor air/fuel ratio, for example) the conversion to carbon dioxide is not completed and carbon monoxide can be formed. This is inefficient and possibly hazardous as carbon monoxide is both highly poisonous, even in small concentrations, and possibly explosive if more air is allowed into the exhaust duct.

The efficiency of a combustion process can therefore be monitored by looking at the concentrations of oxygen, carbon dioxide and carbon monoxide in the exhaust ducting. It is becoming increasingly common for the combustion control strategy to use this stack analysis as part of the control loop. A rise of the oxygen concentration, for example, would indicate that the air/fuel ratio is too lean, and the ratio control would automatically be adjusted to give a richer, more efficient mix. High carbon monoxide levels similarly indicate that the ratio is too rich, and can be used to make the flame leaner.

The techniques are also becoming commonly used in the automotive industry for monitoring and controlling the operation of internal combustion engines. Cheap exhaust sensors are now to be found in the more sophisticated on-board engine management control systems.

10.3.2. Optical analysers

Every chemical compound blocks particular wavelengths of light from passing through it. The atmosphere of the earth blocks most of the sun's radiation and only permits the visible light radiation with wavelengths from 0.38 μm (violet) to 0.78 μm (red) to reach the surface of the earth. Figure 10.4 shows part of the absorption spectrum for carbon dioxide and carbon monoxide. Both of these are

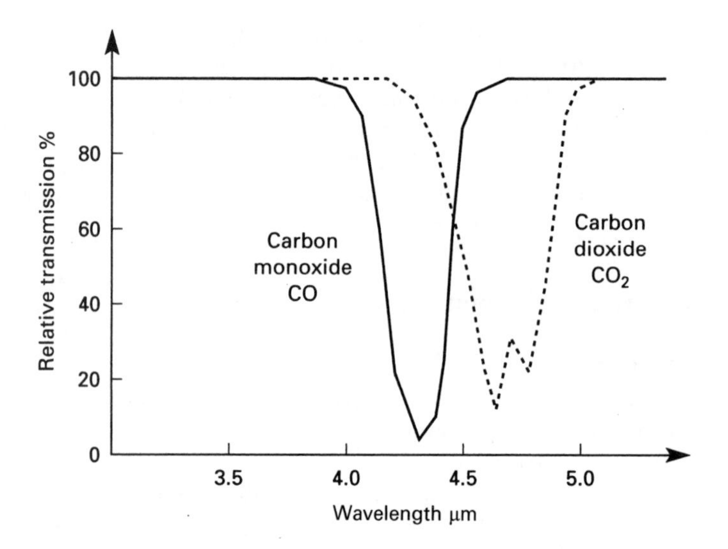

Fig. 10.4 *Spectral response of carbon monoxide and carbon dioxide.*

in the infrared region of the spectrum. The parts of the spectrum where light is blocked are called the absorption bands for the gas.

Different parts of the spectrum are used for different gases. The infrared region is used for carbon monoxide and carbon dioxide, the visible region for nitrogen oxides (the so-called NOX emissions) and the ultraviolet region for sulphur dioxide, a toxic emission formed when sulphur-rich coal or oil is burnt.

The principle of an optical gas analyser is shown on Fig. 10.5. Two wavelengths are chosen. The first, called the reference wavelength, is at a part of the spectrum where there is no absorption either by the gas being monitored or by

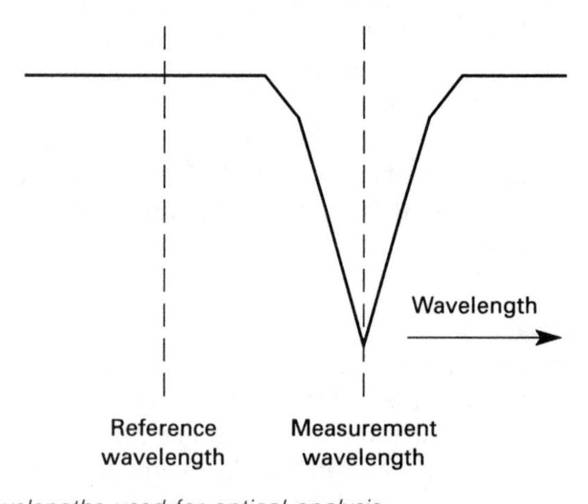

Fig. 10.5 *Wavelengths used for optical analysis.*

any other which may be present. The second wavelength, called the measurement wavelength, is located at an absorption band unique to the gas of interest.

A sample flow of the gas is then passed through a pipe as shown on Fig. 10.6. A single light beam (infrared actually) passes across the pipe to two detectors. The light to the first passes through a filter which only allows the reference wavelength to pass. The second filter passes the measurement wavelength.

The total intensity received by both detectors will depend not only on the gas composition, but also on gas temperature, gas borne dust, the intensity of the light source and the cleanliness of the sample windows on both sides of the pipe. Because both reference and measurement sensors are equally affected, it is simple to compensate for these effects in the display electronics.

Figure 10.6 uses two sensors, so there is still the possibility of error if the two sensors are not matched or age differently. The arrangement of Fig. 10.7 is more commonly used. Here there is just a single source and a single receiver giving total matching between reference and measurement signals and compensation for long-term drift in the sensing electronics. An electric motor rotates two filters, (one reference and one measurement) in front of the sensor. A position measuring device (usually a simple optical sensor) informs the display electronics which filter is in place at any given instant.

With all optical sensors, the vulnerable parts are the windows. Most flue ducts will carry dust particles as well as gas, and the windows consequently become scratched and progressively more opaque. The use of a single source and detector as in Fig. 10.7 can compensate for this to a large extent, but maintenance procedures must be established for cleaning and changing the windows at regular intervals. The windows also limit the maximum pressure that can exist in the sample pipe.

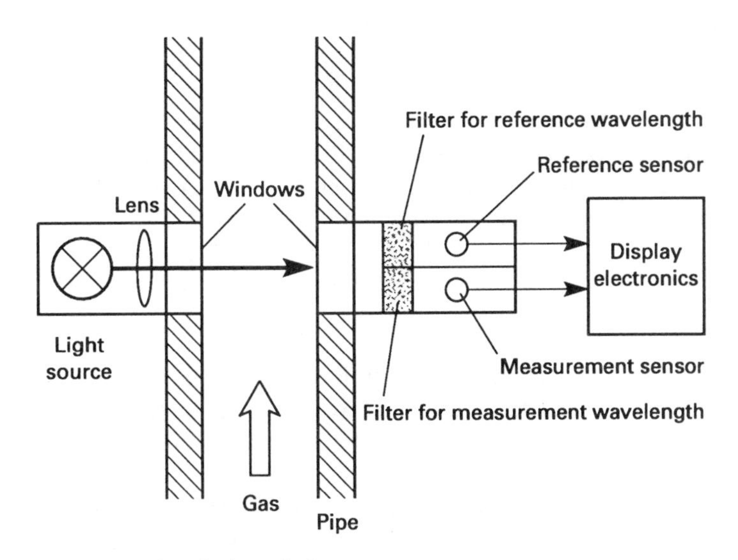

Fig. 10.6 *Principle of optical analysis.*

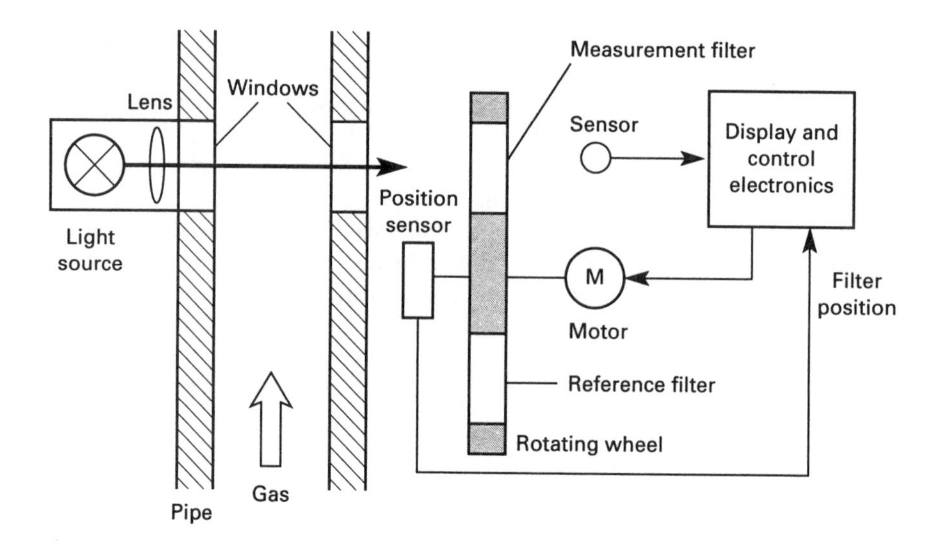

Fig. 10.7 *Optical analyser with automatic compensation.*

10.3.3. Electrometric gas analysers

Electrometric gas analysers use the properties of zirconium oxide to measure the oxygen content in the gas stream from a combustion process. If the combustion is being performed with an accurate air/fuel ratio the oxygen content should be near zero.

The sensor consists of a zirconium oxide disc coated on both sides with porous platinum to allow electrical connections. One side of the disc is exposed to the gas stream of interest, the other side to either a reference gas or the atmosphere. A dry air stream from a normal instrument air supply is usually used as the reference gas. Often the sensor is formed as a tube with the reference gas contained inside as shown on Fig. 10.8.

When the cell is heated to around 800 °C, the cell body behaves as an electrolytic conductor and a voltage is developed across the two faces of the cell. The voltage is given by the Nernst equation:

$$E = AT \ln\left(\frac{P_1}{P_2}\right) + B \tag{10.4}$$

where T is the absolute temperature, P_1 is the partial pressure of oxygen in the reference gas, P_2 the partial pressure of oxygen in the sample and A and B are constants for the cell. The output is of the order of 50 mV per decade and decreases for increasing oxygen content in the sample.

The operation of the cell is very temperature dependent. The cell temperature is measured with a thermocouple and the power to the heater (which is needed to obtain the required 800 °C) set by a temperature control loop.

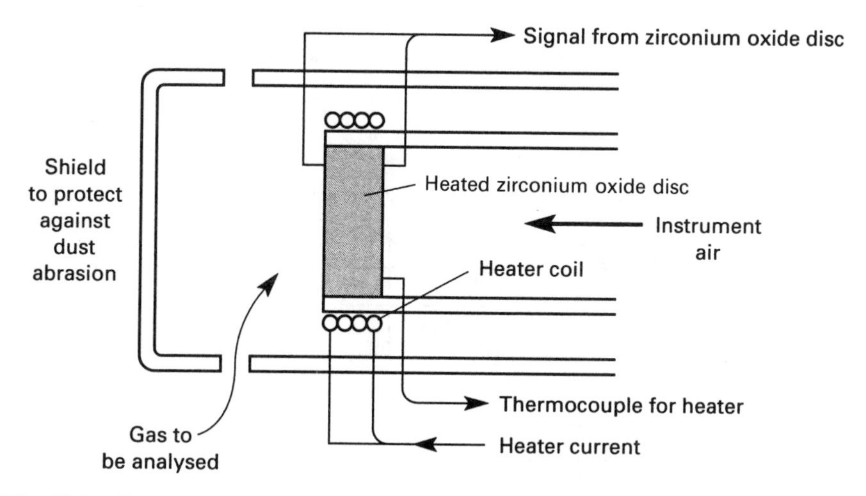

Fig. 10.8 *Zirconium oxide cell for oxygen analysis.*

10.3.4. Electrochemical sensors

An electrochemical oxygen sensor uses a (highly corrosive) strongly basic chemical solution which is exposed to the sample gas stream. Oxygen in the sample gas oxidises the chemical to produce hydroxide molecules and a small electrical potential between two electrodes immersed in the solution.

Although much simpler (and hence cheaper) than the electrometric and paramagnetic sensors, the accuracy is not as good. The corrosive nature of the basic solution and the need for frequent replacement as the chemicals are consumed tend to make these not very attractive for industry.

10.3.5. Paramagnetic sensors

Oxygen has a large, and almost unique, paramagnetic property. The only other commonly found gases exhibiting this property are nitric and nitrous oxide, and for both the effect is about an order of magnitude smaller than oxygen.

A typical instrument is shown in Fig. 10.9. A dumb-bell-shaped non-magnetic sensor is suspended in a non-homogeneous magnetic field. A fine metal or quartz fibre is used for the suspension to reduce stiction. The sensor is exposed to the sample gas stream whose oxygen produces a force on the sensor displacing it towards an area of lower magnetic field.

This force (or rather the resulting motion) is sensed by photocells and a position control loop used to provide a counter-acting rotation to bring the sensor back again to the null position. The force required to keep the sensor in the null position is a measure of the oxygen content in the sample gas stream.

Paramagnetic oxygen instruments are accurate but somewhat delicate. They cannot be operated in areas with high magnetic fields such as close to transformers and electric motors.

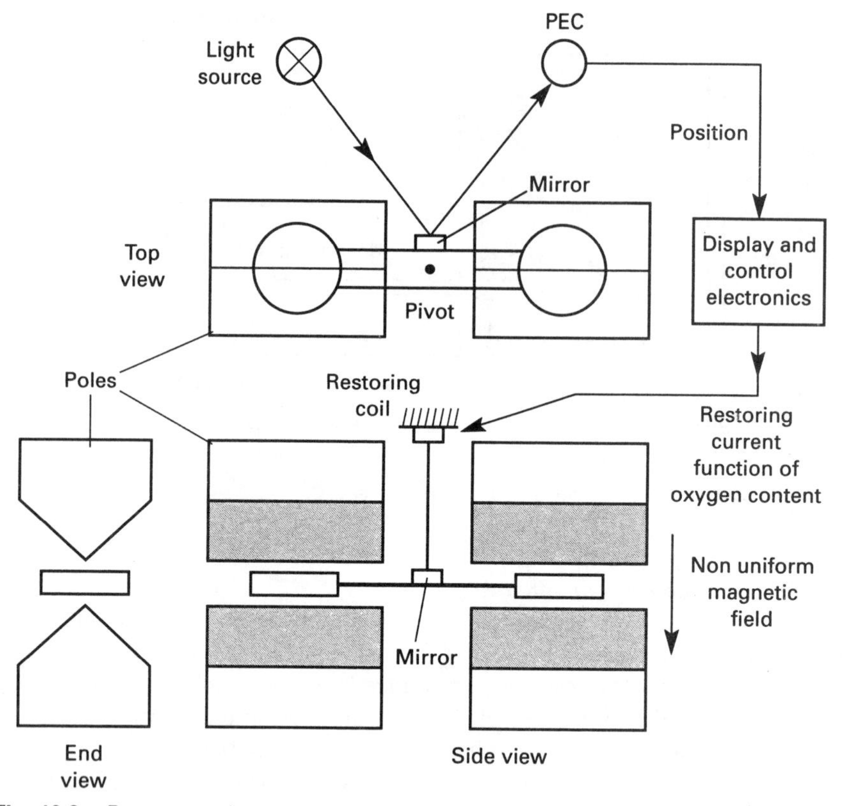

Fig. 10.9 *Paramagnetic oxygen cell.*

10.3.6. Thermal conductivity sensors

Carbon dioxide content is often measured by using the thermal conductivity of the gas. The principle is shown on Fig. 10.10 where two identical small chambers are used, each containing identical heating coils made of platinum wire. The resistance of the coils is measured by connecting each as one leg of a Wheatstone bridge.

Equal volumes of flue gas and atmospheric air are drawn through the chambers by metering pumps. Carbon dioxide removes heat at a markedly different rate than oxygen, nitrogen or carbon monoxide. This removal of extra heat unbalances the bridge and the resultant voltage is a function of the carbon dioxide concentration.

10.3.7. Resistive gas sensors

These sensors are often used to detect the build up of explosive gases such as LPG, methane, propane, butane, etc. They consist of two platinum wires, one coated with a material such as titanium oxide, and one as a compensator for temperature effects. The wires are heated to around 600 °C at which temperature

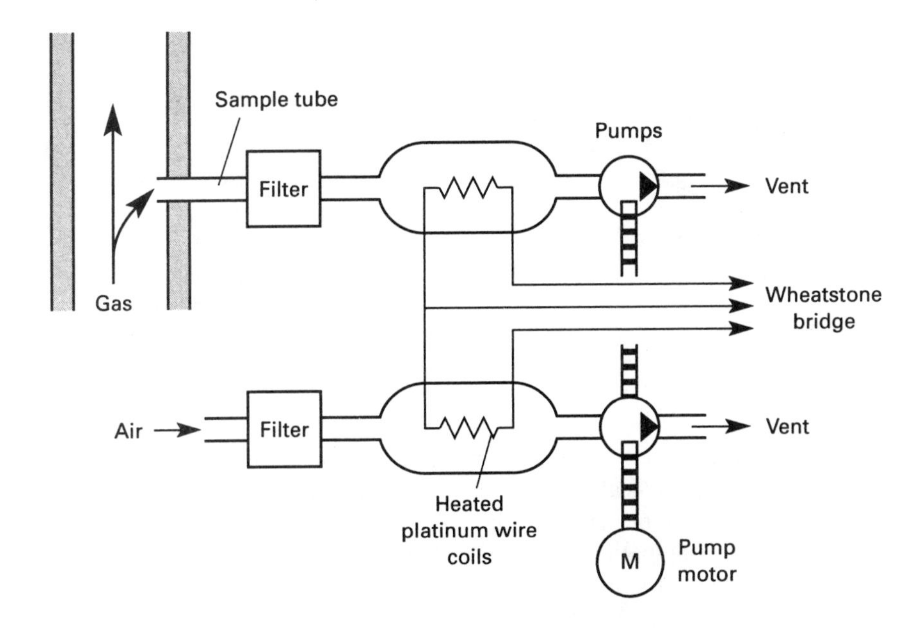

Fig. 10.10 *Thermal conductivity gas analyser.*

any combustible gas will ignite and burn. The wires are enclosed within a strong gauze covering to contain the combustion (identical to the principle of the Davy safety lamp for mines).

The burning gases heat up the wires further causing a change in resistance which can be measured by connecting the sensor and compensation wires in a Wheatstone bridge.

10.4. Spectroscopy

Light is emitted when excited atoms fall to a lower energy state, with the frequency being given by equation 8.12. The energy levels differ from element to element, so it is possible to identify the presence of elements, and their relative proportions, by observing the light emitted when the substance is heated. Light from silicon, for example, contains a distinct spectral line at 2124 Å, and light from sodium contains two very close lines in the orange part of the spectrum (hence the orange colour of sodium street lamps.) These spectral lines can be used for chemical analysis (particularly in metallurgy.) This analytical technique is called *spectroscopy* or spectral analysis.

The basis of a spectral analyser is shown on Fig. 10.11. Atoms of the test sample are excited either by heating or (more commonly) by the striking of a high voltage electric arc. The emitted light is passed through a prism where it is split into its component spectral lines.

Light detection devices (usually photomultipliers) are placed behind thin slits

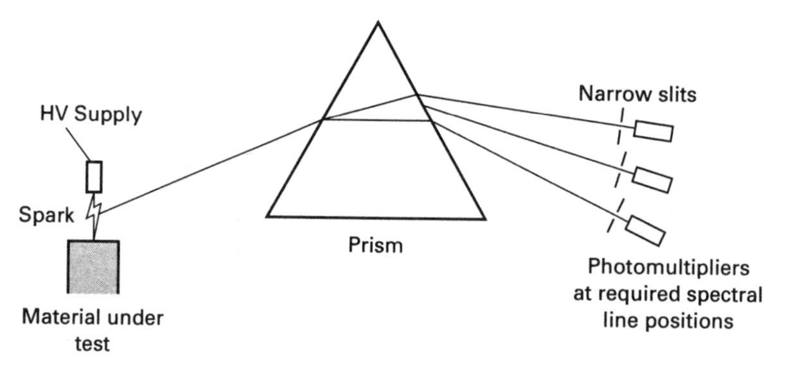

Fig. 10.11 *Principle of spectroscopy.*

at the positions corresponding to the spectral lines of interest. The current from each photomultiplier indicates the amount of each element present in the test sample.

Although the basic principle of Fig. 10.11 is simple, there are many practical difficulties. Very long path lengths (several metres) are needed on the exit side of the prism to give the required separation. These are usually obtained by folding the beam with mirrors. A typical layout is shown in Fig. 10.12. The long path lengths, though, make the instrument susceptible to temperature changes and vibration. A temperature-controlled environment and cushioned mountings are therefore required.

The operation of absorption analysers was described in Section 10.3.2. Any gas between the source and the photomultipliers will absorb light and give errors. The whole of the optical path after the spark must be contained in a high vacuum. Similarly to avoid errors from stray material at the spark, cleanliness must be observed, and the sparking take place in a inert gas such as argon.

One final problem is that whilst it is relatively easy to detect the presence of an element, it is more difficult to obtain quantitative results. When specific proportions are required (the percentage copper in a steel sample, for example) standard samples are used for calibration.

A spectral analyser requires calibration several times a day, and is essentially a batch process. An analysis takes of the order of a minute from preparation of the sample to the results being printed.

10.5. pH and ORP measurement

Knowledge of the acidity or alkalinity of an aqueous solution is often required, not only for chemical processes but also for less obvious applications like food processing, fish farming and gardening (where lime is added if the soil is too acidic).

The acidity (and alkalinity) of a solution is determined largely by the balance of (positive) hydrogen and (negative) hydroxyl ions. The greater the hydrogen ion concentration, the more acidic the solution will be.

At normal temperatures, water partially dissociates into free hydrogen and hydroxyl ions:

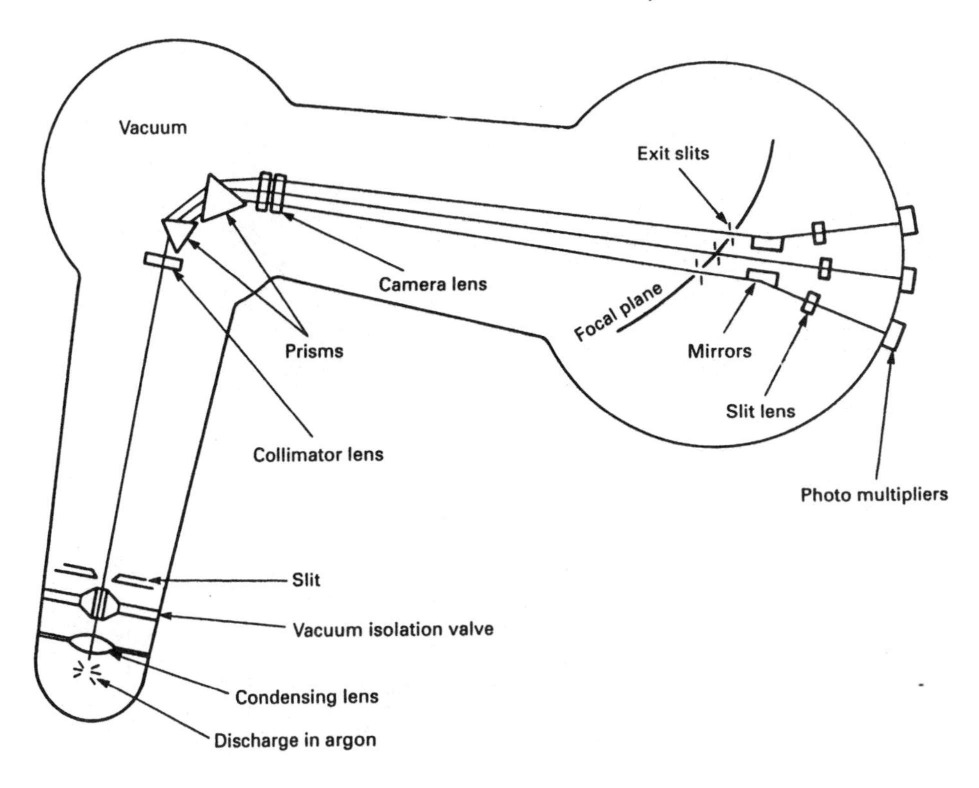

Fig. 10.12 *A spectral analyser.*

$$H_2O \Leftrightarrow H^+ + OH^- \qquad (10.5)$$

The concentration of each type of ion is 10^{-7} gram-molecules per litre. The product of this ion activity is 10^{-14} which is known as the dissociation constant of water.

The pH scale provides a linear measurement scale of hydrogen ion concentration. The 'p' denotes it is a power or logarithmic scale. The pH value is defined as

$$\text{pH} = -\log_{10}(H^+) \qquad (10.6)$$

where H^+ is the hydrogen ion concentration.

The possible range of hydrogen ion concentration is 1 to 10^{-14} which corresponds to a pH range of 0 to 14. Pure water (which is neutral) has a hydrogen ion concentration of 10^{-7} and a pH of 7 which is taken to equate to neutrality. Acids have a pH less than 7 and alkalines (bases) a pH of greater than 7.

Free hydrogen ions will cause a positive charge to build up on an electrode immersed in a solution. The measurement of pH is achieved with two probes, one called a measurement probe, and one a reference probe. Both are immersed in the sample solution as Fig. 10.13.

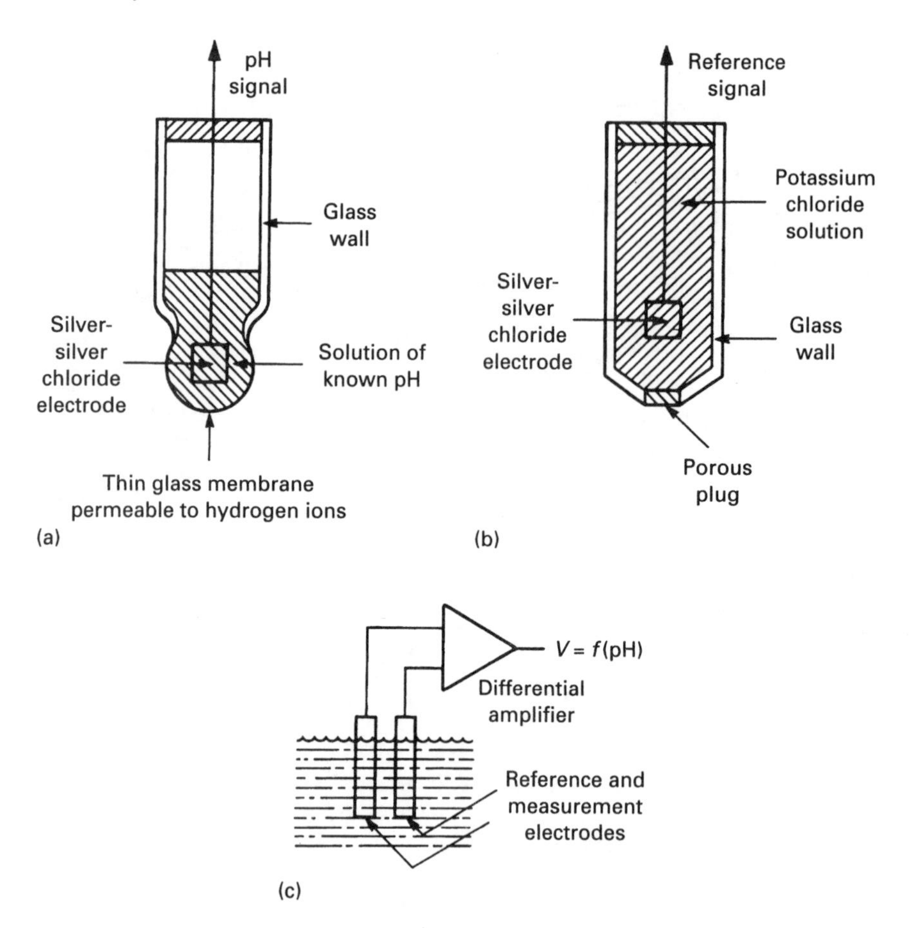

Fig. 10.13 *pH measurement and reference electrodes. (a) pH measurement electrode. (b) Reference electrode. (c) Connection.*

The reference probe generates a small voltage from a silver/silver-chloride electrode immersed in a potassium chloride gel. At the base of the probe a small porous plug makes electrical connection with the sample solution.

The measurement probe again uses a silver/silver-chloride electrode, this time immersed in a gel of known pH. The base of the probe has a very thin glass wall. This wall is permeable to hydrogen ions, causing the voltage from the measurement probe to be affected by the pH level in the solution. The higher the pH compared with the pH of the gel in the measurement probe, the lower will be the voltage across the glass wall.

The net result is that the voltage difference between the leads from the reference and measurement probes is dependent on the pH level in the solution.

Although Fig. 10.13 implies two separate probes are used, in practice the measurement and reference probes are combined in one device as Fig. 10.14.

The voltage from a pH sensor is small (typically a few tens of mV) and is at a

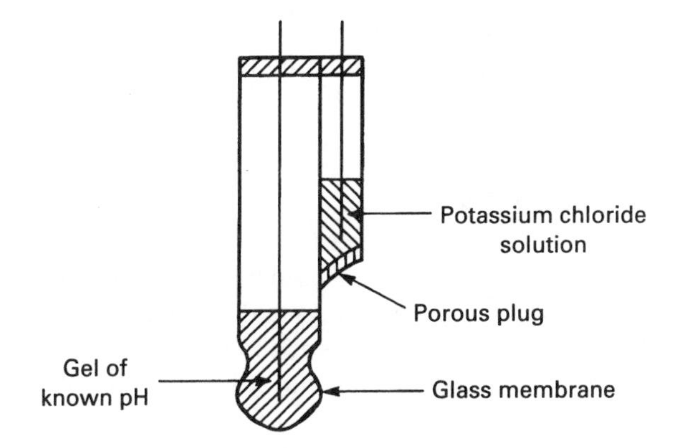

Fig. 10.14 *Combined measurement and reference probe.*

high impedance. It should be connected to a high gain DC amplifier with high input impedance and good CMMR.

The output is also temperature sensitive. To avoid errors the solution should be kept at a constant temperature by a temperature control loop. An alternative approach is to measure the solution temperature (with, say, a PT100) and apply a calculated correction before the reading is displayed.

If the pH probe is to be used as part of a control loop, the user should be aware that they are slow. A typical small pH probe will have a time constant of around two minutes. Large probes can have time constants of tens of minutes.

Calibration of a pH measurement system is performed by using standard calibration solutions. These are normally provided as dry powders in capsules or sealed foil bags and are mixed with precise volumes of de-mineralised or distilled water. Tap water is NOT suitable as its pH can vary widely. A standard solution should only be used on the day it is made and disposed of as soon as possible. Common calibration standards are pH 4 (acid), 7 (neutral) and 9 (alkaline). The outer surface of the probe should be cleaned between calibration checks in each standard fluid to avoid errors from carry-over.

An oxidation reduction potential (ORP) or redox probe is very similar in principle to a pH probe. Oxidation is defined as the loss of electrons by a molecule and reduction as the absorption by another. Both processes occur in equivalent amounts in any chemical reaction.

Liquids that tend to oxidise have a shortage of electrons, and liquids that tend to reduce have an excess of electrons. A solution can thus have a positive ORP (oxidising) or negative ORP (reducing) with respect to a standard voltage from a reference probe.

An ORP reference probe is almost identical to a pH reference probe. An ORP measurement probe is much simpler and is simply a platinum wire exposed to the sample solution. Like the pH probes, the voltage is small and must be connected to a DC amplifier of high input impedance and high CMRR.

pH and ORP probes are not the only sensors for measuring specific ions in solution. By changing the gels in the probes, sensors for any ion can be made. All will use the same basic method of comparing the outputs from a reference and measurement probe.

10.6. Chromatography

If a drop of ink or beetroot juice is placed onto blotting paper and left for an hour or so, the spot will spread and distinct rings of different colours will appear. The ink or juice contains several different compounds, and the rings form because each compound spreads through the paper at a different rate. This is the basis of chromatography.

If a liquid or gas with several component materials is passed through a column of fine powder, the components attach to, and subsequently detach from the powder at different rates. The adhering of one substance to another is called *adsorption*, and the subsequent release again is called *elution*. The effect is to split the component parts into groups separated by time as shown on Fig. 10.15. A peak identifies the presence of a component and the area under the peak the quantity of component present. Detection levels of a few parts per million are easily performed.

Figure 10.16 shows how this principle is used to identify component parts

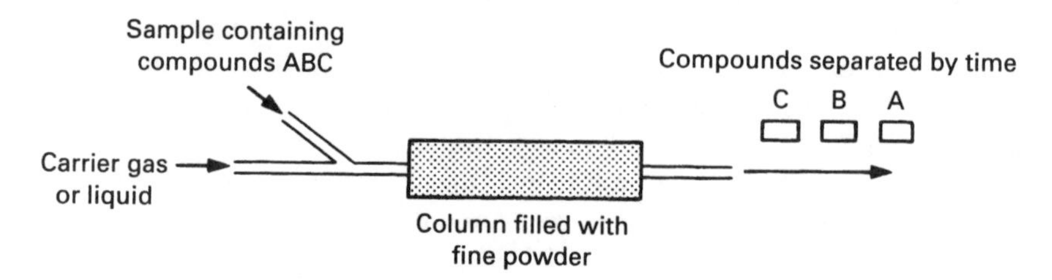

Fig. 10.15 *Principle of gas chromatography.*

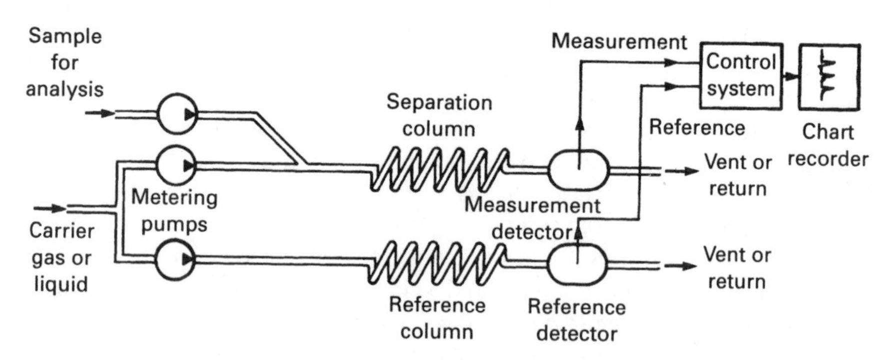

Fig. 10.16 *Chromatograph analysis system.*

present in a gas or liquid. The sample is mixed with a carrier gas or liquid (dependent on the application) and passed through a column filled with material which encourages adsorption and elution for the materials of interest. Column is actually a rather bad descriptive term as it implies something relatively short and vertical. Tube or coil would be better terms as in practice they tend to be narrow (a few mm) and several metres long.

The adsorption/elution separate out the component parts which arrive at a detector on the downstream side of the column. We will return to these detectors shortly, but essentially they simply detect the presence of something in the carrier stream (i.e. *time*, not the sensor, is used to identify which materials are present in the sample.) Each component of interest has a 'gate time' when, if present, it will appear at the detector. The detectors are usually connected to a chart recorder, giving a trace similar to Fig. 10.17.

Figure 10.16 has two identical columns. Chromatography, like all analytical instrumentation, is susceptible to errors from temperature, carrier characteristics and so on. The second column carries purely the carrier flow and is used as a reference.

The commonest type of detector is the thermal conductivity detector. This is identical, in principle, to the technique used to detect carbon dioxide described in Section 10.3.6. The component parts being conveyed by the carrier flow affect the heat loss from heated filaments. The heat loss is detected by resistance changes, with the measuring and reference legs being connected into a Wheatstone bridge.

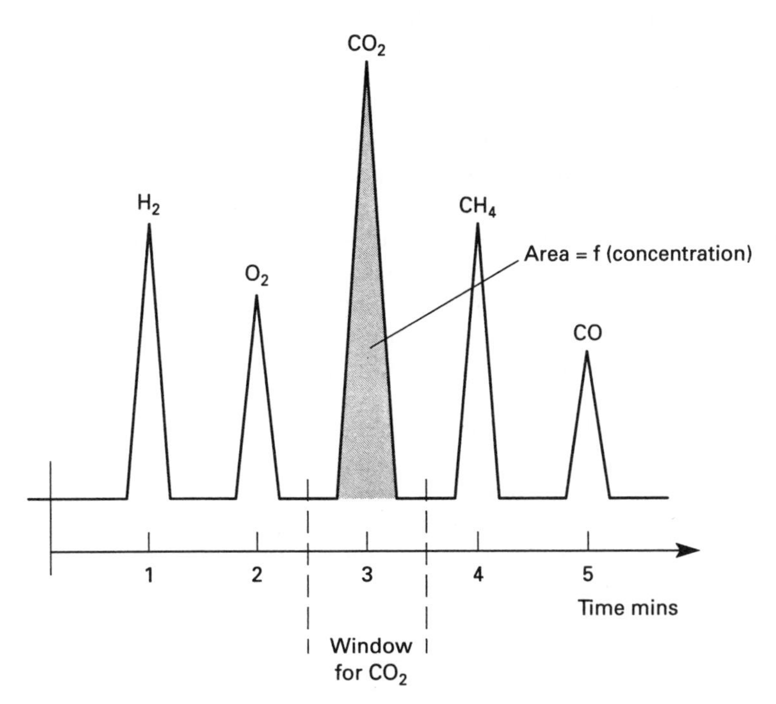

Fig. 10.17 *Typical output from chromatograph.*

The second type of detector, called the flame ionisation detector (FID), passes the flow through a small hydrogen flame. This ionises the components, which are attracted to an electrically charged set of plates close to the flame. Components in the carrier flow cause the (very small) currents from these plates to change.

The final detector, called a photometric detector again uses a flame, but materials are detected by colour changes at the flame (similar in principle to spectroscopy described in Section 10.4). Photocells with suitable filters look for flame colours corresponding to materials of interest. Because of the 'pre-sort' performed by the column, the arrangement is nowhere near as complex or large as the spectroscope described earlier.

Chromatography, like spectroscopy, is essentially a batch process and, by instrumentation standards, very slow. Delays are typically minutes, but hours are not uncommon in high accuracy laboratory instruments. (The longer the columns, the better the separation but the longer the transit delay.) Again like the spectroscope, regular calibration is needed with test samples.

10.7. General observations

Analytical instrumentation has several characteristics which the designer must consider.

They are generally slow by instrumentation standards, which has possible stability implications if they are made part of a closed loop control strategy.

Regular calibration is needed, several times a day for devices such as the spectroscope and chromatograph. The designer must ensure that maintenance procedures are established, and production is not interrupted by calibration checks. During calibration, high standards of cleanliness must be observed. The calibration of pH and similar probes may involve the use of toxic substances with implications for the establishment of safe storage and safe working practices.

In many cases the sensitive parts of sensors will be in direct contact with the process materials (unlike, say, thermocouples which can be protected by thermowells). If these materials are aggressive, sensor life will be short, and there will be safety implications for maintenance technicians who will have to change or repair the sensors.

A decision usually has to be made as to whether sampling is to be made on-line in the process stream, or off-line with samples taken either automatically by pulsing a valve or even manually. The chromatograph and the spectroscope are inherently batch processes. With some sensors (particularly the infrared CO sensor) a bypass route needs to be established with pressure reducing valves and possibly cooling to bring the process pressure or temperature to a level acceptable to the instrument.

The combination of all of these leads to a final conclusion that analytical instrumentation demands a higher level of skill and care from the maintenance technicians.

Chapter 11

DC amplifiers

11.1 Introduction

11.1.1. DC amplifier requirements

Signals in instrumentation and process control are generally represented digitally, pneumatically or as an analog voltage or current. Digital signals are covered in Chapter 13, and pneumatics in Chapter 16. This chapter is concerned with the manipulation of signals represented as an electrical voltage or current.

Instrumentation signals are essentially static for long periods, and as such require amplifiers with predictable characteristics down to 0 Hz, i.e. DC amplifiers. A typical AC audio amplifier, for comparison, would have little gain below about 20 Hz, and would 'droop' on low-frequency signals as Fig. 11.1a. The lower frequency limit in AC amplifiers is usually determined by the impedance of coupling capacitors between stages and emitter decoupling capacitors. In Fig. 11.1b, the impedances of C_1 to C_5 will all increase with

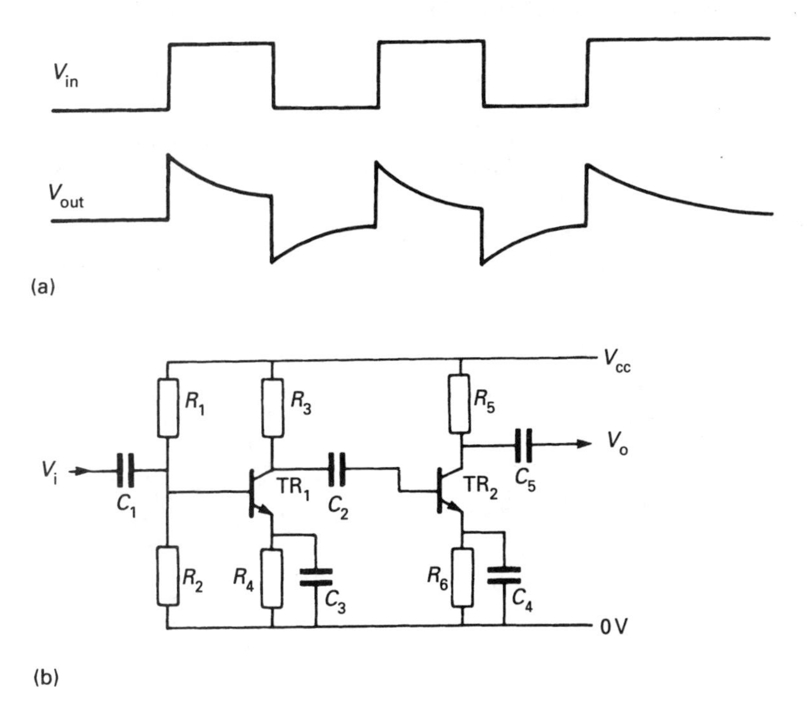

Fig. 11.1 *The effect of AC amplifiers on low frequency signals. (a) The effect of poor low frequency response on a signal. (b) A simple AC amplifier. The low frequency response is determined by capacitors C_1–C_5.*

decreasing frequency, causing the gain to fall. DC amplifiers therefore use direct coupling between stages.

Direct coupling, however, brings its own problems. Figure 11.2a shows a possible design for a simple direct-coupled amplifier. Unfortunately almost all transistor parameters vary with temperature and from device to device. V_{be}, for example, changes by 2 mV per °C, and collector/emitter leakage current doubles every 10 °C. These, and similar effects, will cause the output of Fig. 11.2a to vary in a manner which is indistinguishable from changes caused by the signal itself.

Most DC amplifiers are based on the so-called long-tailed pair circuit of Fig. 11.2b. TR_1 and TR_2 are identical transistors maintained at the same temperature (both conditions being ensured by constructing the circuit on a single integrated circuit–IC–silicon wafer). Resistor R_3 acts as a constant current sink, with the current being split between TR_1/TR_2. Because the two transistors are identical and at the same temperature, leakage current and offsets will cancel and the current split between TR_1, TR_2 will depend solely on V_{in}. The output voltage is therefore an amplified version of V_{in}, and is unaffected by temperature changes. Note that Fig. 11.2b is, in effect, a *differential* amplifier because it amplifies the voltage difference between its two inputs.

11.1.2. Integrated circuit DC amplifiers

It is exceedingly rare nowadays for DC amplifiers to be constructed of individual transistors. The requirements of Fig. 11.2b are identical components and a uniform circuit temperature. These conditions are best met by fabricating the entire circuit as an IC. There are many IC DC amplifiers available with different characteristics, but by far the commonest is the ubiquitous 741, arguably the most successful IC ever designed. Other DC amplifiers are similar in principle, differing only in, say, greater or lesser gain or perhaps frequency response. DC amplifier specifications are described in Section 11.2.

A DC amplifier can be represented by Fig. 11.3a; this has two inputs, two

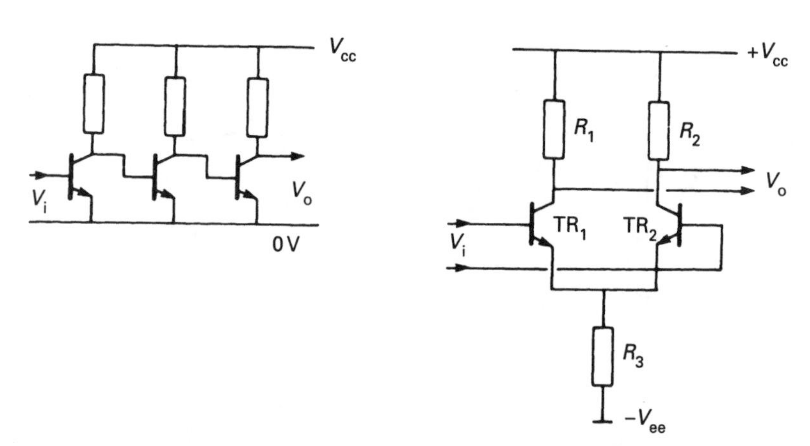

Fig. 11.2 *DC amplifier circuits. (a) Simple DC amplifier. (b) The long-tail pair.*

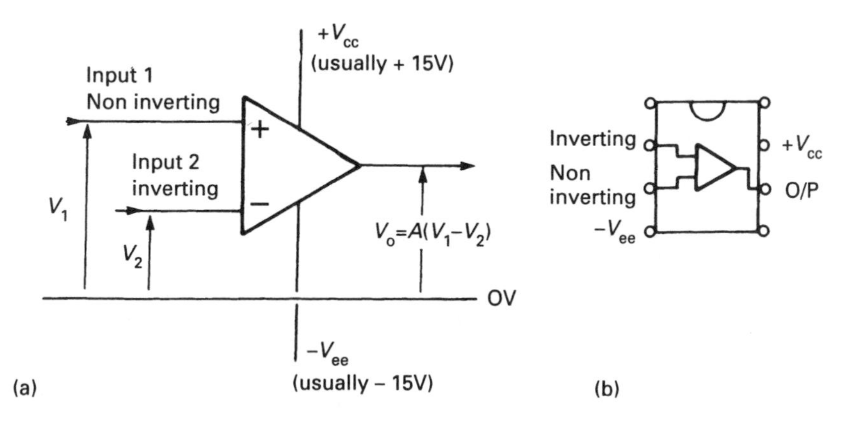

Fig. 11.3 *Practical DC amplifiers. (a) Representation of a DC amplifier. (b) Integrated circuit DC amplifier (741).*

power supply connections (usually symmetrical 15 volts positive and negative) and the output. Note that all voltages (including the power supplies) are referred to the 0 V rail, even though the DC amplifier itself does not need an 0 V connection. The output voltage is given by:

$$V_0 = A(V_1 - V_2) \tag{11.1}$$

Where A is the amplifier gain.

Input 1 is usually called the non-inverting input and input 2 the inverting input, for reasons that can be seen by linking each input to 0 V in turn. With input 2 linked to 0 V, and a signal applied to input 1, $V_0 = AV_1$; the output moves in the same sense as the input signal. With input 1 linked to 0 V, and a signal applied to input 2, $V_0 = -AV_2$; the output moves in the opposite sense to the input signal. Often a + sign is used for the non-inverting input and a – sign for the inverting input; these should not be confused with the power supply connections.

DC amplifiers are usually encapsulated in an eight-pin dual in line (DIL) IC. The arrangement of Fig. 1.3b is widely used, although there are slight variations in the use of pins 1, 5, and 8.

High-gain DC amplifiers were originally used in analog computers where the term operational amplifier, shortened to op amps, was used. This description is usually given to IC DC amplifiers and will be used in the rest of this chapter.

11.2. Op amp specifications

11.2.1. Introduction

There are probably several hundred different op amp ICs. The process control engineer needs to be able to select a device for a specific application. This section describes the relevant items found on a specification sheet. Where typical values are given, the 741 specification has been used.

11.2.2. DC gain (A_{VD})

This is defined as the change in output volts divided by the change in input volts. Referring to Fig. 11.4a:

$$A_{VD} = \frac{\Delta V_0}{\Delta(V_1 - V_2)} \tag{11.2}$$

A typical value for a 741 is around 100 000. Often the gain is given as V/mV, so a gain of 100 000 would be given as 100 V/mV. Sometimes the gain is given in dB:

$$A_{VD} = 20 \log_{10}(\Delta V_0/(\Delta(V_1 - V_2))) \text{ decibels} \tag{11.3}$$

The output of an op amp must, obviously, stay within the supply voltages, so a differential input voltage of less than 1 mV is required to drive the amplifier into saturation. In practice, however, feedback is used to define the gain as described in Section 11.3.

11.2.3. Unity gain bandwidth (BW)

All amplifiers have a frequency response similar to Fig 11.4b. In audio amplifier specifications it is usual to define point X: the 3 dB point. With op amps this point is of little interest because the closed loop gain is usually far lower than A_{VD}. The bandwidth is commonly specified as the frequency at which the gain drops to unity (0 dB), point Y on Fig. 11.4b. Many op amps (including the 741) have a

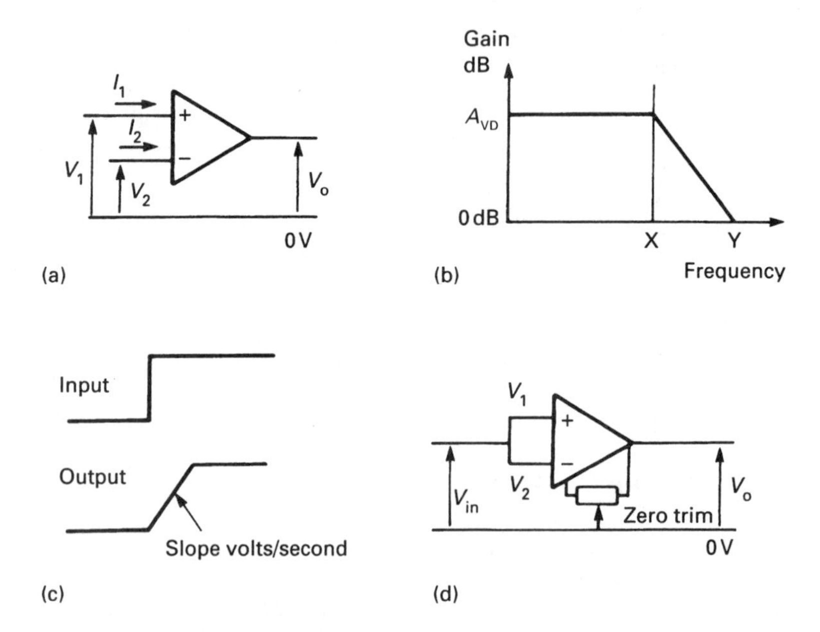

Fig. 11.4 *Definition of DC amplifier terms. (a) DC gain and input bias currents. (b) Unity gain bandwidth. (c) Slew rate. (d) Common mode gain.*

deliberate low-frequency roll of a few hertz to assist stability when feedback is applied. Point X for a 741 is actually 10 Hz, whereas the unity gain bandwidth is 1 MHz. Choosing an op amp with a better frequency response than an application requires can lead to stability problems. (See section 11.3.5.)

11.2.4. *Slew rate (SR)*

In Fig. 11.4c a step input has been applied to an amplifier input, with the amplitude of the step such that it is just sufficient to drive the amplifier into saturation. The output of the amplifier will not be a step, but a ramp as shown. The ramp slope is a measure of the useful frequency range for the amplifier. A 741 has a slew rate of 1 V/μS, but amplifiers are available with slew rates of over 100 V/μS.

11.2.5. *Input offset voltage (V_{IO})*

If V_1, V_2 in Fig. 11.4a are both tied to 0 V, V_o should be zero. In practice, V_o will be either positive or negative by a significant amount. The offset voltage is the differential input voltage that is needed to make V_o zero. The figure for a 741 is 2 mV maximum. In itself, V_{IO} is not particularly important as it can be removed by the zeroing circuits of Section 11.3.4. What is usually more important is how V_{IO} changes with temperature.

11.2.6. *Offset voltage temperature coefficient (αV_{IO})*

This specification states how V_{IO} may change with temperature and is typically 5 μV/°C. This error cannot be zeroed out and can only be controlled by choosing an amplifier with a sufficiently low value of αV_{IO} for the application. Values as low as 0.5 μV/°C are available (at a price).

11.2.7. *Input bias current (I_{IB})*

With $V_1 = V_2 = 0$ V in Fig. 11.4a input currents I_1, I_2 will be non-zero (being the base currents for the input long tail pair). A typical value is 0.1 μA. This will not normally cause any problems provided the source impedances of V_1, V_2 are equal.

11.2.8. *Input offset current (I_{IO})*

This is the difference between I_1 and I_2, and will cause an offset voltage even if the source impedances are equal. A typical value for a 741 is 20 nA. Errors due to I_{IO} can be removed by a zeroing potentiometer.

11.2.9. *Common mode gain (A_{CM})*

In Fig. 11.4d, V_1 and V_2 have been linked and V_{in} set to 0 V. The zeroing potentiometer has been adjusted to set V_o to 0 V. If V_{in} is varied, V_o should stay at zero volts, but in practice will vary slightly. The common mode gain is defined as:

$$A_{CM} = \frac{\text{change in } V_o}{\text{change in } V_{in}} \tag{11.4}$$

The common mode gain is an indication of how well an amplifier will reject common mode noise. Usually, an amplifier specification does not give A_{CM}, but uses the common mode rejection ratio defined below.

11.2.10. Common mode rejection ratio (CMRR)

The DC gain A_{VD} was defined in Section 11.2.2, and the common mode gain A_{CM} above. The CMRR is defined as:

$$\text{CMRR} = \frac{A_{VD}}{A_{CM}} \tag{11.5}$$

The CMRR is large, and is usually given in dB. The 741 has a CMRR of 90 dB.

11.3. Basic circuits

11.3.1. Inverting amplifiers

The commonest DC amplifier circuit is the inverting amplifier of Fig. 11.5. The output voltage V_o will be in the range ± 15 V, so the junction of R_1/R_2 will be in the range $\pm 15/A_{VD}$; typically less than a mV. For all practical purposes, we can assume that both inputs of the amplifier remain at 0 V (the junction of R_1, R_2 being called a virtual earth).

If the input bias current is small compared with I_1, I_2, we can say:

$$I_1 + I_2 = 0 \tag{11.6}$$

$$\frac{V_{in}}{R_1} + \frac{V_o}{R_2} = 0 \tag{11.7}$$

or

$$V_o = -\frac{R_2}{R_1} V_{in} \tag{11.8}$$

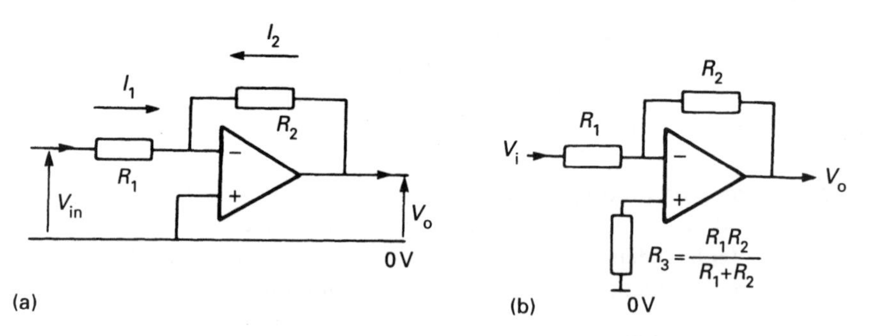

Fig. 11.5 *The inverting amplifier. (a) Theoretical circuit. (b) Practical circuit to eliminate offset from input bias currents.*

The minus sign denotes inversion, i.e. positive V_{in} gives negative V_o. The closed loop gain $(-R_2/R_1)$ is independent of the amplifier gain.

If R_1 and R_2 are large (>100 K) resistor R_3 should be added as Fig. 1.5b to equalise the offset due to I_{IB}. The value of R_3 is chosen to equal the values of R_1, R_2 in parallel (e.g. if $R_1 = R_2 = 330K$, R_3 would be chosen to be 150 K, the nearest preferred value to the theoretical 165K).

Equation 11.8 assumes the amplifier has infinite input impedance, zero output impedance and an open loop gain (A_{VD}) much larger than the closed loop gain (R_2/R_1). This will be true for almost all circuits using IC op amps. If the open loop gain is included, equation 11.8 becomes:

$$V_o = -\frac{R_2}{R_1 + \dfrac{(R_1 + R_2)}{A_{VD}}} V_{in} \tag{11.9}$$

For a typical circuit, the difference between equations 11.8 and 11.9 is less than 0.05%. The effect of finite input impedance (typically 2 MΩ) and non-zero output impedance (about 50 Ω) is even less.

The closed loop input impedance of Fig. 11.5 is simply the value of R_1. This can be unacceptably low if high gain is required. The non-inverting amplifier below has very high input impedance.

11.3.2. Non-inverting amplifiers

The non-inverting amplifier circuit is shown in Fig. 11.6. Resistor R_3 is included purely to minimise the input bias current offset and should equal the value of R_1 and R_2 in parallel. R_3 plays no part in the determination of the gain.

The voltage at the junction of R_1 and R_2 is given by:

$$V_1 = \frac{R_2}{R_1 + R_2} V_o \tag{11.10}$$

As explained previously, the voltages at the two amplifier inputs can be considered equal, so:

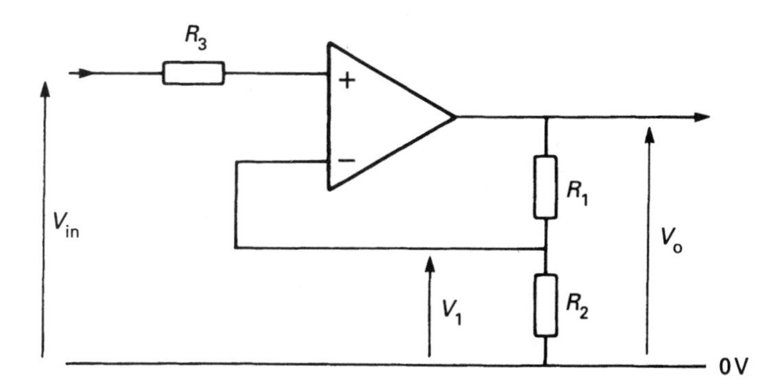

Fig. 11.6 *Non-inverting amplifier.*

$$V_1 = V_{in} \tag{11.11}$$

$$\frac{R_2}{R_1 + R_2} V_o = V_{in} \tag{11.12}$$

$$V_o = \frac{(R_1 + R_2)}{R_2} V_{in} \tag{11.13}$$

The closed loop gain is determined solely by R_1 and R_2 and is independent of the actual amplifier gain (provided the closed loop gain is small compared with the open loop gain).

The input impedance of the non-inverting amplifier is very high, and can be considered infinite for most purposes. If a high input impedance unity gain amplifier is required, the buffer circuit of Fig. 11.7 may be used. Obviously $V_o = V_{in}$.

11.3.3. Differential amplifiers

The circuits of Figs. 11.5 to 11.8 amplify a signal which is referred to the 0 V line. In many applications a true differential input is required. The output from the strain gauge bridge of Fig 11.8a, for example, is a voltage which is not referred to 0 V. Another common requirement is shown on Fig. 11.8b where a thermocouple signal is to be amplified despite common mode noise of several volts being superimposed on the signal lines. In both cases, V_o is to be an amplified version of V_{in} and superimposed input voltages with respect to 0 V are to be ignored.

Figure 11.8c shows a differential output amplifier. If $R_1 = R_2 = R_a$ and $R_3 = R_4 = R_b$, then:

$$V_o = -\frac{R_b}{R_a}(V_1 - V_2) \tag{11.14}$$

The circuit's common mode rejection depends on how close the resistors match. Where a mV signal is to be amplified, precision resistors should be used.

The impedance seen by V_2 is $R_2 + R_4$. The impedance seen by V_1 varies with the signal, but is of the order of R_1. The circuit can handle input voltages outside the supply voltages, provided the actual amplifier inputs stay within the supply range. With unity gain (all resistors equal), the circuit will remove 30 V common mode voltage from a signal of a few mV.

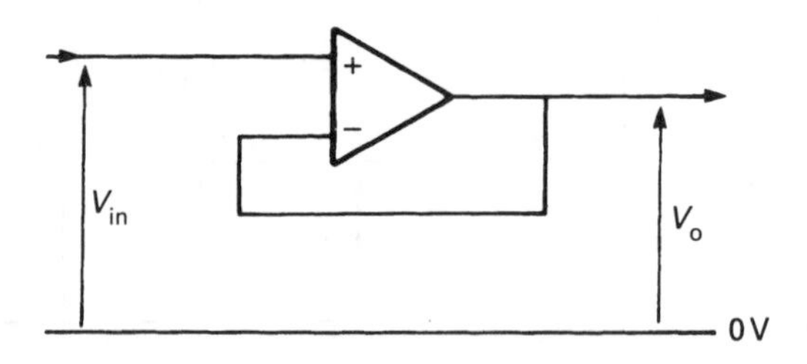

Fig. 11.7 *Unity gain buffer.*

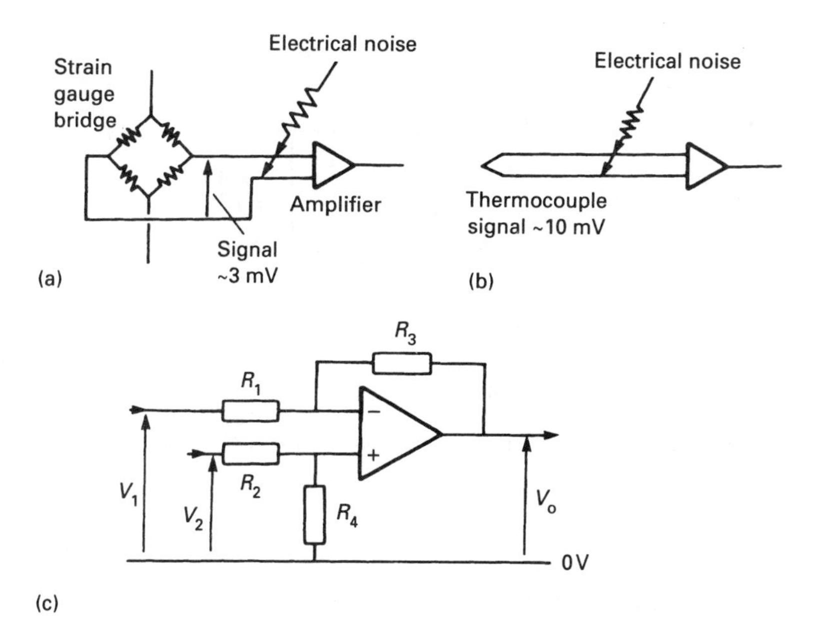

Fig. 11.8 *The differential amplifier. (a) Strain gauge circuit. (b) Thermocouple circuit. (c) Practical circuit.*

11.3.4. Zeroing circuits

The input voltage offset V_{IO} is typically a few mV. In many circuits, where the closed loop gain is small and the signal levels are larger than the offset, the effect of V_{IO} can be ignored. When the signal levels are similar to the offset, a zeroing (or nulling) circuit must be included in the design.

Many ICs, including the 741, have provision for a single zeroing potentiometer to be connected direct to the amplifier, as in Fig. 11.9a. An alternative method for inverting amplifiers is shown in Fig. 11.9b. RV_1 effectively adds an offset current to the virtual earth to produce a voltage output which cancels V_{IO}. The circuit of Fig. 11.9b can also be used to generate an $Ax + B$ function where R_2/R_1 determines A, and the setting of RV_1 determines the value of B.

11.3.5. Stability

Almost all op-amp circuits use feedback and, as shown in Chapter 18, feedback can cause instability.

In general, stability problems only occur where an op amp has high gain at high frequencies, and stray capacitance and lead inductance become significant. Most signal processing circuits deal with frequencies of at most a few kHz. Op amps such as the 741 have a deliberately low high-frequency gain to prevent high-frequency oscillation. Such ICs are said to be unconditionally stable.

If high-frequency operation is required, the gain/frequency response of many op amps can be tailored by the use of external components as shown for the 709 and 308 ICs in Fig. 11.10. Op amps with useful gain up to 500 kHz are available,

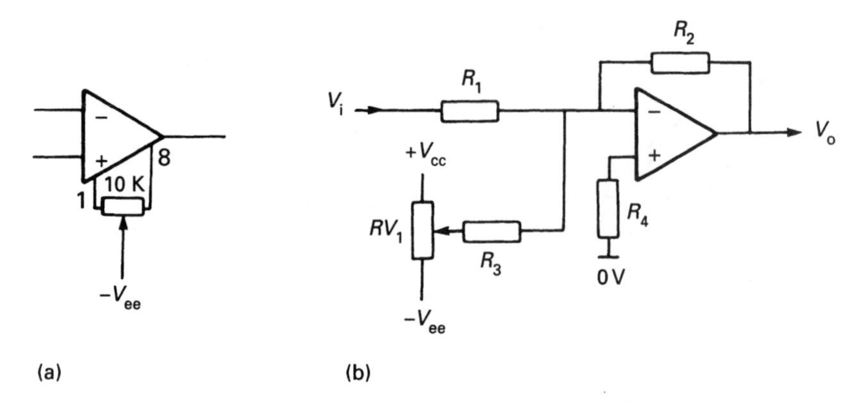

Fig. 11.9 *Zeroing of DC amplifiers. (a) Zero adjust on 741. (b) Zeroing with inverting amplifier.*

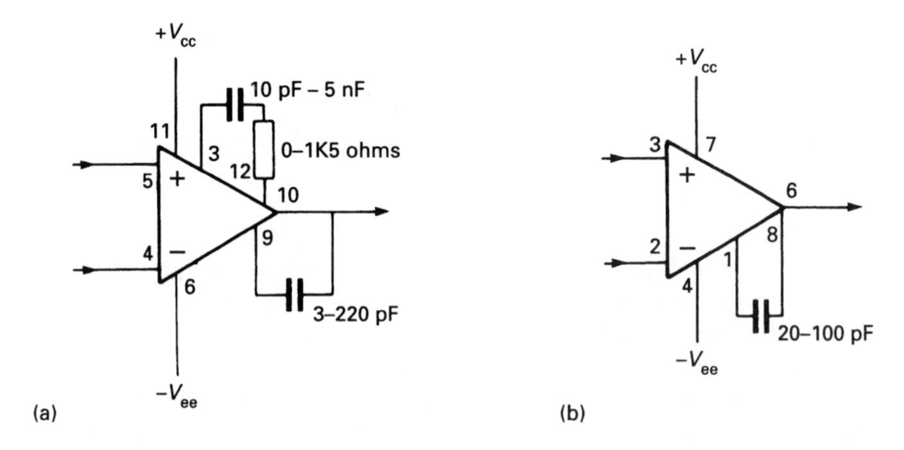

Fig. 11.10 *External frequency compensation. (a) 709 integrated circuit. (b) 308 integrated circuit.*

but care in the circuit layout is required to avoid oscillation. The cardinal rule is choose a frequency response that is just good enough for the application.

11.4. Computing circuits

11.4.1. Addition

Figure 11.11 is used where two or more voltages are to be added. By the reasoning of Section 11.3.1:

$$I_1 + I_2 + I_3 + I_4 = 0 \tag{11.15}$$

$$\frac{V_1}{R_1} + \frac{V_2}{R_2} + \frac{V_3}{R_3} + \frac{V_o}{R_4} = 0 \tag{11.16}$$

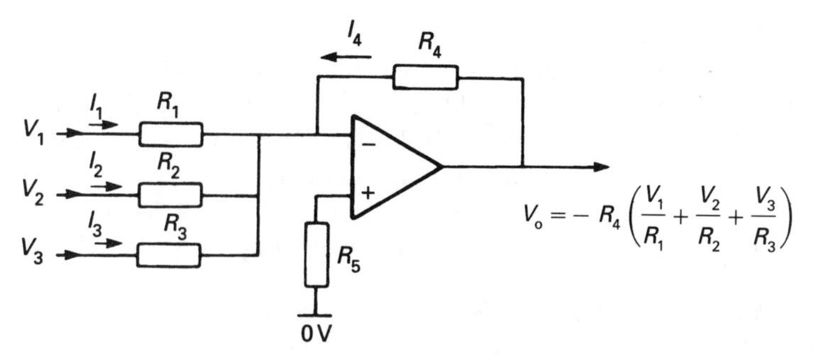

Fig. 11.11 *Summing amplifier.*

or

$$V_o = -\left(\frac{R_4}{R_1}V_1 + \frac{R_4}{R_2}V_2 + \frac{R_4}{R_3}V_3\right) \tag{11.17}$$

if $R_1 = R_2 = R_3 = R$

$$V_o = -\frac{R_4}{R}(V_1 + V_2 + V_3) \tag{11.18}$$

Obviously the circuit can be expanded to any number of inputs. Resistor R_5 is included to remove the effect of I_{IB}. The value of R_5 should equal the parallel combination of the other resistors.

11.4.2. Subtraction

Where one voltage is to be subtracted from another, the differential amplifier of Fig. 11.8c may be used with all resistors equal in value. Where a combination of addition and subtraction (e.g. $V_1 + V_2 - V_3 + V_4$) or a subtraction involving multiplier constants (e.g. $AV_1 - BV_2$ where A and B are constant) is needed, unity gain inverters may be used as shown on Fig. 11.12.

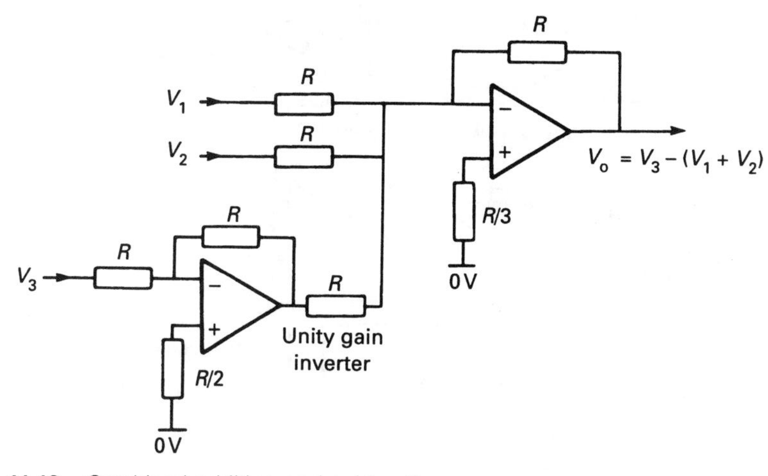

Fig. 11.12 *Combined addition and subtraction.*

11.4.3. Multiplication by a constant

If the constant is less than unity, a simple voltage divider as in Fig. 11.13 should be used. The output voltage should go to an impedance much higher than the parallel combination of R_1 and R_2. If this is not possible, a unity gain amplifier (as shown in Fig. 11.7) should follow the divider.

If the multiplication constant is greater than unity, a fixed gain inverting amplifier (Fig. 11.5) or non-inverting amplifier (Fig. 11.6) should be used.

11.4.4. Multiplication and division of two voltages

Multiplication of two voltages is not widely used because the circuits are complex and not very accurate. In general, input voltages are scaled 0 to 10 volts, and the output is given by:

$$V_o = 0.1 \times V_1 \times V_2 \tag{11.19}$$

With $V_1 = 1.4$ volts and $V_2 = 7.9$ volts, V_o would be 1.106 volts. This keeps the output voltage at reasonable levels.

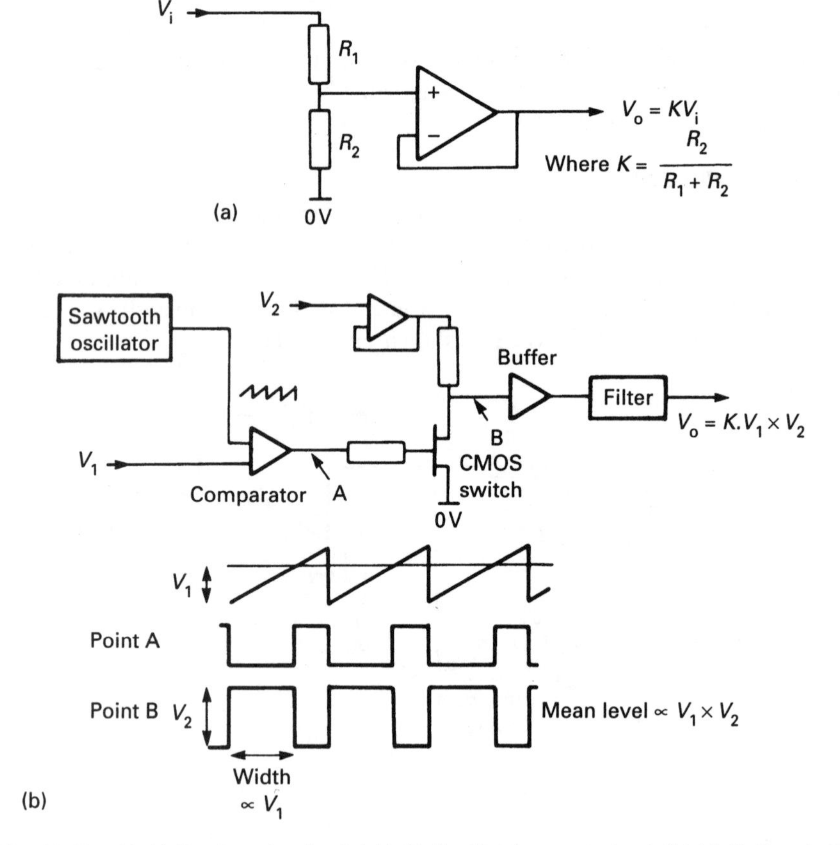

Fig. 11.13 *Multiplication circuits. (a) Multiplication by a constant. (b) Multiplier circuit.*

One possible multiplier circuit, shown in Fig. 11.13, uses a pulse with one voltage controlling the pulse width and the other the pulse height.

A sawtooth voltage is fed to a comparator along with the first voltage V_1. The output of the comparator at point A is a square wave of constant frequency, but pulse width proportional to V_1. This is used to control an electronic CMOS switch connected to input voltage V_2. The pulses at point B have a width proportional to V_1 and height proportional to V_2. If this is passed through a low-pass filter, the output voltage is proportional to the product of V_1 and V_2.

The voltage drop across a semiconductor diode is related to the current through it by the exponential relationship:

$$I = Ae^{BV} \tag{11.20}$$

where A and B are constants. This allows the construction of logarithmic and antilogarithmic amplifiers (by the use of voltage to current and current to voltage amplifiers).

Multiplication of numbers can be performed by the addition of the numbers' logarithms, so a multiplier can be constructed as shown in Fig. 11.14. In practice, ICs such as the AD534 are available which perform multiplication using the log/antilog principle.

Figures 11.13 and 11.14 can only perform multiplication of two positive numbers (called single quadrant multiplication). Correctly signed multiplication

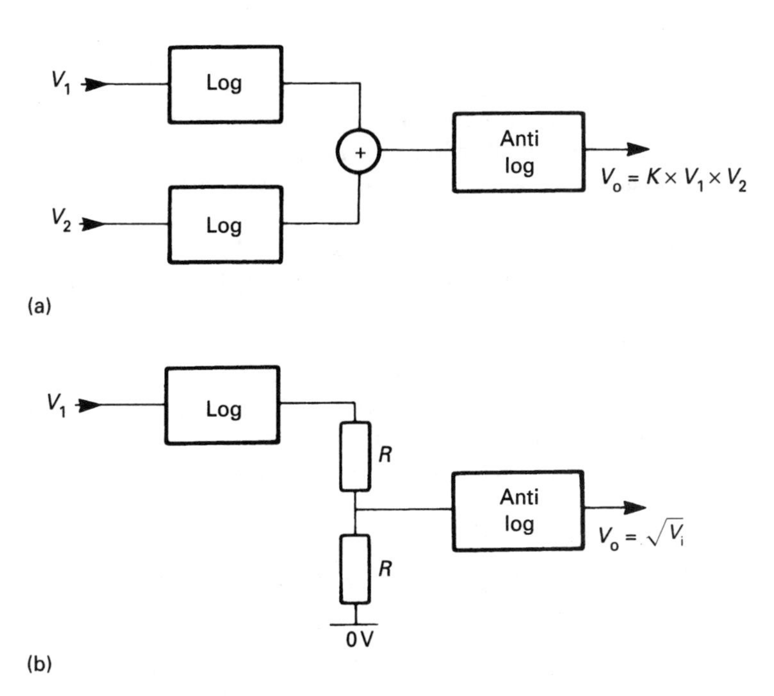

Fig. 11.14 *Use of logarithmic elements. (a) Multiplication using log/antilog amplifiers. (b) Square root extractor.*

of positive and negative numbers (called four quadrant multiplication) is, of course, far more complex but is available in IC form.

Division can also be performed with log/antilog amplifiers by subtracting logarithms. If the adder in Fig. 11.14 is replaced by a subtractor, V_o will become V_1/V_2. Division is rarely performed in practice.

Square root extraction can also be performed with log/antilog amplifiers by taking the log of the input voltage, dividing by 2 and taking the antilog as shown in Fig. 11.14b. Square root circuits are often needed in flow measurement.

11.4.5. Integration

The theoretical circuit for an integrator is shown in Fig. 11.15a; as before, if the amplifier has infinite input impedance, I_1 and I_2 are equal so:

$$\frac{V_i}{R} + C\frac{dV_o}{dt} = 0 \tag{11.21}$$

$$\frac{dV_o}{dt} = -RCV_i \tag{11.22}$$

or

$$V_o = -\frac{1}{RC}\int V_i \, dt + A \tag{11.23}$$

where A is an initialising constant.

A practical circuit is shown in Fig. 11.15b. Difficulties are usually caused by the input bias current of the amplifier (which is integrated by C_1) and leakage current through the capacitor. Low leakage (non-electrolytic) capacitors and FET op amps should therefore be used.

RV_1 is more than a simple zero control. Because V_o is the integral of V_i, the output will not be zero for zero input voltage, but should stay at its current value. RV_1 should be adjusted so that V_o neither rises nor falls with zero input voltage.

Switch SW_1 (which can be a CMOS switch) is used to short out C_1, when the integrator is not in use (to stop integral wind-up when a PID controller is in manual, for example), or to initialise the circuit. R_4 is purely to limit the current through the capacitor when SW_1 is first closed.

Figure 11.15c shows the action of an integration circuit.

11.4.6. Differentiation

Figure 11.16a shows the theoretical circuit for a differentiator. By usual analysis:

$$V_o = -RC\frac{dV_i}{dt} \tag{11.24}$$

A differentiator implies a continually rising gain, which is impossible because of factors such as stray capacitance and the frequency response of the amplifier

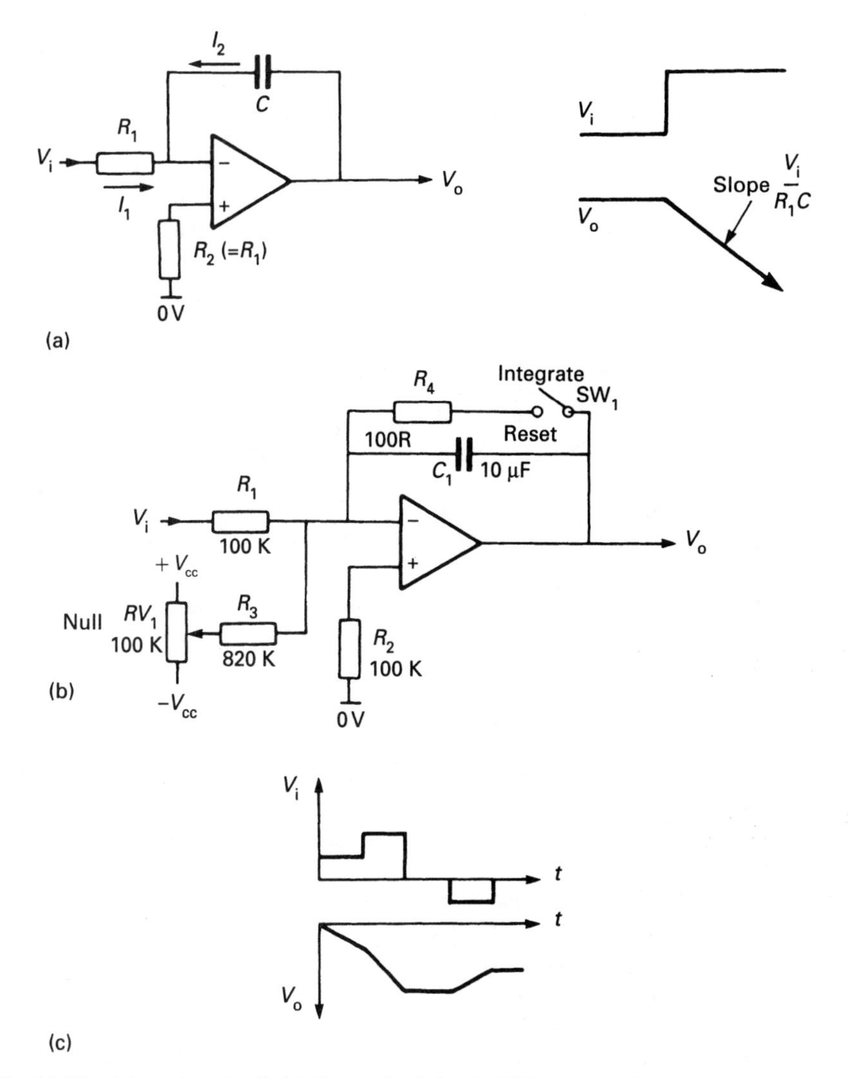

Fig. 11.15 *Integrator circuit. (a) Theoretical circuit. (b) Practical circuit. (c) Typical action.*

itself. To give a predictable high-frequency response, the circuit of Fig. 11.16b is usually used. R_1 and C_1 form the differentiator, and R_2, C_2 are chosen such that:

$$R_1C_1 = R_2C_2 \tag{11.25}$$

The circuit has the response of Fig. 11.16c, and acts as a differentiator up to a frequency given by:

$$f = \frac{1}{2\pi R_1 C_2} \tag{11.26}$$

at which point the gain is R_1/R_2.

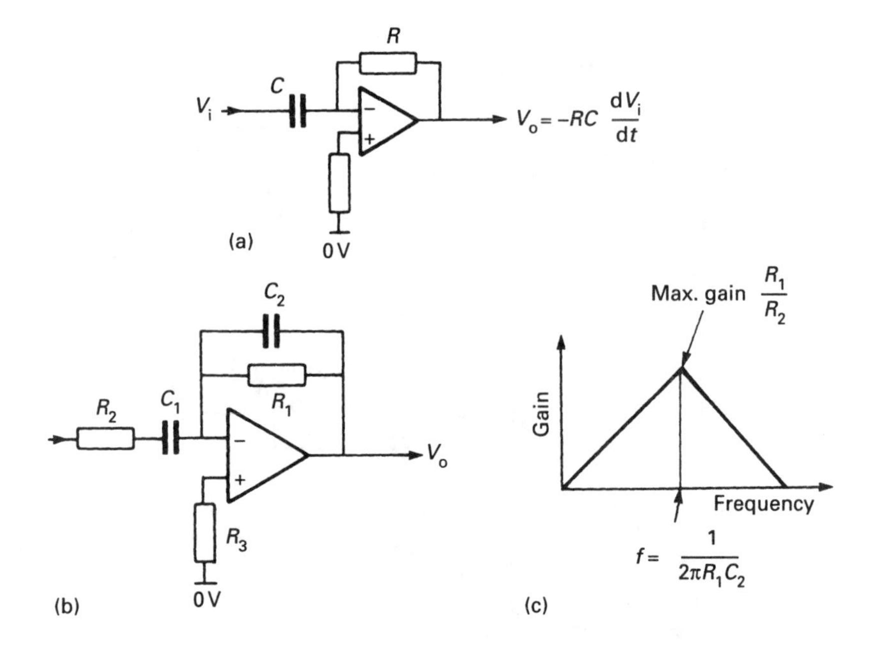

Fig. 11.16 *Differentiator circuit. (a) Theoretical circuit. (b) Practical circuit. (c) Frequency response.*

It should be noted that the increasing gain with frequency characteristic makes differentiation prone to noise pickup. In designing a circuit based on Fig. 11.16b, the frequency at which the gain peaks should be set as low as the application allows.

11.4.7. Analog computers and differential equations

Using the circuits in Sections 11.4.1 to 11.4.6, DC amplifiers can be used to perform complex arithmetical operations and solve simultaneous and differential equations. An analog computer is a collection of DC amplifier circuits which can be interconnected by jumper leads to perform a particular calculation. They are particularly useful for real time simulation (for testing the instrumentation and control strategy for a plant, for example).

Suppose we wish to simulate the equation:

$$5 = \frac{d^2x}{dt^2} + 3\frac{dx}{dt} + 2x \tag{11.27}$$

Figure 11.17a shows that given d^2x/dt^2 we can obtain $-dx/dt$ and x by two integrators. Equation 11.27 can be rearranged:

$$\frac{d^2x}{dt^2} = 5 - 3\frac{dx}{dt} - 2x \tag{11.28}$$

The terms on the right-hand side of the equation are obtained from the two integrators by suitable scaling amplifiers and added to form d^2x/dt^2 as shown in

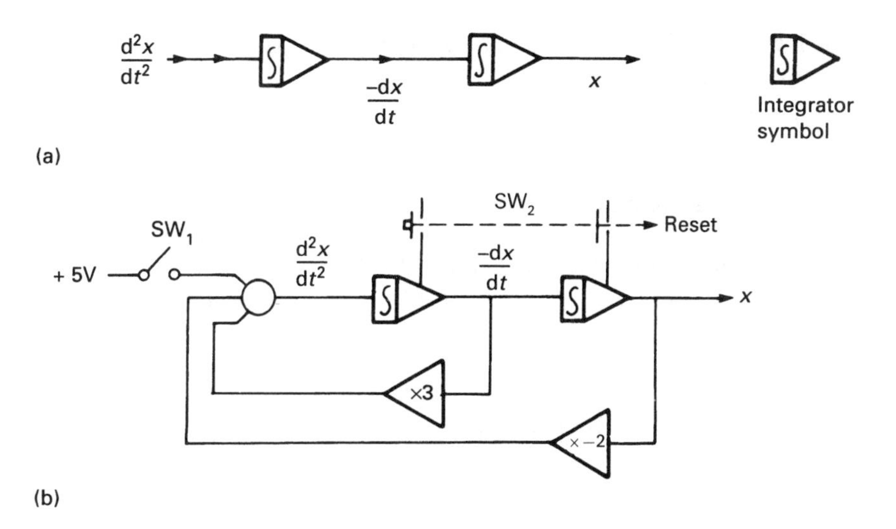

Fig. 11.17 *Analog computing circuits. (a) Basic integrator action. (b) Circuit for solution of a different equation.*

Fig. 11.17b. Switch SW_1 allows the step response of the circuit to be investigated by switching the constant term in and out. Switch SW_2 is used to short out the integrating capacitors to prevent drift when the circuit is not in use.

Circuits similar to Fig. 11.17 can be used to simulate most differential equations, and non-linear effects which are difficult to analyse mathematically (e.g. saturation) can be included by means of the circuits of Section 11.6. Where a process has long time constants (e.g. blast furnaces), the equations can be rearranged to give a scaled time (e.g. 1 second represents 1 minute).

11.5. Filter circuits

11.5.1. Low pass filters

The circuit of Fig. 11.18 simulates a simple exponential rise, and as such is a

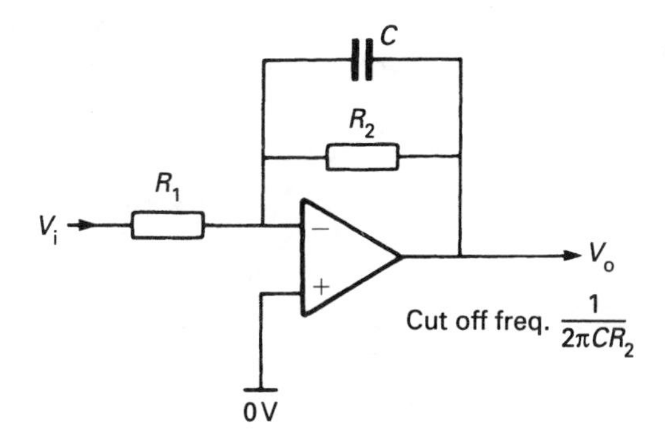

Fig. 11.18 *Simple low pass filter.*

simple low pass filter. The DC gain is, obviously, $-R_2/R_1$ and the 3 dB cut-off frequency is given by:

$$f = \frac{1}{2\pi C R_2} \tag{11.29}$$

The circuit has a roll off of 6 dB/octave above this frequency.

The classical low pass filter is shown in Fig. 11.19a. The circuit has unity gain, and a roll off of 12 dB/octave, and should be designed with $R_1 = R_2 = R$ and $C_2 = 0.5C_1 = C$ (i.e. $C_1 = 2C$). The cut-off frequency is then given by:

$$f = \frac{\sqrt{2}}{4\pi RC} \tag{11.30}$$

If C_1 is not $2C_2$, the cut-off frequency is given by:

$$f = \frac{1}{2\pi R\sqrt{C_1 . C_2}} \tag{11.31}$$

and the damping of the circuit depends on the ratio C_1/C_2 as shown in Fig. 11.19b.

11.5.2. High pass filters

A perfect high pass filter is not feasible because the frequency response of the amplifier itself will ultimately cause high frequency attenuation. The amplifier frequency response is therefore a critical part of the design.

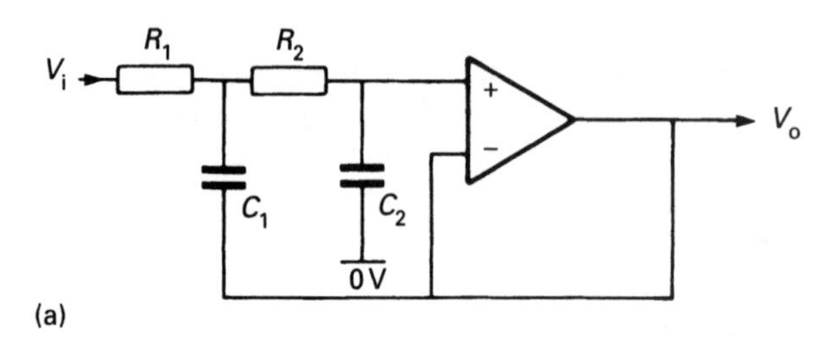

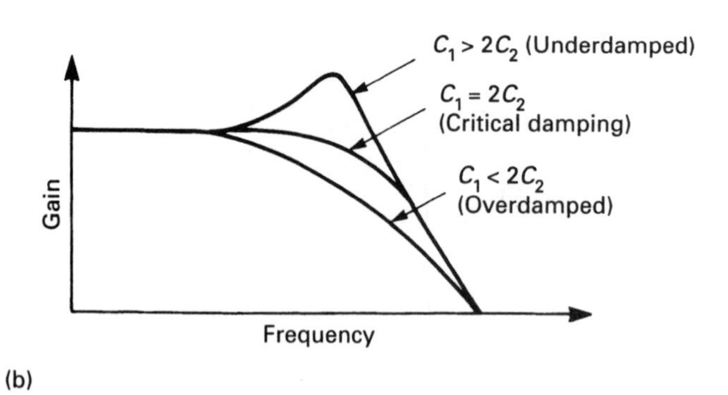

Fig. 11.19 *Two stage low pass filter. (a) Classical low pass filter. (b) Frequency response.*

The simplest high pass filter is shown in Fig. 11.20. This has a high frequency gain of R_2/R_1 (below the point at which the fall off in amplifier gain becomes significant). The cut off frequency is given by:

$$f = \frac{1}{2\pi RC_1} \tag{11.32}$$

An alternative high pass filter is shown in Fig. 11.21a. If the components are chosen such that $R_1 = 0.5R_2 = R$ and $C_1 = C_2 = C$, the cut-off frequency is given by:

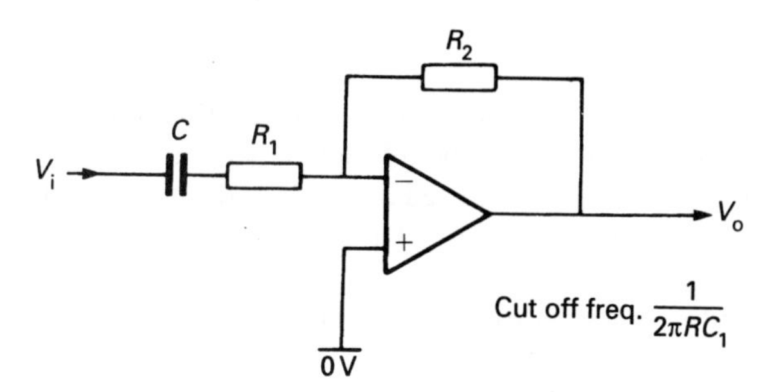

Fig. 11.20 *Simple high pass filter.*

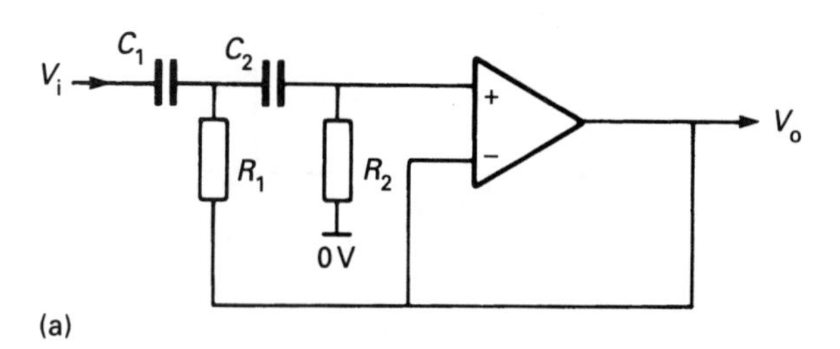

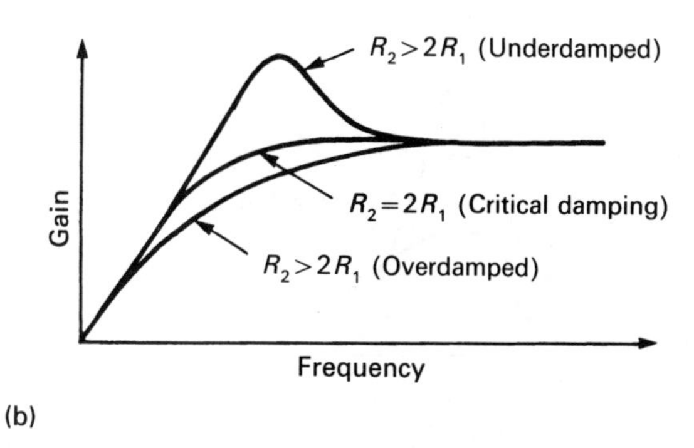

Fig. 11.21 *Two-stage high pass filter. (a) Classical high pass filter. (b) Frequency response. Gain will fall again at high frequencies because of amplifier limitations.*

$$f = \frac{\sqrt{2}}{4\pi RC} \tag{11.33}$$

The response of the circuit, like that of Fig. 11.19, can be adjusted by the ratio of R_1 to R_2. The cut-off frequency is then given by:

$$f = \frac{1}{2\pi C\sqrt{R_1 . R_2}} \tag{11.34}$$

The circuit has unity gain at high frequencies.

11.5.3. Bandpass filters

A bandpass filter passes a specific band of frequencies, and is specified by the centre frequency and the width of the frequency band at the −3 dB point. The latter characteristic is usually expressed as the Q of the circuit, being the ratio of the centre frequency to the bandwidth. The higher the value of Q, the sharper the peak.

A simple circuit is shown in Fig. 11.22. The components should be chosen so

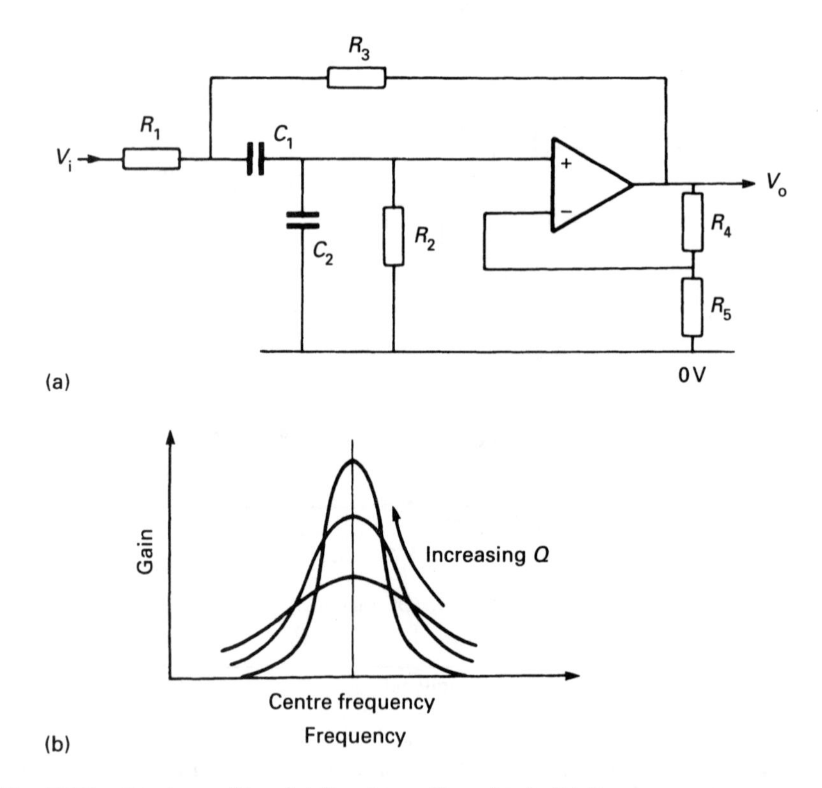

Fig. 11.22 *Bandpass filter. (a) Bandpass filter circuit. (b) Frequency response.*

$R_1 = R_2 = R_3 = R$ and $C_1 = C_2 = C$. The centre frequency is then given by:

$$f = \frac{\sqrt{2}}{2\pi RC} \tag{11.35}$$

and the Q of the circuit by:

$$Q = \frac{R_5\sqrt{2}}{4R_5 - R_4} \tag{11.36}$$

The circuit of Fig. 11.22a is useful where a small bandwidth is required; decreasing Q gives a poorly defined cut-off for a larger bandwidth, as shown in Fig. 11.22b. Where a wide bandwidth with sharply defined cut-off is required, a high pass and low pass filter can be combined as in Fig. 11.23a, which is effectively Figs. 11.19 and 11.21 combined. The upper and lower cut-off frequencies can then be determined separately as in Fig. 11.23b.

11.5.4. Notch filters

The notch filter is used to reject a band of frequencies, usually to suppress 50 Hz (or 60 Hz) mains noise induced onto signal lines. A narrow notch is usually required, which implies a high Q.

The circuit of Fig. 11.24 is commonly used. The values should be chosen such that:

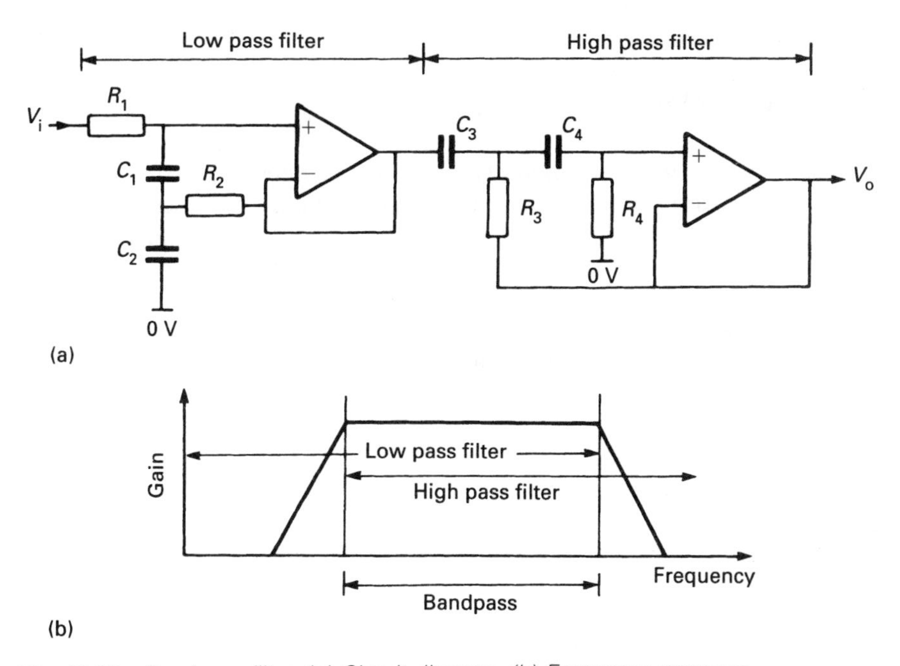

Fig. 11.23 *Bandpass filter. (a) Circuit diagram. (b) Frequency response.*

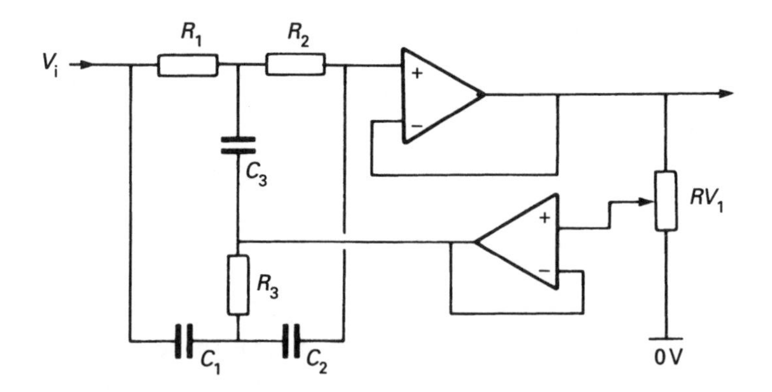

Fig. 11.24 *Notch filter circuit.*

$$R_1 = R_2 = R$$
$$R_3 = 0.5R$$
$$C_1 = C_2 = C$$
$$C_3 = 2C$$

The centre frequency is then given by:

$$f = \frac{1}{2\pi RC} \tag{11.37}$$

The value of Q is determined by RV_1, the maximum value being obtained with the wiper connected to the output.

11.6. Miscellaneous circuits

11.6.1. Voltage-to-current and current-to-voltage conversion

Signals in instrumentation are often transmitted as an electrical current (4–20 mA, for example) because of the high common mode rejection of a current loop. This requires circuits to give voltage-to-current and current-to-voltage conversion.

Voltage-to-current conversion is shown in Fig. 11.25a. The current loop goes through the load, then through the current setting resistor R. The voltage at both amplifier inputs must be equal, so:

$$IR = V_{in} \tag{11.38}$$

or

$$I = \frac{V_{in}}{R} \tag{11.39}$$

which is independent of load provided the amplifier output does not saturate.

Current-to-voltage conversion is achieved by connecting a differential amplifier

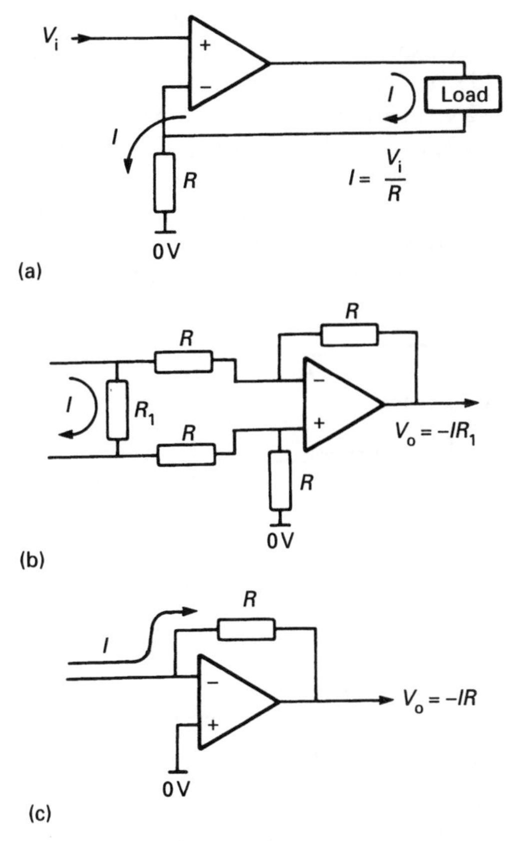

Fig. 11.25 *Voltage–current–voltage conversion. (a) Voltage-to-current conversion. (b) Floating current to voltage conversion. (c) Non-floating current-voltage conversion.*

across the load resistor. The output voltage is then given by:

$$V_o = -IR_1 \tag{11.40}$$

if the amplifier is set for unity gain (all resistors equal). The value of the load resistor should be small compared with the amplifier resistors R.

If the current is not from a current loop, with the return route being via the 0 V or supply rails, the circuit of Fig. 11.25c may be used. The inverting input acts as a virtual earth, and

$$V_o = -IR \tag{11.41}$$

A common application of this circuit is in photometry where the current from a photodiode is to be converted to a voltage.

11.6.2. Ramp circuit

Motor drive circuits require acceleration and deceleration to be limited to prevent mechanical damage to couplings and gearboxes. This can be achieved by the low pass filter of Fig. 11.18, but the exponential rise causes the drive to take an excessive time to reach speed.

The circuit of Fig. 11.26a gives a constant ramp change as shown in Fig. 11.26b. IC1 acts as a comparator (it has no feedback, so the full amplifier gain is used) and compares V_i with V_o. At balance $V_o = -V_i$. Any deviation will cause the output of IC1 to saturate positive or negative.

IC2 is an integrator, whose output will ramp negative at a constant rate when IC1 is saturated positive, and ramp positive when IC1 is saturated negative. The circuit waveforms are shown in Fig. 11.26b. When $V_o = -V_i$ the output of IC1 is

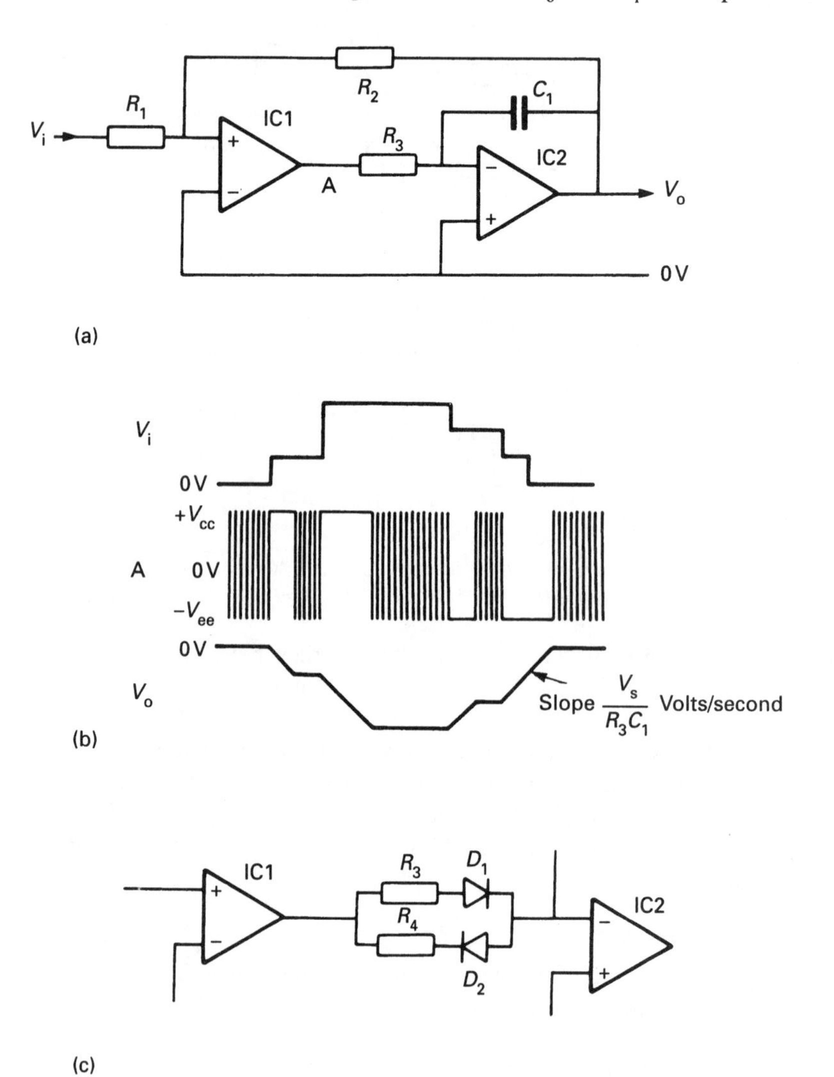

Fig. 11.26 *Ramp circuit. (a) Circuit diagram. (b) Typical waveforms. (c) Modification to give different positive and negative ramp slopes.*

theoretically zero, but in practice small random variations will cause it to dither as shown.

The gain of the circuit is given by:

$$A = -\frac{R_2}{R_1} \tag{11.42}$$

and the ramp slope by:

$$\frac{dV}{dt} = -\frac{V_s}{R_3 C_1} \tag{11.43}$$

where V_s is the saturation voltage of IC1.

If different acceleration and deceleration rates are required, R_3 can be replaced by the two-diode circuit of Fig. 11.26c.

11.6.3. Inverting/non-inverting amplifier

The circuit of Fig. 11.27 has either unity positive gain or unity negative gain dependent on the state of SW_1. With SW_1 closed, the circuit acts as a conventional inverting amplifier with unity negative gain. With SW_1 open the circuit becomes a unity gain buffer, with $V_o = V_i$.

SW_1 can be a mechanical contact (e.g. a relay or a switch) or a CMOS analog switch (e.g. the 4016). If V_i is only positive, an NPN transistor may be used in place of SW_1.

11.6.4. Phase advance

The circuit of Fig. 11.28a delays the feedback signal by the inclusion of capacitor C. This causes the amplifier output to saturate on change of input signal as shown in Fig. 11.28b. The circuit is commonly used with hydraulic proportional valves to give an initial kick to overcome stiction.

11.6.5. Peak picker

A peak picker circuit is used to hold the maximum value of a variable for subsequent measurement. The circuit, and its action, is shown in Fig. 11.29. IC1

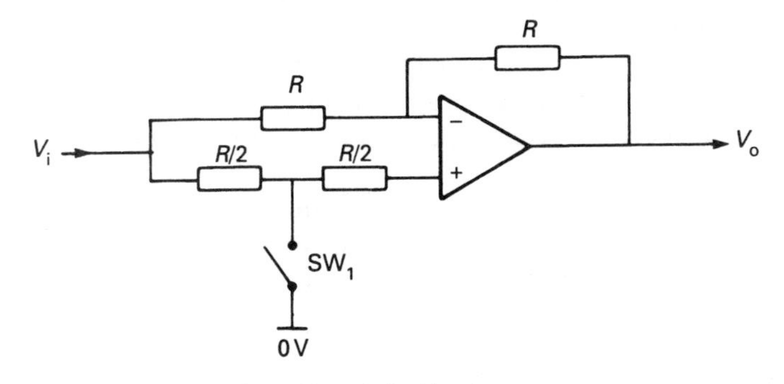

Fig. 11.27 *Unity gain amplifier with switchable sign.*

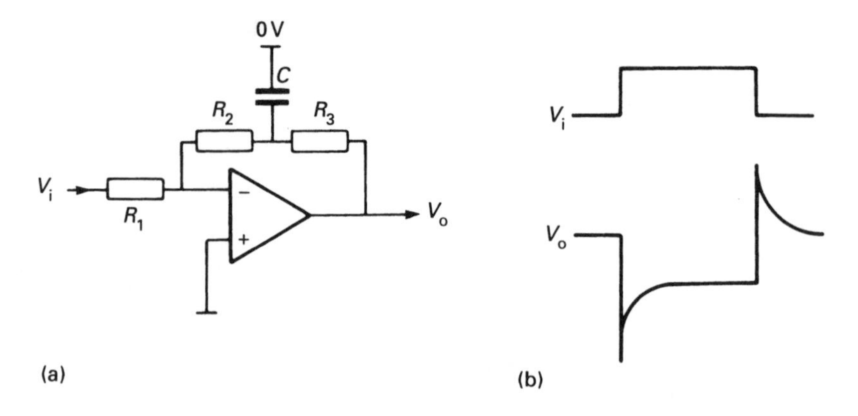

Fig. 11.28 *Phase advance circuit. (a) Circuit diagram. (b) Step response.*

acts as a comparator; if V_o is less than V_i, IC1 output goes positive charging the capacitor via the diode until V_o equals V_i. IC2 simply acts as a buffer to prevent the output load discharging C. Ideally IC2 should be a FET amplifier.

The output voltage therefore follows the maximum values of V_i as shown. Resistor R is included to give a controlled discharge of C if required. If the diode is reversed the circuit follows the minimum value of V_i.

11.6.6. Sample and hold

If the diode in Fig. 11.29 is replaced by a 'switch' SW_1 as Fig. 11.30a, the circuit can be used to 'freeze' a varying signal for measurement. With SW_1 closed, V_o will follow V_i. With SW_1 open, V_o will hold the value of V_i at the instant that the switch was opened as shown in Fig. 11.30b.

Sample and hold circuits are an essential part of analog-to-digital conversion, a topic covered in Section 13.9.2.

11.6.7. Schmitt triggers

Voltage comparison is often required in control systems. This is usually provided by a circuit with built in 'backlash', or 'hysteresis' as it is more commonly called. This is summarised in Figs. 11.31a and b. The circuit is characterised by two voltages, the upper and lower trigger points. The output changes when the input goes above the UTP and below the LTP. The differential between the upper and lower trigger points gives protection against output jitter which can occur when a simple single point comparison circuit is used and the input voltage is near the switching point.

A comparison circuit with hysteresis is usually called a Schmitt trigger, and is shown in its simplest form in Fig. 11.31. Suppose V_{in} is above the UTP; the output will be saturated negative, and the voltage at the junction of R_1 and R_2 will be:

$$V_1 = -V_{sat} \cdot \frac{R_2}{R_1 + R_2} \tag{11.44}$$

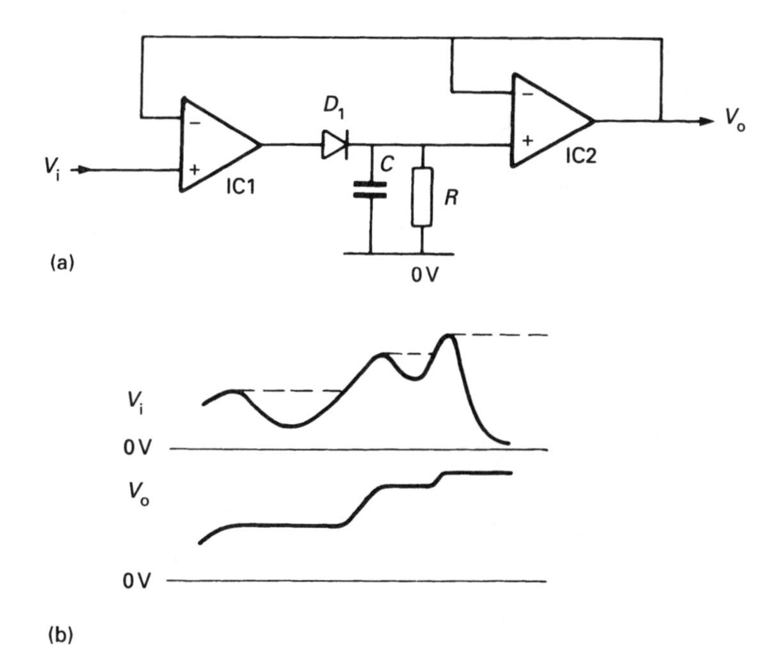

Fig. 11.29 *Peak picker circuit. (a) Circuit diagram. (b) Typical circuit operation.*

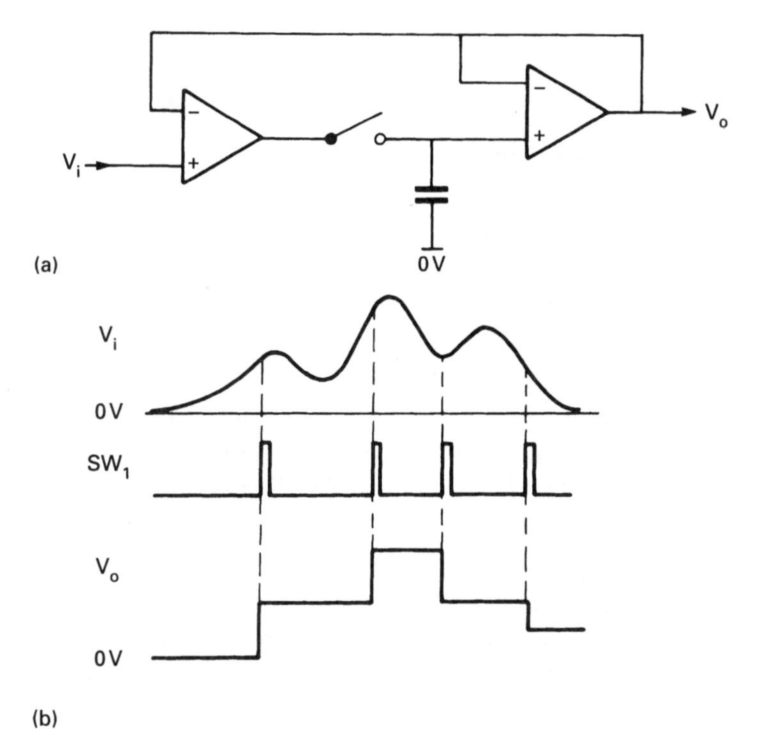

Fig. 11.30 *Sample and hold circuit. (a) Circuit diagram. (b) Typical circuit operation.*

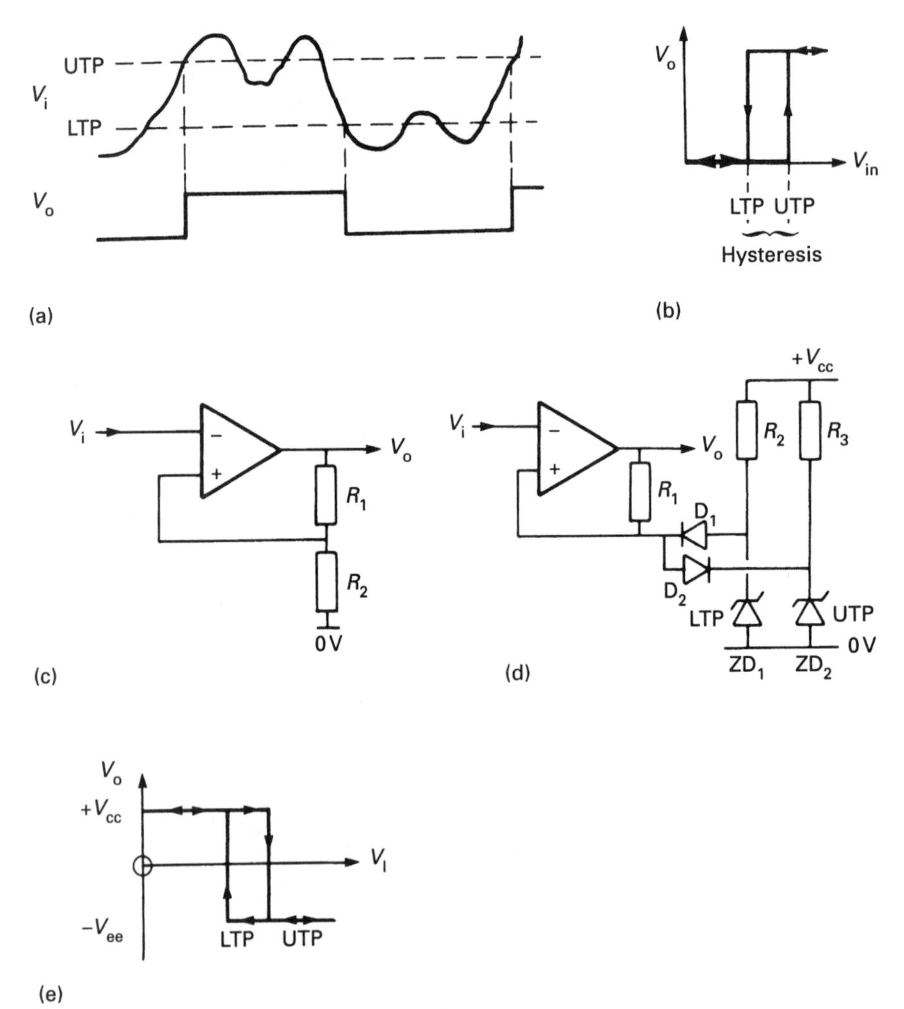

Fig. 11.31 *The Schmitt trigger. (a) Schmitt trigger operation. (b) Circuit hysteresis. (c) Simple Schmitt trigger with UTP and LTP symmetrical about zero. (d) Schmitt trigger with UTP and LTP having same sign. (e) Operation of circuit d.*

This is the lower trigger point LTP because V_{in} will need to go more negative than V_1 before the circuit will switch. The output will now go positive, and the voltage at the junction of R_1 and R_2 will be:

$$V_2 = +V_{sat} \cdot \frac{R_2}{R_1 + R_2} \tag{11.45}$$

This is the upper trigger point UTP because the input must go more positive than V_2 before the circuit will switch. The hysteresis of the circuit is therefore the difference between V_1 and V_2 (remembering that V_1 is negative).

The circuit of Fig. 11.31c has a symmetrical response, with positive UTP and negative LTP. The circuit of Fig. 11.31d has positive UTP and LTP determined by zener diodes. R_1 should be an order of magnitude larger than R_2 and R_3 to avoid the output switching significantly affecting the zener voltages. The circuit has the response of Fig. 11.31e.

11.6.8. Limiting circuits

A limiting circuit is a linear amplifier provided the input signal stays within predefined levels. If these levels are exceeded, the output signal stays constant as shown in Fig. 11.32a. A typical example is error limiting in speed control circuits to prevent excessive motor currents.

Figure 11.32b is a non-inverting amplifier, with input limiting by D_1 and D_2. RV_1 sets the upper limit and RV_2 the lower limit. The value of R_1 should be at least an order of magnitude greater than RV_1 and RV_2.

The inverting circuit is shown in Fig. 11.32c. If R_1 and R_2 are equal, the limiting input voltages will be twice the voltage set on RV_1 and RV_2. Again, the

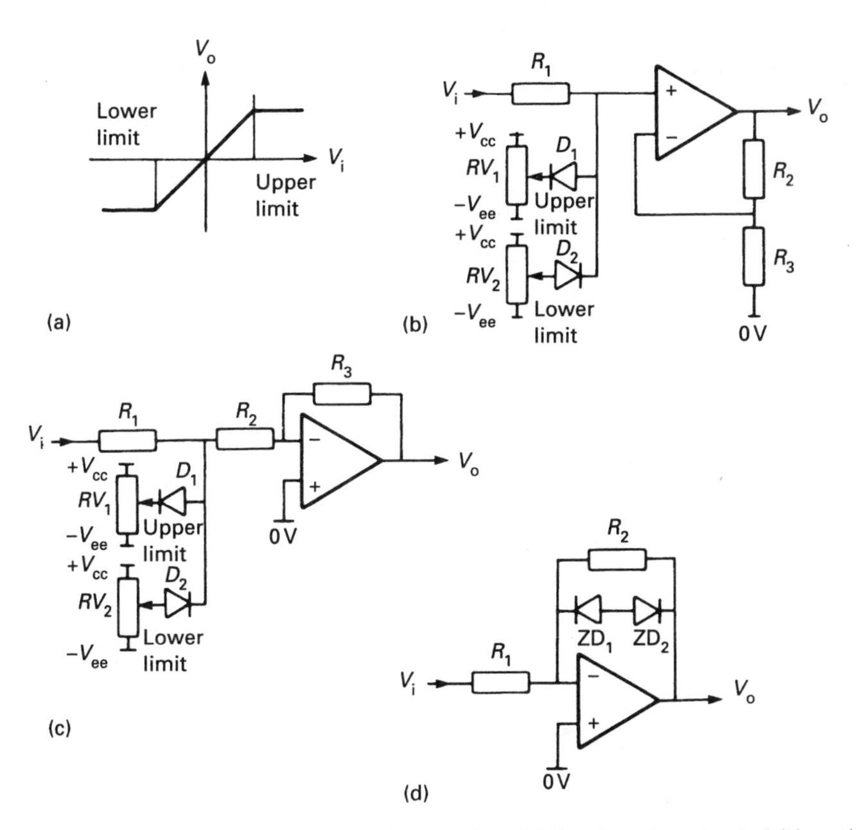

Fig. 11.32 *Limiting circuits. (a) Circuit operation. (b) Non inverting circuit. (c) Inverting circuit. (d) Inverting circuit with output limits set by zener diodes.*

value of R_1 should be at least an order of magnitude greater than RV_1 and RV_2.

Shunt limiting can be provided by zener diodes as shown in Fig. 11.32d. ZD_2 directly sets the output limiting positive voltage, and ZD_1 the negative voltage.

Figures 11.32b and c should be used for variable gain amplifiers when the limiting is to be performed on the *input* signal level and Fig. 11.32d where the limiting is to be performed on the *output* signal level.

11.6.9. Deadband amplifier

A deadband amplifier has the response of Fig. 11.33a, and gives a region of zero sensitivity. The circuit is often used on the error signal in position control systems to prevent 'dither' about the home positions.

With V_{in} at zero, V_1 and V_2 can be set by RV_1 and RV_2. The input voltage has to rise above V_1 or fall below V_2 before the output starts to change. RV_1 therefore sets the positive deadband and RV_2 the negative deadband. Note that R_1 and R_2 allow different slopes to be set for the positive and negative sections.

11.6.10. Linearisation circuits

Many instrumentation circuits require linearisation of a signal; common examples are the square root function needed for many flow transducers and the

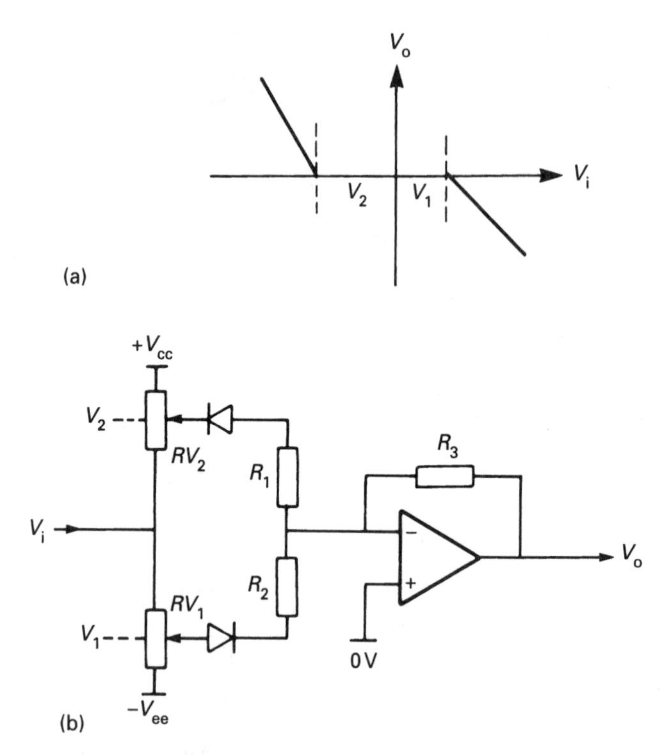

Fig. 11.33 *Deadband amplifier. (a) Circuit operation. (b) Circuit diagram.*

multiterm series needed for thermocouples and PTRs. This linearisation can be performed by operational amplifiers.

The generalised circuit is shown in Fig. 11.34a, where a non-linear element is placed on the feedback of an op amp. The output voltage is related to the input by:

$$V_o = G\left(\frac{V_{in}}{R}\right) \tag{11.46}$$

where $G(V_{in}/R)$ is the inverse function of $F(V_o)$.

In Fig 11.34b, for example, a diode is used in the feedback, which has an exponential current/voltage relationship, i.e.

$$F(V_o) = A\exp(BV_o) \tag{11.47}$$

where A and B are constants.

The inverse function is a logarithm, so:

$$V_o = C\log(DV_i) + E \tag{11.48}$$

where C, D and E are constants. This circuit is the basis of a DC amplifier based multiplier or divider (see Section 11.4.4).

An alternative approach is to approximate the required curves by straight lines. The curve in Fig. 11.35a, for example, is approximated by three straight lines. This can be achieved by the circuit of Fig. 11.35b, with RV_{1-3} to 3 setting the break points and RV_{4-6} the slopes of the individual sections. Practically any function can be achieved by combining amplifiers with various $Ax + B$ characteristics, deadband amplifiers and limiters and summing the outputs.

11.6.11. P + I + D controllers

The classic three-term control algorithm uses three components: error, time integral of error, and time derivative of error. This can be achieved, in a rather sledgehammer way, by using four amplifiers as in Fig. 11.36a, with the integral term being provided by a circuit similar to Fig. 11.15 and the derivative term by one similar to Fig. 11.16.

More elegant circuits are shown in Figs. 11.36b and c. In the latter circuit the proportional gain is determined by R_3/R_1, the integral action mainly by R_1C_2

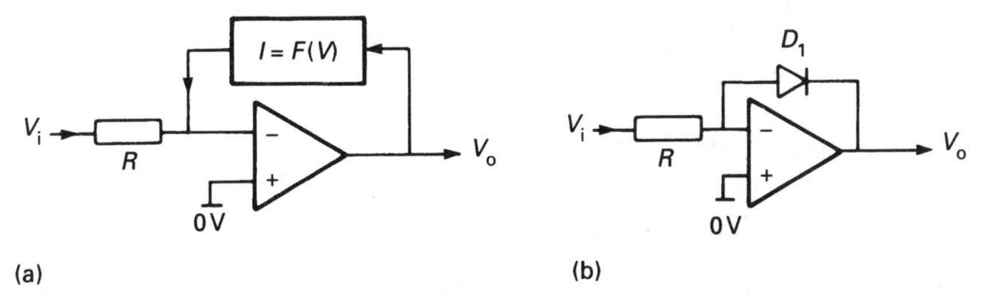

Fig. 11.34 *Linearisation circuit. (a) General principle. (b) Logarithmic circuit.*

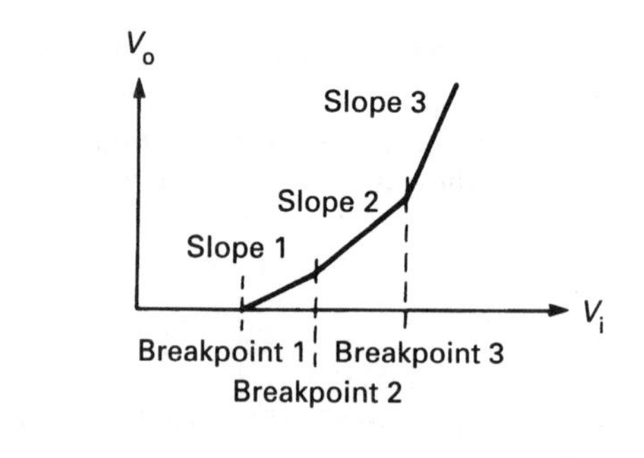

(a)

(b)

Fig. 11.35 *Linearisation by straight line approximation. (a) Circuit operation. (b) Practical circuit.*

and the derivative action by R_2, R_3, C_1. If the resistances are made adjustable, it follows that there will be considerable interaction between the terms. Figures 11.36a and b allow relatively independent adjustments.

11.6.12. Precision rectifiers

Instrumentation signals are often obtained as an AC voltage (LVDT position transducers, for example). These signals must be rectified to give a linear DC signal. Voltage drops across the diodes in a bridge rectifier introduce errors, and make the circuit unusable with voltage below a few volts. Precision error-free rectifiers can be constructed using op amps.

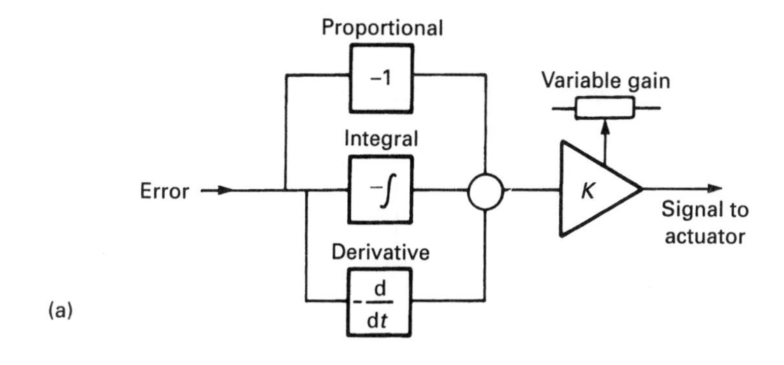

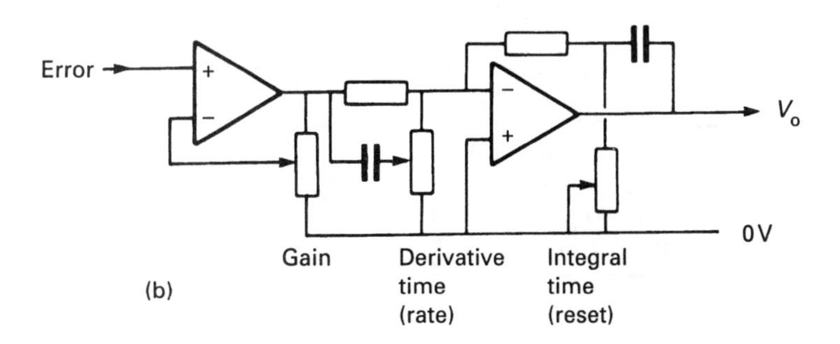

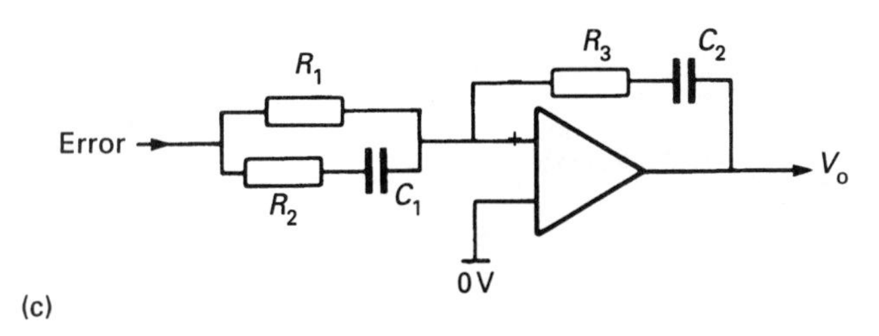

Fig. 11.36 *Three term controller circuits. (a) Black diagram of PID controller. (b) Two amplifier PID controller. (c) Simple interacting PID controller.*

A half wave circuit is shown in Fig. 11.37a. On the negative half cycle, $V_{out} = -V_{in}$ because R_1 and R_2 are equal (i.e. V_{out} is a positive half cycle). Note that the amplifier output, V_x, will be equal to V_{out} plus the voltage drop of D_1. On the positive half cycle D_2 will conduct, limiting V_x to -0.8 V. The junction of R_1 and R_2 will still be a virtual earth, so provided negligible current is drawn through R_2, V_{out} will be zero. If D_1 and D_2 are reversed, a negative half wave rectifier can be constructed.

A full wave rectifier is shown in Fig. 11.37b. IC1 is a negative half wave rectifier with output V_a. IC2 sums V_{in} and $2V_a$ (because of the choice of resistor values). When V_{in} is positive, $V_a = -V_{in}$:

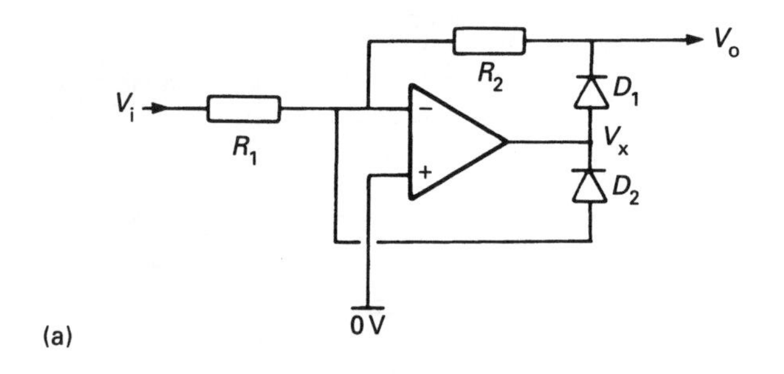

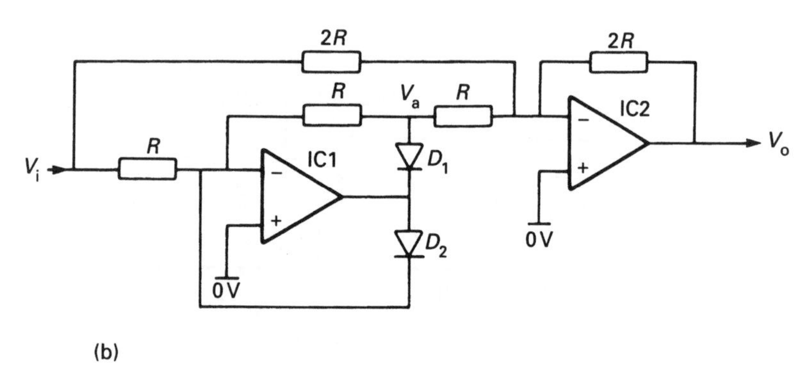

Fig. 11.37 *Precision rectifier circuits. (a) Half wave rectifier. (b) Full wave rectifier.*

$$V_o = -(V_{in} - 2V_{in})$$
$$= V_{in} \text{ (a positive signal)} \qquad (11.49)$$

When V_{in} is negative, V_a is zero, so:

$$V_o = -(V_{in} + 2 \times 0)$$
$$= -V_{in} \text{ (a positive signal again)} \qquad (11.50)$$

Whatever the polarity of V_{in}, the output signal is positive but equal in magnitude to V_{in}.

The circuits will work at fairly high frequencies (several tens of kHz). The limiting factor is the slew rate of the half wave rectifier which has to slew its output from +0.8 V (one forward diode drop) to −0.8 V as the input signal goes through zero.

11.6.13. Isolation amplifiers

Instrumentation signals are often low level and consequently prone to noise. Instrumentation amplifiers therefore need very high common mode rejection. The signals will also probably originate at some distance from the control cubicle, and will probably share cable ducts with high voltage signals. Under fault conditions it is possible for inter-cable shorts to occur, causing high voltages to appear on instrumentation lines. If no precautions are taken, the high voltage will go through all the instrumentation and destroy an entire control system. Both these problems are overcome by the use of isolation amplifiers, shown diagrammatically in Fig. 11.38.

In Fig. 11.38a, the input signal is chopped into AC by a CMOS switch. The resulting AC signal is amplified and passed to the primary of a transformer. The chopper circuit and the amplifier are powered by their own floating power supply derived from a DC to DC inverter. The AC signal on the transformer secondary is rectified and smoothed (see Section 11.6.12) to give the original signal. The use of transformer coupling and the floating power supply gives almost perfect common mode noise rejection and protection against high voltages on the input. An inter-cable fault will still probably damage the transducer and the input stage of the isolation amplifier, but will not spread into the rest of the system.

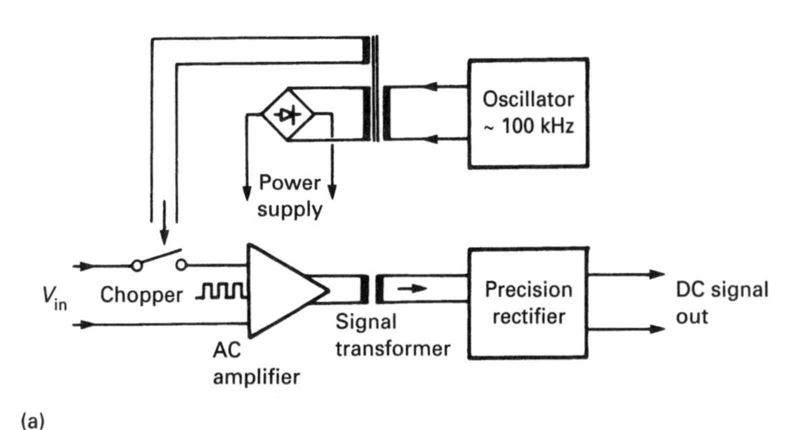

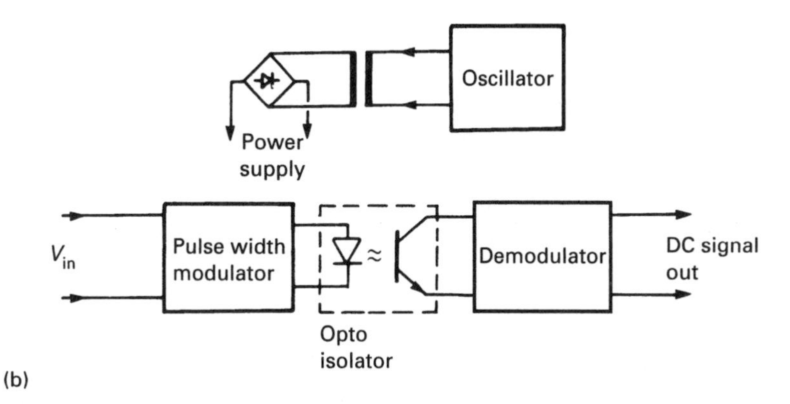

Fig. 11.38 *Isolation amplifiers. (a) Chopper amplifier transformer coupled. (b) PWM amplifier with opto isolation.*

An alternative approach, shown in Fig. 11.38b, uses optical isolation. The input signal is converted to a pulse width modulated signal which drives the light emitting diode in an opto isolator. The input circuit and pulse width modulator are again driven from a floating supply derived from a DC to DC inverter. The pulse width modulated signal from the opto isolator is restored to DC form for use by the rest of the system.

Isolation amplifiers are usually purchased in encapsulated form, and are available with any required gain. A typical unit will give isolation with input voltages up to 1 kV.

Chapter 12

Rotating machines and power electronics

12.1. Introduction

The electric motor is the commonest industrial prime mover, and is often used as the final control in a process control scheme. This chapter describes various types of electric motor. Many applications require the speed of the motor to be accurately controlled. Thyristors and their uses in speed control are also described.

The hydraulic motor, which is an alternative form of rotating machine, is covered in Chapter 15.

12.2. Basic principles

Electric motors and generators are closely related devices, and have a common background theory. Faraday found experimentally that an electrical potential is produced when a conductor is moved relative to a magnetic field. (It does not matter whether the conductor moves relative to a fixed field, or the field moves with respect to a fixed conductor; the resulting potential is the same.)

In Fig. 12.1a, a conductor of length L metres is moving at a uniform velocity v metres per second in a uniform magnetic field B webers per square metre. Faraday's experimental work showed that the induced voltage e is given by:

$$e = \text{flux linked per unit time} \tag{12.1}$$

$$= BLv \text{ volts} \tag{12.2}$$

Equation 12.2 is the basic equation for an electrical generator.

By conservation of energy, work must be done to generate the electrical

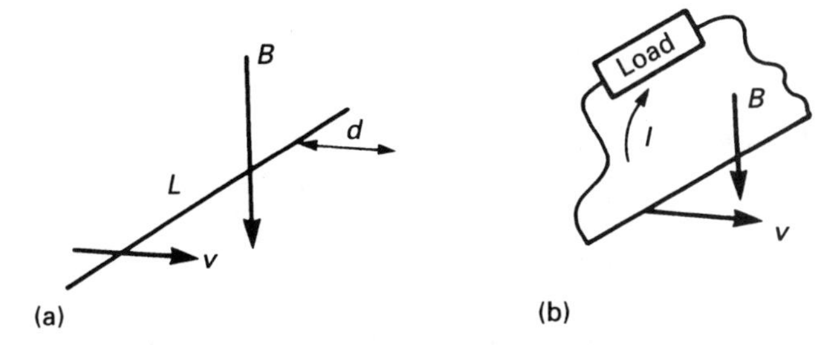

Fig. 12.1 *Principle of electrical motor and generator. (a) Conductor moving in a magnetic field. (b) Conductor connected to a load.*

potential. In Fig. 12.1b, a conductor moving in a magnetic field is connected to an electrical load. When the conductor moves, a current I will flow.

$$\text{power} = eI \text{ watts} \tag{12.3}$$

If the conductor moves at uniform speed for time T secs:

$$\begin{aligned}\text{work done} &= \text{power} \times \text{time}\\ &= eIT\\ &= BLvIT \text{ joules}\end{aligned} \tag{12.4}$$

The generation of electrical energy produces an opposing force, F on the conductor. Mechanical work is done moving the conductor against this force. This work is given by:

$$\begin{aligned}\text{work done} &= \text{force} \times \text{distance}\\ &= FvT\end{aligned} \tag{12.5}$$

Combining equations 12.4 and 12.5 gives:

$$FvT = BLvIT$$

or

$$F = BLI \text{ newtons} \tag{12.6}$$

Equation 12.6 gives the force on a conductor in an electrical generator. It also follows that if a current is driven through a conductor in a magnetic field by an external power source, the conductor will experience a force. This force is also given by equation 12.6, and is the basis of electric motors.

If the conductor is not restrained, the force will cause it to accelerate. Once it starts to move, an emf will be generated. This will oppose the applied voltage (and hence is called the back emf), causing the current to fall. Eventually the conductor will reach a steady speed where the force given by equation 12.6 balances mechanical forces such as friction.

If the mechanical load decreases, the conductor will accelerate because the electrical force no longer balances the reduced mechanical force. The increased speed raises the back emf, reducing the current, and the conductor settles at a higher speed where the electrical force again balances the mechanical force.

12.3. Simple generators

12.3.1. Slip ring generator

In Fig. 12.2a a single turn coil is rotated in a fixed magnetic field. An electrical potential is generated in the coil by equation 12.2, with the instantaneous velocity relative to the field being a sinusoid. The generated voltage is therefore also sinusoidal, as in 12.2b. The frequency is exactly the same as the rotational frequency, and the amplitude depends on the size of coil and the magnetic field strength. Note that speeding up the rotational speed will cause the output voltage frequency and amplitude to increase.

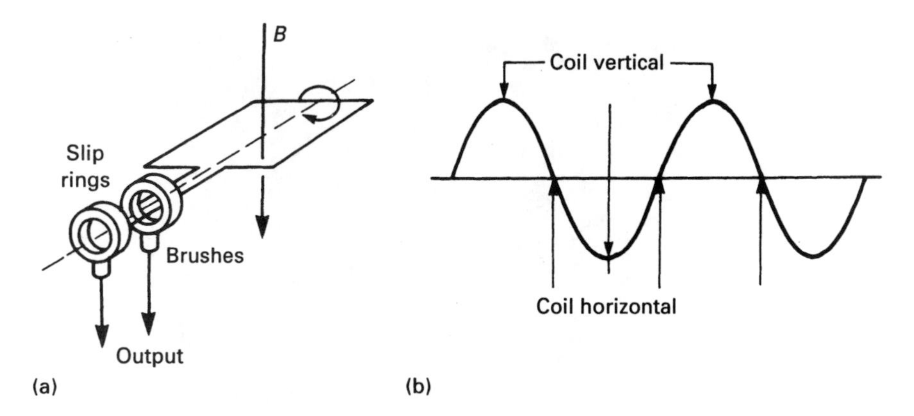

Fig. 12.2 *Slip ring AC generator. (a) Mechanical arrangement. (b) Output voltage.*

The voltage induced in the coil is taken out by stationary carbon brushes rubbing on slip rings on the coil. In a practical generator, the coil will consist of many turns to increase the output voltage. The fixed magnetic field will be derived either from permanent magnets (in small machines) or via electro-magnets (in larger machines).

12.3.2. DC generator

Figure 12.3a is similar to the slip ring generator, but the coil is brought out via a split copper cylinder called a commutator. This causes the coil to brush connections to reverse every 180 degrees of rotation.

The voltage generated in the rotating coil is, again, a sinusoid, but the commutator reverses the coil connections as the voltage passes through zero. The output waveform is therefore pulsating DC, as in Fig. 12.3c.

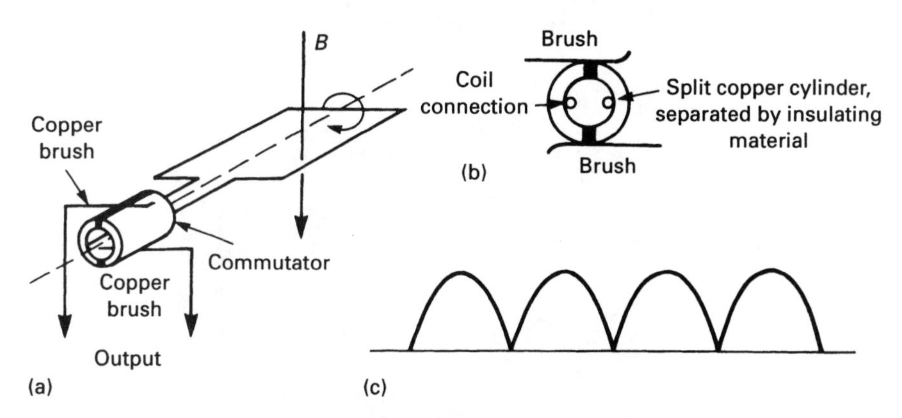

Fig. 12.3 *Commutator DC generator. (a) Mechanical arrangement. (b) Commutator end view. (c) Output voltage. (d) Assembling the rotor on a DC machine (e) A typical commutator and brushgear on a DC machine. (Photos courtesy of ASEA.)*

(d)

(e)

Fig. 12.3 *contd.*

12.4. DC motors

12.4.1. Simple motor

Figure 12.3a can also be used as a simple DC motor. If a voltage is applied to the brush connections with the coil stationary and not horizontal (where the brushes would short the coil), a current I will flow which is determined by the voltage and the coil resistance, by Ohm's law.

By equation 12.6, a force will be produced on the coil which will start it turning. Inertia will take the coil through the commutator changeover point, and it will start to rotate and accelerate. As the coil speeds up, it also starts to act as a generator and a back emf is produced, as described in Section 12.2.

It is common practice to use the symbol V for the applied voltage, and E for the back emf as shown in Fig. 12.4. At any speed these are related by:

$$V = E + IR \tag{12.7}$$

E is directly proportional to motor speed, and for a fixed magnetic field the torque will be proportional to I. As the speed increases, the increased E causes a lower torque. The motor will settle at a speed where the torque exactly balances the load torque.

The motor based on Fig. 12.3a has many shortcomings; it gives a pulsating torque and is non-self-starting if the coil comes to rest in a position where it is shorted by the brushes. The next three subsections describe practical DC motor construction.

12.4.2. Ring wound machines

In Fig. 12.5a, the rotor consists of a laminated core wound with twelve coils. These are connected to twelve commutator sections as shown. The brushes are shown on the inside of the commutator for ease of illustration; in practice, of course, they bear on the outside surface.

A voltage is applied to the brushes, and this causes an equal current I to flow through each half of the coil as shown in Fig. 12.5b. Note that the total current at each brush is $2I$. The current passing through the coils produces a torque on the rotor, causing it to rotate. As it rotates, however, the commutator switches the current to new coil sections to give an almost uniform torque.

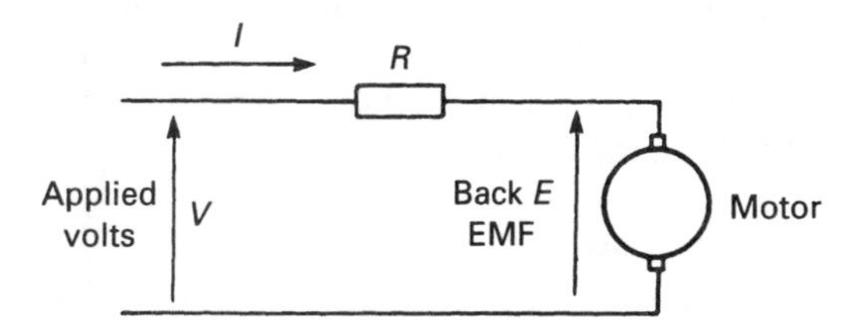

Fig. 12.4 *Basic motor equations.*

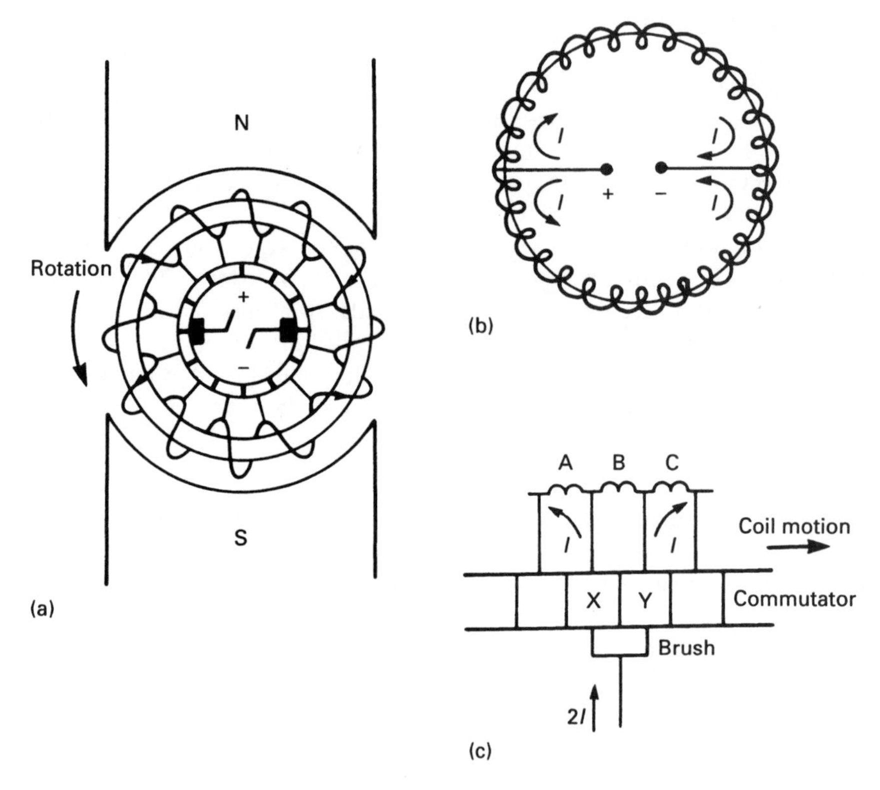

Fig. 12.5 *The ring wound DC motor. (a) Mechanical arrangement. (b) Current flow. (c) Commutator action.*

The action of the commutator is critical, and is more involved than may at first be thought. Consider Fig. 12.5c; the coil is moving to the right. Coils A and C both have current I flowing through them, but the current directions are opposite. As the coil A moves to the right past the brush, its current must first be reduced to zero, then reversed. The coils will inevitably have large inductance, which inhibits fast changes of current. A failure to reduce the current in the coil to zero whilst it is shorted by the brush as in Fig. 12.5c leads to a large induced voltage in the coil, which in turn causes sparking and damage to the brush and commutator. The problem gets worse as speeds and armature current rise, and in large machines interpoles are used (see Section 12.4.5 below).

The ring winding is difficult to construct, and DC machines are more commonly based on the lap or wave winding described below, which lend themselves to more automated assemblies.

12.4.3. Lap wound machines

The lap wound machine is based on prefabricated coils, as in Fig. 12.6a. The ends of the coil are connected to adjacent commutator segments as shown, and the coil laid into slots in a laminated armature as shown in Fig. 12.6b. Coils are overlaid

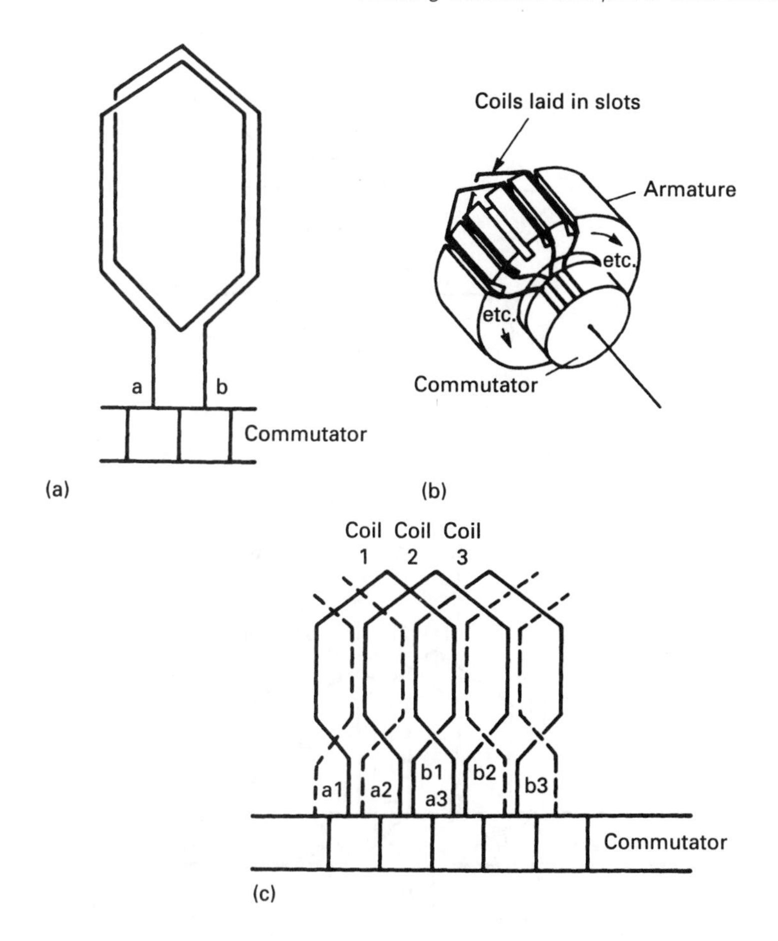

Fig. 12.6 *The lap wound machine. (a) Lap winding. (b) Machine construction. (c) Commutator connections.*

as shown in Fig. 12.6c, with the finish of coil 1 connected (at the commutator) to the start of coil 2, and so on.

To give smooth torque, multiple magnetic poles are used. Figure 12.7a shows a typical four pole arrangement. In a lap wound machine, the number of brushes is equal to the number of poles, and brushes under like poles are parallelled as shown.

Figure 12.7b shows what is happening under two adjacent pole faces for a machine with four poles and twelve coils (and hence twelve commutator segments). Current I_1 enters via brush X, and proceeds via commutator segment a into the start of coil 1. It then flows through coils 2 and 3 (linked at segments b, c) to exit at segment d and brush Y. Because the coil widths and the pole-to-pole distances are equal, the forces on each coil are equal and in the same sense.

Figure 12.7b is not, however, a complete picture. Coil portions 10f, 11f, 12f will also be under the north pole, and an equal current I_2 will also go via brush X, through segment a, coils 12, 11, 10 to brush W. Similarly coil portions 4s, 5s, 6s lie

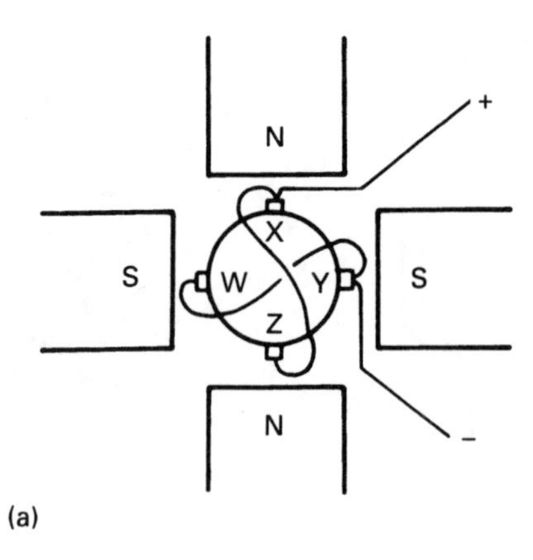

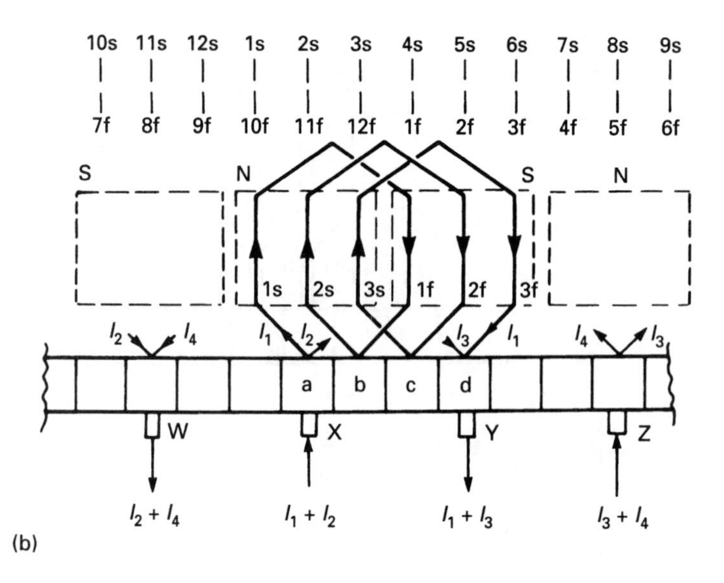

Fig. 12.7 *Coil connections in a lap wound machine. (a) Four pole arrangement. (b) Current flow in a four pole machine.*

under the south pole, and a current I_3 flows from brush Z to brush Y via coils 4, 5, 6. At any instant there are four parallel currents flowing: X to Y, X to W, Z to Y, Z to W.

12.4.4. Wave wound machines

The coil for a wave wound machine is shown in Fig. 12.8a. A wave wound machine has an odd number of coils and commutator segments; Fig. 12.8b has thirteen, for example. Each coil is connected to segments spaced just over 180° apart. In Fig. 12.8b, coil 3 is connected between segments c and j. This

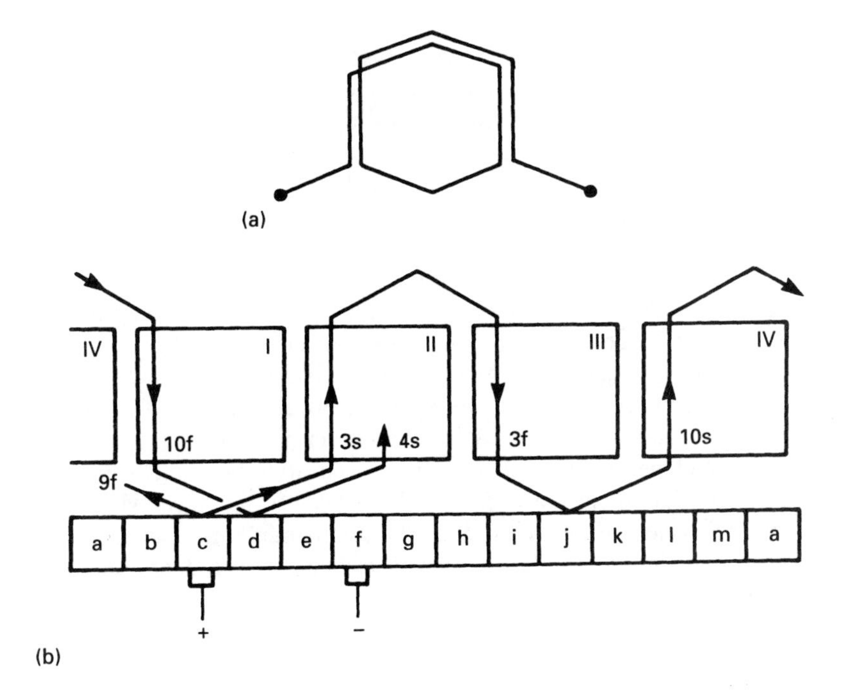

Fig. 12.8 *The wave wound machine. (a) The wave winding. (b) Coil connections.*

arrangement causes the coils to progress in waves and form a continuous series chain of coils. A wave wound machine has two brushes, spaced 90° apart, regardless of the number of poles.

Suppose the positive brush is at segment c; the negative brush is spaced 90° away on segment f. Current enters and passes through coil 3 to segment j, coil 10 to segment d, coil 4 to segment k, coil 11 to segment e, coil 5 to segment 1 and coil 12 to segment f and the negative brush. A parallel path can also be traced via coils 9, 2, 8, 1, 7, 13, 6.

As before, the current direction in front of each pole piece is such that a constant reinforcing torque is produced on the coils, causing the armature to rotate.

12.4.5. Compensating windings and interpoles

As the brushes move from one segment to the next, the current in an armature coil must be reduced to zero and then reversed (see Fig. 12.5c above). This action is called commutation. If the changeover is not completed, excessive sparking occurs at the trailing edges of the brushes. The problem increases as speed or armature current rises.

The problem can be relieved, to some extent, by the use of high resistance brushes which span several commutator segments, thereby increasing the time for commutation. Such techniques are only suitable for small machines, however.

In large machines, another effect becomes apparent. The armature current will itself produce a magnetic field. This has the effect of 'twisting' the main field as shown in Fig. 12.9a. If the brushes were correctly aligned for zero current, they will be misaligned at full load. In old machines a lever for manual brush movement was provided, the operator adjusting the brush position for minimum sparking.

In more modern motors, an additional small pole called an interpole is added before each main pole, as shown in Fig. 12.9b. In motors this opposes the next pole, and is energised by armature current so its field strength depends on load current. The interpoles assist commutation, as their field assists the reversal of current in the coils undergoing commutation. As their field strength increases with current, they automatically compensate for changes in load.

A related problem can arise when rapid load changes occur. With a fast di/dt, transformer action in the armature coils can produce large voltages between adjacent commutator segments. Ultimately a flashover can occur, which will wreck the machine. To prevent this, a compensating winding is added. This consists of conductors in the surface of the pole faces, again carrying armature current. Their field cancels that of the armature conductors. The compensating winding also assists commutation at high currents as it reduces the armature reaction effect of Fig. 12.9a.

12.4.6. Motor equations

The first requirement is to derive equations for power and torque. In Fig. 12.10a a mass M kg is being raised at a constant speed by a rope being wound into a drum.

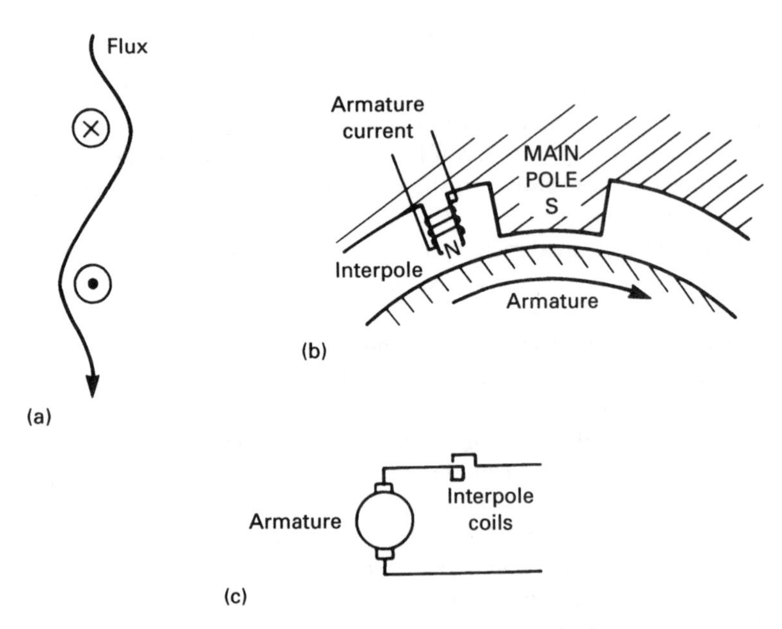

Fig. 12.9 *Interpoles in large machines. (a) Effect of armature current on magnetic field. (b) Interpole arrangement. (c) Interpole connection.*

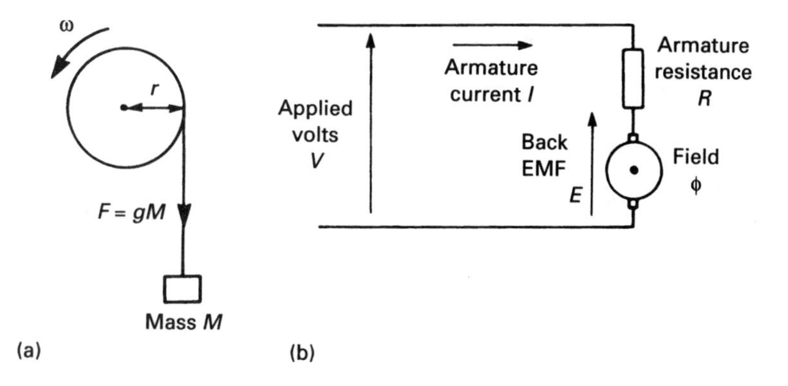

Fig. 12.10 *Fundamental motor principles. (a) Torque, work and power. (b) Motor equations.*

The force being exerted on the mass is given by:

$$F = Mg \text{ newtons} \tag{12.8}$$

The torque at the drum periphery is:

$$T = Fr \text{ metre newtons} \tag{12.9}$$

In one revolution, the mass is raised by $2\pi r$ metres. The work done is given by:

$$w = 2\pi Fr \tag{12.10}$$

$$= 2\pi T \text{ joules (newton metres)} \tag{12.11}$$

Power is work done per second. In one second the drum rotates $\omega/2\pi$ revs, therefore:

$$\text{power} = 2\pi T \times \frac{\omega}{2\pi} \tag{12.12}$$

$$= \omega T \text{ watts} \tag{12.13}$$

The basic motor equation was given earlier in equation 12.7. Figure 12.10b shows the representation of a simple motor. V denotes the applied voltage, E the back emf, I the armature current, R the armature resistance and ϕ the field strength. We have, by simple circuit theory:

$$V = E + IR \tag{12.14}$$

The back emf is given by:

$$E = A\phi\omega \tag{12.15}$$

where A is a constant for the specific machine, and is related to the motor constructional details such as the number of poles, coils, turns on coils, etc.

The useful electrical power is given by:

$$P = EI \text{ watts} \tag{12.16}$$

and by substitution:

$$P = A\phi\omega I \text{ watts} \tag{12.17}$$

The motor torque is given by:

$$T = A\phi I \text{ metre newtons} \tag{12.18}$$

Note that for a fixed ϕ, $T \propto I$ and $P \propto \omega I$.

The power equations above relate useful electrical power. The total power balance is:

$$VI = EI + I^2R \tag{12.19}$$

where VI is the applied power, and I^2R the armature loss which appears as heat. Not all the motor power EI is delivered to the load, however, as frictional losses and losses through windage, etc., must be subtracted. The efficiency of the machines is therefore lower than the equation above would imply.

The above equations indicate that the speed of a motor can be controlled in two ways: by control of armature volts (and hence armature current) or by control of the field strength.

12.4.7. Field circuits

So far it has been assumed that the magnetic field has been produced by permanent magnets. Such an arrangement is only suitable for small motors, an electromagnetically produced field being used on most machines.

There are essentially five different field connections, shown in Fig. 12.11. The first, and most versatile, has a totally separately controlled field. This is the commonest arrangement for applications such as steel rolling mills, mine winders and paper manufacture. The field will have its own control circuit.

The parallel (shunt) field of Fig. 12.11b is used where the armature voltage is not likely to vary widely. The series field of Fig. 12.11c gives a large starting and accelerating torque, as the field strength is directly proportional to armature current. The equations in Section 12.4.6 however, indicate that the speed of a motor rises for a reduced field. Unlike the connections in Fig. 12.11a,b, a series connected motor does not have a clearly defined maximum speed (for the separate and shunt field, E cannot exceed V). If a series connected machine is deprived of a load, its speed can rise to a point where the commutator is destroyed by centrifugal force. It follows that series connected motors should not be used for chain or belt drives.

The final circuits of Fig. 12.11d,e combine the advantages of series and shunt machines and are known as compound machines. The arrangement of Fig. 12.11d is known as a short shunt, and Fig. 12.11e a long shunt.

Field windings, by their nature, inherently have a large inductance. This can

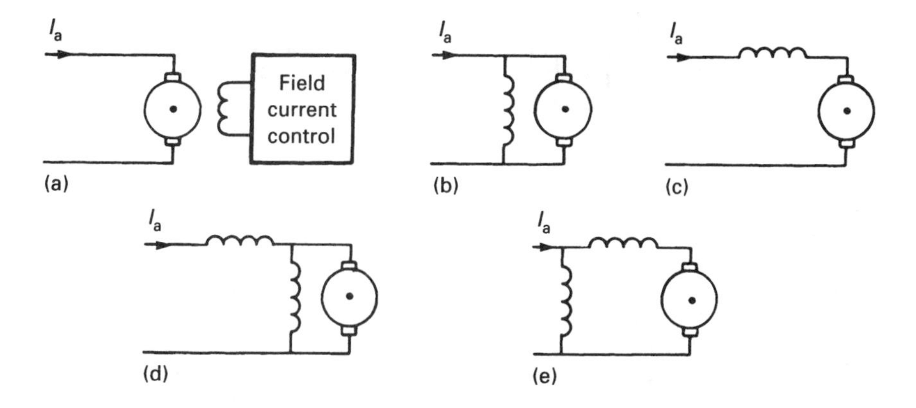

Fig. 12.11 *Field circuits. (a) Separately excited field. (b) Shunt field. (c) Series field. (d) Compound field short shunt. (e) Compound field long shunt.*

cause problems as large voltages can be induced by changes of field current. Later sections deal with the characteristics of circuits containing significant inductance.

12.5. Power semiconductors

12.5.1. Rectifier diodes

Power diodes are used where alternating voltage (from the AC mains supply) is to be converted to DC. Typical examples are electrolysis processes, battery charging, electromagnets and motor field circuits.

Rectifiers based on copper oxide or selenium plates were once common, but these devices are physically large and modern devices are almost entirely constructed from semiconductor materials with a pn junction, as in Fig. 12.12a. Silicon and germanium diodes are used, with silicon being more common. The mechanism of rectification need not concern us, but a typical power diode will have a voltage/current response similar to Fig. 12.12c.

In the reverse direction, a negligible leakage current flows (typically 1 mA for every 5 A of rated forward current). This remains fairly constant until an avalanche voltage is reached, where the current increases dramatically (and the diode is irreversibly damaged).

In the forward direction, the current-to-voltage relationship is approximately exponential, but can be considered as a straight line from a threshold voltage V_{th} and slope $1/r$ where r is the slope resistance. Typically V_{th} is 0.8 V and r is 0.001 ohms. If I_m is the mean current, and I_r the rms current (which may not be the same as I_m), the heat dissipated in the diode is:

$$P = I_m V_{th} + I_r^2 r \tag{12.20}$$

To calculate I_r from I_m, the form factor k for the application must be known, since:

$$I_r = kI_m \tag{12.21}$$

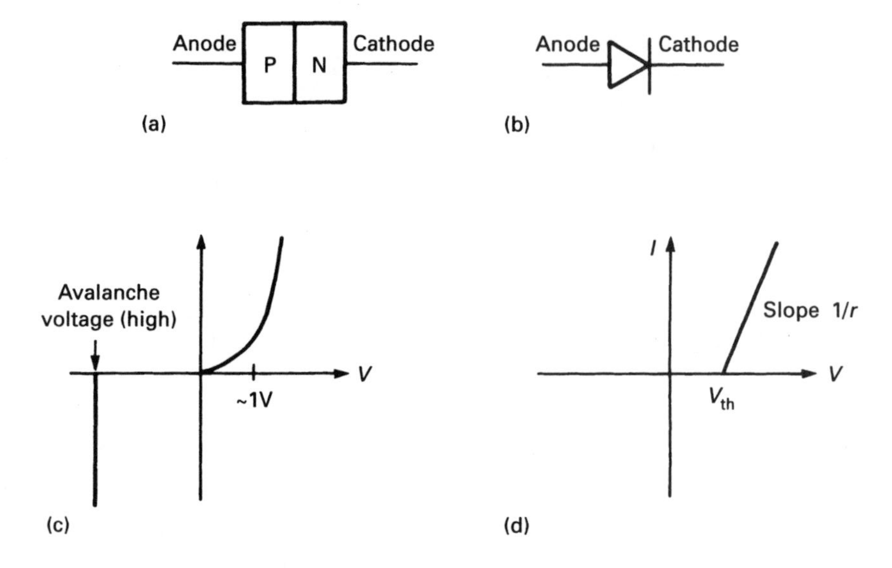

Fig. 12.12 *The power diode. (a) Diode construction. (b) Diode symbol. (c) V/I relationship. Note different voltage scales for positive and negative voltages. (d) Straight line approximation.*

k varies from 1.0 for a pure resistive load to 2.5 for a capacitive load.

As a rough rule of thumb, power diode losses can be approximated as:

0.5 W per A DC per germanium diode
1.2 W per A DC per silicon diode

This is obviously a lot of heat for, say, a 500 A diode, so the construction of the diode encapsulation is made to encourage dissipation via an external heat sink. Typical devices are shown in Fig. 12.13.

Diodes usually fail, however, not from overcurrent or overtemperature, but from overvoltage in the reverse region of Fig. 12.12c. Figure 12.14a shows a simple half wave diode rectifier, and Fig. 12.14b the voltage appearing across the diode. This is small whilst the diode is conducting, but follows the AC whilst the diode is blocking.

There are three voltages of interest. These are, as shown, the peak working voltage, the maximum recurrent voltage and the peak transient voltage. The peak transient voltage must never be exceeded, even briefly, or device failure will result. Transients are difficult to predict, and can arise internally from switching of inductive loads, or 'ringing' of transformer cores at switch-on. Externally generated transients can come from other users (such as electric arc furnaces) or from naturally occurring sources such as lightning strikes on the grid. A voltage margin, defined as PTV/PWV, of 2.5 is usually found to be adequate. If in doubt, protection snubber circuits (resistor and capacitor in series) can be placed across the diodes. Semiconductor manufacturers' data sheets give design criteria for these snubber circuits.

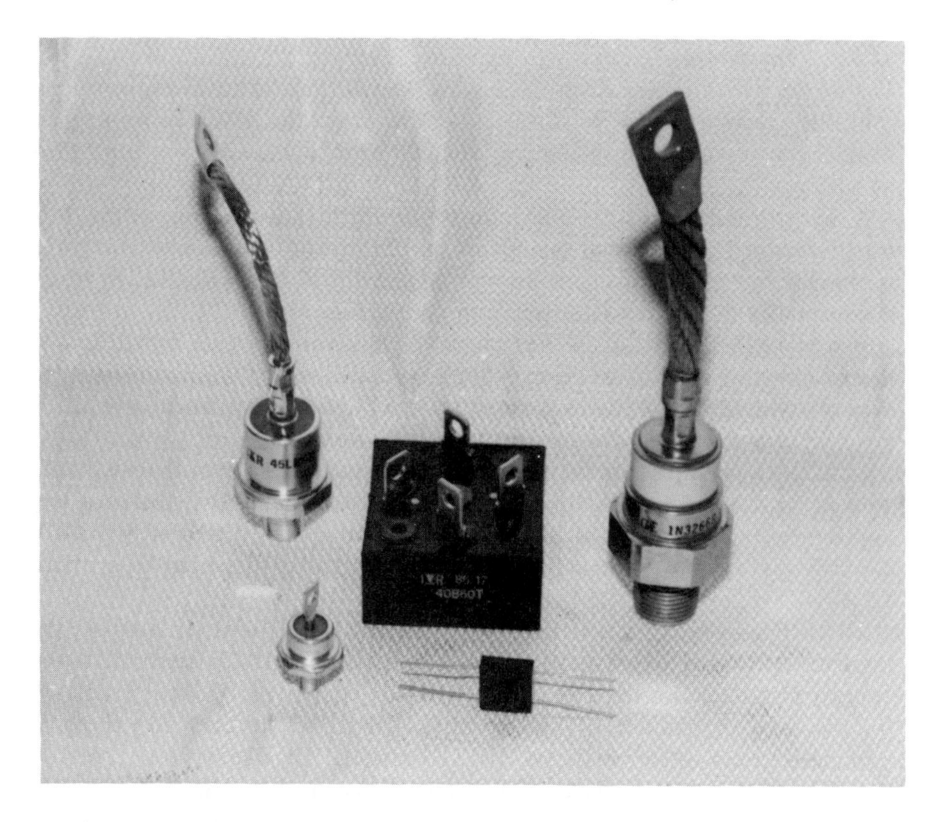

Fig. 12.13 *Typical power diodes.*

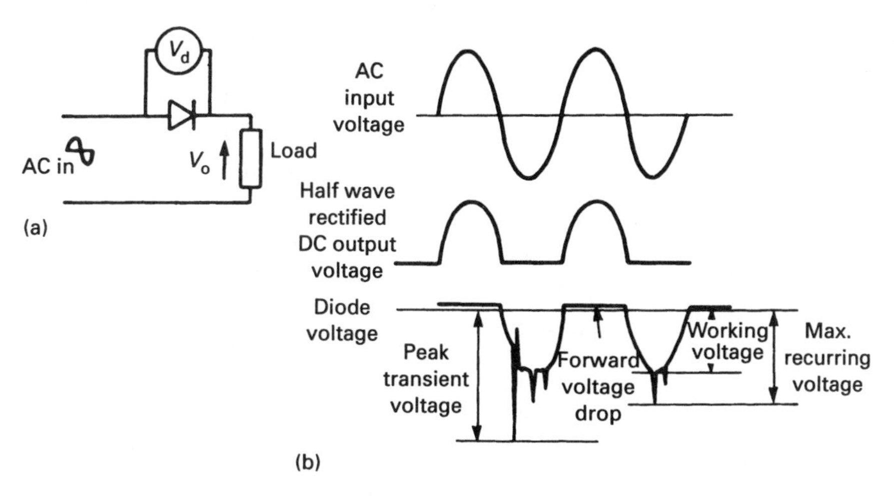

Fig. 12.14 *Voltages in a simple diode rectifier. (a) Half wave rectifier. (b) Voltages in circuit. Transients are not shown in AC in and V_o waveforms for simplicity.*

12.5.2. Thyristors

The workhorse of power electronics is the thyristor, also known as the silicon controlled rectifier or SCR. (This term arises from the device as controlled rectifier constructed from silicon not, as might be thought, a rectifier controlled by silicon.)

A thyristor is a four layer semiconductor device as shown in Fig. 12.15a. It is usually denoted by the circuit symbol of Fig. 12.15b. The operation of a thyristor is complex, but can be considered by rearranging Fig. 12.15a as Fig. 12.15c which approximates to the transistor pnp/npn pair of Fig. 12.15d.

Figure 12.16 shows typical V/I characteristics with no gate signal. In the reverse direction, a thyristor behaves like a normal diode, having a small leakage current (typically 1 mA per 5 A rated current) until the avalanche voltage is reached, at which point the current rises catastrophically.

In the forward direction. minimal leakage current occurs until the breakover voltage, V_{bo}, is reached. At this point the device suddenly switches to a low impedance state with low voltage drop. The conducting portion of Fig. 12.16 corresponds to a conventional diode with a slightly higher threshold V_{th} of about 1.5 V.

Thyristors are rarely, if ever, used without a gate signal, however. Suppose the thyristor is forward biased and sitting at point A on Fig. 12.16. Referring to the two transistor relationships of Fig. 12.15d, the emitter of TR_1 will be biased positive, and the emitter of TR_2 negative. Current is injected into the gate terminal; current will enter the base of TR_2 causing a small collector current to flow. This acts as base current for TR_1 which turns on, giving more base current to TR_2. A regenerative action takes place until both transistors are turned hard on.

The effect on Fig. 12.16 is that the device goes from point A to a point on the conduction region of the characteristic determined by the load and the rest of the circuit. The gate is therefore used to trigger the thyristor into conduction in the forward direction. A typical gate signal is 1.5 V at 100 mA for 100 μs.

Once conduction starts, it will continue until stopped by one of two conditions:

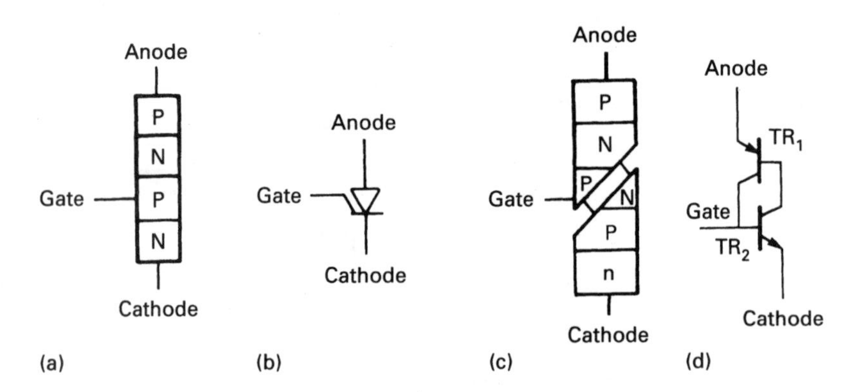

Fig. 12.15 *The thyristor. (a) Thyristor construction. (b) Thyristor symbol. (c) Transistor analogy. (d) Transistor equivalent.*

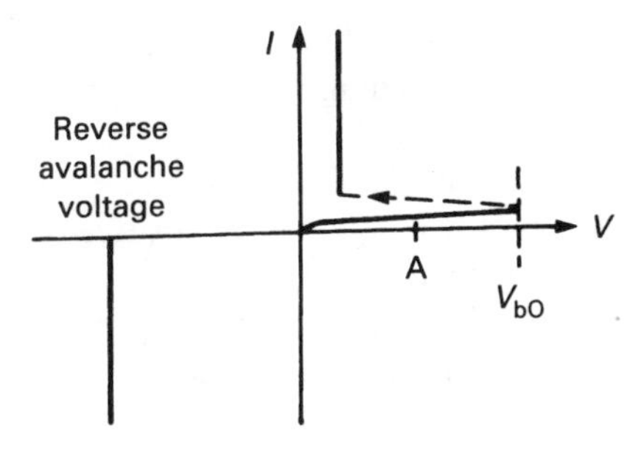

Fig. 12.16 *The thyristor V/I relationship.*

(1) The forward current is reduced below a specified level called the holding current (typically a few mA).
(2) The voltage across the device reverses, taking it into the reverse section of Fig. 12.16.

Once conduction has ceased, it will only restart by application of gate current in the forward section, or by taking the forward voltage above V_{bo}.

Figure 12.17 shows some simple thyristor circuits. In Fig. 12.17a, the operation

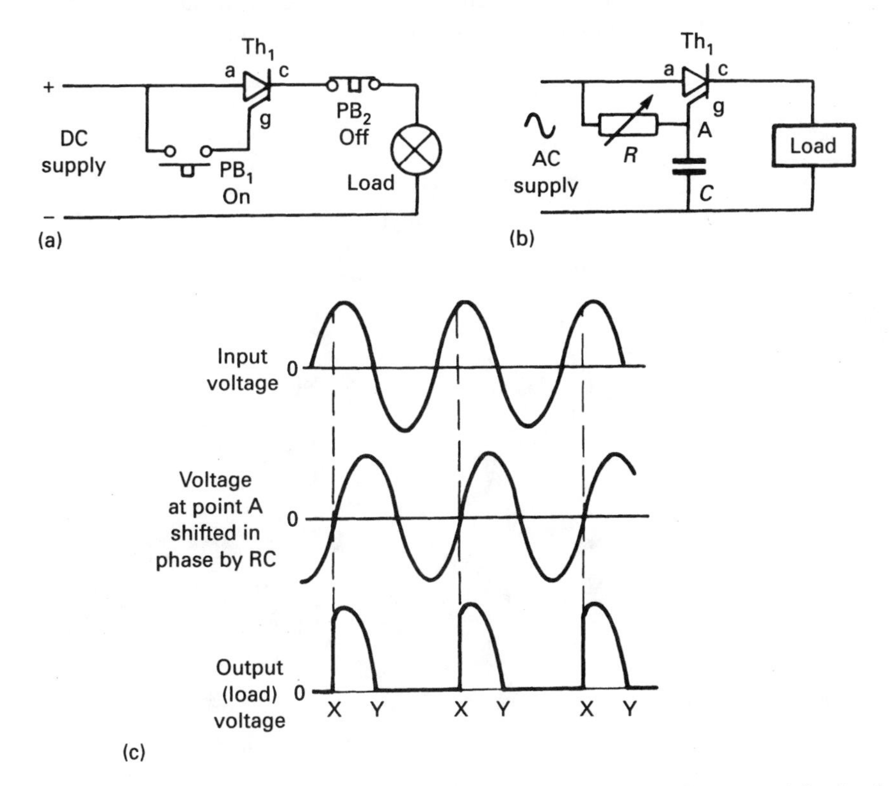

Fig. 12.17 *Simple thyristor circuits. (a) DC on/off circuit. (b) AC phase shift circuit. (c) Voltages in phase shift circuit.*

of PB_1 will inject gate current causing Th_1 to turn on. The bulb will light, and stay on (as the current is above the holding current). Operation of PB_2 breaks the current, causing Th_1 to turn off, and stay off.

Figure 12.17b is an AC circuit. The RC network delays the gate waveform as shown in Fig. 12.17. At point X the gate goes positive causing Th_1 to turn on. Current flows through the load until the AC supply reaches zero at point Y. Load current then falls to zero and the voltage across Th_1 reverses. Conduction ceases until point X again in the next positive half cycle. Load voltage waveforms are shown in Fig. 12.17c; the point of conduction is determined by the value of R and C. Practical versions of this circuit are used for light dimmers and speed control for electric drills.

Thyristors, like diodes, can be considered to have a threshold voltage and slope resistance when conducting. These lead to a loss (heat dissipation) which is typically 1.8 W per A DC per thyristor. The construction of thyristors is again made to facilitate heat removal; typical devices are shown in Fig. 12.18. The gate connection is usually a short tag or lead.

Comments in Section 12.5.1 on reverse voltage protection and current rating apply equally to thyristors. There are other constraints, however, that need to be considered.

The construction of a thyristor creates capacitance between anode and gate as shown in Fig. 12.19. When the anode voltage changes, a current *I* will flow given by:

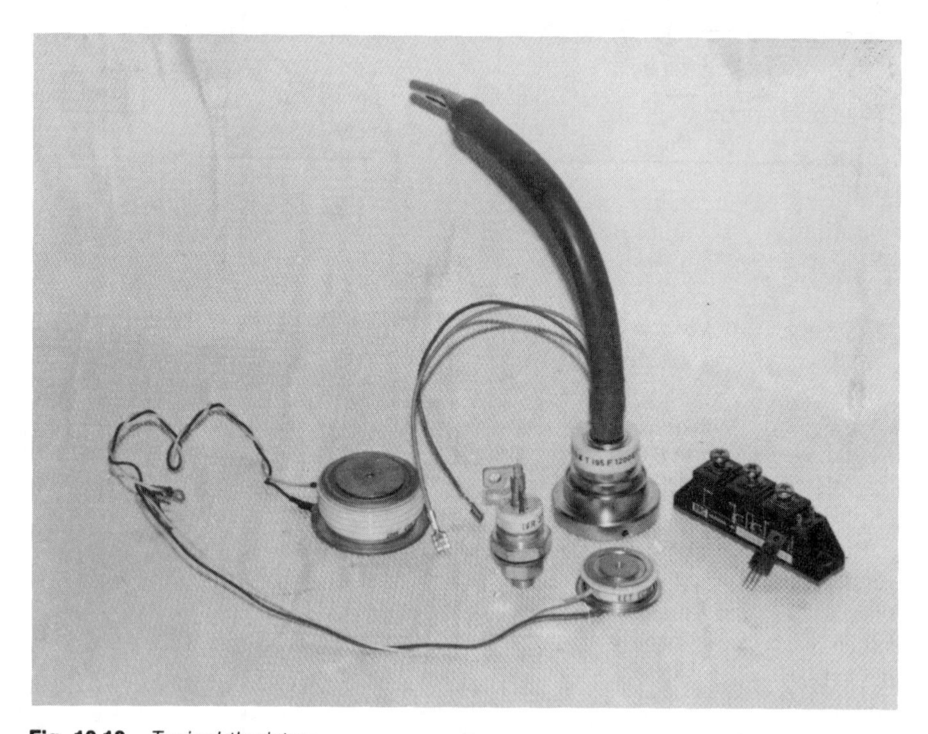

Fig. 12.18 *Typical thyristors.*

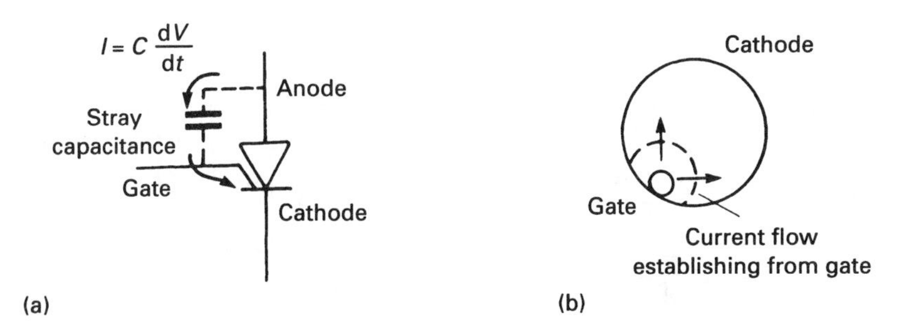

Fig. 12.19 *Thyristor failure modes. (a) False triggering by dV/dt. (b) Local heating by dI/dt.*

$$I = C\frac{dV}{dt} \tag{12.22}$$

If the anode voltage changes sufficiently quickly, I can be large enough to fire the thyristor. Although this does not, in itself, cause a device failure, in most applications harm would result to the thyristor or other components. Manufacturers accordingly specify a maximum dV/dt on their data sheets (typically 2 kV/μs).

Rate of rise of current (dI/dt) must also be limited. When a thyristor is fired, conduction initially starts in a small area by the gate, and spreads across the device as shown in Fig. 12.19b. The speed of propagation is typically 0.1 mm/μs, so for a typical 10 A device with 4 mm slice, some 40 μs must elapse for the thyristor to conduct completely. Too high a dI/dt leads to local high dissipation.

12.5.3. Triacs

The thyristor is a unidirectional device, and as such produces a controlled (albeit pulsating) DC output from an AC supply. The triac is a bidirectional device that can be used to give a controlled AC output. In some respects it can be considered as two thyristors connected as in Fig. 12.20 (although this arrangement is a somewhat incomplete analogy).

A triac is a three terminal device constructed as in Fig. 12.20b with the circuit symbol of Fig. 12.20c. Current flow is between MT_1 and MT_2 (for main terminal) and conduction is possible in either direction. Its characteristic is shown in Fig. 12.20d. Conduction is blocked in either direction until a gate pulse of *any* polarity is applied when the device goes to a low impedance state. Conduction continues until the current falls below a holding current (typically a few mA).

Figure 12.21a shows a controlled AC circuit using a triac (and is a full wave version of Fig. 12.17b). As before, R and C form a phase shifting circuit to delay the gate signal. D_1 is a device called a diac (described in the following section) which is used as a bidirectional voltage switch, only passing the voltage at A to the gate when its amplitude (positive *or* negative) exceeds the diac's breakdown voltage.

Waveforms are shown in Fig. 12.21b. The gate is triggered by the diac at X on

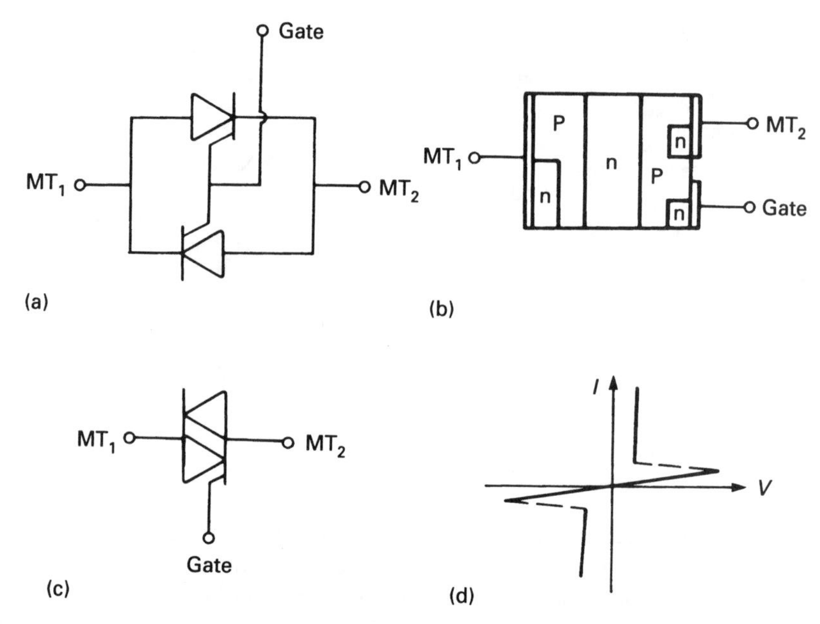

Fig. 12.20 *The triac. (a) Simplified representation. (b) Construction. (c) Symbol. (d) V/I relationship.*

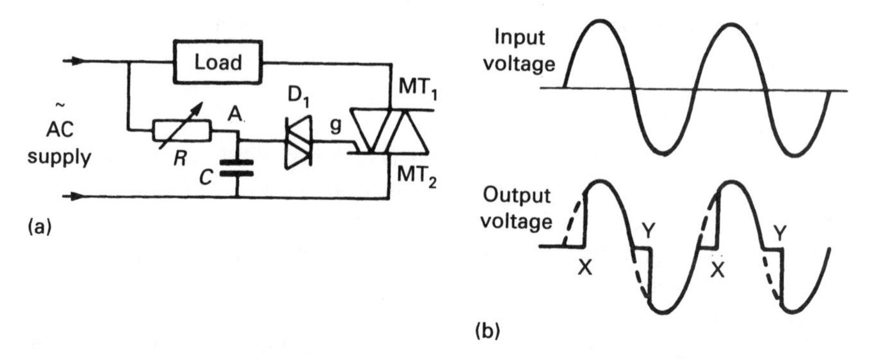

Fig. 12.21 *Simple triac circuit. (a) Circuit diagram. (b) Waveforms.*

the positive half cycle and Y on the negative half cycle. The position of X (and Y) is controlled by the values of R and C. The circuit of Fig. 12.21 gives full wave control of an AC load.

Triacs are subject to the same design constraints (transient voltage rating, dI/dt, etc.) as thyristors.

12.5.4. Diacs

The diac is a two-terminal symmetrical device constructed as in Fig. 12.22a with circuit symbols as in Fig. 12.22b. The device is essentially two thyristors

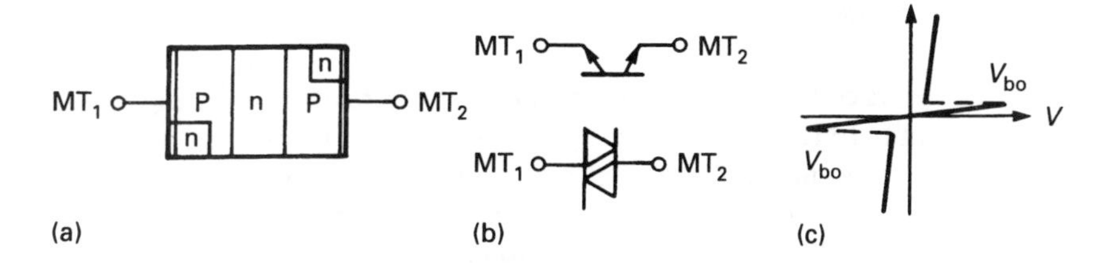

Fig. 12.22 *The diac. (a) Construction. (b) Symbols. (c) V/I relationship.*

connected as in Fig. 12.20a but with no gate connection. The device is therefore characterised by the symmetrical curve of Fig. 12.22c, in particular the two breakover voltages; typically 30 V in each direction.

Diacs are used as a voltage switch, blocking a signal until it reaches some predetermined level. Diacs are typically used as part of the gate triggering circuits for triacs.

12.5.5. Gate turn off (GTO) thyristors

A conventional thyristor blocks in the reverse direction, and can be triggered into conduction by a positive gate pulse (or by exceeding the breakover voltage). Once conducting, it can only be returned to an off state by reducing the current below the holding value, or by reversing the voltage across the device.

A variant is the gate turn off thyristor, whose circuit symbol is shown in Fig. 12.23. This behaves as a conventional thyristor, with the additional feature that it can be taken out of conduction by a negative gate pulse. The gate pulse to turn on the device is similar to a conventional thyristor (typically 1.5 V at about 100 mA for about 100 μs). The turn off pulse is negative (typically -5 V) but the required current is of the same order of magnitude as the anode current, albeit for a few μs. The higher the gate current, the faster the turn off.

The ability to turn off a large circuit has its side effects, however, particularly if the load contains significant inductance when large transients can be generated. A conventional thyristor only turns off at zero current, and tends to suppress inductive transients by itself.

Gate turn off thyristors are commonly used in PWM inverters, described in Section 12.11.3.

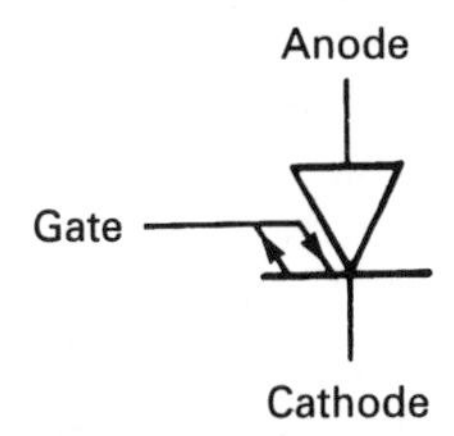

Fig. 12.23 *The gate turn off (GTO) thyristor.*

12.6. Rectifier circuits

Table 12.1 shows common arrangements of rectifier circuits, converting AC from the mains supply to DC (either directly, or via a suitable transformer). The table shows circuit diagrams, and the relationship between AC and DC current and voltages. It should be noted that the PIV quoted is the theoretical value, and in practice a considerable safety margin should be made to allow for transients, etc.

The table is for the most part self explanatory, but there are a few points that may need amplification. In the single phase bridges (a to c) the unsmoothed output is pulsating DC, whereas the three phase circuits have inherently a low ripple content (4% fundamental for the full wave bridge e).

It is also interesting to compare circuits b and c. The biphase circuit gives the same output as the full wave bridge, but is apparently a more economical circuit. Examination, however, shows that the PIV of the diodes for circuit b is twice that for circuit c, and in addition a probably more expensive transformer is required. Similar comparisons apply to circuits d, e and f with the additional observation that circuit e can be used with Star or Delta connected transformers. Note also the difference in ripple frequency between the various three phase circuits; a higher ripple frequency is easier to smooth.

Circuit f is used (albeit rarely) for heavy current low voltage applications, and gives low output ripple. The interphase reactor is used to improve transformer utilisation, and is called a hexaphase rectifier. If the reactor is omitted (and the two star points linked) the circuit is called a diametric connection.

Another consideration is the currents flowing in the AC legs of the rectifier. In general, non-bridge circuits will have a DC component which will reflect into the AC supply (for direct connected rectifiers) or require a larger than necessary transformer because of the resultant magnetising current. The zig-zag connection of the circuit g gives zero DC component whilst minimising the number of rectifiers (albeit at the expense of transformer complexity and cost).

12.7. Burst fired circuits

The simplest thyristor and triac control circuits are those used to control heaters and similar devices. In these the triac (say) is simply used as a switch or relay to turn power on and off. The power on/off ratio controls the power fed to the load.

Figure 12.24a shows a block diagram of a burst fired circuit. A synchronising circuit produces the gate signal to fire the triac. The control signal enables the gate signals and could come from a thermostat (for temperature control with a heater load) or from an oscillator with adjustable mark/space ratio for proportional power control.

Waveforms are shown in Fig. 12.24b. The control signal is present from time A to time B. The synchronising circuit, however, delays firing the triac to the next voltage zero at time X, and the natural action of the triac keeps current flowing after time B to point Y. This switch on at zero voltage, off at zero current,

Table 12.1 *Common rectifier arrangements*

		V_{dc}	Diode current		Diode	Controlled rectifier
			Peak	Mean	PIV	V_{dc} continuous I
(a) Single phase half wave	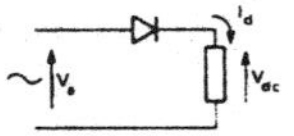	$\frac{V_a\sqrt{2}}{\Pi}$ (0.45 V_a) 0 with continuous current	I_d	I_d	3.14 V_{dc}	Zero
(b) Biphase half wave	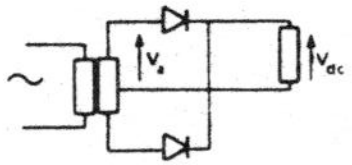	$\frac{V_a 2\sqrt{2}}{\Pi}$ (0.9 V_a)	I_d	$\frac{I_d}{2}$	3.14 V_{dc}	$V_a\frac{2\sqrt{2}}{\Pi}\cos\alpha$
(c) Single phase full wave bridge	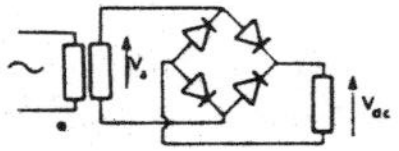	$\frac{V_a 2\sqrt{2}}{\Pi}$ (0.9 V_a)	I_d	$\frac{I_d}{2}$	1.57 V_{dc}	$V_a\frac{2\sqrt{2}}{\Pi}\cos\alpha$
(d) Three phase half wave	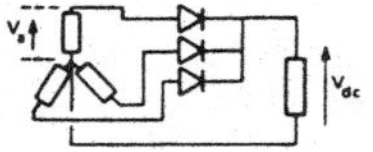	$V_a\frac{3\sqrt{6}}{2\Pi}$ (1.17 V_a)	I_d	$\frac{I_d}{3}$	2.09 V_{dc}	$V_a\frac{3\sqrt{6}}{2\Pi}\cos\alpha$
(e) Three phase full wave	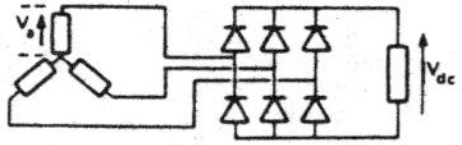	$\frac{V_a 3\sqrt{6}}{\Pi}$ (2.34 V_a)	I_d	$\frac{I_d}{3}$	1.045 V_{dc}	$V_a\frac{3\sqrt{6}}{\Pi}\cos\alpha$
(f) Double star	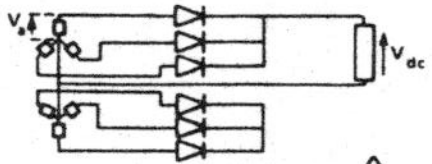	$V_a\frac{3\sqrt{2}}{\Pi}$ (1.35 V_a)	0.5 I_d	$\frac{I_d}{6}$	2.09 V_{dc}	$V_a\frac{3\sqrt{2}}{\Pi}\cos\alpha$
(g) Zig zag	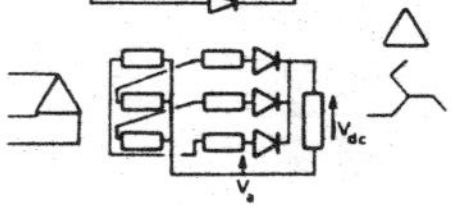	$\frac{V_a 3\sqrt{6}}{2\Pi}$ (1.17 V_a)	I_d	$\frac{I_d}{3}$	2.09 V_{dc}	$V_a\frac{3\sqrt{6}}{2\Pi}\cos\alpha$

No capacitive smoothing is assumed.

Note that AC voltages are given with respect to neutral. For phase to phase AC voltages reduce V_{dc} by $\sqrt{3}$. For example, 240 V phase to neutral, and 415 V phase to phase give the same V_{dc}.

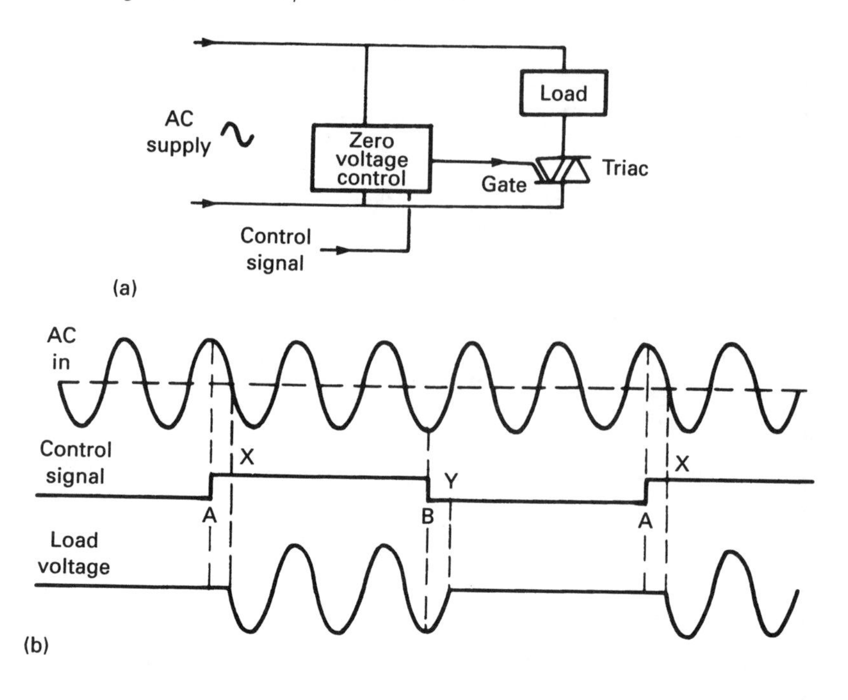

Fig. 12.24 *Burst firing of triacs. (a) Circuit diagram. (b) Waveforms.*

minimises transients fed back into the supply, and prevents interference with radio and television receivers.

Synchronising circuits are, not surprisingly, available in IC form. Figure 12.25a shows the block diagram of a proportional device. The control voltage is compared with a relatively slow ramp (typical period 10 to 100 seconds). When the control voltage is above the ramp pulse (as detected by the comparator) the triac is enabled as shown in Fig. 12.25b.

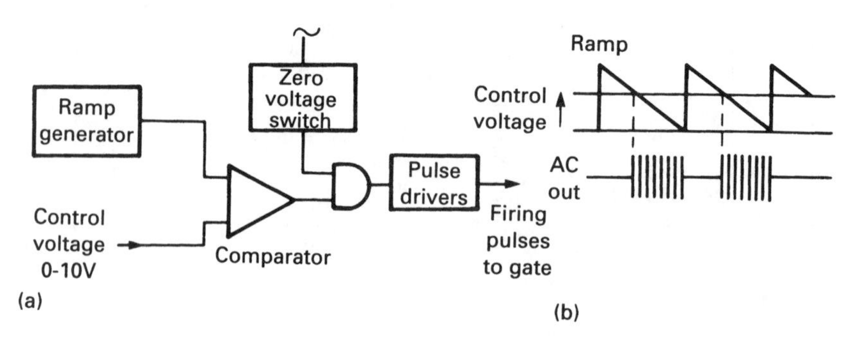

Fig. 12.25 *Proportional control circuit for burst firing. (a) Circuit. (b) Waveforms.*

12.8. Phase shift control

12.8.1. Basic principles

Burst firing is simple, but is only suitable for loads with long time constants such as heaters. For speed control, the pulsating torque would be totally unacceptable. The commonest type of power control varies the power fed to the load by varying the percentage of each cycle of the AC supply.

Figure 12.26a shows a simple biphase rectifier similar to that in Table 12.1 except that the diodes have been replaced by thyristors. These are fired by a synchronising circuit, and the gate pulses can be adjusted over 180° of the cycle; thyristor Th_1 over one half cycle, thyristor Th_2 over the other. The details of the synchronising circuit need not concern us at present.

Waveforms for a resistance load are shown in Fig. 12.26b. During time X, the thyristors are being fired at 30° after the zero crossing point, each thyristor conducts for the rest of the half cycle once fired, and the power fed to the load is nearly the same as that obtained from a simple biphase rectifier.

During time Y, the firing of the thyristors is delayed to 90°, and the power fed to the load is significantly reduced. Finally, during time Z, the firing is delayed to 120°, and the power fed to the load is reduced even further.

The power fed to the load is therefore controlled by the point in the cycle at which the thyristors are fired. The firing angle as described above is called the angle of delay, denoted by α. Maximum power is delivered for $\alpha = 0°$, and minimum power for $\alpha = 180°$ for a resistive load (inductive loads are described later). The firing point can also be described by the angle of advance, denoted by β, as shown in Fig. 12.26c. Obviously $\beta = 180° - \alpha$, and maximum power occurs for $\beta = 180°$.

The average output voltage of Fig. 12.26 is given by:

$$V = \frac{\sqrt{2}E}{\pi}(1 + \cos\alpha) \tag{12.23}$$

12.8.2. Inversion and the effects of inductance

Figure 12.26 was drawn for a purely resistive load. Most industrial loads, particularly motor armatures and fields, have significant inductance and this significantly modifies the behaviour of controlled rectifier circuits.

Figure 12.27a is the simplest controlled rectifier arrangement: a single half wave bridge. In Fig. 12.27b this is fired at an α of 60° at time T_1. At time T_2 the transformer voltage passes through zero (and conduction would cease for a resistive load). The inductance, however, keeps current flowing until time T_3, so the voltage follows the line voltage negative from T_2 to T_3.

During the time T_1 to T_2, power is being transferred from the supply to the load; during T_2 to T_3 power is being returned from the stored energy in the inductor to the supply.

When the controlled rectifier is returning power to the supply it is said to be

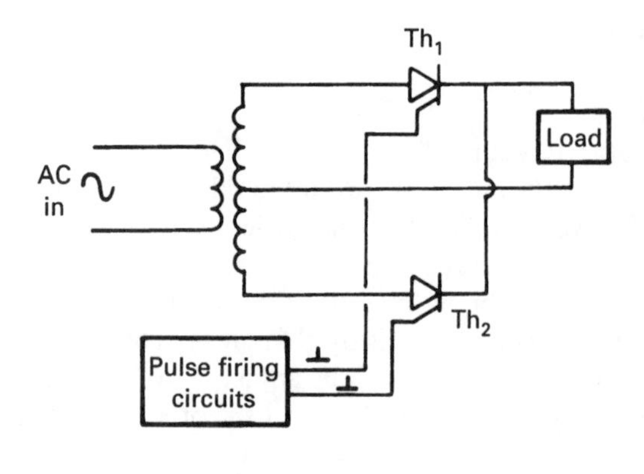

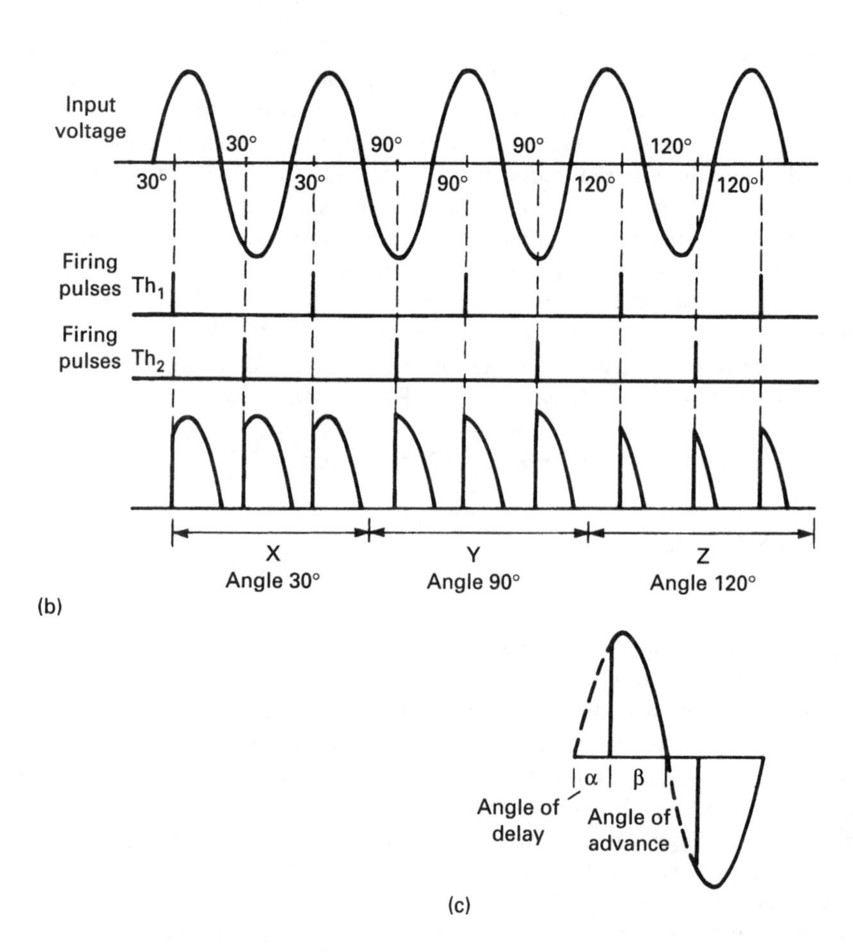

Fig. 12.26 *The phase shift controlled rectifier. (a) Phase controlled rectifier. (b) Waveforms for various angles. (c) Firing angles.*

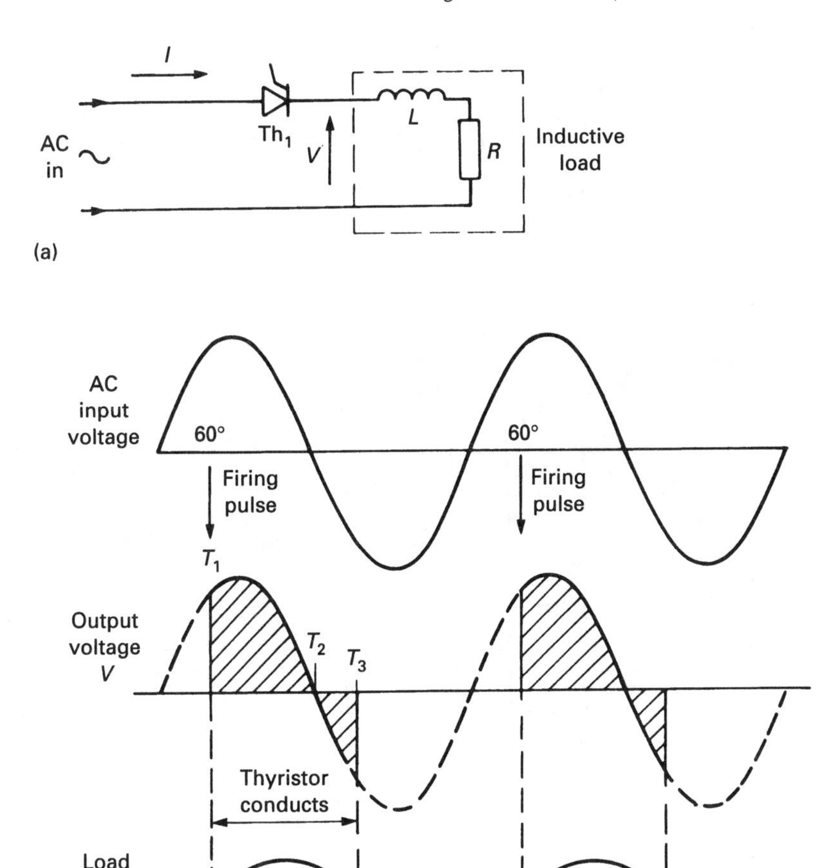

Fig. 12.27 *The effect of load inductance. (a) Circuit. (b) Waveforms.*

acting as an inverter. When power is flowing from the supply to the load, the circuit is said to be acting as a rectifier.

Inversion occurs in Fig. 12.27 because of the load inductance. It will also occur in any situation where the load current is continuous. Later sections will describe dynamic braking of a motor by returning its kinetic energy to the supply via its own back emf and a controlled rectifier circuit acting as an inverter.

Figure 12.28a shows a controlled full wave bridge connected to a load which draws continuous current – a highly inductive load, or a motor acting as a generator, for example. The output voltage for various values of α are shown in Fig. 12.28b. Note that the continuous current causes the output voltage to follow the line voltage negative until the other thyristor pair is fired.

For $\alpha = 30°$ and $\alpha = 60°$ the average voltage is positive and power is, overall, transferred to the load, i.e. the circuit behaves as a rectifier. For $\alpha = 90°$ the

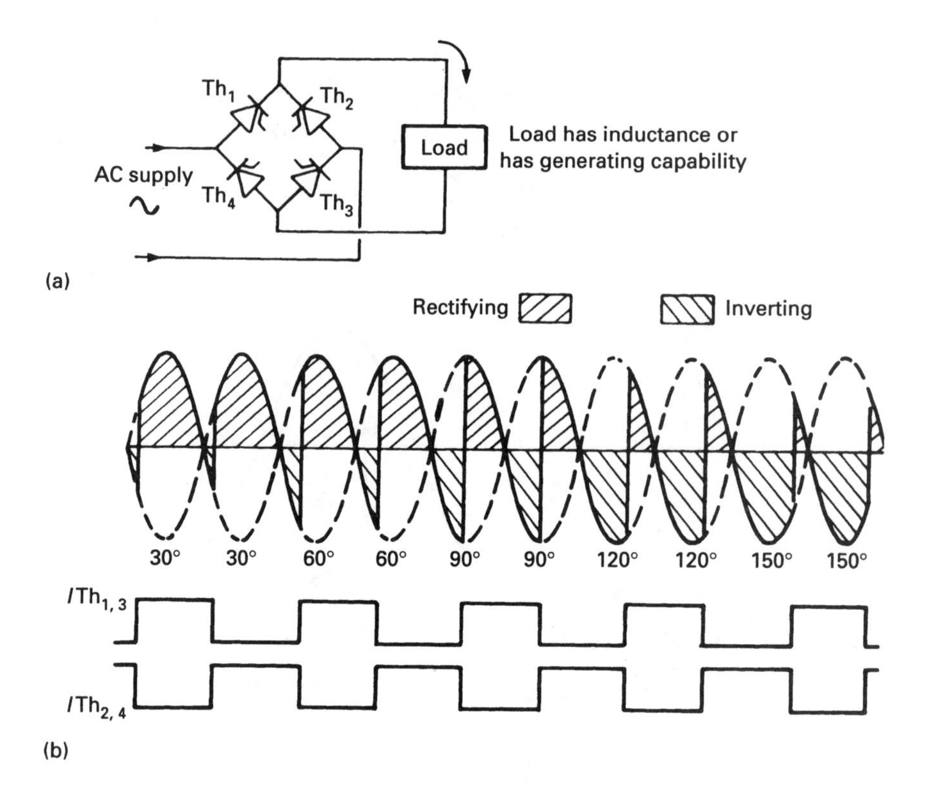

Fig. 12.28 *The fully controlled full wave rectifier with continuous current. (a) Circuit. (b) Waveforms.*

positive and negative portions of the waveforms are equal and there is no net transfer of power. For $\alpha = 120°$ and $\alpha = 150°$ the average voltage is negative and power is being transferred from the load to the supply, i.e. the circuit acts as an inverter.

If the load current is continuous, therefore, the circuit acts as a controlled rectifier from $\alpha = 0°$ to $\alpha = 90°$, and as an inverter from $\alpha = 90°$ to $\alpha = 180°$. The output voltage is given by:

$$V = \frac{2\sqrt{2}E}{\pi}\cos\alpha \tag{12.24}$$

This should be compared with the resistive case of equation 12.23. Both are plotted in Fig. 12.29 for comparison. A circuit containing inductance, but not sustaining continuous current, will follow an intermediate curve similar to the dotted curve in Fig. 12.29.

It is conventional, but by no means universal, to define the firing angle in terms of α for rectification and β for inversion.

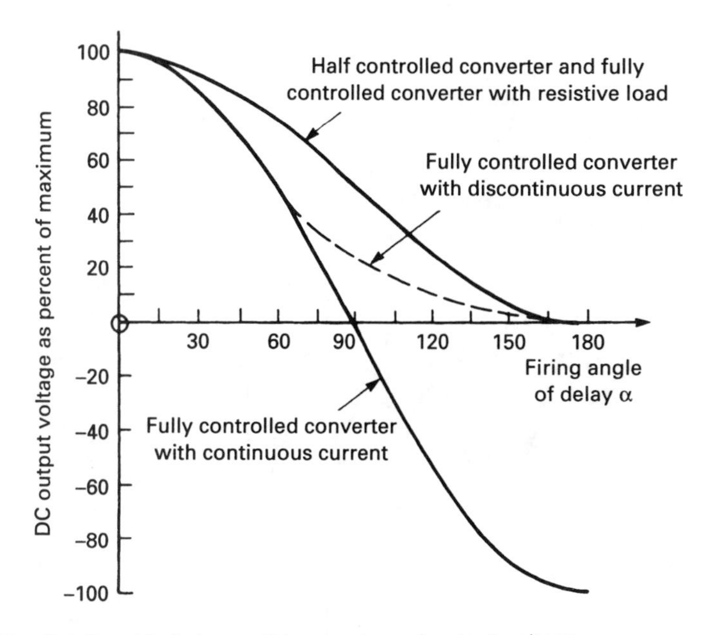

Fig. 12.29 *Relationship between firing angle and output voltage.*

12.8.3. Three phase circuits

Most motor control circuits convert a three phase AC supply to DC. These are usually based on the half wave bridge, d, and the full wave bridge, e, of Table 12.1.

The half wave bridge circuit is shown in Fig. 12.30a. As before, the production of the firing pulses need not concern us at this stage. The firing angle α is defined from the natural commutation point of the bridge, as shown in Fig. 12.30b which gives waveforms for continuous current from 30° to 150°. As we saw for the single phase circuits with continuous current, the circuit rectifies for $\alpha < 90°$, and inverts for $\alpha > 90°$.

Figure 12.31a shows the full wave circuit – probably the commonest controlled rectifier arrangement. Waveforms are again given in Fig. 12.31b with the firing angle defined from the natural commutation point. As before, the circuit rectifies for $\alpha < 90°$ and inverts for $\alpha > 90°$.

12.8.4. Half controlled circuits

Superficially, Fig. 12.32a is a full wave thyristor bridge, but closer inspection will show that devices D_1 and D_2 are simply rectifiers. A circuit with mixed thyristors and diodes is called a half controlled, or mixed, bridge.

Assuming continuous current, let us follow the circuit action from the point where Th_1 fires. Current flows through Th_1 and D_1 until the end of the half cycle. At this point, current switches from D_1 to D_2, the continuous current circulating through Th_1, the load and D_2 with zero transformer current. This state continues until Th_2 fires, when current flows from the transformer through Th_2, the load and D_2. At the end of the half cycle, the current switches from D_2 back to D_1 and

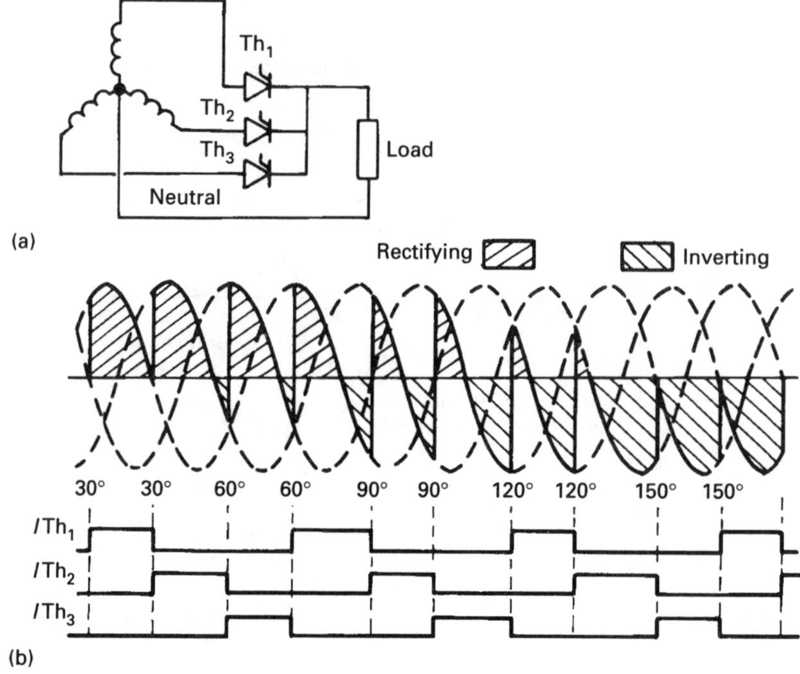

Fig. 12.30 *The three phase half wave controlled rectifier with continuous current. (a) Circuit diagram. (b) Waveforms.*

current circulates through Th_2 and D_1 until Th_1 fires again. Waveforms are shown in Fig. 12.32b.

The action of D_1 and D_2 prevents the bridge output voltage from going negative, and consequently the circuit can only operate as a rectifier, not an inverter. The output voltage is similar to that for a purely resistive load, and the average voltage given by equation 12.23.

Figure 12.32a is not the only half controlled circuit. Figure 12.33a shows an alternative single phase circuit, and Fig. 12.33b and c three phase equivalents. Figure 12.33c is commonly used where inversion is not required from a three phase motor control circuit.

12.8.5. Commutation and overlap

In any practical controlled or uncontrolled rectifier, the AC source will have significant inductance in each phase from the transformers, supply lines and any suppression circuits on the AC supply. This inductance will prevent the instantaneous switch of current from one rectifier to another.

A three-phase half wave rectifier is shown in Fig. 12.34a. Current theoretically switches instantaneously from D_1 to D_2 to D_3 and back to D_1 at the natural commutation points. The effect of inductance is shown in Fig. 12.34b. The natural commutation point is denoted by X, and D_2 starts to conduct at this point. The inductance, however, keeps current flowing in D_1 until point Y.

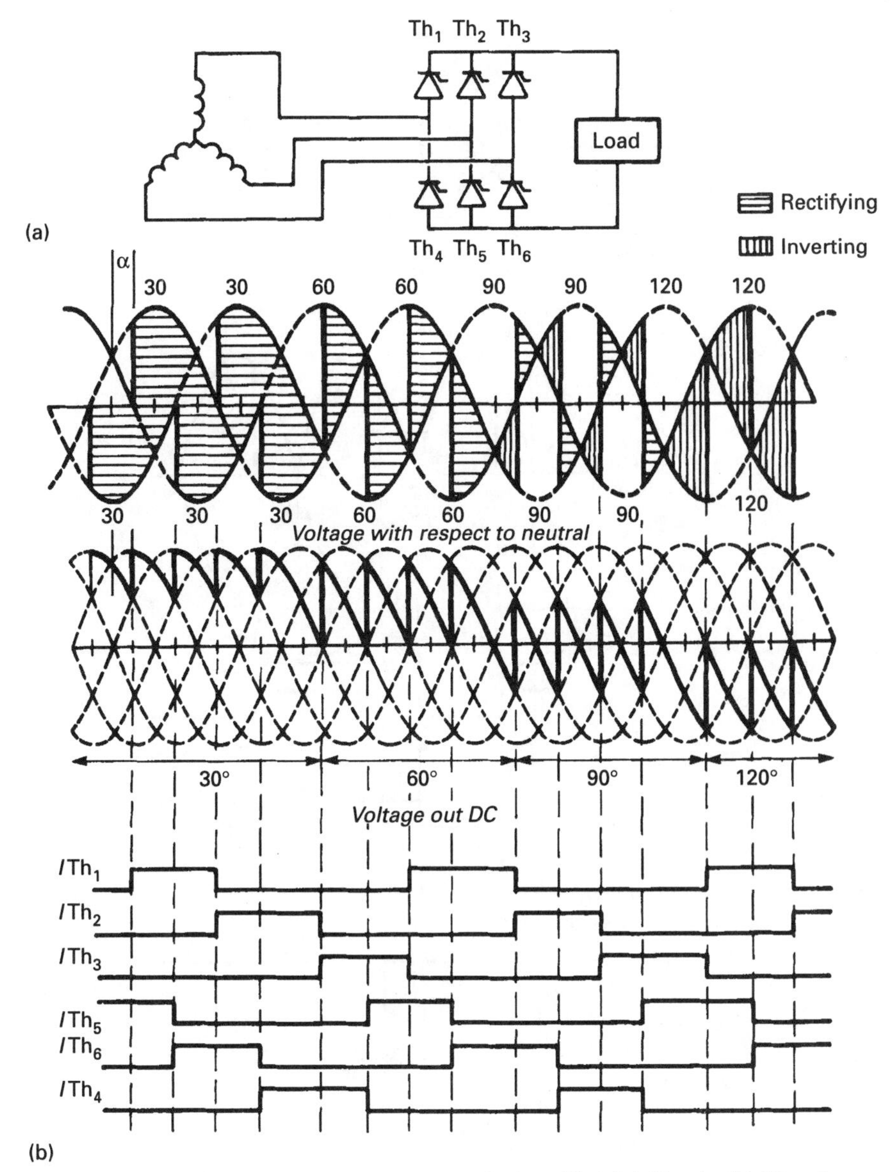

Fig. 12.31 *The three phase full wave controlled rectifier. (a) Circuit. (b) Waveforms.*

Between X and Y D_1 and D_2 both conduct, and the output voltage is the average of the two phase voltages giving a notch in the output. The time between X and Y is called the commutation angle, or overlap, and is typically a few degrees. It is often denoted by the Greek letter δ.

If the diodes are replaced by controlled thyristors, overlap gives the waveform of Fig. 12.34b. Again, a notch appears in the output waveform as two thyristors conduct together.

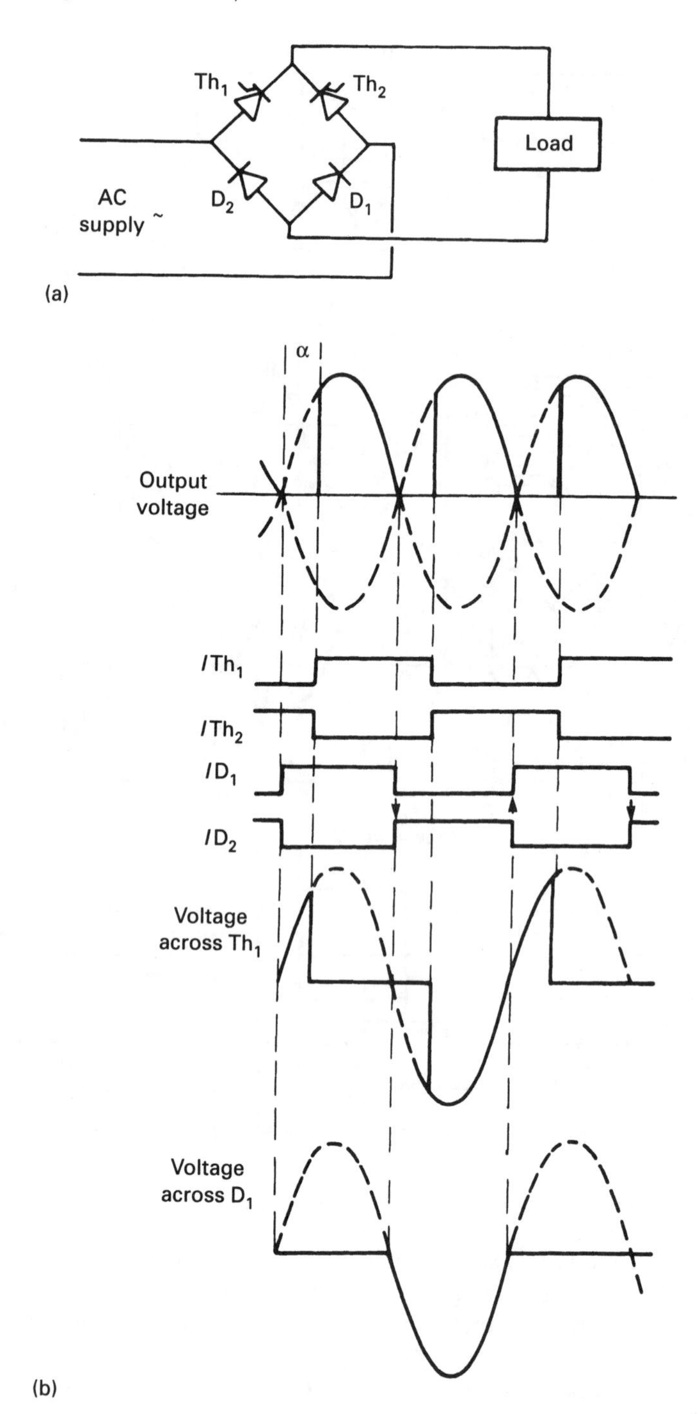

Fig. 12.32 *The half controlled rectifier. (a) Circuit. (b) Waveforms.*

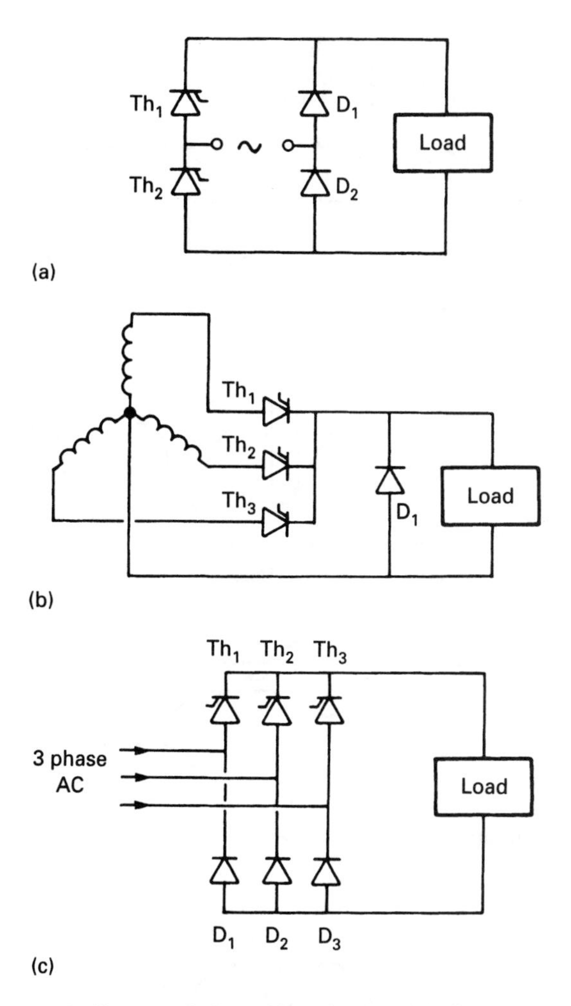

Fig. 12.33 *Common half controlled rectifier circuits. (a) Circulating current flows through D_1, D_2. (b) Three-phase half wave rectifier. Circulating current flows through D_1. (c) Three-phase full wave rectifier. This operates in a similar manner to Fig. 12.32.*

Overlap sets a maximum limit for α (and minimum limit for β) and prevents inversion to the theoretical value of $\alpha = 180°$. If the commutation between successive thyristors has not occurred before 180°, a commutation failure will occur and current transfer will not take place. In practice, a minimum value of β of about 10° must be set to ensure commutation.

12.9. Motor control

12.9.1. Simple single phase circuit

Section 12.4.6 showed that the speed of a DC motor can be controlled either by varying the armature voltage (or current) or the field strength. Figure 12.35a

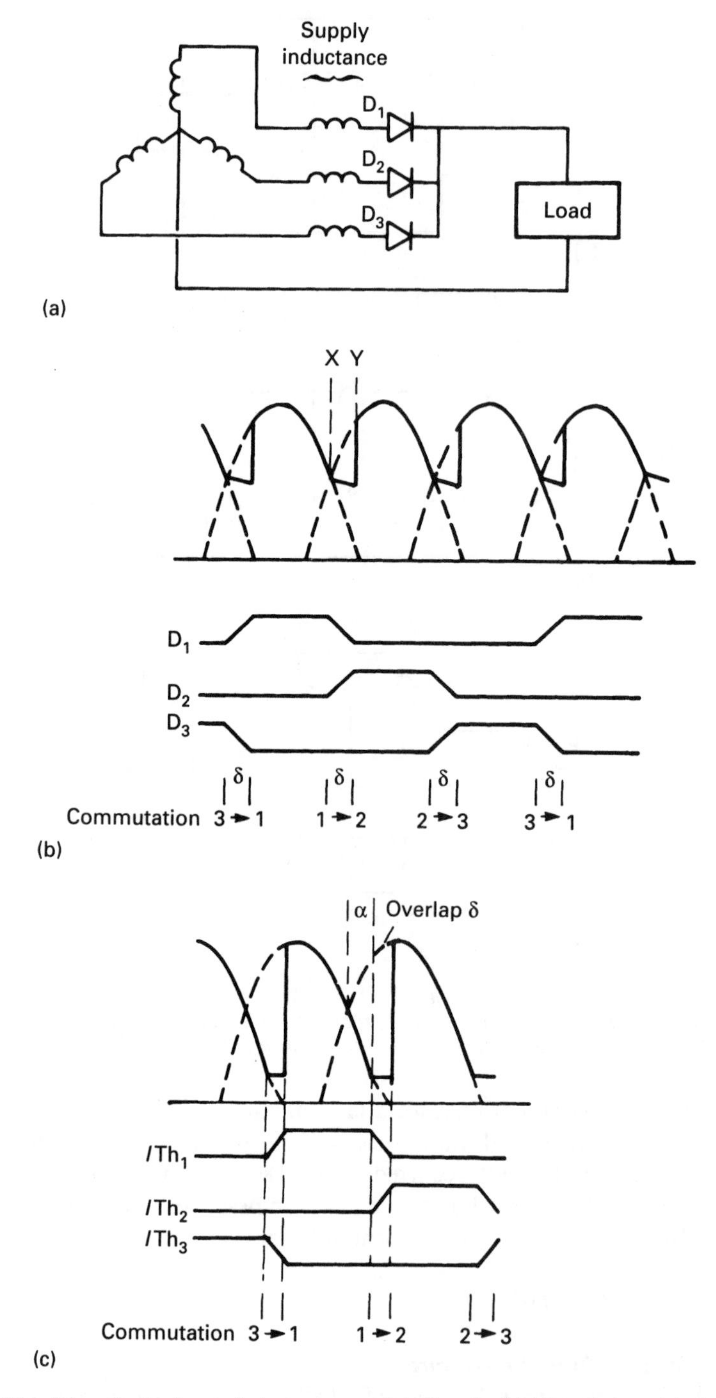

Fig. 12.34 *The effect of supply inductance. (a) Circuit. (b) Waveforms. (c) Controlled rectifier waveforms.*

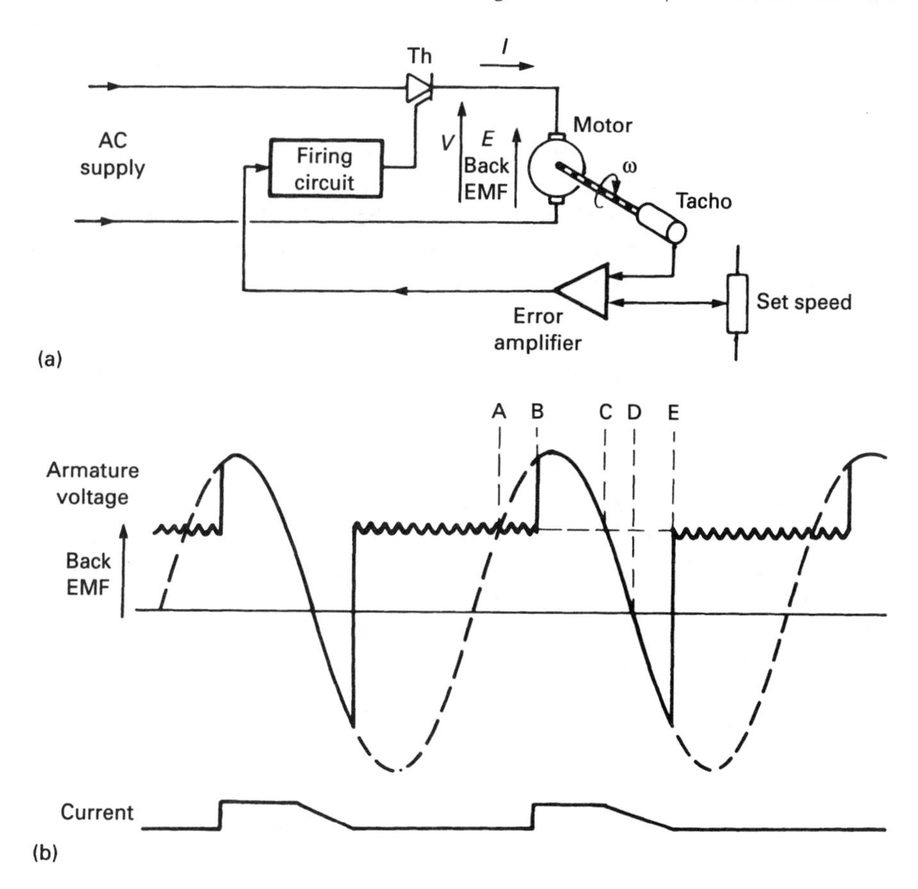

Fig. 12.35 *Simple motor speed control. (a) Circuit. (b) Waveforms.*

shows a single phase half wave controlled circuit, possibly the simplest motor control circuit possible. This operates with a fixed field, and speed is controlled by varying the armature current through Th_1. The current is discontinuous at all times, which limits the effectiveness of the circuit.

The speed is measured by a tacho generator and compared with the set speed. The error is used to advance or retard the firing pulses to Th_1 (usually with P and I control) to make the actual speed equal the set speed.

Waveforms are shown in Fig. 12.35b, and may be different from those expected. Applied volts are denoted by V and the back emf by E (it will be remembered that E is proportional to speed). Thyristor current is denoted by I. The thyristor cannot be fired until point A when the input volts exceeds the back emf. The firing circuit fires the thyristor at point B, and armature current continues until point C, when the input voltage falls below the back emf.

Armature inductance, however, means that the current cannot fall to zero instantly, and Th_1 continues to conduct, and the armature voltage follows the input. At point D the voltage across the armature reverses (and Th_1 becomes an inverter). At point E the current falls to zero, and the armature voltage jumps to the back emf, where it remains to point A in the next cycle. The ripple between points E and A is caused by the commutator segments on the motor.

It is interesting to note that the circuit of Fig. 12.35 has a degree of self-regulation, even without the tacho feedback. If the load increases and the speed consequently falls, the back emf will also be lowered. This will move point C further back and increase the length of the armature current pulse, mitigating the effect of the increased load.

The circuit of Fig. 12.35 cannot provide a braking action in the same direction as it provides motoring action. Although the circuit inverts between points D and E in Fig. 12.35b, it is the inductive energy that is being returned, not the kinetic energy of the motor. If a reduction in speed is required, the gate pulses would be phased right back and the motor allowed to coast down under the action of the load and friction until the new speed was reached. At this time the pulses would be advanced sufficiently to maintain speed.

Figure 12.35 can, however, provide braking in the reverse direction as shown in Fig. 12.36. Reversed direction gives a reversed, i.e. negative, back emf as shown. If Th_1 is fired at point A in Fig. 12.36a, current will be returned from the kinetic energy of the motor to the supply (V is negative, I is positive, so energy is flowing from the motor to the supply) until point B, where the supply volts goes more negative than the back emf (neglecting the effects of armature inductance). The circuit is acting as an inverter, albeit in a limited manner because of the discontinuous current.

If the firing point is advanced further as in Fig. 12.36b, the braking becomes more efficient. The circuit, however, is now acting as a rectifier from points A to B forcing current through the armature. Inversion still takes place from B to C, but on average the power transfer is from supply to motor (and the motor kinetic energy is converted to heat, not returned to the supply).

12.9.2. Single, two and four quadrant operation

Figures 12.35 and 12.36 are said to be two quadrant circuits. Let us examine what this means and the implications for motor control. Figure 12.37a shows a block representation of a controlled rectifier connected to a DC load. Assuming the load is capable of sustaining continuous current (either via inductance or, say, a generator), there are four possible operating conditions, shown in Fig. 12.37a. These are called the four operating quadrants. A simple uncontrolled rectifier operates in quadrant 1 for example.

In quadrants 1 and 3, power is transferred from the AC supply to the load. In quadrants 2 and 4, power is transferred from the load back to the supply.

No single controlled rectifier bridge can operate in more than two quadrants. Fully controlled rectifiers (with thyristors in each bridge position), such as Figs. 12.28, 12.30 and 12.31, can operate in quadrants 1 and 2 (or 3 and 4 by reversing the connections from the rectifier to the load).

Half controlled bridges (with mixed thyristors and diodes) such as Fig. 12.33 can only act as rectifiers and hence only operate in quadrant 1 (or quadrant 3 by

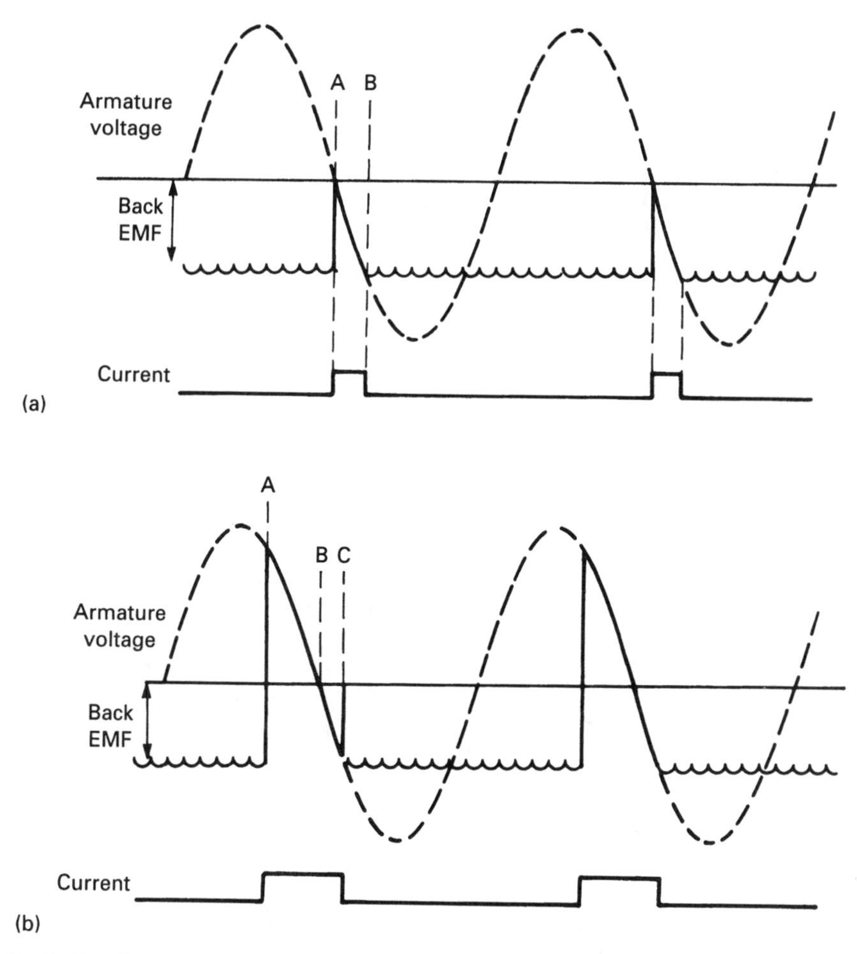

Fig. 12.36 *Dynamic braking of a motor. (a) Firing at zero crossing point. (b) Firing before zero crossing point.*

reversal of connections). Such circuits cannot invert.

The four quadrants of Fig. 12.37 are redrawn as Fig. 12.38 in terms of motor rotation and motor torque. Note that for motoring quadrants, motor torque and rotation have the same sign, and $M_t > L_t$ and $V > E$. Similarly for generating quadrants, torque and rotation are of opposite sign, and $M_t < L_t$ and $V < E$.

It follows that a single quadrant converter can only motor, and a fully controlled (two quadrant) converter can only operate as a motor in one direction, and as a generator in the opposite direction. There are few applications for this type of control, the only common ones being shown in Fig. 12.39. The first is a bidirectional drive where material is wound back and forth through some process between coils (e.g. papermaking). The motors alternately drive and provide back tension (e.g. A motoring and B generating). The second, and more common, is a winch, where the load falls with the motor generating and rises with the motor driving.

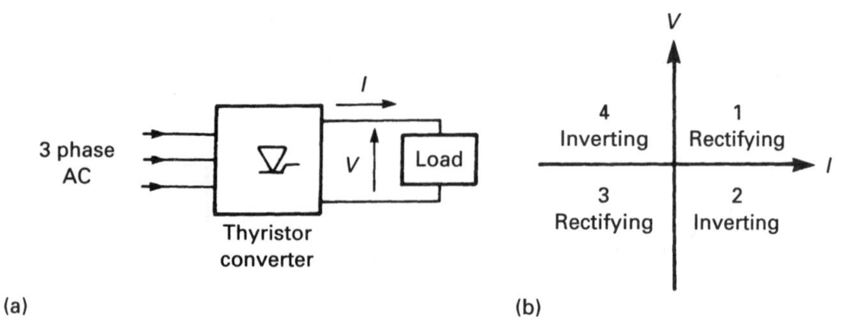

Fig. 12.37 *Four quadrant converters (a) Circuit diagram. (b) Four quadrants of operation.*

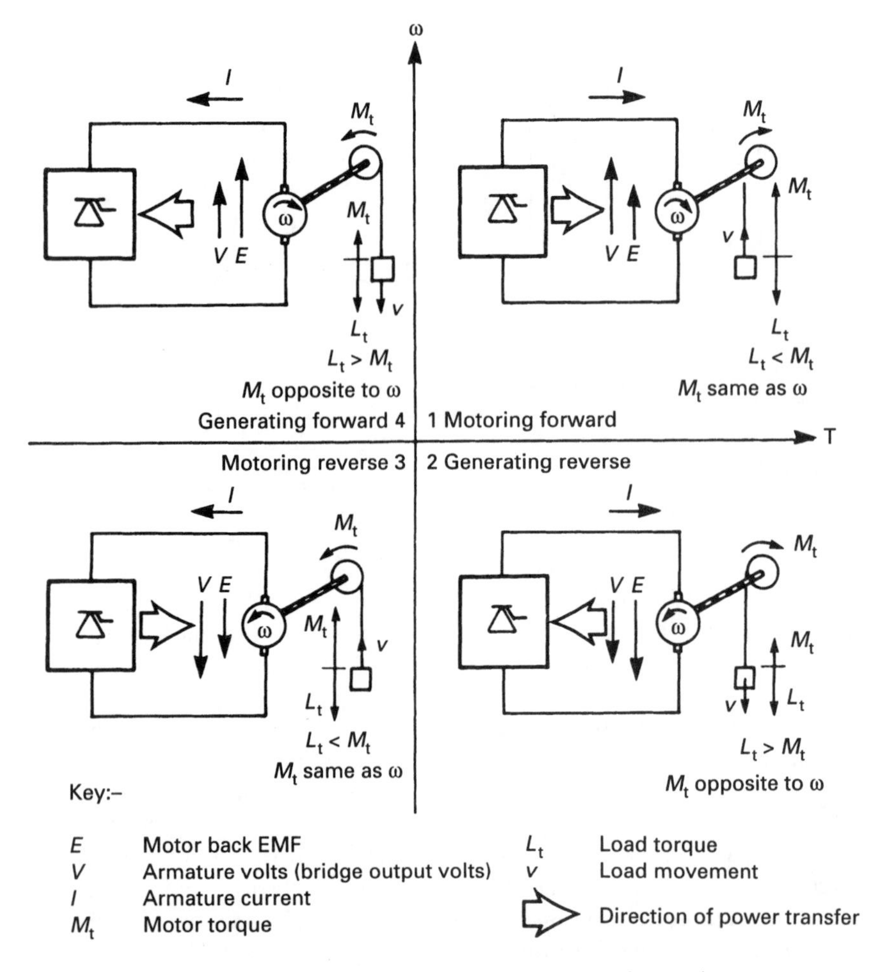

Fig. 12.38 *Four quadrant operation in terms of motor rotation and torque.*

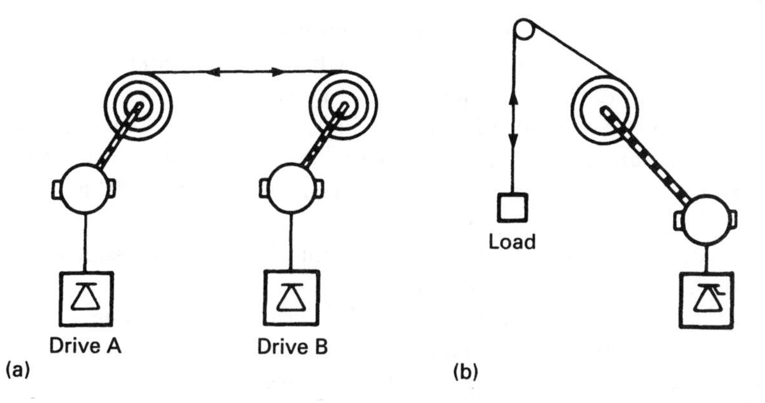

Fig. 12.39 *Two quadrant applications. (a) Back tension control. Transfer left to right; A inverts, B drives. Transfer from right to left; A drives, B inverts. (b) Winch drive. Load rises with motor driving; falls with motor inverting.*

Most applications require operations in all four quadrants. This can be achieved by any of the schemes of Fig. 12.40. Circuit a uses a contactor to reverse the armature and circuit b the field. Both employ a two quadrant converter, and require external sequencing to control the contactor, and in particular ensure that switching only takes place at zero armature current.

Figure 12.40c uses two separate converters: A operating in quadrants 1 and 2, say, and converter B in quadrants 3 and 4. This configuration is discussed further in Section 12.9.4.

12.9.3. Plugging

Quadrants 2 and 4 are often labelled 'braking'. Whilst it is true that generating does act as a brake, it is only efficient at high speeds, and the net torque reduces as the motor speed falls.

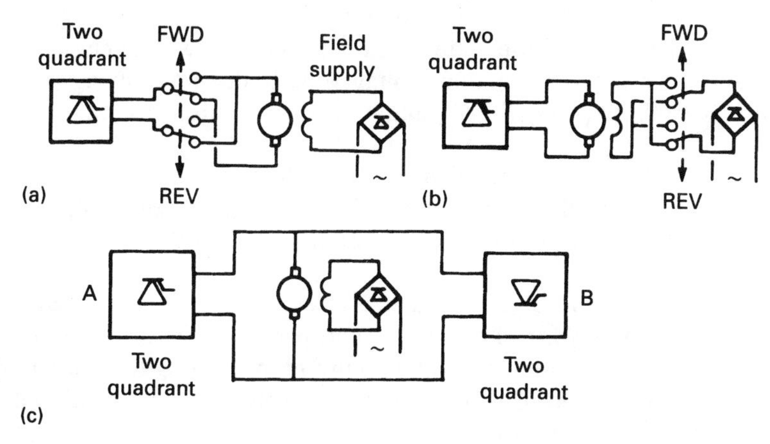

Fig. 12.40 *Various four quadrant arrangements. (a) Armature changeover. (b) Field changeover. (c) Dual converter system.*

Quadrants 1 and 3 can also be used to perform very efficient braking, but the motor kinetic energy is dissipated as heat rather than being returned to the supply. Suppose, adopting the notation of Fig. 12.38, we have a motor spinning clockwise (positive ω) and we switch it into quadrant 1. V and E will be of opposite sense, and the current will rapidly bring the motor to a standstill (and even reverse its direction if the rectifier does not return α to 90° as the motor stops).

In practice, current control is essential to limit I. Normally the rectifier would start braking in quadrant 2, and shift into quadrant 1 as the motor speed falls. This action is sometimes called plugging. A typical situation occurs in the winch drive of Fig. 12.39b where the load is being lowered at a slow controlled speed. The converter will be rectifying to provide the necessary torque (and the motor will significantly heat up).

12.9.4. Dual converter systems

The principle of a dual converter system is shown in Fig. 12.41. Each of the converters can be considered to be a variable voltage source in series with a diode. Both converters can achieve a positive or negative output voltage, but the series diodes limit the direction of current flow. Both converters are externally controlled so they always have the same voltage.

Let us assume we are motoring forward with an armature voltage of +50 volts, and this gives clockwise rotation. Both V_a and V_b are set at 50 V but bridge A is rectifying (via D_a) and bridge B is blocked by D_b. We now reduce speed by dropping V_a and V_b to +25 V. D_a blocks bridge A, and the motor generates into bridge B until its speed falls to the point where bridge A conducts and starts rectifying, and bridge B is blocked.

If we wish to reverse direction, V_a and V_b will be reduced to zero, then increased negative, and bridge B will at first invert, then plug, then rectify until the motor achieves the required speed. For a reduced reverse speed, the motor inverts into bridge A until the speed falls to the required level.

Effectively the two converters are operating up and down the curve of Fig. 12.41c, and for any given armature voltage one bridge is rectifying and one inverting. Note that $\alpha_A + \alpha_B = 180°$, and for zero armature volts both inverters have $\alpha = 90°$.

It is not possible to make the voltages from the two bridges identical. The imbalance can be handled in two ways. A circulating current bridge allows current to flow between bridges, usually with inductors in the DC legs of each bridge.

A more common arrangement, though, is to use external logic to ensure only one bridge fires at once. This is known as an anti-parallel suppressed half (APSH) bridge. The bridge selection logic requires current monitoring to ensure bridge selection changes only take place at zero current. There is inevitably a small delay during which time the motor coasts.

Suppose we are motoring forward on bridge A, and a speed reversal is called. A typical sequence could be:

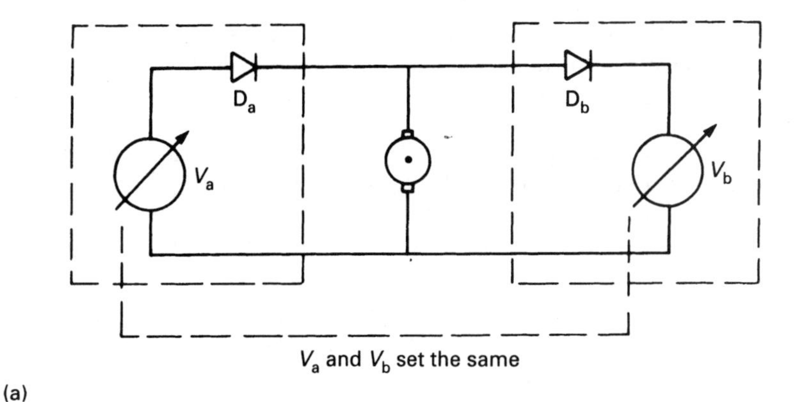

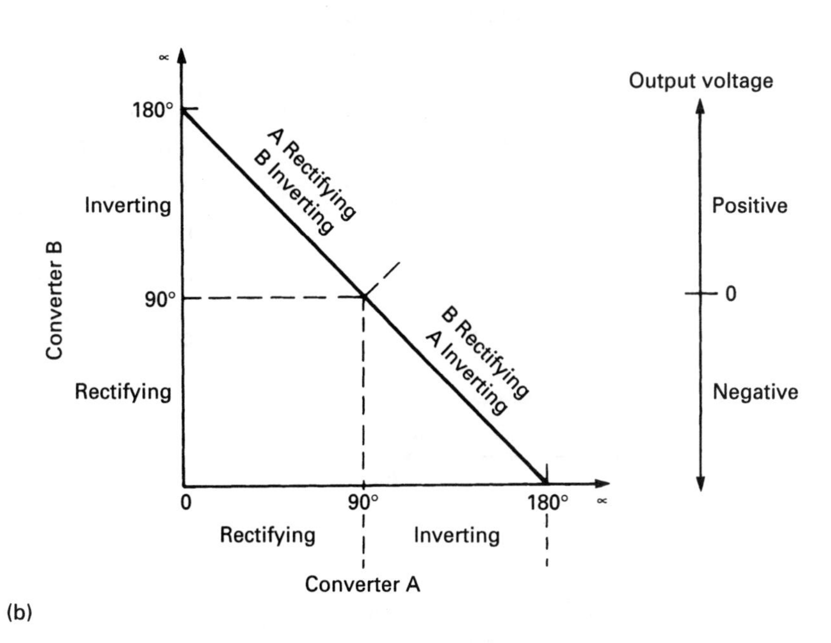

Fig. 12.41 *Operation of dual converter system. (a) Representation of dual converter system. (b) Relationship between the two converters.*

(a) A phases right back (large α) and inverts until armature current falls to zero (energy stored in armature inductance). Motor is still travelling forward.
(b) Logic sees zero current; blocks bridge A, releases gate pulses on bridge B.
(c) B phases forward to invert from motor back emf. Note that current direction is now reversed. Motor deceleration is controlled by bridge B. As motor speed falls, B phases further forward with $\alpha < 90^\circ$ near zero speed.
(d) Motor stops, and accelerates in reverse up to required speed on bridge B.

It is also instructive to follow the sequence where a speed reduction is called without a reversal of direction. Again let us start motoring forward on bridge A, and call for a speed reduction:

(a) A phases back and inverts ($\alpha > 90°$) to reduce the armature current to zero.
(b) Logic sees zero current, blocks bridge A and releases bridge B. Current flow through motor is reversed.
(c) Motor back emf inverts into bridge B; motor speed falls under control of bridge B.
(d) When correct speed is reached bridge B decreases α to give zero current. When zero current is detected bridge B is blocked and A released.
(e) Bridge A now drives motor at new speed.

12.9.5. Armature voltage control and IR compensation

Equations 12.14 and 12.15 show that the back emf of a motor is proportional to speed, and the applied armature volts approximate to the back emf if the IR term is small. From this a simple speed control can be derived as shown in Fig. 12.42. The armature voltage is compared with the set speed. The resultant error is used

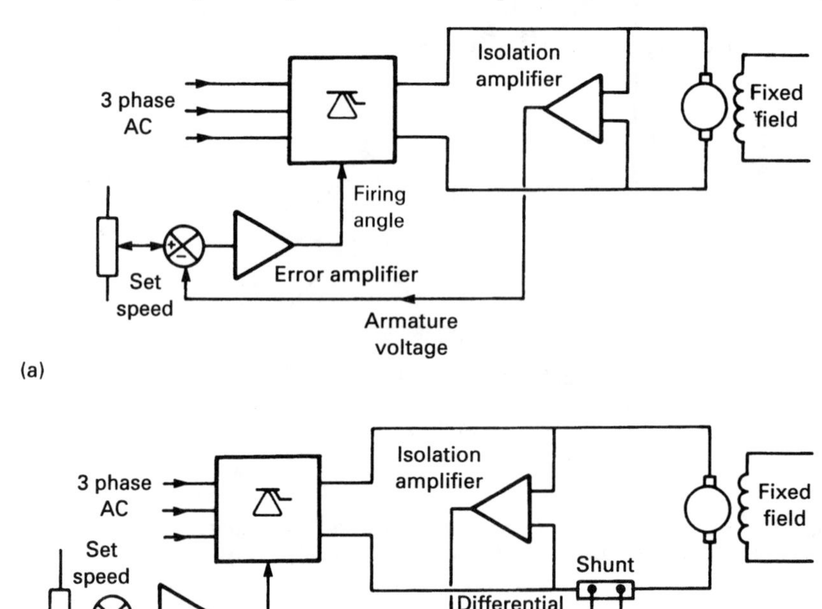

Fig. 12.42 *Speed control without a tachometer. (a) Armature voltage control. (b) IR Compensation.*

to advance, or retard, the firing angle of the thyristor bridge and hence control the speed. Usually a P + I controller is used.

Figure 12.42a does not, however, compensate for load changes. Figure 12.42b measures both armature volts and current (V and I in Equation 12.14) and computes E from the relationship $E = V - IR$. This can be done with a few DC amplifiers. E (and hence ω) are compared with the set speed and used to control the bridge firing angle. This, in theory, gives true speed control with load changes and is called IR compensation. In practice, if the IR term is too large the control becomes unstable, so it is normal to set the drive for about 5% speed 'droop' from no load to full load. Motor windings also experience severe temperature changes, which cause the value of R to change significantly with load and time. This further reduces the effectiveness of IR compensation. The technique is, however, simple and cheap and is widely used with small motors and reasonably constant loads.

12.9.6. Tacho feedback and current control

Figure 12.43 shows the commonest arrangement for a reversing thyristor drive, and represents an APSH circuit with tacho speed feedback. The set speed is represented by a voltage in, say, the range -10 V to $+10$ V. This is passed through a ramp circuit to limit the drive acceleration (a possible circuit is Fig. 11.26 in Chapter 11).

The actual speed comes from a tachogenerator (typically 1 V per 100 rpm) and the resultant speed error presented to a P + I amplifier. The output from this amplifier is a current set point (limited to safe values by a limit circuit). The actual current is measured either by a shunt in the motor armature circuit or, as shown, by rectifying the output from current transformers on the AC supply. The latter is cheaper as it avoids the need for isolation amplifiers, but adds a lag into the current measurement via the filters on the rectifier.

Bridge selection logic enables the A/B firing circuits as described above, and controls the polarity of the current feedback signal (the output of the current feedback rectifier being unipolar). Gating is sequenced by the output of zero current detection circuit.

12.9.7, Field weakening

Equation 12.15 shows that a motor speed can also be controlled by the field strength; for a given armature voltage a reduced field would cause an increase in speed. Any motor has a maximum armature voltage determined by the insulation and similar factors. This armature voltage will correspond to a no-load speed called the base speed at the specified field current.

Armature voltage/current can only be used to control the motor up to base speed, but the speed can be increased further by reducing the field strength once maximum armature voltage is reached. The circuit to achieve this is shown in Fig. 12.44.

The armature voltage is reduced to a reasonable level by an isolation amplifier, and rectified to give a unipolar DC signal. This is compared with the maximum armature voltage set on RV_1 (often called the spillover voltage). As long as V_a is

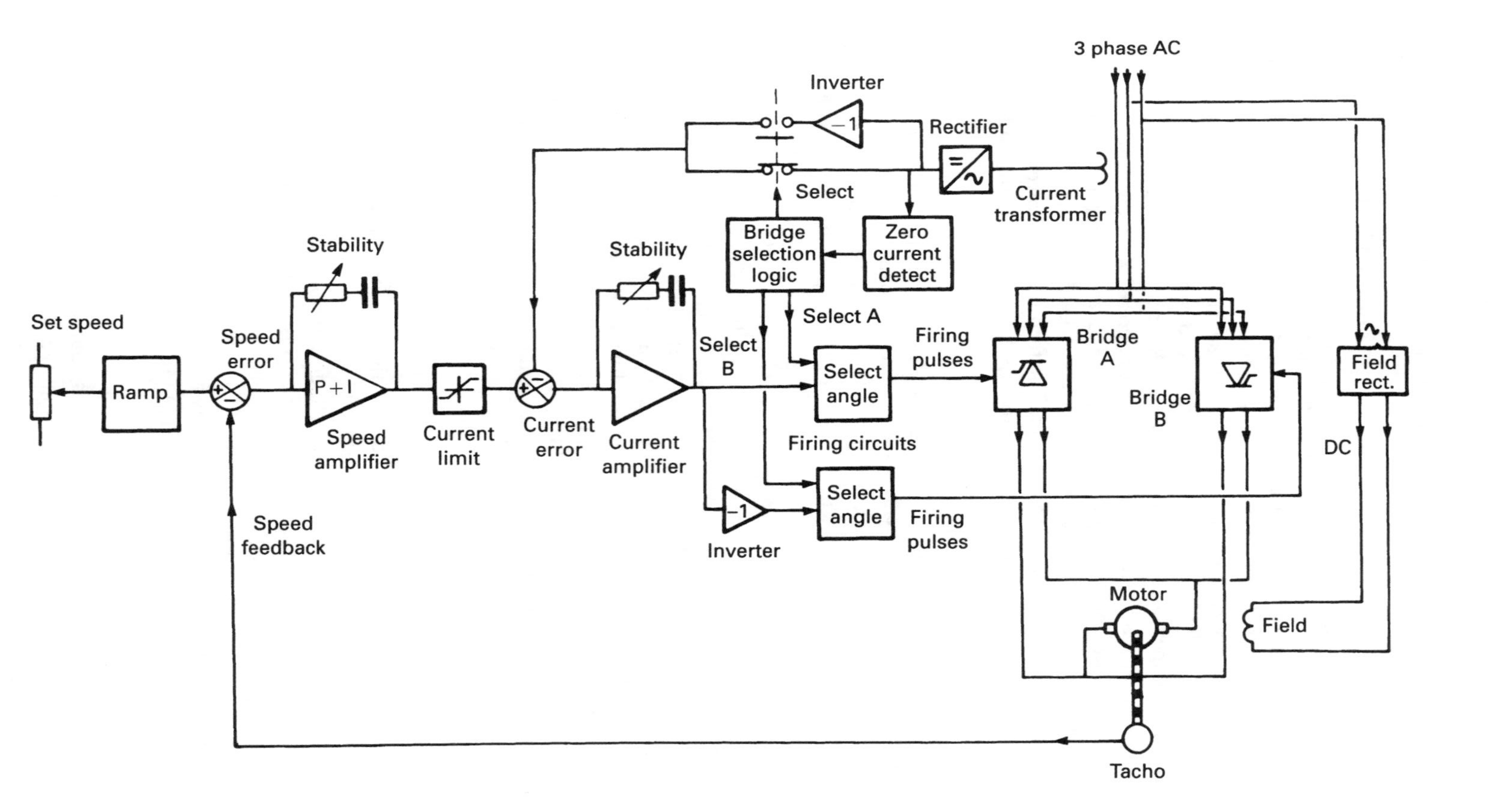

Fig. 12.43 *Block diagram of APSH drive.*

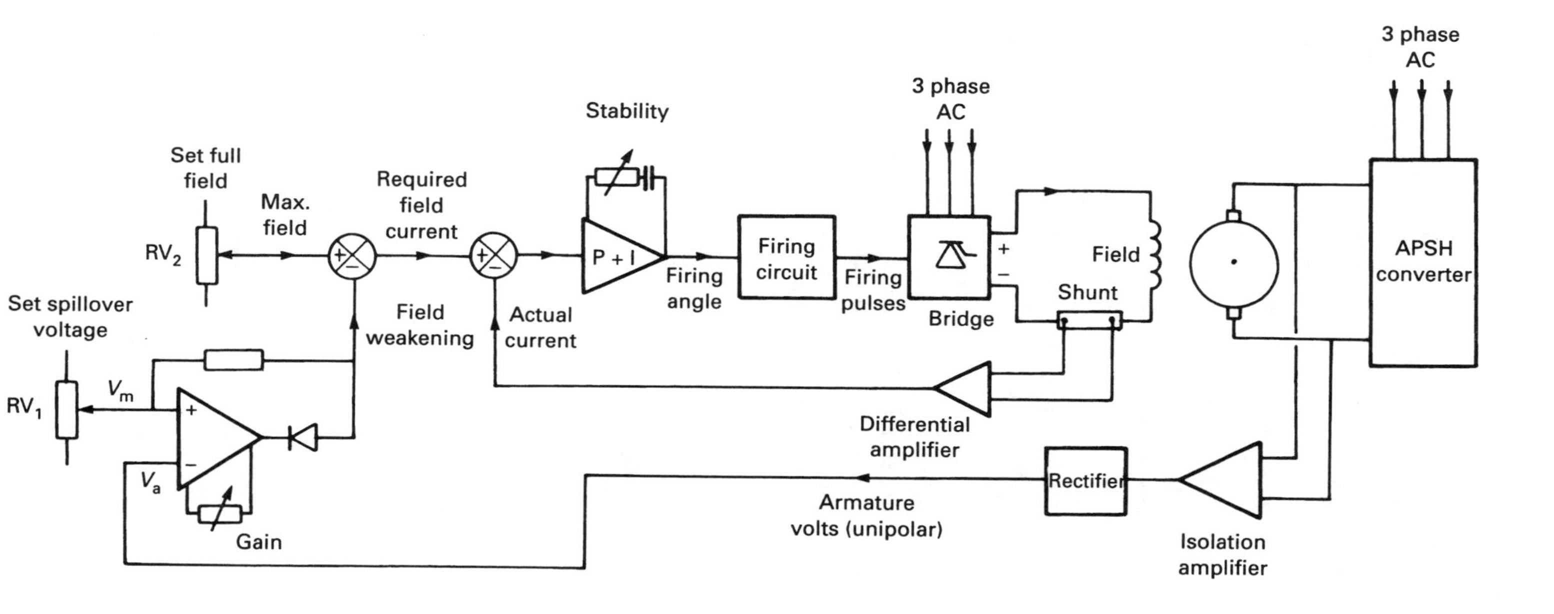

Fig. 12.44 *Field converter with field weakening.*

less than V_m, the output from the spillover amplifier will saturate positive, and the reference for the field controlled rectifier will be purely set by RV_2. If V_a starts to rise above V_m, however, the spillover amplifier will reduce the current reference, causing the field current to reduce.

The reduced field will cause an increase in speed, and the main speed loop will reduce the armature voltage. The net result is that V_a stays nearly equal to V_m, and speed is controlled by field current above base speed.

Below base speed, the motor is current limited and the maximum available torque is fixed (as in Fig. 12.45). A motor accelerating in current limit, for example, is accelerating with constant torque. Above base speed, the maximum available power is fixed, and the maximum torque falls. Ultimate speed is limited either by centrifugal forces on the armature or commutator segments, or by commutation failures (leading to excessive sparking).

Commutation can be improved by having a current limit which reduces as speed rises above base speed.

12.9.8. Gate firing circuits

The firing circuits in a thyristor drive are required to convert a control signal, say 0 to 10 volts, into firing pulses over a 0° to 180° range for each bridge thyristor. Figure 12.46 shows the firing pattern for a typical three phase bridge, and it can be seen that each thyristor requires two pulses 60° apart to ensure operation with

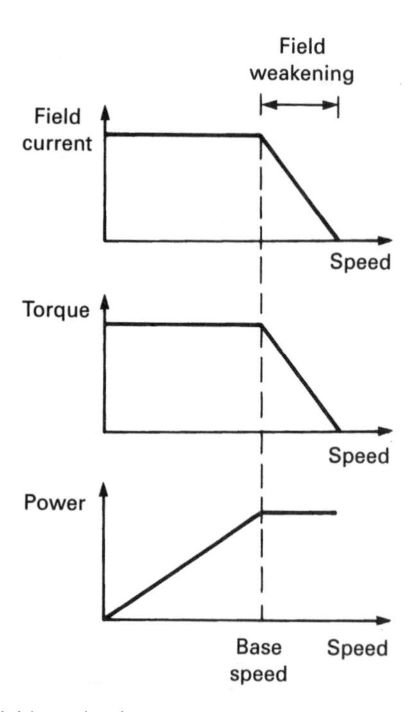

Fig. 12.45 *Effect of field weakening.*

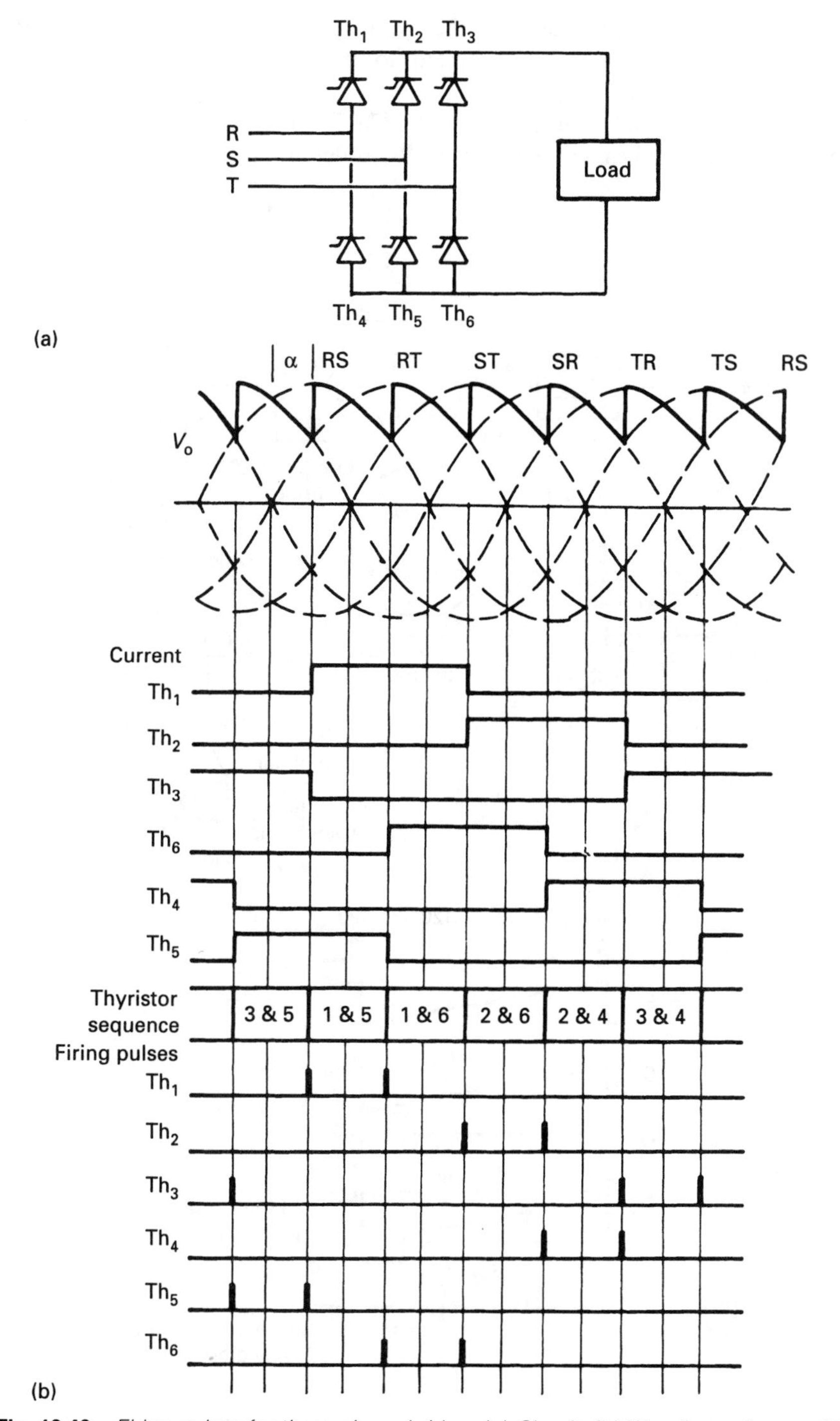

Fig. 12.46 *Firing pulses for three phase bridge. (a) Circuit. (b) Waveforms for $\alpha = 30°$.*

discontinuous current. Although, in theory, the pulses can be moved over a 180° range, in practice α is usually limited to the range 7° to 150° to avoid commutation failures.

One common circuit is shown in Fig. 12.47a. Zero crossing pulses are derived from one phase, and these are set by an RC differentiator to be 7° wide. These

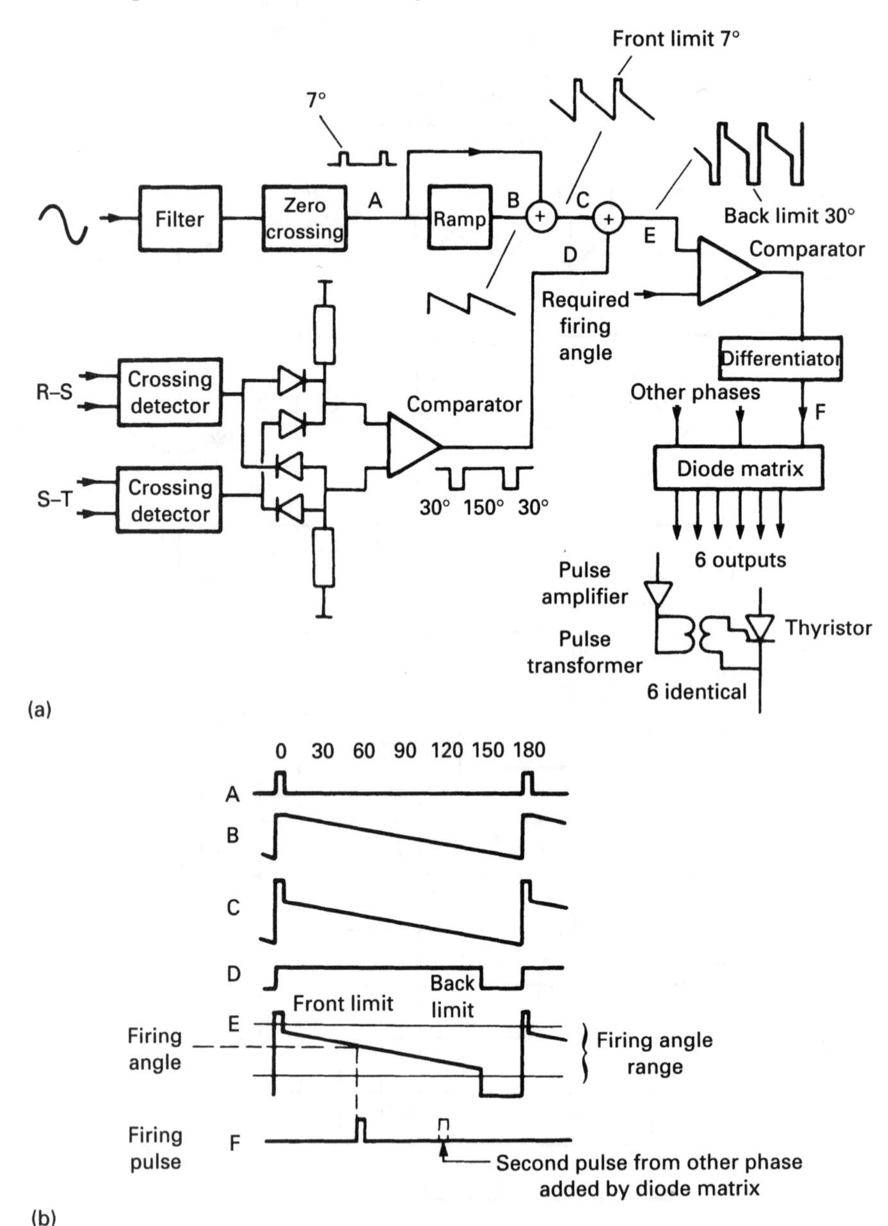

Fig. 12.47 *Typical gate firing circuit. (a) Circuit diagram. (b) Waveforms.*

trigger a ramp, and the pulses are added to the ramp to set the front 7° limit. The 30° back limit is derived by detecting crossover points between phases, the resultant square wave also being added to the ramp.

The composite ramp is now compared with the control voltage to give a firing pulse at the point where the ramp and control voltage are equal. Pulses from two identical circuits on the other phases are combined in a diode matrix to give the six double pulses for the six thyristors. Pulse amplifiers fire the thyristors via pulse transformers. Waveforms are shown in Fig. 12.47b.

The circuit of Fig. 12.47 has a non-linear response between control voltage and bridge output (as shown by equations 12.23 and 12.24). A linear response can be obtained by replacing the ramp of Fig. 12.43 with a cosine curve. This is simply obtained by phase shifting one supply phase with an RC network. Front and back stop pulses are added in a similar way to Fig. 12.47.

12.9.9. Ward Leonard control

The classical motor control scheme is the Ward Leonard system shown in Fig. 12.48, although this is only found nowadays on large motors with sudden load changes (such as mine winders or steel mill drives).

A DC generator is run at fixed speed by an AC motor, with a flywheel to absorb load changes. The output of the generator is attached directly to the motor armature, and controlled by the generator field. This is set by the low power signal from the speed error amplifier. The motor field is controlled by a spillover field weakening circuit similar to Fig. 12.44.

The only electronics in Fig. 12.48 are the low power field circuits, so the arrangement is very robust. The flywheel prevents impact loads being reflected into the supply. The disadvantages are the extra expense of the motor generator set, and a relatively slow response caused by the highly inductive field windings.

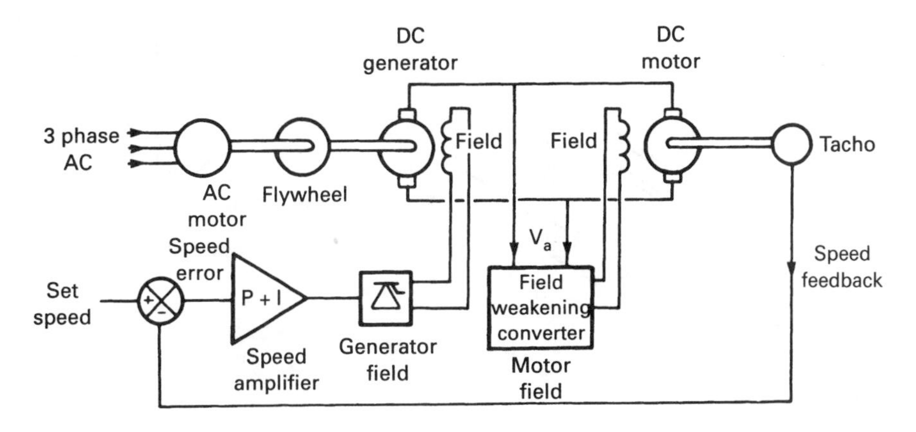

Fig. 12.48 *Ward Leonard control.*

12.10. AC machines

12.10.1. Rotating magnetic fields

DC motors have excellent controllability, but are expensive and need regular inspection and maintenance; particularly the brushgear and commutator. In many applications, motors are required to run at a fixed speed (or have adjustments over a limited range), and accurate speed holding is not required. In these circumstances there is an overwhelming cost advantage in using an AC motor.

Almost all industrial AC machines are fed from three phase AC at around 415 V, and work by producing a rotating magnetic field. Consider first, for simplicity, the two phase circuit of Fig. 12.49. The two coils, denoted A and B, are fed with voltages displaced 90°, as shown in Fig. 12.49a. The resulting magnetic fields are shown in Fig. 12.49b,c and d.

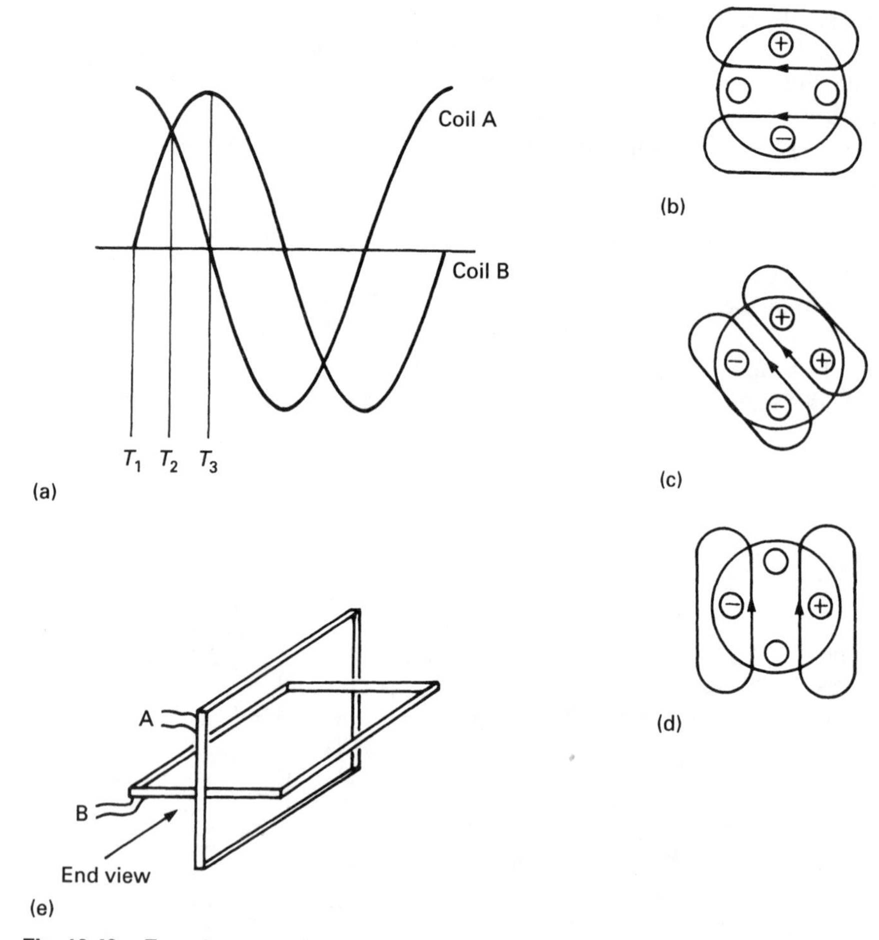

Fig. 12.49 *Two phase rotating magnetic field. (a) Coil voltages. (b) Time T_1. (c) Time T_2. (d) Time T_3. (e) Coil arrangement.*

At time T_1, there is only current in the A coil, producing a magnetic flux with direction parallel to the B coil. At time T_2, 45° later, the currents in the two coils are equal, causing the flux direction to move as shown. Similarly, at time T_3 the current in A coil has fallen to zero and the flux direction rotates further. Examination of further points in time would show that the flux would rotate through 360° (one revolution) for one complete cycle of the AC supply.

Similar considerations apply to the three-phase circuit of Fig. 12.50. Here we have three coils spaced 120° apart fed with three-phase AC as in Fig. 12.50a. The arrangement again produces a rotating magnetic field, examples of which are shown at 30° intervals in Fig. 12.50b, c and d.

It can be seen that, for Fig. 12.50, the angle of rotation of the magnetic field is the same as the phase in Fig. 12.50a. At T_3, for example, the three phases have advanced by 60°, and the field rotated by 60°. In one full mains cycle, the flux rotates through one revolution, i.e.:

$$n = f \tag{12.25}$$

where n is the rotational speed of the flux, and f the supply frequency.

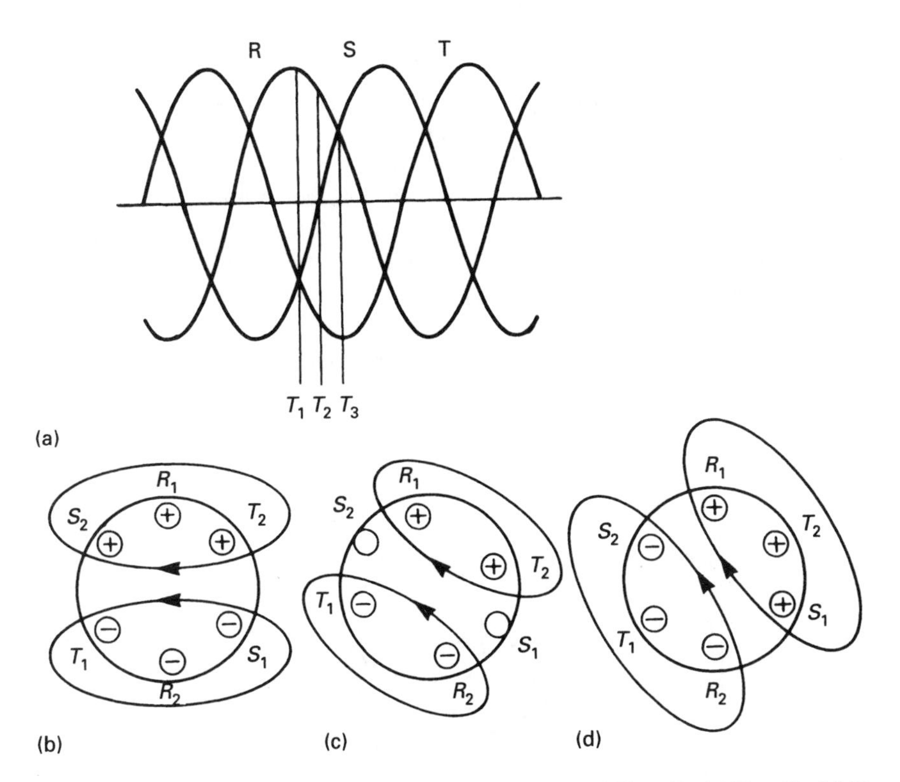

Fig. 12.50 *Three phase rotating field. (a) Coil voltages. (b) Time T_1. (c) Time T_2. (d) Time T_3.*

The stator in Fig. 12.50 is wound for one pole pair. If p is the number of pole pairs, equation 12.25 can, by similar reasoning, be amended to:

$$n = \frac{f}{p} \tag{12.26}$$

Values of n, in rpm, for 50/60 Hz supplies and different number of pole pairs are:

Pole pairs	50 *Hz*	60 *Hz*
1	3000	3600
2	1500	1800
3	1000	1200
4	750	900

It can be seen that the flux rotational speed decreases with the number of poles.

12.10.2. Synchronous machines

In Fig. 12.51 a permanently magnetised rotor is under the influence of a rotating magnetic field produced by a three phase stator, as described above. The rotor magnetisation can be achieved either by a permanent magnet or from DC windings connected by slip rings.

The interaction between the stator and rotor fields produces a rotational torque which causes the rotor to rotate at the same speed as the field, but lagging by an angle which is a function of the torque required to rotate the rotor and load.

Figure 12.51 is called a synchronous motor because it rotates in exact synchronism with the stator field (and hence some multiple of the supply frequency according to the number of pole pairs). A four pole (two pole pairs) synchronous motor, for example, will run at 1500 rpm, regardless of load up to the point where the motor cannot supply the required torque, at which point it will stall.

Synchronous machines, however, are only widely used for clocks and timer applications where a permanently magnetised rotor is used. Large synchronous

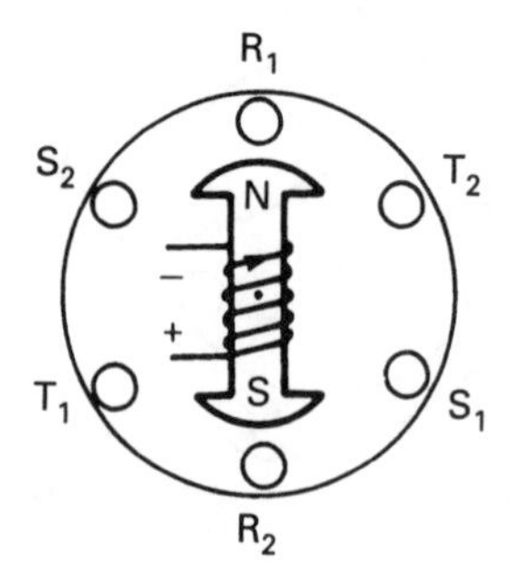

Fig. 12.51 *The synchronous motor.*

machines with DC rotors (upwards of 300 to 10 000 kW) are also used for some pumps and ID fans. The need for slip rings and heavy rotor windings, however, makes the induction motor, described below, more attractive for most industrial applications.

12.10.3. Induction motors

In Fig. 12.52, a stationary conductor is under the influence of a moving magnetic field. From Fleming's right-hand rule (generators), a current will be induced in the conductor. This current, however, produces a magnetic field which interacts with the external field to produce a force on the conductor (Fleming's left-hand rule, motors).

If the conductor is free, it will accelerate in the direction of the field movement. As its speed increases, however, its velocity relative to the field reduces and the induced current falls, reducing the accelerating force. The conductor will settle at a constant speed, less than the velocity of the field, where the force on the conductor balances the force required to propel the conductor and any external load.

In the simplest induction machine, called the squirrel cage motor, the rotor consists of copper, or aluminium, conductors shorted by end rings as in Fig. 12.53. A laminated iron rotor core increases the magnetic efficiency. The rotor is free to rotate inside a stator similar to Fig. 12.50 which produces a rotating magnetic field.

Induced currents from the field cause the rotor to rotate as described above, with the rotor rotating at a lower than synchronous speed to produce torque. If the rotor speed is ω_r and the supply frequency ω, the fractional slip is defined as:

$$s = \frac{\omega - \omega_r}{\omega} \quad \text{with} \quad 0 \le s \le 1 \tag{12.27}$$

Slip increases with load torque, and will be typically 5% for a normal machine under design operating conditions.

At start-up, the rotor is stationary and the induced currents are large. As the motor speeds up, the induced currents (and frequency) fall. The induction motor

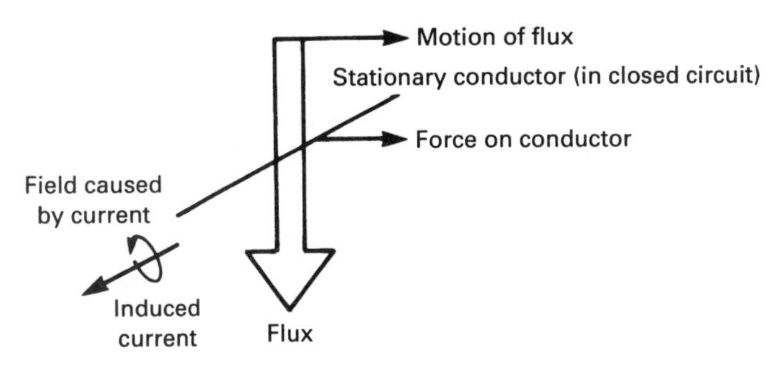

Fig. 12.52 *The principle of the induction motor.*

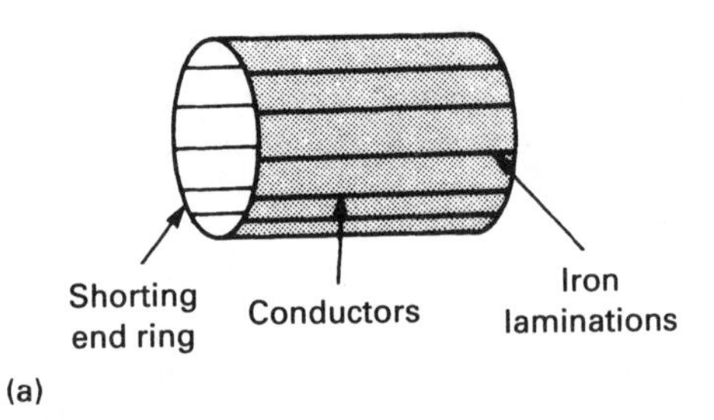

Fig. 12.53 *The squirrel cage induction motor. (a) Rotor construction. (b) Construction of a typical induction motor. (Photo courtesy of GEC small machines.)*

has a large starting torque, and a correspondingly (sometimes embarrassingly) high starting current.

At a slip s, the induced rotor current has a frequency, sf, where f is the supply frequency. The rotor circuit impedance is therefore

$$Z = \sqrt{R^2 + (sX)^2} \tag{12.28}$$

where X is the inductive impedance at standstill (where the rotor current frequency and supply frequently match) and R the rotor resistance.

The rotor input power is ωT. Neglecting windage losses, this is split between useful output power, $\omega_r T$, and rotor iron losses, $3I^2R$ (for a three phase machine), i.e.:

$$\omega T = \omega_r T + 3I^2R \tag{12.29}$$

$$\mathrm{T}\,(\omega - \omega_r) = 3I^2R \tag{12.30}$$

$$\frac{(\omega - \omega_r)}{\omega} . \omega T = 3I^2R \tag{12.31}$$

or

$$s\omega T = 3I^2R \tag{12.32}$$

i.e. slip × (input power) = rotor I^2R loss.

The electrical power dissipated in the rotor is $3I^2R$

$$= \frac{3s^2E^2}{R^2 + (sX)^2} \tag{12.33}$$

where E is the induced rotor emf at standstill. Substituting into equation 12.30 gives:

$$T(\omega - \omega_r) = \frac{3s^2E^2R}{R^2 + (sX)^2} \tag{12.34}$$

$$\frac{T(\omega - \omega_r)}{\omega} = \frac{3s^2E^2R}{\omega(R^2 + (sX)^2)} \tag{12.35}$$

$$sT = \frac{3E^2}{\omega} \cdot \frac{s^2R}{R^2 + (sX)^2} \tag{12.36}$$

or

$$T = K\frac{sR}{R^2 + (sX)^2} \tag{12.37}$$

where K is a constant for the machine $(3E^2/\omega)$.

For small values of slip, as is usually the case, the term $(sX)^2$ can be neglected, and:

$$T = \frac{Ks}{R} \tag{12.38}$$

i.e. for small slip, the torque is proportional to the slip.

The torque/slip relationship, equation 12.37, is shown for different relationships between R and X in Fig. 12.54a. Note that slip = 1 corresponds to a stationary rotor, and slip = 0 corresponds to synchronous speed.

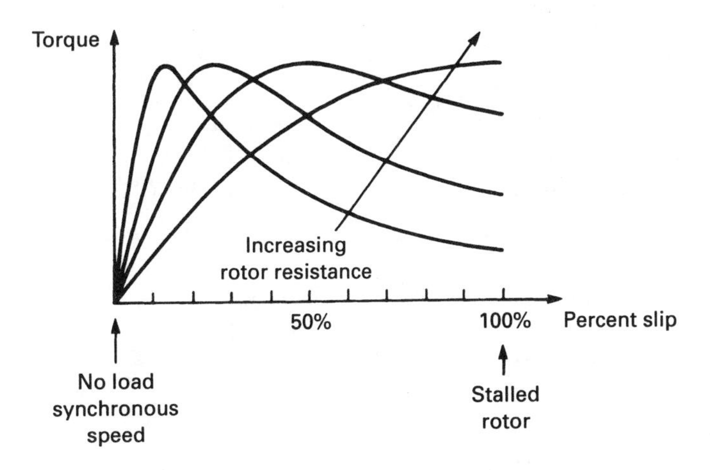

Fig. 12.54 *The performance characteristic of the induction machine. Torque/slip relationship.*

For practical machines, $X > R$, giving a speed/torque relationship similar to Fig. 12.55. The maximum torque occurs for a slip of about 10%, and is roughly twice the minimum torque. The machine is normally operated in the region AB. The motor's nominal torque is the minimum torque, corresponding to point C. This also occurs at point D where the slip is about half that at point A. Note that the starting torque is roughly 150% of the minimum torque. Operation at speeds below point A is not recommended (other than for starting and stopping).

The normal running current of a three phase induction motor is given by:

$$I = \frac{KW \times 1000}{\sqrt{3} \times \text{Voltage} \times \text{efficiency} \times \text{power factor}} \tag{12.39}$$

where the efficiency is expressed in the range 0 to 1, e.g. 0.85 for an 85% efficient motor. For a typical 415 V 11 kW motor with 88% efficiency and power factor ($\cos\phi$) of 0.9, the running current at full load will be 19.3 A.

12.10.4. Starting induction motors

An induction motor has a very high initial starting current because the slip is 100% when the rotor is stationary. In a typical motor this can be five times the normal running current, (over 450 A for a 55 kW motor with a normal running current of about 90 A). This high current is accompanied by a large shock load into the motor, its couplings and the load.

The large current spike will also be fed back into the supply causing light flicker and possible trips of upstream breakers. A restriction is normally applied to the size of motor that can be started direct on line (DOL). This will vary according to

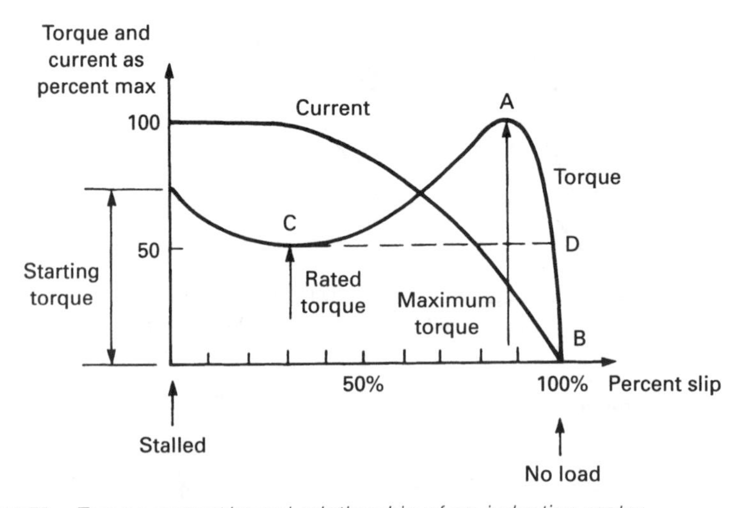

Fig. 12.55 *Torque-current/speed relationship of an induction motor.*

the user and the locality. A small machine shop in a residential area may be limited to 4 kW. At the steel plant where the author works, however, motors up to 90 kW are commonly started DOL.

A common method of reducing the current surge is the star-delta starter shown on Fig. 12.56. The motor is wound with all six coil ends brought out. These are connected initially in star (C_1 and C_2 energised) to provide a reduced voltage for the start. After a short delay (usually a few seconds) set by the timer T, C_2 is de-energised and C_3 energised to bring the motor to full speed and torque. There are two current surges; a much reduced starting pulse and a large (but very short) pulse as the motor changes from star to delta. Ready-built star-delta starters are available.

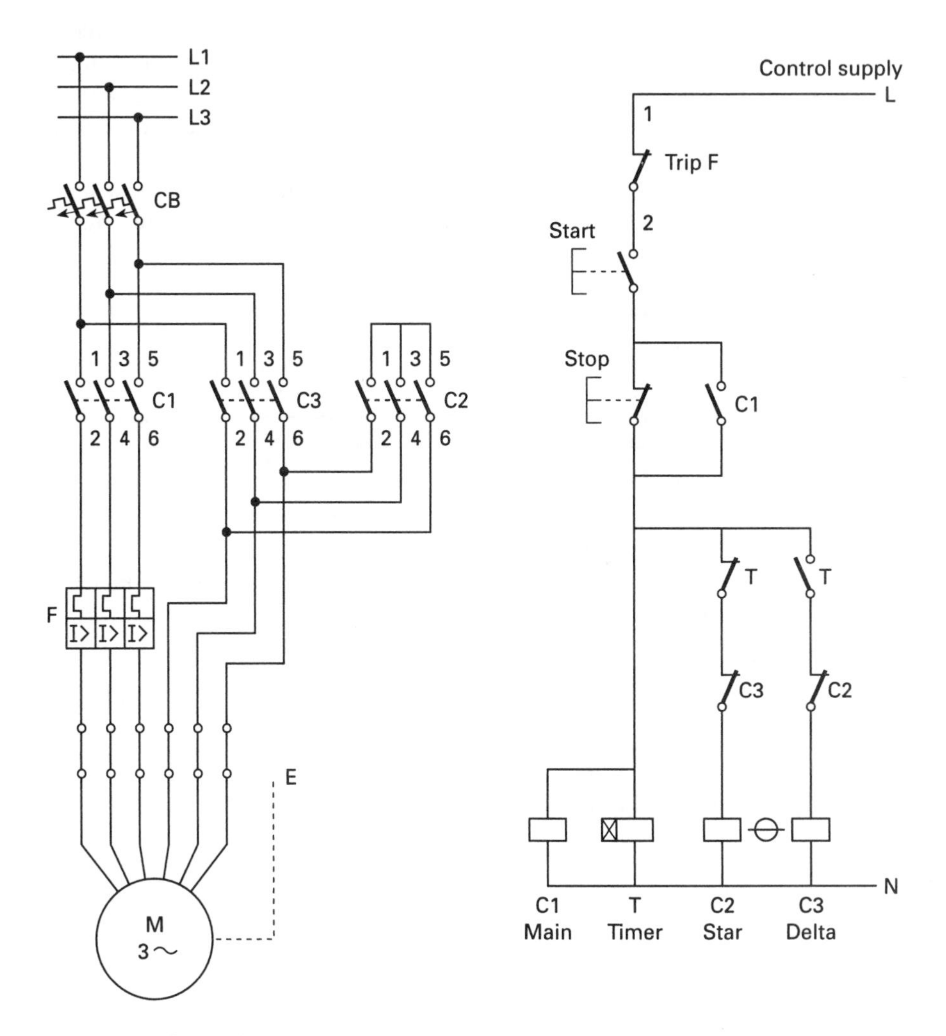

Fig. 12.56 *Star-delta starter.*

Soft starters, described in the following section are a more modern method of reducing the starting current of an induction motor.

12.10.5. Soft start units

Soft starters control the starting current by slowly increasing the voltage applied to the motor over a period of time (usually between ten and twenty seconds.) The power electronics is very simple, just six thyristors connected as antiphase parallel pairs as Fig. 12.57. At start-up these are phased right back, giving zero volts, then the phase angle slowly decreased to raise the voltage and give a smooth start. The ramp rate controls how fast the phase angle is reduced.

A simple linear voltage ramp as Fig. 12.58a can cause problems with a high stiction load. There is no movement (and hence a high current) until a certain voltage is reached when the load suddenly kicks off. This effect can be overcome by either a quick pulsed start before the ramp as Fig. 12.58b or by the provision of an initial starting current (torque) usually called the pedestal as Fig. 12.58c. The duration of the kickstart duration and the height of the pedestal are adjusted by the user.

Some soft starts also have a constant current acceleration mode. As its name implies, the phase angle is adjusted to keep the current constant during

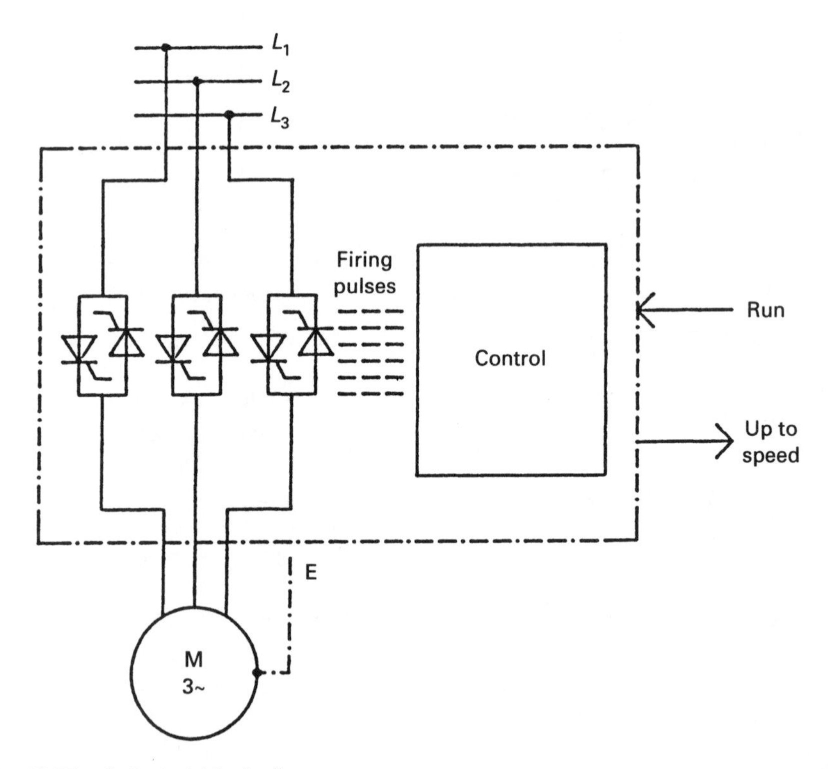

Fig. 12.57 *Soft start block diagram.*

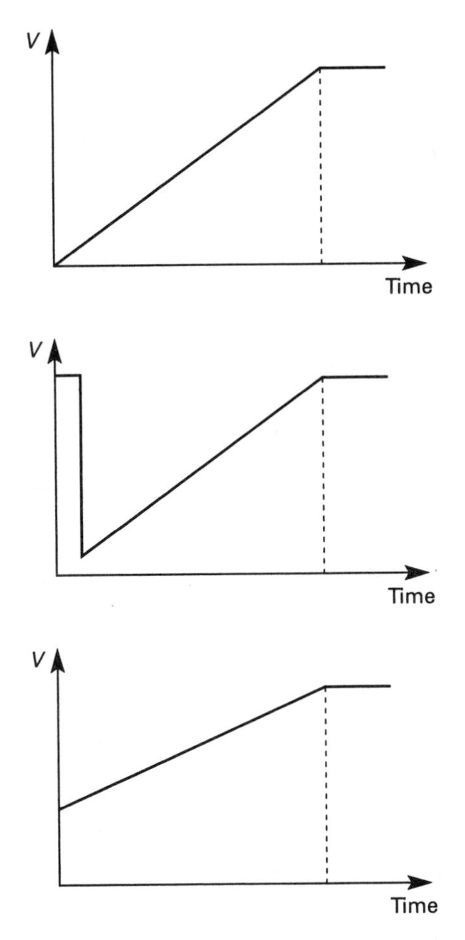

Fig. 12.58 *Soft start characteristics. (a) Linear ramp. (b) Kickstart. (c) Pedestal.*

acceleration. This mode is useful if there is considerable backlash in the load or the supply feeding the motor has little spare capacity. A constant current acceleration can bring a motor up to speed in the shortest possible time for a given supply constraint.

Care must be taken with constant current ramps, however. If the current is set too low the motor can hang at an intermediate speed and be unable to get to full speed. The motor then suddenly kicks to high speed (with the usual shock load and current surge) when the soft start ramp finishes.

Figure 12.59 shows a typical arrangement. A three phase contactor is still needed for safety reasons, an emergency stop circuit cannot be applied to the run command of a soft start unit as the control electronics may have failed. The contactor switches the supply and an auxiliary contact enables the soft start. Because the contactor contacts do not have to close onto the DOL starting

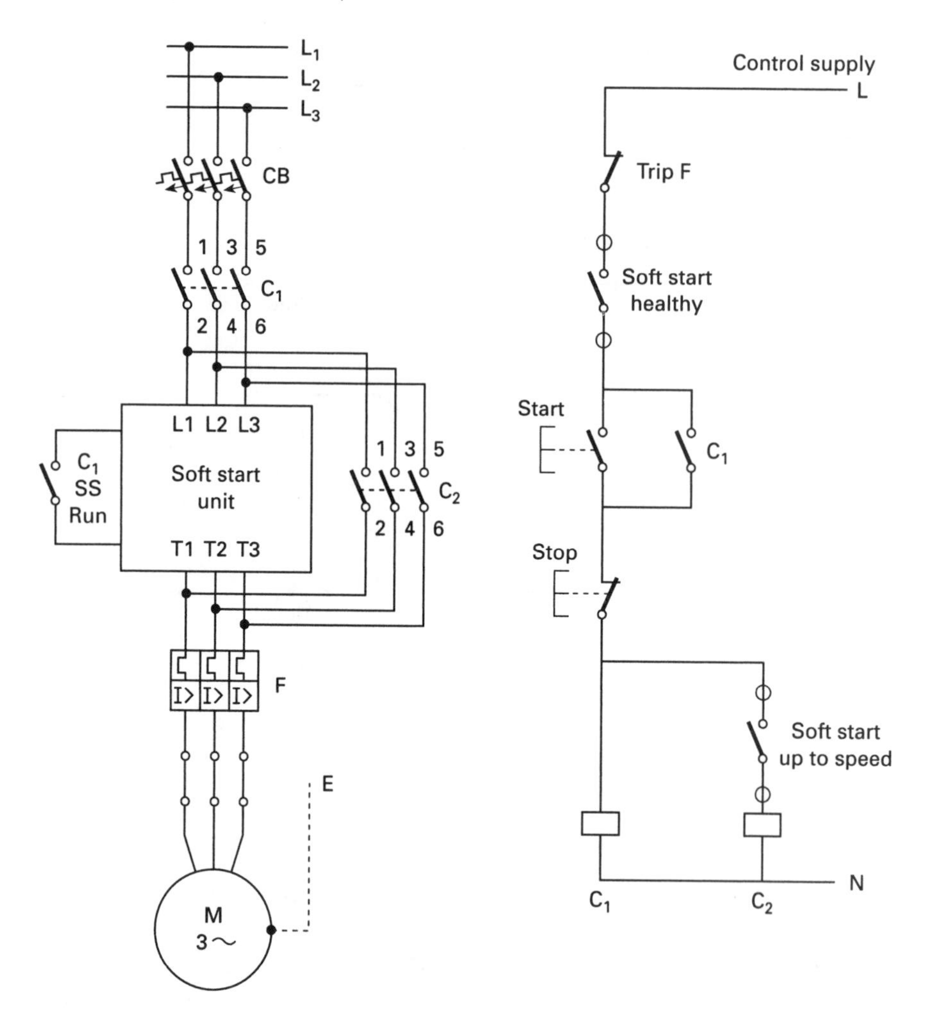

Fig. 12.59 *Soft start with bypass.*

current their life will be considerably extended. Most soft starts provide an 'up to speed' contact output. This can be used to bring a bypass contactor in around the soft start. A bypass is normally only needed if the heat produced by the current flowing through the thyristors in the soft start is a problem.

It is possible to use a soft start to obtain a rather crude form of closed loop speed control by running the motor at a reduced voltage and large slip. The speed feedback from the tacho reduces the motor voltage. This is really only feasible with fans and pumps which have an exponential load characteristic. Care must be taken with motor cooling, the increased slip will cause extra heat and the lower speed will reduce the motor fan's cooling ability. Variable voltage speed control can be unpredictable, and works best with a motor having high resistance coils.

Motors are manufactured with coils designed specifically for this type of control.

Crude open loop speed control is also possible with high inertia loads by pulsing (burst firing) the thyristors. Again care must be taken to avoid overheating the motor.

Sophisticated soft starts also incorporate a soft stop facility. This is based on DC injection braking where DC is passed through one coil. The DC field and the AC field from the rotating motor produce a braking force determined by the magnitude of the current.

Another available facility is energy saving, although this is really only effective with fans and pumps. Energy-saving soft starts sense the slip angle and reduce the voltage until the slip is just within acceptable limits. A by-pass contactor obviously cannot be used with energy saving. When energy saving is taking place the thyristors are running with a firing angle which may cause harmonic problems on the supply.

12.10.6. Direction reversal

The direction of rotation of an induction motor is determined by the direction of rotation of the magnetic flux. Reversal of direction can simply be achieved by interchanging any two rotor connections. Figure 12.60 shows a simple DOL reversing starter. Note that both mechanical and electrical interlocking are provided to prevent both contactors being energised at the same time.

As a motor changes direction there will again be a large current surge similar to that described in the previous section. If there is a temptation for the operator to go forward, reverse and forward again quickly, anti-plug timers can be used to ensure that the motor comes back to a standstill before starting in the opposite direction.

12.10.7. Slip power recovery

If power is extracted from the rotor of an induction motor, the speed will reduce. Figure 12.61a shows a fairly crude way of achieving this. A wound rotor is used, whose connections are brought out via slip rings. If these are shorted, the motor acts as a normal induction motor.

In Fig. 12.61a, the rotor connections are linked via three variable resistances. The higher the resistance, the higher the I^2R loss and the lower the speed. The speed will, however, be also load dependent, and as such is best suited for simple applications such as crane hoist control. The rotor resistors also dissipate a fair amount of heat.

Figure 12.61b is a more sophisticated version of Fig. 12.61a, which can be adapted to give closed loop control over a limited speed range. The output from the rotor (which is low frequency AC) is rectified and fed back to the supply via a naturally commutated controlled converter. This arrangement is called slip power recovery. It can, of course, only be used to reduce the speed in an induction motor.

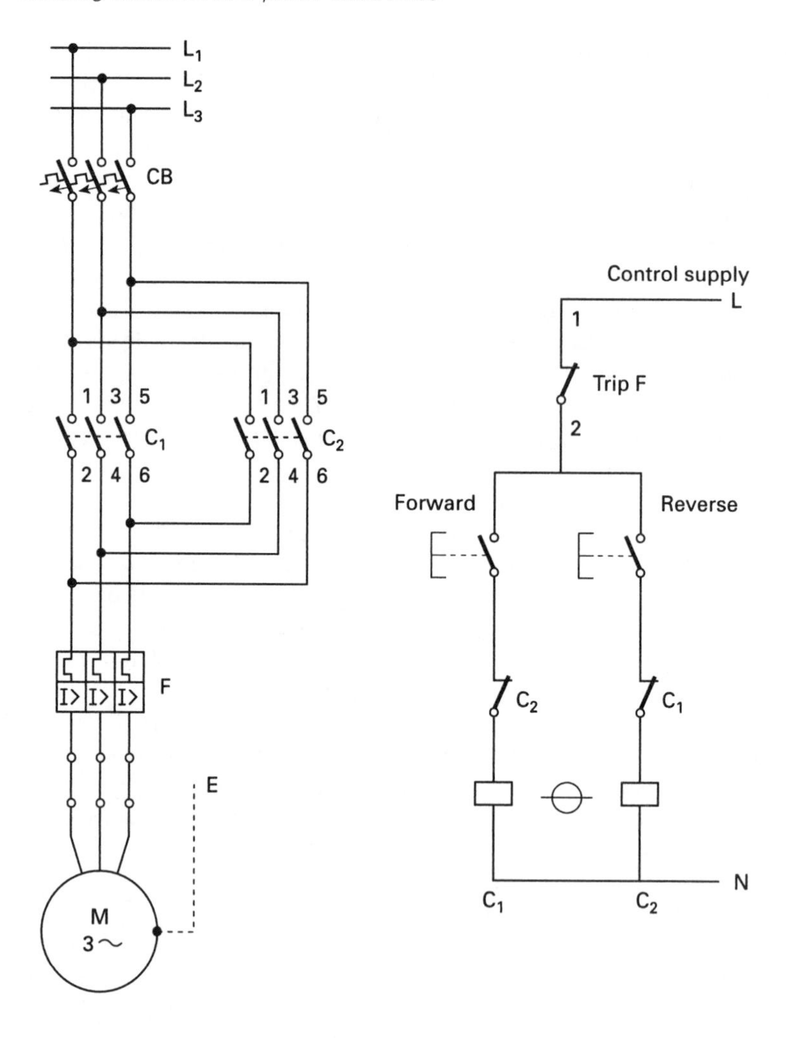

Fig. 12.60 *Reversing an induction motor.*

12.11. Speed control of AC motors

12.11.1. Introduction

Ignoring slip, the speed of an induction motor is given by

$$\text{speed} = 60f/p \text{ rpm} \tag{12.40}$$

where f is the supply frequency in Hz and p is the number of pole pairs. A four-pole 50 Hz motor has a no load speed of 1500 rpm.

For a normal motor and load there are two common ways the speed can be

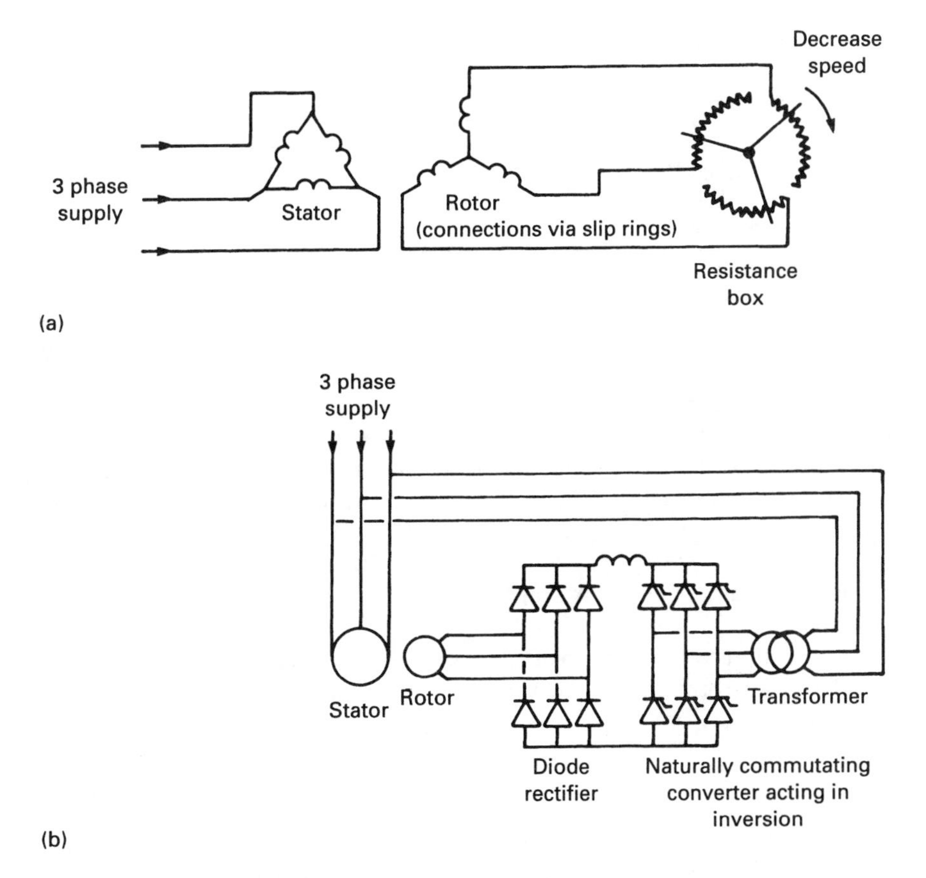

Fig. 12.61 *Slip power recovery speed control. (a) Speed control via resistance box. (b) Speed control via rectifier/converter.*

varied; changing the number of poles or changing the frequency. There is actually a third method, varying the voltage which changes the slip, but this is limited to fans and similar loads and ideally needs special motors with high resistance coils. Voltage speed control is described in Section 12.10.5 as it is a type of soft start.

Pole switching can vary the number of poles in an integer ratio. If all the coil connections of a four-pole NSNS motor are brought out (six wires) it can be connected by contactors as a two-pole NNSS motor running at 3000 rpm or at 1500 rpm in the original four-pole arrangement, a 2 : 1 ratio. A better method called pole amplitude modulation (PAM, not to be confused with pulse amplitude modulation VF drives described in Section 12.11.2) uses non-uniform windings in the stator to give two distinct (but still integer ratio) speeds. There are six terminal connections, three for one speed and three for the other. Two contactors, one for each speed, are required.

Common arrangements are:

Poles	*Speeds (rpm)*
4:6	1500:1000
6:8	1000:750
8:10	750:600

although speed ranges as wide as 3000 : 250 rpm (2 : 24 poles) can be obtained. The author rather likes the labelling of terminals used in some motors, where the high speed is shown as a hare, and the low speed as a tortoise.

If true speed control is required, the only solution is to vary the frequency. Early ways of achieving this used a variable speed DC motor controlling the speed of an AC alternator which supplied the induction motors whose speed needed to be varied. This method still has some advantages where the motors are subject to very high shock loads or exposed to water and high humidity. The commonest method, though, is to provide a variable frequency supply from a power electronics inverter. A range of near zero to around 90 Hz is normally available.

An induction motor needs to operate with a constant amplitude magnetic flux, but the flux is proportional to supply voltage and inversely proportional to supply frequency, i.e.:

$$\text{flux} = \frac{KV}{f} \tag{12.41}$$

where K is a constant for the motor. If the frequency is reduced below the motors designed frequency (50 Hz in the UK) the voltage must be reduced in proportion. The motor will then operate at constant torque below its designed speed.

If the motor is run higher than its design speed the voltage should be increased to maintain constant flux. This is not advisable, as the voltage rating of the motor insulation might be exceeded (and in any case is not possible with the DC link design described below). The voltage is therefore held at the rated voltage for higher speeds. This leads to a fall-off in torque and constant power. The relationships between voltage, power, torque and speed are shown on Fig. 12.62.

At low frequencies the rotor resistance has an increasing effect, leading to a fall-off in torque. This can cause problems starting, and a common user selectable option is to boost the voltage at low frequencies as shown on Fig. 12.62d.

All inverters have the block diagram of Fig. 12.63. The incoming AC (which can be single phase or three phase) is rectified to DC which is then 'chopped' by normal thyristors, GTOs (gate turn off thyristors) or high voltage IGBT (insulated gate bipolar transistors).

There are three distinct ways in which an inverter can be operated. It can supply varying frequency voltage (the PAM and PWM inverters), varying frequency current (the CSI inverter) or flux vector where the inverter uses a model

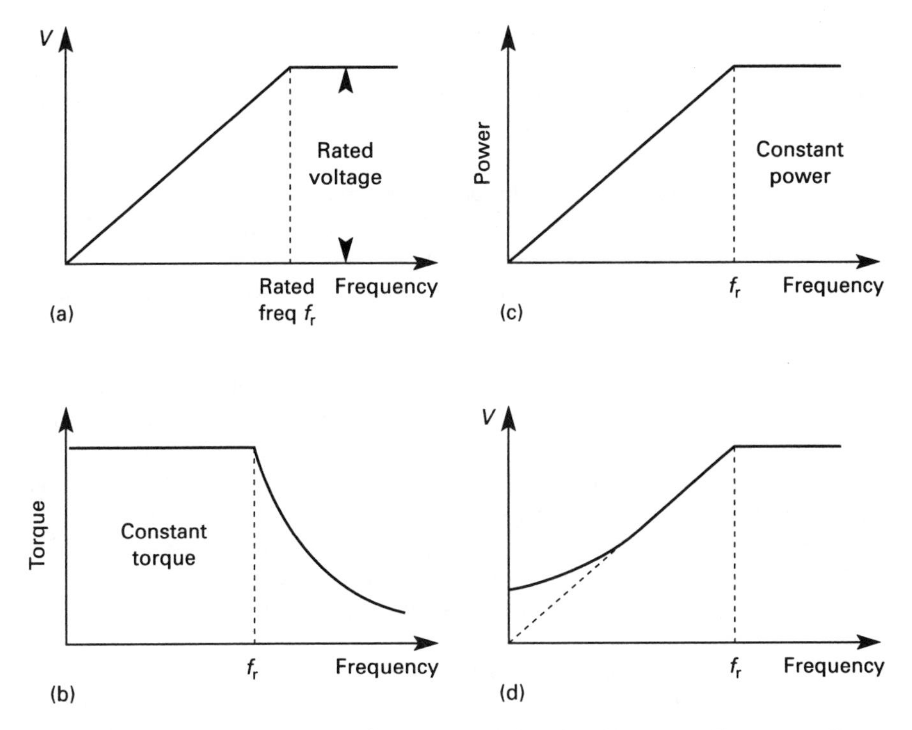

Fig. 12.62 *Variable frequency drive characteristics. (a) Voltage. (b) Torque. (c) Power. (d) Stator resistance compensation.*

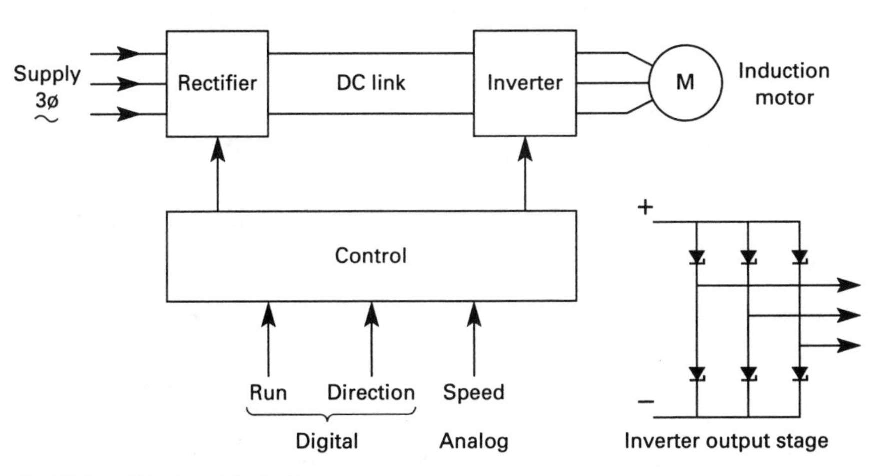

Fig. 12.63 *VF drive block diagram.*

of the motor to control the position of the magnetic flux. Voltage source inverters must be used where several motors are operated in parallel. The flux vector is a high performance drive which gives similar dynamic performance to a DC drive. These are now described in more detail.

12.11.2. Pulse amplitude modulation (PAM) inverters

These vary the voltage of the DC bus either by means of a chopper (Fig. 12.64a) or a controlled rectifier (fig. 12.64b). The voltage is set as required by the voltage/frequency relationship described in the previous section. The frequency of the output voltage is then controlled by the firing of the thyristors in the output inverter stage.

The firing pattern is fixed, with the time duration being set by the frequency. The crudest firing pattern is the six-pulse waveform shown on Fig. 12.65a for 50 Hz and 25 Hz operation. These seem to bear little relationship to a sine wave, but are phase-to-phase voltages. The phase-to-neutral voltages, shown on Fig. 12.65b are better (but by no means perfect).

The eighteen-pulse waveform of Fig. 12.65c is more commonly used. This gives a better approximation to a sine wave, reducing harmonics interference back into the supply and reducing motor heat dissipation and losses.

12.11.3. Pulse width modulation (PWM) inverters

The PWM approach is probably the commonest design of inverter at present. It uses a fixed voltage DC bus (i.e. an uncontrolled rectifier) and the switching of the output thyristors is used to control both voltage and frequency as shown on Fig. 12.66.

A reference sine wave of the desired frequency and amplitude corresponding to the voltage frequency ratio needed by the motor is compared with a 'delta' triangular waveform as shown. The comparison produces firing pulses for the inverter thyristors.

The ratio of the delta frequency to the reference frequency is called the '*n* ratio'.

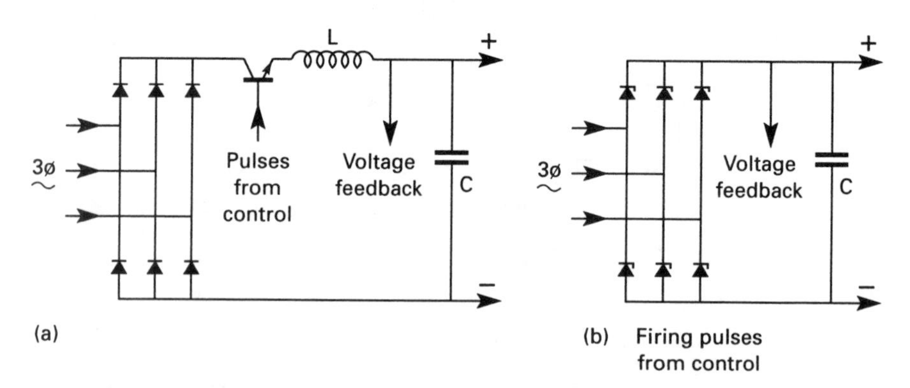

Fig. 12.64 *DC Link voltage control in PAM VF. (a) Uncontrolled rectifier with chopper. (b) Controlled rectifier.*

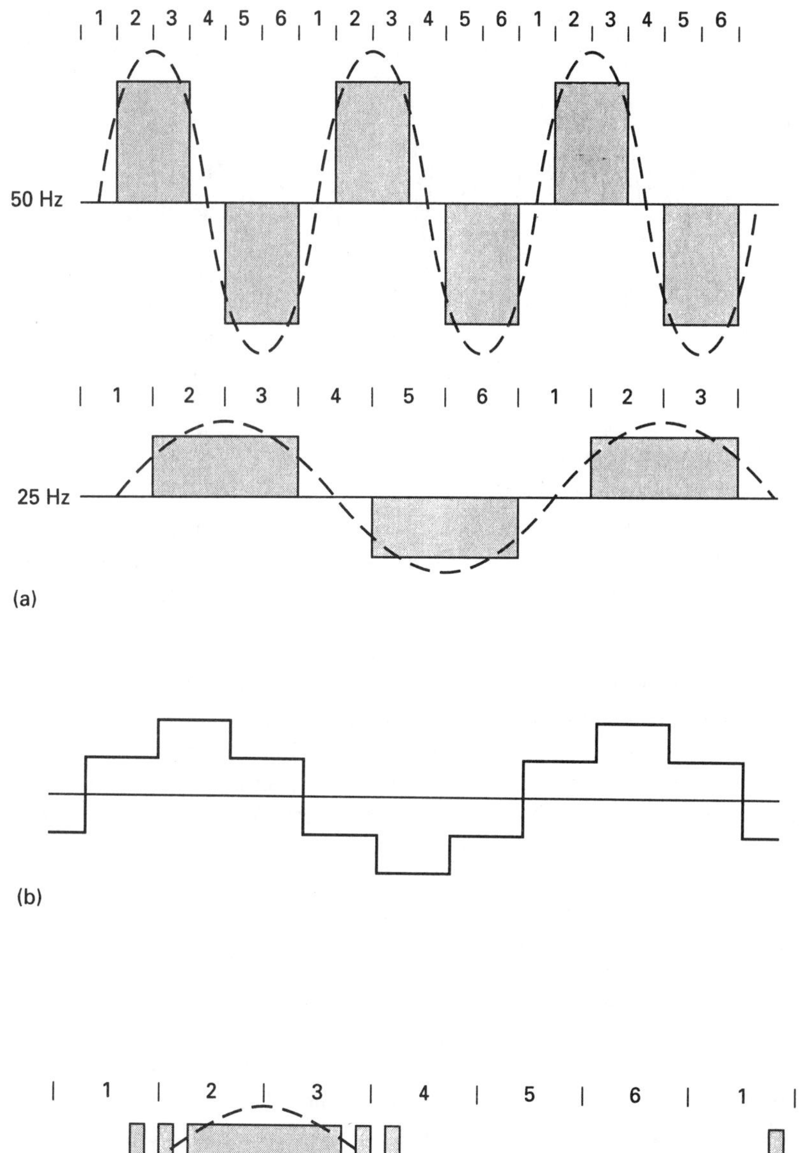

Fig. 12.65 *Pulse Amplitude modulation (PAM) VF. (a) Phase-to-phase voltage (6-pulse). (b) Phase-to-neutral voltage. (c) Phase-to-phase voltage (18-pulse).*

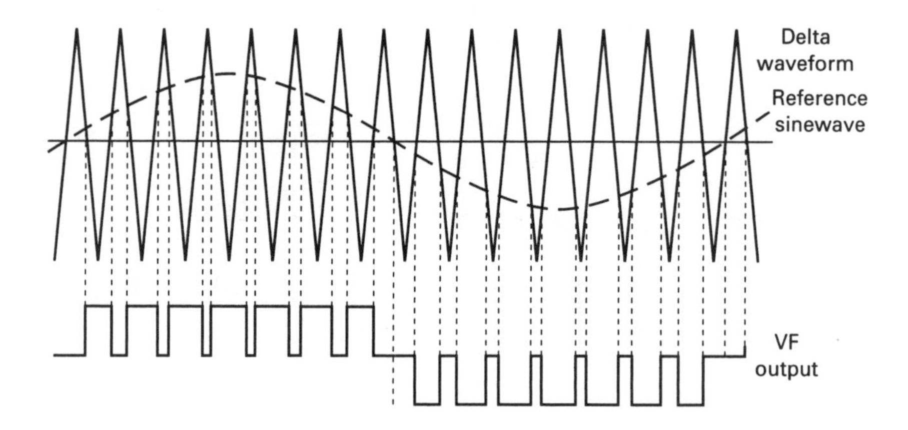

Fig. 12.66 *Pulse width modulation (PWM) VF.*

Theoretically it should be as high as possible to simulate the best sinusoidal waveform and to aid the reactance of the motor coils in minimising the harmonic effect. If there is a multiple of three to one ratio between the desired frequency and the delta frequency the three phases will cancel out all harmonics divisible by three.

Unfortunately there are limits on the speed at which a thyristor can be switched, and this is compounded by the need to force commutate the thyristors from an 'On' state to an 'Off' state which is wasteful of energy. High voltage transistors and field effect transistors (FETs) have a better response.The *n* ratio is normally made high at low frequencies and reduces as the frequency rises. Above mains frequency a PWM inverter normally operates with a fixed *n* ratio which gives either a six-pulse or eighteen-pulse PAM waveform. The *n* ratio does not continually vary with frequency, but switches at predetermined points. This produces a marked 'gear changing' sound which is characteristic of PWM inverters.

12.11.4. Current source inverter (CSI)

The current source inverter, as its name implies, switches current pulses rather than voltage pulses to the motor. A controlled rectifier feeds a DC link which has a large inductance to maintain continuous current. The thyristor inverter stage has additional components (diodes and capacitors) to maintain the current whilst allowing commutation as shown in Fig. 12.67.

With the PAM and PWM inverters the voltage is determined by the inverter. With a CSI the voltage is determined purely by the load, and will vary as the frequency or load changes.

The main advantage of a CSI is that it is short circuit proof, and the motor runs smoothly and quietly at all speeds. There are, however, major disadvantages. A CSI must be matched to the load (PAMs and PWMs are universal) and cannot be run with parallelled motors. There must always be a load connected, so no downstream switching (isolators or contactors) is allowed. It is not good practice

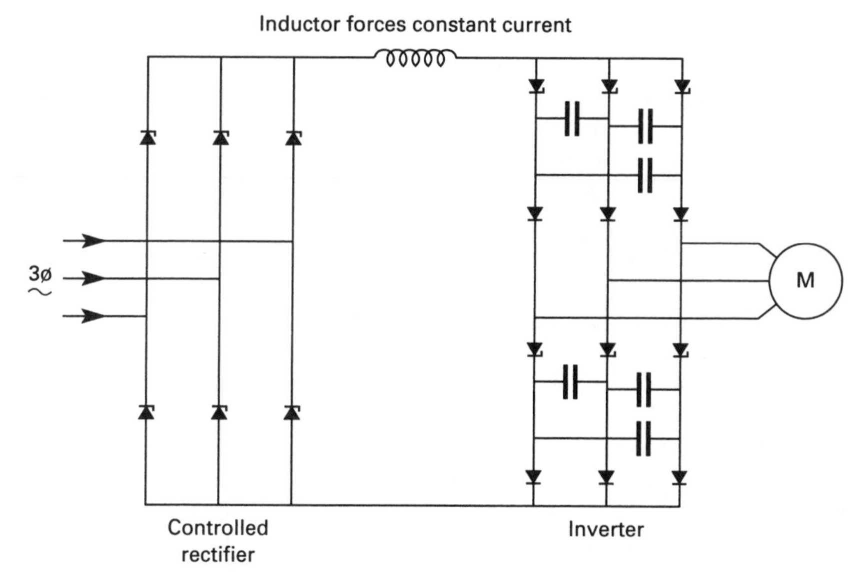

Fig. 12.67 *Current source inverter (CSI) VF.*

to open circuit the output on any inverter, but it will cause damage and danger with a CSI. A CSI cannot be used with devices such as slip rings which can give brief breaks of connection in normal operation.

12.11.5. The flux vector drive

The inverters described so far are relatively cheap and are used for simple speed variation. In most applications the speed control is open loop; a speed reference is set and the motor will run at the set frequency minus the load determined slip. Closed loop control is possible, even common, but the nature of the induction motor gives a dynamic response (to load or reference changes) which is poor compared to a DC drive.

The flux vector inverter aims to give improved *dynamic* performance by directly controlling not the voltage or current, but the magnetic flux within the motor itself. To achieve this, the control needs feedback of motor current, rotor speed and rotor position, (or more strictly the change of rotor position relative to the stator field) plus a model of the motor. The last two feedback items are provided by an incremental encoder which must be connected to the rotor shaft.

The equations are non-linear and complex so fast powerful computational ability is required in the drive. The result is an induction machine drive with dynamic performance as good as a DC drive.

It is very important to recognise the applications that a flux vector drive is intended for. It is not in competition with the PWM/PAM and CSI drives. Its complexity, cost and the disadvantages of needing a motor model and a shaft mounted encoder totally rule it out from simple applications (but this does not, of course, stop salesmen trying to sell them as 'the latest technology'). Simple applications such as fans, conveyors and material feeders only need PAMs or PWM drives.

The flux vector drive is in direct competition with the DC drive. An expensive flux vector drive coupled to a very cheap, robust, maintenance free induction motor can be very competitive when compared with a moderately expensive DC drive coupled to a very expensive DC motor needing regular maintenance. Flux vector drives are appearing in paper making and rolling mill drives where they do have distinct cost benefits.

12.11.6. Braking and four quadrant operation

If the inverter frequency is reduced, the motor will regenerate and cause the DC link voltage to rise. In extreme cases the rise of voltage can be sufficient to cause the drive to trip. If the inverter has a controlled rectifier the drive can regenerate back into the mains giving full four quadrant operation. In cheaper drives the rise in DC voltage is detected by a voltage comparator which switches in a high wattage resistor as in Fig. 12.68. The regenerated energy from the drive is lost as heat, so the resistor is nearly always mounted outside the drive cubicle. In many drives this is called a braking unit and is supplied as an optional extra.

12.11.7. General comments

Most VFs on the market at present are based on the PWM idea. CSI drives have almost vanished but may still be encountered where their few advantages outweigh their disadvantages.

Speed control is of great benefit with fans and pumps. Fluid (water and gas) movement has a square law torque and cube law power curves. Running a fan or pump at 90% of its rated speed will reduce the torque to 81% and the power to 73% of the full speed values. A fan or pump running at 50% of its rated speed uses just 12.5% of its rated power. It is therefore very inefficient to control fluid flow by running a fan or pump at fixed full speed and restricting the output with a vane or damper. The energy savings from using a VF are tremendous (pay back times of under a year usually), so it is not surprising that the major purchasers of VF drives are water boards.

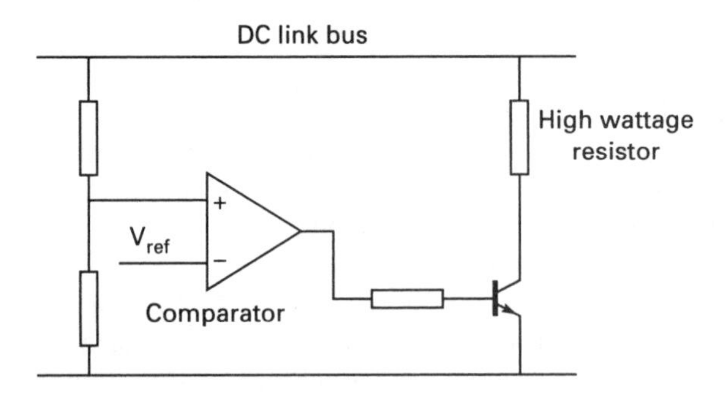

Fig. 12.68 *Simple VF braking circuit.*

The main disadvantages of VF drives are cooling and de-rating. A standard induction motor has a shaft-mounted cooling fan. As the speed of the motor is reduced the fan efficiency falls dramatically (see previous paragraph). If a motor is to be operated continuously at low speeds a separate fixed speed cooling fan may need to be added.

The waveform in even the best VF drive is not sinusoidal, and the resultant harmonics will cause additional motor losses. These cause heating which aggravates the problems with the fan outlined in the previous paragraph. Even at rated speed a VF-fed motor will need de-rating by about 10%. Ideally thermistors in the motor windings should be used as protection against overheating.

Motors with integral brakes (common on long travel and winch drives) often catch people out. These brakes are connected across two phases and lift automatically when a supply is applied to the motor. Often the existence of these brakes is unknown to the user, and when a VF is retrofitted the motor exhibits odd characteristics: running fine at high speed, but running rough, drawing high current and tripping or refusing to run at low speed. The reduced voltage from the VF, of course, allows the brake to come back on progressively. Fortunately most integral brake motors allow the brake to be disconnected and driven by a separate supply.

12.11.8. Variable speed couplings

Altering the speed of a motor is not the only way to alter the speed of a load. In many applications, notably fans and pumps, it is often cheaper to run an AC motor at fixed speed and use a variable speed coupling.

Gearboxes are, of course, one solution, particularly with infinitely variable ratios as in Fig. 12.69a. Other solutions are the eddy current couplings shown in Fig. 12.69b where a low power field current controls the coupling between input and output and the hydraulic coupling shown in Fig. 12.69c where a vane angle controls the coupling.

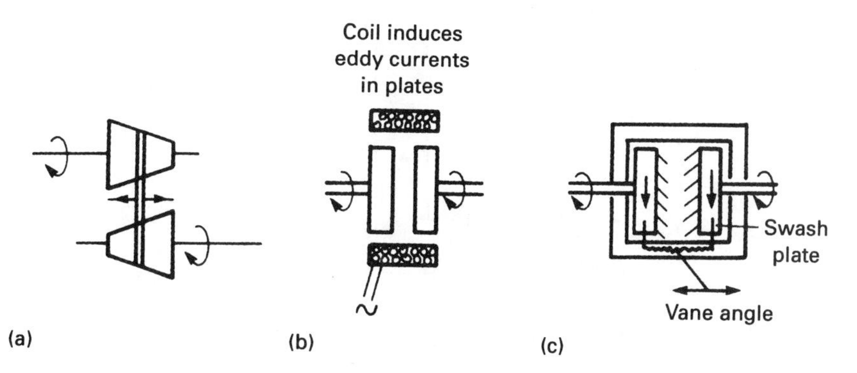

Fig. 12.69 *Other methods of speed control. (a) Infinitely variable coupling. (b) Eddy current coupling. (c) Hydraulic coupling.*

12.12. Current measurement

AC and DC drives both require measurement of the motor current for control and protection. The commonest method is the current transformer shown in Fig. 12.70. A large AC current passes through the primary of a transformer. The secondary, which has a large number of turns, is short-circuited. To achieve flux cancellation:

$$\frac{I_s}{I_p} = \frac{N_p}{N_s} \tag{12.42}$$

Usually there is just one primary turn, i.e. $N_p = 1$. The primary current is then reduced by the number of secondary turns (e.g. a 200 : 5 A current transformer would have forty turns). The AC current output is usually passed through a small resistor (usually around one ohm) and the resultant AC voltage amplified and rectified to give a DC voltage proportional to current.

A current transformer (CT) must always have a load connected. With no load a very high voltage (in the ratio N_s to N_p) will occur across the secondary terminals which is hazardous and could lead to arcing and damage.

The CT measures current on the AC side, and the rectification and filtering required can introduce a small delay of a few milliseconds which may be undesirable, particularly in DC drives where a fast response is needed.

The simplest DC current measurement method is a shunt in the DC lines. The current flow through the shunt will produce a small (few millivolt) potential which can be amplified by an isolated DC amplifier to give a usable current signal with good dynamic response.

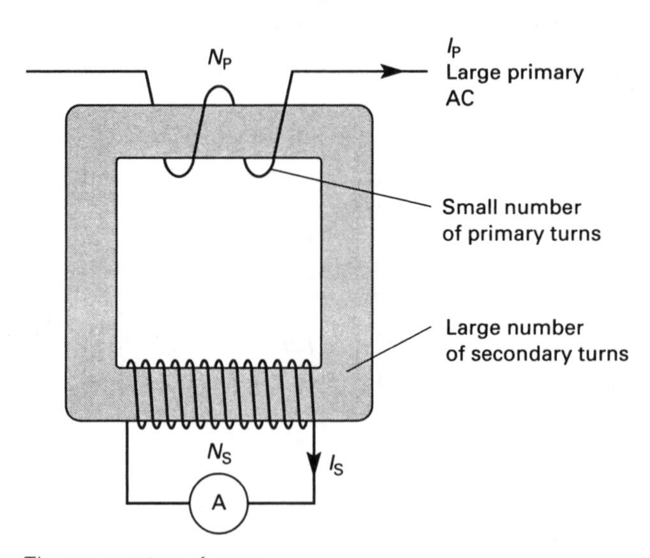

Fig. 12.70 *The current transformer.*

If a shunt cannot be mounted, a close relation of the CT, the DC-CT can be used. A single cable (or bus bar) with the DC current to be measured passes through a toroidal transformer core as Fig. 12.71. The transformer primary comes from a fixed high frequency oscillator, and induces voltages in the secondary winding which can be rectified. Increasing DC current will induce a DC flux into the core which reduces the coupling between primary and secondary. Although rectification and filtering is still required, the higher frequency of the oscillator significantly reduces the delay and improves the response.

A final, but much less common, method uses Hall effect transducers. The Hall effect occurs in some materials when a magnetic field is applied to a conductor carrying a DC current. This causes a voltage to appear across the faces of the material at 90° to both the field and the current as shown in Fig. 12.72. In a Hall

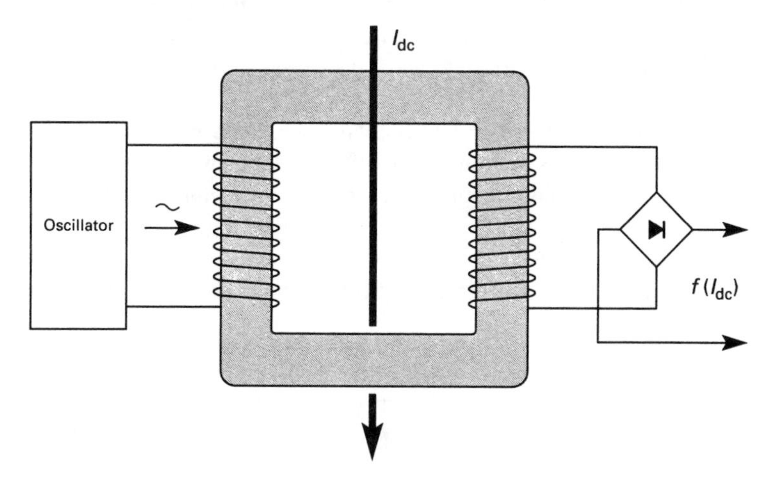

Fig. 12.71 *The DC current transformer (DCCT).*

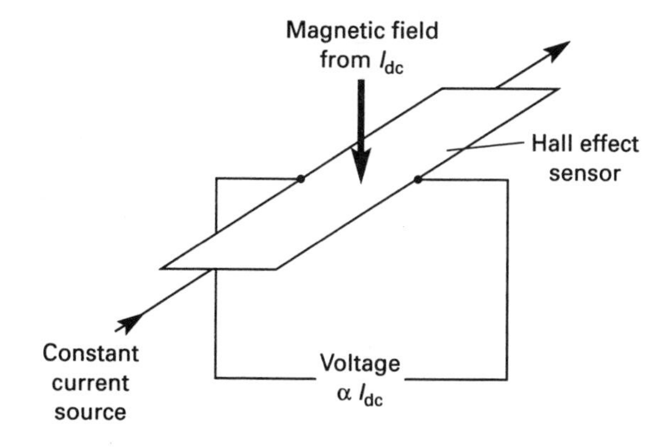

Fig. 12.72 *Hall effect current measurement.*

effect current sensor, a constant current is passed through the Hall effect material and the magnetic field is derived from the current to be measured.

The direction of the current is provided directly by the shunt and the Hall effect transducer. In both cases the voltage changes polarity if the current reverses direction. With the CT and the DC-CT the direction has to be inferred by other means.

12.13. The stepper motor

If an AC or DC motor is used for position control, some form of separate position transducer and control system must be used. The stepper motor combines the functions of both an output angular actuator and an angular position transducer. It is therefore possible to drive a stepper motor and know where the load is at all times without a separate position transducer. The torque available from stepping motors is, however, small, and they can only be used in applications such as small robots.

One form of stepper motor, called the variable reluctance motor, is shown in Fig. 12.73. The rotor has a number of soft iron teeth which is unequal to the number of stator teeth. The stator has a series of coils which are driven as three (or more) separate groups from the controlling logic. This is sometimes (incorrectly) called a three phase supply.

When a group is energised (phase 1 in the diagram), the rotor will align itself

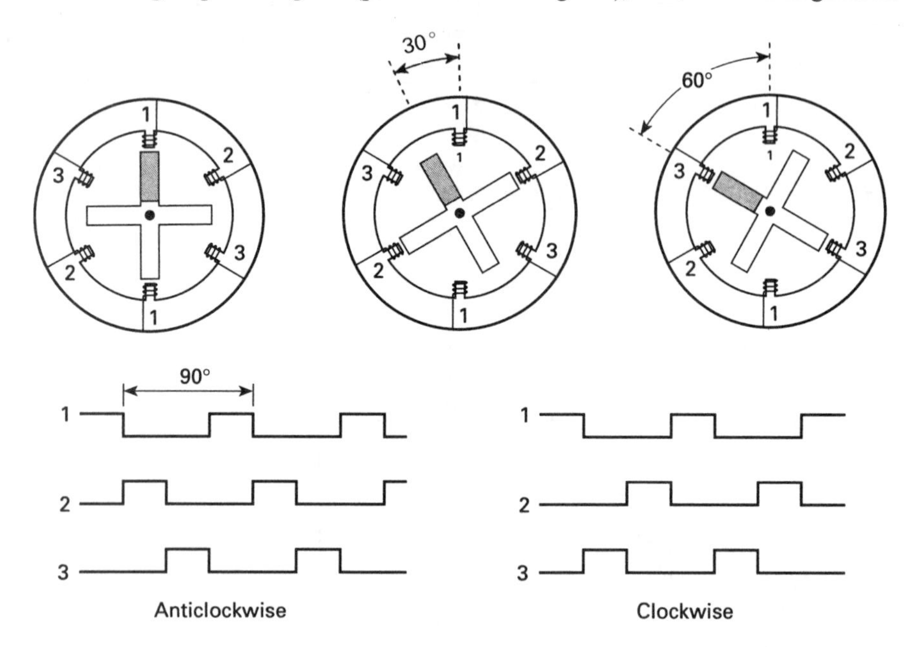

Fig. 12.73 *Simple 30° stepper motor. (a) Simple stepper motor moving in 30° steps anticlockwise. (b) Pulse trains for rotation.*

with the energised coils to give least magnetic reluctance. To move the rotor, the current group is de-energised, and the new group energised. The rotor rotates to a new position of minimum reluctance. If phase 2 were energised after phase 1, the rotor would rotate one step to the left, for example. Phase 3 after phase 1 would cause it to rotate one step to the right.

Figure 12.73b shows the pulse sequences required to drive the rotor in each direction. The step angle is given by:

$$\text{Step angle} = \frac{360^\circ}{S \times R} \qquad (12.43)$$

where S is the number of stator phases and R is the number of rotor teeth. For Fig. 12.73, with three phases and four teeth, this gives a step angle of 30°. The step angle can be reduced by using more teeth and multiple phases as Fig. 12.74. This has three phases and eight teeth giving a step angle of 15°. A common arrangement uses 50 teeth and four rotor phases which gives a step angle of 1.8°.

The step size can also be halved by energising two phases at once. If the phase sequence of Fig. 12.73b was changed to that of Fig. 12.75a, the step size would halve to 15°. If applied to Fig. 12.74 steps of 7.5° would be obtained. Note that the sequence is one phase, two phases, one phase and so on. Not surprisingly this is known as *half-stepping*.

The idea can be developed further to give even finer resolution by varying the amplitude of the current in each phase as Fig. 12.75b. This is called *micro-stepping*.

Dedicated ICs are available for producing the phase sequences needed to drive stepper motors, and microprocessors can easily be programmed to produce the phase outputs directly.

The phase windings have significant inductance. For best response, step current changes (rather than step voltage changes) are required. The easiest circuit is the simple transistor of Fig. 12.76a. This must be equipped with a spike suppression diode D to kill the large voltage spike generated when the transistor

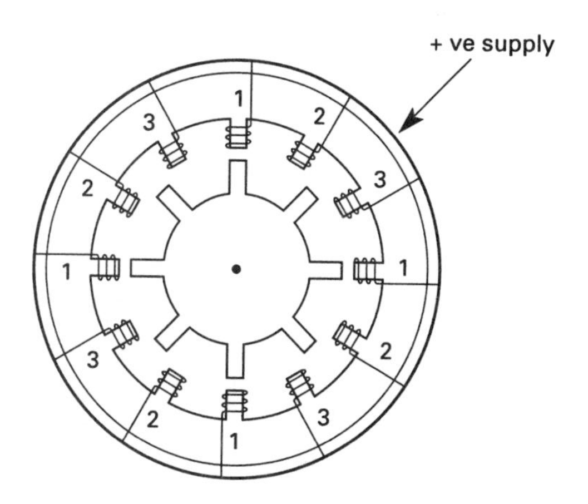

Fig. 12.74 *15° stepper motor.*

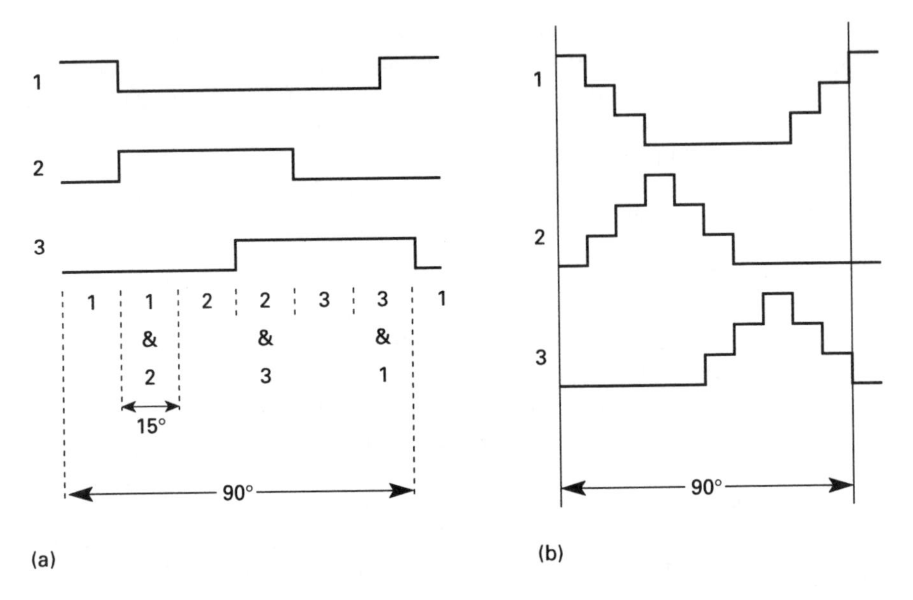

Fig. 12.75 *Increasing the resolution. (a) Half-stepping a 30° stepping motor. (b) Microstepping with 10° steps.*

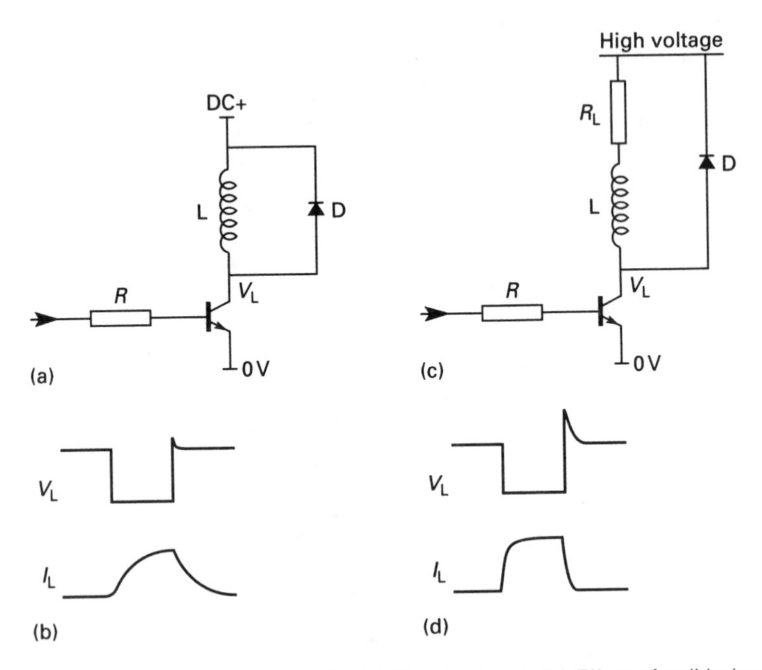

Fig. 12.76 *Driving stepper motor coils. (a) Simple circuit. (b) Effect of coil inductance. (c) Resistor acting as current source. (d) Improved speed.*

turns off. Unfortunately the coil inductance causes an exponential rise and fall of current as Fig. 12.76b which limits the maximum rotational speed. An improved response can be obtained by increasing the supply voltage and adding a series resistor *R* as shown on Fig. 12.76c. The value of *R* is chosen to prevent the coil being overvoltaged at low speeds, but acts like a constant current source at high speeds.

Where optimum performance is required, the coil current is controlled by a current feedback loop with a transistor pair as Fig. 12.77 (known as a closed loop chopper). The circuit is fed from a high voltage source. Transistor TR_1 is turned on all the time a phase is energised, and transistor TR_2 is under the control of the current feedback loop, acting as a switched high impedance current source. Again dedicated ICs are available for controlling a closed loop chopper.

A stepping motor application will have some restrictions placed on its performance. There will be a maximum rotational speed and a maximum acceleration. If these are exceeded, the rotor will slip out of lock, and the load position will be lost. A similar event will occur if the maximum load torque on the rotor is exceeded.

Unlike an AC or DC motor, a stepper motor does not rotate smoothly, but moves in small steps, even at high speed. The small steps will pulse the load, and if the load structure has a resonant frequency in the stepper motor pulse range severe vibrations may result. The mechanical designers of a system using stepper motors must ensure that the structure does not have a marked resonant frequency inside the stepping motor phase pulse range.

12.14. Microelectronics and rotating machines

The microprocessor is appearing in an increasing number of drive control systems, along with a shift from analog to digital control systems. Figure 12.43 showed a typical APSH drive. Conventionally this would be implemented with an analog tacho, and a collection of DC amplifiers for the various operations.

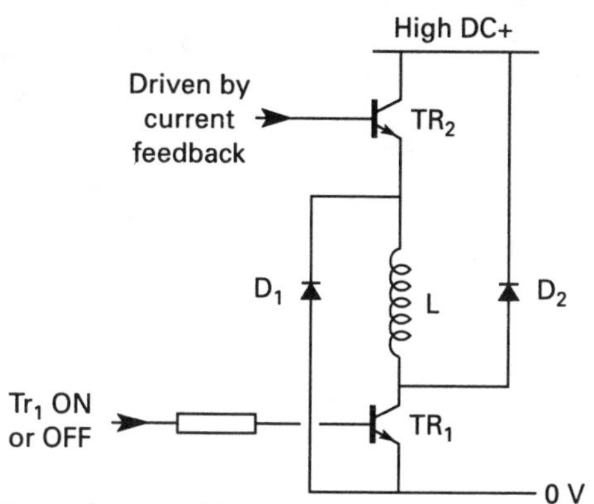

Fig. 12.77 *Closed loop chopper driver.*

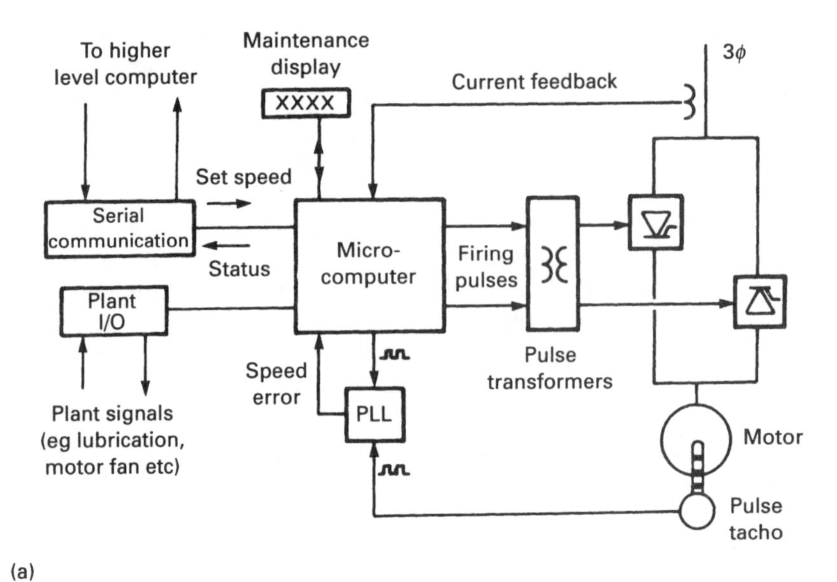

(a)

(b)

Fig. 12.78 *Microelectronics and thyristor drives. (a) Block diagram of digital drive. (b) Modern digital drive; the GEC Gemdrive Micro. Digital techniques are used for speed/current control, thyristor firing and interfacing to higher level computers/controllers. Drive set up and diagnostics are achieved via an operator's keypad. The interior view shows how little equipment is required for digital drives. (Photo courtesy of GEC Electrical Projects, Kidsgrove.)*

In a digital drive, all the functions would be performed by a dedicated microcomputer, which would interface to the outside world via the usual analog and digital conversion circuits. Increasingly, analog tachos are being replaced by pulse tachos, giving speed holding characteristics of better than 0.01%. Speed can be determined by counting pulses for a fixed time, timing between pulses, or determining speed error directly from a phase locked loop (PLL) IC.

Figure 12.78a shows a block diagram for a typical device. Note that this is part of a distributed computer system, receiving its speed reference and start commands via a serial link, and returning its status in a similar way. Figure 12.78b is the complete electronics for a digital drive. The only other equipment in the drive are the thyristor pulse transformers, power components (such as isolators) and the thyristor stacks themselves. All the functions of Fig. 12.43 are performed in software.

Digital drives can also provide powerful diagnostic aids for maintenance and fault finding, such as a 'replay' of events leading to a drive trip, along with auto tuning of the speed and current control functions.

Current feedback in large drives is usually obtained from current transformers on the AC side of the thyristor stack. The signals from these must be rectified and smoothed to give a current signal that can be used by the current amplifier. This operation introduces a significant lag into the current loop. Figure 12.79 shows a digital current sensing method which introduces minimal lag, and can measure the current in one bridge leg. The outputs from the CTs are taken to voltage controlled oscillators. The oscillator outputs feed counters which are read and reset by the control microcomputer in synchronism with the supply frequency. The count is proportional to the mean current during the previous mains cycle.

Large scale integration (LSI) ICs are available for many drive functions. Increasingly, DC and VF drives are constructed from a small number of specialist

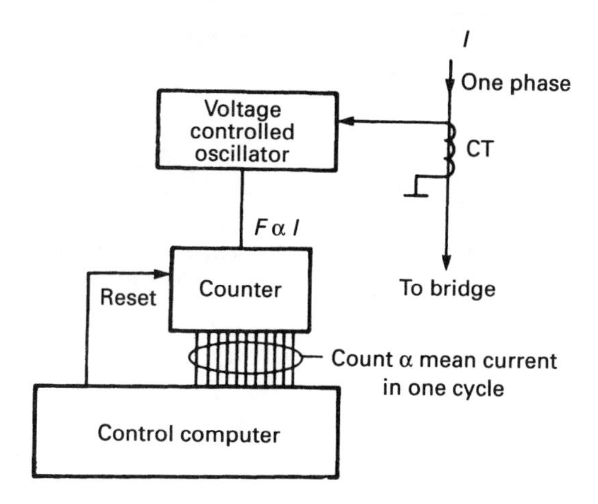

Fig. 12.79 *Digital current measurement.*

ICs. This in turn leads to smaller, cheaper and more reliable drives, and the possibility of treating drives of up to about 200 amp rating as replaceable items. This tendency has been helped by the arrival of compact thyristor stacks with insulated mounting blocks, allowing the drive electronics to be neatly and compactly mounted directly on to the heatsink as in Fig. 12.80.

Fig. 12.80 *Modern thyristor drives. Note how insulated encapsulated thyristors allow the heatsink to be used as an integral part of the chassis. (Photo courtesy of GEC Electrical Projects, Kidsgrove.)*

Chapter 13

Digital circuits

13.1. Introduction

13.1.1. Analog and digital systems

Signals in process control are conventionally transmitted as a pneumatic pressure or electrically as a voltage or current. A pneumatic signal of 3 to 15 psi, for example, could represent a liquid flow from 0 to 600 litres per minute, or an electric current of 4 to 20 mA could represent a temperature from $-100\,°C$ to $+400\,°C$.

These signals are said to be continuously variable in that they can take any value between the two extreme limits. In the above temperature measurement, for example, a reading of 250 °C would be represented by a current of 15.2 mA (given by $350 \times 16/500 + 4$). Similarly a flow of 540 litres per minute would be represented by a pressure of 13.8 psi in the flow measurement above.

In each of these examples an electrical or pneumatic signal is used as an analog of the process variable and the signal follows the process variable within the accuracy limits of the system. Such systems are called analog systems.

Digital systems are concerned with signals that can only take certain values. Most digital systems deal with electrical signals that can only have two values; 5 V or 0 V, for example. Many systems are inherently of this type: a light can be on or off, a valve open or shut, a motor running or stopped.

Figure 13.1 shows two possible approaches to liquid level measurement. In Fig. 13.1a a weighted float is mechanically linked to a potentiometer which gives an output voltage proportional to level. Within the resolution limits of the

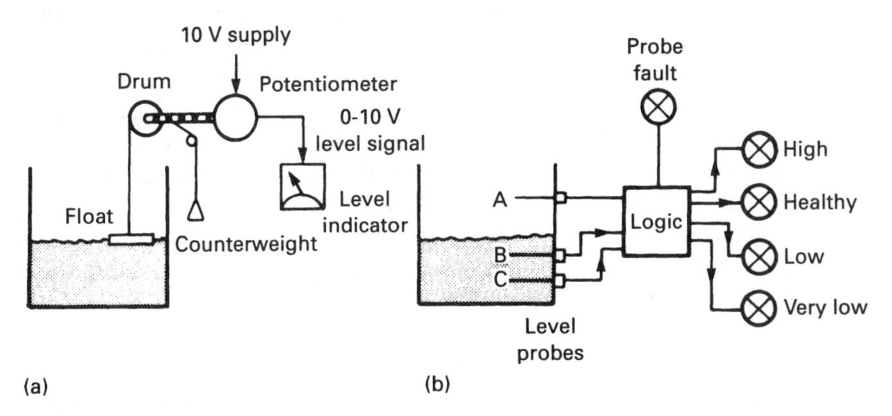

Fig. 13.1 *Comparison of analog and digital systems. (a) Analog system. (b) Digital system.*

potentiometer, the voltage can take any value between 0 and 10 V, so this is an example of an analog system.

Figure 13.1b uses three level switches A, B, C. These can be ON or OFF (ON being defined as the switch being submerged).

There are four possible conditions:

State	*A*	*B*	*C*
Very low level	OFF	OFF	OFF
Low level	OFF	OFF	ON
Healthy	OFF	ON	ON
High level	ON	ON	ON

There are actually four other possible combinations of A, B, C; these all indicate a switch failure (e.g. A on, B off, C on). Figure 13.1b uses on/off signals and is an example of a digital system.

13.1.2. Types of digital circuit

Digital applications can, in general, be classified into three types. The simplest of these are called combinational logic (or static logic), and can be represented by Fig. 13.2a. Such systems have several digital inputs and one or more digital outputs. The output states are uniquely defined for every combination of input states, and the same input combination always gives the same output states. Figure 13.2a is a combinational logic system.

A sequencing logic system is superficially similar to Fig. 13.2a, but the output states depend not only on the inputs but also on what the system was doing last (its previous state). Sequencing systems therefore have memory and storage elements. A very simple example is the motor starter of Fig. 13.2b. The start input causes the motor to start running and keep running even when the start signal is removed. The stop input stops the motor. Note that with neither signal present the motor could be running or stopped dependent upon which signal occurred last; the output state is not defined solely by the present input states.

Another sequencing example is shown in Fig. 13.2d. The digital circuit has three inputs – start and two limit switches – and two outputs – extend and retract. On a start signal the hydraulic ram extends until LS_1 is made, then retracts until LS_2 is made, at which point the ram remains until another start signal is received. With no inputs present, the ram can be travelling out or in.

The final group of digital systems uses digital signals to represent, and manipulate, numbers. Such systems cover the range from simple counters and digital displays to complex arithmetic and computing circuits.

13.1.3. Logic gates

The simplest digital device is the electromagnetic relay, and it is useful to describe some of the fundamental ideas in terms of relay contacts. In Fig. 13.3a, the coil Z

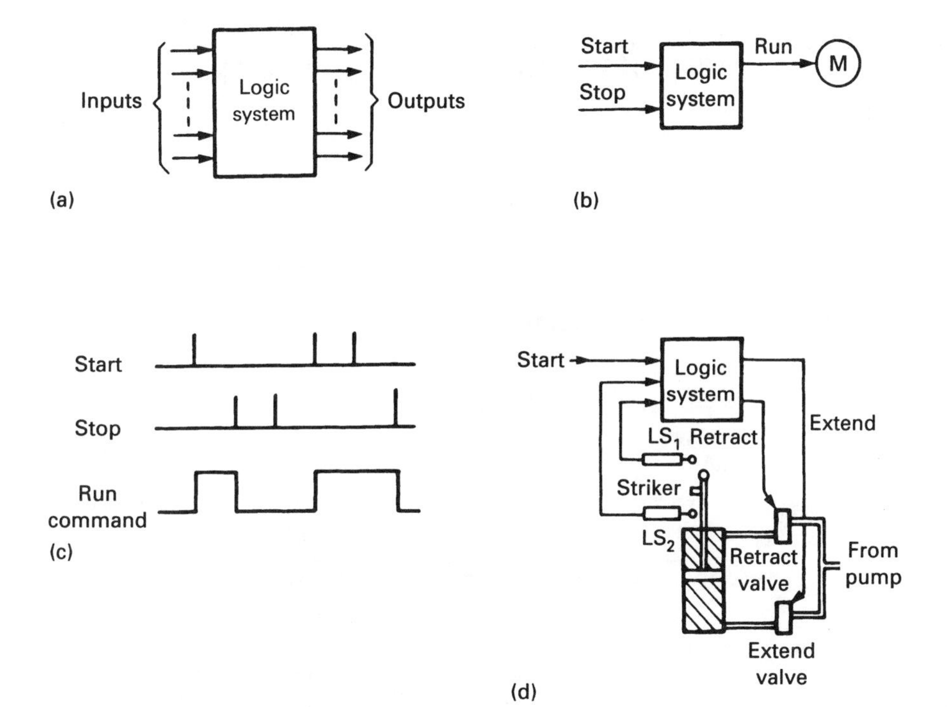

Fig. 13.2 *Types of digital system; combinational and sequencing logic. (a) Representation of combinational logic system: output states solely defined by input states. (b) A simple sequencing system, a motor starter. (c) Operation of motor starter. (d) Sequencing system.*

will energise when contact A AND contact B AND contact C are made. The series connection of contacts performs an AND function.

Similarly, in Fig. 13.3b the coil Z will energise when contact A OR contact B OR contact C are made. The parallel connection of contacts performs an OR function.

In Fig. 13.3c, coil Z is energised when the push button is pressed. A normally closed contact of Z controls coil Y. When Z is energised Y is de-energised, and vice versa. The normally closed contact can be said to invert the state of its coil.

Combinational logic circuits are built round combinations of AND, OR and INVERT circuits. In Fig. 13.4, for example, Z will be energised for:

(A not energised) AND (B energised OR C energised)

Such verbal descriptions are impossibly verbose for more complex combinations. Circuit operations are more conveniently expressed as an equation. Normally closed contacts are represented by a bar over the top of the contact name (e.g. $\bar{A}$, verbalised as A bar). The circuit of Fig. 13.4a can then be represented as:

$$Z = \bar{A} \text{ AND } (B \text{ OR } C)$$

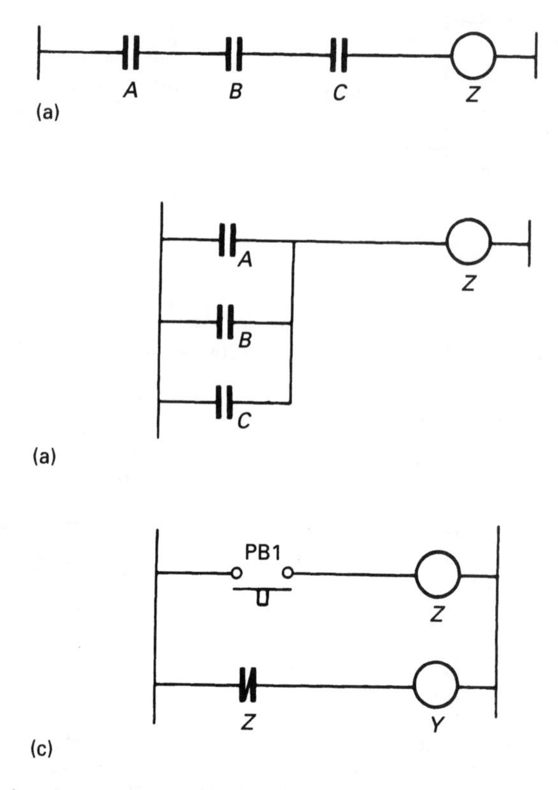

Fig. 13.3 *Simple relay logic. (a) AND combination. (b) OR combination. (c) Inversion.*

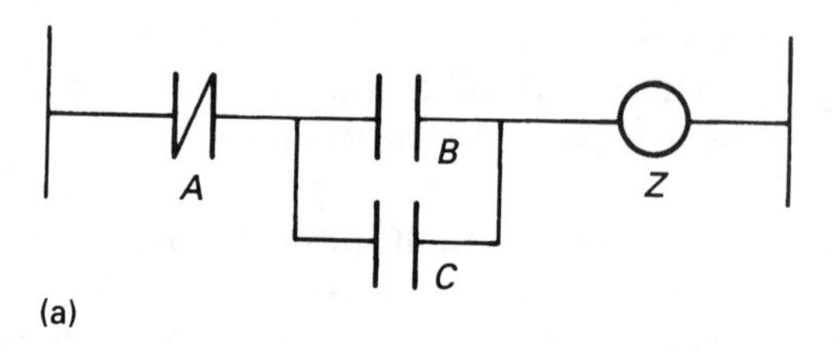

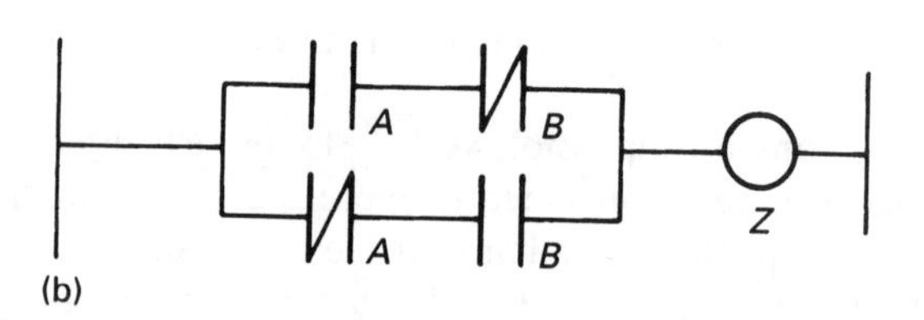

Fig. 13.4 *Further relay logic. (a) Z is energised when (A is not energised) AND (B energised or C energised). (b) Stairwell lighting circuit.*

Similarly the circuit of Fig. 13.4b (commonly used for stairwell lighting) can be represented by:

$$Z = (A \text{ AND } \bar{B}) \text{ OR } (\bar{A} \text{ AND } B)$$

These are known as Boolean equations, a topic discussed further in Section 13.3.3.

Relays can perform all logic functions but are slow (typically 20 operations per second), bulky and power hungry. Electronic circuits performing similar functions are called logic gates. These work with signals that can only have two states. A signal in CMOS logic, for example, can be at 12 V or 0 V and could represent a limit switch made or open.

The two logic states can be called high/low, on/off, true/false and so on. The usual convention, however, is to call the higher voltage 1 and the lower voltage 0. For a CMOS gate, therefore, 12 V is 1 and 0 V is 0.

Figure 13.5a shows the circuit of a simple AND gate. Neglecting diode drops, the output Z will be equal to the lower of the two input voltages. In other words, it will be a 1 if, and only if, both inputs are 1. This can be represented by Fig. 13.5b (which is called a truth table).

On circuit diagrams it is clearer to use logic symbols rather than the actual circuit diagram. The symbol for an AND gate is shown in Fig. 13.5c; the output Z is 1 when A AND B are both 1.

In Fig. 13.6a the output Z will be equal to the higher of the two inputs (again neglecting diode drops). Z will therefore be 1 if either input is 1, giving the truth

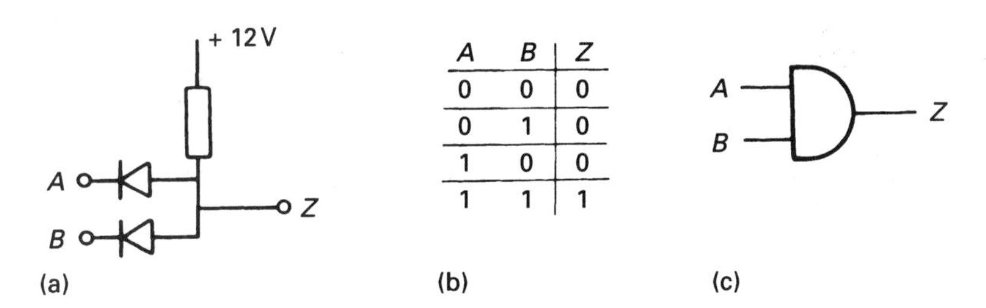

A	B	Z
0	0	0
0	1	0
1	0	0
1	1	1

Fig. 13.5 *The AND gate. (a) Circuit. (b) Truth table. (c) Logic symbol.*

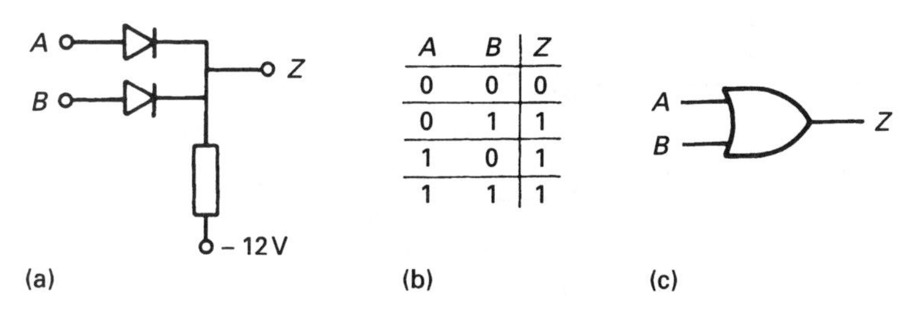

A	B	Z
0	0	0
0	1	1
1	0	1
1	1	1

Fig. 13.6 *The OR gate. (a) Circuit. (b) Truth table. (c) Logic symbol.*

table of Fig. 13.6b. The logic symbol for an OR gate is shown in Fig. 13.6c.

The invert function is given by the simple saturating transistor of Fig. 13.7a. When A is 0, the transistor is turned off and the output Z is pulled to a 1 state by the collector load resistor. When A is 1, the transistor is saturated on taking Z to 0 V; a 0. The circuit behaves as the truth table of Fig. 13.7b and has the logic symbol of Fig. 13.7c.

Combinational logic circuits can be drawn purely in terms of AND gates, OR gates and inverters (although other gate types are more commonly used for reasons given later in Section 3). The stairwell lighting circuit of Fig. 13.4b is drawn with logic symbols in Fig. 13.8a. This behaves as the truth table of Fig. 13.8b which shows that Z is 1 if only one input is 1. This circuit is known as an exclusive OR and is sufficiently common to merit its own logic symbol, shown in Fig. 13.8c. This is called an XOR gate.

If an inverter is used after an AND gate as in Fig. 13.9a, the truth table of Fig. 13.9b is produced. This arrangement is called a NAND gate (for NOT-AND) and has the logic symbol of Fig. 13.9c. The NAND gate is probably the commonest logic gate.

Adding an inverter to an OR gate as in Fig. 13.10a gives the truth table of Fig.

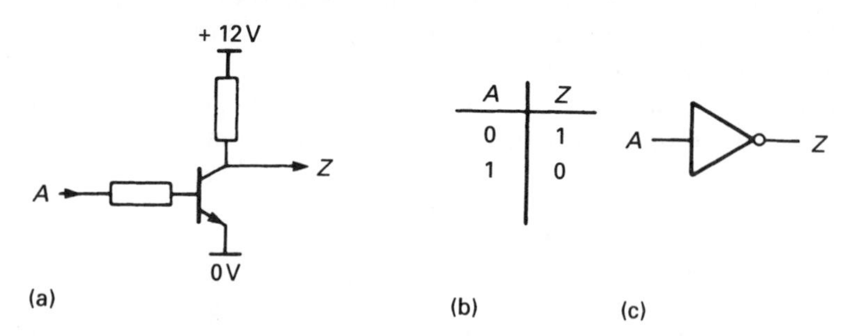

A	Z
0	1
1	0

Fig. 13.7 *The inverter. (a) Circuit. (b) Truth table. (c) Logic symbol.*

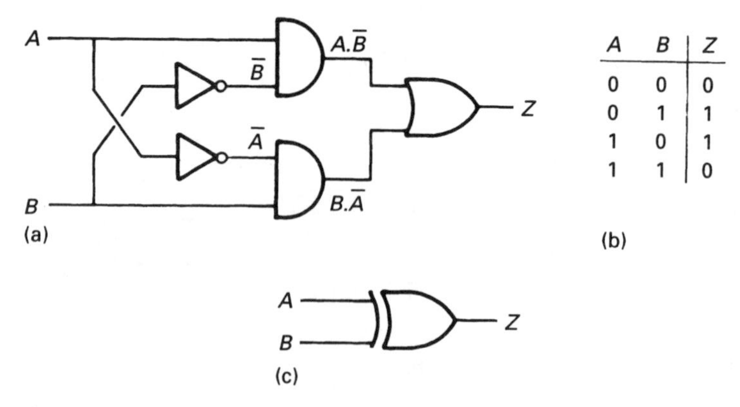

A	B	Z
0	0	0
0	1	1
1	0	1
1	1	0

Fig. 13.8 *The exclusive OR gate. (a) Logic diagram. (b) Truth table. (c) Logic symbol.*

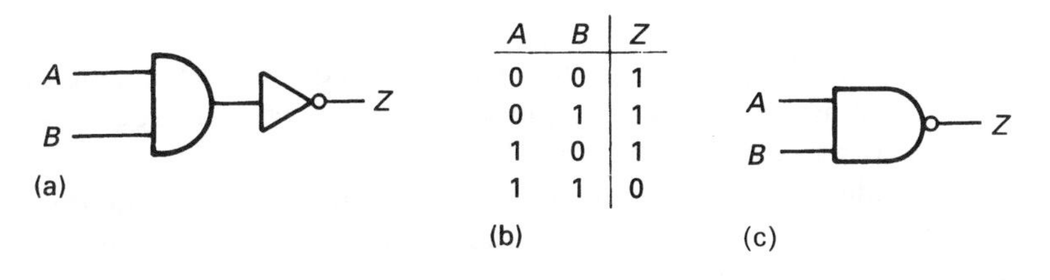

A	B	Z
0	0	1
0	1	1
1	0	1
1	1	0

Fig. 13.9 *The NAND gate. (a) Logic diagram. (b) Truth table. (c) Logic symbol.*

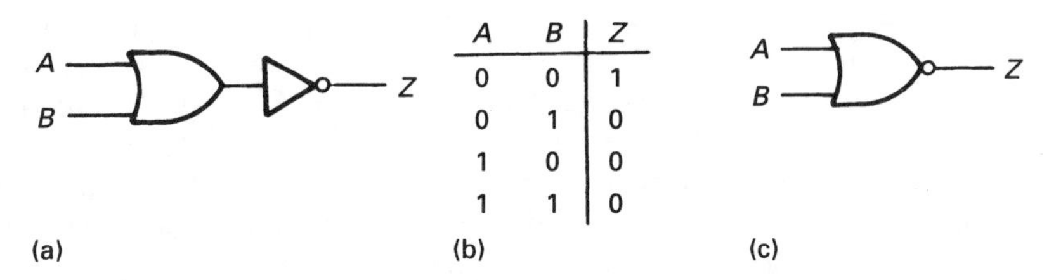

A	B	Z
0	0	1
0	1	0
1	0	0
1	1	0

Fig. 13.10 *The NOR gate. (a) Logic diagram. (b) Truth table. (c) Logic symbol.*

13.10b. This is known as a NOR gate (for NOT-OR) and is given the logic symbol of Fig. 13.10c.

Note that the logic symbols for NAND/NOR gates are similar to those for AND/OR gates with the addition of a small circle on the output. The circle denotes an inversion operation.

The illustrations for AND/OR/NAND/NOR gates show two inputs. In reality these gates can have any required number of inputs (up to eight being readily available in commercial logic families). The exclusive OR gate inherently has only two inputs and the inverter, of course, has only one.

13.2. Logic families

13.2.1. Introduction

The circuits of Figs. 13.5 to 13.7, whilst illustrating the principles of logic gates, have many shortcomings. The voltage drops across the diodes would lead to severe degradation of the voltage levels after several gates, and the operating speed would be limited by the time constant formed by the load resistors and stray capacitance.

Most logic circuits are constructed from integrated circuits, and have high operating speed and well defined levels. Two logic families (TTL and CMOS) are widely used in industrial applications and a third family (ECL) may be encountered where very high speed is required. Before these are described, we must first examine how the various factors of a logic gate's performance are specified.

13.2.2. Speed

A logic gate such as the inverter of Fig. 13.11a does not respond instantly to a change at its input. For infinitely fast input signals the output will be delayed and the edges slowed, as shown in Fig. 13.11b.

The delay is called the propagation delay and is defined from the mid point of the input signal to the mid point of the output signal. Typical values are around 5 ns for TTL.

The edge speeds are defined by the rise time (for the 0 to 1 edge) and the fall time (for the 1 to 0 edge). These are measured between the 10% and 90% points of the output signal. Typical values are 2ns for TTL.

Propagation delays and rise/fall times determine the maximum speed at which a logic family can operate. TTL can operate in excess of 10 MHz, CMOS around 5 MHz and ECL at over 500 MHz (although considerable care needs to be taken with board layout at speeds over 10 MHz).

Power consumption is related to speed, as increased speed is obtained by reducing RC time constants formed by stray capacitance, and by using non-saturating transistors. CMOS, for example, has a power consumption of about 0.01 mW per gate compared with ECL's figure of 60 mW/gate.

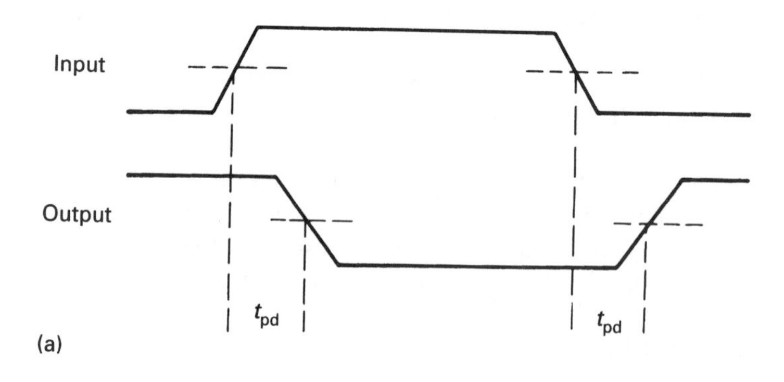

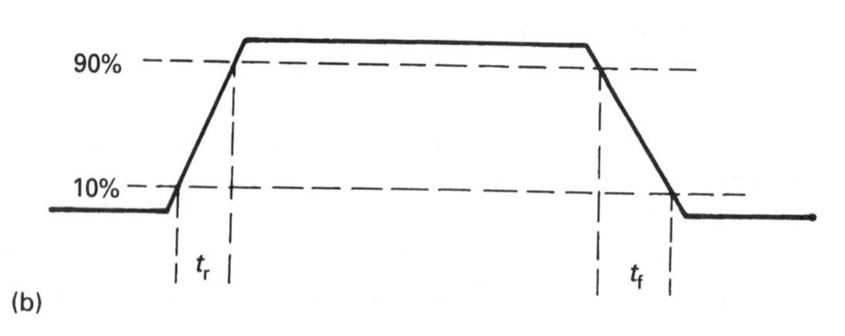

Fig. 13.11 *Speed definitions. (a) Propagation delay (t_{pd}). (b) Rise and fall times (t_r and t_f).*

13.2.3. Fan in/fan out

The output of a logic gate can only drive a certain load and remain within specification for speed and voltage levels. There is therefore a maximum number of gate inputs a given gate output can drive.

A simple gate input is called a standard load, and is said to have a fan in of one. A gate output's drive capability is called its fan out, and is defined in unit loads. A TTL gate output, for example, can drive ten standard gate inputs and correspondingly has a fan out of ten.

Some inputs appear as a greater load than a standard gate. These are defined as a fan in of an equivalent number of standard gate inputs. An input with a fan in of three, for example, looks like three gate inputs.

Obviously the sum of all the fan in loads connected to a gate output must not exceed the gate's fan out.

13.2.4. Noise immunity

Electrical interference may cause 1 signals to appear as 0 signals, and vice versa. The ability of a gate to reject noise is called its noise immunity.

Defining noise immunity is more complex than it might at first appear, but the method usually adopted is that shown in Fig. 13.12a. The voltages given are those for a TTL gate which has a nominal 1 voltage of 3.5 V and a nominal 0 voltage of 0 V.

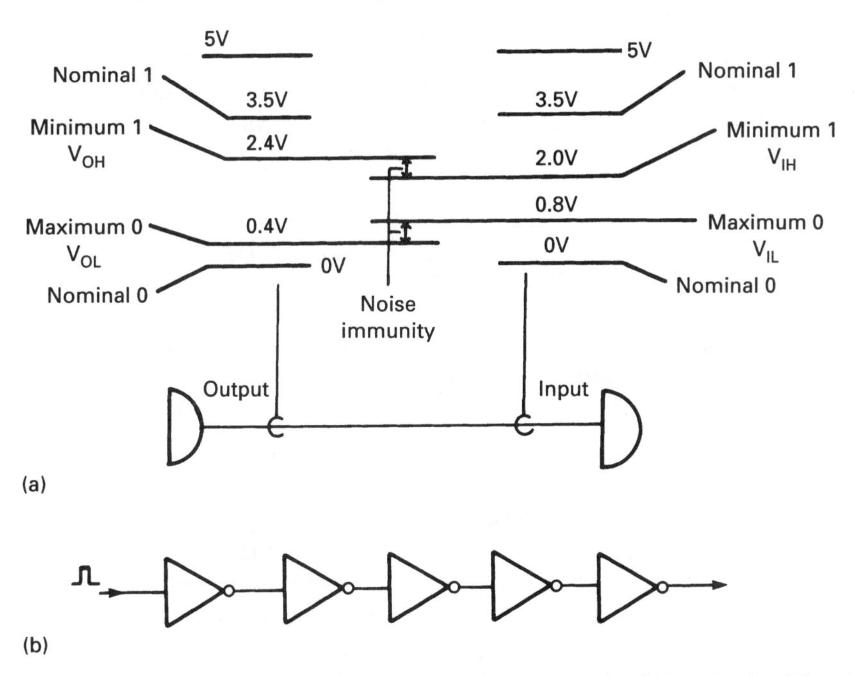

Fig. 13.12 *Definitions of noise immunity. (a) DC noise margin. (b) Testing for AC noise immunity.*

Next we define how an output 1 can fall to (2.4 V) and a 0 rise to (0.4 V). These are respectively termed V_{OH} and V_{OL}. Finally we define how low a gate's input 1 can fall and an input 0 rise without allowing its output to go between V_{OH} and V_{OL}. These volts are called V_{IH} (2.0 V) and V_{IL} (0.8 V). The noise immunity is then the smaller of $V_{OH} - V_{IH}$ or $V_{IL} - V_{OL}$. For TTL the figure is 0.4 V. This is a worse case value, a more typical noise immunity being about 1.2 V.

A figure sometimes quoted is the AC noise margin. This is defined as the largest pulse that will not propagate down a chain of gates similar to Fig. 13.12b. This gives a more favourable figure than Fig. 13.12a, but is a more realistic test.

13.2.5. *Transistor transistor logic (TTL)*

TTL is probably the most successful logic family. TTL is NAND based logic, the circuit of a two input NAND gate being shown in Fig. 13.13. The rather odd-looking dual emitter transistor can be considered as two transistors in parallel or three diodes, as shown.

If both inputs are high, Q_2 is turned on by current from R_1 supplying base current to Q_3. The output is therefore nominally 0 V. With either input low, Q_1 is turned on, Q_2 turned off and Q_4 pulls the output high to a nominal 3.5 V.

The output transistors Q_3, Q_4 are called a totem pole output and play a significant part in increasing the operating speed. When the output is a 0, Q_3 acts as a saturated transistor. When the output is a 1, Q_4 acts as an emitter follower. Both states have low output impedances which reduce RC time constants with stray capacitance.

There are at least six versions of TTL with differing speeds and power consumption. Schottky versions use Schottky diodes within the gate to reduce hole storage delays. The six common types are:

Name	*Suffix*	*Prop delay(ns)*	*Speed (MHz)*
Normal	None	10	25
Schottky	S	3	45
Low power Schottky	LS	9.5	25
Advanced low power Schottky	ALS	5	35
Fast	F	2	100
Advanced fast	AF	2	105

All TTL is part of the so-called 74 series (originally conceived by Texas Instruments) having the same pin arrangements on all the ICs. They can also be intermixed although care must be taken because of the different input loadings and output capabilities (an LS gate input, for example, looks like 0.5 of a normal gate input). All run on a 5 V supply and use logic levels of 3.5 V and 0 V. The suffix in the above table appears as part of the device identification; a 74LS06, for example, is a low power Schottky gate.

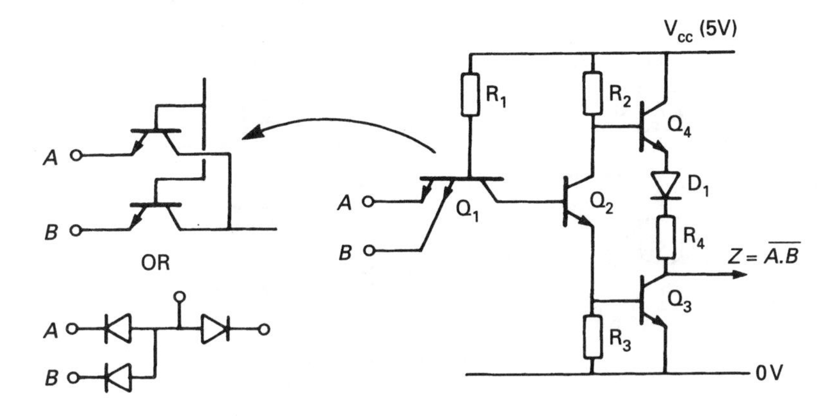

Fig. 13.13 *Transistor transistor logic (TTL).*

13.2.6. Complementary metal oxide semiconductor (CMOS) logic

CMOS is virtually the ideal logic family. It can operate on a wide range of power supplies (from 3 to 15 V), uses little power (approximately 0.01 mW at low speeds), has high noise immunity (about 4 V on a 12 V supply) and very large fan out (typically in excess of 50). It is not as fast as TTL or ECL but its maximum operating speed of 5 MHz is adequate for most industrial purposes (too high a maximum speed can actually be a disadvantage as it makes a system more noise prone).

CMOS is built around the two types of field effect transistors shown in Fig. 13.14. From a logic point of view these can be considered as a voltage operated switch. These switches can be used to manufacture logic gates.

Figure 13.15a shows how an inverter can be implemented. With A low, Q_1 is turned on and Q_2 off. With A high, Q_2 is turned on and Z is low.

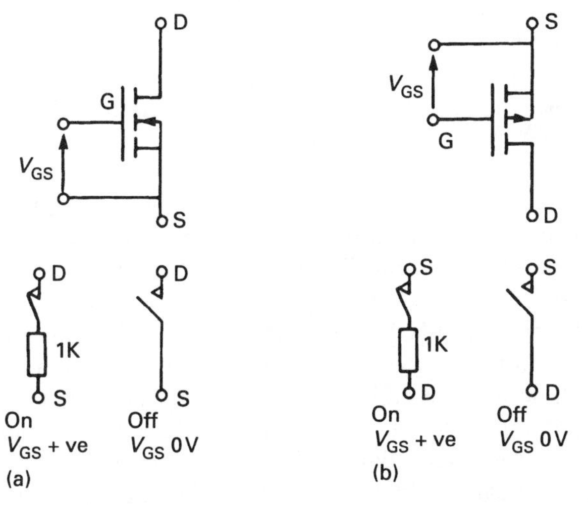

Fig. 13.14 *Metal oxide semiconductor (MOS) transistors. (a) n channel. (b) p channel.*

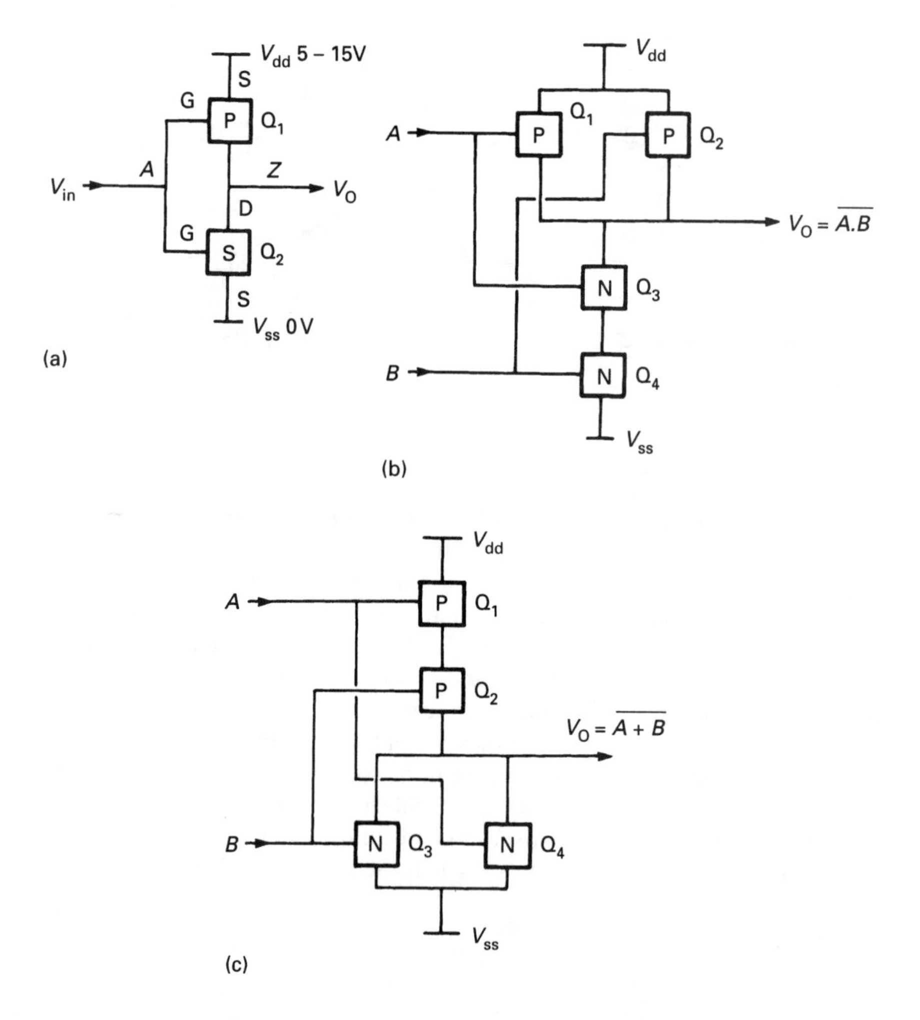

Fig. 13.15 *Complementary metal oxide semiconductor (CMOS) logic gates. (a) CMOS inverter. (b) CMOS NAND gate. (c) CMOS NOR gate.*

Similarly a NAND gate can be constructed as in Fig. 13.15b. If A or B is low, Z will be high because one of the parallel pair Q_1, Q_2 will be on, and one of the series pair Q_3, Q_4 will be off. The output Z will be low only when both A and B are high when Q_1, Q_2 are both off and Q_3, Q_4 are both on.

Figure 13.15c shows a CMOS NOR gate. If A or B is high, one of Q_3 or Q_4 will be on, taking the output Z low (with one of Q_1, Q_2 off). When both A and B are low, Q_1, Q_2 will both be on and Q_3, Q_4 off, taking the output high.

The high input impedance of FETs can present handling problems, and early devices could be irreparably damaged by static electricity from, say, nylon clothing or leakage currents from unearthed soldering irons. Modern CMOS now includes protection diodes and can be treated like any other component.

Another effect of the high input impedance is the tendency for unused inputs to charge to an unpredictable voltage. All CMOS inputs must go somewhere; even unused inputs on unused gates on multigate packages must go to a supply rail (thereby forcing a 1 or 0 state).

CMOS was originally sold in the so-called 4000 series which is a rationalisation of the original RCA COSMOS and Motorola McMOS ranges. A B suffix denotes buffered signals and improved protection; needless to say, the B devices are better suited for industrial systems. Further variations are discussed in Section 13.2.9.

13.2.7. Emitter coupled logic (ECL)

ECL is the fastest commercially available logic family, and with care it can operate at 500 MHz. At such speeds, however, extreme care needs to be taken with the circuit board layout to avoid crosstalk and power supply induced noise.

ECL acquires its speed from the use of non-saturating transistors and high power levels (around 60 mW per gate compared with the CMOS figure of 0.01 mW). Figure 13.15 shows an ECL NOR gate, which is based superficially on a long tailed pair.

The logic levels in ECL are -0.8 V and -1.6 V (giving a rather poor noise immunity of 0.25 V). Q_1 sets a bias voltage of -1.2 V at the base of Q_2.

If A or B is at -0.8 V, Q_3 or Q_4 will pass current (but not saturate) and the voltage at Q_5 base will fall, taking the output low to -1.6 V. Q_5 is an emitter follower, giving a low impedance output. When both A and B are at -1.6 V, Q_2 will pass current and Q_3, Q_4 will be off. R_3 takes Q_5 base high giving an output voltage of -0.8 V.

ECL is very fast, but its odd voltage levels, strict wiring and power supply requirements, and poor noise immunity preclude its use in industrial applications except where very high speed is needed.

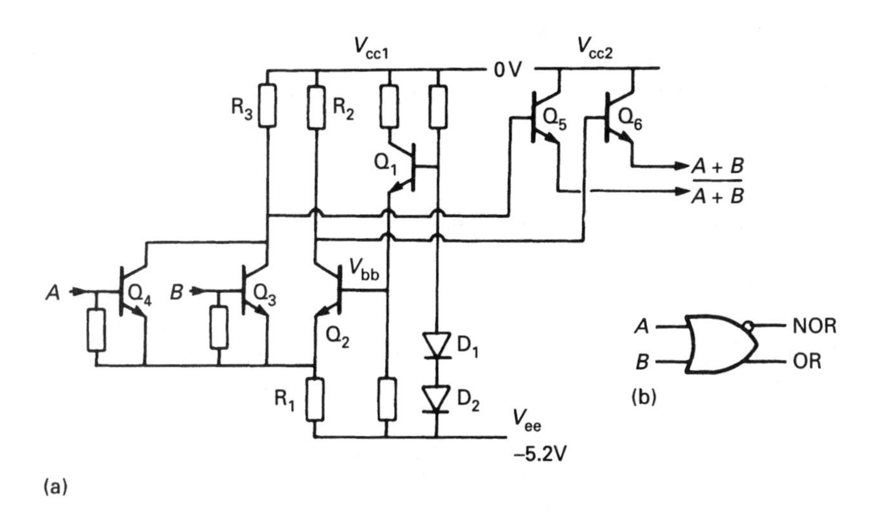

Fig. 13.16 *Emitter coupled logic. (a) Circuit diagram. (b) Symbol.*

13.2.8. Industrial logic families

Many sequencing applications only require operating speeds of at most 100 Hz. Industrial logic families, such as Mullard's NORBIT or the German Sigmatronic, are designed to replace relay panels. These devices are not based on integrated circuits, but are usually constructed from discrete components encapsulated in epoxy resin.

Although physically larger than IC logic such as TTL or CMOS, industrial logic families are virtually indestructible (usually able to withstand 240 V AC on inputs and power supplies), and the slow speed gives very high noise immunity. They are well suited to applications where the technical expertise of maintenance staff is low because devices can usually be changed solely with a screwdriver or by unplugging snap connectors.

13.2.9. Choosing a logic family

Until the latter part of the 1980s the designer really had to choose between TTL (with the LS family being the popular choice) and CMOS, which was slower and had a much smaller range of devices, but had the advantages of very low power consumption, better noise immunity and a wide supply tolerance. Since then, though, there has been a tendency for the families to merge.

The trend started with the 74C series of CMOS which provided CMOS devices with the same pinning as TTL, but with CMOS B series electrical characteristics (slower than TTL, but with 3 to 15 V supply). These were useful, but the major impact was the introduction of the 74HC, 74AC, 74HCT and 74ACT families. These use improved technologies, were as fast as TTL, and (as their name implies) they follow the 74 series pinning.

Taking them in turn:

74AC is the high speed member of the family, capable of operating at speeds of 125 MHz. The voltage supply range is 3 to 6 V (essentially TTL with a wider tolerance), and the transfer characteristic is the standard CMOS near ideal symmetrical curve of Fig. 13.17a.

74HC is a near replacement for LS TTL with an operating speed of 30 MHz. Other characteristics are similar to 74AC.

As mentioned earlier, the output levels of TTL, shown on Fig. 13.17b, are approximately 0.5 V in the 0 state, and 3.5 V in the 1 state. Standard forms of TTL can, just, connect to 74AC or 74HC devices, but the resultant noise immunity is poor. Two further forms of CMOS were developed with a transfer function whose input side mimicked a TTL device. These are known as 74ACT (high speed version) and 74HCT (practically a direct replacement for LS TTL). These should not be viewed as a family to be used for a complete project, as to do so would give the poorer TTL noise immunity. They are, though, exceedingly useful when a circuit has to mix TTL and CMOS devices.

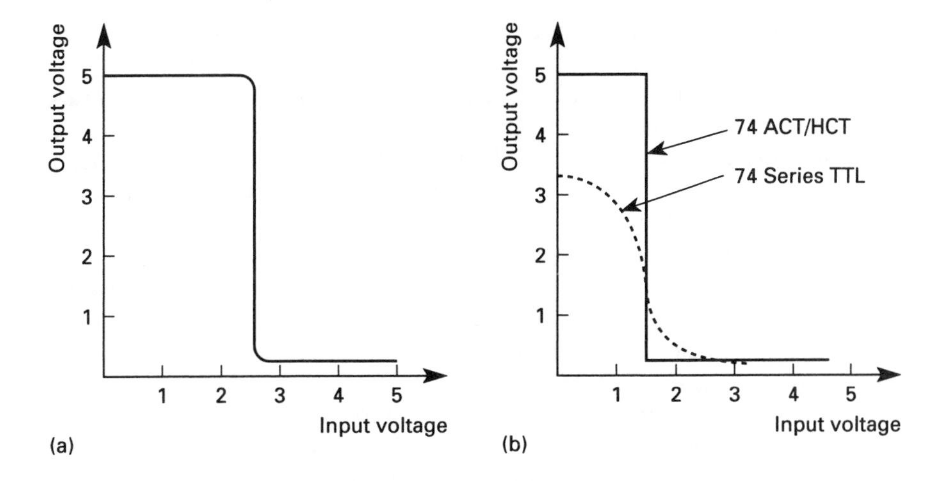

Fig. 13.17 *Transfer functions of digital families. (a) CMOS. (b) 74 series TTL and 74ACT/HCT CMOS.*

In CMOS, therefore, there is now:

Family	*Supply*	*Speed (MHz)*	*Comments*
4000B	3 to 15 V	2	Useful for battery circuits, slow. Seems unlikely to develop further.
74C	3 to 15 V	2	As 4000B with TTL pinning.
74AC	2 to 6 V	125	Very fast, TTL pinning, power consumption rises with speed, so normal CMOS low power may not be relevant. Care needed with layout (as with all high speed circuits)
74HC	2 to 6 V	30	Designed as direct replacement for LS TTL.
74ACT	2 to 6 V	125	As 74AC with TTL input levels.
74HCT	2 to 6 V	30	As 74HC with TTL input levels.

An interesting development is the view that the corner supply pins on TTL are not the best arrangement for power supply and ground noise, and there seems to be a move toward centre pinning on some high speed CMOS circuits.

There are four TTL families in common current use; LS, ALS, F and AS. It is worth listing these in tabular form:

Family	*Speed (MHz)*
74LS	25
74ALS	35
74F	100
74AS	105

All require a 5 V +/−0.25 V supply.

Choosing a device is quite straightforward. First where there is little choice; for out and out speed use ECL (but remember the precautions needed to avoid noise). For high speed, use 74AC (but again take care with the layout).

For logic circuits in industrial control, needing slow speed and robustness and also subject to regular change and modification, DIN rail mounted logic blocks are best. This does, though, seem to be a dying market as programmable controllers (PLCs, see Section 14.4) provide the same functionality, are cheap and give more flexibility.

Battery circuits are best designed with 4000B or 74C devices, the supply is less critical, and both will run on a 9 V battery until it is flat without the need of a regulator circuit. Being slower they are also less prone to noise.

For 'cooking' logic, 74HC seems best suited with a reasonable speed, lower power and better noise immunity than LS TTL. The only problem is an incomplete coverage of the TTL family at present (the useful 7490/92 counters are missing, for example), so the odd LS or ALS TTL circuit may be needed, with 74HCT devices being used as interfaces between TTL and CMOS.

Figure 13.18 shows a comparison between these families.

13.3. Combinational logic

13.3.1. Introduction

Combinational logic is based around the block diagram of Fig. 13.19a. Such systems have several inputs and one or more outputs. The output states are uniquely defined for each and every combination of inputs and the 'block' does not contain any device such as storage, timers or counters. We therefore have n inputs I_1 to I_n and Z outputs Q_1 to Q_z. In systems with multiple outputs it is

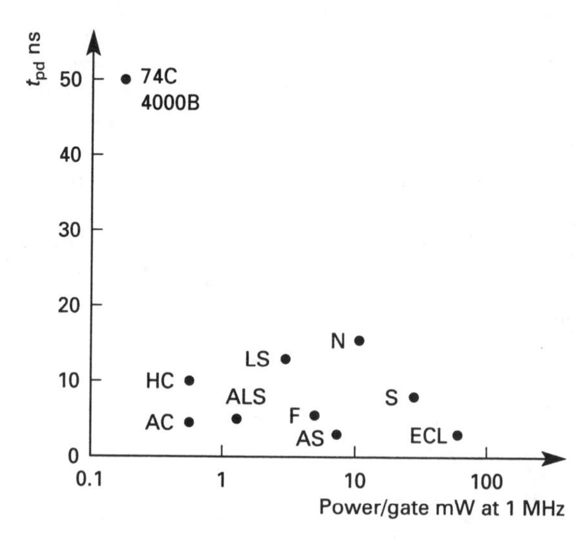

Fig. 13.18 *Comparison of speed/power for digital families. Note the power axis is logarithmic.*

usually easier to consider each separately as in Fig. 13.19b, allowing us to consider the circuit as Z blocks, each different but represented by Fig. 13.19c.

The number of possible input states depends on the number of inputs:

for two inputs there are four input combinations
for three inputs there are eight input combinations
for four inputs there are sixteen input combinations

and so on. Not all of these may be needed. There are frequently only a certain number of input combinations that may occur because of physical restrictions elsewhere in the system.

The design of combinational logic systems first involves examining all the input states that can occur and defining the output states that must occur for each and every input state. A logic design to achieve this is then constructed from the gates described in Section 13.1.3. In many systems the design can be done in an intuitive manner, but the rest of this section describes more formal design procedures.

Few real-life systems need pure combinational logic; most need storage and similar dynamic functions. Such systems can be analysed and designed by considering them as smaller subsystems linked together. The design of dynamic systems is discussed in Section 13.8.

13.3.2. Truth tables

A truth table is a useful way of representing a combinational logic circuit, and can be used to design the circuit needed to achieve a desired function.

Suppose we have three contacts monitoring some event (overpressure in a chemical reactor, for example) and we wish to construct a majority vote circuit. If the three switches are called A, B, C and the majority vote Z, this would have the truth table:

A	B	C	Z	
0	0	0	0	
0	0	1	0	
0	1	0	0	
0	1	1	1	←
1	0	0	0	
1	0	1	1	←
1	1	0	1	←
1	1	1	1	←

It can be seen that Z is 1 for:

$\bar{A}$ and B and C
or A and $\bar{B}$ and C
or A and B and $\bar{C}$
or A and B and C

The desired logic function can then be constructed directly from the truth table as

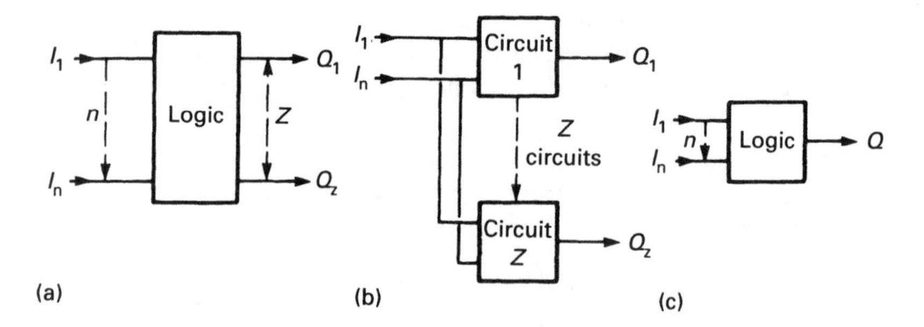

Fig. 13.19 *Combinational logic block diagrams (a) Generalised problem. (b) Separate circuits. (c) One circuit.*

in Fig. 13.20. In general, the circuit derived from a truth table will consist of a set of AND gates whose outputs are OR'd together.

A truth table always gives a design which works and is logically correct, but does not always give a circuit which uses the minimum combination of gates. To do this we need one of the other techniques described below.

The form of a logic design derived from a truth table is always a series of AND gates whose outputs are OR'd together (Fig. 13.20 is a typical example). This form of circuit is known as a sum of products (see Section 13.3.3 below), and one of the reasons for the popularity of NAND gates is that a sum of products expression can be formed purely with NAND gates.

Consider the expression:

$$Z = (A \text{ and } B) \text{ OR } (C \text{ and } D)$$

This has the simple circuit of Fig. 13.21a, which obviously fulfils the logic function. Consider, however, the totally NAND based circuit of Fig. 13.21b.

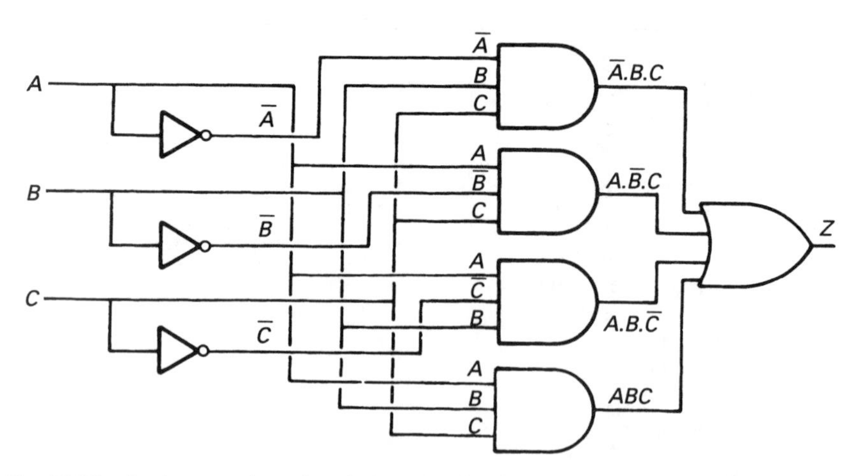

Fig. 13.20 *Implementation of majority vote circuit direct from truth table.*

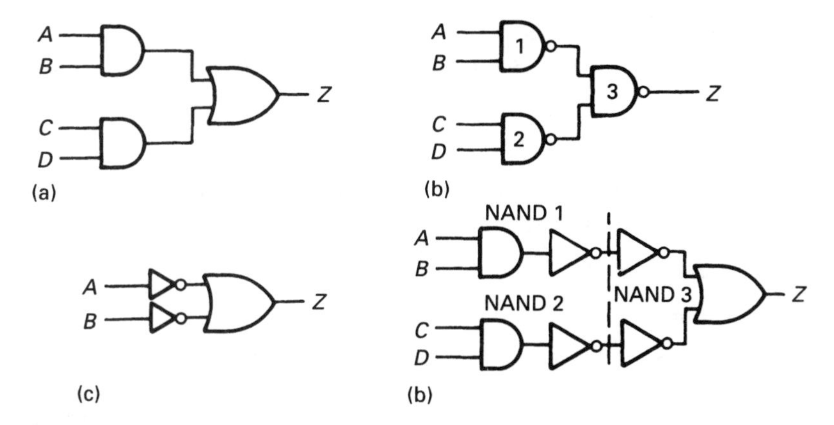

Fig. 13.21 *Logic circuits based solely on NAND gates. (a) Required logic. (b) NAND gate equivalent. (c) Representation of a NAND gate. (d) Operation of circuit (b).*

Straightforward, if laborious, testing of all sixteen possible input states will show that it behaves identically to Fig. 13.21b. In some mysterious way, the right-hand NAND gate is behaving as an OR gate.

This rather surprising fact is a result of De Morgan's theorem, described in the next section. Intuitively, however, we can see the reason by drawing up the truth table for the OR gate preceded by inverters as in Fig. 13.21c:

A	B	Z
0	0	1
0	1	1
1	0	1
1	1	0

This is the same as a NAND gate, so a NAND gate can, with legitimacy, be drawn as Fig. 13.21c.

The circuit of Fig. 13.21b could now be drawn as Fig. 13.21d with the ingoing NANDs drawn as ANDs followed by an inverter, and the outgoing NAND by the arrangement of Fig. 13.21c. Obviously the intermediate inverters cancel, leaving the equivalent circuit of Fig. 13.21a.

13.3.3. Boolean algebra

In the nineteenth century a Cambridge mathematician and clergyman, George Boole, devised an algebra to express and manipulate logical expressions. His algebra can be used to represent, design and minimise combinational logic circuits.

The AND function is represented by a dot (.), so:

$$Z = A \, . \, B$$

means Z is 1 when A is 1 AND B is 1. Often the dot is omitted, e.g. $Z = AB$.

The OR function is represented by an addition sign (+), so:

$$Z = A + B$$

means Z is 1 when A is 1 OR B is 1.

The invert function is represented by a bar $\bar{}$, so

$$Z = \bar{A}$$

means Z takes the opposite state to A.

Boolean algebra allows complex expressions to be written in a concise manner. Figure 13.4b, for example, is:

$$Z = (A \,.\, \bar{B}) + (\bar{A} \,.\, B)$$

and Fig. 13.20 is:

$$Z = (\bar{A} \,.\, B \,.\, C) + (A \,.\, \bar{B} \,.\, C) + (A \,.\, B \,.\, \bar{C}) + (A \,.\, B \,.\, C)$$

Boolean algebra can also be used to simplify expressions. To achieve this, a series of rules are used. The first eleven of these are self-obvious (or can be visualised by considering the equivalent relay circuits):

(a) $A \,.\, A = A$
(b) $A + A = A$
(c) $A \,.\, 1 = A$
(d) $A \,.\, 0 = 0$
(e) $A + 1 = 1$
(f) $A + 0 = A$
(g) $\bar{\bar{A}} = A$
(h) $A \,.\, \bar{A} = 0$
(i) $A + \bar{A} = 1$
(j) $A + B = B + A$
(k) $A \,.\, B = B \,.\, A$

The next two laws allow us to group brackets around variables with the same operator:

(l) $(A + B) + C = A + (B + C) = A + B + C$
(m) $(A \,.\, B) \,.\, C = A \,.\, (B \,.\, C) = A \,.\, B \,.\, C$

The next two laws are called the absorptive laws, and tell us what happens if the same variable appears with AND and OR operators:

(n) $A + A \,.\, B = A$
(o) $A \,.\, (A + B) = A$

The above two laws are not immediately obvious, and are shown in relay form in Fig. 13.22.

The next laws (called the distributive laws) tell us how to factorise Boolean equations:

(p) $A + B \,.\, C = (A + B) \,.\, (A + C)$
(q) $A \,.\, (B + C) = A \,.\, B + A \,.\, C$

In general, Boolean expressions can be expressed in two forms. The first form,

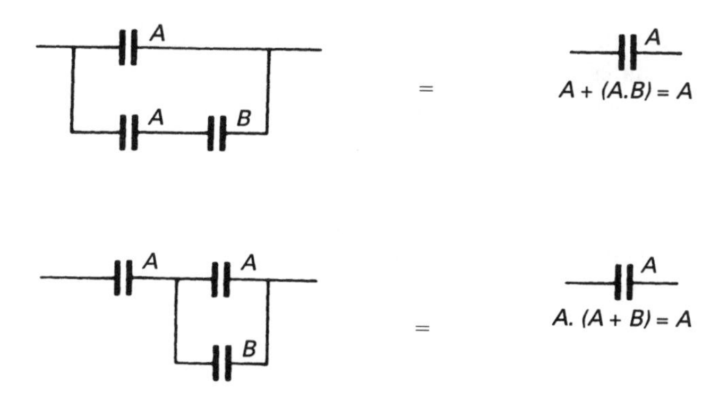

Fig. 13.22 *Relay demonstration of absorptive laws.*

called product of sums (or P of S), brackets OR terms and ANDs the results, e.g.:

$$Z = (A + \bar{B}) . (B + C + D) . (\bar{A} + \bar{D})$$

The second form, called sum of products (S of P), groups AND terms and ORs the results, e.g.:

$$Z = AB\bar{D} + \bar{B}C + A\bar{D}$$

Truth tables, described above in Section 13.3.2, inherently give an S of P result.

The complementary function of a Boolean expression yields the inverse of the expression (i.e. where the expression yields 1, the complement yields 0). The expressions $A + B$ and $\bar{A} . \bar{B}$, for example, can be shown to be complementary by simply constructing their truth tables.

The last two laws, known as De Morgan's theorem, show how to form the complement of a given expression (and give one way to interchange S of P and P of S forms):

(r) $\overline{A . B . C \ldots . N} = \bar{A} + \bar{B} + \bar{C} + \ldots \bar{N}$
(s) $\overline{A + B + C + \ldots . + N} = \bar{A} . \bar{B} . \bar{C} \ldots . \bar{N}$

In its formal representation, De Morgan's theorem appears rather daunting. It can be more easily expressed:

'To form the complement of an expression:

(1) Replace each ' + ' in the original expression with '.' and vice versa.
(2) Complement each term in the original expression.'

For example, the expression $\bar{A} + B . C$ is complemented as below:

Step 1: replace ' + ' by '.' and '.' by ' + ' giving $A . (B + C)$
Step 2: complement each term $A . (\bar{B} + \bar{C})$

which is the complement of $\bar{A} + B.C$ (as can be verified by trying all eight possible input states).

De Morgan's theorem explains the behaviour of Fig. 13.21b. The output Z is given by:

$$Z = \overline{\overline{(A.B)}.\overline{(C.D)}}$$

i.e. the complement of $\overline{(A.B)}.\overline{(C.D)}$.

Applying De Morgan's theorem to $\overline{(A.B)}.\overline{(C.D)}$ gives the complement form to be $(A.B) + (C.D)$, hence:

$$Z = (A.B) + (C.D)$$

which is the required expression.

Boolean algebra can be used to minimise logical expressions, but the method is rarely obvious, and it is easy to make errors with double bars and swapping of '.' and '+'. Minimisation by Boolean algebra makes good examination questions, but is rarely used in practice. One example will suffice. Consider the expression:

$$Z = ABC + A\bar{B} + (\overline{\bar{A}\bar{C}})$$

Applying De Morgan's theorem to the right-hand term gives:

$$Z = ABC + A\bar{B}(\bar{\bar{A}} + \bar{\bar{C}})$$
but $\bar{\bar{A}} = A$ and $\bar{\bar{C}} = C$ giving
$$Z = ABC + A\bar{B}(A + C)$$
$$Z = ABC + AA\bar{B} + A\bar{B}C$$
$$Z = ABC + A\bar{B} + A\bar{B}C$$

we observe $ABC + A\bar{B}C = AC(B + \bar{B}) = AC$

hence $\quad Z = AC + A\bar{B}$

which is the minimal form.

An easier way to achieve the same minimisation result is to use the graphical Karnaugh map, described below.

13.3.4. Karnaugh maps

A Karnaugh map is an alternative way of presenting a truth table. The map is drawn in two dimensions; two, three and four variable maps are shown in Fig. 13.23.

Each square within the map represents one line on the truth table. For example:

square X represents $A = 1$, $B = 0$ which can be written $A\bar{B}$
square Y represents $A = 0$, $B = 1$, $C = 1$ which can be written $\bar{A}BC$
square Z represents $A = 1, B = 0, C = 1, D = 0$ which can be written $A\bar{B}C\bar{D}$

The essential feature of a Karnaugh map is the way in which the axes are labelled. It will be seen that only one variable changes between horizontally adjacent squares in any row, and only one variable changes between vertically adjacent squares in any column.

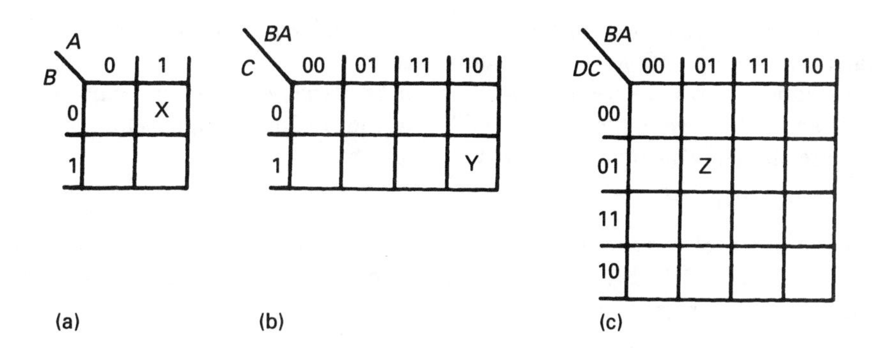

Fig. 13.23 *Karnaugh maps. (a) Two variable map. (b) Three variable map. (c) Four variable map.*

The use of this feature is not immediately apparent, but consider Fig. 13.24a. The truth table contains four terms giving a 1 output. These are:

$$A\bar{B}C\bar{D},\ ABC\bar{D},\ A\bar{B}CD,\ ABCD$$

so we could write (quite correctly)

$$Z = A\bar{B}C\bar{D} + ABC\bar{D} + A\bar{B}CD + ABCD$$

Examination of the map, however, shows that the D variable and B variable change without affecting the output. The circled squares, in fact, represent AC, so the above expression can be simplified to:

$$Z = AC$$

This result could, of course, also have been obtained (with great effort) by Boolean algebra.

In a three variable map such as Fig. 13.24b, each cell represents some

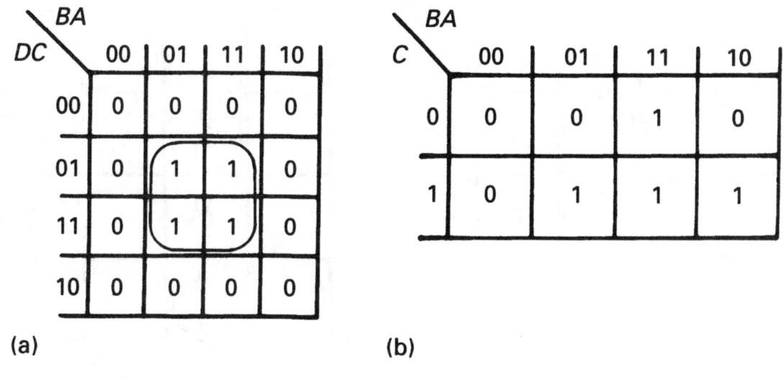

Fig. 13.24 *Four and three variable maps. (a) Representation of $Z = AC$ on four variable map. (b) Plot of $Z = A.B.C + A.B.\bar{C} + A\bar{B}C + \bar{A}BC$.*

combination of the three variables. On a four variable map such as Fig. 13.24b, each cell represents some combination of the four variables.

Groups of two adjacent cells on a three variable map represent some combination of *two* of the three variables. In Fig. 13.25a, groupings for AB and $C\bar{B}$ are shown. This map represents:

$$Z = AB + C\bar{B}$$

Two adjacent cells on a four variable map represent some combination of three of the four variables. In Fig. 13.25b, groupings for $\bar{A}BC$, $B\bar{C}\bar{D}$, $A\bar{B}\bar{D}$ and $\bar{B}\bar{C}D$ are shown. This map represents:

$$Z = \bar{A}BC + B\bar{C}\bar{D} + A\bar{B}\bar{D} + \bar{B}\bar{C}D$$

Groups of four adjacent cells on a three variable map represent a single variable. The group in Fig. 13.26 represents the variable A, hence:

$$Z = A$$

Groups of four adjacent cells on a four variable map represent some combination of two of the four variables. The groups in Fig. 13.26b represent $\bar{B}\bar{D}$ and BD. The map represents

$$Z = \bar{B}\bar{D} + BD$$

A group of eight adjacent cells on a four variable map represents a single variable. The groups in Fig. 13.27 represent C and $\bar{B}$, so:

$$Z = C + \bar{B}$$

It is important to realise that top and bottom edges are considered adjacent, as are right and left sides. Grouping can therefore be made around the tops and sides, as in Fig. 13.28 which represents:

$$Z = \bar{A}C + A\bar{C}$$

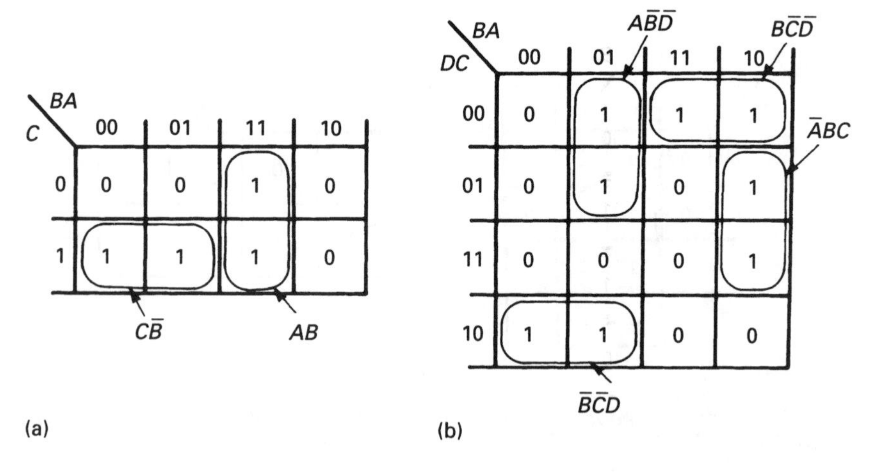

Fig. 13.25 *Grouping of 2 adjacent cells. (a) Three variable. (b) Four variable.*

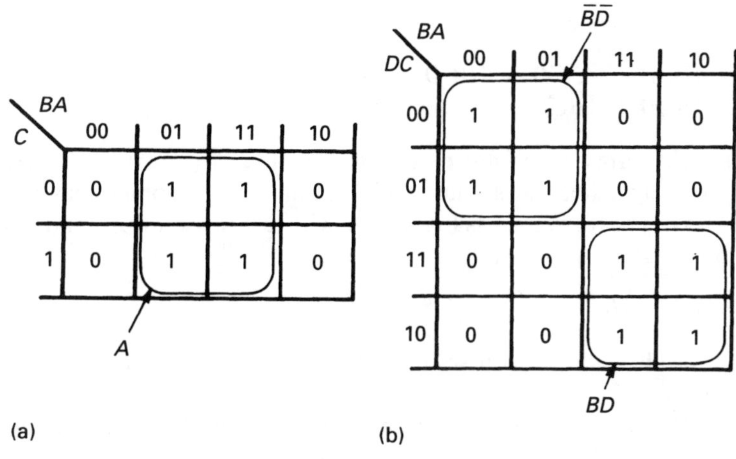

Fig. 13.26 *Grouping of 4 adjacent cells. (a) Three variable. (b) Four variable.*

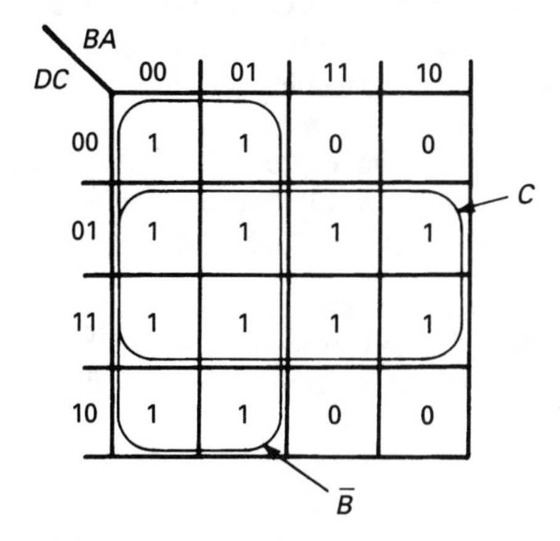

Fig. 13.27 *Grouping of 8 adjacent cells.*

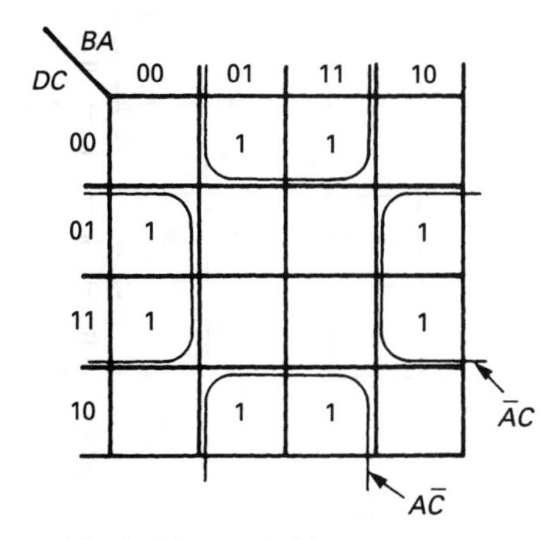

Fig. 13.28 *Adjacency of top/bottom and sides.*

Consider the expression:

$$Z = \bar{A}B\bar{C}\bar{D} + \bar{A}BC\bar{D} + ABC\bar{D} + A\bar{B}C\bar{D} + A\bar{B}CD + ABCD + \bar{A}BCD + \bar{A}B\bar{C}D$$

This has eight terms which are plotted in Fig. 13.29a (equally, this could be derived from eight lines of a truth table). Using the grouping ideas outlined above, these can be regrouped as Fig. 13.29b with just two terms AC and $\bar{A}B$. The expression for Z above becomes:

$$Z = AC + \bar{A}B$$

The same result could, of course, have been obtained by lengthy analysis by Boolean algebra.

The rules for minimisation using Karnaugh maps are simple and straightforward:

(1) Plot the Boolean expression or truth table on to the Karnaugh map.
(2) Form new groups of 1s on the map. Groups must be rectangular and contain one, two, four or eight cells. Groups should be as large as possible and there should be as few groups as possible. Do not forget overlaps and possible round-the-edge groupings.
(3) From the map, read off the expression for each group. The minimal expression is then obtained in S of P form, and can be directly implemented in AND/OR gates (as in Fig. 13.20) or NAND gates (as in Fig. 13.21b).

In section 13.3.2 we designed a majority vote circuit (Fig. 13.20) direct from a truth table. This is plotted on to a Karnaugh map in Fig. 13.30a, and grouped on Fig. 13.30b. It will be seen that this has three terms giving the (simpler) circuit of Fig. 13.30c.

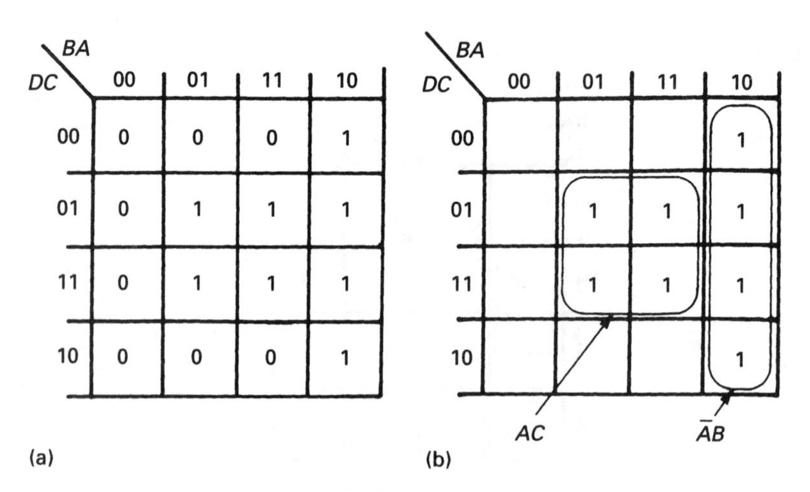

Fig. 13.29 *Grouping for minimal expression. (a) Four variable. (b) Minimal grouping.*

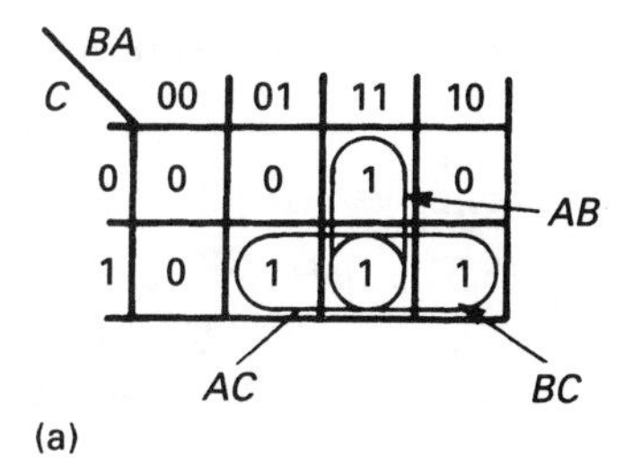

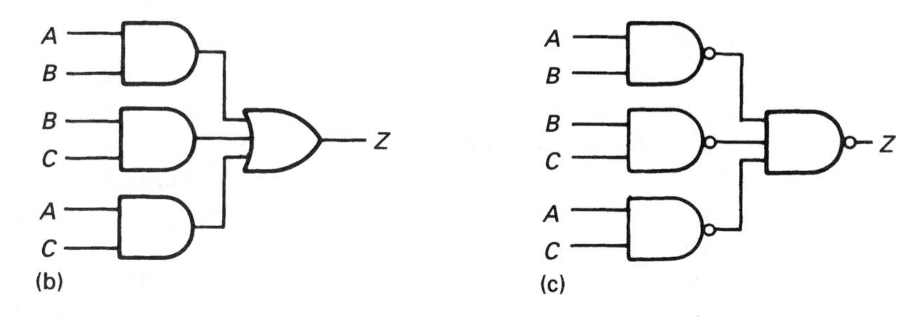

Fig. 13.30 *Design of majority vote circuit. (a) Karnaugh map of majority vote circuit. (b) AND/OR implementation. (c) NAND implementation.*

13.3.5. Integrated circuits

Many complex functions are available in IC form, and a circuit designer should aim to minimise cost and the number of IC packages rather than the number of gates. A minimisation exercise, whether by Boolean algebra or Karnaugh map, should always be preceded by a search of an IC catalogue for a suitable off-the-peg device.

13.4. Storage

13.4.1. Introduction

Most logic systems require some form of memory. A typical relay circuit is the motor starter circuit of Fig. 13.31 which 'remembers' which of the two operator

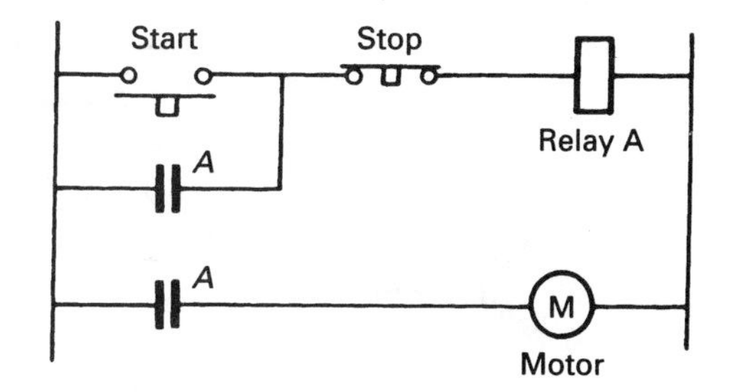

Fig. 13.31 *A relay storage circuit, a motor starter.*

push buttons was pressed last. The memory is achieved by the latching contact of relay A.

13.4.2. Cross-coupled flip flops

The logical equivalent of Fig. 13.31 is the cross-coupled NOR gate circuit of Fig. 13.32a. Assume both inputs are 0, and output Q is at a 1 state. The output of gate a will be 0, and the two 0 inputs to gate b will maintain Q in its 1 state. The circuit is therefore stable.

If the reset input is now taken to a 1, Q will go to a 0 and $\bar{Q}$ to a 1. Similar analysis to that above will show that the circuit is stable in this state, even when the reset input goes back to a 0.

The set input can be used now to switch the Q output to 1 and the $\bar{Q}$ back to 0. The set and reset inputs cause the output to change state, with the outputs indicating which input was last at a 1 state, as summarised by Fig. 13.32b. If both inputs are 1 together both outputs go to a 0, but this condition is normally disallowed.

The cross coupled NOR gate circuit is called an RS flip flop, and is shown on logic diagrams by the symbol of Fig. 13.32c.

It is also possible to construct a cross coupled flip flop from NAND gates, as in Fig. 13.33a. Analysis will show that this behaves in a similar way to Fig. 13.32, but the circuit remembers which input last went to a 0, as shown in Fig. 13.33b. The logic symbol for a NAND-based RS flip flop is shown in Fig. 13.33c; the small circles on the input show that the flip flop responds to 0 inputs.

13.4.3. The transparent latch

The transparent latch (known also as a hold/follow latch) is used to freeze digital data. It is constructed as in Fig. 13.34 (and is usually obtained in IC form, such as the TTL 7475). With the enable input at 1, the output Q follows the input A. When the enable input goes to a 0, Q indicates the state of A at the instant the enable went from 1 to 0.

Transparent latches are typically used to transfer data from, say, a fast counter to a display.

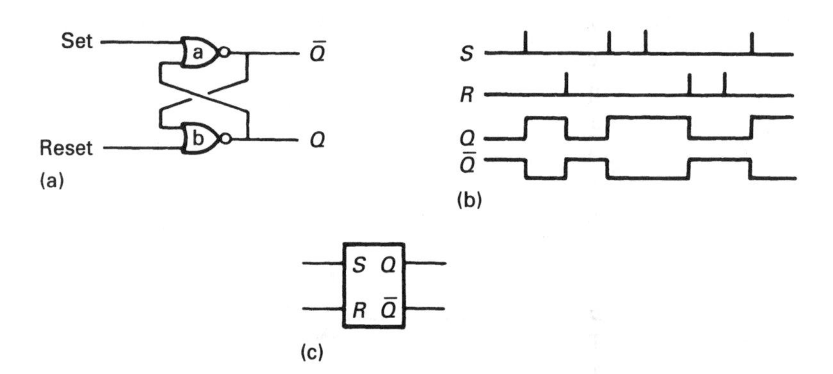

Fig. 13.32 *The NOR gate RS flip flop. (a) Logic diagram. (b) Operation (c) Logic symbol.*

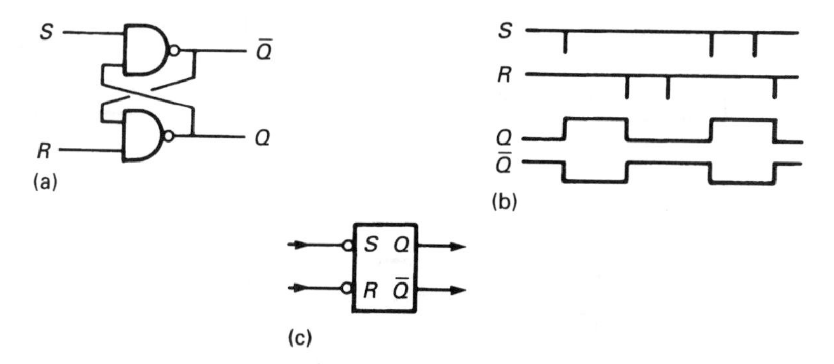

Fig. 13.33 *The NAND gate RS flip flop. (a) Logic diagram. (b) Operation. (c) Logic symbol.*

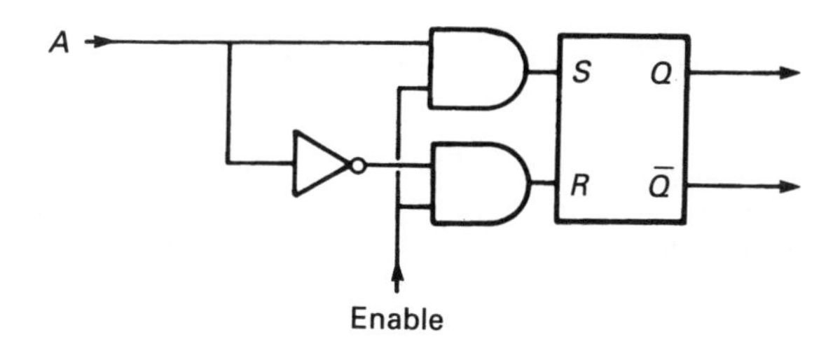

Fig. 13.34 *The transparent latch.*

13.4.4. The D-type flip flop

The D-type flip flop shown in Fig. 13.35a has a single data input (D), a clock input and the usual Q and $\bar{Q}$ outputs. Superficially this is similar to the latch memory above, but the clock operates in a more subtle way. The operation of a typical D-type flip flop is shown in Fig. 13.35b. The clock samples the D input when the clock input goes from 0 to 1, but the output changes state when the clock goes from 1 to 0. The significance of this is explained below in Section 13.4.6.

There are several ways in which a D-type flip flop can be implemented. A common circuit uses the master/slave arrangement of Fig. 13.35c. When the clock input is 1, the D input sets, or resets, the master flip flop. When the clock input is 0, the state of the master flip flop is transferred to the slave flip flop (and the outputs take up the state of D when the clock input was 1). Note that the master flip flop is isolated from the D input whilst the clock is 0.

Although it would be feasible to construct a master/slave flip flop from discrete gates, IC D-types (such as the TTL 7474 or the CMOS 4013) are readily available.

13.4.5. The JK flip flop

In Section 13.4.2 the NOR based RS flip flop was described, and it was stated that the input state $R = S = 1$ was normally disallowed. The JK flip flop is a clocked RS flip flop with additional logic to cover this previously disallowed state. The

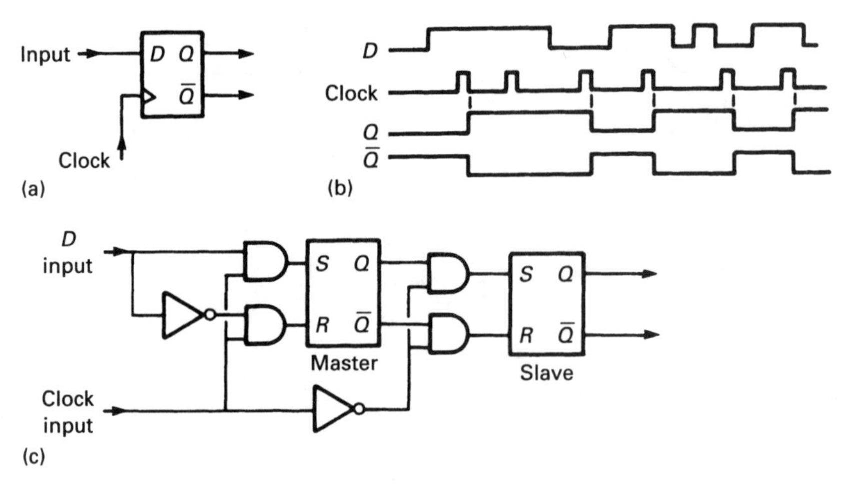

Fig. 13.35 *The D-type flip flop. (a) Logic symbol. (b) Operation. (c) Logic diagram for master/slave circuit.*

clock input acts as described above for the D-type flip flop, i.e. sampling the inputs on one edge, and causing the outputs to change on the other.

The outputs after a clock pulse for $J = 1, K = 0$; $J = 0, K = 1$; $J = 0, K = 0$ are as would be expected for a clocked RS flip flop. If $J = K = 1$, the outputs toggle; that is, the states of the Q and $\bar{Q}$ interchange. This action is summarised in Fig. 13.36b and in the table below.

		Output after clock pulse		
J	K	Q	$\bar{Q}$	*Comment*
0	0	No change		Q=old Q, $\bar{Q}$=old $\bar{Q}$
0	1	0	1	Reset
1	0	1	0	Set
1	1	Toggle		Q=old $\bar{Q}$, $\bar{Q}$=old Q

In data sheets, the above would be represented:

J	K	CK	Q_{n+1}
0	0	⎍	Q_n
0	1	⎍	0
1	0	⎍	1
1	1	⎍	$\bar{Q}_n$

The toggle state is the basis for counters, described in Section 13.7.

13.4.6. Clocked storage

The D-type and JK flip flop described above are examples of clocked storage. The advantages and implications of this are probably not immediately obvious.

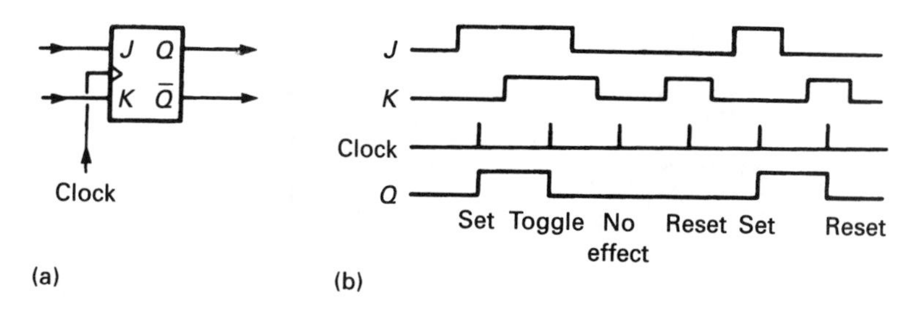

Fig. 13.36 *The JK flip flop. (a) Logic symbol. (b) Operation.*

In all bar the simplest systems, data is often required to be moved around from one storage position to another. In Fig. 13.37, for example, data is to be moved through stores A, B, C in an orderly manner. If transparent latches were used along with a single enable as shown, the data would shoot straight through all the stages. If clocked storage is used, the data will sequence from A to B to C, moving one position for each clock pulse.

13.5. Timers and monostables

Control systems often need some form of timer; a gas igniter might operate under a pilot flame for 5 seconds, say, before a flame failure detector is enabled, or a nitrogen purge of a reactor vessel undertaken for 2 minutes before the reagents are admitted. Timing functions in logic circuits are provided by devices called monostables or delays.

There are many types of delay, although all can be considered, as Fig. 13.38a, to consist of an input, Q and $\bar{Q}$ outputs, and an RC network which determines the delay period.

The commonest timer, often called the one shot or monostable, gives an output pulse, of known duration, for an input edge. The user can select which edge (0–1 or 1–0) triggers the circuit. In Fig. 13.38b a 0–1 edge is used. Monostables are the basis of all other delay circuits and are widely available (74121, 74122 in TTL, 4047, 4098 in CMOS).

Pure delays are shown in Fig. 13.38c, d and e, and these can be constructed by adding gates to monostable outputs. Figure 13.38f,g shows the circuit for a delay off.

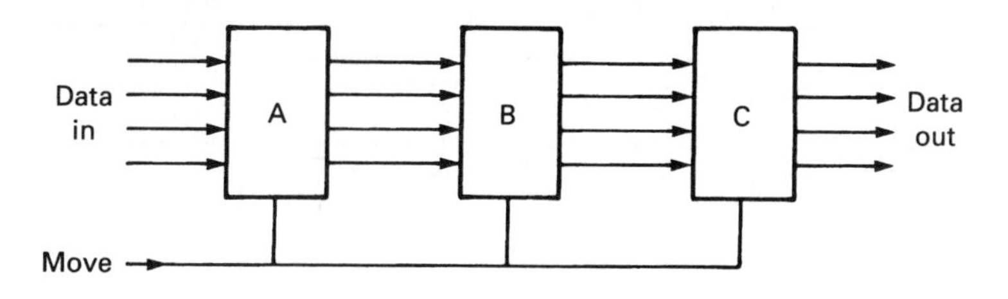

Fig. 13.37 *Clocked storage.*

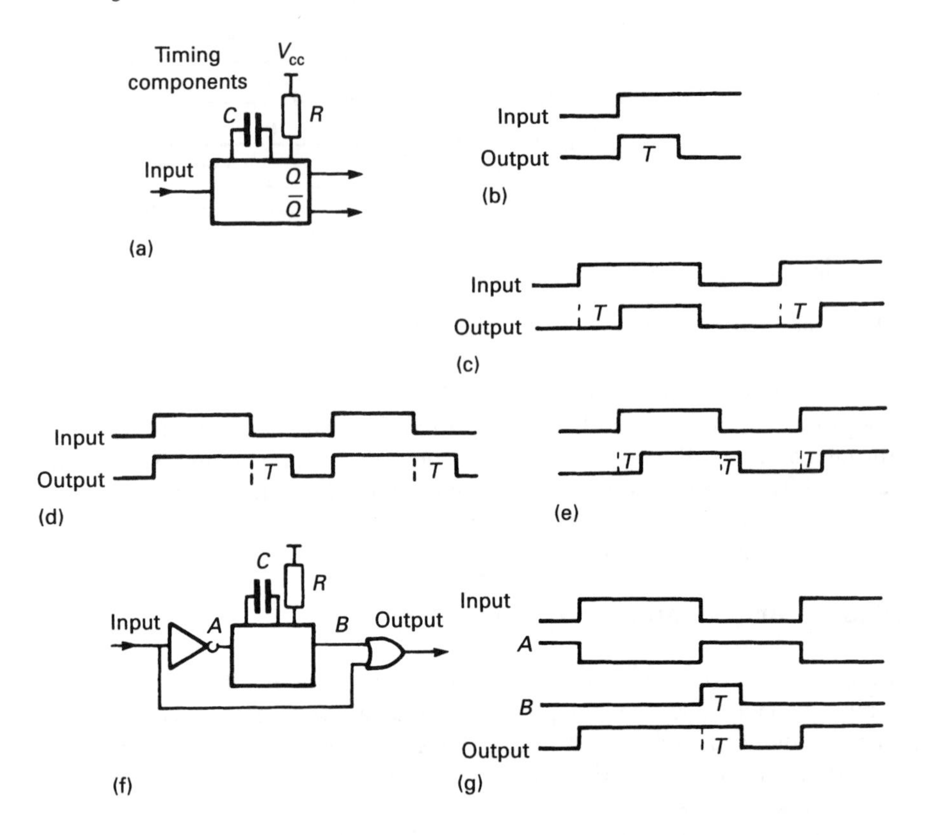

Fig. 13.38 *Timers and monostables. (a) General representation of a timer. (b) One shot timer. (c) Delay on. (d) Delay off. (e) Delay on and off. (f) Delay off circuit using a monostable. (g) Waveforms for circuit (f).*

A variation of the monostable is the retriggerable monostable. In most monostable circuits the timing logic ignores further input edges once started. In a retriggerable monostable each edge sets the timing circuit back to the start again. The actions of a retriggerable and normal monostable are compared in Fig. 13.39. Retriggerable monostables are commonly used as a low speed alarm with

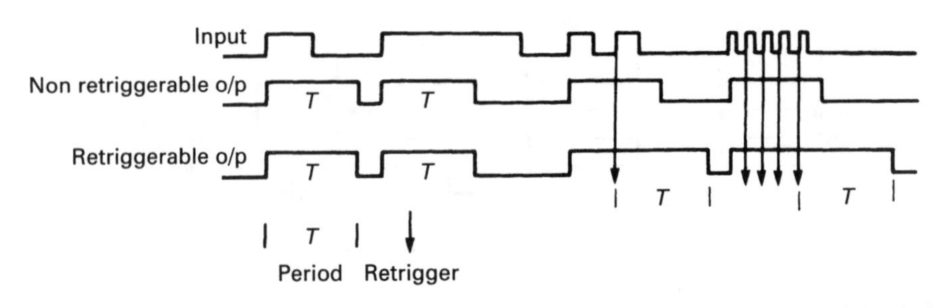

Fig. 13.39 *Retriggerable monostables.*

pulse tachos or as a watchdog protection for process control computers.

Monostables can also be used to construct digital oscillators (called clocks). Two monostables connected as in Fig. 13.40 will give a stable easy-to-adjust pulse train.

13.6. Arithmetic circuits

13.6.1. Number systems, bases and binary

In previous sections, logic signals have been assumed to represent events such as printer ready, or low oil level. Digital signals can also be used to represent, and manipulate, numbers.

We are so used to the decimal number system that it is hard to envisage any other way of counting. Normal everyday arithmetic is based on multiples of ten. For example, the number 9156 means:

9 thousands	$= 9 \times 10 \times 10 \times 10$
plus 1 hundred	$= 1 \times 10 \times 10$
plus 5 tens	$= 5 \times 10$
plus 6 units	$= 6$

Each position in a decimal number represents a power of ten. Our day-to-day calculations are done to a base of ten because we have ten fingers. Counting can be done to any base, but of special interest are bases 8 (called octal), 16 (called hex for hexadecimal) and two (called binary).

Octal uses only the digits 0 to 7; the octal number 317, for example, means

$3 \times 8 \times 8$	= decimal 192
plus 1×8	= decimal 8
plus 7	= decimal 7
Total	= decimal 207

Hex uses the letters A–F to represent decimal ten to fifteen, so hex C52, for example, means

$12 \times 16 \times 16$	= decimal 3072
plus 5×16	= decimal 90
plus 2	= decimal 2
Total	= decimal 3164

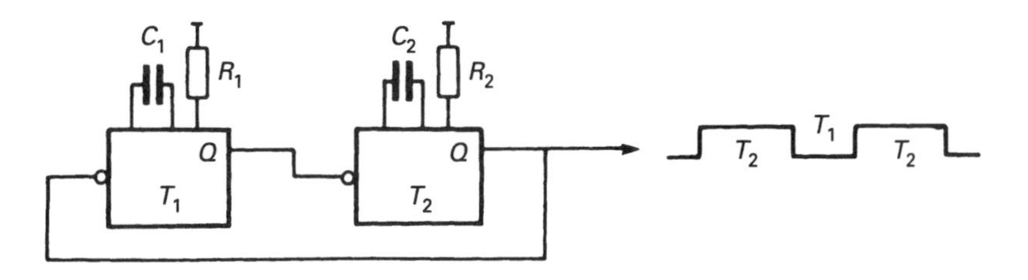

Fig. 13.40 *Oscillator built from two monostables.*

Binary needs only two symbols, 0 and 1. Each position in a binary number represents a power of two and is called a bit (for BInary digiT), the most significant to the left as usual, so 101101 is evaluated:

$$\begin{array}{lr} 1 \times 2 \times 2 \times 2 \times 2 \times 2 & = 32 \\ \text{plus } 0 \times 2 \times 2 \times 2 \times 2 & = 0 \\ \text{plus } 1 \times 2 \times 2 \times 2 & = 8 \\ \text{plus } 1 \times 2 \times 2 & = 4 \\ \text{plus } 0 \times 2 & = 0 \\ \text{plus } 1 & = 1 \\ \text{Total} & = \text{decimal } 45 \end{array}$$

Similarly 1101011 is evaluated:

$$\begin{array}{l} 1 \times 64 = 64 \\ 1 \times 32 = 32 \\ 0 \times 16 = 0 \\ 1 \times 8 = 8 \\ 0 \times 4 = 0 \\ 1 \times 2 = 2 \\ 1 = 1 \\ \text{Total} = \text{decimal } 107 \end{array}$$

Conversion from decimal to binary is achieved by successive division by two, noting the remainders, Reading the remainders from the top (LSB–least significant bit) to bottom (MSB–most significant bit) gives the binary equivalent, e.g. decimal 23;

23
11 r 1 (LSB)
5 r 1
2 r 1
1 r 0
0 r 1 (MSB)

Decimal 23 is binary 10111.

Similarly decimal 75:

75
37 r 1 (LSB)
18 r 1
9 r 0
4 r 1
2 r 0
1 r 0
0 r 1 (MSB)

Decimal 75 is binary 1001011.

Octal and hex give a simple way of representing binary numbers. To convert a binary number to octal, the binary number is written in groups of three (from the LSB) and the octal equivalent written underneath; for example, 11010110:

grouped in threes	11	010	110
octal	3	2	6

Hex conversion is similar, but groupings of four are used. Taking the same binary number 11010110:

grouped in fours	1101	0110
hex	D	6

The octal number 326 and the hex number D6 are both representations of the binary number 11010110.

13.6.2. Binary arithmetic

Consider the decimal sum:

$$\begin{array}{r} 345 \\ +\ 272 \\ \hline 617 \end{array}$$

This is evaluated in three stages:

$5 + 2 = 7$ no carry
$4 + 7 = 11$ one down (as result) plus carry
$3 + 2 + \text{carry} = 6$

At each stage we consider three 'inputs': two digits and a possible carry from the previous stage. Each stage has two outputs: a sum digit and a possible carry to the next, more significant stage. A single digit adder can therefore be considered, as in Fig. 13.41a. Several single digit adders can be cascaded, as in Fig. 13.41b, to give an adder of any required number of digits. Note the carry out of the most significant stage becomes the most significant digit.

Binary addition is similar, except that there are only two possible values for each digit. If Fig. 13.41a is a binary adder, there are eight possible input combinations:

Inputs			*Outputs*	
Digit 1	*Digit* 2	*Carry*	*Sum*	*Carry*
0	0	0	0	0
0	1	0	1	0
1	0	0	1	0
1	1	0	0	1
0	0	1	1	1
0	1	1	0	1
1	0	1	0	1
1	1	1	1	1

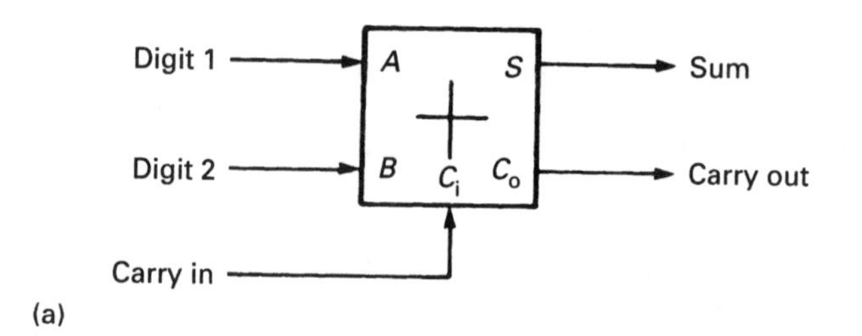

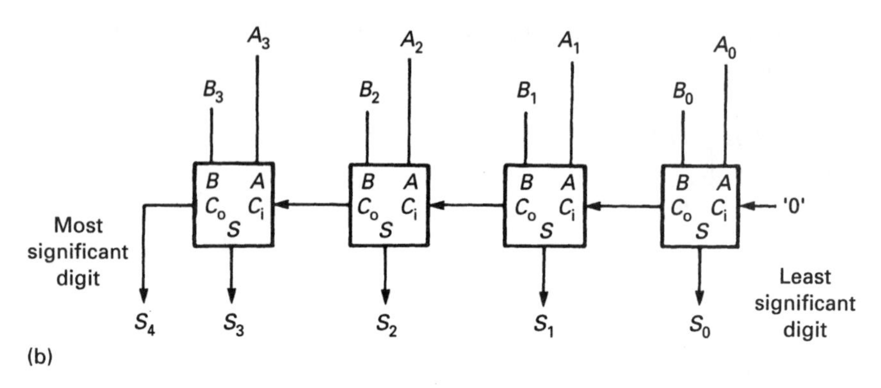

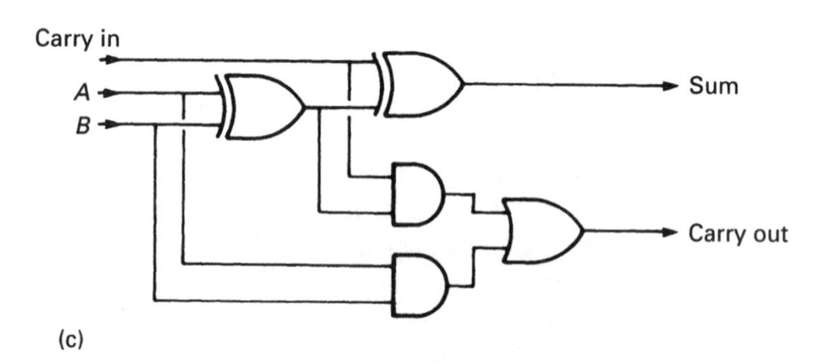

Fig. 13.41 *Digital adder circuits. (a) Representation of one-digit adder. (b) Four-digit adder constructed from four one-digit adders. (c) One-bit dinary adder.*

An example of binary arithmetic is:

```
 1 0 1 1 0 1 0
 0 1 0 1 0 1 1
 -------------
1 0 0 0 0 1 0 1    Sum (result)
 1 1 1 1 0 1 0     Carry
```

The implementation of the adder truth table is a simple problem of combinational logic; one possible solution is shown in Fig. 13.41c. In practice, of course, adders such as the TTL 7483 are readily available in IC form.

Negative numbers are generally represented in a form called twos complement. The most significant digit represents the sign, being 1 for negative numbers and 0 for positive numbers. The value part of a negative number is complemented and 1 added. For example, +12 in twos complement is 01100 (the MSB 0 indicating a positive number).

To get the twos complement for −12 we complement 1100 giving 0011, set the MSB to 1 giving 10011, then add 1 giving 10100 which is the twos complement representation of −12. Similarly:

+43	0101011
Complement	1010100
Add 1	1010101, which is −43

In each case, addition of the positive and negative number will give the result zero, e.g.:

```
 +43    0 1 0 1 0 1 1
 −43    1 0 1 0 1 0 1
       ______________
      1 0 0 0 0 0 0 0
```

The top carry is lost, giving the correct result of zero.

Twos complement representation allows subtraction to be done by adding a negative number, for example 12 − 3:

```
  0 1 1 0 0    +12
  1 1 1 0 1    −3
 ___________
1 0 1 0 0 1
```

The top bit is lost, giving the correct result of +9.

13.6.3. Binary coded decimal (BCD)

A single decimal digit can take any value between 0 and 9. Four binary digits are therefore needed to represent one decimal digit. In BCD, each decimal digit is represented by 4 bits. For example:

9	4	0	7	6
1001	0100	0000	0111	0110

BCD is not as efficient as pure binary. In pure binary 12 bits can represent 0 to 4095, compared with 0 to 999 in BCD. BCD, however, has advantages where decimal numbers are to be read from decade switches or position measuring encoders.

13.6.4. Unit distance codes

Figure 13.4.2 shows a possible application of binary coding. The position of a shaft is to be measured to 1 part in 16 by means of an optical grating moving in

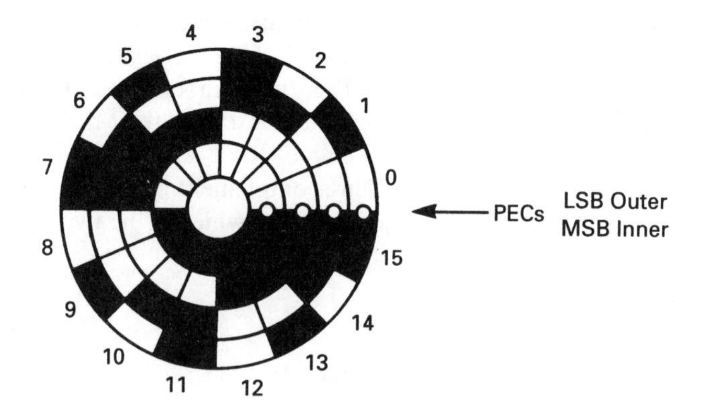

Fig. 13.42 *Shaft encoder.*

front of four photocells. The photocell outputs give a binary representation of the shaft angular position.

Consider what may happen as the shaft goes from position 7 (0111) to position 8 (1000). It is unlikely that all the cells will switch together, so we could get:

0111→0000→1000 or
0111→1111→1000

or any other combination of 4 bits. These possible incorrect intermediate states can be avoided by using a code in which only 1 bit changes between adjacent positions. Such codes are called unit distance codes.

The commonest unit distance code is the Gray code, shown in 4 bit form below. It will be noted that the code is reflected about the centre. Sometimes the term 'reflected' code is used for unit distance codes.

Decimal	*Gray*		
0	0000		
1	0001		
2	0011		
3	0010		
4	0110		
5	0111		
6	0101		
7	0100	decimal	↑
8	1100	cyclic	symmetrical ↓
9	1101		
10	1111		
11	1110		
12	1010		
13	1011		
14	1001		
15	1000		

A unit distance code can be constructed to any even base by taking an equal number of combinations above and below the centre point of a Gray code. A decimal version (called the XS3 cyclic BCD code) is shown above. In this code 0 is 0010, 1 is 0110, 2 is 0111 and so on to 9 which is 1010.

Conversion between binary and Gray code is straightforward, and is achieved with XOR gates as shown in Fig. 13.43a and b.

13.7. Counters and shift registers

13.7.1. Ripple counters

Counters are used for two basic purposes. The first, and obvious use is the counting, or totalising, of external events. Batch counters, traffic recorders, frequency meters and such devices all use counters for totalisation. The second use of counters is the division of a frequency to give a new, lower frequency. A visual display unit (VDU), for example, is built around a timing chain which produces frequencies from several MHz down to 50 Hz from a single oscillator.

The 'building block' of all counters is the toggle flip flop which changes state each time its clock input is pulsed. Usually the toggling occurs on the negative edge, as shown in Fig. 13.44a. A toggle flip flop can be constructed from JK or D-type flip flops, as shown in Fig. 13.44b,c.

If the Q output of a toggle flip flop is connected to the clock input of the next stage as shown in Fig. 13.45a, a simple binary counter can be constructed to any desired length. Figure 13.45 is a three bit counter, with A the LSB and C the MSB. This counts:

Pulse	*C*	*B*	*A*
0	0	0	0
1	0	0	1
2	0	1	0
3	0	1	1
4	1	0	0
5	1	0	1
6	1	1	0
7	1	1	1

Another pulse will take it to state 0 again. It can be seen that Fig. 13.45 is counting up.

To count down, the $\bar{Q}$ outputs are connected to the following stage as in Fig. 13.46a, and the signal outputs taken from the Q lines. Examination of Fig. 13.46b will show that this counter is counting down.

There are two limitations to the speed at which a counter chain similar to Figs. 13.45 and 13.46 can operate. The first is the maximum speed at which the first (fastest) stage can toggle. This is typically 20 MHz for a TTL device. The second restriction is not so obvious.

Consider the case of an 8-bit counter going from 01111111 to 10000000. The LSB toggling causes the next to toggle, and so on to the MSB. The change has to

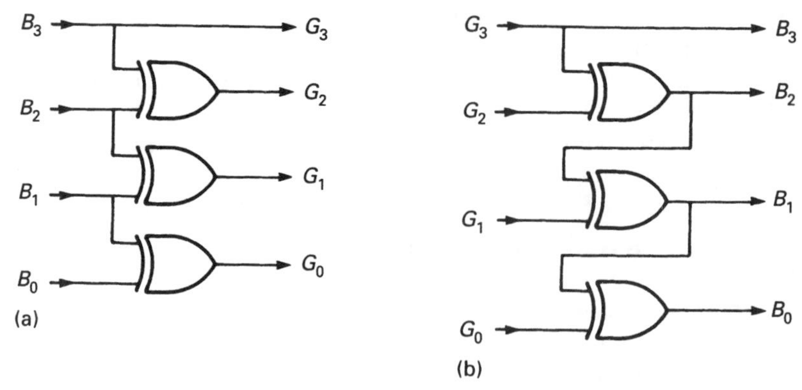

Fig. 13.43 *Binary/Gray conversion. (a) Binary to Gray. (b) Gray to binary.*

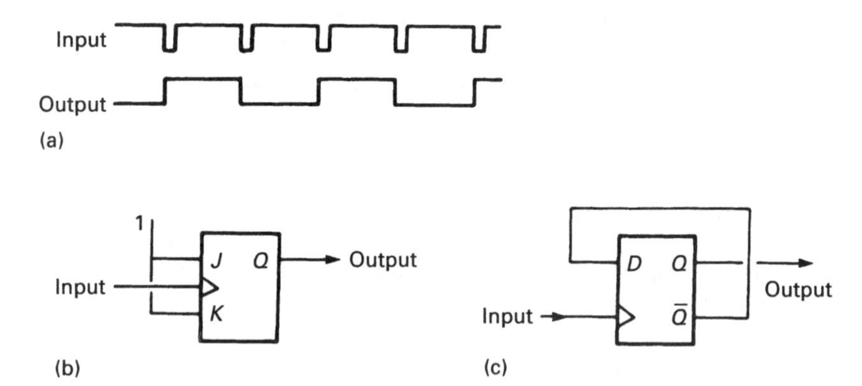

Fig. 13.44 *The toggle flip flop. (a) Counter operation. (b) JK based toggle flip flop. (c) D-type based toggle flip flop.*

propagate through all 8 bits of the counter, so circuits similar to Figs. 13.45 and 13.46 are called ripple counters.

During the 'ripple' the counter will assume invalid states and cannot be sensibly read. Obviously the propagation delay through all the stages should be considerably less than the input period. High speed applications use synchronous counters, described below.

In both Figs. 13.45 and 13.46 the frequency of output C is precisely one eighth of the input frequency. A simple ripple counter can act as a frequency divider. If we define:

$$N = f_{in}/f_{out}$$

then $N = 2^m$ for m binary stages.

It will also be seen that the output of any stage of a binary counter has equal mark space ratio regardless of the input mark/space, provided the input frequency is constant.

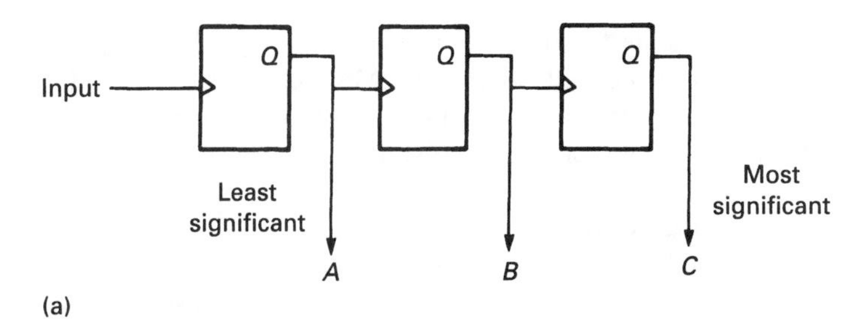

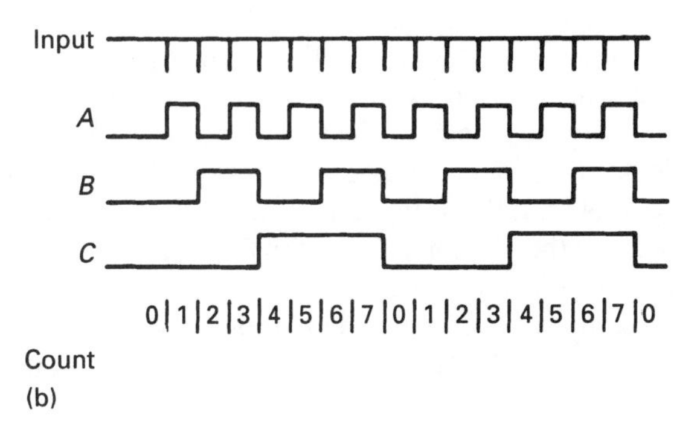

Fig. 13.45 *Simple three-bit binary counter. (a) Three-bit binary counter constructed from toggle flip flops. (b) Counter operation.*

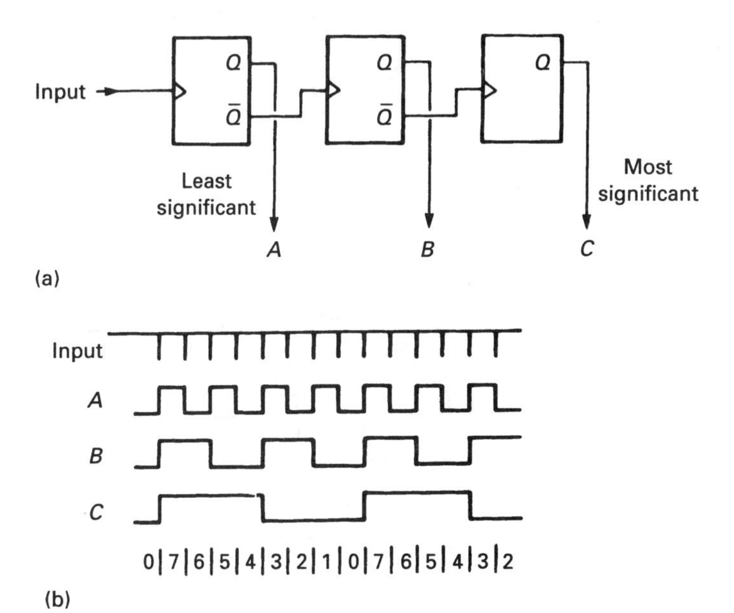

Fig. 13.46 *Three-bit down counter. (a) Down counter logic diagram. (b) Counter operation.*

Although it is feasible to construct ripple counters with D-type and JK flip flops, it is usually more cost effective to use MSI ICs such as the TTL 7493 4-bit counter or the CMOS 4024 7-bit counter. These incorporate features such as a reset line to take the counter to a zero state.

13.7.2. Synchronous counters

Ripple counters are limited in both speed and length by the cumulative ripple through propagation delay and also temporarily exhibit invalid outputs. Although these limitations are not important in slow speed applications, they can cause difficulties in high speed counting.

These restrictions can be overcome by the use of a synchronous counter where all required outputs change simultaneously. There is no ripple propagation delay through the counters and no transient false count stages. The only speed restriction is the toggling frequency of the first stage.

The building block of a synchronous counter is the JK flip flop/AND gate arrangement of Fig. 13.47a. If the T input is 1, the JK flip flop will toggle on the receipt of a clock pulse. If the T input is 0, the flip flop will not respond to a clock pulse. The carry output is 1 if T is 1 and Q is 1.

A synchronous up counter is constructed as in Fig. 13.47b, which is simply the circuit of Fig. 13.47a repeated. Note that the clock input is common to all stages, and the carry from one stage is the T input of the next.

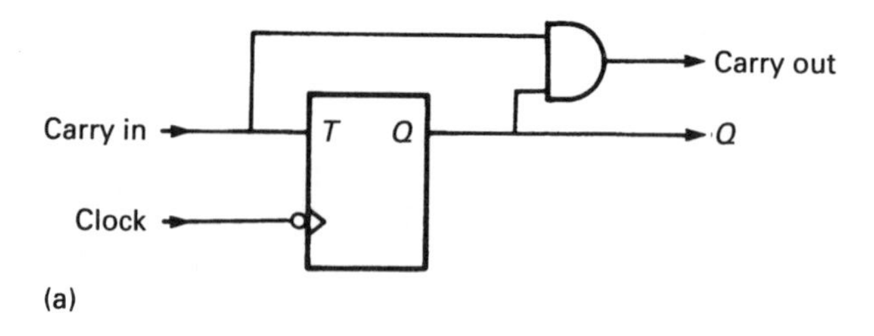

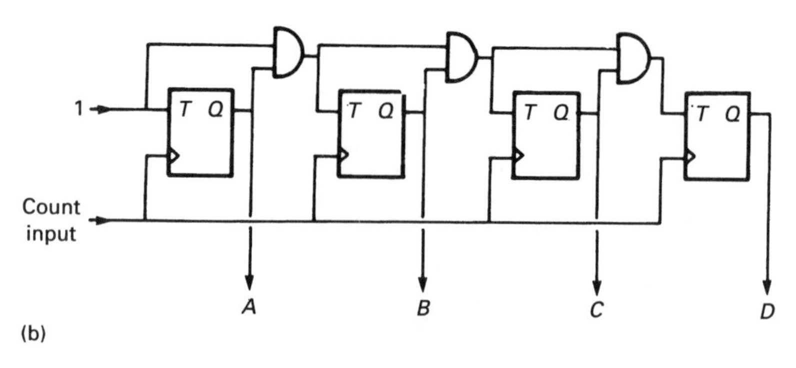

Fig. 13.47 *Synchronous counters. (a) Basic circuit for synchronous counter. (b) Series connected synchronous counter.*

It will be seen that the T inputs T_b, T_c, T_d will be 1 when all the preceding outputs are 1. T_c will be 1, for example, when A and B are both 1. This is the condition when a counter stage should toggle, taking $DCBA$ from, say, 0011 to 0100.

It is also possible to construct a synchronous down counter by counting the AND gate input of Fig. 13.47a to the $\bar{Q}$ output rather than the Q, and observing the counter state on the Q output (superficially similar to Fig. 13.46). A synchronous up/down counter with selectable direction can be constructed as in Fig. 13.48. If the direction line is a 1, gates 1, 2, 3 are enabled, the Q outputs pass to the next stage and the counter counts up. If the direction line is a 0, gates 4, 5, 6 are enabled, the $\bar{Q}$ outputs pass to the next stage and the counter counts down.

13.7.3. Non-binary counters

Counting to non-binary bases is often required; a BCD count is probably the most common requirement. When the required count is a subset of a straight binary count (as BCD is) the circuit of Fig. 13.49a can be used. The counter output is decoded by external logic. When the counter reaches the desired maximum count the decoder output forces the counter to its zero state (which is 0000 for a BCD counter, but need not be for other counters).

A single BCD stage constructed on these principles is shown in Fig. 13.49b. The circuit shown is a ripple counter, but could equally well be a synchronous counter. The gate detects a count of ten (binary 1010) and resets the counter to zero via direct reset inputs on the JK flip flops. Waveforms are shown in Fig. 13.49c.

Where a non-binary count is needed (e.g. a Gray code count), it is best to use synchronous counters and an arrangement similar to Fig. 13.50. This is drawn for D-type flip flops, but JK-based design is similar.

A combinational logic network looks at the counter outputs and sets the D inputs for the next state. If the counter, say, was required to step from 1101 to 0011, the combinational logic output to the D inputs would be 0011 for an input of 1101. Effectively there are four combination circuits in the network, one for each D input.

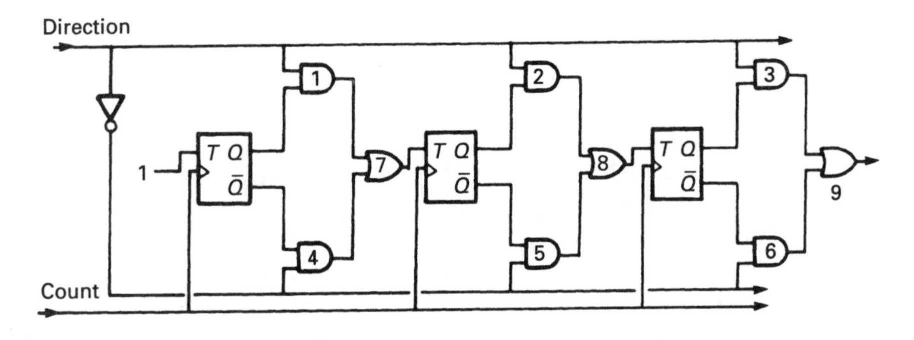

Fig. 13.48 *Up/down synchronous counter.*

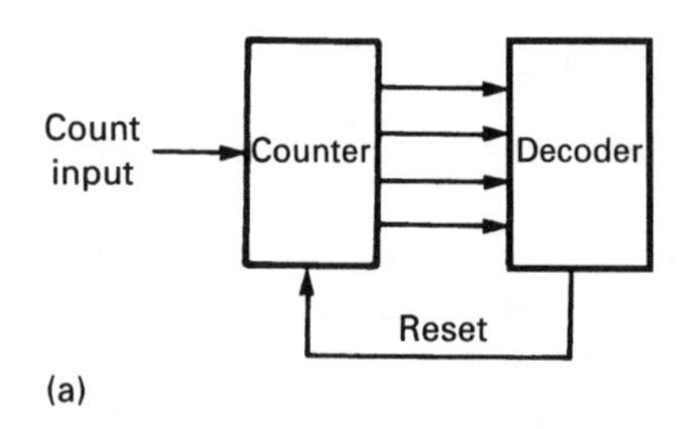

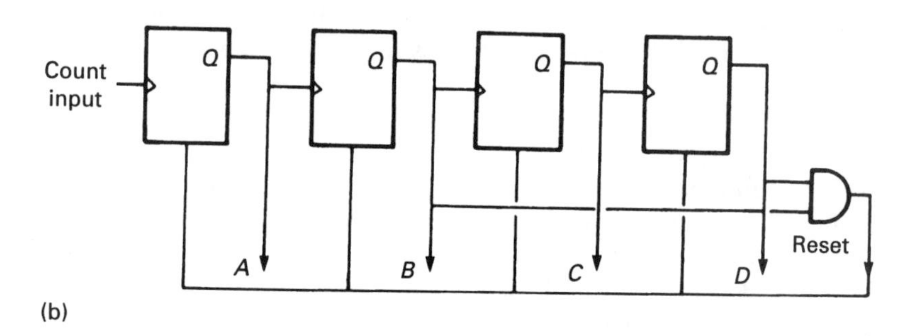

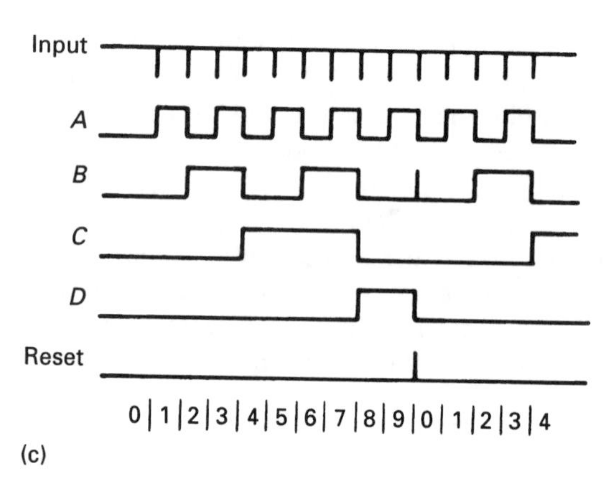

Fig. 13.49 *BCD counter. (a) Principle of operation. (b) Logic diagram. (c) Counter operation.*

13.7.4. Shift registers

A simple shift register is shown in Fig. 13.51a. Data applied to the serial input, S_{in}, will move one place to the right on each clock pulse as shown on the timing diagram of Fig. 13.51b.

Shift registers are used for parallel/serial and serial/parallel conversions. They are also the basis of multiplication and division circuits as a shift of one place towards the MSB is equivalent to a multiplication by 2, and one place towards the LSB an integer division of 2.

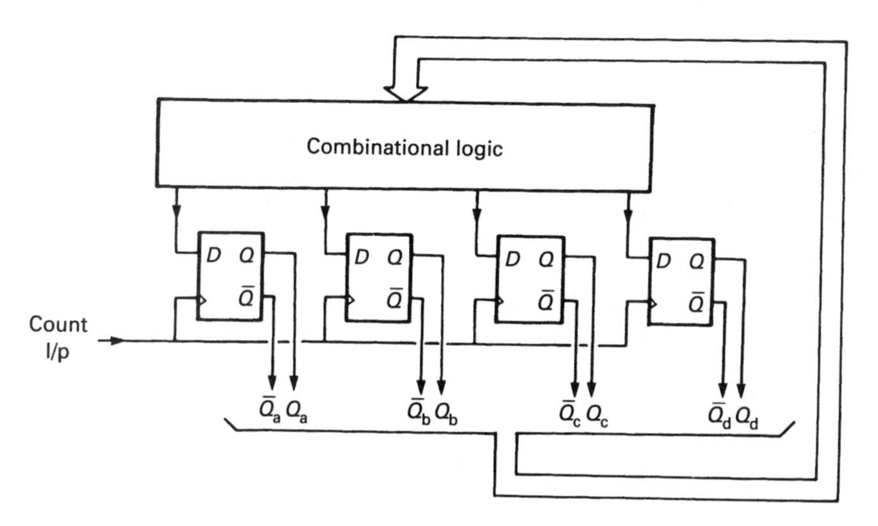

Fig. 13.50 *Generalised counter design with D-type flip flops.*

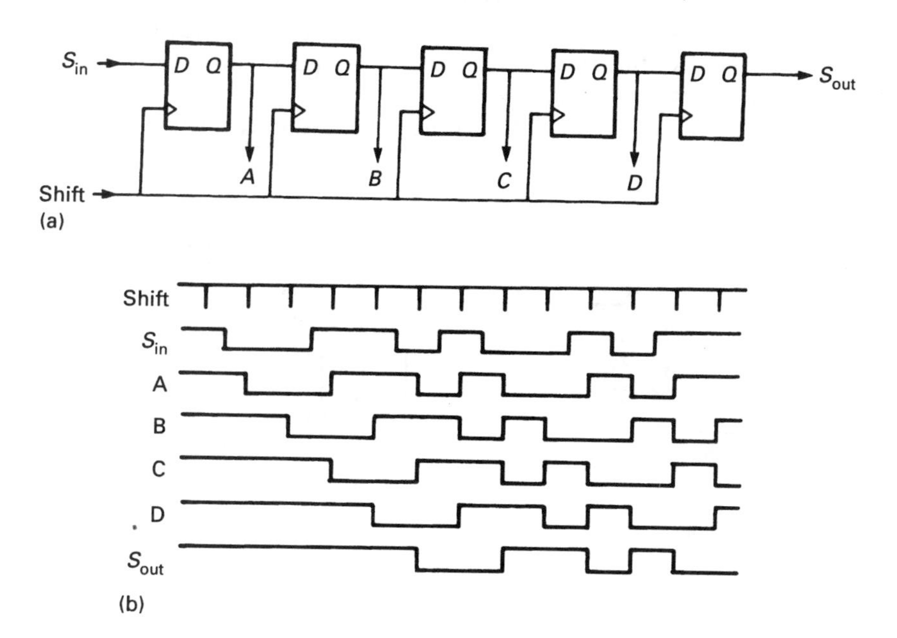

Fig. 13.51 *Simple shift register with D-type flip flops. (a) Circuit diagrams. (b) Waveforms.*

13.8. Sequencing and event driven logic

Many logic systems are driven by randomly occurring external events, and follow a sequence of operations. In such systems, the output states do not depend solely on the input states, but also on what the system was doing last. These types of systems are said to be sequencing or event driven logic; simple examples were the motor starter and the hydraulic ram of Fig. 13.2b and d.

Sequencing logic is designed using a state diagram. This shows the possible conditions the system can be in, the signals that are required to move from one state to the next and the outputs in each state. Figure 13.52a shows the state diagram for the motor starter and Fig. 13.52b the state diagram for the hydraulic ram.

A more complex example is shown in Fig. 13.53, which is a state diagram for a gas burner control. When the start PB is pressed, a 15 second air purge is given (set by timer 1). The pilot valve is opened, and the igniter started for 4 seconds (timer 2). If, at the end of this time, the flame detector shows the flame to be lit, the main gas valve is opened. At any time the stop button terminates the sequence. A non-valid signal from the flame detector (i.e. flame present in states 1 and 2 or no flame in state 4) puts the system to an alarm state, as does the incorrect signal from the air flow switch. Note that these are checked for being 'unfrigged' at the start of the sequence.

Event driven logic is built around flip flops, usually one for each state. The flip flop corresponding to state 4 is shown in Fig. 13.54a, and is set by the required conditions from state 3 and reset by the possible next states (1 and 5). Outputs are simply obtained by ORing the necessary states. The pilot output, shown in Fig. 13.54b, is simply state 3 OR state 4.

It is possible to minimise event driven circuits to use fewer flip flops, but such

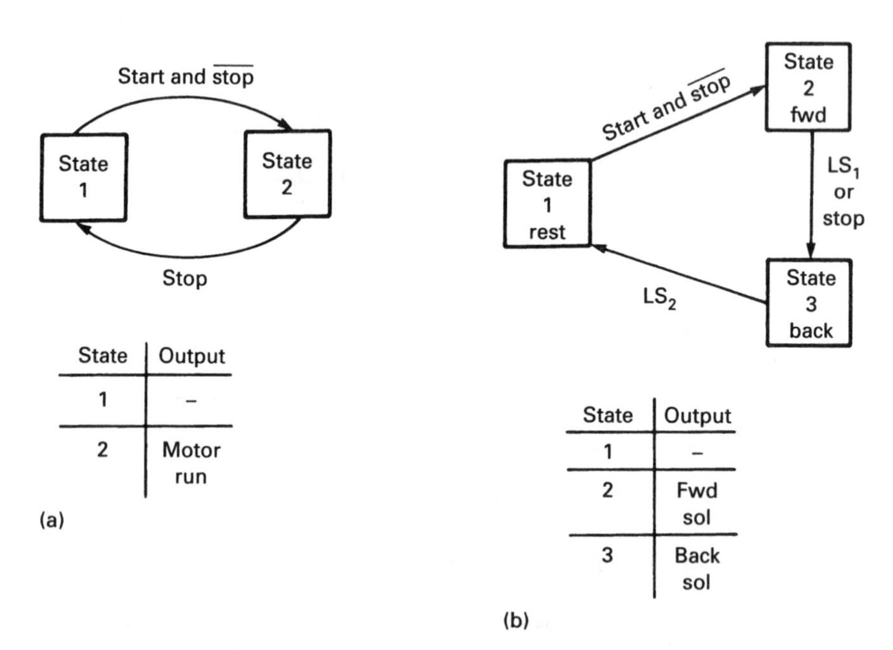

Fig. 13.52 *Simple state diagrams. (a) State diagram for motor starter. (b) State diagram for hydraulic ram of Fig. 13.2(d).*

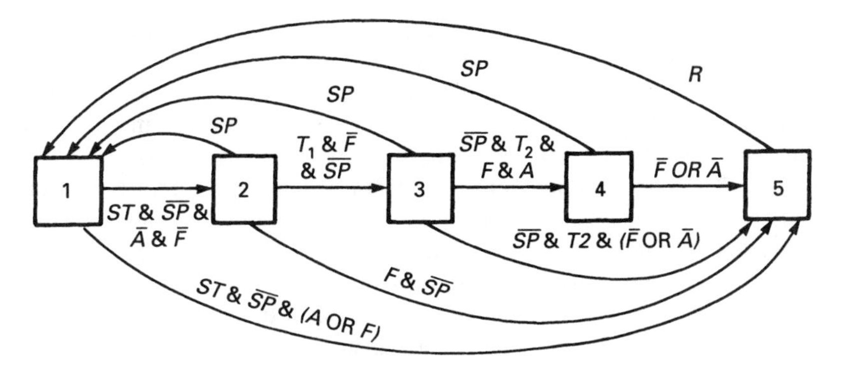

Inputs: Start PB *(ST)*, Stop PB *(SP)*, Flame present *(F)*, Reset PB *(R)*, Timer 1 complete (T_1), Timer 2 complete (T_2), Air flow SW *(A)*

Outputs:

State	Description	Air	Pilot valve	Ignition	Gas valve	Start timer 1	Start timer 2	Alarm bell
1	Off	0	0	0	0	0	0	0
2	Air purge	1	0	0	0	1	0	0
3	Ignition	1	1	1	0	0	1	0
4	On	1	1	0	1	0	0	0
5	Alarm	1	0	0	0	0	0	1

Fig. 13.53 *State diagram for gas burner.*

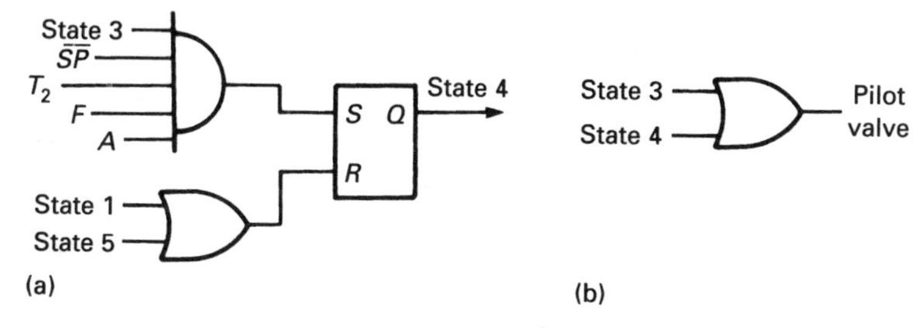

Fig. 13.54 *Circuit implementation of state diagram. (a) Circuit for state 4. (b) Circuit for pilot valve output.*

an approach is usually not required in industrial applications. A straightforward state diagram similar to Fig. 13.53 is easy to design, understand and modify, and simplifies fault finding for maintenance personnel. State diagrams can also be used to write programs for programmable controllers.

State diagrams are being formalised by the International Electrotechnical Commission (IEC) and the British Standards Institute (BSI), and already exist with the French Standard Grafset. These are basically identical to the approach outlined above, but introduce the idea of parallel routes which can be operated at the same time. Figure 13.55a is called a divergence; state 0 can lead to state 1 for condition 's' OR to state 2 for condition 't' with transitions 's' and 't' mutually exclusive. This is the form of the state diagrams described so far.

Figure 13.55b is a simultaneous divergence, where state 0 will lead to state 1 and state 2 simultaneously for transition 'u'. States 1 and 2 can now run further sequences in parallel.

Figure 13.55c again corresponds to the state diagrams described earlier, and is known as a convergence. The sequence can go from state 5 to state 7 if transition 'v' is true OR from state 6 to state 7 if transition 'w' is true.

Figure 13.55d is called a simultaneous convergence (note again the double horizontal line) state 7 will be entered if the left-hand branch is in state 5 AND the right-hand branch is in state 6 AND transition 'x' is true.

The state diagram is so powerful that most medium size PLCs include it in their programming language in one form or another. Telemecanique give it the name Grafcet (with a 'c'); others use the name Sequential Function Chart (SFC) (Allen Bradley) or Function Block (Siemens). The IEC have adopted state diagrams as one of their formalised methods of PLC programming in IEC 1131–3. We will return to this in Section 14.3.6.

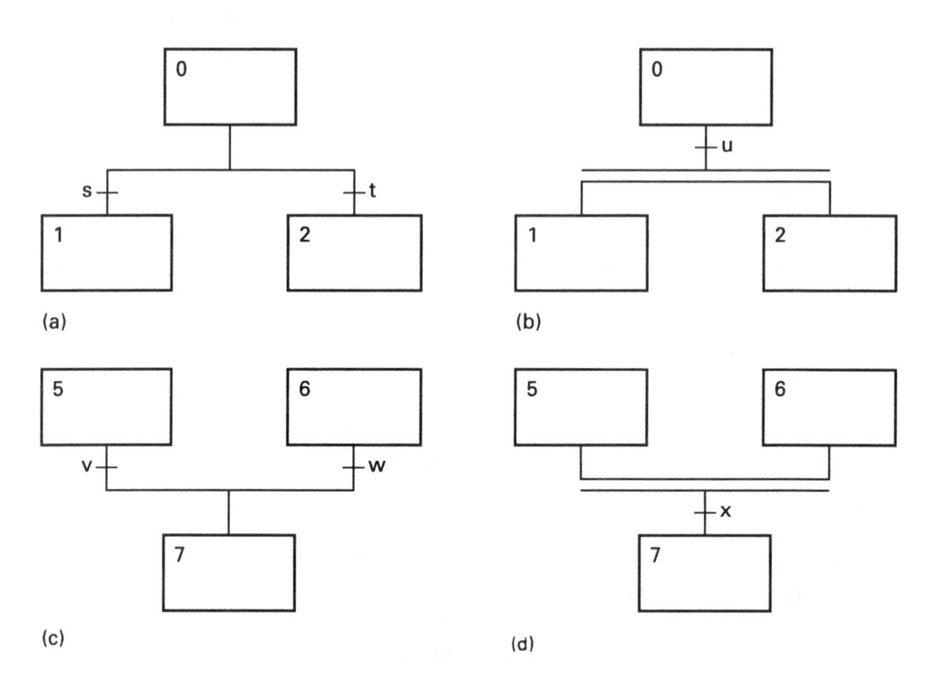

Fig. 13.55 *State transition diagram symbols. (a) Divergence. (b) Simultaneous divergence. (c) Convergence. (d) Simultaneous convergence.*

13.9. Analog interfacing

13.9.1. Digital-to-analog converters (DACs)

A binary number can represent an analog voltage. An 8-bit number, for example, represents a decimal number from 0 to 255 (or −128 to +127 if twos complement representation is used). An 8 bit number could therefore represent a voltage from 0 to 2.55 volts, say, with a resolution of 10 mV. A device which converts a digital number to an analog voltage is called a digital-to-analog converter, or DAC.

Common DAC circuits are shown in Fig. 13.56; in each case the output voltage is related to the binary pattern on the switches. In practice, FETs are used for the switches, and usually an IC DAC is used. The R–2R ladder circuit is particularly well suited to IC construction.

13.9.2. Analog-to-digital converters (ADCs)

There are several circuits which convert an analog voltage to its binary equivalent. The two commonest are the ramp ADC, shown in Fig. 13.57, and the

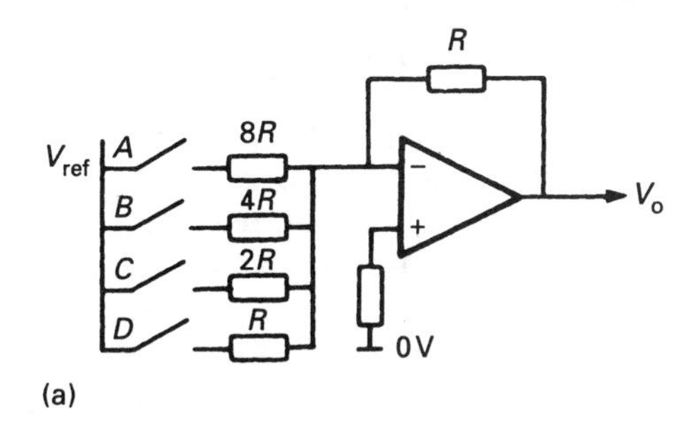

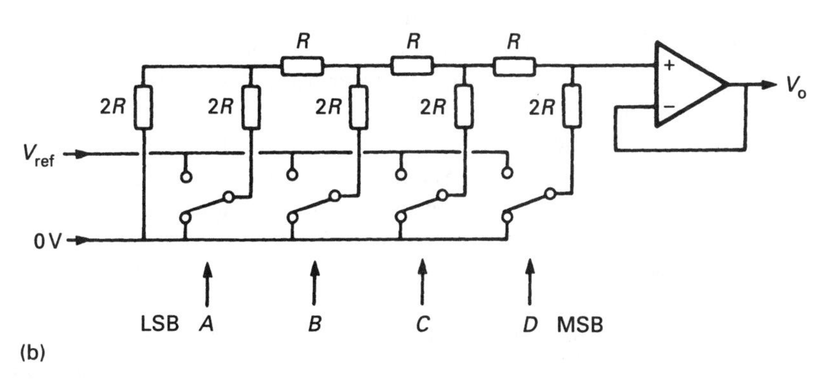

Fig. 13.56 *Digital to analog conversion (DAC). (a) Simple digital to analog converter. (b) R-2R ladder DAC.*

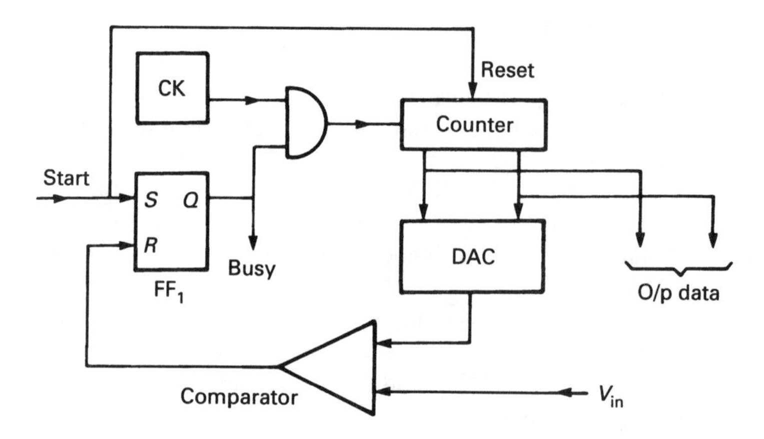

Fig. 13.57 *Block diagram of ramp ADC.*

successive approximation ADC of Fig. 13.58. Both of these compare the output voltage from a DAC with the input voltage.

The operation of the ramp ADC commences with a start command which sets FF_1 and resets the counter to zero. FF_1 gates pulses to a counter. The counter output is connected to a DAC whose output ramps up as the counter counts up. The DAC output is compared with the input voltage, and when the two are equal FF_1 is reset, blocking further pulses and indicating that the conversion is complete. The binary number in the counter now represents the input voltage. A variation of the ramp ADC, known as a tracking ADC, uses an up/down counter that follows the input voltage.

The ramp ADC is simple and cheap, but relatively slow (typical conversion time >1 ms). Where high speed or high accuracy is required, a successive approximation ADC is used. The circuit shown in Fig. 13.58 uses an ordered trial and error process.

The sequence starts with the register cleared. The MSB is set, and the comparator output examined. If the comparator shows the DAC output is less than, or equal to, V_{in}, the bit is left set. If the DAC output is greater than V_{in}, the bit is reset. Each bit is similarly tested, in order from MSB to LSB, causing the DAC output to home in on V_{in} as shown.

Successive approximation ADCs are fast (conversion times of a few μS) and accurate (0.01%). Unlike the ramp ADC, the conversion time is constant. They are, however, more complex and expensive than the simpler ramp ADC.

13.10. Practical details

Real-life digital systems have to connect to the outside world, and this can often bring problems when noise and effects such as contact bounce are encountered. Precautions also need to be taken against inadvertent introduction of high voltages into logic systems via inter-cable faults on the plant.

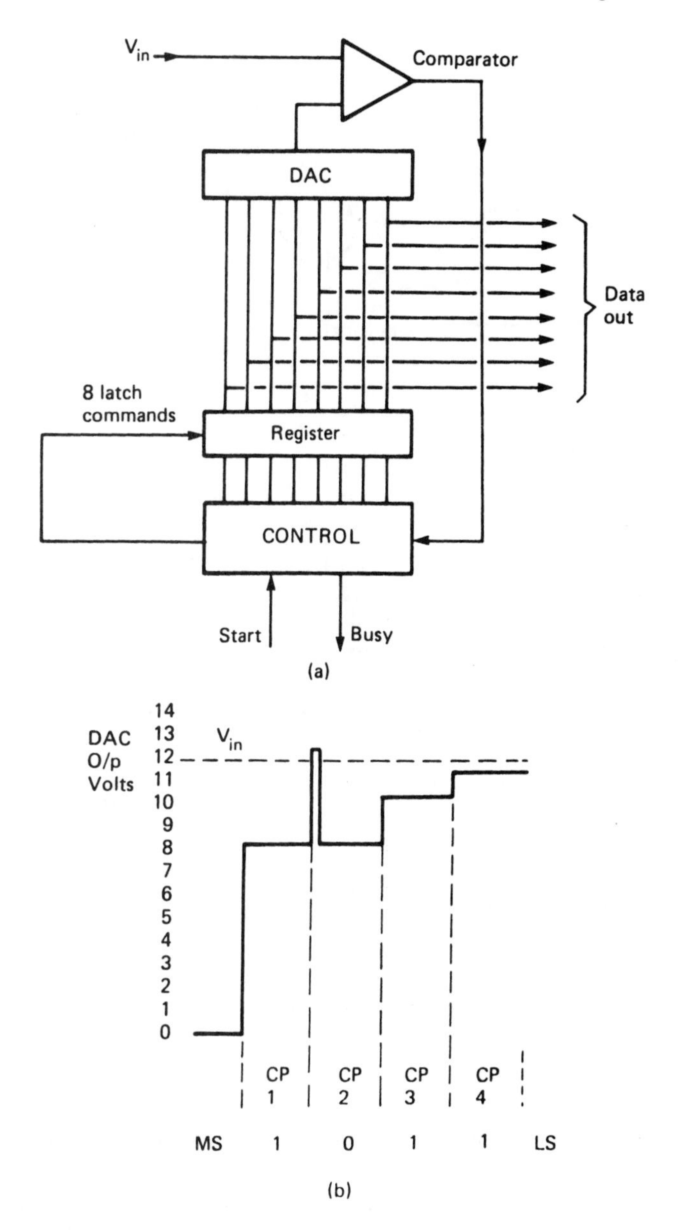

Fig. 13.58 *Successive approximation ADC. (a) Block diagram. (b) Operation.*

All signals between a logic system and the outside world should use a technique called opto isolation when cable lengths are longer than a few metres. Figure 13.59 shows typical input and output circuits. In both, the signal is electrically isolated by using a coupled LED and phototransistor. Because the plant side power supply and digital power supply are totally separate, the system will

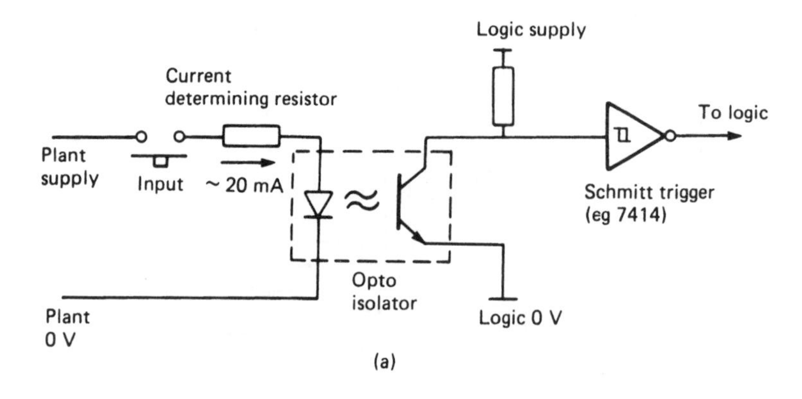

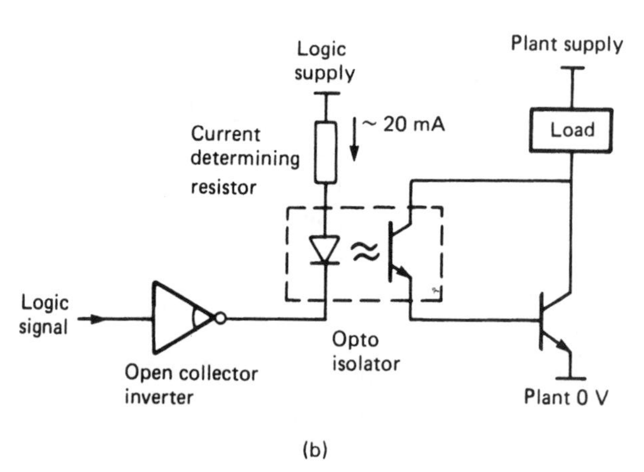

Fig. 13.59 *Optical isolation of DC inputs and outputs. (a) Input circuit. (b) Output circuit.*

withstand voltages of up to 1 kV without damage to the digital equipment (although such voltages would probably damage the plant side components, of course). The absence of ground loops and relatively high current levels (around 20 mA) also give excellent noise immunity. Isolated AC input and output circuits are shown on Fig. 13.60.

Opto isolators (such as the TIL 107) are usually constructed in a six pin IC, and characterised by a current transfer ratio. This is defined as the ratio between the phototransistor collector current to the LED current. A typical value is 0.3, so 20 mA input current will give 6 mA output current. If Darlington phototransistors are used, transfer ratios as high as 1 : 2 can be obtained.

Noise can also enter digital systems via the power supply rails so filtering is necessary, both on the DC side and (with LC filter) on the AC supply side. It is particularly important to adopt a sensible segregation of 0 V rails such that digital logic, relays/lamps and analog circuits have separate 0 V returns to some common earth point. Under no circumstances should high currents flow along logic 0 V lines, or the logic 0 V be taken outside its own cubicle.

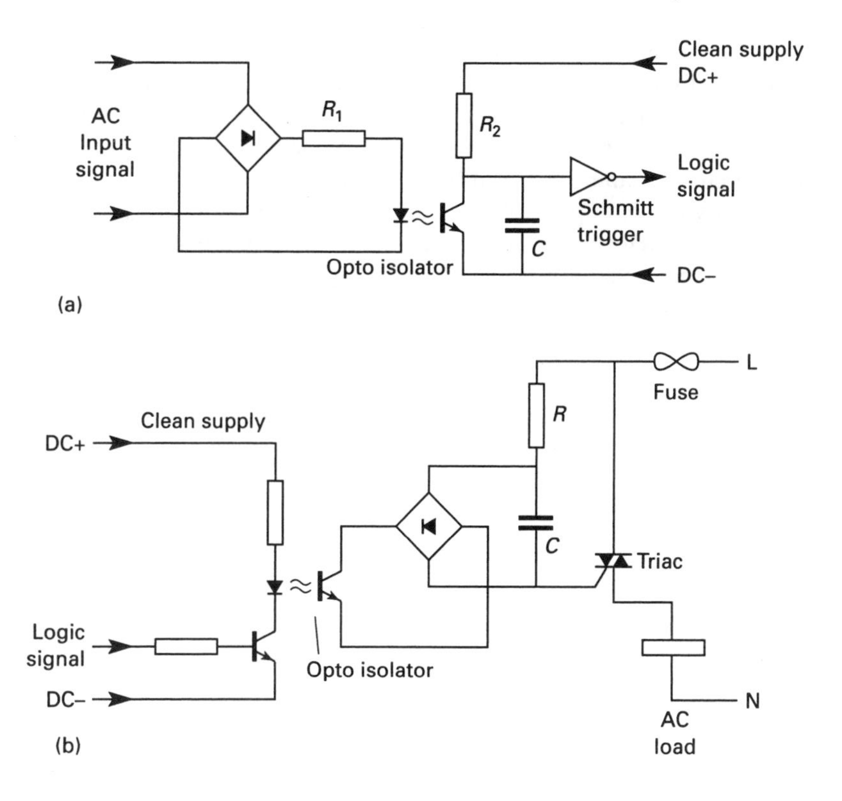

Fig. 13.60 *Isolated input and output circuits for AC signals. (a) Isolated input. (b) Isolated output.*

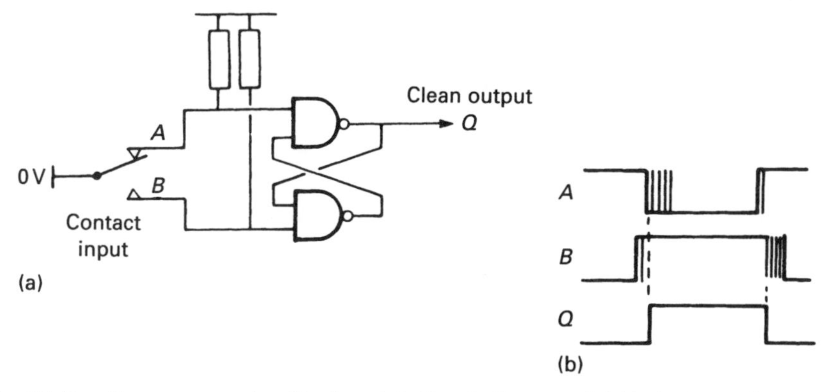

Fig. 13.61 *Bounce removing flip flop. (a) Circuit diagram. (b) Typical waveform.*

A digital IC can also generate its own noise on power supplies (TTL is particularly troublesome). It is therefore highly desirable to provide each IC with its own local 0.01 μF capacitor. A single large value electrolytic has no effect as the noise is caused by rapid di/dt and the PCB track inductance.

Mechanical contacts from switches, relays, etc., do not make instantly but 'bounce' rapidly for 1 to 4 ms due to dirt and the uneven contact surfaces. In many combinational logic systems this does not matter, but where counting, sequencing or arithmetic circuits are used, trouble can ensue.

Contact bounce can be removed by RC filters, but the best solution is to use a bounce removing flip flop, as in Fig. 13.61. Provided break before make contacts are used, the circuit gives totally bounce-free true and complement outputs. If the contacts are some distance from the digital system, opto isolation should, of course, be used before the flip flop.

Chapter 14

Computers and industrial control

14.1. Industrial computers

14.1.1. Introduction

A computer is a device that follows predetermined instructions to manipulate input data in order to produce new output data. A computer can therefore be represented by Fig. 14.1a. For a computer used for payroll calculations the input data would be employees' names, salary grades and hours worked. This data would be operated on by instructions written to include current tax and pension rules and the result would be output data in the form of wage slips (or in this day and age more likely direct transfer to bank accounts).

Early computer systems tended to be based on commercial functions: payroll, accountancy, banking and similar activities. The operations tended to be batch processes: a daily update of stores stock, for example.

The computer from Fig. 14.1a can be used as part of a control system as shown in Fig. 14.1b. Note that the operator's actions (e.g. start Widget process) are not instructions, they are part of the input data. The instructions will define what action is to be taken as the input data (from both the plant and the operator) changes. The output data are control actions to the plant and status displays to the operator.

Early computers were large, expensive and slow. Speed is not that important for batch-based commercial data processing, but is of the highest priority in

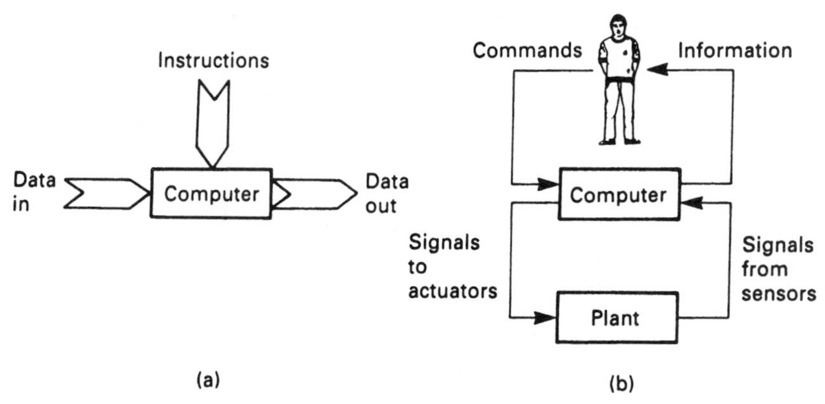

Fig. 14.1 *The computer in industrial control. (a) A simple overview of a computer. (b) The computer as part of a control system.*

industrial control which has to be performed in 'real time'. Many emergency and alarm conditions require action to be taken in fractions of a second.

Commercial (with the word 'commercial' used to mean 'designed for use in commerce') computers were also biased towards receiving data from punched cards and keyboards and sending output data to printers. An industrial process requires possibly hundreds of devices to be read in real time and signals sent to devices such as valves, motors, meters and so on.

There was also an environmental problem. Commercial computers were designed to exist in an almost surgical atmosphere: dust free and an ambient temperature that can only be allowed to vary by a few degrees. Such conditions can be almost impossible to achieve close to a manufacturing process.

The first industrial computer application was probably a system installed in an oil refinery in Port Arthur, USA in 1959. The reliability and mean time between failure of computers at this time meant that little actual control was performed by the computer, and its role approximated to the monitoring subsystem description in Section 14.2.2.

14.1.2. Computer architectures

It is not essential to have intimate knowledge of how a computer works before it can be used effectively, but an appreciation of the parts of a computer is useful for appreciating how a computer can be used for industrial control.

Figure 14.1a can be expanded to give the more detailed layout of Fig. 14.2. This block diagram (which represents the whole computing range from the smallest home computer to the largest commercial mainframe) has six portions.

(1) An input unit where data from the outside world is brought into the computer for processing.

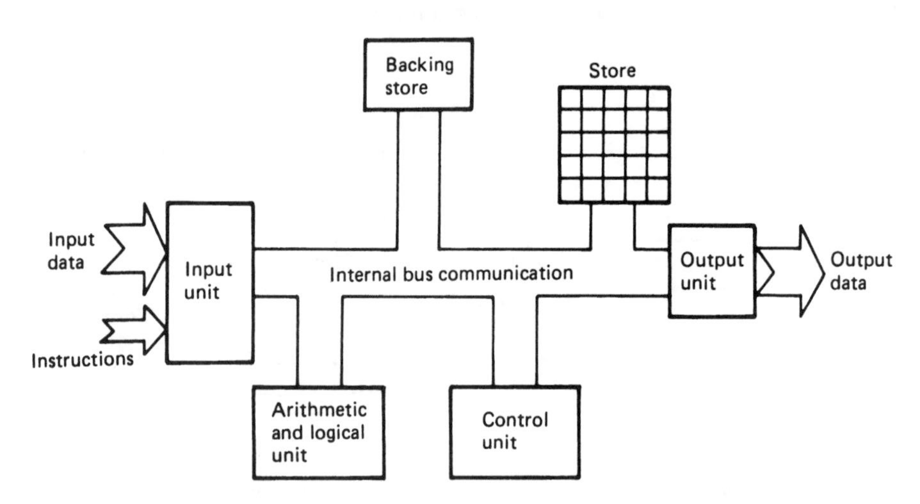

Fig. 14.2 *Computer architecture.*

(2) A store, or memory which will be used to store the instructions the computer will follow and data for the computer to operate on. This data can be information input from outside or intermediate results calculated by the machine itself. The store is organised into a number of boxes, each of which can hold one number and is identified by an address as shown in Fig. 14.3. Computers work internally in binary (see Section 13.6.1 for a description of binary, hexadecimal (hex) and other number systems) and the store does not distinguish between the meanings that could be attached to the data stored in it. For example, in an 8-bit computer (which works with numbers 8 bits long in its store) the number 01100001 can be interpreted as:

(a) The decimal number 97.
(b) The hex number 61 (see Section 13.6.1).
(c) The letter 'a' (see Table 19.1).
(d) The state of eight digital signals such as limit switch states.
(e) An instruction to the computer. If the machine was the old Z80 microprocessor, hex 61 moves a number between two internal stores.

A typical desk-top computer will use 16-bit numbers (called a 16-bit word) and have over a million store locations. A typical 'Windows'-based PC requires 4 Mbyte words of memory. Some industrial computers called programmable controllers (or PLCs, described below in Section 14.4) are far more frugal in their memory usage, about 16 000 (16K) locations, but even smaller machines with just 1000 store locations are common.

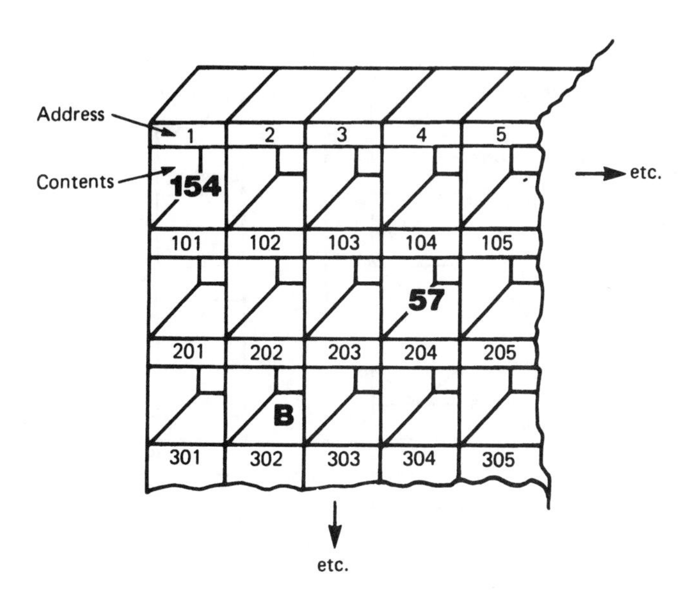

Fig. 14.3 *The store represented as pigeon holes. Location 104 contains 57 and location 202 contains 'B'.*

(3) Data from the store can be accessed very quickly, but commercial computers often need vast amounts of storage to hold details such as bank accounts or names and addresses. This type of data is not required particularly quickly and is held in external storage. This external storage is usually magnetic discs or tapes and is called secondary or backing storage. Modern PCs tend to use hard discs with around 400 Mbyte storage capacity or floppy discs with 1.44 Mbyte capacity.
(4) An output unit where data from the computer is sent to the outside world.
(5) An arithmetic and logical unit (called an ALU) which performs operations on the data held in the store according to the instructions the machine is following.
(6) A control unit which links together the operations of the other five units. Often the arithmetic and logical unit and the control unit are known together as the Central Processor Unit or CPU. A microprocessor is a CPU in a single integrated circuit.

The instructions the computer follows are held in the store and, with a few exceptions we will consider shortly, are simply followed in sequential order as Fig. 14.4a.

The control unit contains a counter called an Instruction Register (or IR) which says at which address in the store the next instruction is to be found. Sometimes the name Program Counter (and the abbreviations PC) is used.

When each instruction is obeyed, the control unit reads the store location whose address is held in the IR. The number held in this store location tells the control unit what instruction is to be performed.

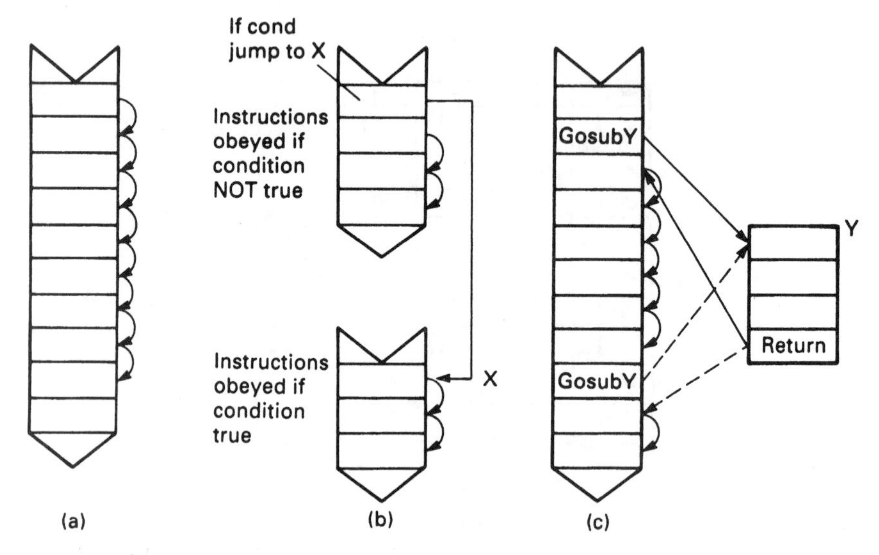

Fig. 14.4 *Program flow in a computer. (a) Simple sequential flow. (b) Conditional jump. (c) Subroutine call.*

Instructions nearly always require operations to be performed on data in the store (e.g. add two numbers) so the control unit will bring data from the store to the ALU and perform the required function.

When the instruction has been executed, the control unit will increment the IR so it holds the address of the next instruction.

There are surprisingly few types of instruction, the ones available on most microprocessors are variations on:

(a) MOVE data from one place to another (e.g. input data to a store location, or move data from a store location to the ALU).
(b) ALU operations on two data items, one in the ALU and one in a specified store location. Operations available are usually add, subtract, and logical operations such as AND, OR.
(c) Jumps. In Fig. 14.4a we implied that the computer followed a simple sequential list of instructions. This is usually true, but there are occasions where simple tests are needed. These usually have the form

```
IF   (some condition) THEN
           Perform some instructions
 ELSE
           Perform some other instructions
```

To test a temperature, for example, we could write:

```
IF   Temperature is less than 75 °C THEN
           Turn healthy light on
           Turn fault light off
ELSE
           Turn healthy light off
           Turn fault light on
```

Such operations use conditional jumps. These place a new address into the IR dependent on the last result in the accumulator. Conditional jumps can be specified to occur for outcomes such as result positive, result negative or result zero, and allow a program to follow two alternative routes as shown on Fig. 14.4b.
(d) Subroutines. Many operations are required time and time again within the same program. In an industrial control system using flows measured by orifice plates, a square root function will be required many times (flow is proportional to the square root of the pressure drop across the orifice plate, see Section 5.2). Rather than write the same instructions several times (which is wasteful of effort and storage space) a subroutine instruction allows different parts of the program to temporarily transfer operations to a specified subroutine, returning to the instruction after the subroutine call as shown on Fig. 14.4c.

14.1.3. Machine code and assembly language programming

The series of instructions that we need (called a 'program') has to be written and loaded into the computer. At the most basic level, called machine code programming, the instructions are written in the raw numerical form used by the machine. This is difficult to do, prone to error, and almost impossible to modify afterwards.

The sequence of numbers

```
16 00 58 21 00 00 06 08 29 17 D2 0E 40 19 05 C2 08 40 C9
```

genuinely are the instructions for a multiplication subroutine starting at address 4000 for a Z80 microprocessor, but even an experienced Z80 programmer would need reference books (and a fair amount of time) to work out what is going on with just these nineteen numbers.

Assembly language programming uses mnemonics instead of the raw code, allowing the programmer to write instructions that can be relatively easily followed. For example, with:

```
LOAD         Temperature
SUB          75
JUMP POSITIVE to Fault_Handler
```

It is fairly easy to work out what is happening.

A (separate) computer program called an *assembler* converts the programmer's mnemonic based program (called the source) into an equivalent machine code program (called the object) which can then be run.

Writing programs in assembly language is still labour intensive, however, as there is one assembly language instruction for each machine code instruction.

14.1.4. High level languages

Assembly language programming is still relatively difficult to write, so ways of writing computer programs in a style more akin to English were developed. This is achieved with so called 'High level languages' of which the best known are probably Pascal, C, Fortran and the ubiquitous BASIC. (There are many, many languages: RPG, FORTH, LISP and CORAL to name but a few, each with its own attractions).

In a high level language, the programmer writes instructions in something near to English. The Pascal program below, for example, gives a printout of a requested multiplication table:

```
program multtable (input,output);
var number, count : integer;
begin
     readln ('Which table do you want?',number);
     for count=1 to 10 do
     writeln (count, 'times', number, 'is', count*number);
end. {of program}
```

Even though the reader may not know Pascal, the operation of the program is clear; if asked to change the table from a ten times table to a twenty times table, for example, it is obvious which line would need to be changed.

A high level language source program can be made to run in two distinct ways. A compiler is a program which converts the entire high level source program to a machine code object program offline. The resultant object program can then be run independent of the source program or the compiler.

With an interpreter, the source program and the interpreter both exist in the machine when the program is being run. The interpreter scans each line of source code converting them to equivalent machine code instructions as they are obeyed. There is no object program with an interpreter.

A compiled program runs much faster than an interpreted program (typically five to ten times as fast because of the extra work that the interpreter has to do) and the compiled object program will be much smaller than the equivalent source code program for an interpreter. Compilers are, however, much less easy to use, a typical sequence being:

(a) A Text Editor is loaded into the computer.
(b) The source program is typed in or loaded from disc (for modification).
(c) The resultant source file is saved to disc.
(d) The compiler is loaded from disc and run.
(e) The source file is loaded from disc.
(f) Compilation starts (this can take several minutes). If any errors found go back to step (a).
(g) Object program produced which can be saved to disc and/or run. If any runtime errors found, go back to step (a).

These steps are summarised in Fig. 14.5.

An interpreted language is much easier to use, and for many applications the loss of speed is not significant. BASIC is usually an interpreted language; Pascal, C, and Fortran are usually compiled.

14.2. Types of control strategies

14.2.1. Introduction

The term SCADA is often applied to applications of computers in process control. The letters stand for supervisory control and data acquisition which rather neatly describes the functions that a computer can perform. Another common term is CIM for computer integrated manufacturing.

If a computer is to be used for industrial control, the areas it can be applied to must be identified. It is very easy to be confused and overwhelmed by the size and complexity of large industrial processes. Most if not all, can be simplified by considering them to be composed of many small sub-processes which fall into three distinct areas.

14.2.2. Monitoring subsystems

These display the process state to the operator and draw attention to abnormal or fault conditions which need attention. The plant condition is measured by suitable sensors which are connected to input cards.

Digital sensors measure conditions with distinct states. Typical examples are Running/Stopped, Forward/Off/Reverse, Fault/Healthy, Idle/Low/Medium/High,

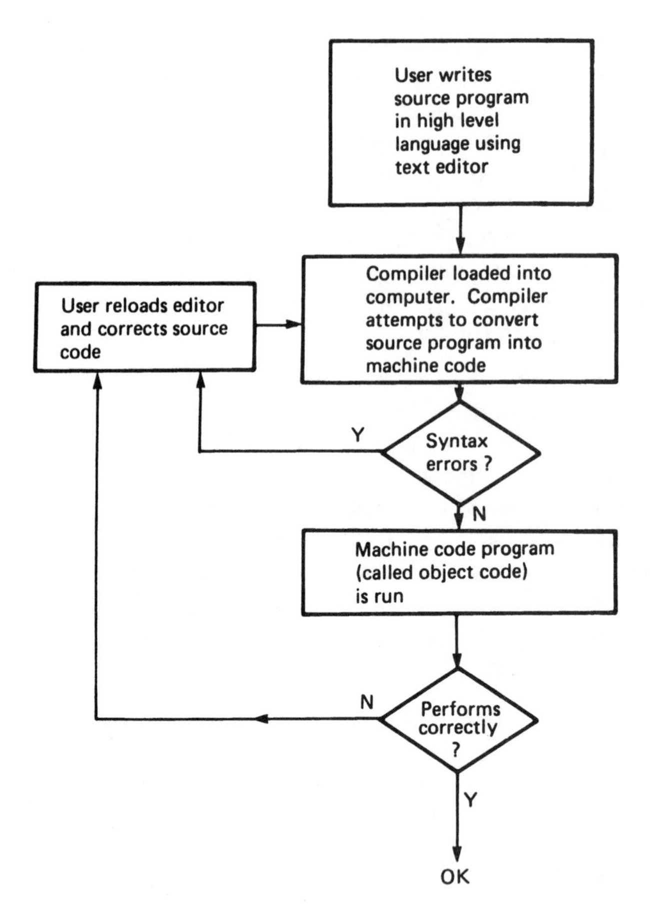

Fig. 14.5 *Using a compiler.*

High Level/Normal/Low Level.

Analog sensors measure conditions which have a continuous range such as temperature, pressure, flow or liquid level.

The results of these measurements are displayed to the operator via indicators (for digital signals) or by meters and bargraphs for analog signals.

The signals can also be checked for alarm conditions. An overtravel limit switch or an automatic trip of an overloaded motor are typical digital alarm conditions. A high temperature or a low liquid level could be typical analog alarm conditions. The operator could be informed of these via warning lamps and an audible alarm.

A monitoring system often keeps records of the consumption of energy and materials for accountancy purposes, and produces an event/alarm log for historical maintenance analysis. A pump, for example, may require maintenance after 5000 hours of operation.

Because the plant can still operate without the computer (albeit with reduced operator information and poor performance) early industrial computer systems tended to be of this type.

14.2.3. Sequencing subsystems

Many processes follow a pre-defined sequence. To start a gas burner, for example, the sequence could be:

(1) Start button pressed, if sensors are showing sensible states (no air flow and no flame) then sequence starts.
(2) Energise air fan starter. If starter operates (checked by contact on starter) and air flow established (checked by flow switch) then
(3) Wait two minutes (for air to clear out any unburnt gas) and then
(4) Open gas pilot valve and operate igniter. Wait 2 seconds and then
(5) If flame present (checked by flame failure sensor) open main gas valve.
(6) Sequence complete. Burner running. Stays on until stop button pressed OR air flow stops OR flame failure.

The above sequence works solely on digital signals, but sequences can also use analog signals; (e.g. Open valve V1 until 250 kg of product A have been added.)

14.2.4. Closed loop control subsystems

In many analog systems, a variable such as temperature, flow or pressure is required to be kept automatically at some preset value or made to follow some other signal. In a batch sequence, for example, the temperature of the reagents could be required to follow some heating/cooling profile during the process.

Such systems can be represented by the block diagram of Fig. 14.6. Here a particular characteristic of the plant (e.g. temperature) denoted by P_V (for process variable) is required to be kept at a preset value S_P (for set point). P_V is measured by a suitable sensor and compared with the S_P to give an error signal. If, for example, we are dealing with a temperature controller with a set point of 80 °C and an actual temperature of 78 °C the error is 2 °C.

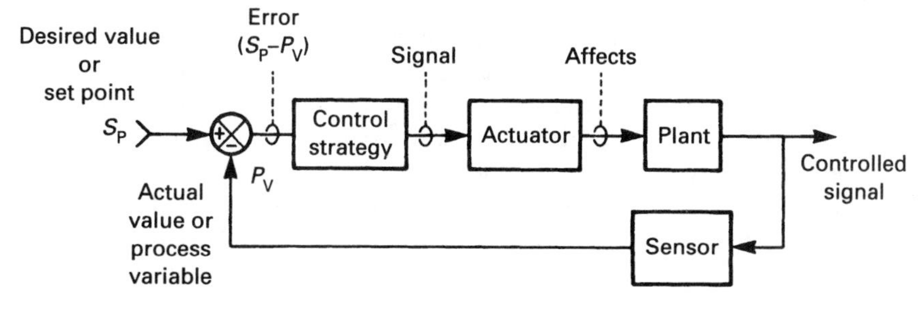

Fig. 14.6 *A closed loop control system.*

This error signal is applied to a control algorithm. There are many possible control algorithms, and this topic is discussed in detail in chapter 18, but a simple example for a heating control could be 'If the error is negative turn the heat off, if the error is positive turn the heat on'.

The output from the control algorithm is passed to an actuator which affects the plant. For a temperature control, the actuator could be a heater, for a flow control the actuator could be a flow control valve. The control algorithm will adjust the actuator until there is zero error, i.e. the process variable and the setpoint have the same value.

In figure 14.6, the value of P_V is fed back to be compared with the setpoint, leading to the term 'feedback control'. It will also be noticed that the block diagram forms a loop, so the term 'closed loop control' is also used.

If a computer is to be used for closed loop control, there are two possible approaches. The computer itself can perform the control algorithm. This is known as direct digital control (or DDC). An alternative method uses closed loop controllers (described in Chapter 18) with the computer providing the setpoints. This is known as supervisory control. Although more expensive supervisory control does remove the possibility of a total plant shutdown in the event of a computer failure.

14.2.5. Requirements for industrial control

Industrial control has rather different requirements than other applications. It is worth examining these in some detail.

A conventional computer takes data, usually from a keyboard, and outputs data to a VDU screen or printer. The data being manipulated will generally be characters or numbers (e.g. item names and quantities held in a stores stock list).

An industrial control computer is very different. Its inputs come from a vast number of devices. Although some of these are numeric (flows, temperature, pressures and similar analog signals) most will be single bit, On/Off, digital signals.

There will also be a similar large amount of digital and analog output signals. A very small control system may have connections to about twenty input and output signals; figures of over two hundred connections are quite common on medium sized systems. The keyboard, VDU and printer may exist, but they are not necessary and the function will probably be different to those on a normal desktop or mainframe computer.

Although it is possible to connect this quantity of signals into a conventional machine, it requires non-standard connections and external boxes. Similarly, although programming for a large amount of input and output signals can be done in Pascal, BASIC or C, the languages are being used for a purpose for which they were not really designed, and the result can be very ungainly.

In Figure 14.7a, for example, we have a simple motor starter. This could be connected as a computer-driven circuit as Fig. 14.7b. The two inputs are identified by addresses 1 and 2, with the output (the relay starter) being given the address 10.

If we assume a program function bitread(N) exists which gives the state (on/off)

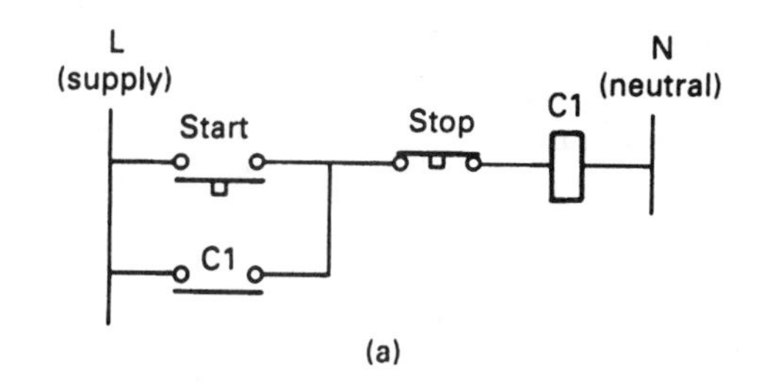

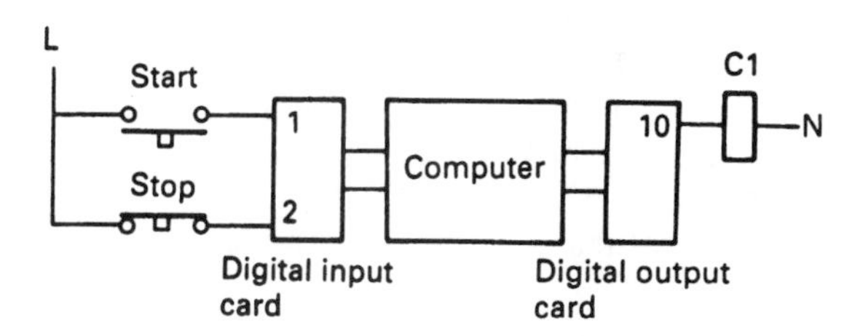

Fig. 14.7 *Comparison of hardwire and computer-based schemes. (a) Hardwire motor starter circuit. (b) Computer-based motor starter.*

of address N, and a function bitwrite(M,var) which sends the state of program variable var to address M, we could give the actions of Fig. 14.7 by

```
repeat
      start: =bitread(1);
      stop: =bitread(2);
      run: = ((start) or (run)) & stop;
      bitwrite (10,run);
until hellfreezesover
```

where start, stop and run are one-bit variables. The program is not very clear, however, and we have just three connections.

An industrial control program rarely stays the same for the whole of its life. There are always modifications to cover changes in the operations of the plant. These changes will be made by plant maintenance staff, and must be made with minimal (preferably none) interruptions to the plant production. Adding a second stop button and a second start button into Fig. 14.7 would not be a simple task.

In general, computer control is done in real time, i.e. the computer has to respond to random events as they occur. An operator expects a motor to start (and more important to stop!) within a fraction of a second of the button being pressed. Although commercial computing needs fast computers, it is unlikely that the difference between a one second and two second computation time for a spreadsheet would be noticed by the user. Such a difference would be unacceptable for industrial control.

Time itself is often part of the control strategy (e.g. start air fan, wait 10 seconds for air purge, open pilot gas valve, wait 0.5 second, start ignition spark, wait 2.5

seconds, if flame present open main gas valve). Such sequences are difficult to write with conventional languages.

Most control faults are caused by external items (limit switches, solenoids and similar devices) and not by failures within the central control itself. The permission to start a plant, for example, could rely on signals involving cooling water flows, lubrication pressure and temperatures within allowable ranges. For quick fault finding the maintenance staff must be able to monitor the action of the computer program whilst it is running. If, as is quite common, there are ten interlock signals which allow a motor to start, the maintenance staff will need to be able to check these quickly in the event of a fault. With a conventional computer, this could only be achieved with yet more complex programming.

The power supply in an industrial site is shared with many antisocial loads: large motors stopping and starting, thyristor drives which put spikes and harmonic frequencies onto the mains supply. To a human these are perceived as light flicker; to a computer they can result in storage corruption or even machine failure.

An industrial computer must therefore be able to live with a 'dirty' mains supply, and should also be capable of responding sensibly following a total supply interruption. Some outputs must go back to the state they were in before the loss of supply, others will need to turn off or on until an operator takes corrective action. The designer must have the facility to define what happens when the system powers up from cold.

The final considerations are environmental. A large mainframe computer generally sits in an air-conditioned room at a steady 20 °C with carefully controlled humidity. A desk-top PC will normally live in a fairly constant environment because human beings do not work well at extremes. An industrial computer, however, will probably have to operate away from people in a normal electrical substation with temperatures as low as −10 °C after a winter shutdown, and possibly over 40 °C in the height of summer in a substation at the steel plant where the author works. Even worse, these temperature variations lead to a constant expansion and contraction of components which can lead to early failure if the design has not taken this factor into account.

To these temperature changes must be added dust and dirt. Very few industrial processes are clean, and the dust gets everywhere (even with IP55 cubicles, because an IP55 cubicle is only IP55 when the doors are shut and locked; maintenance staff note well! IP ratings are discussed in Section 21.7.) The dust will work itself into connectors, and if these are not of a highest quality, intermittent faults will occur which can be very difficult to find.

In most computer applications, a programming error or a machine fault can often be humorous (bills and reminders for 0 p) or at worse expensive and embarrassing. When a computer controlling a plant fails, or a programmer misunderstands the plant's operation, the result could be injuries or fatalities. Under the all powerful UK Health and Safety at Work Act, prosecution of the design engineers could result. It behoves everyone to take extreme care with the design.

Our requirements for industrial control computers are very demanding, and it is worth summarising them:

(a) They should be designed to survive in an industrial environment with all that this implies for temperature, dirt and poor quality mains supply.
(b) They should be capable of dealing with bit form digital input/output signals at the usual voltages encountered in industry (24 V DC to 240 V AC) plus analog input/output signals. The expansion of the I/O should be simple and straightforward.
(c) The programming language should be understandable by maintenance staff (such as electricians) who have no computer training. Programming changes should be easy to perform in a constantly changing plant.
(d) It must be possible to monitor the plant operation whilst it is running to assist fault finding. It should be appreciated that most faults will be in external equipment such as plant-mounted limit switches, actuators and sensors, and it should be possible to observe the action of these from the control computer.
(e) The system should operate sufficiently fast for realtime control. In practice, 'sufficiently fast' means a response time of around 0.1 second, but this can vary dependent on the application and the controller used.
(f) The user should be protected from computer jargon.
(g) Safety must be a prime consideration.

14.3. The programmable controller

14.3.1. Introduction

In the late 1960s the American motor car manufacturer General Motors was interested in the application of computers to replace the relay sequencing used in the control of its automated car plants. In 1969 it produced a specification for an industrial computer similar to that outlined at the end of the previous section.

Two independent companies, Bedford Associates (later called Modicon) and Allen Bradley responded to General Motors specifications. Each produced a computer system similar to Fig. 14.8 which bore little resemblance to the commercial minicomputers of the day.

The computer itself, called the central processor, was designed to live in an industrial environment, and was connected to the outside world via racks into which input, or output cards could be plugged. In these early machines there were essentially four different types of cards:

(1) DC digital input card
(2) DC digital output card
(3) AC digital input card
(4) AC digital output card

Each card would accept 16 inputs or drive 16 outputs. A rack of 8 cards could thus be connected to 128 devices. It is very important to appreciate that the card allocations were the user's choice, allowing great flexibility. In Fig. 14.8b the user has installed one DC input and one DC output card, three AC input cards, and

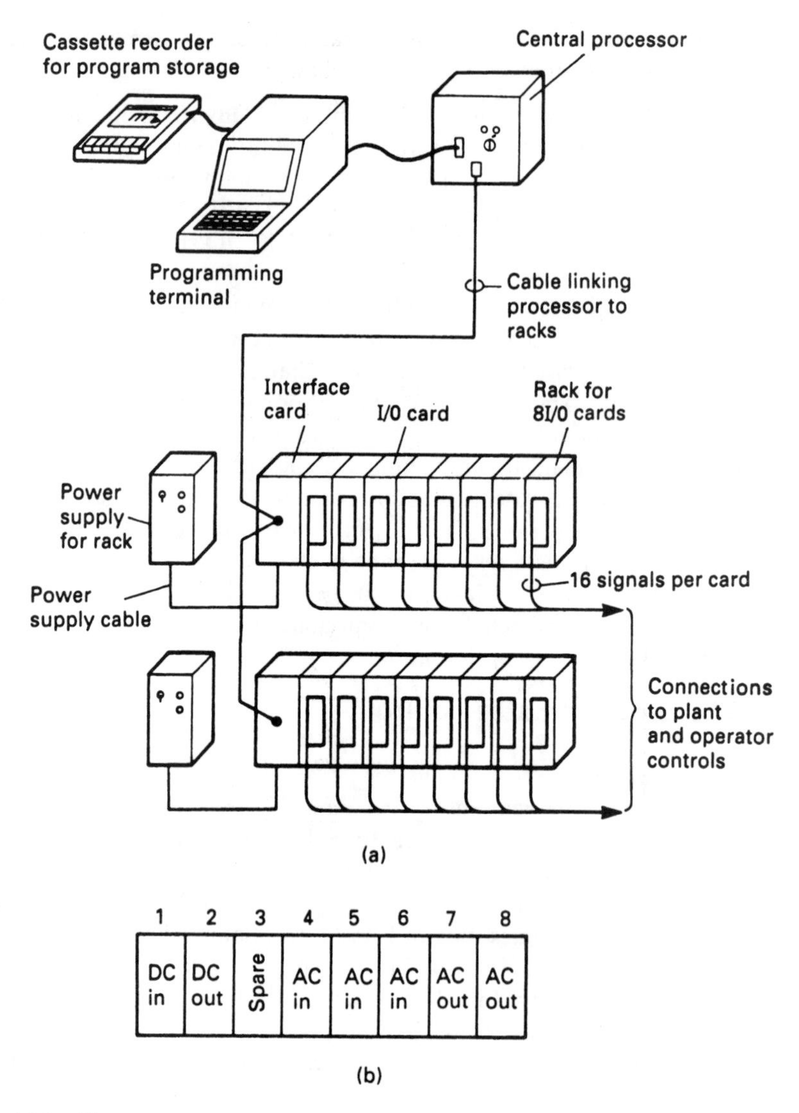

Fig. 14.8 *The component parts of a PLC system. (a) An early PLC system. (b) A typical rack of cards.*

two AC output cards leaving one spare position for future expansion. This rack can thus be connected to:

16 DC input signals
16 DC output signals
48 AC input signals
32 AC output signals.

Not all of these, of course, need to be used.

The most radical idea, however, was a programming language based on a relay schematic diagram, with inputs (from limit switches, pushbuttons, etc) represented by relay contacts, and outputs (to solenoids, motor starters, lamps, etc) represented by relay coils. Figure 14.9 shows a simple hydraulic cylinder which can be extended or retracted by pushbuttons. Its stroke is set by limit switches which open at the end of travel, and the solenoids can only be operated if the hydraulic pump is running. This could be controlled by the computer program of Fig. 14.9b which is identical to the relay circuit needed to control the cylinder. These programs look like the rungs on a ladder, and were consequently called 'ladder diagrams'.

The program was entered via a programming terminal with keys showing relay symbols (normally open/normally closed contacts, coils, timers, counters, parallel branches, etc), with which a maintenance electrician would be familiar. Figure 14.10 shows the programmer's keyboard for an early PLC. The meaning

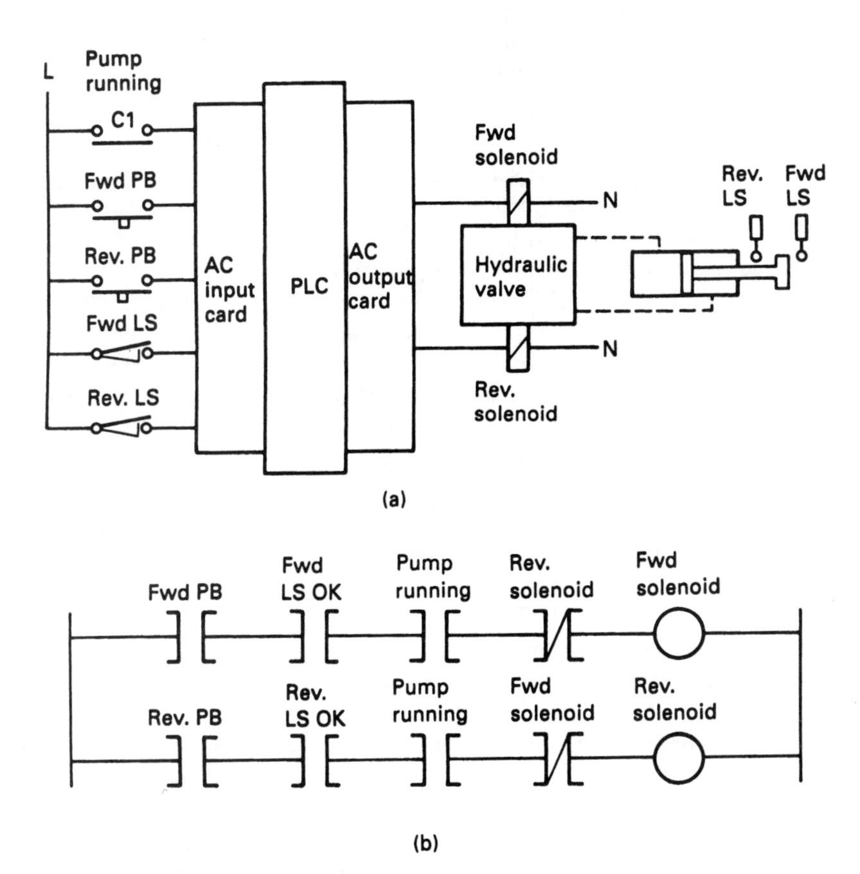

Fig. 14.9 *A simple PLC application. (a) A simple hydraulic cylinder controlled by a PLC. (b) The 'ladder diagram' program used to control the cylinder. This is based on American relay symbols. —] [—means that signal is present, and —]/[—means that signal is not present.*

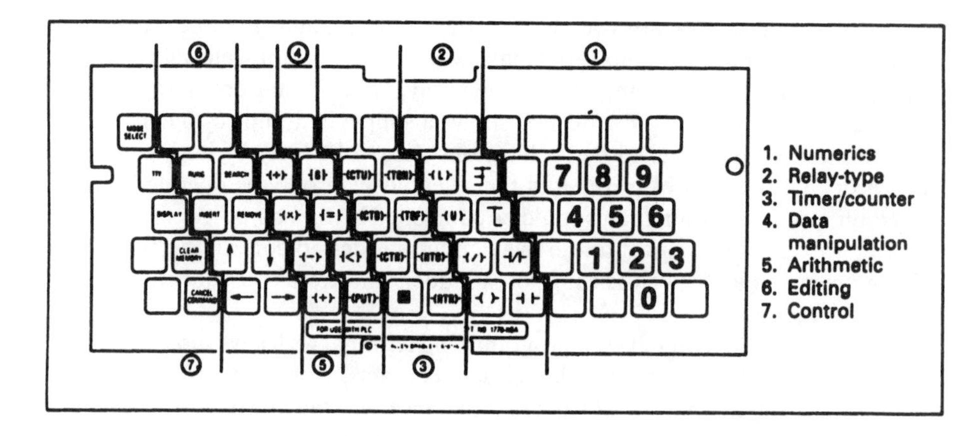

Fig. 14.10 *The programming terminal keypad for an early Allen Bradley PLC. The relay nature of the programming language, coils, contacts, timers, etc. can be readily seen. (Reproduced by permission of Allen Bradley.)*

of the majority of the keys should be obvious to any maintenance electrician. The program, shown exactly on the screen as Fig. 14.9b, would highlight energised contacts and coils allowing the programming terminal to be used for simple faultfinding.

The name given to these machines was programmable controllers or PCs. The name programmable logic controller or PLC was also used, but this is, strictly, a registered trade mark of the Allen Bradley Company. Unfortunately in more recent times the letters PC have come to be used for personal computer, and confusingly the worlds of programmable controllers and personal computers overlap where portable and lap-top computers are now used as programming terminals. To avoid confusion, we shall use PLC for a programmable controller and PC for a personal computer.

Figure 14.11 shows a range of modern PLCs from Allen Bradley, Siemens, ABB and CEGELEC. A modern programming terminal, using a standard portable PC is shown on Fig. 14.12.

14.3.2. Input/output connections

Internally a computer usually operates at 5 V DC. The external devices (solenoids, motor starters, limit switches, etc) operate at voltages up to 110 V AC. The mixing of these two voltages will cause severe and possibly irreparable damage to the PLC electronics. A less obvious problem can occur from electrical 'noise' introduced into the PLC from voltage spikes, caused by interference on signals lines, or from load currents flowing in AC neutral or DC return lines. Differences in earth potential between the PLC cubicle and outside plant can also cause problems.

The question of noise is discussed at length in Chapter 21, but there are obviously very good reasons for separating the plant supplies from the PLC

(a)

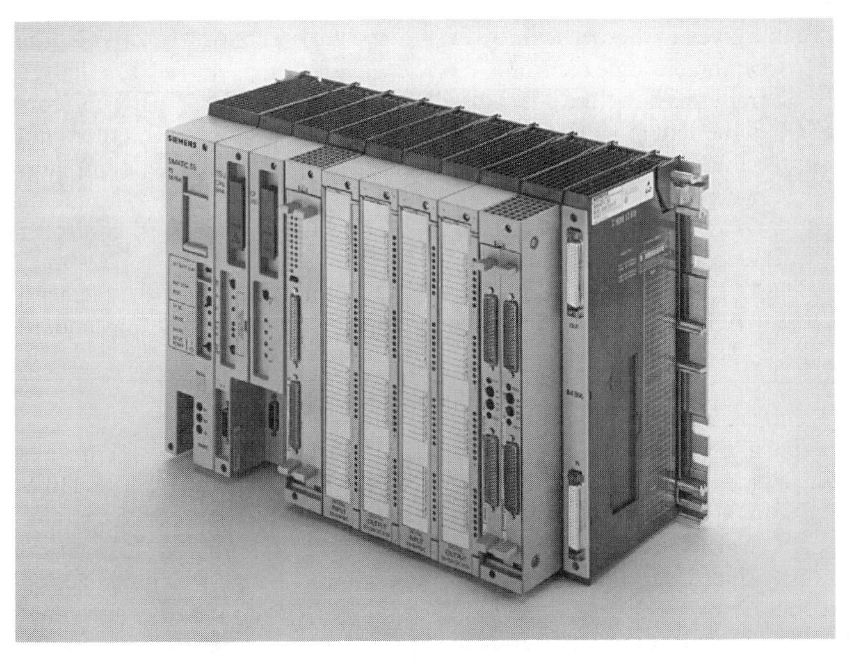

(b)

Fig. 14.11 *Four medium-sized PLCs. (a) The Allen Bradley PLC–5. (b) The Siemens SIMATIC S5–115. (c) The CEGELEC GEM–80. (d) The ABB Master. (Photos courtesy of the manufacturers.)*

(c)

(d)

Fig. 14.11 *contd.*

Fig. 14.12 *Modern programming terminal for Siemens SIMATIC S5 range of PLCs. Like many programming terminals this is based on an industrialised portable IBM compatible PC. (Photo courtesy of Siemens.)*

supplies with some form of electrical barrier. This ensures that the PLC cannot be adversely affected by anything happening on the plant. Even a cable fault putting 415 V AC onto a DC input would only damage the input card; the PLC itself (and the other cards in the system) would not suffer.

This separation is usually achieved by optical isolators, a light emitting diode and photoelectric transistor linked together as shown earlier in Fig. 13.60.

14.3.3. Input/output identification

The PLC program must have some way of identifying inputs and outputs. In general, a signal is identified by its physical location in some form of mounting frame or rack, the card position in this rack, and the card connection to which the signal is wired.

In Fig. 14.13, a relay is connected to output 5 on card 6 of rack 2. In Allen Bradley notation, this is signal

O: 26/05

The limit switch is connected to input 2 on card 5 of rack 3, and (again in Allen Bradley notation) is

I: 35/02

Most PLC manufacturers use a similar scheme.

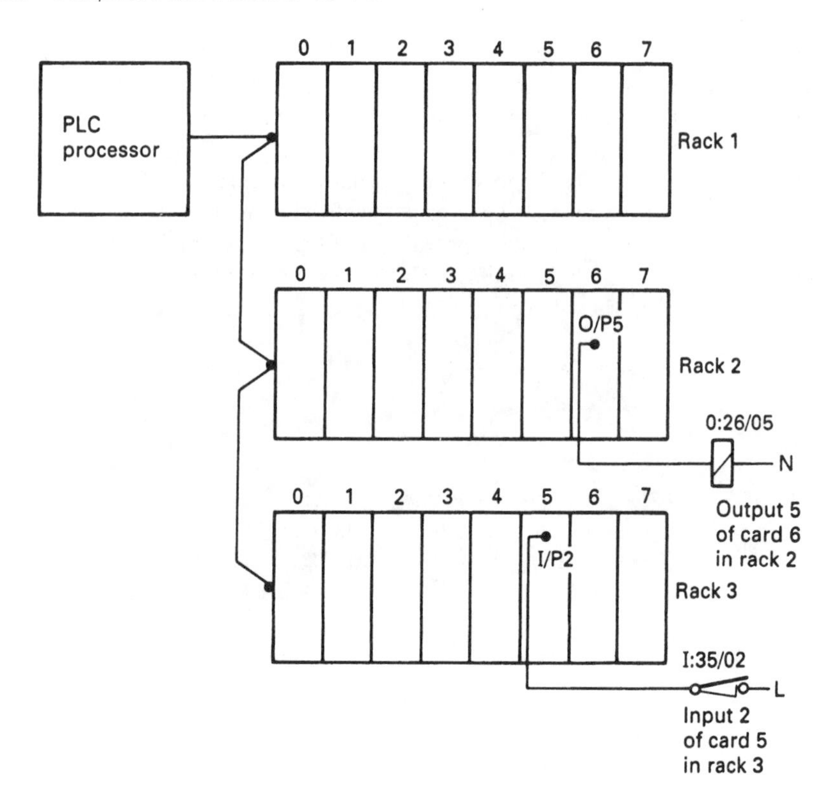

Fig. 14.13 *Identification of plant signals connected to an Allen Bradley PLC–5 rack.*

14.3.4. Remote I/O

So far we have assumed that a PLC consists of a processor unit and a collection of I/O cards mounted in local racks. Early PLCs did tend to be arranged like this, but in a large and scattered plant, all signals would have to be brought back to some central point in expensive multi-core cables. It can also make commissioning and fault finding rather difficult, as signals can only be monitored effectively at a point that may be some distance from the device being tested.

In all bar the smallest and cheapest systems, PLC manufacturers therefore provide the ability to mount I/O racks remote from the processor, and link these racks with simple (and cheap) screened single pair or fibre optic cable. Racks can then be mounted up to several kilometres away from the processor.

There are many benefits from this. It obviously reduces cable costs as racks can be laid out local to the plant devices and only short multi-core cable runs are needed. The long runs will only need the communication cables (which are cheap and only have a few cores to terminate at each end) and hardware safety signals (which should not be passed over remote I/O cable, or even through a PLC for that matter, a topic discussed further in Section 14.8).

Less obviously, remote I/O allows complete units to be built, wired to a built-in rack, and tested off-site prior to delivery and installation. A control pulpit, for example, could be delivered to site all pre-tested minimising expensive site installation and testing work.

If remote I/O is used, provision should be made for a program terminal to be connected local to each rack. It negates most of the benefits if the designer can only monitor the operation from a central control room several hundred metres from the plant. Fortunately, manufacturers have recognised this and most allow programming terminals to also connect to the processor via similar screened twin cable.

We will discuss serial communication further in Chapter 19.

14.3.5. The program scan

A PLC does not control a plant continuously (as is commonly thought) but works by taking repeated 'snapshots' of the plant input state, working on this picture then updating the outputs as required. This operation of read inputs, obey instructions, update outputs summarised on Fig. 14.14a is called the program scan, and will typically operate every few tens of milliseconds. With a remote I/O system driven by serial communications as Fig. 14.14b, a further

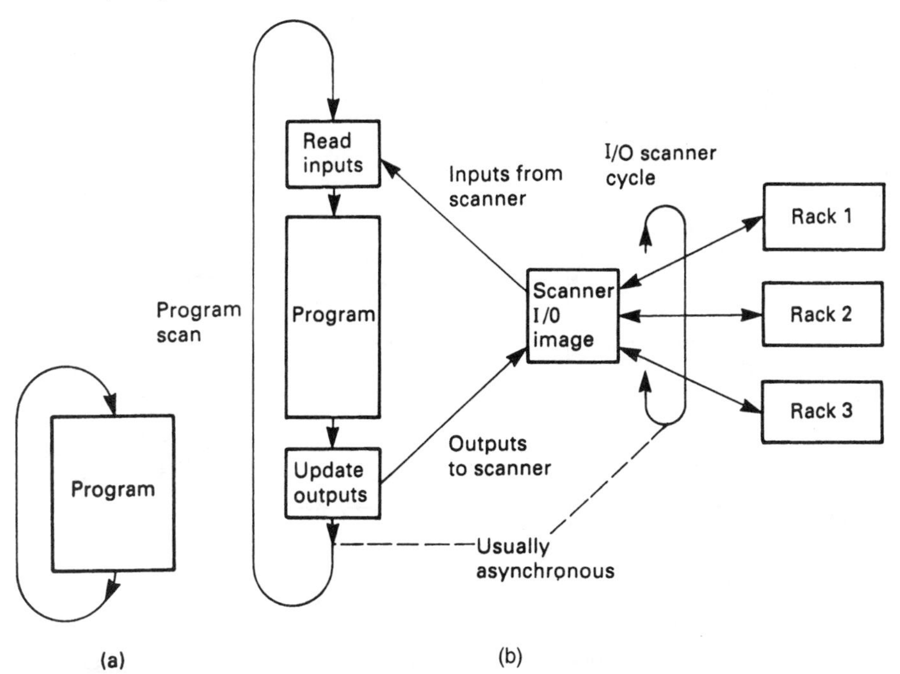

Fig. 14.14 *The PLC program scan which introduces an uncertainty in the response time of the PLC. (a) The PLC scan cycle; read inputs, obey program, update outputs. (b) The additional effect of the remote I/O scan, which is usually not synchronised with the program scan.*

uncertainty in timing will be added as the remote I/O scan, again a few tens of milliseconds, is not usually synchronised with the program scan. As a result there is always an uncertainty, shown on Fig. 14.15, in the response time of the PLC between one program scan (minimum) and two program scans plus two remote I/O scans (maximum).

In most applications this uncertainty of around 50 ms is unimportant except where material travelling at speed is involved (50 ms at 15 m/s is 0.75 m) but the programmer can compound this delay if care is not taken. Most PLCs start at the first instruction, and work in strict sequence to the end. If the program logical flow is written against this it will take one program scan for each backward written instruction to have an effect. The effect often occurs when the program is written in modular blocks, then the blocks are placed into the program without thought as to how the signals flow between them. It does not take many backward flowing instructions to raise the delay to around a second.

The program scan also limits the maximum rate at which a PLC can count signals. Figure 14.16 shows the effect of different frequency signals. In theory, a PLC can just handle a maximum frequency going at half the scan rate, but in practice filters on input cards tend to aggravate the situation. A good rule of thumb is one quarter of the scan rate. If pulse rates of more than about 5 Hz are to be handled by a PLC the designer should check the scan time. Pulses shorter than the scan time can also be missed for similar reasons.

14.3.6. PLC programming and IEC 1131/3

PLCs can be programmed in several different ways. In recent years the International Electrotechnical Commission (IEC) have been working towards

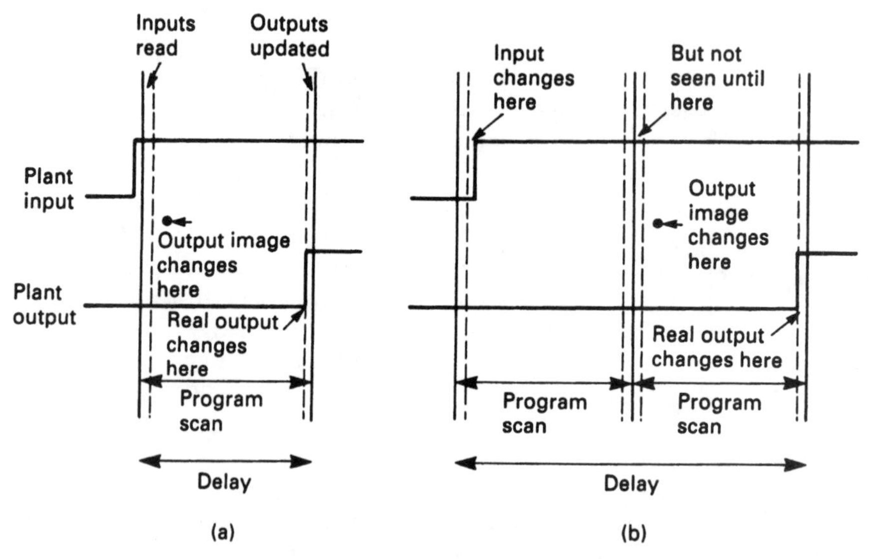

Fig. 14.15 *The effect of program scan on response time. (a) Best case. (b) Worst case.*

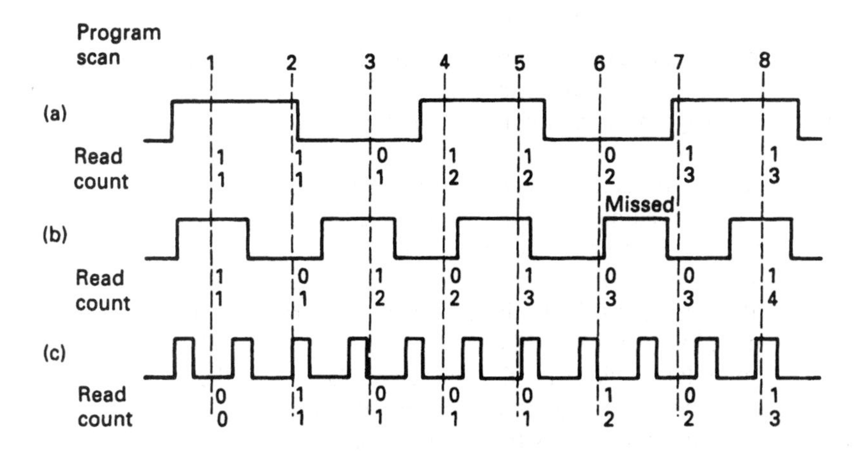

Fig. 14.16 *The effect of program scan on fast pulses.*

defining standard architectures and programming methods for PLCs. The result is IEC 1131, a standardised approach which will help at the specification stage and assist the final user who will not have to undergo a mind-shift when moving between different machines.

The earliest, and probably still the commonest, programming method described is the ladder diagram (or LD in IEC 1131,) shown earlier in Fig. 14.9.

Function block diagrams (FBDs) use logic (AND/OR etc) for digital signals and numeric function blocks (arithmetic, filters, controllers, etc.) for numeric signals. FBDs are similar to PLC programs for the ABB Master and Siemens SIMATIC families (shown on Fig. 14.17). There is a slight tendency for digital programming to be done in LD, and analog programming in FBD.

Many control systems are built around state transition diagrams, and IEC 1131–3 calls these sequential function charts (SFCs). The standard is based on the French Grafcet standard shown earlier on Fig. 13.55.

Finally are text-based languages. Structured Text (ST) is a structured high level language with similarities to Pascal and C. Instruction List (IL) contains simple mnemonics such as LD, AND, ADD, etc. IL is very close to the programming method used on small PLCs where the user draws a program up in ladder form on paper, then enters it as a series of simple instructions.

Figure 14.18 illustrates all of these programming methods.

A given project does not have to stick with one method – they can be intermixed. A top level, for example, could be an SFC, with the states and transitions written in ladder rungs or function blocks as appropriate.

It will be interesting to see the effect of IEC 1131–3. Most attempts at standardisation fail for reasons of national and commercial pride. MAP, and latterly Fieldbus, have all had problems in gaining wide acceptance. A standard will be useful at the design stage, and could be accepted by the end user if programming terminals presented a common face regardless of the connected

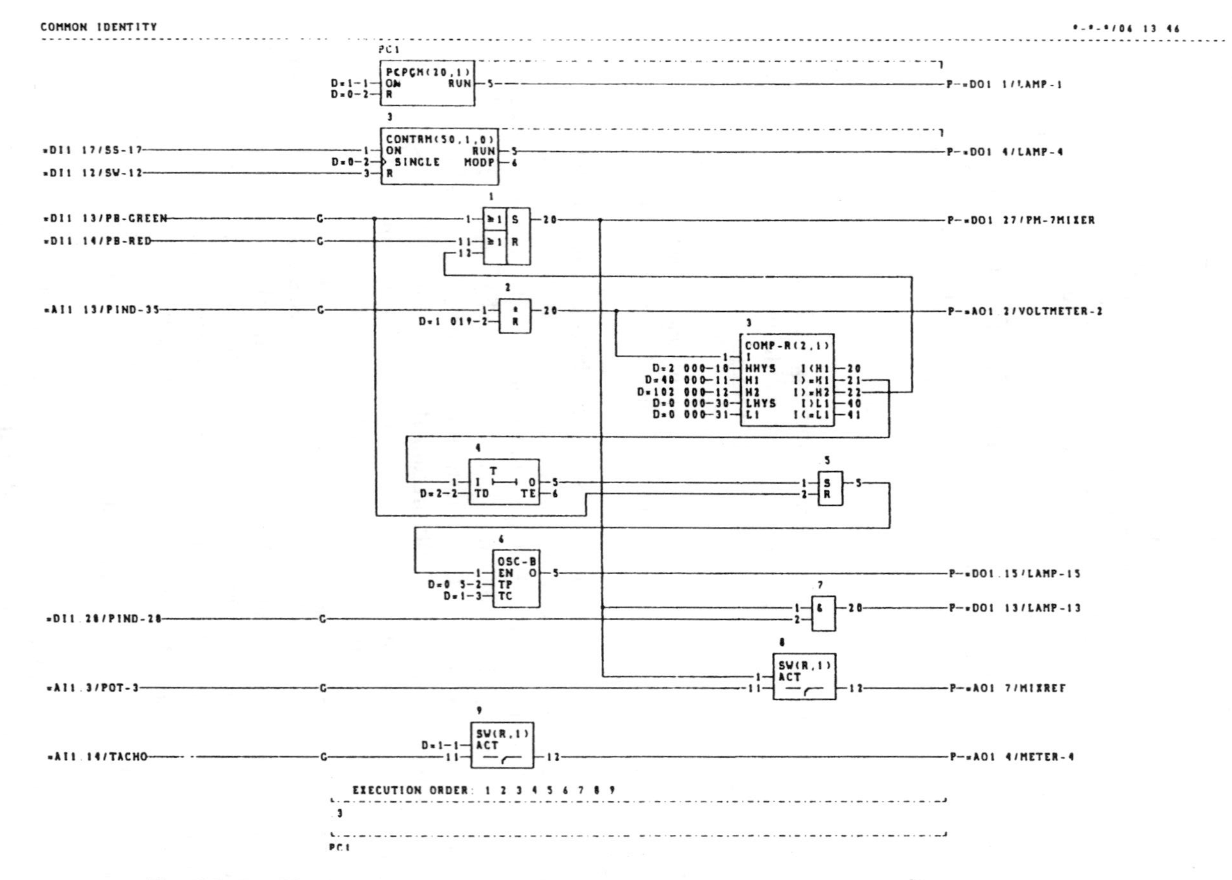

Fig. 14.17 *Block/logic symbols used by ABB for their Master series of controllers.*

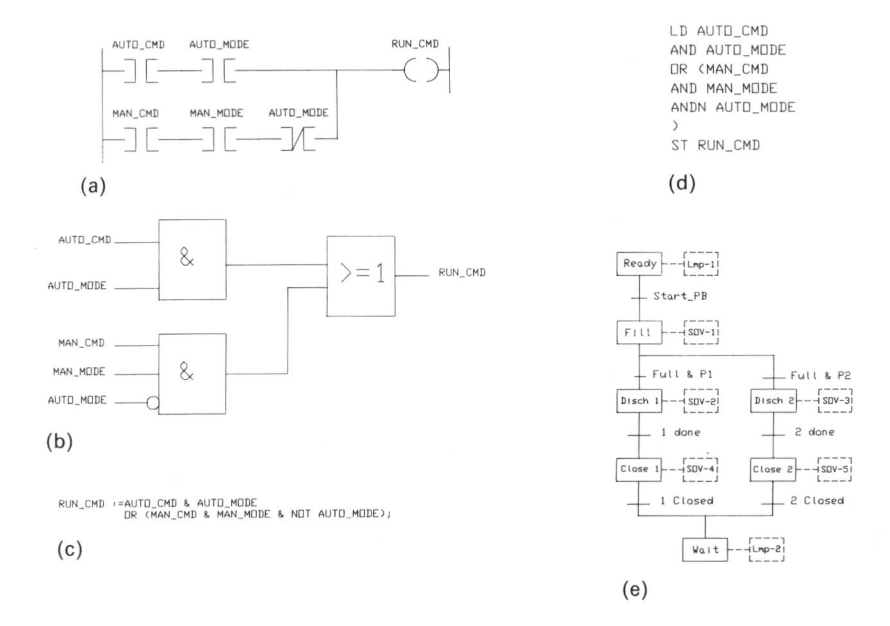

Fig. 14.18 *Programming methods from IEC 1131–3. (a) Ladder rung (LD) language. (b) Function block (FBD) language. (c) Structured Text (ST) language. (d) Instruction list (IL) language. (e) Sequential function chart (SFC) language.*

machine. It is to be hoped it doesn't act as a brake on design ingenuity and inhibit development.

14.3.7. The advantages of PLC control

Any control system goes through several stages from conception to a working plant. We will return to these stages in Section 14.5, but a PLC system brings advantages at each stage.

The first stage is *design* when the required plant is studied and the control strategies decided. With conventional systems every 'i' must be dotted before construction can start. With a PLC system all that is needed is a possibly (usually!) vague idea of the size of the machine and the I/O requirements (so many inputs and outputs). The cost of the input and output cards are cheap at this stage, so a healthy spare capacity can be built in to allow for the inevitable omissions and future developments.

Next comes *construction*. With conventional schemes, every job is a 'one-off' with inevitable delays and costs. A PLC system is simply bolted together from standard parts.

The next stage is *installation*, a tedious and expensive business as sensors, actuators, limit switches and operator controls are cabled. A distributed PLC system (discussed in Chapter 19) using serial links and pre-built and tested desks can simplify installation and bring huge cost benefits. The majority of the PLC program is usually written at this stage.

Finally comes *commissioning*, and this is where the real advantages are found. No plant *ever* works first time. Human nature being what it is, there will be some oversights. ('We need a limit switch to only allow feeding when the discharge valve is shut' or 'Whoops, didn't we say the loading valve is energised to UNLOAD on this system' and so on). Changes to conventional systems are time consuming and expensive. Provided the designer of the PLC systems has built in spare memory capacity, spare I/O and a few spare cores in multi-core cables, most changes can be made quickly and relatively cheaply. An added bonus is that all changes are inherently recorded in the PLCs program and commissioning modifications do not go unrecorded as is often the case.

There is an additional fifth stage called *maintenance* which starts once the plant is working and is handed over to production. All plants have faults, and most tend to spend the majority of their time in some form of failure mode. A PLC system provides a very powerful tool for assisting with fault diagnosis.

A plant is also subject to many changes during its life to speed production, ease breakdowns or because of changes in its requirements. A PLC system can be changed so easily that modifications are simple and the PLC program will automatically document the changes that have been made.

14.4. Conventional computers

14.4.1. Introduction

We have seen how powerful PLCs are, so an obvious question is why use a conventional computer at all. The main advantage is brute mathematical computing power, high speed and ease of connection to printers, keyboards, disc drives and the like. Usually the programs require specialist knowledge to change. This can be an advantage or a disadvantage dependent on the application. A PLC program is easy to understand and easy to modify. It can be quite difficult for engineering management to control and keep track of program changes. All PLCs have some form of access control via keys and password, but these are actually little protection. Access has to be provided for maintenance staff, who must be in possession of keys or password. Inevitably there will be 'midnight programmers'.

A computer with a program written in 'C' which is compiled and stored in ROM is as secure as a bank vault (Changing the program is, however, another story). If an application has few real I/O, needs a lot of mathematical operation, is unlikely to change, has a need for several printers or graphics monitors, or security is important, a conventional computer may be a sensible choice.

In general, computers at this level fall into three categories: bus–based systems, industrialised clones of the ubiquitous IBM PC family, and mini-computers such as the DEC Vax. The rest of this section briefly describes industrial computers belonging to these classes.

14.4.2. Bus–based machines

14.4.2.1. Introduction

The architecture of any computer (be it PLC, personal computer, mini-computer, games machine or company mainframe) can be represented by Fig. 14.19 and consists of a central processor (CPU), memory store and input/output (I/O) linked to the outside world. These are linked by a bus system (for BUSbar or OmniBUS depending on what source you are first exposed to) which has three components. The data bus carries data between the various elements; I/O to store, store to CPU and so on. The address bus carries the address of the store or I/O port concerned with the data movement, such as bring data from I/O port 17 to register C or store the contents of register D in store location address hex E147. The final bus is the control bus. This carries timing and direction signals.

This structure allows the idea of an expandable DIY computer to be implemented. The bus is laid out on a printed circuit backplane with established connections for the data, address, and control signals. The designer can then plug his own CPU, memory, video and I/O cards to build the computer needed to perform the required task. There are several bus standards, and the commonest

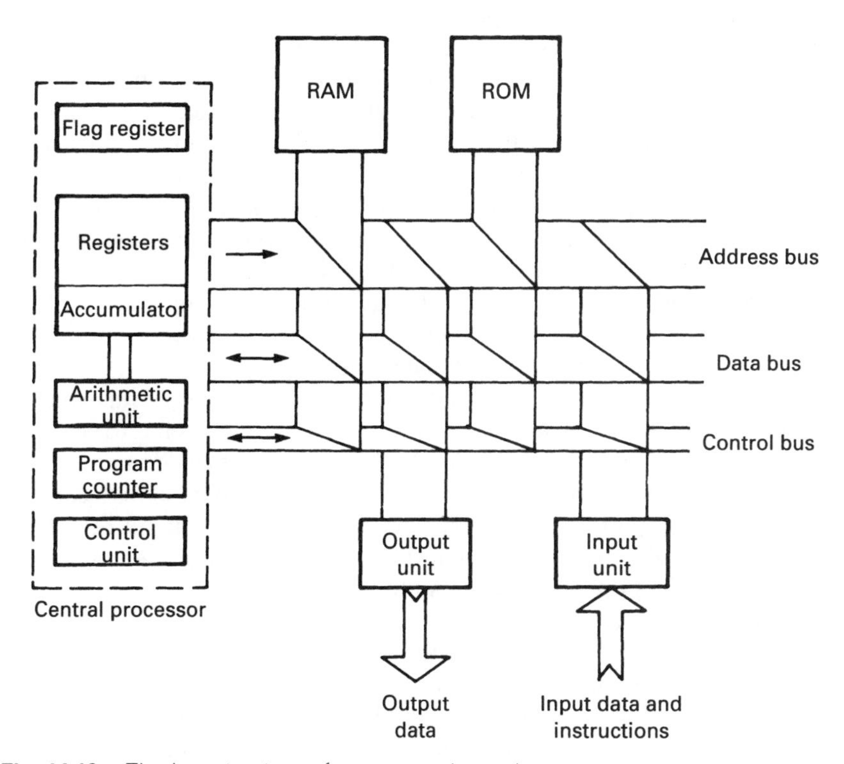

Fig. 14.19 *The bus structure of any computer system.*

are described briefly later. The IBM PC clone has this form as well, but the designer has less choice in the selection of the CPU.

There are actually two uses of the term 'bus-based machine'. The first of these is constructed as above, in the second, a master computer is linked to several external devices via a ribbon cable. Data can be read from, or written to, these external instruments. There is a fundamental difference between these two approaches, shown by the dotted line which represents the boundary of 'the computer'. We will first look at the common GPIB (IEEE-488) bus which is of this second type.

14.4.2.2. IEEE–488 parallel interface bus

This system was originally developed by Hewlett Packard to link HP computers to HP instruments. In its original form it was known as HP-IB (Hewlett Packard Instrumentation Bus). In 1975, the standard was formulated by the American Institute of Electrical and Electronic Engineers as standard IEEE-488 (also popularly known as GP-IB for General Purpose Interface Bus). This allows the linking of up to 15 devices and a computer with a total transmission length of 20 m.

The IEEE-488 bus can support three types of device: listeners, talkers and controllers. Listeners accept data from the bus; typical examples are a printer or a display. Talkers place data, on request, onto the bus; a measuring instrument is a typical talker. A controller assigns the role of any other devices on the bus but only one controller can be active at any one time. The designations listener, talker, controller are attributes of a unit (rather than a description of a unit's function) and many devices can fill more than one role. A computer, for example, can act as all three.

Signals on the bus can be grouped into a bi-directional data bus (which serves the three roles of data transfer, address selection and control selection), transfer control, interface management and grounds/shields.

Signalling is done at TTL levels, with 0 V representing '1' and 3.5 V a '0' (the inverse of normal TTL signals). Open collectors are used to allow the bi-directional data bus and bi-directional control signals to operate.

The data bus is used for several purposes. It can obviously carry data to one (or more) listeners or data from talkers. It can be used as an address bus to enable or disable one (or more) devices. Up to 15 primary devices addresses and 16 secondary addresses can be supported. Normally a secondary address controls an auxiliary function within a primary device. For an analog input device, for example, the channel would be selected with a secondary address, and the input value read with the primary address. Address 31 has a special function, being used to disable all active listeners.

The data bus action is determined by the active controller which uses the ATN line to signal whether the data bus is carrying data or control (address) information.

The major attraction of the IEEE-488 bus is its ease of use, with the operation being totally transparent to a programmer working in a high level language. For example, the instruction

OUTPUT 702, Setpoint

sends the data in the variable 'Setpoint' in the computer (address 7 represented as 700 and acting as a Talker) to the Listener with address 02. The actual bus operation corresponding to this instruction has four steps:

(1) Deselect all Listeners
(2) Select talker (7)
(3) Select listener (02)
(4) Perform data transfer.

IEEE–488 interface cards are available for many devices, both instruments and computers such as IBM PC clones.

14.4.2.3. Backplane bus systems

Backplane bus systems are built in the form of Fig. 14.20. A backplane provides the data, address and control bus signals, and various cards can be plugged onto the bus to configure the system as required. A typical bus computer is shown on Fig. 14.21.

The advantages of a bus system are obvious, standardisation, use of off-the-shelf cards, ease of expansion and a DIY, bolt it together yourself, approach. Unfortunately each and every microprocessor manufacturer and many equipment manufacturer devised their own standards, with different sized data words (8 bit, 16 bit or 32 bit) different address ranges and, of course, different edge connectors and pin layouts. Fortunately some common standards do seem to be emerging, notably the VME and STE Bus standards.

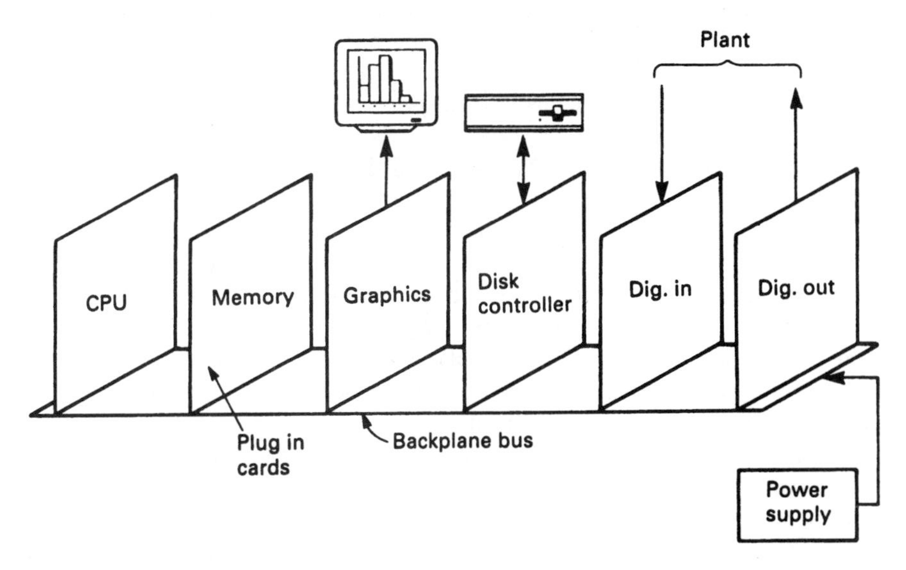

Fig. 14.20 *Bus-based computer.*

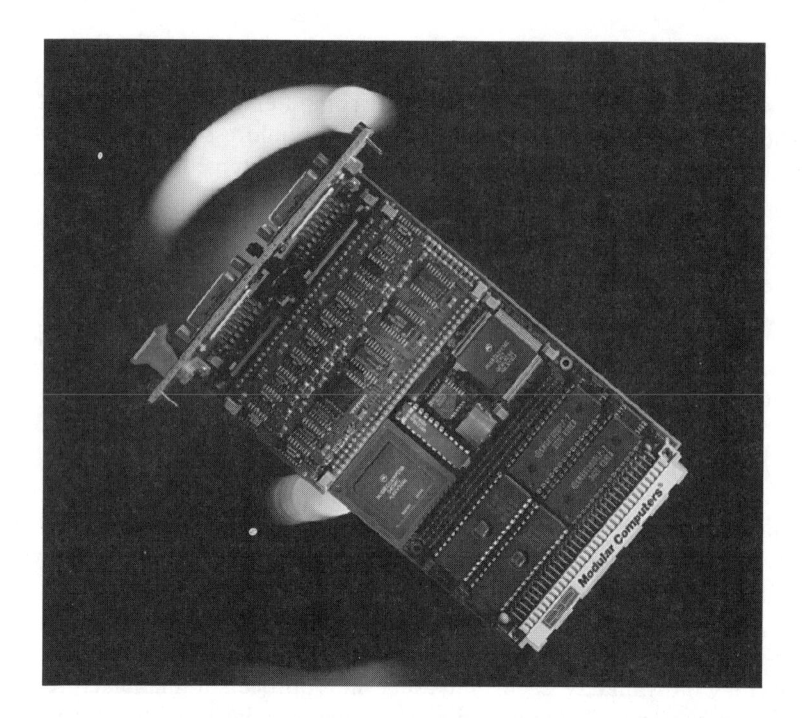

Fig. 14.21 *A single board VME bus computer. (Photo courtesy of PEP Modular Computers of Shoreham, West Sussex.)*

VME bus is a system designed originally for 16-bit machines and based on an earlier 16-bit bus system known as Versabus. A 24-bit address bus gives a large address range. The introduction of 32-bit microprocessors such as Intel's 80386 or Motorola's 68020 led to the upgrade of the VME bus to handle 32-bit data with a second connector. The bus thus exists in 16-bit (single 96 way DIN 41612 connector) or 32-bit (two 96-way connector) forms. In 32-bit form, the address bus was also extended to 32 bit. VME bus cards are generally compatible with either form.

With a high speed clock (24 MHz data transfer rate), 32-bit data bus and an enormous address range, the VME bus is arguably the most powerful industrial control system available. If speed of performance is critical it cannot be beaten. It is, however, sophisticated and expensive and, as one user said, 'You don't go shopping in a formula 1 racing car, you walk, go by bike or take the family estate car'.

In this latter category comes the STE bus (and the PLCs discussed earlier). The STE bus is an 8-bit data bus system with origins in the earlier STD bus. It has 20 address lines (allowing over 1 Megabyte of memory space) and 4 kilobytes of addressable I/O. It has been formalised under the IEEE-1000 standard. This ensures compatibility between different manufacturers, and has led to it becoming a general purpose accepted standard where the higher performance of the VME bus is not required.

It has many attractive features. The cards are based on the compact Eurocard size (100 × 160 mm) and connections between the cards and the backplane are made by a robust two-part connector (DIN 41612) which is resistant to vibration and shock loads.

STE bus has been designed around an interface between a master processor and a wide range of I/O boards. The microprocessor type is not defined, and any microprocessor can be used providing the interface between the CPU board and the backplane meets the defined standards. The bus can support up to three master CPU boards on the bus, although, of course, only one of these can be active at any one time. A well-defined procedure is laid down for the selection of control of the bus when contention occurs between the different masters.

14.4.3. IBM PC clones

Up to the early 1980s, the world of personal desk-top computers (PCs), was very varied, with a wide range of different machines and no common standard between them. This was matched by a plethora of operating systems, with usually each manufacturer having their own.

In 1981 the major mainframe manufacturer IBM entered the PC market. The effect of this was dramatic. IBM dominate the commercial computer market, and their timing for entering the PC market was, through design or luck, immaculate. Personal computers were falling in price and had reached a level where their widespread purchase could be justified by most companies. Although the specification for the PC was not particularly noteworthy (and the graphics ability of the early machines was poor), IBM's reputation ('Nobody ever got fired for buying IBM') ensured that the IBM family of PCs rapidly dominated the market.

IBM chose the software company Microsoft to provide the operating system, known originally as PCDOS. This operating system had a loose relationship to an earlier Z-80 based operating system called CP/M. To its credit, IBM were very open about the hardware and software of the PC and designed a bus system which allowed easy expansion. A vast market of add-on cards appeared, along with cheap IBM-clone computers. For these Microsoft provide an operating system known as MSDOS which is identical to PCDOS for all practical purposes.

Since its introduction in 1981 the IBM PC family has undergone a steady development. The original machine, with floppy disc storage and known simply as the IBM PC was based on the Intel 8088 microprocessor, a 16-bit version of the ubiquitous Intel 8080 (and the Zilog Z80) family. This was followed rapidly in 1982 by the IBM XT, again an 8088 based machine with hard disc for storage.

The next step occurred in 1984 with the introduction of the IBM AT. This was based on the more powerful Intel 80286 micro which could handle a large amount of memory (16 Mbytes compared with the 1 Mbyte of the 8088 and 64 kbytes of earlier micros such as the Z-80). Unfortunately MSDOS (and PCDOS) were designed around 1 Mbyte memories and could not directly use all the memory capability of the 286; and the 640K memory limit of MSDOS still remains a limitation. The 80286 could also support multitasking where the processor can operate more than one task at once.

The PC and XT had been constructed with a well-documented bus system allowing the addition of cards such as modems, interface cards, etc. The AT introduced a bus with additional capabilities (which was still compatible with the earlier standard). We will describe these bus systems shortly.

In 1987 IBM introduced a new family, the PS/2 (for Personal System 2) computers with a variety of number suffixes (PS/2-30, PS/2-50, etc). These are based on a range of Intel processors from the 286 (in the PS/2-30, PS/2-50, PS/2-60), to the 32-bit 80386 (used in the PS/2-55, PS/2-70 and PS/2-80) and 80486 (used in the PS/2-486). The PS2 machines also introduced improved graphics facilities and a totally new bus system called MCA (for Micro Channel Architecture).

This new bus system was incompatible with the earlier bus standards and has resulted in two divergent systems. The early bus (in the original PC and XT) was based on a 62-way edge connector with an 8-bit data bus and a 19-bit address bus plus control, timing and power supply lines (+12 V, +5 V, 0 V and −12 V). The 8-bit data bus was a restriction and IBM increased the data bus to 16 bits in the IBM AT bus with the addition of an extra 36-pin connector. Both this (and the 62-way connector) use printed circuit board edge connectors which are far less robust than the two-part connectors used on the STE bus. The two common standards, usually known as the PC-bus, and AT-bus, appear as Fig. 14.22 along with smaller half card formats. These standards are also known as Industry Standard Architecture (or ISA, which has nothing to do with the Instrument Society of America). Although IBM had originated the ISA bus, it never actively pursued legal action against clone and interface board manufacturers, and a flourishing industry in machines and add-ons developed.

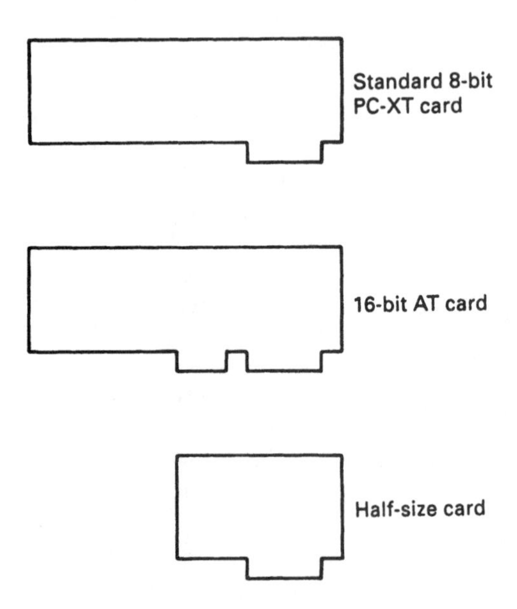

Fig. 14.22 *Expansion cards for IBM PC family.*

IBM's approach to the PS/2 MCA bus has, however, been markedly different. The MCA is superior to, but incompatible with the ISA bus, introducing features such as automatic configuration, higher speed and more tolerance to electrical interference. IBM has protected its rights to the MCA with strict licensing fees and a far less open approach.

Many of the clone manufacturers chose to co-operate against IBM and jointly developed an improved version of ISA called EISA (for Extended Industry Standard Architecture) as an alternative. There are thus four different bus systems for the IBM family: the PC bus, the AT bus, the EISA bus and the MCA. Future developments are not clear at the time of writing, but the ISA bus will be available for the foreseeable future.

Although the original PC had poor graphics, these have improved through the XT (CGA), AT (EGA), and PS/2 (VGA). The poor graphics capability of the early IBM-PC led to the development of graphics cards by external manufacturers, such as the monochrome Hercules Graphics Adaptor. These have largely been superseded by the EGA and VGA standards. These latter standards are useful for industrial graphics terminals.

Industrial PCs are based on clones and clone adaptor boards, usually to the AT bus standard. Using serial communications they often act as programming terminals or operator stations with PLCs.

14.4.4. Minicomputers

The most powerful computers used in industrial control tend to be the minicomputers such as the powerful DEC Vax of Fig. 14.23. These usually act to

Fig. 14.23 *Digital Equipment's range of VAX and MicroVAX computers with a VAX station terminal. (Photo courtesy of Digital Equipment Co. Ltd.)*

coordinate and report on the activities of the lower-based PC and PLC-based control systems. Minicomputers have a large amount of disc and memory storage for holding plant data for management reporting.

14.4.5. Programming for real-time control

The use of conventional programming languages was briefly discussed in Section 14.1.4. Languages such as BASIC, Fortran, Pascal and C were generally designed for general purpose, or scientific, computing and do not normally provide functions for real-time control. These are exceptions, however, with real-time variations on the standard language. MACBASIC, for example, is a version of BASIC, with instructions such as AIN(M,N) which gets an analog input from channel N on card M. Most of the single board and bus board computers described earlier operate with non-standard additional instructions to BASIC or C.

The programmer has to ensure that the computer program responds to plant actions and operator inputs in a reasonable time. One way of achieving this is simply to write a version of the PLC program scan of Section 14.3.4 which would have the form:

```
Begin
      Repeat
            Read Plant Inputs
            Work out Required Actions
            Write Plant Outputs
      Until Hellfreezesover
End. {of program}
```

This is possibly satisfactory for small schemes, but could be wasteful of computing time in large schemes because manual operations would expect a response in under 0.5 second, but the level of water in, say, a large storage tank would only need to be examined, perhaps at once a minute. Combined in a single scan, activities with widely differing speed requirements would be difficult to manage.

An alternative is to have the required action broken down into a series of tasks which are controlled by a common executive as shown on Fig. 14.24. The executive can call tasks at different intervals. Task 1, for example, is an auto/manual changeover run at 0.5 second intervals, task 2 is checking the level in a water tank, at 60 second intervals, task 3 is checking the oil level, pressure and filter state in a hydraulic system every 2 seconds and so on. This helps to streamline the process by having no wasted time, and aids programming as each task is totally separate from every other task and can be written independently.

Specialist real-time control languages are available, such as ICI's RTL, (for real time language) the US defence language ADA and the CEGB's CUTLASS (designed initially for power station control).

CUTLASS, for example, is a compiled language, originally written for DEC minicomputers, in which a control scheme is broken down into tasks which are

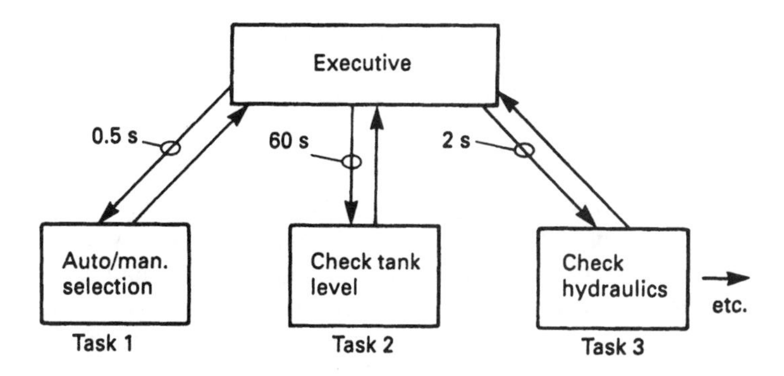

Fig. 14.24 *A computer program broken down into separate tasks under the control of an executive program.*

activated at preset time intervals. A task program starts with a definition giving its name, priority (in the event of a clash, tasks with highest priority are run first) and its run rate, for example

```
TASK AUTOSLEW PRIORITY = 236 RUN EVERY 600 ms
```

Next comes the definition of variables. These can be *global* (for the whole program and usable in any task) or *local* to the task. CUTLASS supports the usual forms of reals, integers and booleans (the latter being called *logic*) but introduces the concept of good and bad data. Any data coming from the outside world has the possibility of being erroneous due to plant failures. In CUTLASS, data can have a value or the state *bad*. Reals, for example, can have a numeric value or bad. Logic (Boolean) can be true, false, or bad. This state is carried through operations, for example

```
Aver: = (temp1 + temp2)/2
```

will yield the average temperature if both temp1 and temp2 are good values, but bad if either temperature readings are faulty. Some operations can produce good outputs with some bad data. A majority vote, for example, of 2 out of 3 will give a good output with one bad value.

Forth is a language also designed for real-time control. Most languages come from academic and research backgrounds but Forth was designed by an astronomer for the control of a telescope at Kitts Peak in the USA. It is an unconventional language in many respects, but once its peculiarities are learned it is ideally suited for industrial use.

Forth uses the idea of a pushdown stack, which can be considered similar to the spring-loaded piles of plates seen in cafeterias. As a plate is added, the pile moves down. Numbers in Forth are treated in a similar way with most operations concerned with numbers at the top of the stack. Polish notation is used, with the arithmetic symbol or operation following the data. The addition 273 + 28 is written

```
273 28 +
```

which leaves the result 301 on the top of the stack.

In Forth, the programmer adds to the language by defining a series of instructions and giving them a name. For example, we will write a series of instructions to convert a temperature in Fahrenheit to Centigrade. It assumes that the temperature in degrees Fahrenheit is at the top of the stack and it leaves the corresponding temperature in Centigrade in its place on the stack. The definition of a new instruction called F_TO_C goes

```
:     F_TO_C   {: means this is a definition}
      32       {goes to top of stack, pushing degrees F down}
      -        {subtract top of stack from next down}
      5*
      9/       {Degrees C now on stack}
      ;        {; means end of definition}
```

We can now write

```
68 F_TO_C
```

and 20 will be left on the stack.

Suppose we want to control a batch process where two chemicals are added to a vat, mixed, heated to some preset temperature, mixed for some time again, and then drained ready for a new batch. We could define a new word BATCH

```
:     BATCH
      ADD_1 ADD_2 MIX_1 HEAT MIX_2 DRAIN;
```

These are all new words, with new definitions, for example

```
:     ADD_1
      OPENV1
            BEGIN TSTLEVEL1 UNTIL {This is a Forth Loop}
      SHUTV1;
```

and

```
:     MIX_1
      MOTORON
      BEGIN TIME1UP UNTIL
      MOTOROFF
```

Again new words are introduced (OPENV1, TSTLEVEL1) which are again defined until eventually the ‘built-in’ Forth words can be used. OPENV1 uses standard words.

```
:       OPENV1
        1          {state bit 1 = 'ON']
        3          {channel number}
        5          {card number}
        DIGOUT     {standard word in TCS Forth}
;
```

and TSTLEVEL1 is simply

```
:       TSTLEVEL1
        4          {channel number}
        2          {card number}
        DIGIN      {leaves state 1 or 0 of Digital input 4 of card 2
                   on the stack}
;
```

Analog inputs and outputs are handled in a similar manner.

When all the user-defined words have been broken down to original Forth words, the sequence is run with the one word BATCH.

Forth programs are perfect examples of top-down programming, where a requirement is split into smaller tasks, which are split into sub-tasks and so on until units of small size and minimal complexity are created.

Where speed or minimal memory size is of the absolute importance, the programmer has little choice but to work in machine code. Normally programs are written in assembly code and turned into machine code by an assembler provided by the suppliers of the target computer system. The resulting program will be compact and fast, but can be difficult to change or maintain if the documentation is not good. The ability to monitor the running program (standard in all PLCs) will not be available unless fault-finding procedures have been written in as part of the specification.

14.5. Software engineering

Any project goes through six stages during its life. The first of these, *analysis*, is the most difficult as the project requirements are usually unclear. Most projects that come unstuck do so because this first stage has been cut short or overlooked.

Next comes *specification*, which is documenting the analysis so everyone concerned can agree what is to be done and what the end result should be. If you can't produce a specification, how can you sensibly design it? Never say 'We'll sort that out later' because later becomes 3 a.m. as the plant starts up. The final testing procedures must also be defined at the specification stage; again, if you don't know how you will test it, how will you know if it's working properly? Defining testing procedures in the cold light of day several months before the final frantic rush to meet a deadline also helps the poor commissioning engineer to resist the pleas for a premature start up.

The importance of these two first stages cannot be over-emphasised, too often the users do not know, or do not say, what they want, but once the project is complete they are sure it wasn't *that*. With these first two difficult stages over, the rest of the project becomes much easier!

The *design* stage can now start (simple with a good specification), followed by *installation*. Next comes *testing/commissioning*. This can also be a difficult time, as in any project the control engineer ends up collecting everybody's delays and comes under pressure to 'get the plant away'. It is here that the advantage of the test schedule from the specification stage will be invaluable.

It is not generally understood that commissioning involves both positive and negative testing. Positive testing is obvious; it is ensuring that when the 'Firkling' button is pressed the plant 'Firkles'. Practically everyone sees the need for this. Negative testing is less obvious; it is ensuring that the control system deals correctly with all the unlikely circumstances and fault conditions. Negative testing takes far longer, because there are far more fault modes than healthy modes. It is very common for people to say 'It works, let's go' when only the positive testing has been done. Try to resist this pressure, at best it can lead to damaged plant a few years hence, at worse some safety features could be overlooked.

Finally the plant is handed over to the maintenance department. In commercial software it is generally thought that over 50% of the effort goes into *maintenance* as changes are made to meet new requirements or correct the inevitable bugs. For easy maintenance all the documentation must be complete and up-to-date.

14.6. Computer graphics and the operator interface

14.6.1. Introduction

Chapter 17 discusses individual operator devices such as push buttons, switches, indicators, analog meters and digital displays. Increasingly all of these functions are being provided by computer graphic screens. These can be a display device designed specifically for a particular range of PLCs (for example, the Allen Bradley Panelview and the CEGELEC Imagem) and general purpose graphic display devices (such as ABB/ASEA's excellent Tesselator) or graphics software running on conventional industrial MSDOS computers.

It is useful to first consider the merits, and disadvantages of using computer displays. If everyone was totally honest, often the main reason for the choice is that they *do* look good and impress visitors. Too often the result is stunning colourful flashy graphics that are impossible to view for more than a few minutes without acquiring a headache, have to be searched for useful information and have an update speed of several seconds. At the author's site several plants are controlled by screen, and there are usually spectacular 'visitor's screens' and more mundane, restful (and useful) working screens.

The major advantages are simplicity of installation and flexibility. A graphics terminal has just two connections to the outside world, a serial link connection (see Chapter 19) and a power supply. If it is used to replace a desk full of switches and indicators there are obvious cost savings. A good quality switch occupying about 60 × 40 mm of desk space costs about £20 at the time of writing, to which must be added 1/16th of the cost of an input card, a share of a PLC rack, about three connectors, one core in a multi core cable plus labour for building the desk and PLC cubicle, pulling the cable, and ferruling the cable cores. A single device can be *very* expensive when all the costs are considered. There are software costs and a large capital cost for a graphics terminal, but these generally work out significantly cheaper.

The designer of desks or control stations often has to deal with changes and modifications (another example of the 'Didn't we tell you' syndrome which usually manifests itself as a retrofitted 30 mm pushbutton with dymotape label in a desk originally fitted with 20 mm controls). Constructing a desk is always a fine balance of time, choosing between waiting until all the requirements are clear, and the minimum time needed to make it. Modifications at the commissioning stage rarely look neat.

The displays on a graphical terminal can be modified relatively easily, and, more importantly, the modifications leave no scars. If the design of a normal desk can only start when the desk contents are 95% finalised (which is about right) a graphic screen can be started at 75% finalised. This flexibility is of great assistance as no job is ever right first time.

There are disadvantages, though. The most important of these is the limited amount of information that can be displayed on a single screen. It is very easy to overcrowd a screen (giving a screen similar to a page full of text on a word processor) making it difficult for the operator to identify critical items. A useful rule of thumb is not to use more than 25 to 30% of the screen. For a typical 80 × 25 character screen this means about 500 available positions which includes both identifying text and data. 'Motor Speed NNN rpm', for example, uses 16 characters excluding spaces.

The effect of this is often a need to build up a hierarchy of screens; the top screen showing an overview, lower screens showing more and more detail. The problem with this is the time delay needed to shift through the screens. Direct screen to screen movement is possible by calling for a page number (which needs a good human operator memory, or a directory piece of paper, or wasted screen space) or by making all screen changes via an intermediate directory page (with additional delay). These time delays are small (less than a second typically) but the cumulative annoyance is large.

The time taken to update screen data can also be problematical, particularly where a machine to machine link is involved. Again a response time of around one second is typical, but several seconds is by no means uncommon. The use of a graphic terminal for fault finding on a fast moving plant is not really feasible.

If the amount of data to be seen at any time is large, or a fast update speed is required, a large wall-mounted mimic such as Fig. 14.25 should be considered. These use tiles on a grid structure which permits relatively simple modifications.

There are generally two types of graphic terminal. The simplest, known as block graphics, has one store location for each character position on the screen. An

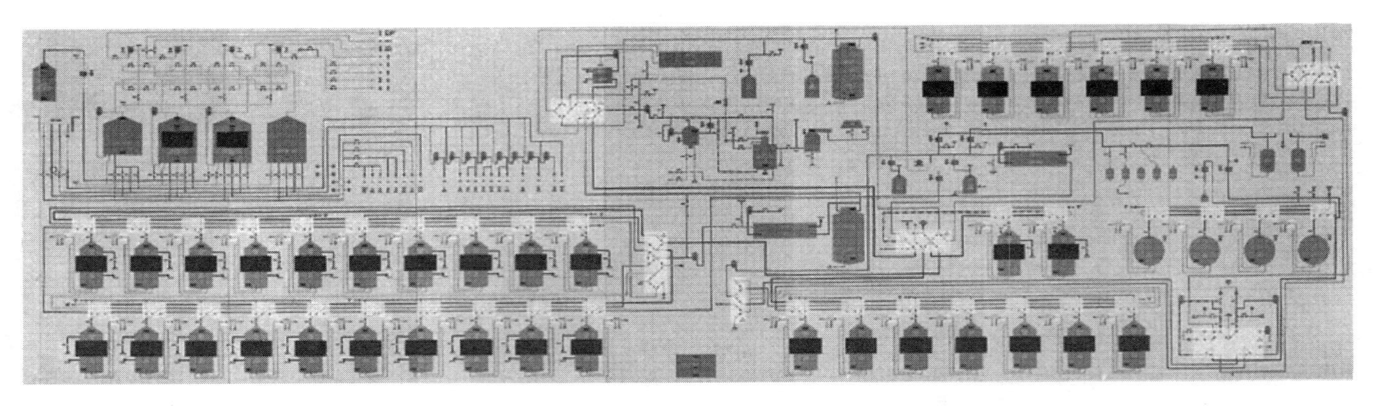

Fig. 14.25 *Plant mimic display built from tiles for a chemical plant. The amount of data on display here would be impossible to show on computer screens. (Photo courtesy of TEW instruments.)*

80 × 25 character display will thus have 2000 store locations. Each location will commonly have two bytes (one 16-bit word) arranged as Fig. 14.26. The first of these holds the character to be displayed, a single byte giving 256 possibilities. Standard ASCII (see Table 19.1) provides 128 alphanumeric characters, the other 128 being assigned to useful semigraphics characters. Figure 14.27 shows some of the block graphic symbols available on IBM PC clones, and the Allen Bradley Panelview. The second byte determines the colour, using 3 bits for foreground colour (giving 8 colours) 3 bits for background colour (again 3 bits) leaving 2 bits for functions such as flash, double height or bright/dim.

Block graphics terminals cannot usually show moving graphical displays trend charts or meters (although vertical and horizontal bargraphs can be easily built.) Figure 14.28 shows a typical block graphics screen from an Allen Bradley Panelview.

The second type of display deals not with individual *characters*, but with individual *points* on the screen called 'pixels'. A typical medium-resolution screen will have 640 (horizontal) by 350 (vertical) pixels, a total of 224 000 points. High-resolution screens for computer-aided design (CAD) use even more pixels. Each of these can be accessed individually, allowing lines to be drawn at any angle, fill patterns of any type to be used and trend graphs of plant variables to be displayed. Each individual pixel can have its own colour (from over 256 possible colours in some displays) and intensity. The result is an almost photographic resolution. Figure 14.29 shows a pixel-built screen.

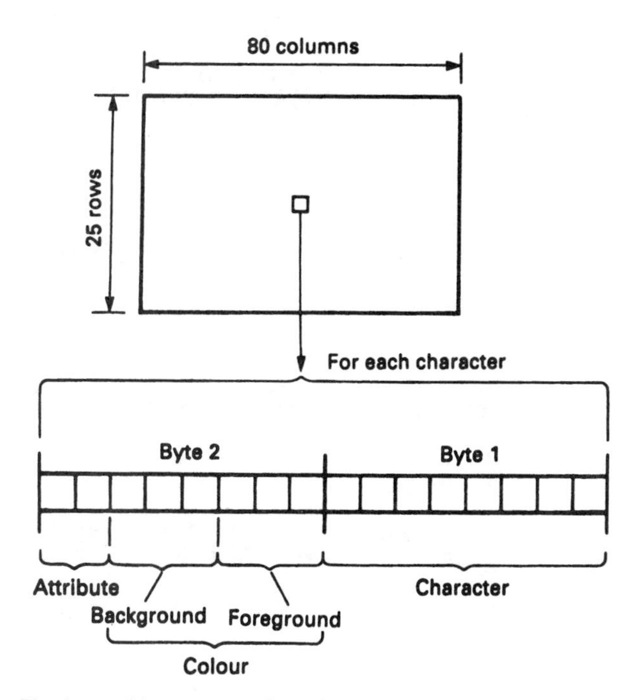

Fig. 14.26 *Block graphic memory allocation.*

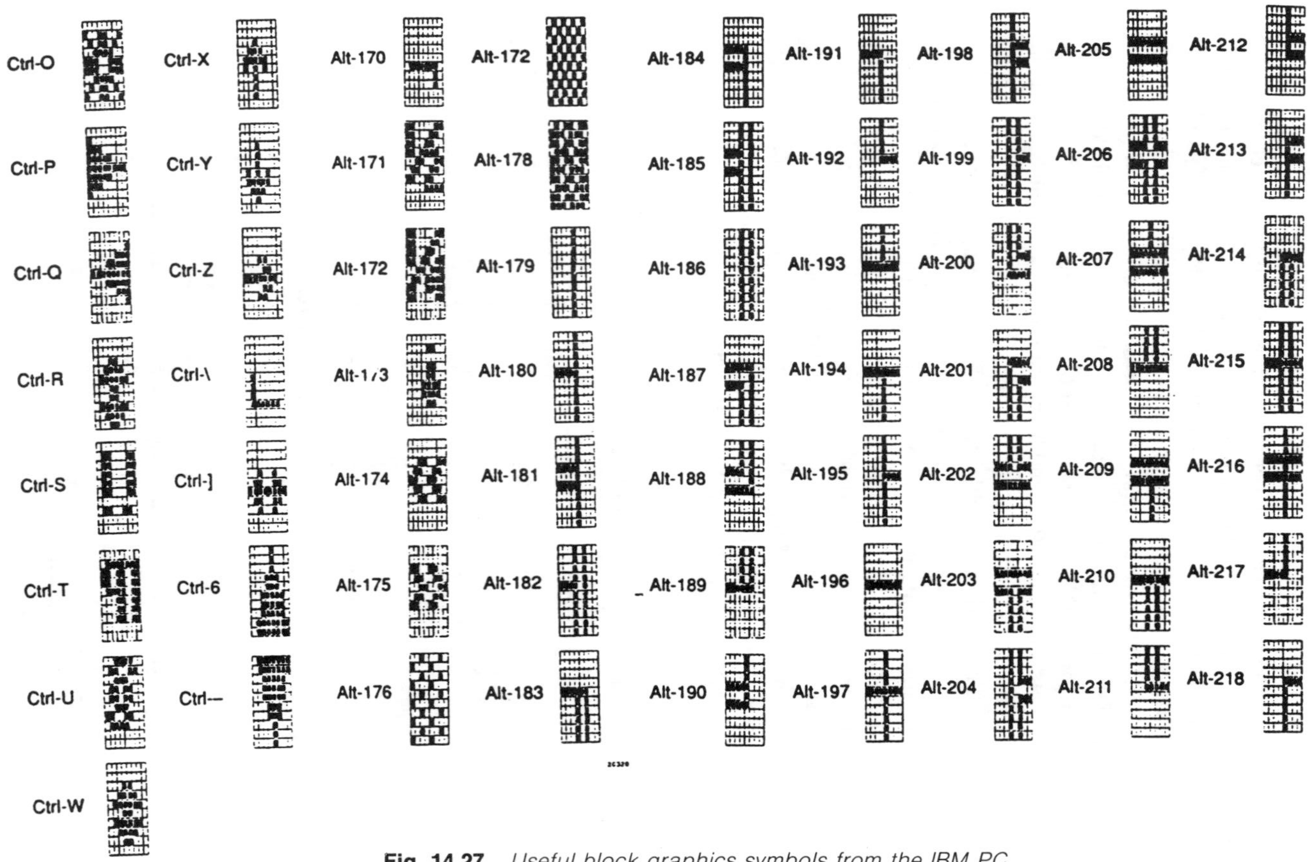

Ctrl-O Ctrl-X Alt-170 Alt-172 Alt-184 Alt-191 Alt-198 Alt-205 Alt-212
Ctrl-P Ctrl-Y Alt-171 Alt-178 Alt-185 Alt-192 Alt-199 Alt-206 Alt-213
Ctrl-Q Ctrl-Z Alt-172 Alt-179 Alt-186 Alt-193 Alt-200 Alt-207 Alt-214
Ctrl-R Ctrl-\ Alt-173 Alt-180 Alt-187 Alt-194 Alt-201 Alt-208 Alt-215
Ctrl-S Ctrl-] Alt-174 Alt-181 Alt-188 Alt-195 Alt-202 Alt-209 Alt-216
Ctrl-T Ctrl-6 Alt-175 Alt-182 Alt-189 Alt-196 Alt-203 Alt-210 Alt-217
Ctrl-U Ctrl-– Alt-176 Alt-183 Alt-190 Alt-197 Alt-204 Alt-211 Alt-218
Ctrl-W

Fig. 14.27 *Useful block graphics symbols from the IBM PC.*

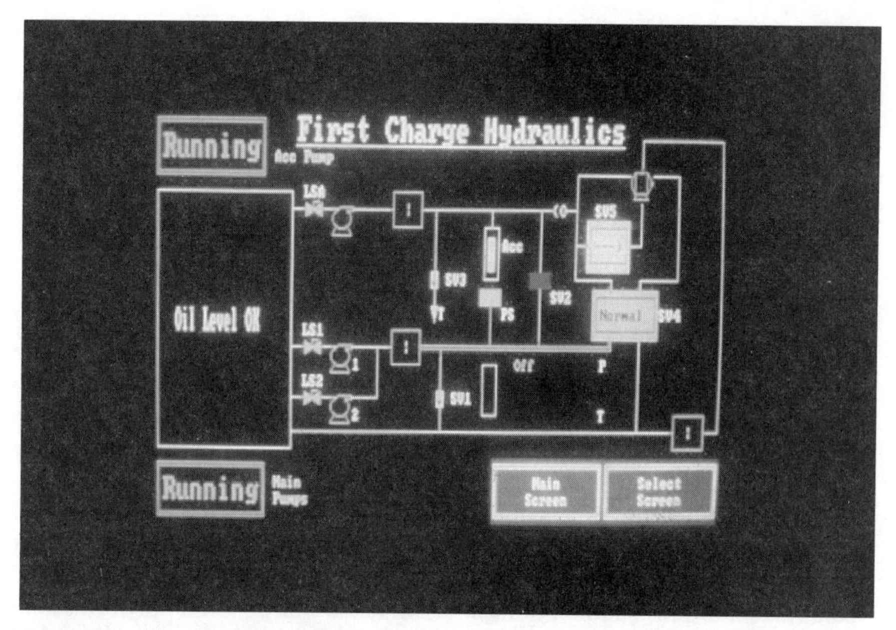

Fig. 14.28 *Allen Bradley Panelview screen based on block graphics. The Panelview is a touchscreen, and the double bordered blocks represent pushbuttons. The screen shows a diagnostic display for a hydraulic unit.*

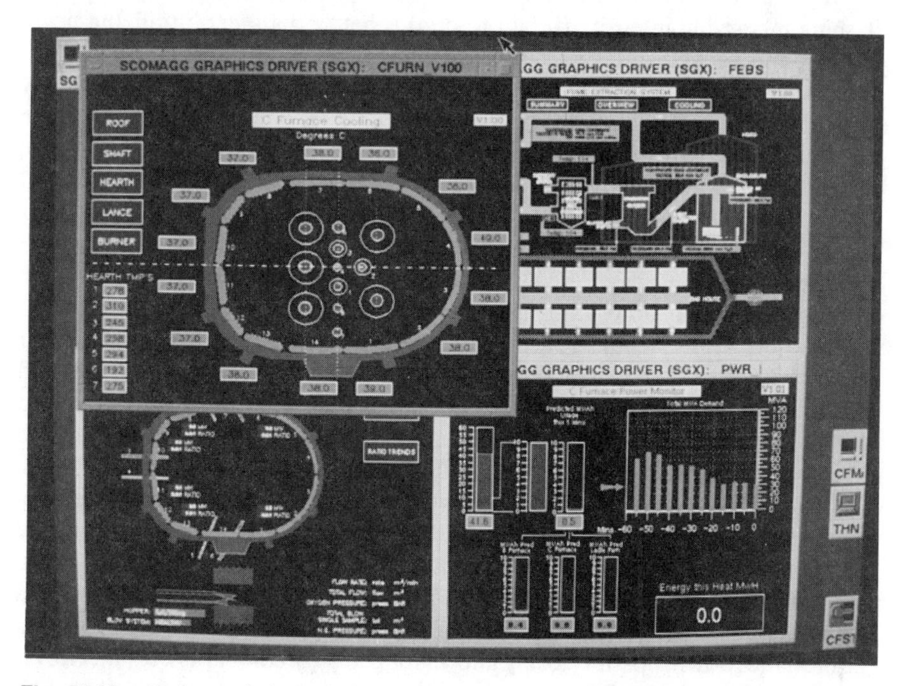

Fig. 14.29 *High resolution pixel-addressed graphics using DecWindows Motif on a VAXstation. (Photo courtesy of CoSteel Sheerness and Scomagg.)*

There are additional costs, the most obvious of which is a large store requirement. The system hardware and software is more complex (and hence more expensive but, perhaps surprisingly, this is not apparent to the user; pixel graphics displays are often easier to program than block graphics units. Recent home computers (from the era of the BBC and the Spectrum) have all used pixel graphics. Programming for these screens is surprisingly simple with instructions using keywords like DRAW FROM < > TO < >.

14.6.2. Practical considerations

One major advantage of graphical displays is that mistakes can be rectified without leaving scars, albeit with some cost. None the less it is far better not to make mistakes in the first place. Perhaps the most important consideration is to realise that the system is being designed for an operator who will be sitting in front of it for about 8 hours a day, and not for the managing director and his visitors. By all means include impressive 'visitors' screens', but remember the poor operator. Above all, remember you are *not* designing a video game.

A typical mistake is over-use of flashing displays (coupled with similar problems from flashing lights on desks). A video display (or a desk) in normal operating conditions should not flash or blink. Flashing should only be used to attract an operator's attention (an unaccepted alarm, for example) and should go steady when the operator acts. On a screen, a flashing small box alongside fixed text is much more friendly and easier to assimilate than flashing text. Text flashing in varying intensity (bright to dim) is better than text on/off and never use text and background switching colours which is almost impossible to read.

Another common problem is SCREENS WHICH SHOUT AT YOU IN UPPER CASE ALL THE TIME. All modern screens can support upper case and lower case. Road transport research found many years ago that lower case text is far easier to read, as can be seen on any road sign. Careful use of initial capital letters Can be Used to Emphasise Text and Draw the Operators Attention. Think about layout.

Screens should be uncluttered and consistent. The 25–30% usage rule is a good starting point as it allows an operator to quickly scan a screen for relevant information. Consistency ensures similar operations are performed in similar ways, with colours having the same meaning on different screens. Pumps should not, for example, be run with separate start/stop pushbuttons on one screen, and with push on/push off single buttons on another. If an 'End of Travel' limit switch is yellow on one screen, it should be yellow on all (and not red, green or blue). Consistency problems normally arise where more than one person has been involved, and can generally be overcome by laying down standards at the start of a project.

Bright colours (yellow/white) tire the eye, and should not be used in large areas (and be avoided for background colours). Overuse will cause the operator to turn the brilliance down, possibly losing information in dark colours as a result. Grey is a much more restful background colour. Blue characters on a black background are particularly vulnerable to vanishing if the contrast or brilliance is turned down.

Good colour combinations are black on green, black on yellow, black on red, red on white, blue on white, green on white, red on black, green on black and white on blue (the latter is very good for large areas of text and is often used for word processors).

Colour combinations to avoid are yellow/green and yellow/white (which merge together) and blue on black for fine detail (the visibility is very dependent on the setting of the contrast and brilliance control). Cyan/blue is also poor to the point of vanishing into an unreadable bluish block.

Colour should only be used to emphasize values and not really be part of the control. A significant part (about 10%) of the population is colour blind. If colour is important, colour-blindness tests will have to be implemented.

With touchscreens, a useful standard is colour on black for an unactivated pushbutton, and black on the colour for an activated state. With start and stop buttons, for example, in the stopped state the start button would show 'Start' in green on black, and the Stop button show 'Stopped' in black on red. When the start button is pressed, it changes to 'Running' in black on green, whilst the stop button changes to 'Stop' in red on black. Note the text changes from allowed action to state, and the total intensity changes from bare text on dark background to black text on light background to allow a colour-blind operator to use the screen.

The environment around a display needs to be carefully considered. Most screens are mounted angled up, and are prone to annoying reflections from overhead lights and windows. Bright lighting (and above all, direct sunlight) can make a display impossible to read.

Displays are also adversely affected by magnetic fields. Close proximity to electric motors, transformers or high-current cables will cause a picture to wobble and the colours to change. The effect can be distinctly bilious. The effect can be overcome by screening the monitor with a mu-metal cage (normal steel or iron does not work).

The size and weight of the monitors are often overlooked making them difficult to mount neatly, and even more difficult to change. Access should be made as easy as possible; trying to hold a 25 kg display in place with one hand whilst undoing interminably long mounting screws is not much fun.

Displays fail, and the implication of this needs to be considered in the design. If all the plant control is performed by screens, what will happen during the ten or so minutes it will take to locate a spare and change the faulty unit, (see the previous paragraph)? Often dual displays (main and standby) are used to overcome this problem.

14.6.3. Data entry

The operator will obviously need to input data and initiate actions. Keyboards are one approach, but many people are nervous of them (home computers help here) and the cable connecting the keyboard always seems prone to damage. In dirty environments keys can become blocked with dirt and membrane keypads with tactile (feel) feedback should be used.

Another useful approach is softkeys. Here a set of buttons (often 10) is positioned on the keyboard below a set of (software driven) blocks on the screen. The push button can thus change their meaning as the screen changes as shown on Fig. 14.30.

If the operator has to access points anywhere on the screen, a tracker ball is a useful device. Rather like an upside-down mouse it controls the movement of a cursor on the screen. All normal actions can be performed with three buttons on the trackerball and a numerical keypad. Trackerballs work surprisingly well in dirty environments as they are open underneath and dirt seems to fall straight through. Mice perform a similar function but are vulnerable to damage and dirt and seem more suited to an office environment.

Touchscreens have already been mentioned briefly. Combining a display area with operator controls they provide a very compact interface, but their use should be tempered with care. There is absolutely no tactile feedback for the operator to sense a button, so their use is not recommended for an application where the operator has to look at the plant (and not the screen) when operating controls. It is also easy (far too easy!) to operate buttons by mistake. Dangerous buttons should be removed from the main screens or protected by a delay which means a button has to be touched for, say 1.5 seconds before it has an effect. A similar effect can occur when a touchscreen is cleaned; a blank screen should always be provided for this purpose. The continued touching of the screen leads to a build up of greasy fingermarks accentuating this problem.

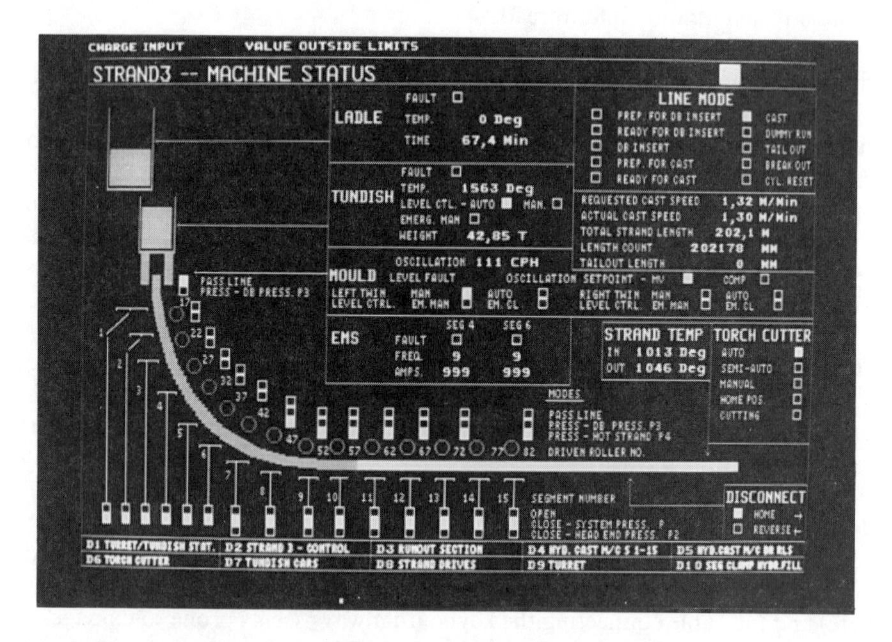

Fig. 14.30 *ABB Tesselator with ten softkeys at the bottom of the screen corresponding to ten function buttons on the operator's keypad. (Courtesy of ABB.)*

14.6.4. Human computer interaction

Interesting psychological effects occur when a human being and a computer have to communicate. The implications of this are discussed, along with the general subject of ergonomics and the man-machine interface, in Section 17.7. Recent legislation concerning the use of screens is also discussed.

14.7. Computer-aided design and manufacture (CADCAM)

A manufactured product goes through many stages from conception to the first sale. It must be designed, the parts and raw materials ordered, and manufacturing process established. Computerisation is playing an increasing role in these areas of traditional manufacturing processes which are the bedrock of a country's economy.

Many computer-aided design packages are available, from simple layout packages which check that components will fit within a case with adequate clearance to full finite element analysis software which will calculate and display the effect of applied loads on three dimensional objects.

Computer aided draughting (CAD), shown on Fig. 14.31, is now very common. With packages such as Autosketch and TurboCAD available for very small amounts of money the advantages of CAD can be used even by one-man operations. There are many advantages of CAD, but the main one is the time saving that comes from repetitive drawings. The main rule of CAD is don't draw what you can copy or modify from somewhere else. As most organisations tend to manufacture variations on the same basic ideas, there is a great time saving from

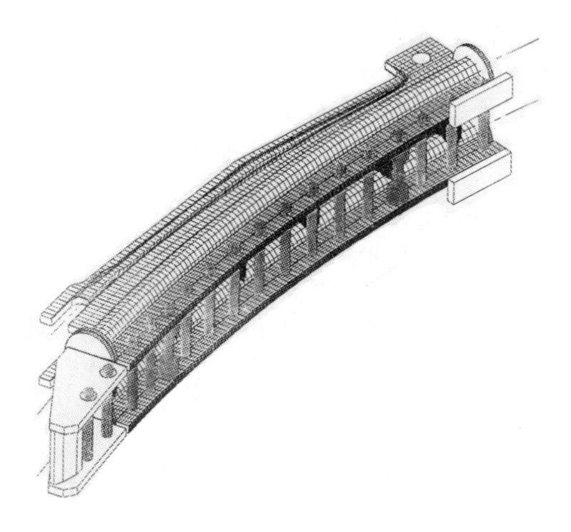

Fig. 14.31 *A three-dimensional model built on a CAD system with solid modelling and hidden line removal. The object can be viewed from any angle. (Drawing courtesy of CoSteel Sheerness.)*

having standard symbol libraries from which parts can be picked.

Higher level CAD systems include features such as three-dimensional modelling which allows 3-D representations of an object or complete plant to be built and viewed from any angle. It is used for checking views and, with animation, assembly procedures and clashes between components. It is likely that virtual reality technology will be used with 3-D modelling allowing designers to 'walk round' a model of an object which could be as large as a multi-storey plant or as minute as a complex molecule.

Manufacturing can be performed by robots (fig. 14.32) or numerically-controlled machine tools (NC tools). These can easily be re-programmed for the manufacture of different products. The programming language used is usually APT which is built around geometry definitions such as

```
P1 = POINT/0,0
P2 = POINT/45*COSF(60),45*SINF(60)
C1 = CIRCLE/CENTER,P1,RADIUS,20
L1 = LINE/LEFT,TANTO,C1,P2
```

and cutting tool movement definitions such as

```
CUTTER/20
GOFWD/C1,TANTO,L1
GOFWD/L1,PAST,L2
```

The first of these instructions says to load the cutter with 20 mm radius which is used as an offset in the movement instructions (i.e. we move the cutter face, not the shaft). The cutter then follows the circumference of circle C1 until it reaches the tangent with line L1 which is then followed until the cutter is past intersecting line L2 and so on.

NC machines can be programmed manually, but there is increasing use of Flexible Manufacturing Systems (FMS) which coordinate design, production control and manufacturing in a site-wide computer network linked with LANS such as Ethernet, MAP and Fieldbus described in Chapter 19. With FMS the production schedule and the cutting program can be directly loaded into the NC machine tool.

CADCAM needs planning to work smoothly. There is a considerable waste of money in having raw material, spares and, worse still, finished products sitting around on shelves. The 'just in time' philosophy aims to save money by reducing material sitting idly around in stores. To achieve this the whole manufacturing process needs to be coordinated and production schedules planned for each machine along with the parts and raw materials needed. In a complete flexible manufacturing system all the production will be controlled by production planning and control software on a high-level computer.

14.8. Safety

In the early 1980s the Health and Safety Executive became concerned about the safety aspects of programmable devices and published a consultative booklet

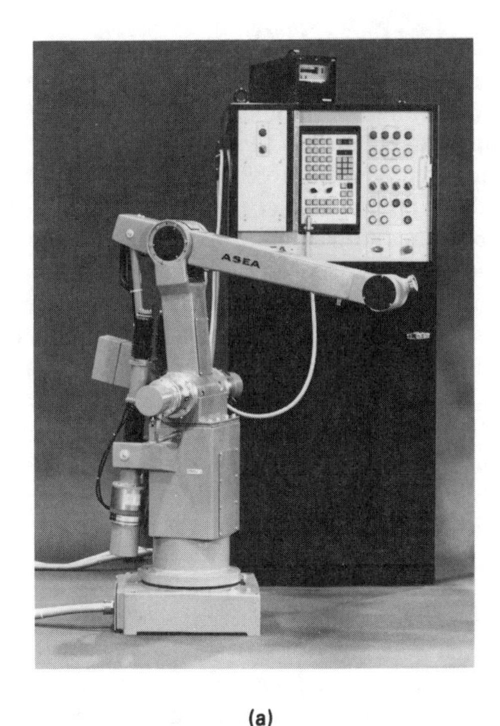

(a)

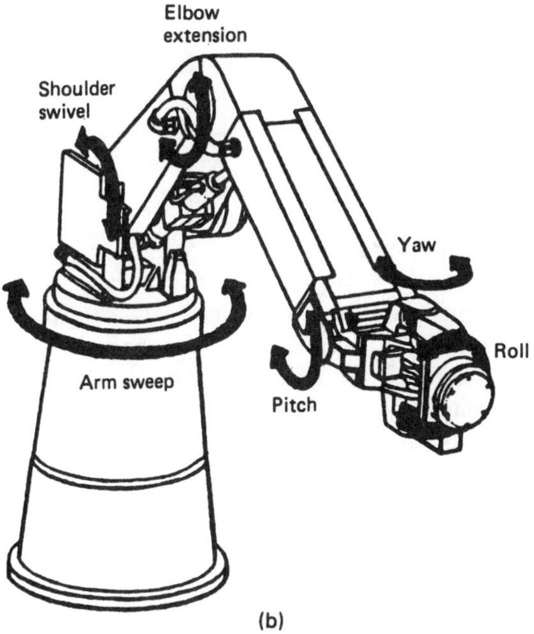

Elbow
extension
Shoulder
swivel
Yaw
Roll
Arm sweep
Pitch

(b)

Fig. 14.32 *Robotics. (a) An industrial robot. (Photo courtesy of ASEA, Vasteras, Sweden.) (b) A robot with 6 axes (degrees of freedom).*

called 'Microprocessors in Industry, safety implications of the uses of programmable electronic systems in factories'. This was followed in 1987 by the two-volume *Programmable Electronic Systems in Safety Related Applications*. Volume 1 is a general introduction, and Volume 2 is Technical Guidelines.

These two volumes are often held to state that you cannot use PESs (as the HSE calls PLCs and computers) in safety applications. This is not true. What is stated is that PESs bring additional failure modes into the system which the user should recognise and deal with to produce a system which is at least as safe as could have been achieved by conventional means. PLCs have been widely used in public transport and public lifts, but the designers have examined the reliability and probability of failure to show that the result is safer than conventional systems.

There are two major problems to be addressed. The failure mode of a PES is not predictable, both inputs and outputs can fail in an on- or off-state. Simply taking a safety signal through a PLC has therefore acquired two additional failure modes and has made the system more unsafe.

The second problem concerns the program itself. It is a most impossible to check if there is a logical failure in the program design. A bug may lurk for years and surface just when it becomes important. Testing can reduce, but not eliminate, this problem. The difficulty is compounded by later modifications which are never as well designed or tested as the original program.

The subject of producing safe computer programs is complex, but usually is based on the idea of redundancy of signals and programs. Simple duplication does not bring safety, a program bug in one program will obviously exist in a duplicate copy. The redundancy must ensure that different programs designed by different people are used. This is known as diverse redundancy and eliminates so-called common mode failures.

For simple sequencing PLCs, it is usually easier to perform safety functions (e.g. Emergency stops, safety gates) outside the computer program. For complex systems the designer should be aware of, and follow, the HSE guidelines.

14.9. Application programs

Increasingly, as computers become more widespread, many programs have been written which allow the user to define the tasks to be performed without worrying unduly about how the computer achieves it. These are known as application programs and are typified by spreadsheets such as Lotus 123 and Visicalc and databases such as DBase and DataEase. In these the user is defining complex mathematical or database operations without 'programming' the computer in a conventional sense.

Amongst packages of interest to engineers are software for analysing and predicting the behaviour of closed-loop control systems, Critical Path Analysis (CPA) for project control and design software for anything from PCB layouts to steam pipe calculations. The difficulty is knowing what is available! If you want to do some tedious operation there is probably software specifically written to make the task easier.

Chapter 15

Hydraulics

15.1. Introduction

Electric motors and solenoids are widely used where an object is to be moved or a force applied. Electricity is not the only prime mover, however. Hydraulics (the use of liquids) and pneumatics (the use of air) offer a viable alternative in many applications. They are particularly useful where small size is important, precise control is required, a force (as opposed to movement) is needed or explosion hazards preclude the use of electrical machines.

This chapter discusses the use of hydraulics in industrial control. Pneumatic systems are described in Chapter 16.

15.2. Fundamentals

15.2.1. Pressure

Hydraulics and pneumatics are largely concerned with pressure in closed systems. It is therefore important to appreciate the precise meaning of the term 'pressure'. It is defined as force per unit area i.e.

$$P = \frac{F}{A} \tag{15.1}$$

where F is the force, A the area acted on, and P the resultant pressure.

In Fig. 15.1 a ram is supporting a weight of 200 kg. The ram cross-sectional area is 50 cm^2 (about 4 cm radius) so a pressure of 200/50 or 4 kgf per cm^2 is transmitted into the fluid.

In the SI system, the unit of pressure is the pascal, defined as 1 newton per square metre. For historical reasons, however, most hydraulic systems in the UK and USA work in pounds per square inch (psi). The bar (100 kPa; approximately 1 kgf/cm^2 or 14.5 psi) is also becoming common. Pressure in hydraulic systems is measured with respect to atmosphere (called gauge pressure). Strictly a g suffix (e.g. psig) should be used. Units of pressure are discussed further in Chapter 3.

The pressure of 4 kgf/cm^2 in Fig. 15.1 is transmitted uniformly through the confined fluid, and acts with equal force on every unit area of the surface of the system. The base, area 600 cm^2, experiences a downward force of $600 \times 4 =$ 2400 kgf. The top of the side chamber, area 300 cm^2, experiences an upward force of 1200 kgf.

Very large pressure can be developed by applying relatively small forces to small areas. A 10 kg force applied to the 3 cm^2 cork in a filled bottle will develop a pressure of 3.3 kgf/cm^2. A typical bottle has a bottom of area 30 cm^2 and

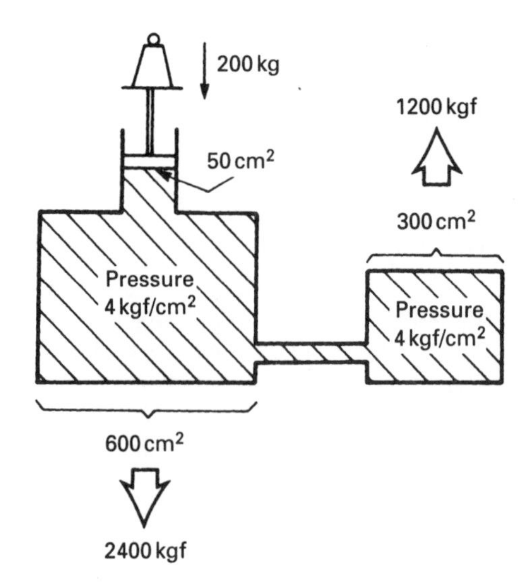

Fig. 15.1 *Pressure and force.*

experiences a force of 100 kgf – which could easily shear the bottom off the bottle.

The equality of pressure in a confined fluid is dignified by the name 'Pascal's law', and is the basis of hydraulics.

In Fig. 15.2a, a piston of surface area 2 cm^2 is linked hydraulically to a larger piston of area 500 cm^2. If a force of 8 kgf is applied to the smaller piston, a pressure of 4 kgf/cm^2 will develop in the fluid. This will be transmitted to the larger piston, which will experience an upward force of $4 \times 500 = 2000$ kgf. The input force of 8 kgf has been magnified to an output force of 2000 kgf. This force magnification, which is the basis of hydraulic presses and jacks, is the ratio of the areas of the output and input pistons.

Energy must, however, be conserved. Suppose the force causes the smaller piston to move down 10 cm. A volume of fluid $10 \times 2 = 20$ cm^3 is transferred from left to right. This causes the large piston to rise just $20/500 = 0.04$ cm. Although there is a force magnification, there is a movement reduction by the same factor.

Work is defined as the product of force and distance moved, and energy as a capacity for performing work (see Section 15.2.3). As the force magnification factor and the displacement reduction factor are equal, there is no net system gain or loss of energy (other than that lost by friction). There is an obvious analogy to the pulley of Fig. 15.2b, the lever of Fig. 15.2c and the gears of Fig. 15.2d. In each case a small input force is translated to a larger output force, but displacement is reduced by the same factor, giving conservation of energy.

15.2.2. Volume, pressure, flows and movement

Hydraulic pumps (described further in Section 15.4) deliver a constant flow regardless of output pressure (and are accordingly known as positive displacement

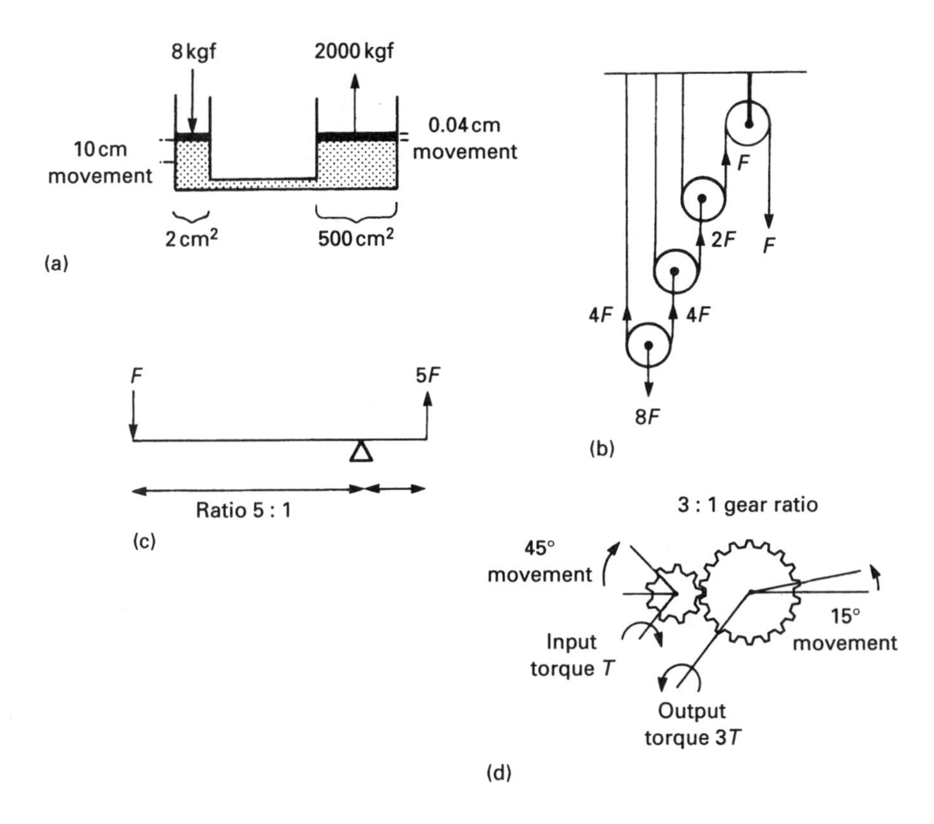

Fig. 15.2 *Examples of mechanical advantage. (a) Hydraulic. (b) Pulleys. (c) Lever. (d) Gear train.*

devices). In electrical engineering terms they are constant current rather than constant voltage sources. A pump specification, for example, could quote a delivery of 30 litres per minute, with a maximum working pressure of 75 bar. This does *not* mean that the pump will develop a pressure of 75 bar in a given system. As we will shortly see, the system pressure is determined by other components. A pump driving fluid into a closed system without some form of pressure regulation will cause a continuing pressure rise until the pump or the piping fails. The pump *creates* the pressure, but does not directly determine the value provided that the pump output meets the system's needs.

Figure 15.3 is a simple hydraulic system, where a hydraulic ram is lifted by a hydraulic pump driven by an electric motor (it could, for example, be a car jack in a garage). Suppose there is a load of F kgf, and the piston area is A cm^2. With the motor stopped, the pressure indicator will show F/A kgf/cm^2.

The pump is rated at Q cm^3/s. If the motor is started, Q cm^3 of fluid will be transferred to the ram each second. As a practical aside, it is not desirable to start a hydraulic pump directly on load; this is a 'thought experiment' we are considering. The speed of advance is simply:

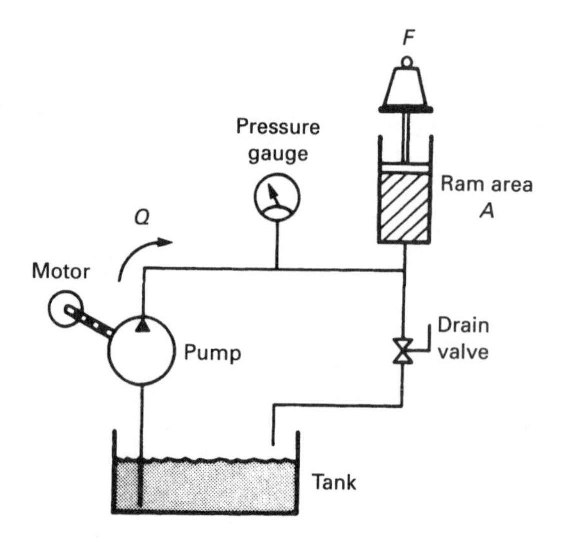

Fig. 15.3 *Pressure, volume and velocity.*

$$\text{velocity} = \frac{\text{volume transferred in unit time}}{\text{ram area}} \tag{15.2}$$

The ram therefore moves at Q/A cm per second.

The ram, however, is still supporting the load F, so the system pressure remains constant at F/A kgf/cm^2. There is no change of pressure when the pump is running (neglecting pressure drops caused by fluid flow in the connecting pipes). Similarly if the pump is stopped and the drain valve opened, the ram and load will fall but the fluid pressure will still be F/A kgf/cm^2.

Let us suppose the drain valve is left open accidentally, and the pump is started. Let us assume also that R cm^3/s flows out of the valve. Regardless of this leak the pump will still deliver its Q cm^3/s, giving a net transfer to the ram of $(Q - R)$ cm^3/s. The system pressure will still, however, be F/A kgf/cm^2 and if $Q > R$ the ram will still rise, at a reduced speed of $(Q - R)/A$ cm per second. If $R > Q$ the load will fall, but the pressure will be unchanged.

Fluid flow in hydraulic pipes should be non-turbulent (called laminar flow). Turbulent flow is undesirable on several counts: it creates wear, it is noisy and it causes friction which in turn lowers the system efficiency. The lost energy appears as heat which must be removed by oil coolers.

In general, hydraulic fluid flows should be kept around 3 ms^{-1} (10 fps) to avoid turbulence. Flow velocity is given by flow rate/pipe area. Note that the area is the *inside* area of the pipe, and the area is proportional to the square of the inside pipe radius.

Flow, particularly through restrictions, causes a pressure reduction as illustrated in Fig. 15.4a. The pump and associated (unshown) pressure regulating

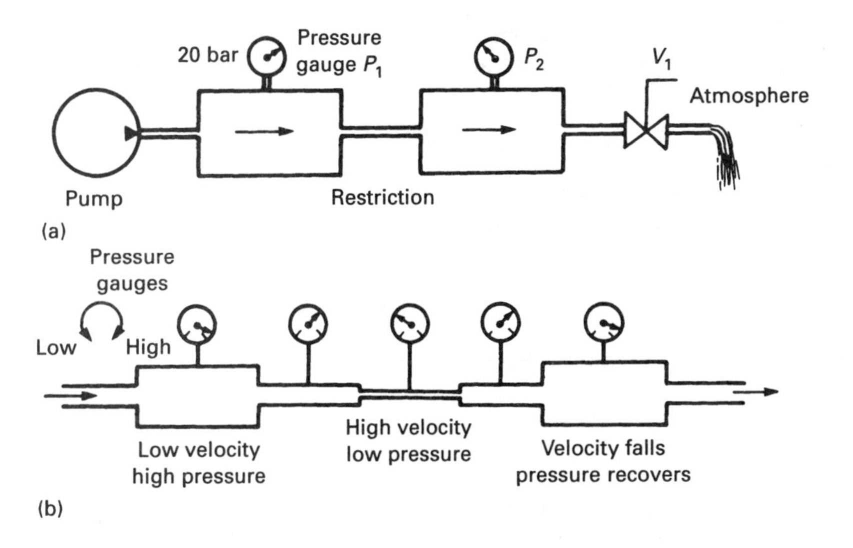

Fig. 15.4 *Pressure drops in a system. (a) Pressure drop caused by a restriction. (b) Velocity/pressure relationship.*

components maintain a constant pressure in the left-hand chamber of 20 bar. With the valve V_1 closed there is no flow, so by Pascal's law P_2 must also be 20 bar and there is equal pressure at all points.

If V_1 is cracked open, fluid will flow from the pump, and a pressure drop – say, 3 bar – will be created across the restriction. P_1 will remain at 20 bar, but P_2 will fall by 3 bar to 17 bar. As the valve is opened further, the flow and the pressure drop will increase, lowering P_2 further. Calculation of pressure drops across restrictions is analogous to orifice plate calculations described in Section 5.2. The calculations are, however, complicated by the laminar flow requirements for hydraulic fluid.

Energy in a hydraulic system is conveyed as a mixture of potential energy (pressure) and kinetic energy (mass flow). As flow velocity increases, pressure falls. In a system not venting directly to atmosphere and with changing diameters as in Fig. 15.4b, pressure will rise and fall as a complex function of the inverse of flow.

Work is done whenever a force is exerted through a distance, i.e.:

$$\text{work} = \text{force} \times \text{distance moved} \tag{15.3}$$

The SI unit of work is the joule (J) and the British (fps) unit is the foot pound.

Mechanical systems are more concerned with the *rate* of doing work (how fast a load is lifted by a crane, for example). The rate of doing work is called power, which is defined as:

$$\text{power} = \frac{\text{work}}{\text{time}} = \frac{\text{force} \times \text{distance}}{\text{time}} \tag{15.4}$$

The SI unit is the watt (joules per second). The British unit is the horsepower (hp), defined as 550 foot pounds per second. One horsepower is 746 watts (0.746 kW).

It follows that power in a hydraulic system is related to flow rate and system pressure, i.e.:

$$\text{power} = K \times \text{flow rate} \times \text{pressure} \tag{15.5}$$

where K is a constant dependent on the units being used. If British units are used, with flow rate in gallons per minute (gpm) and pressure in psi:

$$\text{power} = \frac{\text{gpm} \times \text{psi}}{1714} \text{hp} \tag{15.6}$$

In SI units with flow rate in litres per minute (lpm):

$$\text{power} = \frac{\text{lpm} \times \text{bar}}{600} \text{kW}$$

15.3. Pressure regulation

Figure 15.3 controlled the ram extension by starting and stopping the pump. This is not a particularly convenient (or desirable) control system. Figure 15.5a is one possible attempt at a solution. The pump is kept running all the time, and V_1 is opened to raise, and V_2 to lower, the piston. With V_2 open the piston falls under gravity and fluid is returned to the tank. Unfortunately, when the piston is stationary or being lowered, the pump is driving into a closed system. As explained earlier, this will lead to a theoretically infinite pressure rise; in practice either the pump or the piping will fail.

Figure 15.5b shows a more practical solution. A pressure regulating device is connected from pump output back to the tank. This device will be closed for low pressure and opens when some preset pressure is reached. If Figure 15.5b requires a pressure of 40 bar to raise the load, a suitable setting for the regulator could be about 50 bar.

With V_1 closed, the regulator will open and the pressure indicator will show 50 bar. With V_1 open, the system pressure will fall and the indicator will show the 40 bar necessary to move the load. Note that with V_1 closed the *entire* pump flow will go through the pressure regulator. The pressure regulator is called a relief valve which can be considered to act as in Fig. 15.5c. Under low pressure conditions the spring keeps the ball seated, blocking the flow. When the pressure rises sufficiently high the ball lifts off its seat, allowing fluid to pass. The spring tension (usually adjustable) and the ball cross section determine the pressure at which the valve starts to open.

15.4. Pumps

15.4.1. Introduction

The function of a hydraulic pump is to provide an adequate volume of fluid to the rest of a hydraulic system. The pump is usually driven by a constant speed electric

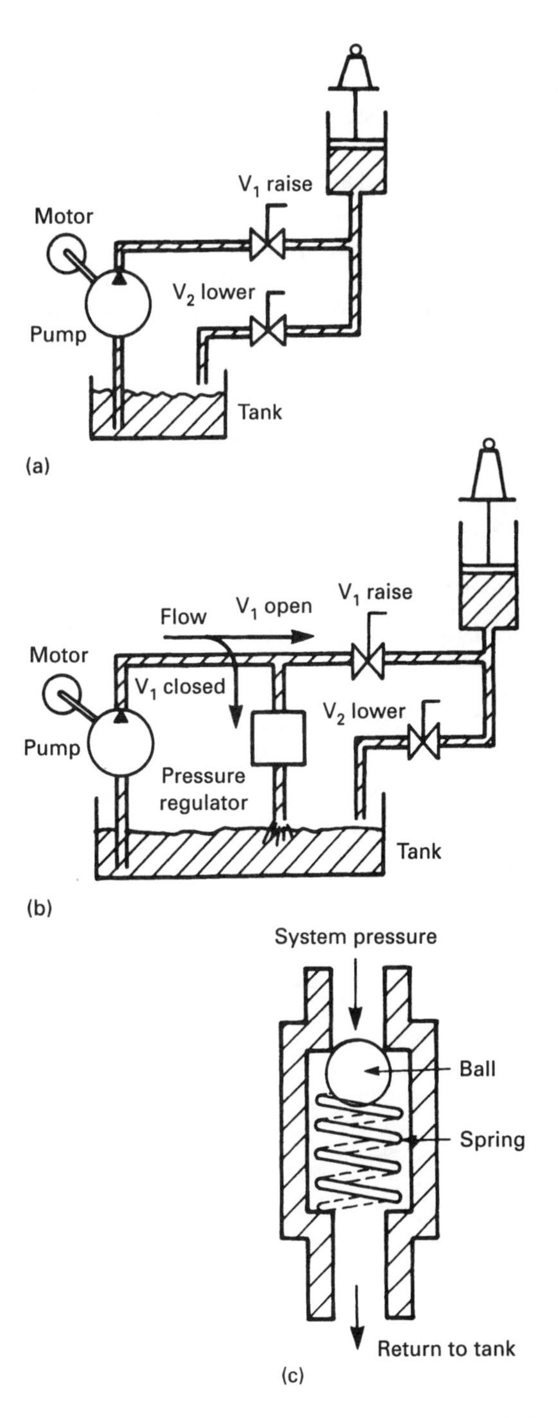

Fig. 15.5 *Pressure regulation. (a) System without pressure regulation. (b) System with pressure regulation. (c) Simple pressure regulator.*

motor. As explained earlier, pumps for hydraulic systems move a fixed volume of fluid per revolution regardless of outlet pressure, and are consequently known as positive displacement, or hydrostatic, pumps. Non-positive displacement devices such as the centrifugal pump of Fig. 15.6a and the propeller of Fig. 15.6b are called hydrodynamic pumps and are used purely to shift fluids from one location to another.

The symbols for a pump, motor and tank are shown in Fig.15.7. The arrow shows the direction of flow (compare this with the rotary motor symbol of Fig. 15.31). Figure 15.7a also illustrates the fact that in many applications a pump must lift fluid a height '*h*' from the tank to the pump inlet. To achieve this, the pump must be capable of creating a partial vacuum at its inlet.

There are, not surprisingly, limits to *h*. In theory, if the pump could create a perfect vacuum, atmospheric pressure could support a column of oil about 9 metres (30 feet) high. In practice far lower lifts are used. Liquids tend to produce vapour as pressure falls below atmospheric, creating bubbles in the fluid

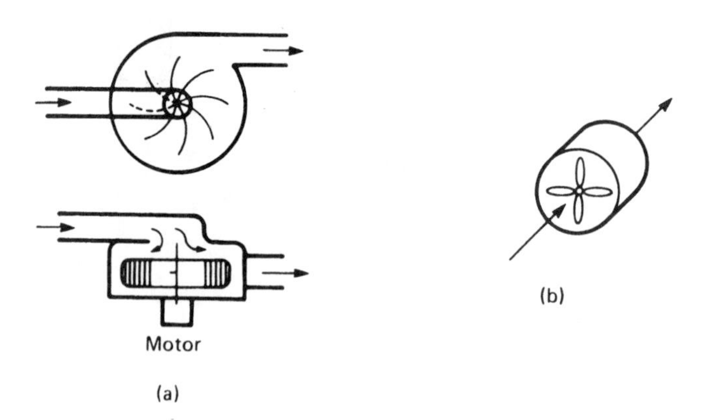

Fig. 15.6 *Non-positive displacement pumps. (a) Centrifugal blower. (b) Propeller.*

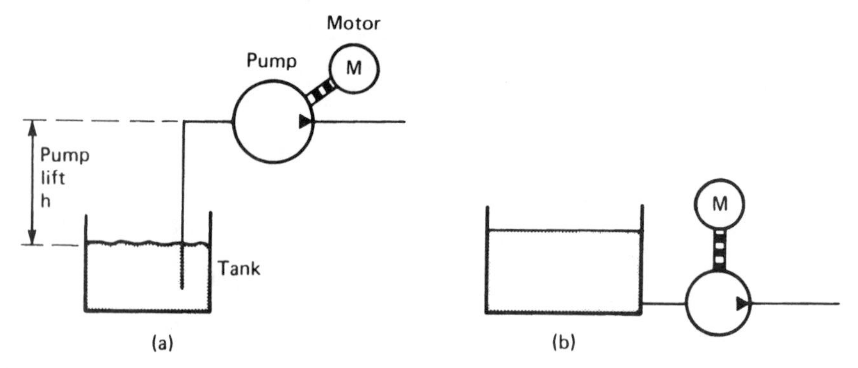

Fig. 15.7 *Pump to tank connection. (a) Definition of pump lift. (b) Self-priming pump.*

which collapse abruptly at the transition from low to high pressure inside the pump. This is known as cavitation and causes severe wear inside the pump. Similar effects occur from entrapped air and water vapour in the fluid. In general lifts of less than 2 metres (about 6 feet) are recommended, and the self-priming arrangement of Fig. 15.7b is often used (provided care is taken to avoid passing any sludge from the tank bottom to the pump).

Cavitation is also induced by high flow velocities. Pump inlet lines will always handle total pump flow, and consequently have the maximum flow in any hydraulic system. Inlet lines should be as large as possible to reduce flow velocity and minimise cavitation and frictional losses.

Pumps are specified by their delivery (in gallons per minute, say) and their maximum working pressure at some fixed input shaft speed (usually 1500 rpm to allow use of a simple 3-phase 4-pole induction motor). Delivery rate is proportional to shaft speed, and is sometimes given in terms of the displacement for one rotation of the pump shaft. Pumps are often sold as a pump/electric motor assembly.

15.4.2. The gear pump

The gear pump is the simplest and most robust hydraulic pump, as it has only two moving parts and these are rotating at uniform speed. The principle is shown in Fig. 15.8.

A partial vacuum is created at the inlet as the gear teeth come out of mesh at the centre. This partial vacuum causes fluid to enter the inlet chamber. Fluid is now

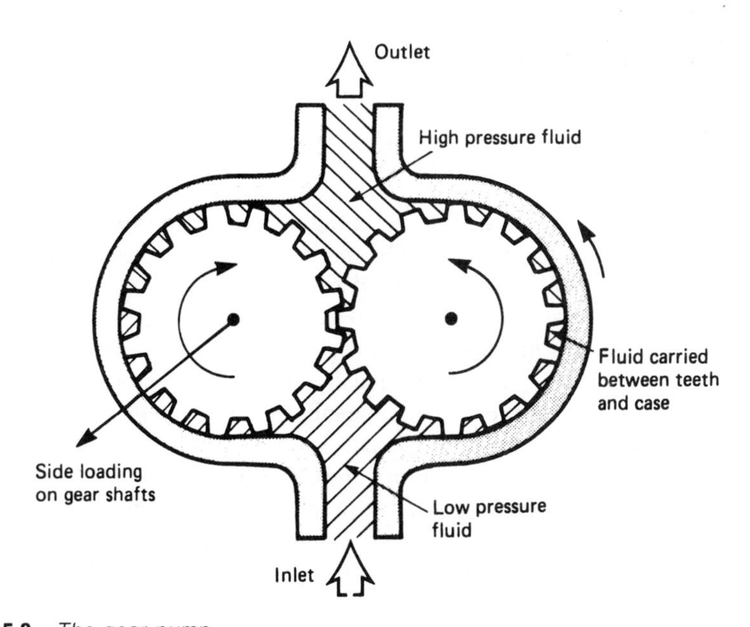

Fig. 15.8 *The gear pump.*

entrapped between the outer teeth and the housing and carried round to the outlet port where it is discharged. Displacement is determined by the volume between teeth and the number of teeth.

A close machined fit between the teeth ensures no oil leaks back where the teeth mesh at the centre. Close machined side plates are also required to avoid leakage over the gear faces. These are often designed as replaceable wear plates.

The difference in pressure between the inlet and outlet port causes a large side load to be applied to the gear shafts (at 45° to the centre line of the pump).

There are many variations on the gear pump principle; Fig. 15.9a is called an internal gear pump and Fig. 15.9b a lobe pump. Gear pumps are, in general, simple, robust and reliable. The side loading is usually the limiting factor.

15.4.3. Vane pumps

The basic principle of the vane pump is shown in Fig. 15.10a. A slotted rotor is fitted with several free moving vanes, and turns inside a cam ring. Hydraulic or spring pressure keeps the vanes in contact with the cam ring at all times.

The rotor and cam ring centres are offset, so oil from the inlet chamber enters the compartments between vanes at the top of the diagram, and is carried to the outlet port. The displacement is determined by the throw of the vanes and the rotor thickness.

As before, the difference in pressure between inlet and outlet ports causes a severe side loading on the shaft. This can be overcome by the balanced arrangement of Fig. 15.10b. Here two inlet and outlet ports are used with an elliptical cam ring and chamber. Both pairs of ports induce side loads, but they are equal and opposite, and hence cancel.

Vane pumps are obviously more complex than gear pumps, with many moving parts. Vane tips and the cam ring are also prone to wear, but this is mitigated by the hydraulic fluid itself acting as a lubricant. Seals are also required between the face plates and the rotors/vanes.

15.4.4. Piston pumps

Piston pumps are based on the principle of Fig. 15.11a. As the input shaft rotates the piston oscillates up and down. On the downward stroke, fluid is drawn through the check valve CV_1 to fill the space above the piston. On the upward stroke, fluid is discharged from above the piston, through check valve CV_2 to the outlet port. This simple arrangement, however, delivers a pulsing flow, and is unbalanced mechanically and hydraulically. Practical piston pumps utilise several pistons to even out the flow and balance the rotating components. Pump designs also eliminate the need for the check valves in Fig. 15.11b.

Figure 15.11b is called a radial piston pump. The cylinder block carries several pistons and rotates, off centre, inside the pump casing. The pistons are kept in contact with the casing at all times by springs, hydraulic pressure or mechanical linkage. The inlet and outlet ports are at the centre of the pump, and open on to chambers.

As the cylinder block rotates, pistons are moving out in the region of the inlet

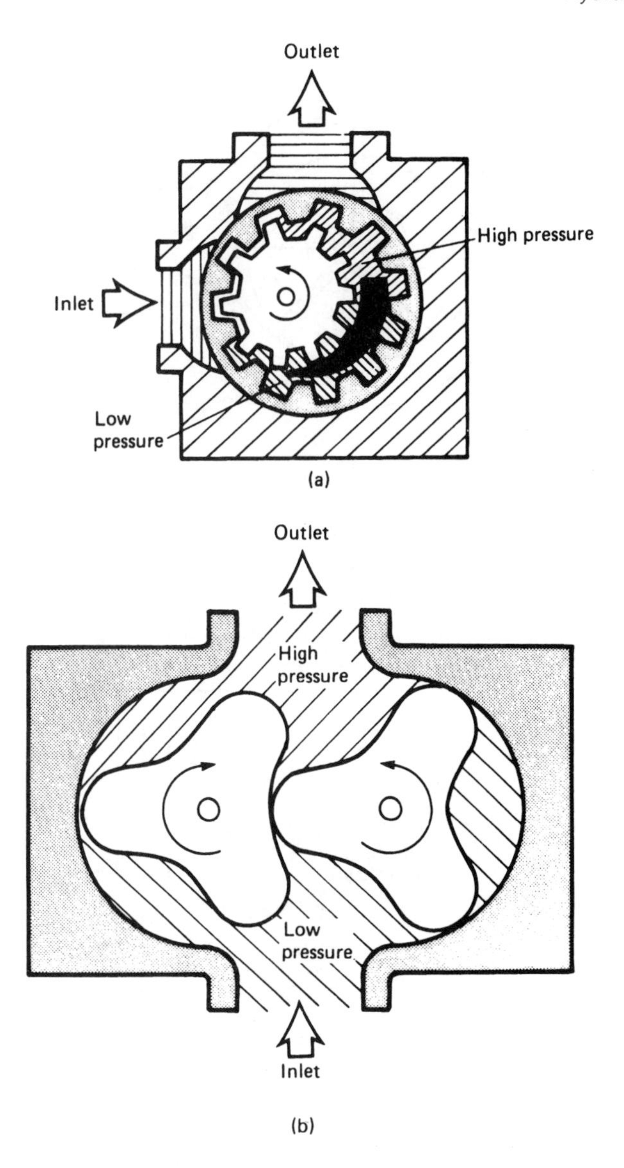

Fig. 15.9 *Variation on the gear pump. (a) Internal gear pump. (b) Lobe pump. Rotors are linked by external gears.*

chamber, and in adjacent to the outlet chamber. Fluid is thus conveyed from inlet to outlet port, without the need for separate check valves. Displacement is dependent on the number of pistons and their bore/stroke.

An alternative piston pump, called a swash plate or axial pump, is shown in Fig. 15.12a. The cylinder block is directly connected to the input shaft. The pistons are attached through ball and socket joints to a shoe plate which revolves

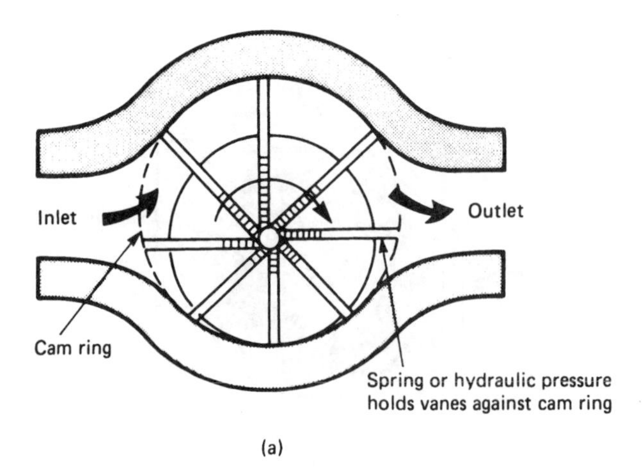

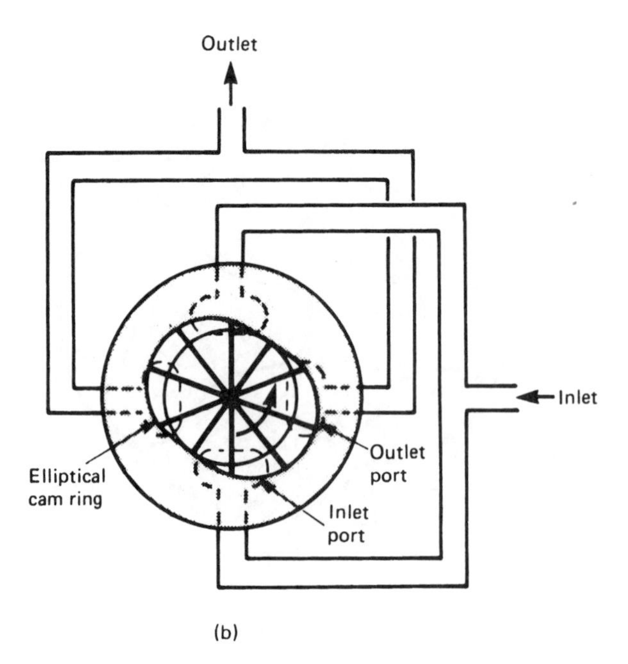

Fig. 15.10 *The vane pump. (a) Unbalanced vane pump. (b) Balanced vane pump.*

on a fixed, angled plate (called the swash plate). As the input shaft (and the cylinder block and pistons) rotates, the angled swash plate causes the pistons to reciprocate, transferring fluid from the inlet to the outlet port as before.

The angle of the swash plate determines the piston stroke giving displacement control from zero to maximum flow. If the swash plate goes beyond the vertical (zero flow) position the pump direction is reversed. The swash plate angle can be adjusted manually by a lever, or via electric or hydraulic actuators. Such pumps are said to be variable displacement.

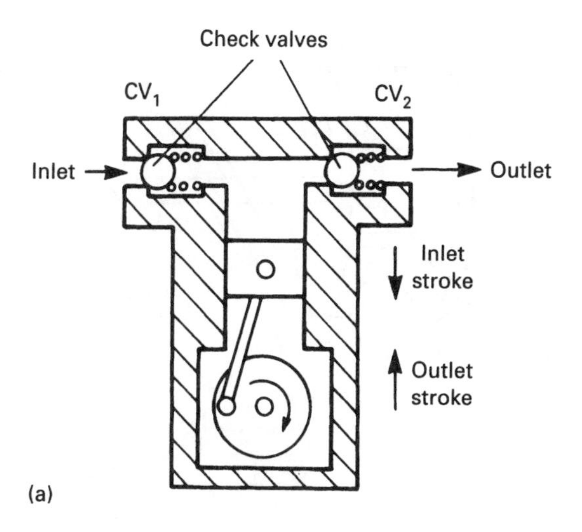

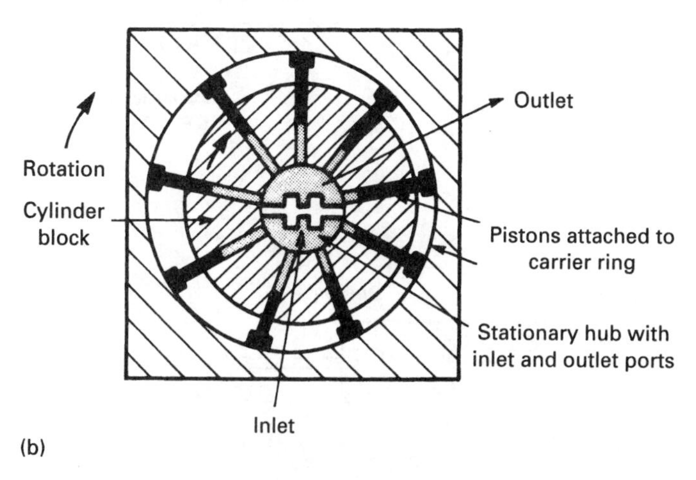

Fig. 15.11 *Piston pumps. (a) Single cylinder piston pump. (b) Radial piston pump.*

The bent axis pump, shown in Fig. 15.12b, is a variation on the piston pump. Here reciprocation of pistons is achieved by the angular difference between the drive shaft and the axis of the cylinder block. The cylinder block is driven by a universal joint to avoid side loads on the pistons. Again pump displacement can be controlled by varying the shaft angle.

15.4.5. Pump unloading

Figure 15.5b used a pressure relief valve to protect the system when no flow was required from the pump. This arrangement keeps the pump output at a high pressure, and by equation 15.6 the pump inlet power will remain high, even when no useful work is being done in the system. Apart from consuming electricity

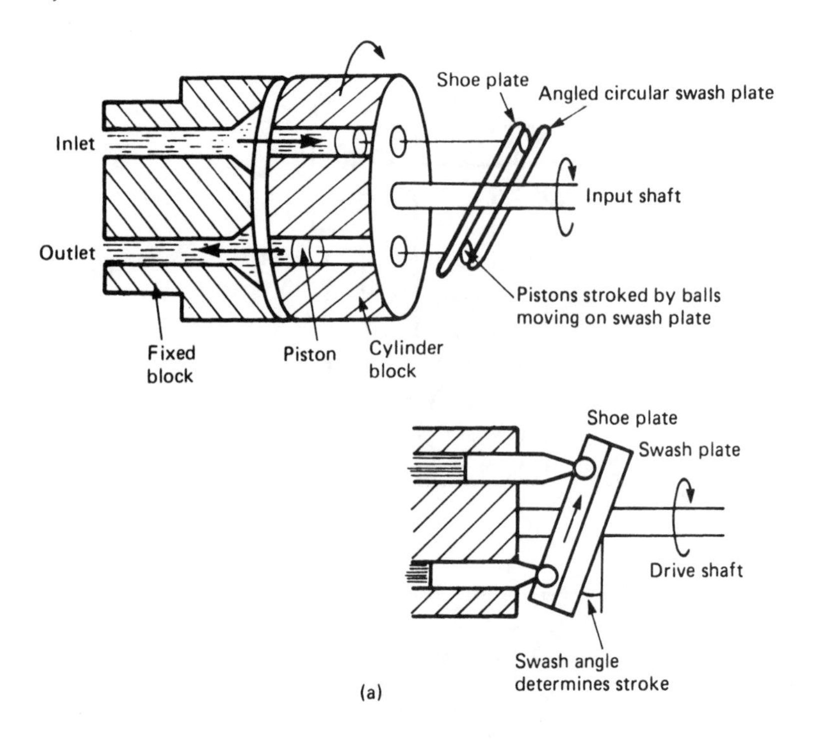

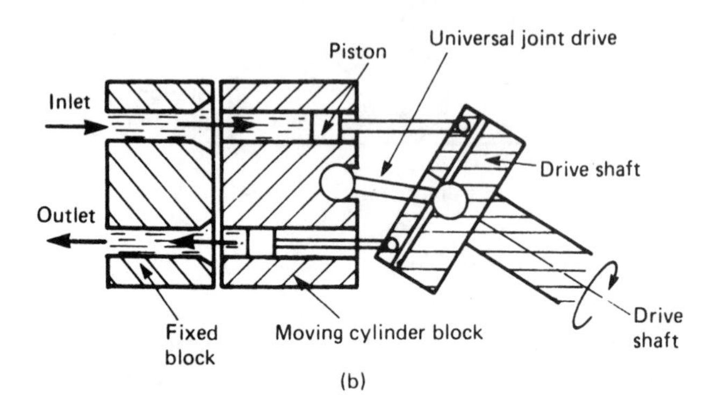

Fig. 15.12 *Variations on the piston pump. (a) Inline swash plate pump. (b) Bent axis pump.*

unnecessarily, the input power is converted to heat and leads to a rapid rise in fluid temperature.

If the hydraulic pump is required to be kept running at all times, but is only called upon to supply fluid intermittently, the arrangement of Fig. 15.13 may be used. A separate valve (called a loading or unloading valve, according to the sense of the control signal) goes from pump output back to the tank. When this

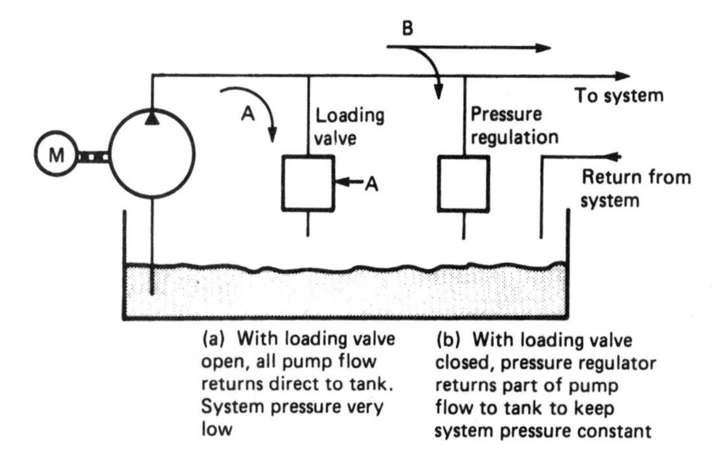

Fig. 15.13 *Pump loading valve.*

valve is opened by the control signal (electrical or hydraulic) the entire pump output is returned direct to the tank. The pump output pressure is minimal, and input power to the motor greatly reduced. When the valve is closed, the relief valve operates and sets systems pressure as normal.

Often the relief valve and unloading valve are combined by utilising a relief valve with remotely adjustable setting. Unloading is then obtained by setting a low relief pressure.

15.4.6. Combination pumps

Many hydraulic systems require two separate operating conditions, usually high flow/low pressure and minimal flow/high pressure. An example is a clamping vice. High flow/low pressure will be required to bring the vice jaws from their open position until they touch the object to be clamped. Once in contact, the jaws require high pressure but no flow.

These requirements can be provided by the two pump arrangement of Fig. 15.14. Pump P_1 is a high pressure low volume pump, and P_2 a high volume pump. RV_1 is a normal relief valve, and RV_2 a relief valve operated by a remote pressure (shown dotted and called a pilot line). RV_1 is set at, and RV_2 set lower than, the high pressure needed by the system.

In the high volume mode system, pressure is low as only friction needs to be overcome. RV_1 and RV_2 are both shut, and both P_1 and P_2 deliver fluid. When higher pressure is needed (e.g. when the vice jaws clamp) system pressure rises, causing RV_2 to open fully. Pump P_2 now unloads via RV_2, and check valve CV_1 isolates it from the rest of the system. The system pressure is determined by the relief valve RV_1 which is set at the required level.

The arrangement of Fig. 15.14 is very common, and complete assemblies called combination pumps (complete with two pumps, relief and check valves) are manufactured.

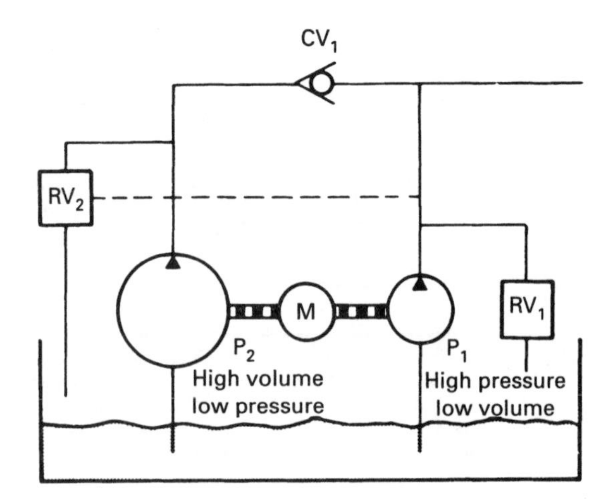

Fig. 15.14 *Combination pump.*

15.5. Hydraulic valves

15.5.1. Graphical symbols

Process control valve symbols have been used so far, but these are inadequate to describe the wide range of hydraulic valves (check valves, proportional valves, changeover valves, relief valves, etc.). A range of graphical symbols for hydraulic valves has evolved.

There are two basic types of valves. Infinite position valves (of which a relief valve is an example) can take any position between fully closed and fully open. Finite position valves (of which a directional valve is an example) can only be fully open or fully closed. Finite position valves generally switch flows between different ports.

The basic valve symbol is a square. Infinite position valves have a single arrow, as shown in Fig. 15.15a. The arrow shows flow direction and is generally drawn in the non-operated position. Control of the valve is shown by symbols on the side of the square. Figure 15.15b incorporates a spring push to right, and pilot pressure push to left. Pilot pressure increases the flow, and spring pressure reduces the flow as the pilot pressure is reduced.

Figure 15.15c therefore is the symbol for a relief valve (inlet pressure increases flow). The arrow on the spring shows adjustable tension.

Finite position valves generally have four ports, as in Fig. 15.16a. The pressure port P is connected to the pump, and the tank port T to the tank. The A and B ports are connected to the device being controlled. In Fig. 15.16b a reversing valve is connected to a ram. In the extend position, P and B are linked as are A and T (to return fluid to the tank from the space above the piston). In the retract position, P and A are linked and B and T.

Finite position valve symbols are constructed from squares, one for each

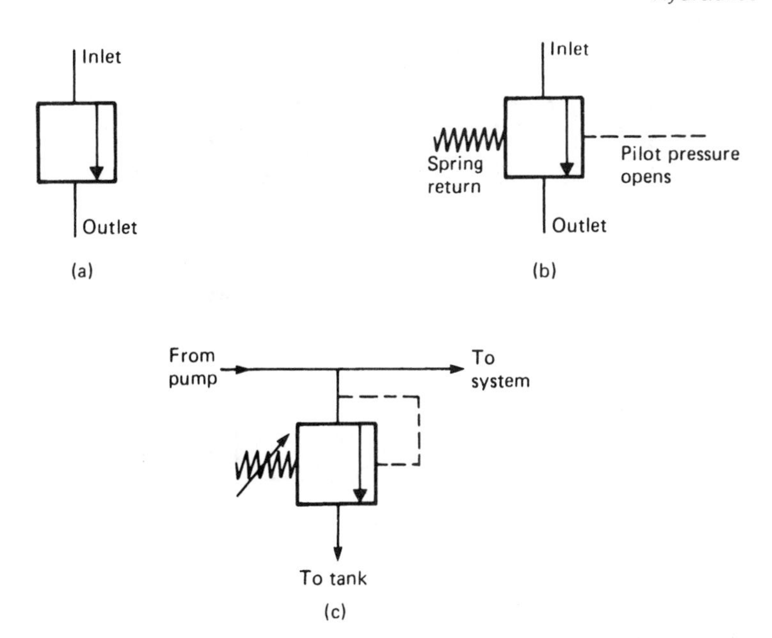

Fig. 15.15 *Valve graphical symbols. (a) Infinite position valve symbol. (b) Pilot pressure open, spring close, (c) Pressure relief value.*

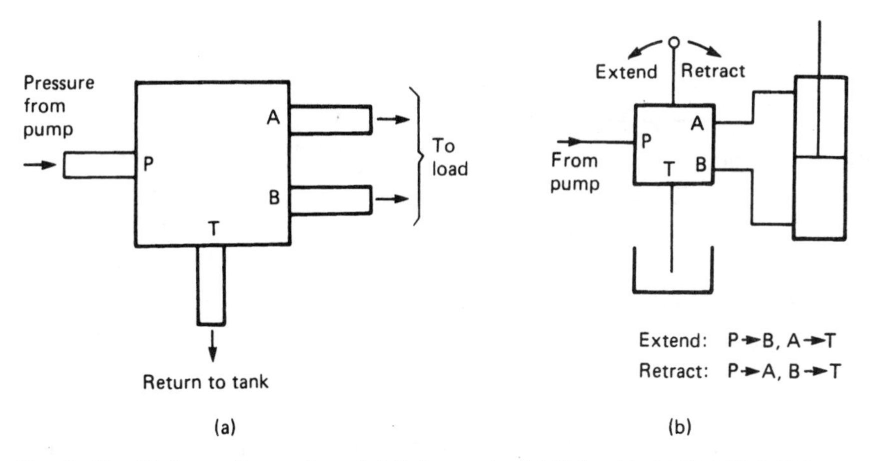

Fig. 15.16 *Finite position valves. (a) Valve ports. (b) Valve connected to a ram.*

possible valve position. Lines and arrows inside the squares show the port linking in each position. Figure 15.17 shows various examples: Fig. 15.17a is a reversing valve with two positions; Fig. 15.17b is a three-position reversing valve with centre off.

Actuating control is shown by symbols at the ends of the valve. Figure 15.18 shows various options. These are for the most part self-explanatory, with the

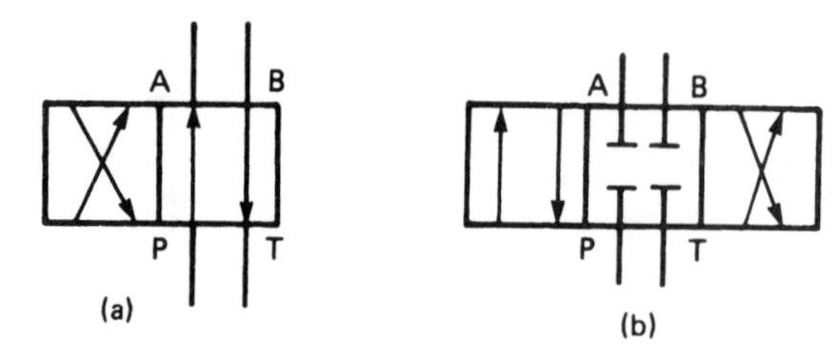

Fig. 15.17 *Finite position valves graphical symbols. (a) Two-position reversing valve. (b) Three-position reversing valve, centre off.*

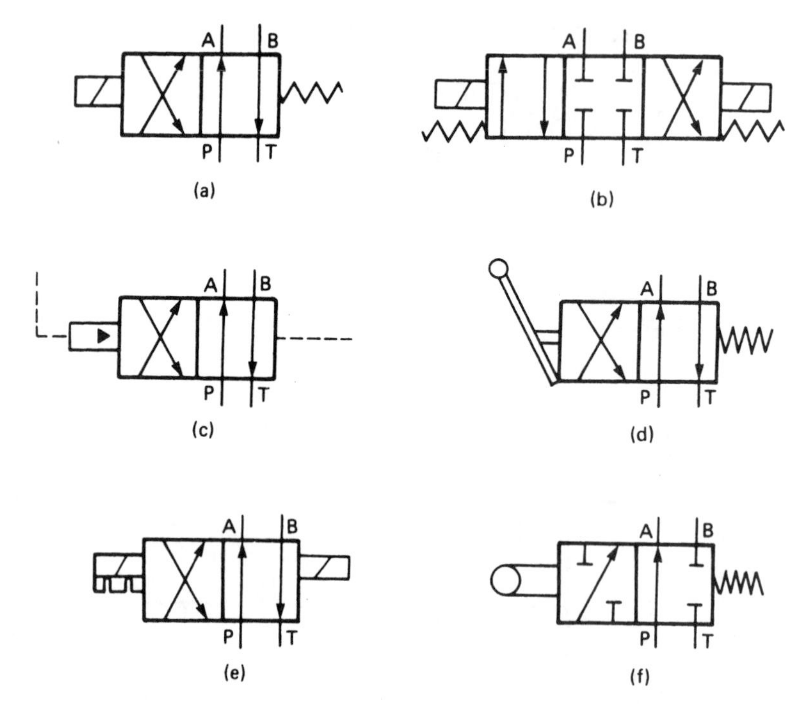

Fig. 15.18 *Actuator graphical symbols. (a) Solenoid operated spring return. (b) Three-position solenoid operated spring return to centre. (c) Pilot operated direct pressure return. (d) Hand operated spring return. (e) Solenoid operated with detent. (f) Cam operated, spring return, changeover valve.*

exception of Fig. 15.18e, with detents. Detent valves hold the last position. The solenoid in Fig. 15.18a must be continually energised to reverse the AB lines. The A solenoid in Fig. 15.18e need only be pulsed. The AB lines will then be reversed until the B solenoid is pulsed.

15.5.2. Check valves

The check valve only allows flow in one direction and as such is analogous to the electronic diode. It is used to block unwanted flow. In its simplest form it is an in-line ball and spring, as in Fig. 15.19a. Pressure from the left lifts the ball off its

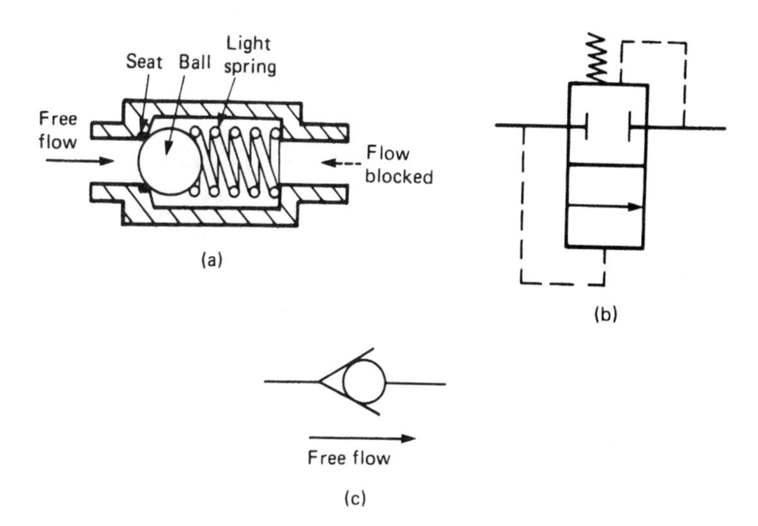

Fig. 15.19 *The check valve. (a) Construction. (b) Functional symbol. (c) Conventional symbol.*

seat, and unimpeded flow is permitted. Pressure from the right pushes the ball tight on to the seat, and flow is blocked. Graphically the valve is represented by Fig. 15.19b, which is rather unnecessarily complex. Usually the representation of Fig. 15.19c is used. There are many variations of check valves, but most are similar in principle to 15.19a.

In lifting ram applications similar to Fig. 15.20a, a changeover valve is used to raise or lower a load. Pump pressure is used to raise the load, and gravity to lower, with oil being returned to the tank. In the valve centre (stop) position, any leakage in the control valve will cause the load to creep down. A check valve in the pipe to the ram will prevent the creep, but prevent lowering when required. In this type of application a pilot operated check valve, shown in Fig. 15.20b, is required.

With no pilot pressure, the valve functions as a normal check valve, flow from A to B being permitted, and blocked from B. If pilot pressure is applied, the valve is open at all times, and flow is allowed freely in both directions. The symbol for a pilot operated check valve is shown in Fig. 15.20c. This device would be inserted in the line from the control valve to the ram. When raised the check valve permits flow to the ram. In the stop position, the check valve prevents creep. To lower, pilot pressure is applied to the check valve and 'down' selected on the control valve.

15.5.3. Relief valves

Relief valves are used for protection and pressure regulation in every hydraulic system. As explained earlier in Section 15.3, hydraulic pumps are positive displacement devices, and as such some external means is needed to prevent over-pressuring a hydraulic system.

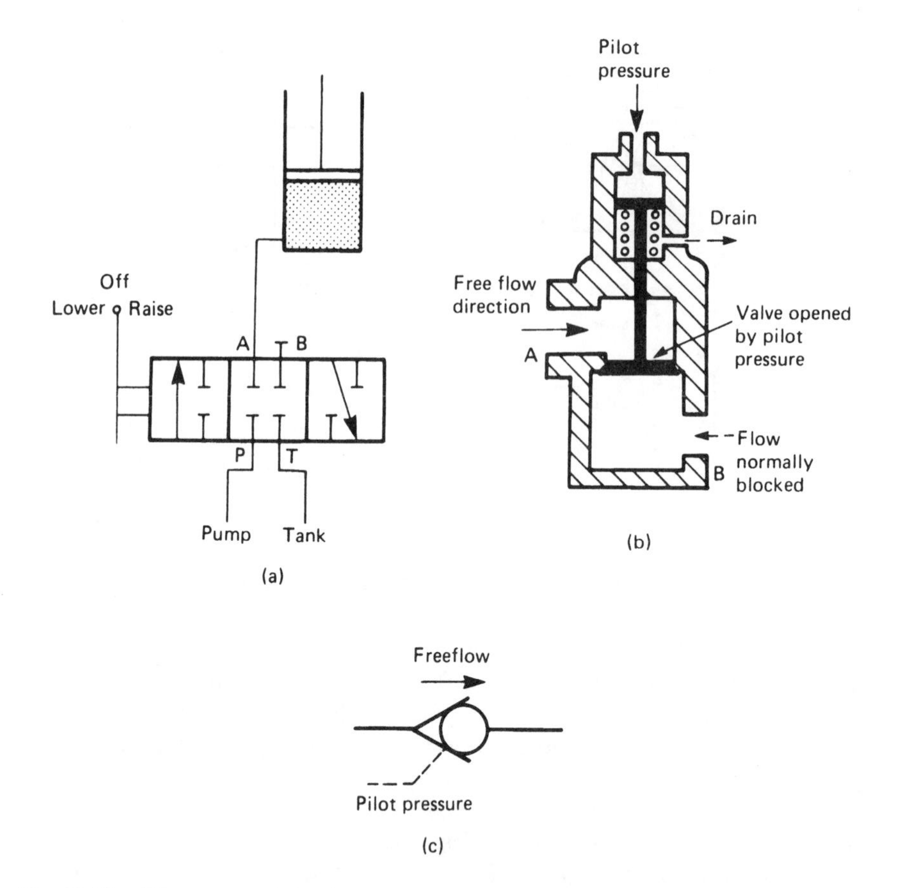

Fig. 15.20 *Pilot operated check valve. (a) Application requiring check valve. (b) '2C' type check. (c) Symbol.*

The simplest relief valve is the spring and ball arrangement of Fig. 15.21a. The spring holds the valve shut until a preset pressure (set by the adjustable spring tension) is reached, when the valve cracks open. If pressure increases further, flow increases further. The valve is therefore an infinite position valve, and has the symbol of Fig. 15.21b.

The pressure at which flow commences is called, not surprisingly, the 'cracking pressure'. There will also be, for a given valve, a pressure when full flow is obtained. Again not surprisingly, this is called the 'full flow pressure'. The difference is called the 'pressure override' and is an indication of the pressure regulation the valve will provide.

Pressure override is related to the spring tension in a simple relief valve. When a small, or precisely defined, override is required, a balanced piston relief valve (shown in Fig. 15.22) is used.

The piston in the valve is free moving, but is normally held in the lowered position by a light spring, blocking flow to the tank. Fluid is permitted to pass to

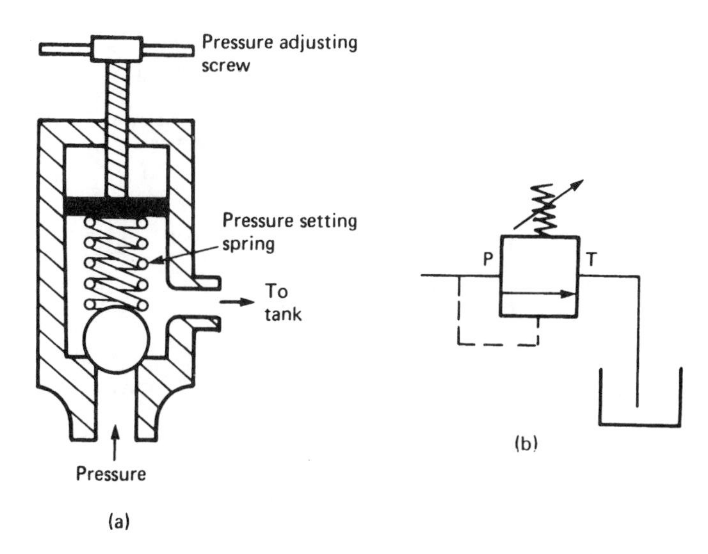

Fig. 15.21 *The relief valve. (a) Construction. (b) Symbol.*

the upper chamber through a small hole in the piston. The upper chamber is sealed by an adjustable spring loaded poppet. In the low pressure state, there is no flow past the poppet, so pressure on both sides of the piston is equal and spring pressure keeps the valve closed.

When the fluid pressure rises, the poppet cracks and a small flow of fluid passes from the upper chamber to the tank via the hole in the piston centre. This fluid is replenished by fluid flowing through the hole in the piston. With fluid flow, there is now a pressure differential across the piston, which is acting only against a light spring. The whole piston lifts, releasing fluid around the valve stem until a balance condition is reached. Because of the light restoring spring, a very small override is achieved.

The balanced piston valve can also be used as an unloading valve. The plug X is a vent connection, and if this is removed fluid will flow from the main line through the piston. As before, this will cause the piston to rise and flow to be dumped to the tank. Controlled loading/unloading can be achieved by the use of a finite position valve connected to the vent connection, as in Fig. 15.22b. When the solenoid is energised, pump output is directed direct to the tank.

15.5.4. Finite position (changeover) valves

The basic principle of finite position valves was described earlier in Section 15.5.1 and Fig. 15.16. Almost all have a pressure P connection, a tank T connection, and plant connected ports denoted A, B. On/off valves are achieved by blocking one port.

The commonest type is the spool valve shown in Fig. 15.23. The spool moves inside the valve body, and raised portions called 'lands' block ports to give the required operation. The valve illustrated changes over pressure between the A and B ports, the A port being selected with the spool to the left and the B port with

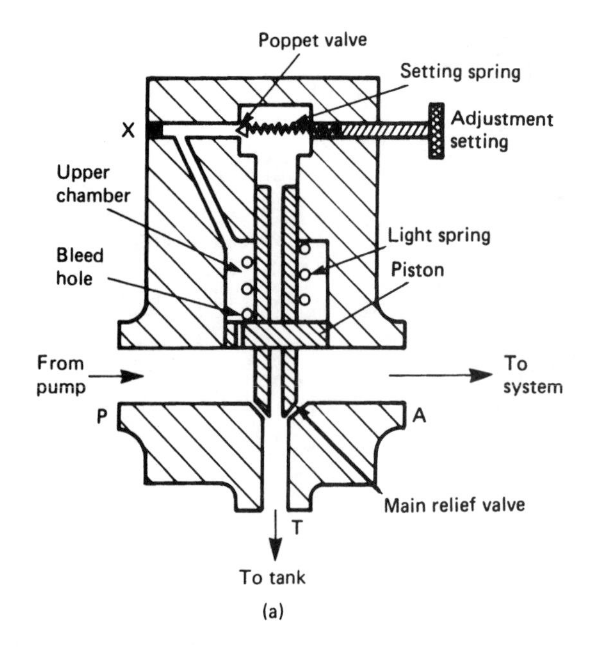

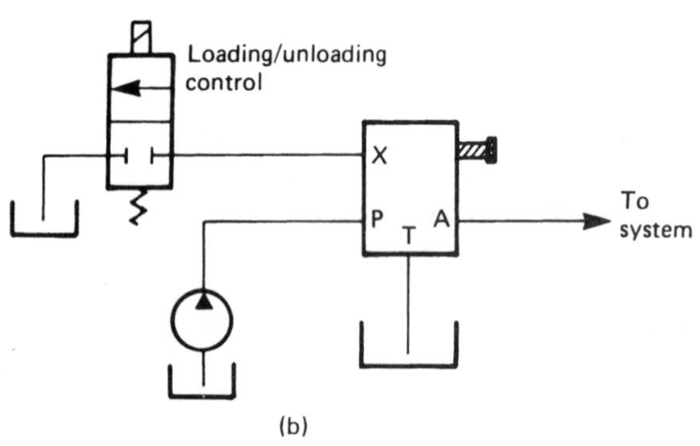

Fig. 15.22 *Balanced piston relief value. (a) Construction. (b) Used as unloading/regulation valve.*

the spool to the right. The tank connection is not used, and only serves to drain leakage from the valve.

Changes in valve operation are achieved by utilising spools with different land patterns, whilst maintaining the same valve body. Figure 15.24 shows a changeover valve using the same valve body as in Fig. 15.23. This approach obviously simplifies valve manufacture.

Three-position valves are constructed in a similar manner. Figure 15.25a shows a three-position changeover valve with centre off position. Three-position valves can be obtained with many different centre positions, some of which are shown in Fig. 15.25c. These different valve configurations are obtained by

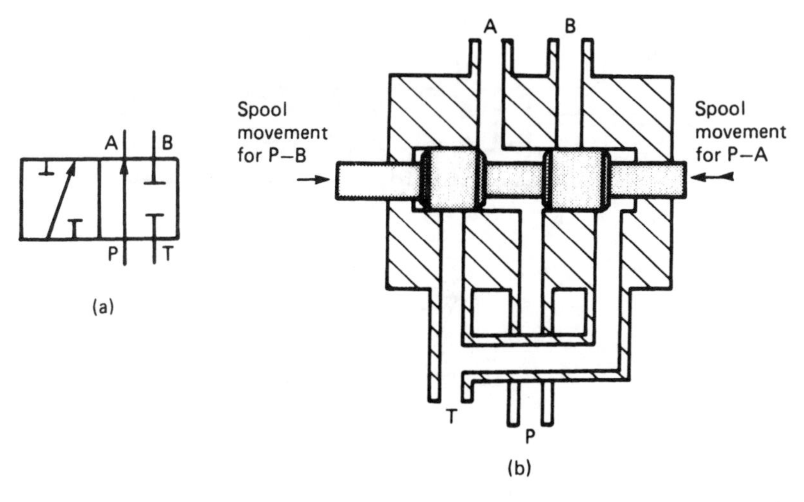

Fig. 15.23 *Two-way spool valve. (a) Symbol. (b) Construction.*

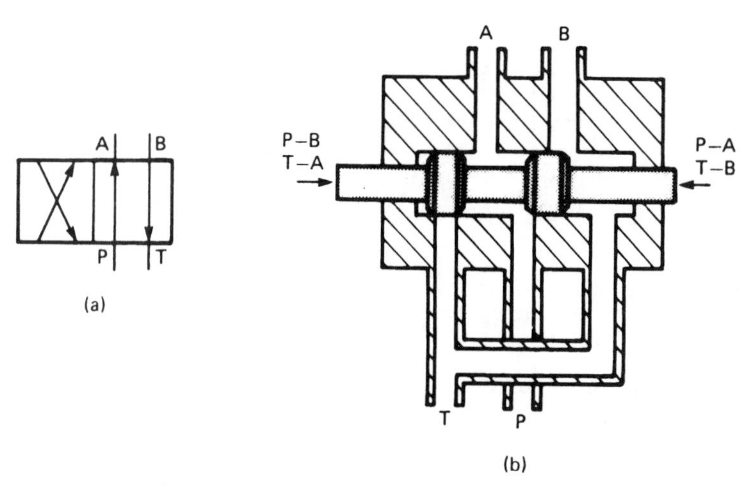

Fig. 15.24 *Four-way spool valve. (a) Symbol. (b) Construction.*

utilising the same valve body with different land patterns on the spools.

Spool movement can be achieved manually by a lever or striker, electrically by solenoids at each end of the spool, or hydraulically by pilot pressure applied to the spool end.

Valves for use at high pressure or high flows require a higher spool force than can be reasonably obtained from a solenoid. In these cases a two-stage valve is used, as in Fig. 15.26a. Solenoids operate a small valve which in turn applies pilot pressure to shift the large spool in the main valve. This arrangement is called pilot operation and is shown schematically in Fig. 15.26b.

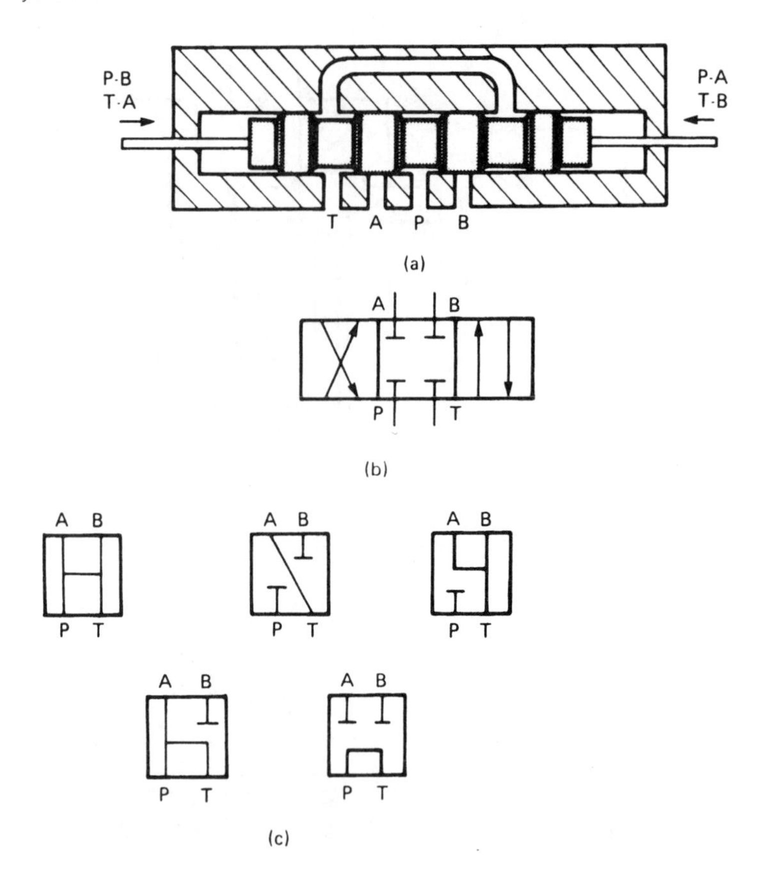

Fig. 15.25 *Three-position four-way valves. (a) Construction of centre off valve. (b) Symbol. (c) Common centre position connections.*

15.6. Linear actuators

Actuators are the output components of a hydraulic system. Linear actuators, which produce a straight line motion, are described in this section. Rotary actuators (motors) are described in Section 15.7.

All linear actuators are based on Fig. 15.27 and consist of a ram which is moved by the introduction of fluid to the enclosed space below the piston. Figure 15.27a is called a 'single-acting' cylinder; it is extended by hydraulic pressure, but returns under gravity which pushes fluid back to the tank. Single-acting cylinders must be mounted vertically unless spring return is employed.

Figure 15.27b uses hydraulic pressure to extend and retract the cylinder. This arrangement is known as a 'double-acting' cylinder, and has a power stroke in each direction. Examination of Fig. 15.27c shows that the area of piston available for the extend stroke is greater than for the return stroke by the area of the connecting rod. The extend stroke is therefore slower than the return stroke (for

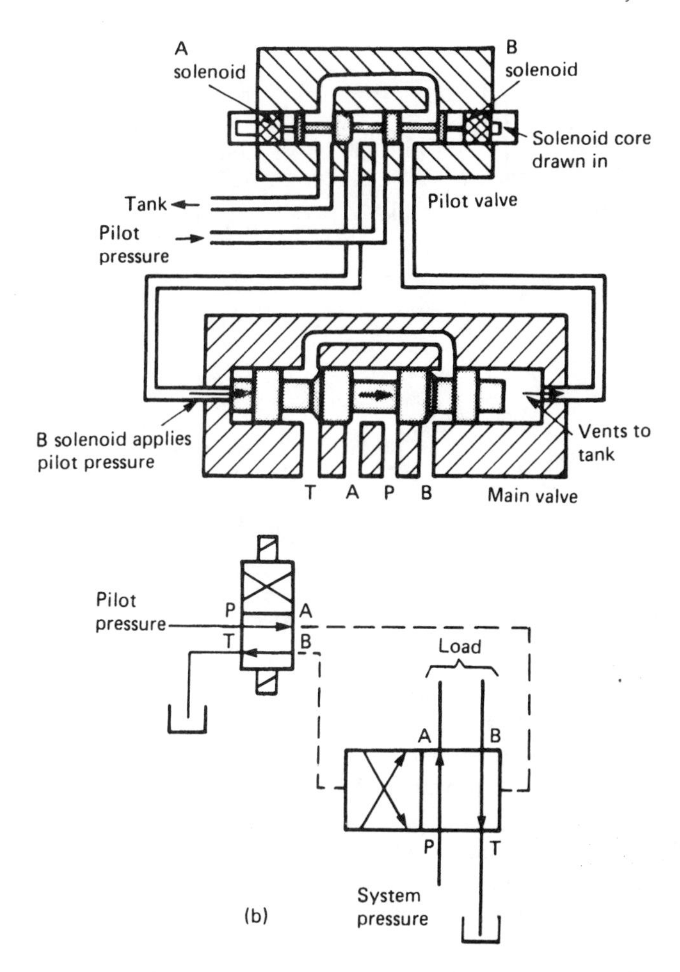

Fig. 15.26 *Pilot operated valve. (a) Construction. Power applied to B solenoid has moved pilot spool to left. This applies pilot pressure to left hand end of main spool, shifting spool to right. (b) Symbol.*

constant flow rate) but is capable of exerting a greater force. Such an actuator is often called a 'differential' cylinder.

The double rod cylinder of Fig. 15.27d has equal areas for extend and retract strokes, and is consequently called a 'non-differential' cylinder.

Cylinders are as simple as Fig. 15.27 suggests, consisting of a barrel, piston, rod, caps, glands and piston seals. They are classified by available stroke, piston area and maximum operating pressure. Force, speed, pressure and flow rate are related by:

$$\text{speed} = K \times \frac{\text{flow rate}}{\text{piston area}} \tag{15.7}$$

$$\text{force} = L \times \text{pressure} \times \text{piston area} \tag{15.8}$$

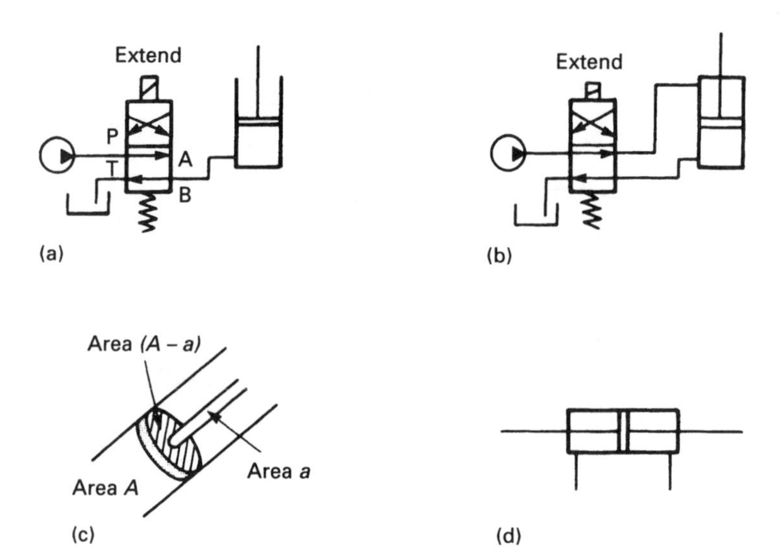

Fig. 15.27 *Linear actuators. (a) Single-acting cylinder. (b) Double-acting cylinder. (c) Effect of piston rod. (d) Non-differential cylinder.*

where K and L are scaling factors.

Stroke is obviously less than the overall piston length, and is often further reduced by internal bushing if side loads are expected. The rod pivots around the end cap, and if the piston travels to the end of the bore considerable side load magnification results. The internal bush reduces this magnification as shown in Fig. 15.28, but also restricts the stroke.

Where long strokes and minimal cylinder size are needed, telescopic cylinders as in Fig. 15.29 can be used. Unfortunately these are single acting and have poor tolerance to side loads.

Cylinder cushions as in Fig. 15.30 are often employed to absorb the shock as the piston reaches the end of the stroke. Progressive deceleration occurs as the plug enters the end cap and reduces the outlet flow. The check valve allows free flow and full pressure to start reverse movement.

15.7. Rotary actuators (hydraulic motors)

15.7.1. Introduction

A rotary hydraulic actuator is usually called a hydraulic motor, and can be used in the same applications as an electric motor. It does, however, have several notable advantages. Hydraulic motors are physically smaller than equivalent electric motors (albeit with the need to have a hydraulic power pack located somewhere close) and give far better control of speed and torque at low speeds.

The symbol for a hydraulic motor is shown in Fig. 15.31. The inverted arrow denotes a motor (compared with the pump symbol of Fig. 15.3). A motor accepts hydraulic fluid at a certain flow rate and pressure, and produces output shaft

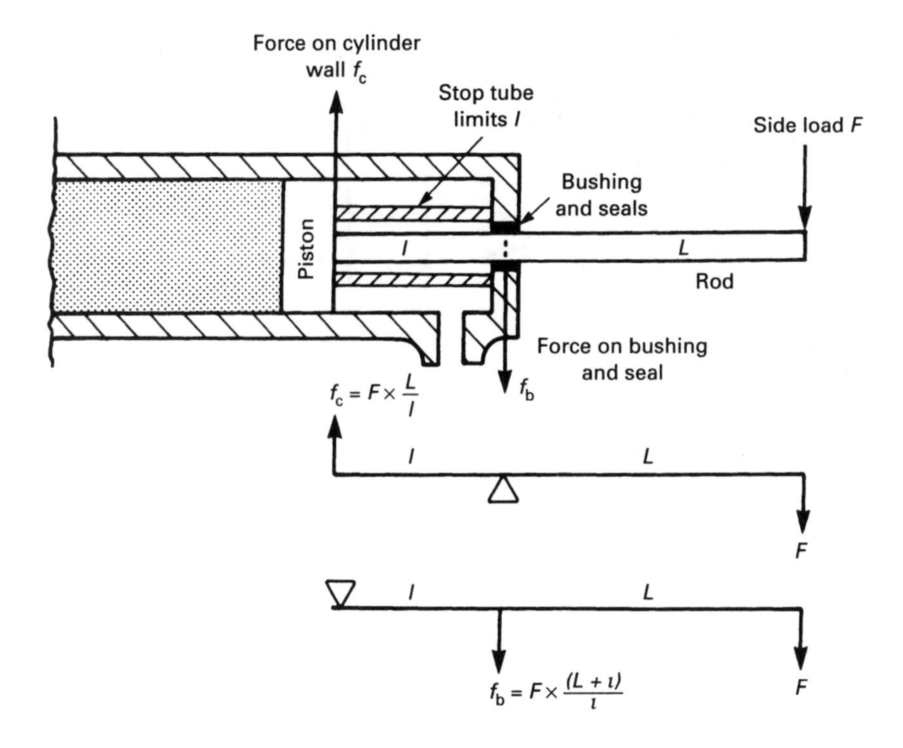

Fig. 15.28 *The effect of side loads.*

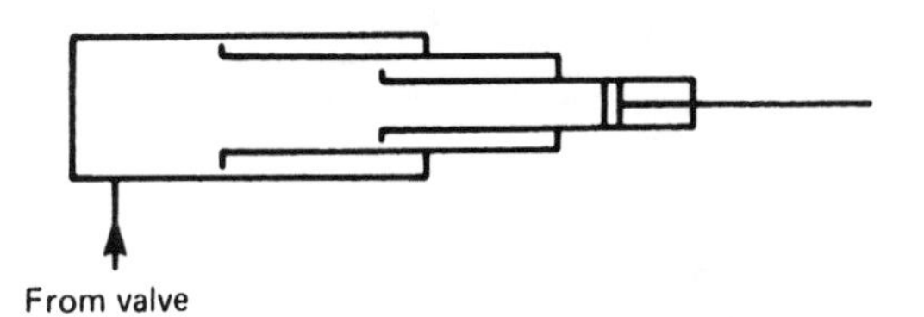

Fig. 15.29 *Telescopic cylinders.*

rotation and torque. Hydraulic motors are rated according to their displacement (i.e. the volume of fluid required for one rotation of the output shaft) torque rating and maximum working pressure. They are all related for any given motor:

$$\text{speed} = \text{flow rate} \times \text{displacement} \tag{15.9}$$

There is no relationship between speed and torque or pressure provided the mechanical load is constant.

$$\text{torque} = K \times \text{pressure} \tag{15.10}$$

where K is a constant for the motor called the 'torque rate' (e.g. a given motor could have a torque rating of 50 Nm per 10 bar).

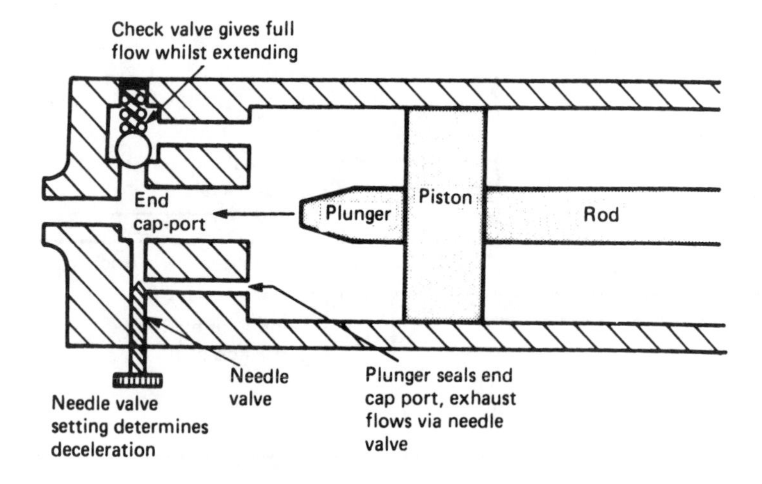

Fig. 15.30 *Cylinder cushioning.*

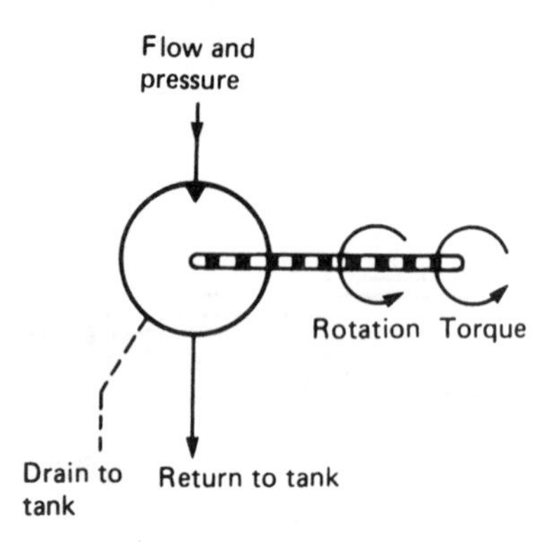

Fig. 15.31 *Graphical symbol for a rotary actuator.*

The 'torque rate' is related to the motor size, and motors with larger displacements require a lower operating pressure for a given torque.

It follows from equations 15.9 and 15.10 that speed is controlled by flow rate, and torque by pressure. These equations are, of course, theoretical, and in any practical system an allowance for losses (typically 30%) must be made. Flow control is discussed later in Section 15.8.1.

15.7.2. Practical motors

Electrical motors and generators have, in many cases, almost identical construction, and a given machine could function as a generator (mechanical power applied to shaft, electrical power obtained from terminals) or as a motor (electrical power applied to terminals, mechanical power obtained from shaft). It is not, therefore, surprising to find that hydraulic pumps and motors have almost identical construction.

Figure 15.32, for example, shows the gear and vane pumps rearranged as motors. The piston pumps of Fig. 15.12 can also be used as the basis of

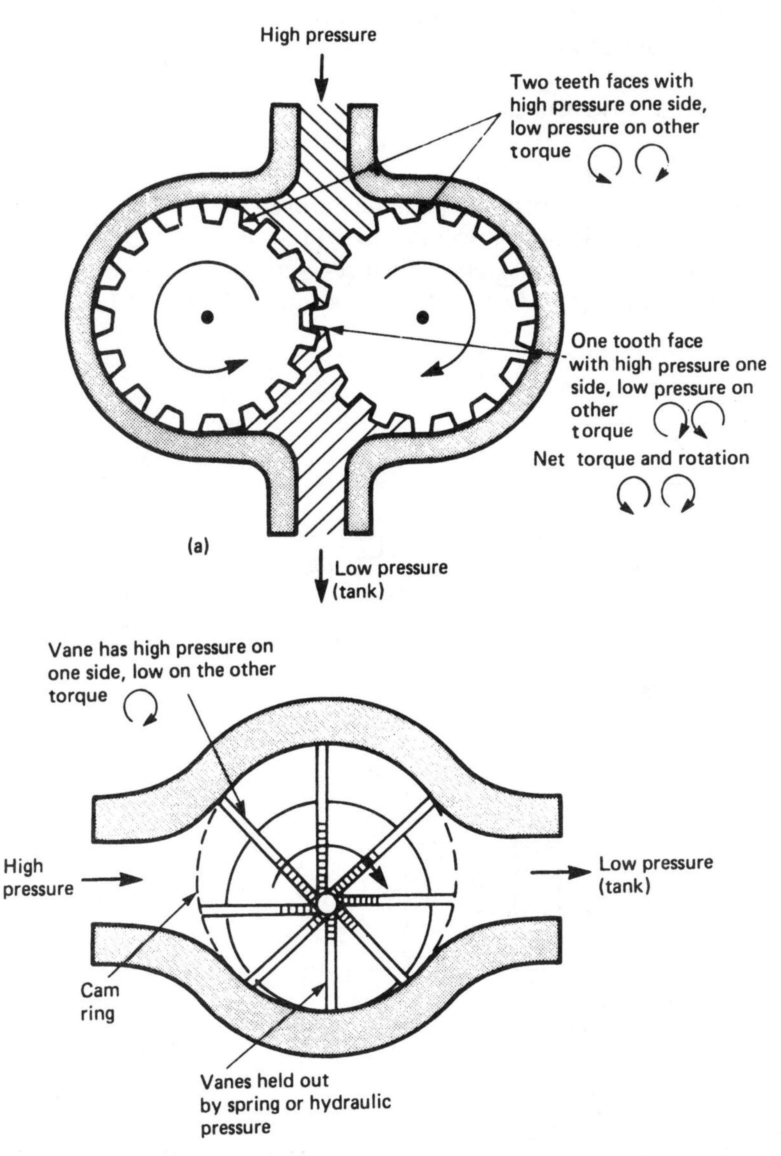

Fig. 15.32 *Hydraulic motors. (a) The gear motor. (b) The vane motor (usually the balanced construction of Fig. 15.10b is used).*

mechanically speed variable motors where the swash plate shaft angle varies the displacement and hence the speed (note that from equation 15.9 decreasing the displacement increases the speed for a given flow rate, but the decreased displacement reduces the available torque).

All hydraulic motors experience some internal leakage of oil. As this oil is essentially static, the pressure inside the casing would eventually build up to full line pressure with the possibility of internal damage to the motor. A drain line, shown dotted on Fig. 15.31, is therefore included to allow leakage oil to return to the tank, preventing internal pressure build-up.

15.8. Miscellaneous topics

15.8.1. Flow control

With linear and rotary actuators, speed is controlled by flow rate, and force (or torque) by pressure. Relief valves, described in Section 15.3, can be used for pressure regulation. This subsection discusses methods of flow control.

There are essentially three types of flow control, illustrated in Fig. 15.33.

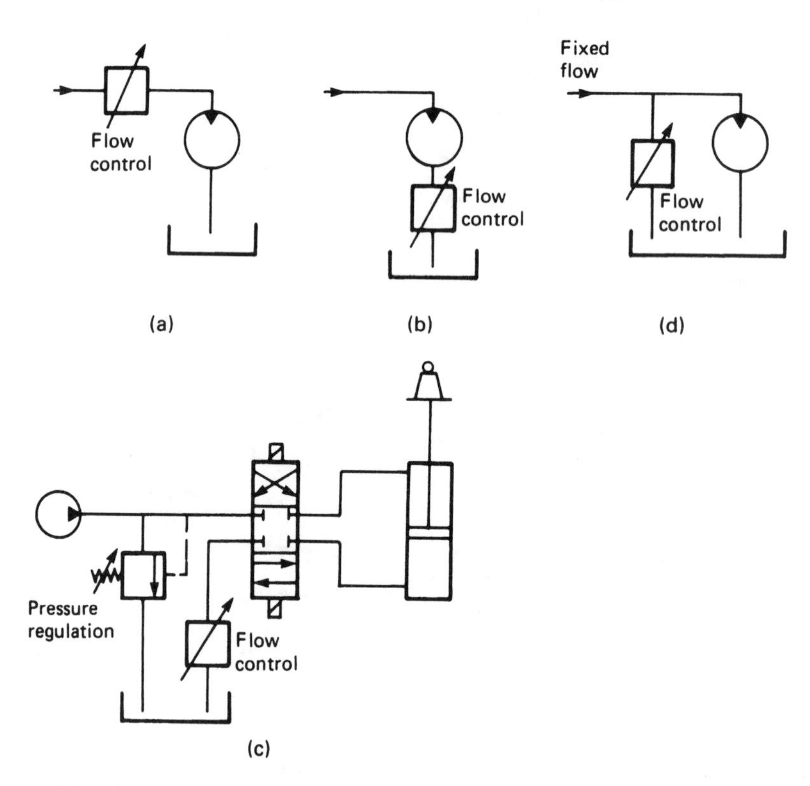

Fig. 15.33 *Flow control. (a) Meter in circuit. (b) Meter out circuit. (c) Piston speed control using meter out. (d) Variable bleed.*

'Meter in' controls the flow of fluid to the actuator (and is possibly the most obvious method of speed control). Meter in can only be used, however, when the load opposes the actuator. It cannot be used where the load could run away (controlling the descending speed of the cylinders on a car inspection ramp, for example). In these circumstances the 'meter out' circuit of Fig. 15.33b is used.

Where directional valves are used, the flow control can be placed on the pump, or tank, side of the directional valve. Figure 15.33c shows a circuit with meter out flow control connected to a lifting cylinder.

An alternative circuit sometimes encountered is the bleed off circuit of Fig. 15.33d where the flow to the actuator is controlled by bleeding off excess fluid. This method of speed control is not as accurate as meter in/out circuits. Where a single actuator and pump are linked, speed control can also be achieved by varying the pump displacement.

Flow control is achieved by applying a restriction in a pipe, and consequently has the symbol of Fig. 15.34a. Simple flow control devices are just an adjustable

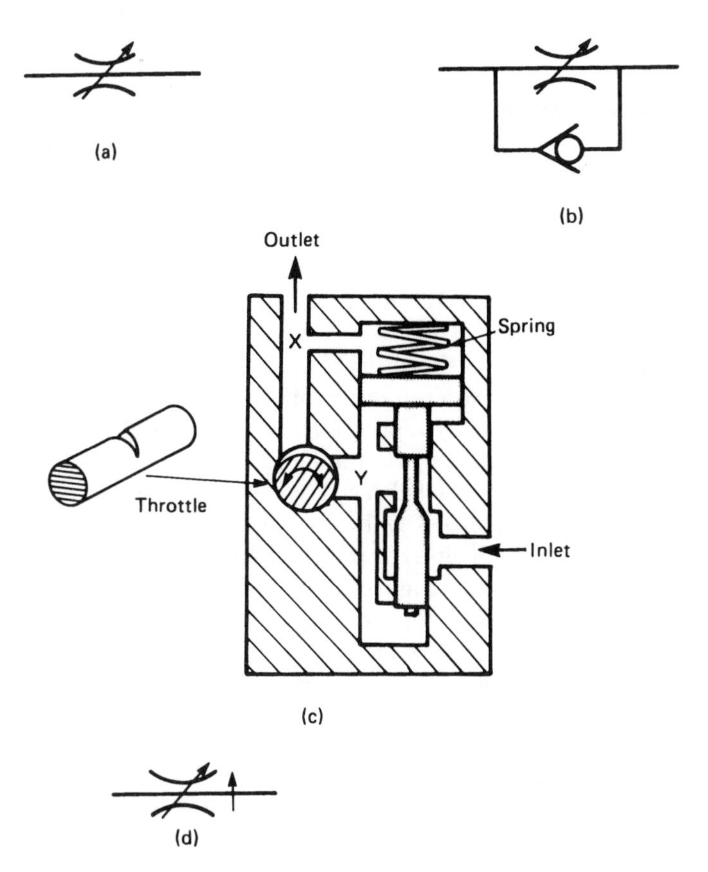

Fig. 15.34 *Flow control valves. (a) Flow control graphical symbol. (b) Uni-directional flow control. (c) Pressure compensated restrictor flow control valve. (d) Symbol for pressure compensated flow control valve.*

obstruction (e.g. a needle valve) and often incorporate a check valve to give unobstructed flow in the opposite direction, as in Fig. 15.34b. Such valves, however, exhibit a pressure drop which varies as the square of the flow, and consequently causes the actuator speed to vary with load.

An ideal flow controller would control flow whilst exhibiting a constant pressure drop. Such a device is shown in Fig. 15.34c. The throttle is a notch in a rotatable shaft. A movable piston controls the inlet of fluid to the valve by the raised land. The piston experiences a downward force due to the spring and the fluid pressure at point X, and an upward force from fluid pressure at point Y (the lower chamber makes the upper and lower piston areas equal). The piston will move up or down until the differential pressure between X and Y matches the spring compressive force.

The throttle therefore controls the flow through the valve, and the spring tension the pressure drop across the valve. This arrangement is known as a pressure compensated flow control valve, and is denoted by the symbol of Fig. 15.34d. Usually a check valve is incorporated to allow free flow in the reverse direction.

15.8.2. Servo (proportional) valves

Variable flow control can be achieved by the use of movable spool valves, similar to Fig. 15.25 except that the spool can take an infinite range of positions, not just two or three. There is little problem positioning the main spool with direct manual action, but remote electrical or pilot operation requires some form of feedback of main spool position to ensure a linear relationship between input signal and output flow. Such valves are generally called proportional, or servo, valves.

A common arrangement is shown in Fig. 15.35. The input electrical signal shifts the spool of a pilot valve a distance proportional to the input current (the restoring force being provided by a spring). The pilot valve sleeve is mechanically linked to the main spool, and pivots at a fulcrum.

Pilot pressure can be applied to either end of the main spool, but because of the difference in areas equal pressure will cause the spool to move to the right. Movement of the pilot spool either applies equal pressure to both ends (main spool moves to right, pilot sleeve moves to left) or pressure to end B, end A to tank (main spool moves to left, pilot sleeve moves to right). Only when the sleeve exactly matches the centre land on the pilot spool does the main spool stop.

This following action causes the main spool to follow precisely the movements of the pilot spool regardless of load and pressure variations. Note that the pilot and main spools move in opposite directions, with the ratio of relevant movements being set by the fulcrum position.

Figure 15.35 shows an electrical input signal (sometimes called a torque motor pilot). Obviously the movement of the pilot spool could equally well be controlled by an applied variable pilot pressure to the end of the pilot spool.

Figure 15.36a is a fully position-controlled valve. The main spool position is measured (usually by an LVDT) and compared with the input signal. Any error is

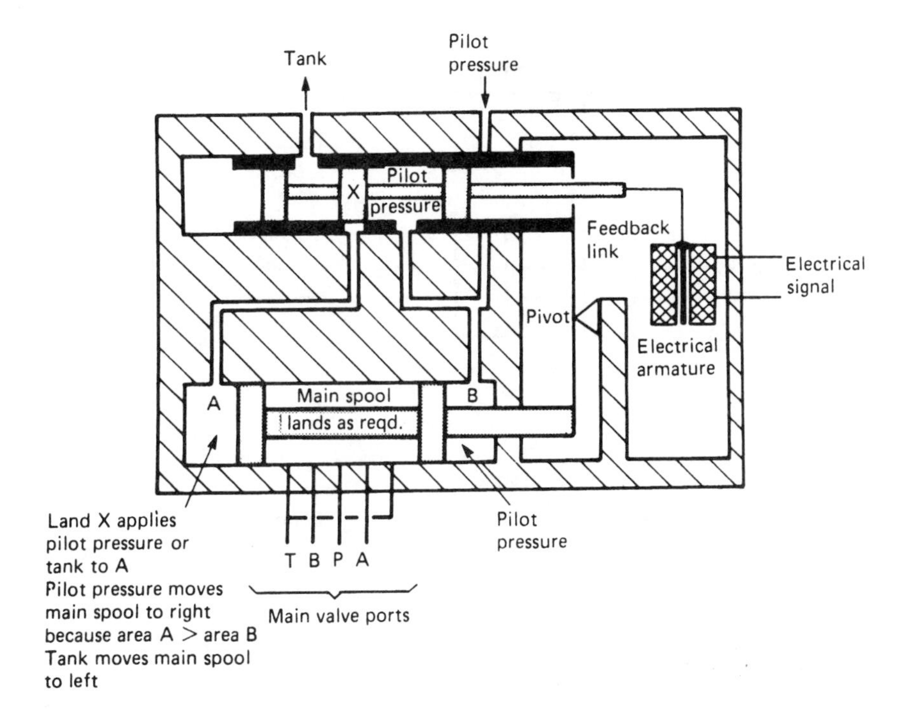

Fig. 15.35 *Two-stage servo valve.*

amplified and used to shift the pilot spool until the positional error is zero.

Servo valves are precision devices, and operate with very small displacement of the pilot spool. They are particularly prone to sticking and erratic operation resulting from dirty or 'gummy' oil. A common practice with electrically controlled pilot valves is the addition of a 'dither' signal to the applied input, as in Fig. 15.36b. This is simply a small sinusoidal signal (at 50 or 60 Hz) which does not affect the mean valve position, but serves to keep the valve permanently in slight oscillation. This helps displace dirt and overcome stiction.

15.8.3. Accumulators

Accumulators are used to 'store' pressure in a hydraulic system, and are used in several circumstances. The first is where a large volume of fluid is required for very short periods. In this application the accumulator allows a pump to be used of lower capacity than the peak, thereby saving on initial pump and piping (as well as running) costs.

Accumulators also are used to reduce running costs where instant response is needed. If a pump/unloading valve arrangement is used (as in Fig. 15.13) there is a short delay whilst the pressure builds up after the pump has been put on load. An accumulator linked unloading circuit, described later, gives instant response combined with the power saving of a pump unloading valve.

Finally an accumulator can act as a 'buffer' to absorb shocks and transient

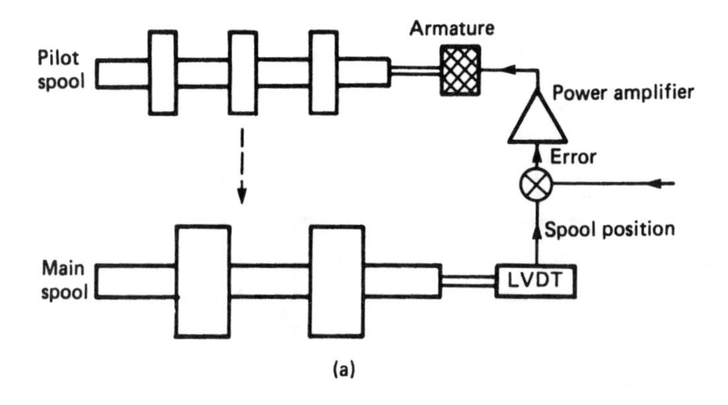

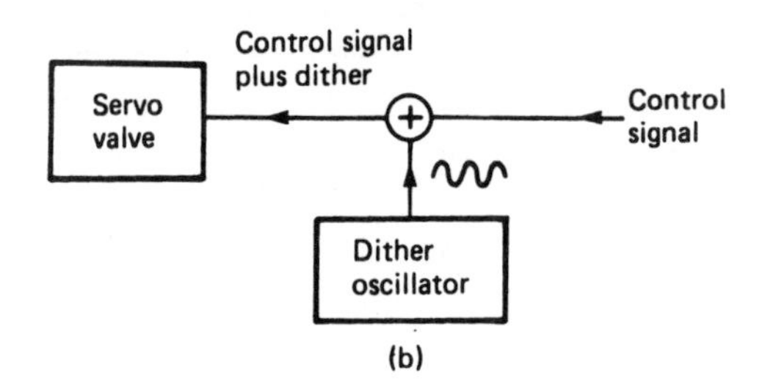

Fig. 15.36 *Electrical modifications to basic servo valve. (a) Position-controlled servo valve. (b) Overcoming stiction with dither.*

pressure peaks which occur when fluid is stopped or reversed. In a non-accumulator system these shocks manifest themselves as loud bangs and hammering noises, which can often be severe enough to cause failures of piping or valves.

There are two types of accumulator, shown in Fig. 15.37. The piston/spring uses the compressive force of the spring as the pressure store, and the gas filled a pressurised inert gas. The gas is usually separated from the hydraulic fluid by a

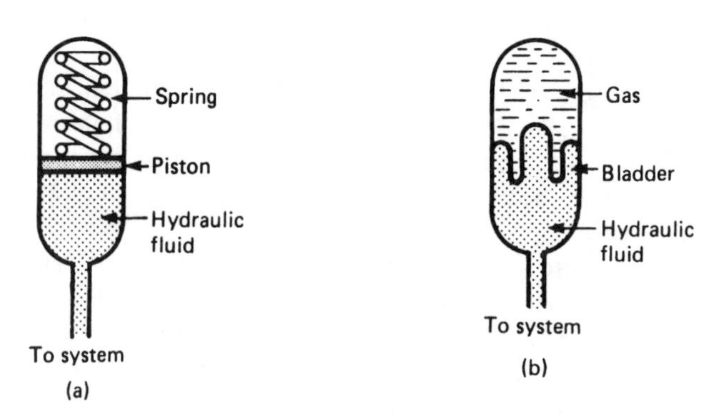

Fig. 15.37 *Accumulators. (a) Piston/spring. (b) Gas filled.*

flexible bladder. In both cases the stored pressure and fluid are available, on demand, to the rest of the system.

Figure 15.38a shows a typical accumulator pressure control system. Two pressure switches are connected to the system, set for a slight differential. The higher set switch opens the unloading valve, and the lower set switch closes it with the circuit of Fig. 15.38b. The system pressure thus cycles between the two pressure switch settings. The pump only comes on load when required to deliver fluid, but the accumulator gives instant pressure and flow on demand. Instant response with no wasted power is thereby achieved.

Accumulators can stay fully charged for days, or even months, and this brings a possible hazard to hydraulic systems. It is *very* dangerous to work on a pressurised hydraulic system; oil at high pressure can easily blind or maim, or even cause a fatality. An accumulator based system must include a mechanism by which the stored pressure can be released, and indicating devices which allow system pressure to be checked before any maintenance work is commenced. The author writes from personal experience of having been totally covered in oil during work on a system which was thought, incorrectly, to be depressurised.

A common arrangement is the automatic blow down circuit of Fig. 15.38c. The

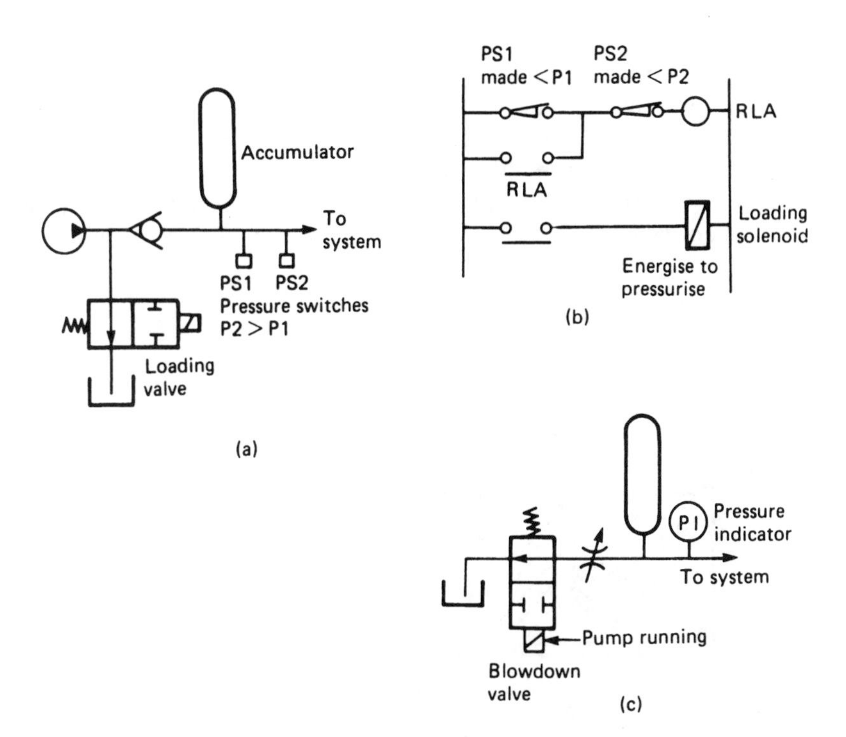

Fig. 15.38 *Practical accumulator circuits. (a) Automatic accumulator charging. (b) Electrical circuit. (c) Safety considerations.*

relief valve is held shut by the volts applied to the coil of the pump motor contactor. When the pump stops, or trips, the spring return on the relief valve causes accumulator stored fluid, and pressure, to be released to tank. The flow control allows the blow down rate to be adjusted. The pressure indicator should be checked to ensure blow down has been achieved before any work on the system is started.

15.8.4. Tanks and reservoirs

Hydraulic systems utilise a closed system of fluid, and as such require a tank or reservoir to hold the fluid between its return from the system and the pump intake. The reservoir is more than a simple tank, however, as examination of Fig. 15.39 will show. The tank must hold sufficient volume to allow for volume changes caused, for example, by single acting cylinders and temperature effects. The depth must be such that the pump inlet does not cause a whirlpool effect at the surface which would draw air into the system. A common rule of thumb is a tank volume of three times the volume delivered by the pump per minute.

The reservoir serves to cool the oil, and to assist this the inlet and outlet tank halves are separated by a baffle to make the oil take a circuitous route round the walls. If extra heat removal is necessary a water cooled heat exchanger can be included in the return line or the tank itself. The baffle also serves to reduce turbulence, and allows contaminants to settle out.

Inevitably a layer of sludge will form in the tank bottom. A drain plug and access plate should therefore be provided for regular cleaning. It is essential that this sludge is not drawn into the system, so a coarse strainer is provided on the pump inlet line. The return line is also angled so that the flow is directed away from the bottom and does not stir up the sludge.

If the tank level falls (as it will, because no system is leak free!) there is a danger that a whirlpool will form, drawing air into the pump. This will cause poor

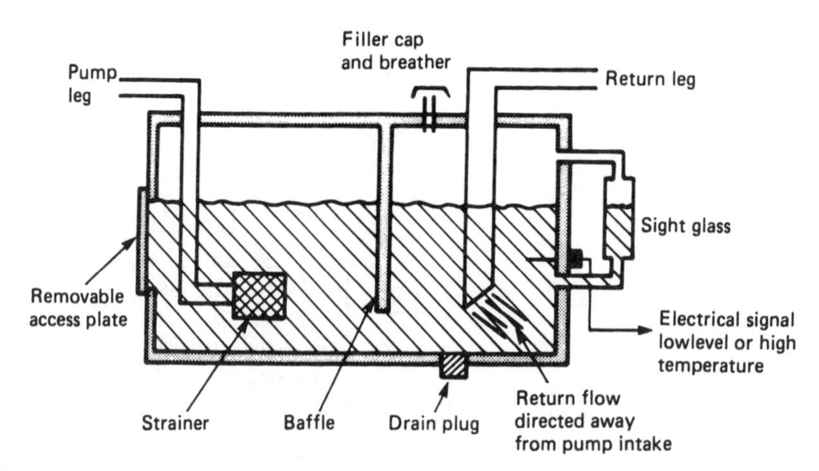

Fig. 15.39 *A hydraulic tank.*

performance and even damage. A visual sight glass and low level float switch are provided to allow the level to be examined and a remote alarm given. Two level switches are sometimes provided. The first indicates a low level alarm, and the second, lower, switch stops the pump. A fluid overtemperature alarm is often incorporated with the level switch.

15.8.5. Filters

Cleanliness is next to godliness in hydraulics. It is generally dirt that causes sticking valves, failure of seals and premature wear. Particles of dirt as small as 20 microns can cause problems (a micron is one millionth of a metre; the naked eye can just resolve 40 Microns). Filters are specified in microns or meshes per linear inch (sieve number).

The inlet line in the tank will be fitted with a strainer, but this will be a coarse wire mesh element for removing relatively large metal particles and similar contaminants. A separate filter is needed to remove the finer particles. This filter can be installed in three places as shown in Fig. 15.40a, b and c.

Inlet line filters protect the pump, but must be designed for low pressure drop or the pump will not be able to raise fluid from the tank. Low pressure drop implies a coarse filter or a large physical size. Pressure line filters which protect the valves and actuators can be finer and smaller, but must be able to withstand the full system operating pressure. Return line filters can be very fine and, paradoxically, serve to protect the pump by limiting the size of particle returned to the tank.

Filters can also be classified as full or proportional flow. In Fig. 15.41a, all the flow passes through the filter. This is obviously efficient in terms of filtration, but will incur a large pressure drop. This pressure drop will increase as the filter

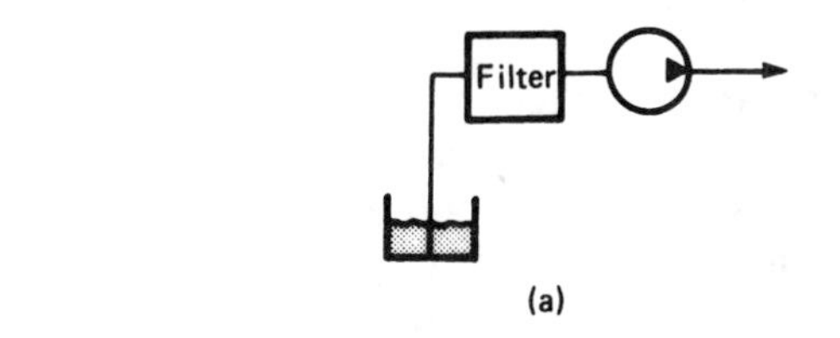

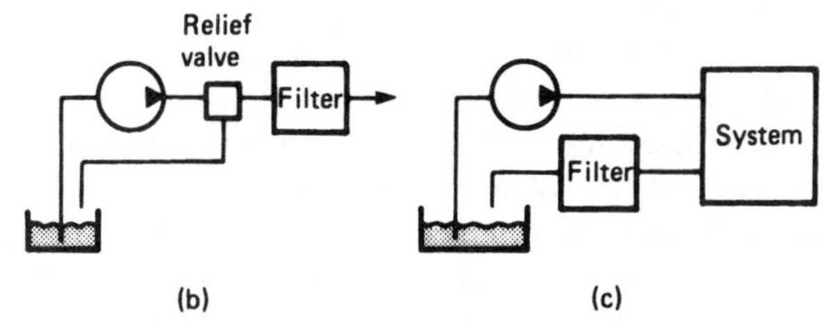

Fig. 15.40 *Filter positions. (a) Inlet line filter. (b) Pressure line filter. (c) Return line filter.*

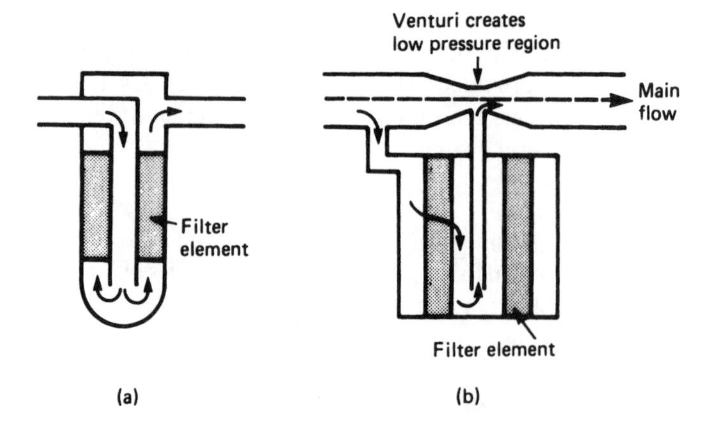

Fig. 15.41 *Filter types. (a) Full flow filter. (b) Proportional flow filter.*

becomes increasingly polluted, so a full flow filter usually incorporates a relief valve which cracks when the filter becomes unacceptably blocked. The filter should, of course, have been changed before this state was reached.

In Fig. 15.41b the main flow passes through a venturi, creating a localised low pressure area. The pressure differential across the filter element draws a proportion of the fluid through the filter. This design is accordingly known as a proportional flow filter, as only a proportion of the main flow is filtered. It is characterised by a low pressure drop, and does not need the protection of a pressure relief valve.

The pressure drop across the filter element is an accurate indication of its cleanliness, and many filters incorporate a differential pressure meter calibrated with a green (clear)/amber (warning)/red (change overdue) indicator. Such types are called indicating filters.

The filtration material can be mechanical or absorbent. Mechanical filters are relatively coarse, and utilise fine wire mesh or a disk/screen arrangement as in Fig. 15.42. Absorbent filters are based on porous materials such as paper, cotton or cellulose. The filtration size can be made very small as the filtration is done by the pores in the absorbent material. Mechanical filters can usually be removed, cleaned and refilled, whereas absorbent filters are usually replaceable items.

15.8.6. Hydraulic fluids

Hydraulic fluid is used as the transmission medium, and its characteristics determine the performance and reliability of any hydraulic system. The fluid must meet several, often conflicting, requirements. To eliminate losses it must flow freely and it should be incompressible so that actuators respond instantly to valves.

The fluid also acts as a lubricant. In pumps, valves, actuators, etc., a thin film of fluid is employed where moving parts slide over each other. This thin film must also act as a seal around the lands of a valve spool, for example, stopping high pressure fluid leaking to the adjacent chamber.

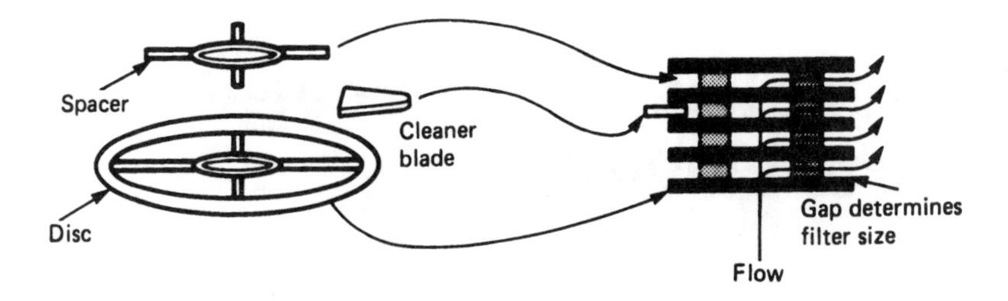

Fig. 15.42 *Edge-type filter.*

In operation, hydraulic fluid experiences a wide range of temperature. At start-up, after a long weekend shutdown, temperature near freezing may be encountered. In operation, heat is generated from pipe friction and the action of pressure regulators, valves, actuators, etc., causing a significant temperature rise, possibly to the extent that cooling is necessary. The fluid must not notably change its characteristics with temperature changes. The nature of the fluid should also encourage heat loss to the pipe and reservoir walls.

A major cause of hydraulic problems is the sticking of valve spools. This is caused by dirt and a sticky, gummy deposit which can form when some oils stand or undergo temperature changes. Hydraulic fluid must have stable characteristics with age.

Water is present in all hydraulic systems, entering via the tank breather and mixing with the fluid as condensation forms with temperature changes. When water and oil are mixed, a white emulsion can form which, again, leads to sticking valves. Water can also lead to foaming, and cause cavitation and erosion damage. Hydraulic fluid should not change its characteristics in the presence of small amounts of water.

Hydraulic fluid, like all oils, is defined by its viscosity. This is a measure of its resistance to flow. Fluidity is the measure of how easily a fluid flows. Treacle has high viscosity and low fluidity. Petrol has low viscosity and high fluidity. Selection of the viscosity for hydraulic fluid is a compromise between sealing and lubrication (which requires high viscosity) and friction losses and speed of response (which require low viscosity).

Viscosity can be defined in absolute terms in poise or centistokes, or in relative terms in SUS (Saybolt universal seconds) or SAE (Society of Automobile Engineers) numbers which were developed to specify the viscosity of motor oil over a range of temperatures. SAE numbers are the most common.

Most hydraulic fluids are petroleum-based oils. These are, however, inflammable and a hazard in industries with open flames or where welding is common. Fire resistant fluids are based on water/oil emulsions, or water/glycol mixes. These require regular testing as evaporation reduces the water content. Some synthetic fluids (e.g. phosphate esters) are also fire resistant but rather expensive. All fluids have additives to stabilise the characteristics with age and temperature and to reduce the formation of gummy deposits.

Chapter 16

Pneumatics and process control valves

16.1. Basic principles

16.1.1. Introduction

Chapter 15 described the use of hydraulics as an alternative to electrically powered actuators. Hydraulic systems utilise a liquid (usually oil or water) as the power transmission medium. It is also feasible to use a gas for transmitting force. Systems using gas are called pneumatic systems (derived from the Greek words for hidden gas). Industrial pneumatic systems are usually based on air.

Pneumatic applications can be loosely split into three groups. The first is linear and rotary actuators, where pneumatic devices are employed to operate devices such as control valves, rams, cylinders, etc. The second group is the use of pneumatics for process control signals. A 4 to 20 mA current is commonly used to represent process variables – a liquid flow from 0 to 3000 litres per minute say. In a similar way, a process variable can be represented as a pressure change. The commonest standard is 0.2 to 1 bar (or the equivalent 3–15 psi). Many pneumatic transducers are available which give a pneumatic 0.2 to 1 bar output signal as a representation of a process variable. Pneumatic three-term controllers (with pneumatic pressure set point, pneumatic process variable and pneumatic output signal) are also available, allowing closed-loop control with non-electrical signals. Such schemes are particularly attractive in applications with explosive atmospheres.

Pneumatics can also provide an alternative to digital logic and relay sequencing. Such systems, called fluidics, are not particularly common at the time of writing, but again are attractive in hazardous areas.

16.1.2. Fundamentals

In many respects there are similarities between hydraulics and pneumatics. In particular, both use the pressure of an enclosed fluid as a force transmission medium. Equation 15.1 relates the force produced by a fluid of pressure P acting on an area A, and can be rearranged as

$$\text{force} = P \times A \tag{16.1}$$

which is the fundamental equation for pneumatic systems.

In most pneumatic systems, gauge pressure measurements are used (i.e. with respect to atmosphere). This allows direct calculation of forces where cylinders are open to atmosphere on one side. In Fig. 16.1, for example, the piston has an

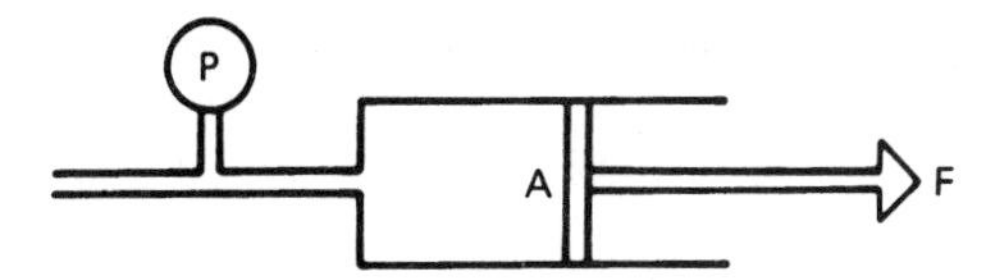

Fig. 16.1 *Relationship between force and pressure.*

area of 20 square inches, and is experiencing a force of 30 psig on one side and atmospheric pressure on the other. The net force is 30 psi × 20 square inches = 600 pounds.

The concepts of work, energy and power were defined in Section 15.2. These apply equally to pneumatic and hydraulic systems.

16.1.3. Gas laws

The major differences between pneumatic and hydraulic systems arise from the compressibility of the fluid used as the transmission medium. Hydraulic fluid is, for all practical purposes, incompressible, whereas a gas can easily be compressed.

The behaviour of a gas subject to pressure, volume or temperature changes is described by three gas laws.

The first, known as Boyle's law, is illustrated in Fig. 16.2a, where a fixed enclosed mass of gas can be compressed by a movable piston. The temperature of the gas is kept constant by a surrounding heat sink (this is known as adiabatic conditions). Boyle's law states that the pressure and volume are related by:

$$PV = K \tag{16.2}$$

where K is a constant. Equation 16.2 can be rearranged to describe conditions before (subscript b) and after (subscript a) compression:

$$\frac{P_a}{P_b} = \frac{V_b}{V_a} \tag{16.3}$$

Halving the volume doubles the pressure. It should be noted that equations 6.2 and 6.3 use absolute pressure. As most pneumatic systems use gauge pressure, conversions are necessary before and after the application of Boyle's law. For

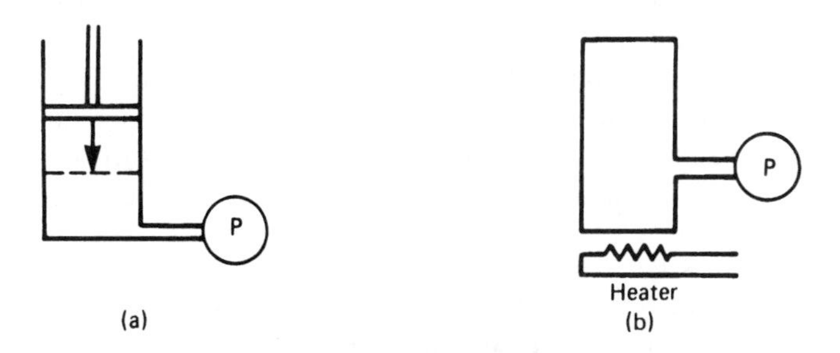

Fig. 16.2 *Gas laws. (a) Boyle's law. (b) Charles' law.*

example, reducing the volume of gas at 30 psi to one third of its initial volume results in a pressure of:

$$3 \times (30 + 14.7) \text{ psia}$$

i.e. 134.1 psia or 109.4 psig.

Charles' law is concerned with the relationship between the pressure and the temperature of a fixed volume of gas, as in Fig. 6.2b. Increasing the temperature causes a pressure rise which is given by:

$$\frac{P}{T} = L \tag{16.4}$$

where L is a constant. As before, this can be more conveniently written to relate the pressure at temperature T_a, T_b as:

$$\frac{P_a}{T_a} = \frac{P_b}{T_b} \tag{16.5}$$

Both pressure and temperature must be given in absolute terms: i.e. temperature in degrees Kelvin (or Rankine). For example, a cylinder containing gas at 30 psig is heated from 20 °C to 100 °C. The resulting pressure is:

$$(30 + 14.7) \times (100 + 273)/(20 + 273) \text{ psia}$$

which is 56.9 psia or 42.2 psig.

The compression of a gas is usually accompanied by a rise in temperature – commonly experienced when a bicycle pump gets warm in use. Figure 16.2a was a special case where the container was surrounded by a heat sink to maintain constant temperature. The generalised equation, called the ideal gas law, relates pressure, volume and temperature by:

$$\frac{P_a V_a}{T_a} = \frac{P_b V_b}{T_b} \tag{16.6}$$

Absolute units must, again, be used for pressure and temperature.

Equation 16.6 does not directly describe what happens to pressure and temperature, say, when the volume of a mass of gas changes. The work done will be converted partly to a change in pressure, partly to a change in temperature. The resulting conditions will be determined by factors such as rate of heat transfer, which are outside the scope of equation 16.6.

It should be noted that although equations 16.2 and 16.6 are usually exemplified by increases in pressure and temperature and decrease in volume, the equations are valid for all changes. A decrease in pressure or volume is accompanied by a fall in temperature, for example; a phenomenon used in refrigeration.

16.1.4. Differences between hydraulic and pneumatic systems

Figure 16.3 shows how a simple extend/retract system would be implemented in hydraulics (Fig. 16.3a) and pneumatics (Fig. 16.3b). The hydraulic system pump

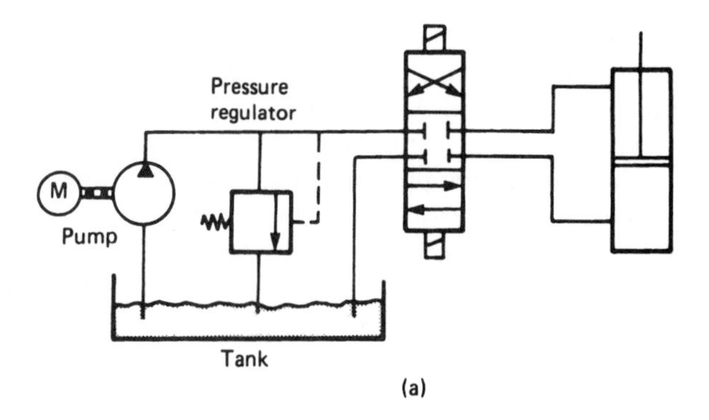

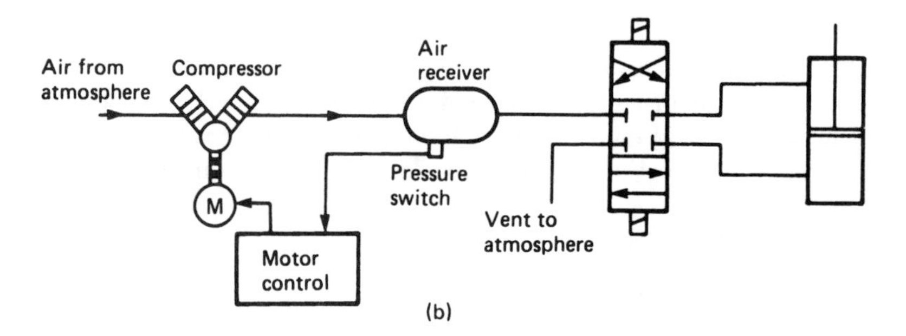

Fig. 16.3 *Comparison between hydraulics and pneumatics. (a) Hydraulic system. (b) Pneumatic system.*

takes oil from a tank, with the pump outlet pressure being determined by a pressure relief valve which returns excess fluid to the tank. The direction control valve connects pressure to one side of the piston and drains oil back to the tank from the other side of the piston. A hydraulic system is essentially a closed system, the oil being recycled between the tank and the plant. The pump runs continuously, with pressure regulation being determined by the pressure relief valve.

A pneumatic system uses compressed air, and this is usually stored in a pressure vessel called an air receiver. Air, from the atmosphere, is delivered to the air receiver by a motor-driven compressor. Unlike a hydraulic pump, the compressor is controlled by a pressure switch on the air receiver and either starts/stops on demand or vents to atmosphere when the receiver is charged. The cylinder movement is again controlled by a directional valve, but air returned from the cylinder is simply vented to atmosphere. A pneumatic system is an open system, the fluid being obtained from and returned to the atmosphere.

Pneumatic systems also require clean dry air (the Three Mile Island nuclear incident was initiated by water in pneumatic lines).

A practical pneumatic system has additional air treatment elements not present in a hydraulic system.

16.1.5. Elements of a pneumatic system

Figure 16.4 is a more detailed version of Fig. 16.3b, and contains all the elements normally found in a pneumatic system. Air is drawn to the inlet side of the compressor via a filter. This is necessary to remove dust and insects which could damage the seals and valves of the compressor. As the air is compressed to decrease the volume and increase the pressure, its temperature rises according to equation 16.6. The compressor is therefore followed by a cooler to reduce the air temperature. Water vapour in the air tends to condense out at this point. Water can cause corrosion and line blockage, so the aftercooler is followed by a drier (sometimes called a separator or primary air treatment).

The air receiver stores the compressed air, and its pressure is controlled by a pressure switch acting directly on the starter of the electric motor for the compressor. The air receiver usually has a safety relief valve which will act if the pressure switch fails.

Ideally, the air should contain a slight oil mist to lubricate the system components. This is provided by secondary air treatment which incorporates further filtration, water removal and the introduction of the oil mist.

The items of Fig. 16.4 are discussed in detail in the sections which follow.

16.2. Compressors

16.2.1. Introduction

A compressor is used to compress atmospheric air to give the desired air pressure. Like hydraulic pumps, compressors can be classified into positive displacement devices which move a fixed volume of air per revolution of the input shaft, and dynamic devices which accelerate the air velocity. Large volume low pressure compressors used in pneumatic conveying, for example, are often dynamic devices and are called 'blowers'.

A plant pneumatic system is usually dealt with as a works-wide central service (in the same way as electricity, gas, steam, water, etc.) and is served by a compressed air distribution system fed from a central compressor station.

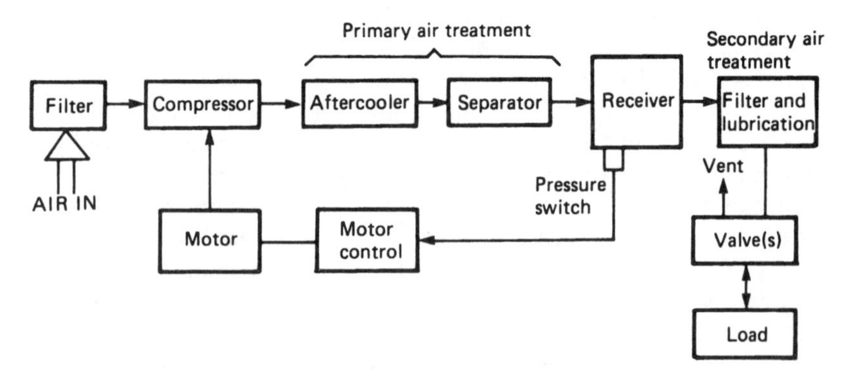

Fig. 16.4 *Components of a pneumatic system.*

Usually several compressors are installed in parallel to allow for servicing and give redundancy against failure.

16.2.2. Reciprocating piston compressors

The reciprocating piston compressor has a superficial resemblance to a motorcar engine (Fig. 16.5a). The crankshaft is rotated by an external electric motor, causing the piston to move cyclically up and down the cylinder bore. An inlet and outlet valve are mounted in the cylinder head. Unlike the valves in an internal combustion engine, these are not operated mechanically by cams but are operated directly by pressure variation above the piston.

As the piston falls, a partial vacuum is formed in the bore above the piston. Air pressure causes V_1 to open, and air is drawn in from the atmosphere via the inlet filter and silencer. The partial vacuum keeps V_2 firmly closed.

As the piston reaches its lowest position, the cylinder is filled with air at approximately atmospheric pressure. On the upward stroke, this air is compressed causing V_1 to close. When the pressure inside the cylinder head exceeds the outlet receiver air pressure P_o, valve V_2 will open and air is delivered to the receiver.

There is therefore an inlet and outlet stroke per cycle, and a mass of air whose volume is the swept cylinder volume at atmospheric pressure is delivered per cycle.

Ball valves are shown in Fig. 16.5a for simplicity, but in practice valves are constructed of 'feathers' of spring steel or disks seating on to annular inlet ports.

A modification of the basic reciprocating compressor is the double-acting arrangement of Fig. 16.5b. This has two inlet and two exhaust strokes per revolution. The crosshead and guide serve to keep the piston rod parallel to the cylinder bore.

Two or more cylinders can be connected in parallel in a variety of ways – V, in line, horizontally opposed, etc., as shown in Fig. 16.5c. Such arrangements are called single stage compressors, and can be used up to about 7 bar.

Where higher pressures are required, a multistage compressor is used, as in Fig. 16.6a or b. When air is compressed, its temperature rises in accordance with equation 16.6. This temperature rise increases the power required to drive the compressor. In a multi-stage compressor, therefore, it is usual to cool the air between stages by a device called an intercooler. This can be a tubed water-cooled heat exchanger, as in Fig. 16.6c, or an air-cooled finned pipe or radiator. The bodies of large compressors are also water cooled or finned to assist heat removal. The effects of cooling on power consumption are dramatic. To provide 5 m^3 of air at 7 bar per minute requires a shaft input power of about 20 kW for a single stage compressor and will result in a temperature rise of 200 °C. If a multistage compressor with cooling is used, the shaft power is only about 10 kW. Cooling must not, however, be overdone or water vapour will condense out and cause considerable damage to the second stage of the compressor.

16.2.3. Rotary compressors

Rotary compressors do not use reciprocating pistons, and as such are smaller, quieter and easier to maintain than reciprocating compressors. The air supply

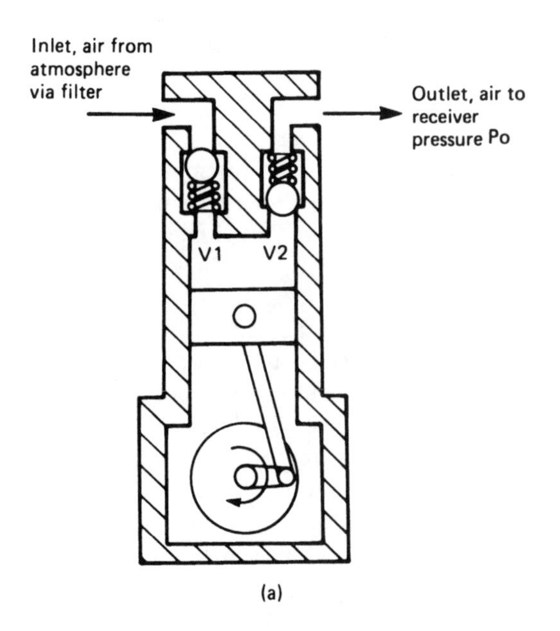

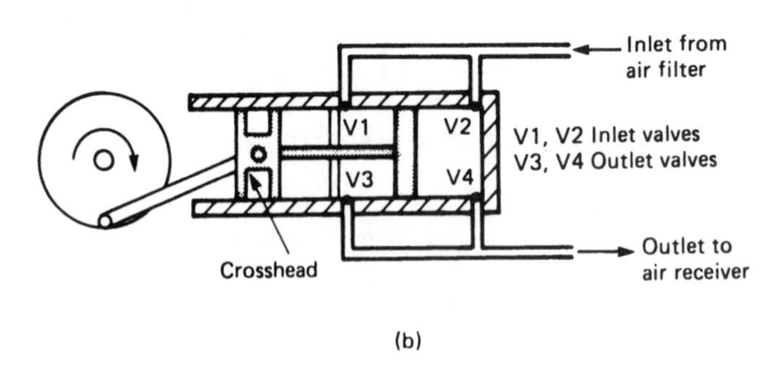

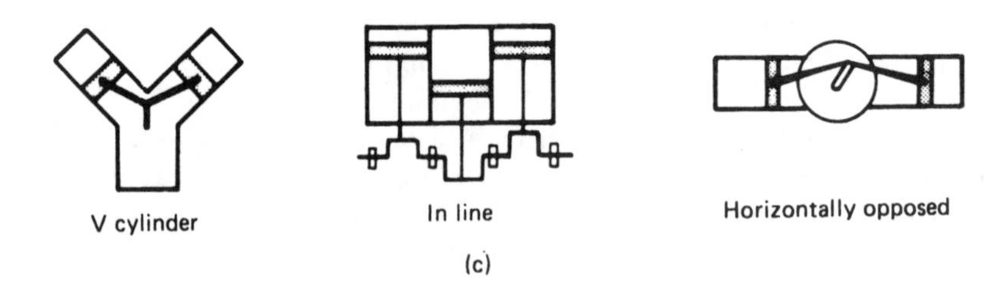

Fig. 16.5 *Reciprocating compressors. (a) Piston compressor. (b) Double-acting compressor. (c) Cylinder arrangements.*

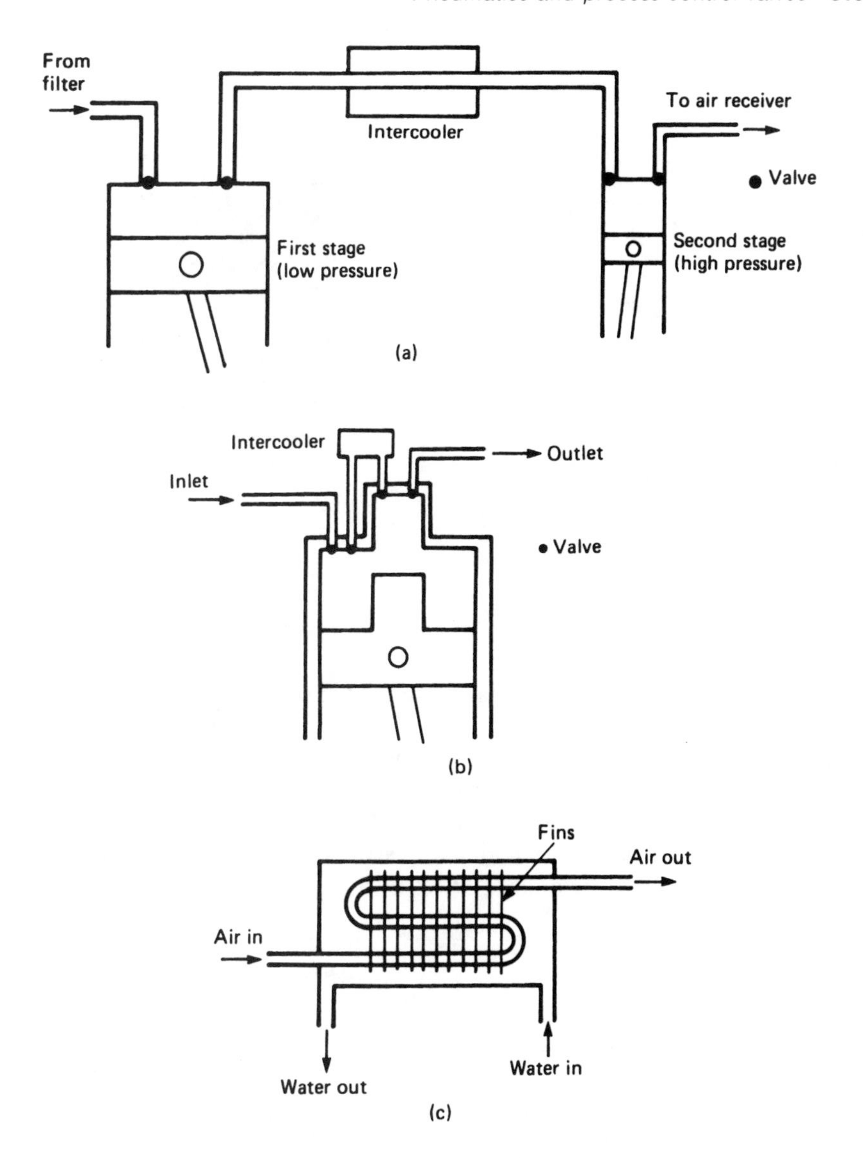

Fig. 16.6 *Multistage compressors. (a) Two-stage compressor. (b) Combined two-stage compressor. (c) Intercooler.*

from a rotary compressor is also smooth and non-pulsating. These benefits are gained, however, at the expense of slightly reduced efficiency and lower operating pressures. This section describes positive displacement rotary compressors.

The vane compressor of Fig. 16.7a operates in a similar manner to the hydraulic vane pump of Section 15.4.3. The rotary screw compressor of Fig. 16.7b uses two intermeshing counter-rotating screws. These mesh with a clearance of a few thousandths of an inch, and are driven by timing gears. As the

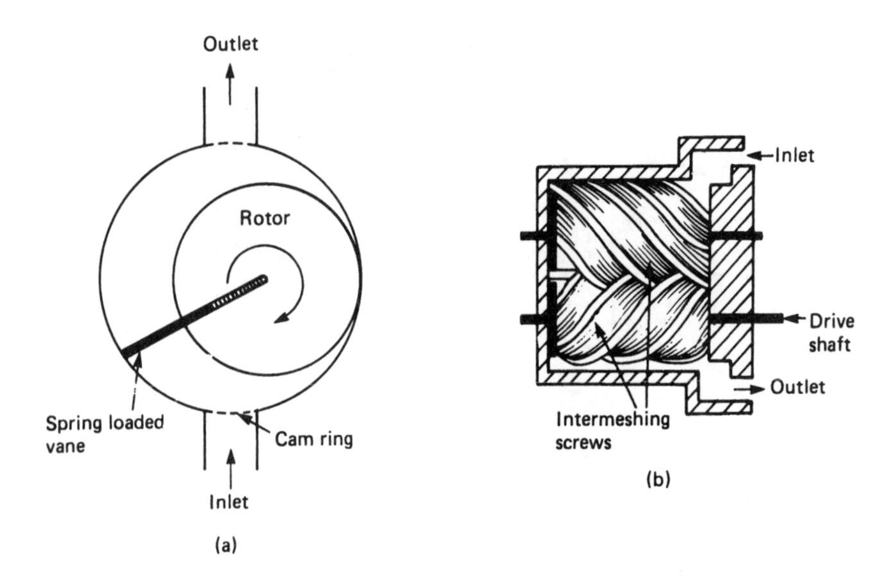

Fig. 16.7 *Various compressors. (a) The vane compressor. (b) The screw compressor.*

screws rotate, pockets of air are carried from the inlet port to the outlet port.

The liquid ring compressor of Fig. 16.8a is a variation on the vane compressor. This device uses many vanes which rotate inside an eccentric casing. The casing is filled with a liquid, usually water, which is flung out by centrifugal force to form a ring which follows the contour of the casing. The volume entrapped by the liquid between vanes therefore increases and decreases as the shaft rotates, delivering air from the inlet ports to the outlet ports.

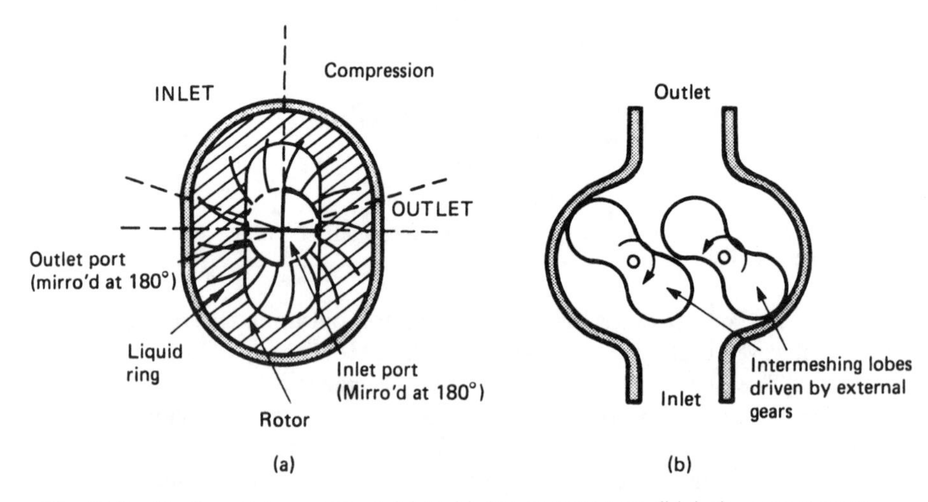

Fig. 16.8 *Further compressors. (a) Liquid ring compressor. (b) Lobe compressor.*

The final type of rotary compressor is the lobe (impeller) compressor of Fig. 16.8b. This operates in a similar manner to the hydraulic gear pump of Section 15.4.2, sweeping pockets of air between the impeller blades and the casing. The lobe compressor is a high volume low pressure device, operating typically up to 2 bar (30 psig). The operating pressure is mainly limited by leakage between blades and between blades and the casing.

Rotary compressors can, like reciprocating compressors, have their output pressure range increased by the use of multi-stages with intercoolers to increase efficiency.

16.2.4. Dynamic compressors

Many applications, such as gas/air burners and manufacturing processes require a large volume low pressure air supply. The reciprocating and rotary compressors described so far are essentially low volume high pressure devices. Dynamic compressors are non-positive displacement devices, and as such there is a direct shaft route from the load to the supply side; if the drive shaft stops, the pressure can decay back from the load through the compressor.

Dynamic compressors can be classified into two types, shown in Fig. 16.9. The centrifugal types use centrifugal force to transfer air from an axial inlet port to a peripheral outlet port. An axial compressor is essentially a series of in-line fans or turbine blades.

Dynamic compressors, often called blowers, operate at high speed (up to 10 000 rpm) and can deliver well in excess of 100 000 m^3 of air per minute.

16.2.5. Practical considerations

Because a compressor can be used over a range of pressures, manufacturers usually specify a compressor by its 'free air capacity'. This is the volume of air

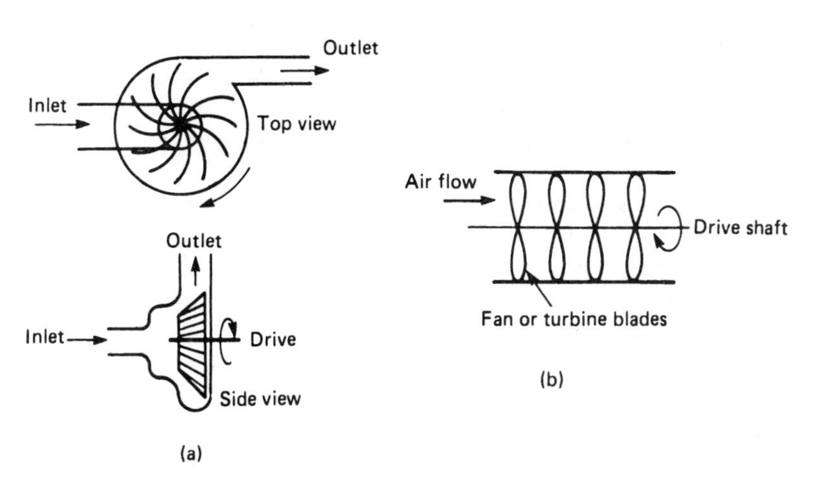

Fig. 16.9 *Non-positive displacement compressors (blowers). (a) Centrifugal type. (b) Axial type.*

drawn into the compressor at atmospheric pressure per unit time (e.g. cubic metres or cubic feet per minute). A maximum outlet pressure will also be specified.

The air consumption rate on the plant can be calculated from the volume of actuators and their rate of use. Knowing the working pressure, the volume of air used (with a healthy allowance for losses) equation 16.3 gives the required free air capacity for the compressor. Sizes should always be chosen conservatively.

Small compressors are usually stopped and started on demand from a pressure switch as shown in Fig. 16.4. Large compressors are often run continuously with an outlet valve being controlled by the air receiver pressure switches. Compressors are usually started off load to limit the starting current in the electric drive motor.

Input air filtering is essential to long compressor life. Filter cartridges can use replaceable paper or gauze elements, or cleanable mesh filters. The elements should never be cleaned in petrol; this can turn compressors into diesel internal combustion engines, with dire results! Filter elements should be examined as part of regular maintenance schedules.

16.3. Air treatment

Air always contains a certain amount of moisture in the form of water vapour. The amount of water present in the air depends on the temperature, pressure and atmospheric conditions. For a given volume of air at a certain temperature and pressure there is a maximum amount of water that can be held in the form of vapour. Air in this state is said to be saturated. The amount of water vapour in a volume of air expressed as a percentage of the maximum water vapour that could be held (i.e. saturated air) is called the relative humidity. Obviously the relative humidity of saturated air is 100%.

We intuitively feel the relative humidity of air by describing days with high relative humidity as 'heavy', 'humid', or 'sticky', the latter description arising because sweat does not readily evaporate from the skin when the relative humidity is high. Low relative humidity is experienced as dry, crisp weather and (in housewives' terms) a good drying day as moisture easily evaporates from clothes.

The amount of water vapour that can be held in a given mass of air is dependent on temperature and pressure, increasing with increasing temperature and decreasing with increasing pressure. Effectively, the mass of water vapour that can be held in a given volume at a certain temperature is independent of pressure. Figure 16.10 shows what happens when the mass of water vapour held in 10 m^3 of saturated air at atmospheric pressure is taken to various temperatures and gauge pressures. At 1 bar and 40 °C, for example, the air will hold about 0.26 kg of water vapour. If this volume of saturated air is cooled to 20 °C, still at 1 bar, it can only hold just over 0.1 kg of water vapour, the excess of 0.16 kg appearing as moisture in the form of condensation.

Figure 16.10 also allows us to predict what happens with non-saturated air undergoing temperature and/or pressure changes. At 50% relative humidity at atmospheric pressure and 20 °C, 10 m^3 of air will contain about 0.08 kg of water vapour. If this air is compressed to 3 bar and allowed to cool back to 20 °C, it can

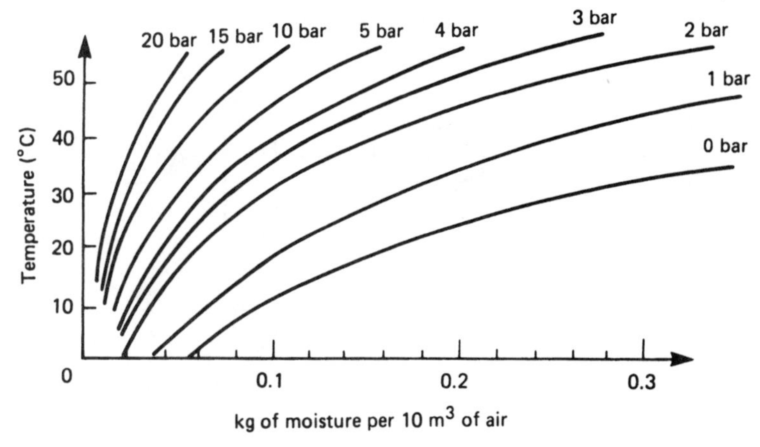

Fig. 16.10 *Moisture curves.*

only hold just over 0.04 kg of water vapour; the remaining 0.04 kg will condense out.

If the pressure of a volume of non-saturated air is held constant whilst cooling takes place, the relative humidity of the air will increase until at some temperature the air becomes saturated and condensation occurs. The temperature at which condensation occurs is called the dew point. From Fig. 16.10, for example, 10 m^3 of air containing 0.1 kg of water vapour at 0 bar will have a dew point of about 10 °C.

When air is compressed in a pneumatic system, the increased pressure results in a large fall in the amount of water vapour that can be held in the air. In the immediate region of the compressor the air temperature is sufficiently raised to prevent condensation, but as the air cools in the receiver and the rest of the piping, moisture will appear. In extreme cases, the moisture will appear as the air pressure drops suddenly which, by equation 16.6, is accompanied by a temperature fall that can be sufficient to form ice particles.

Water particles cause severe problems in pneumatic systems, leading to rust, rapid wear and pitting, and the formation of a sticky oil/water emulsion that can jam valve spools and block fine orifices in process control devices. There is little that can be done to prevent the condensation, and air treatment consists of condensing and removing the water after the compressor but before it can cause problems, as in Fig. 16.11.

In many systems, all that is required is a simple aftercooler following the compressor which cools the air sufficiently for the water to condense out. The water vapour, which now exists as a mist or droplets, is removed in a strainer or separator. The aftercooler is a simple shell and tube heat exchanger, with water, chilled brine or ethylene glycol coolant. In small systems, finned air coolers may suffice.

Pneumatic systems can be roughly split into bulk air systems, for pneumatic driven tools, actuators, etc., and instrument air supplies for controllers. An aftercooler and moisture separator will suffice for bulk air supplies, but instrument air often requires very dry air.

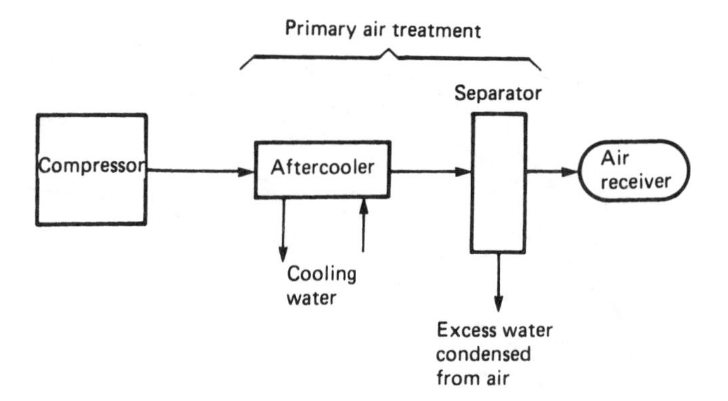

Fig. 16.11 *Air dryer.*

The moisture content can be reduced further by the use of a refrigerated dryer, shown in Fig. 16.12. This chills the air to just above 0 °C in the refrigerator heat exchanger which condenses almost all the water vapour. The air leaving the refrigerator pre-chills the incoming air. The water droplets are removed in the moisture separator.

Where absolutely dry air is required, chemical dryers are used. Various chemicals absorb water from the air by deliquescence or absorption. A deliquescent dryer is shown in Fig. 16.13a. As the dryer chemical removes the water vapour it turns to a liquid which collects at the bottom of the vessel for periodical draining. Fresh desiccant chemical must be added from time to time.

An absorption dryer uses chemicals which exist in a hydrated and dehydrated

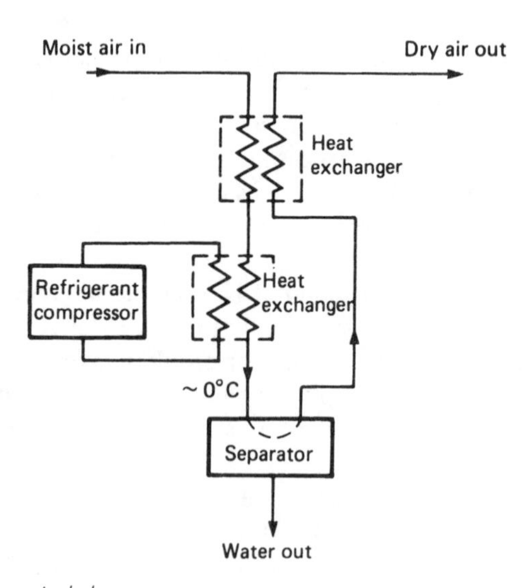

Fig. 16.12 *Refrigerated dryer.*

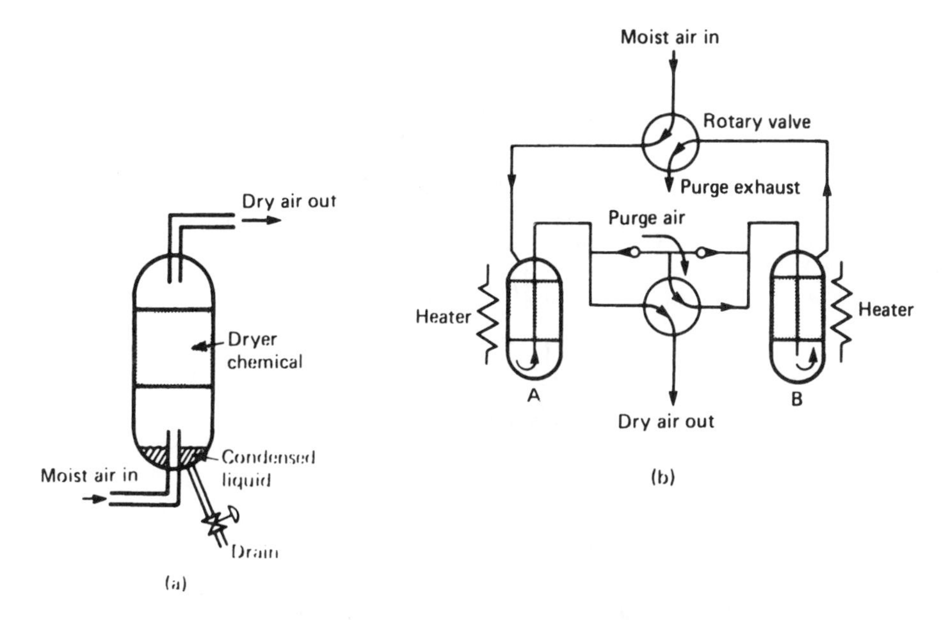

Fig. 16.13 *Chemical dryers. (a) Deliquescent dryer. (b) Absorption dryer.*

state (e.g. copper sulphate). These absorb water vapour, which can be released again by heating. Absorption dryers, an example of which is shown in Fig. 16.13b, use two columns which are sequenced between drying and regeneration. As shown, column A is drying the air and column B is being regenerated by the heater.

Aftercoolers and refrigerated dryers condense the water vapour to a mist or droplets which must be removed by a moisture separator. These are simply vessels in which the air is caused to suddenly reverse direction or swirl, as in Fig. 16.14. The heavier water particles are flung out and down, to collect in the bottom of the trap where they are drained manually periodically or automatically by a float-operated drain valve. Similar separators can also be used to remove oil and other contaminants.

Air treatment may also include the introduction of a carefully controlled amount of oil mist into the air, although process control pneumatics usually require oil-free air. This oil lubricates and protects main parts, but can only be added after the air has been thoroughly dried and cleaned or the oil and water will form a trouble-causing sticky emulsion.

Figure 16.15 shows a typical lubricator. This operates on a similar principle to the petrol/air mixing in the carburettor of an internal combustion engine. As the air passes through the lubricator, its velocity is increased by a venturi ring. This causes a fall in pressure and a partial vacuum in the upper chamber, drawing oil up the riser tube. The oil emerges from a jet to mix with the air. The pressure drop, and hence the oil flow, depends on the air flow rate giving a consistent air/oil ratio over a wide flow range. The needle valve adjusts the pressure difference across the

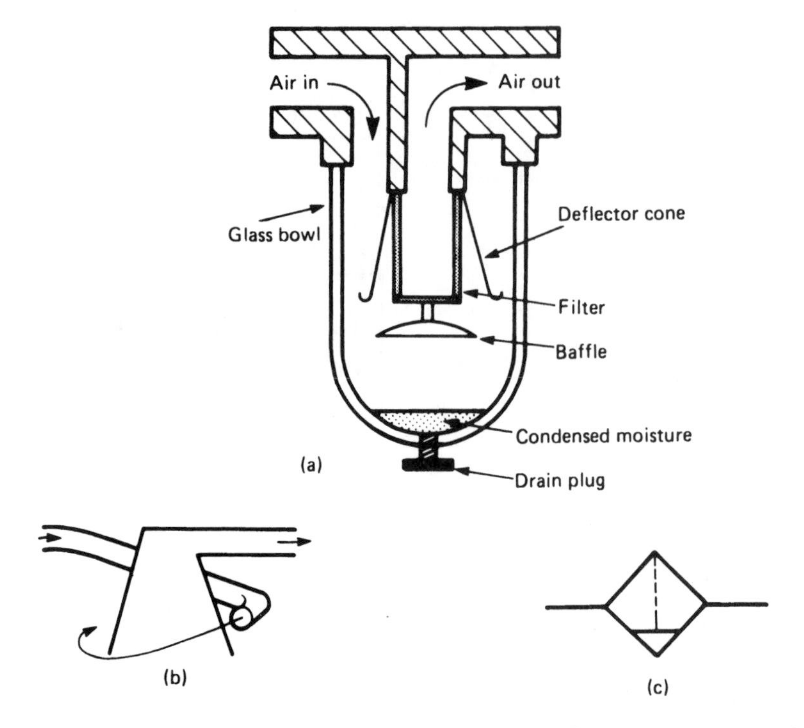

Fig. 16.14 *Air filter and water trap. (a) Construction. (b) Swirl introduced by deflector cone. (c) Symbol.*

oil jet, and hence the air/oil ratio. On leaving the mixing region the air follows a circuitous route to remove any large oil particles, as described for the moisture separator.

Moisture separators, filters and lubricators are frequently combined in one unit along with a pressure indicator and pressure regulator, as shown with the symbolic representation of Fig. 16.15b. Such assemblies are frequently called service units.

16.4. Pressure regulation

16.4.1. Introduction

Pneumatic systems, like hydraulic systems, require pressure regulation. In hydraulic systems, pressure regulation is mainly achieved by relief valves which bypass excess fluid back to the tank. Pressure regulation in pneumatic systems takes many forms.

The pressure in the air receiver is determined by a pressure switch which controls the compressor output, either by starting and stopping the compressor drive motor or by operating a loading valve. The air receiver is protected against failure of the main pressure switch by a relief valve. This is set higher than the main pressure switch, and is sized to be capable of handling the full compressor capacity.

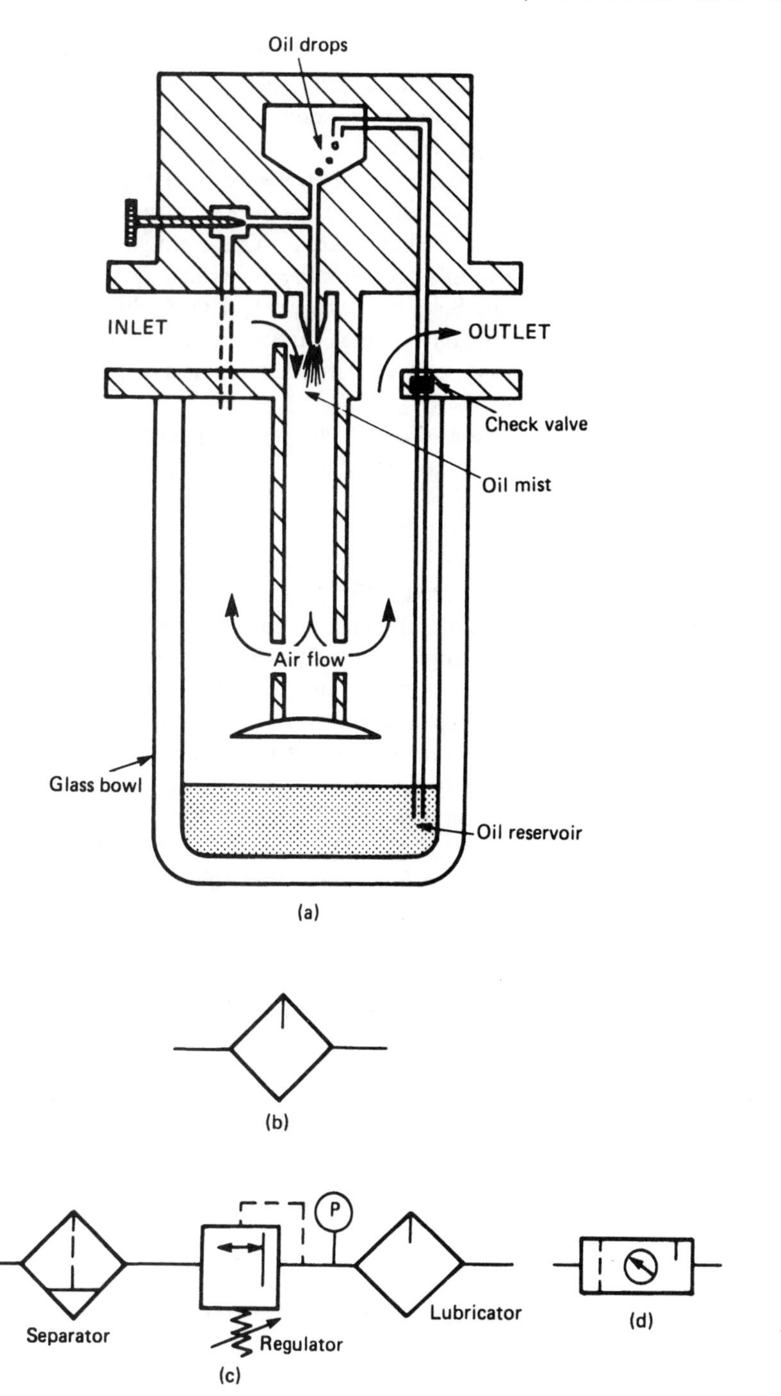

Fig. 16.15 *Lubricators and service unit. (a) Construction. (b) Symbol for lubricator. (c) Service unit. (d) Composite symbol.*

Flow velocities in pneumatics can be quite high, leading to substantial pressure drops. Pneumatic distribution systems usually operate at a higher than required pressure, with local pressure regulation as shown. This pressure regulation can take three forms, shown in Fig. 16.16.

Load A uses a regulator which controls pressure by a variable restriction, essentially controlling the flow to maintain constant pressure. This arrangement requires some minimum flow. If the load takes no air the pressure downstream of the regulator will rise to the supply pressure. Pressure can only be reduced by air being passed through the load. Such devices are called 'non-relieving regulators'.

Load B uses a three-port regulator which can vent air from the load to reduce pressure when required. This type of regulator can handle a 'dead-end' load, and is called a 'relieving regulator'.

The final load uses a large air volume beyond the capacity of a simple in-line regulator. This requires a pressure regulator valve controlled by a separate regulator.

16.4.2. Relief valves

At first sight a relief valve, illustrated in Fig. 16.17, appears similar to a hydraulic (or pneumatic) check valve. The difference in each case is the strength of the spring. In a check valve the spring is relatively weak and serves only to seat the valve. In a relief valve the spring is strong and determines the pressure at which the valve cracks. The spring tension is adjustable to set the relief pressure.

Once cracked, the flow is a function of the excess pressure, an increase in pressure leading to an increase in flow. As stated earlier, a relief valve must be capable of handling the full line flow.

Safety valves fulfil a similar purpose to relief valves, but behave slightly differently. A safety valve goes to fully open once the set pressure is reached

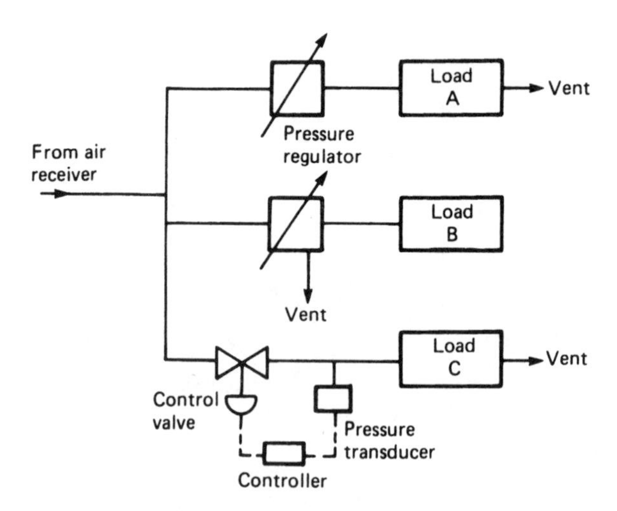

Fig. 16.16 *Types of pressure regulator.*

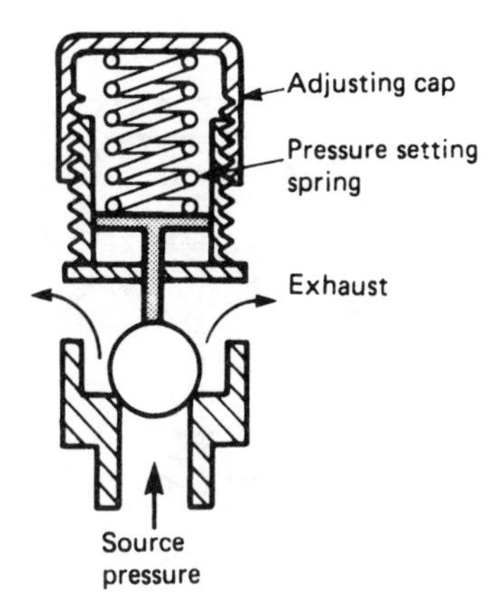

Fig. 16.17 *Relief valve.*

(unlike the flow/pressure relationship of the relief valve), and it is usual for a safety valve to remain open until reset manually.

16.4.3. Pressure regulating valves

Figure 16.18 shows the construction of a non-relieving pressure regulator, as connected to load A in Fig. 16.16. The flow through the valve is controlled by a poppet connected to a spring-tensioned diaphragm. The outlet pressure is applied to the lower face of the diaphragm.

If the outlet pressure falls the spring forces the poppet down, increasing the air flow and hence the load pressure. Similarly a rise in outlet pressure results in the diaphragm moving up and reducing the flow. The valve balances when the outlet pressure acting on the diaphragm balances the spring force.

A relieving pressure regulator is shown in Fig. 16.19. This can deal with dead-end loads similar to load B in Fig. 16.16. If the outlet pressure falls, the

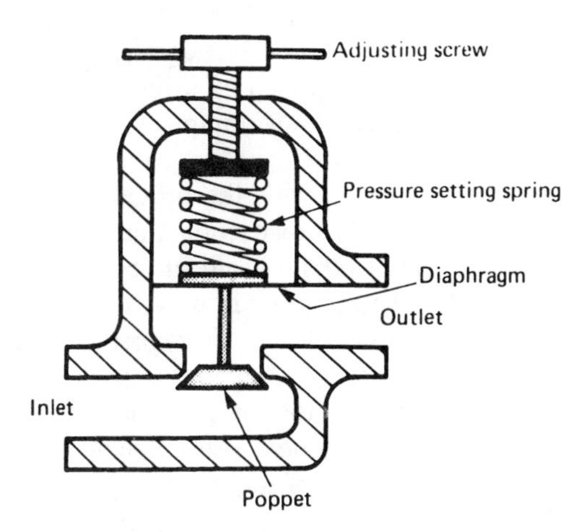

Fig. 16.18 *Non-relieving pressure regulator.*

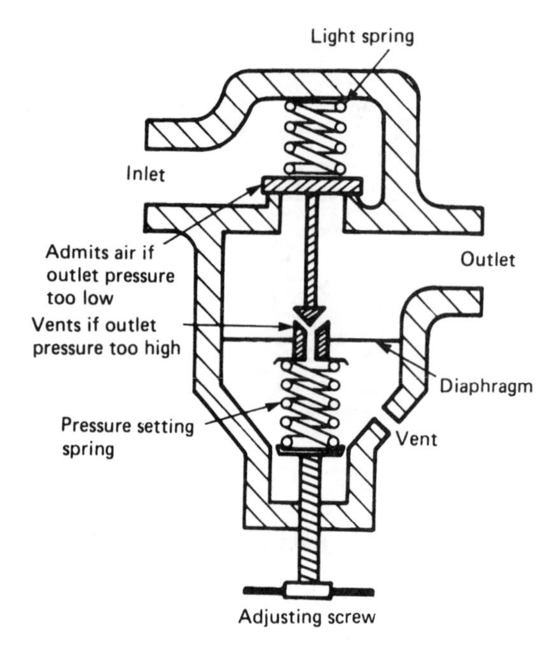

Fig. 16.19 *Relieving pressure regulator.*

poppet valve is pushed up by the adjustable spring, admitting air to the load. The valve at the base of the valve stem is closed, blocking off the vent.

If the outlet pressure rises, the diaphragm is forced down by the increased pressure. This causes the poppet valve to close and the valve between the stem and diaphragm to open, allowing air to pass from the load to the vent port and thereby reducing the outlet pressure.

The valve will balance with the diaphragm just admitting sufficient air to keep the load at the pressure set on the tensioning spring.

Improved performance can be obtained by using a pilot operated regulator. Figure 16.20 shows a relieving regulator for simplicity, but the technique can also be used for a non-relieving regulator.

The outlet pressure is compared with the preset spring force at the pilot diaphragm. Inlet air is bled through a restriction, and either applied to the main diaphragm or vented according to the movement of the pilot diaphragm. If the outlet pressure falls the pilot diaphragm will move down, sealing the ball vent valve. The supply pressure now is applied to the main diaphragm, allowing more air to flow to the load.

If the outlet pressure rises, the pilot diaphragm is lifted and the ball valve vents the air bleed. The main diaphragm now rises, closing the poppet valve and opening the vent through the centre of the poppet spool.

Figures 16.18, 16.19, and 16.20 use spring compression to set the required pressure. If remote pressure setting is required, a pilot pressure can simply be

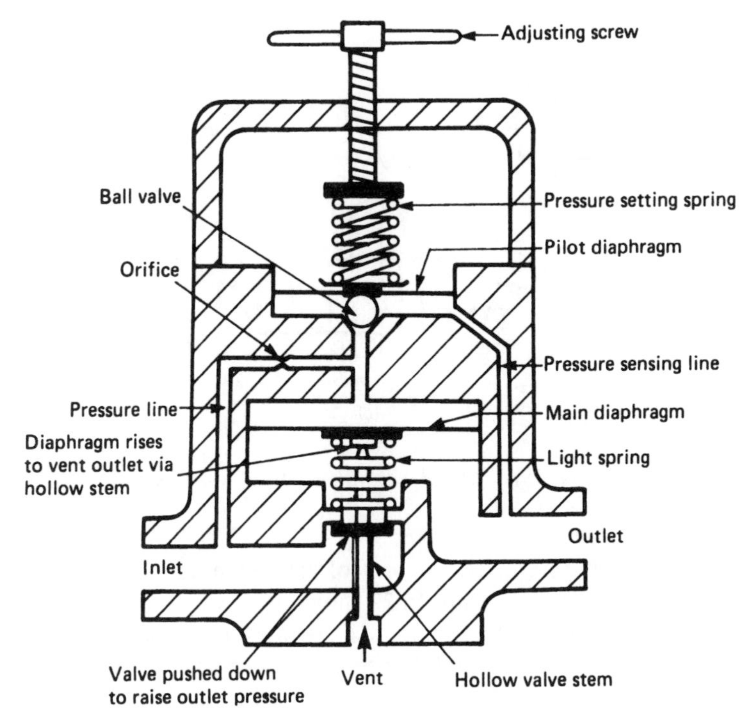

Fig. 16.20 *Relieving pilot operated regulator.*

applied to the sensing diaphragm. The pilot pressure then determines the regulation pressure.

16.5. Control valves

16.5.1. Introduction and symbols

Pneumatic control valves are used to control the flow of air to and from actuators and other devices driven by compressed air. These valves are similar to the hydraulic valves described in Chapter 15, but obviously differ in detail of construction, seal, material, etc.

A valve is described by its number of connections (called ports), and the number of control positions. Symbols for pneumatic valves are similar to those described for hydraulic valves in Section 15.5.1. Figure 16.21a is therefore a two-port, two-position valve (written 2/2-way valve), Fig. 16.21b a three-port, two-position valve (3/2-way) and Fig. 16.21c a 4/3-way valve.

The supply port is usually labelled P (for pressure), and vent ports R, S, etc. Working ports are designated A, B, C, etc., and control ports (e.g. pilot lines) Z, Y, X, etc. The control positions are sometimes labelled a, b, c, etc., with the normal de-energised position denoted 0.

Actuator symbols (solenoid, spring, button, etc.) are identical to those outlined in Section 15.5.1.

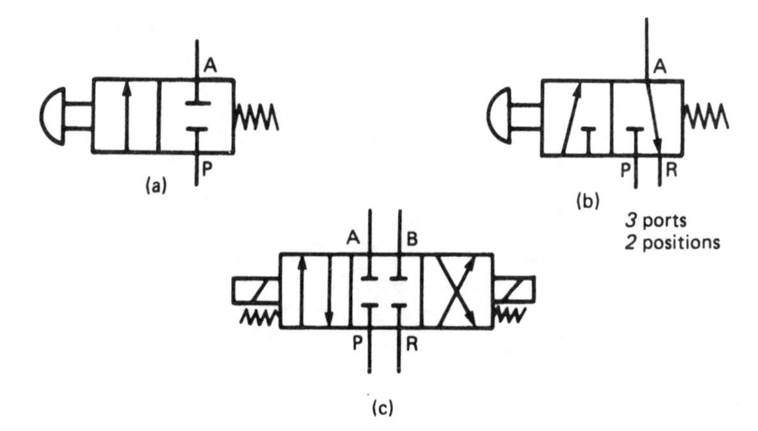

Fig. 16.21 *Valve descriptions. (a) 2/2-way value. (b) 3/2-way valve. (c) 4/3-way valve.*

16.5.2. Valve types

There are essentially three types of control valve: the poppet valve, the spool valve and the rotary valve. Figure 16.22 shows a 3/2-way poppet valve. Button actuation is shown, but it could equally well operate by solenoid or pilot pressure. In the de-energised state, the A and R ports are connected by the hole through the centre of the plunger, and the P port is blocked. When the button is pressed, the plunger descends and contacts the valve disk, sealing off the A and R ports. Further movement pushes the valve disk off its seat, connecting the P and A ports. The valve thus acts as in Fig. 16.21b.

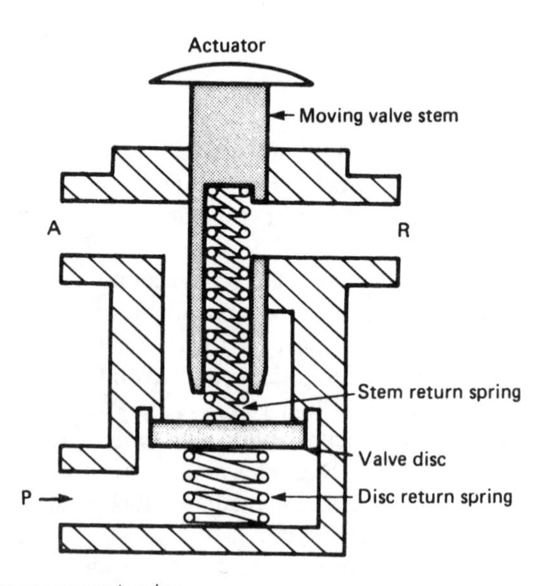

Fig. 16.22 *3/2-way poppet valve.*

Spool valves utilise a moving spool with raised lands which block or uncover the required ports. Figure 16.23a shows the spool valve equivalent of the poppet valve of Fig. 16.22. In the de-energised state, as shown, ports A and R are connected. When the button is pressed, the spool moves over to connect ports A and P and block port R.

Spool valves require less operating force than poppet valves, as pressure forces are equal and opposing on the land faces. The only force the actuator has to overcome is the restoring spring force. On the poppet valve, the actuator has to overcome the spring force plus the air pressure acting on the valve disk.

The action of a spool valve can also be interchanged easily. The valve of Fig. 16.23a can be converted to normally energised by swapping the P and R connections. If this were tried on the poppet valve of Fig. 16.22, the air pressure would open the valve disk, connecting all three ports simultaneously.

Poppet valves are limited to relatively simple operations, whereas spool valves

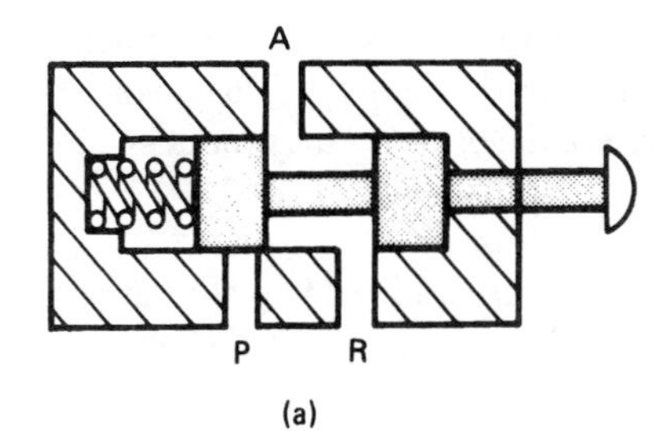

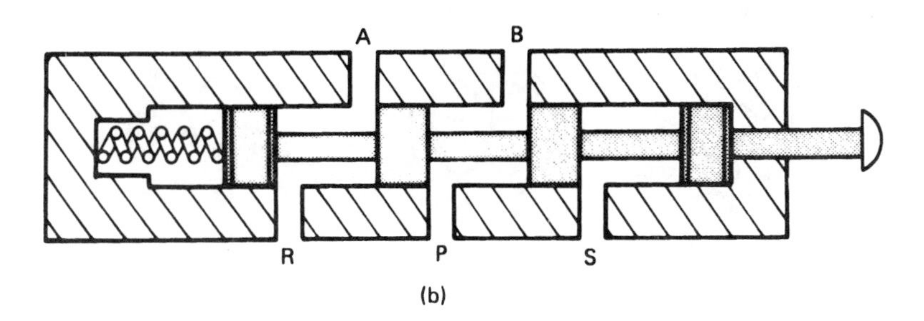

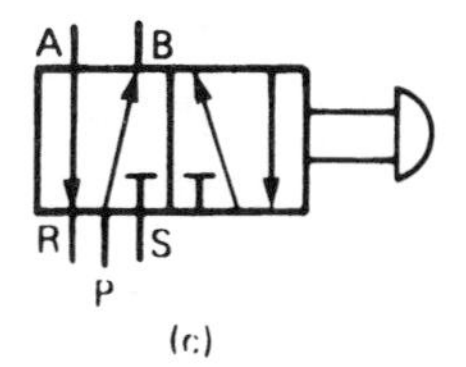

Fig. 16.23 *Spool valves. (a) 3/2-way valve. (b) 5/2-way valve. (c) Symbol for 5/2-way valve.*

can be constructed to almost any desired complexity. Figure 16.23b shows a 5/2-way valve with the operations denoted by the symbol of Fig. 16.23c.

The final valve type, the rotary valve, is shown in Fig. 16.24. These utilise a rotating spool with drilled passages which align with ports in the valve casing to give the required action. The valve shown is a 4/3-way valve.

Where the valve construction results in uneven forces on the valve disk or spool (as is inherent in poppet valves) a pilot valve can be used. A simple application is shown in Fig. 16.25. The actuator causes the small pilot valve to open. This puts full pressure on to the control spool. As this has larger area than the main valve disk, the valve opens. Releasing the actuator vents the space above the pilot spool, causing the valve to close. Operating force is the product of the line pressure and the (small) pilot valve disk.

16.6. Actuators

16.6.1. Linear actuators

A pneumatic cylinder, or ram, is used where a pneumatically controlled linear motion or force is required. Figure 16.26a shows a simple cylinder. Air is introduced to the right of the piston. This produces a force on the piston given by:

$$F = P \times A \tag{16.7}$$

where P is the gauge pressure and A the piston area. The force available at the shaft is slightly less than the force at the piston because of the opposing force from the return spring.

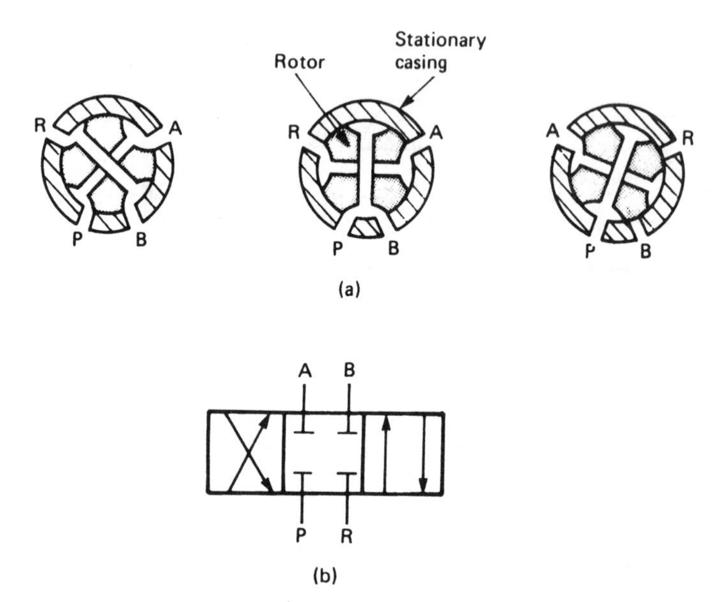

Fig. 16.24 *The rotary valve. (a) 4/3-way valve. (b) Operation of rotary valve.*

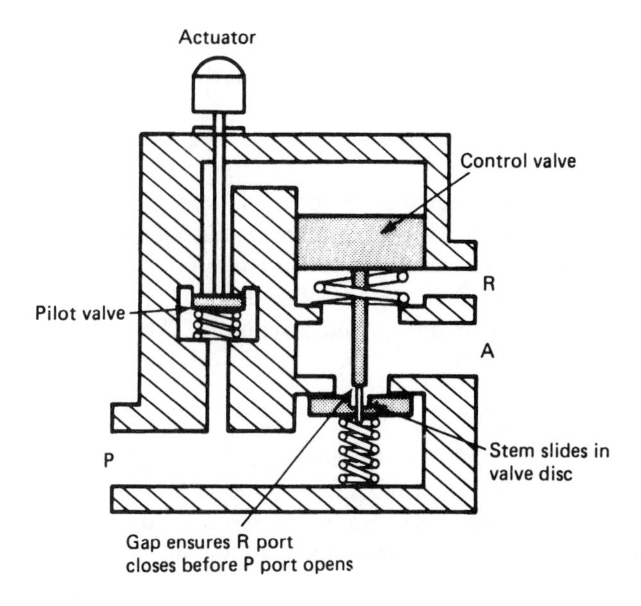

Fig. 16.25 *3/2-way pilot operated poppet valve.*

If the opposing force from the load is less than the shaft force the piston will move to the left with a velocity:

$$V = \frac{Q}{A} \tag{16.8}$$

where Q is the volume of air delivered to the piston per unit time. This flow rate is determined by the flow capacity of the valve controlling the cylinder.

If the space to the right of the piston is vented to atmosphere, the restoring spring will move the piston to the right. Normally the force available from the spring is small (to avoid reducing the force from equation 16.7 significantly) so a cylinder similar to Fig. 16.26a (called a single acting cylinder) can only deliver force in one direction. Typical applications are cylinders for clamping work under a machine tool or a lifting cylinder which returns under gravity. Single acting cylinders with spring return are, of necessity, longer than the stroke to give space for the spring to compress.

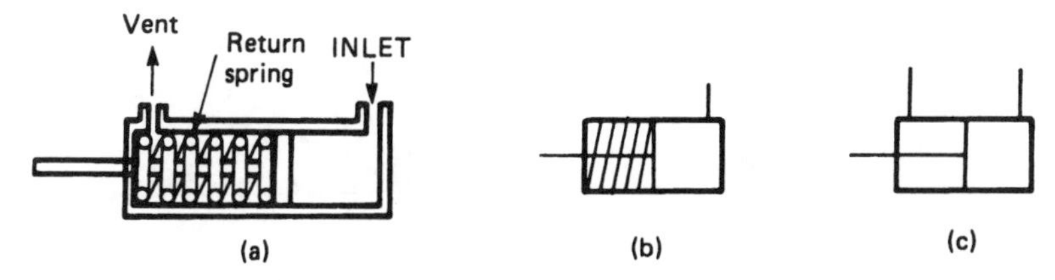

Fig. 16.26 *Linear actuators. (a) Single acting cylinder. (b) Symbol. (c) Double acting cylinder.*

Double acting cylinders, as in Fig. 16.26c, are used where force is required in both directions of motion. It should be noted that the available force is not equal in both directions because the area on one side of the piston is reduced by the output shaft. Control of a double acting cylinder requires a 4/2-way or 4/3-way valve (see Fig. 16.21).

The construction of pneumatic cylinders is generally similar to that of hydraulic cylinders, described in Section 15.6. Details such as stroke restrictors to reduce side loads and cylinder cushions are frequently found in pneumatic cylinders.

Valve actuators are a specialised form of actuator where a linear displacement proportional to input pneumatic pressure is required. These are described further in Section 16.6.4.

16.6.2. Rotary actuators

Rotary actuators, or pneumatic or air motors as they are more commonly called, convert the motion and pressure of air flow to mechanical torque and rotational motion. In general the available torque is determined by the supply pressure, typically 5–10 bar, and the rotational speed by the air flow rate.

Figure 16.27 shows the torque and power curves for a typical motor. It can be seen that the torque is maximum in the zero speed (stalled) state. Like a hydraulic motor (but unlike an electric motor) a pneumatic motor can be stalled indefinitely without damage. The power curve, Fig. 16.27a, is a maximum at approximately half the no-load maximum speed.

Power, torque and rpm are related by:

$$\text{power} = K \times \text{torque} \times \text{rpm} \tag{16.9}$$

where K is a scaling constant dependent on the units used. Imperial units are still commonly used in pneumatics, for which:

$$\text{HP} = \frac{\text{torque (inch-pounds)} \times \text{rpm}}{63025} \tag{16.10}$$

Rotary actuators use a considerable volume of air, typically 1 to 2 cubic metres per minute per kW at an operating pressure of 5 to 6 bar. Care must be taken to

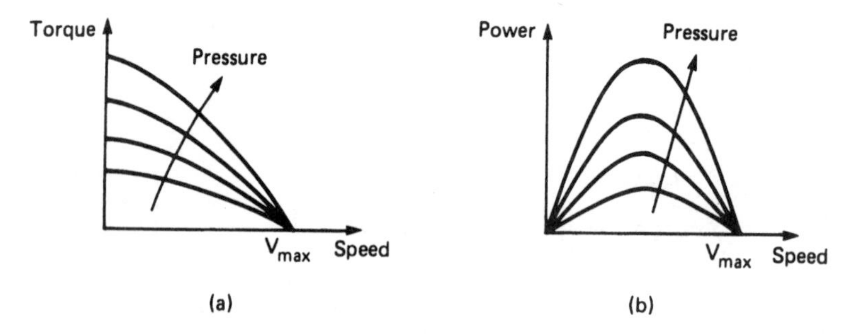

Fig. 16.27 *Torque/power speed curves for rotary actuator. (a) Torque. (b) Power.*

ensure that the compressor can supply an adequate volume of air and the piping does not introduce pressure losses.

The construction of pneumatic motors is generally similar to that of hydraulic motors, described in Section 15.7. The commonest types are vane motors and axial or radial piston motors. Most are positive displacement motors, the exception being high speed turbines used in some hand tools.

16.6.3. Flow control valves

In a large number of pneumatic systems, the final actuator controls the flow of some fluid-liquid, gas or steam, for example. The following section describes valve positioning actuators, but it is first useful to describe the operation of common flow control valves.

All control valves operate by inserting a variable restriction in the flow path. Figure 16.28 shows the three commonest arrangements. The plug, or globe, valve operates by moving a tapered plug, thereby varying the gap between the plug and

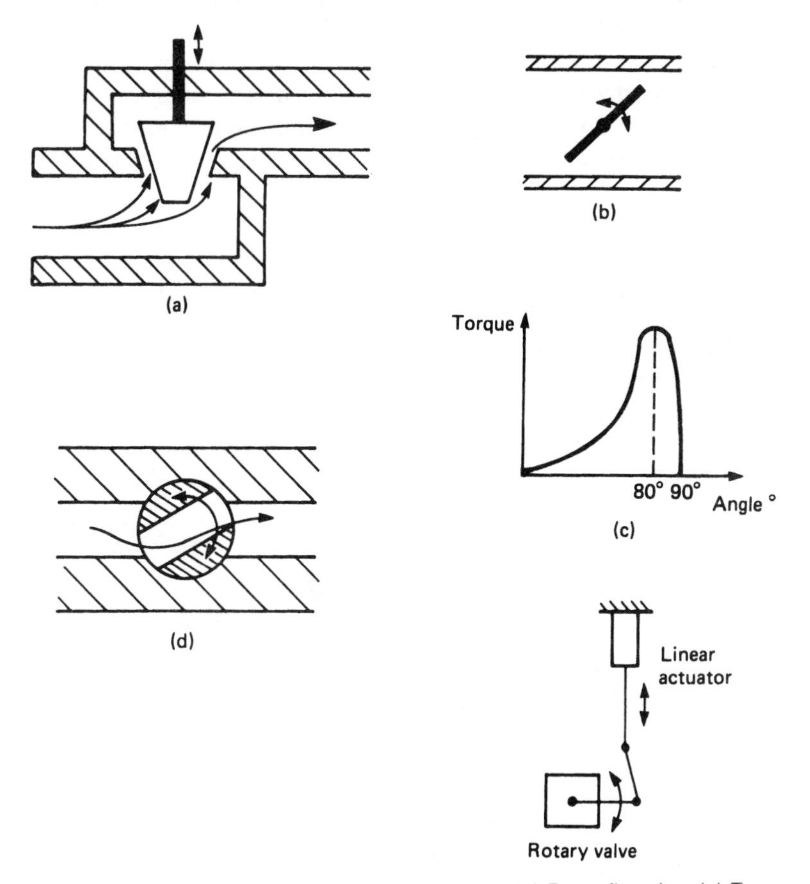

Fig. 16.28 *Flow control valves. (a) Plug (globe) valve. (b) Butterfly valve. (c) Torque on butterfly valve. (d) Ball valve. (e) Use of linear actuator for valve positioning.*

the valve seat. The flow is controlled by linear movement of the valve stem. Normally the plug is guided by a cage (not shown for simplicity) to prevent sideways movement.

Butterfly valves, such as in Fig. 16.28b, utilise a circular disk which is rotated to vary the restriction. The leakage of a butterfly valve in the shut-off position is not as good as that which can be obtained with globe valves. Butterfly valves can, however, be constructed to almost any required size. Dynamic torque effects on the disk limit the travel to about 60 °C from the closed position. Figure 16.28c shows the torque acting on a butterfly valve related to valve position.

The ball valve of Fig. 16.28d uses a variable restriction obtained by rotating a ball with a through hole which moves within an accurately machined seat. Often a V-notch hole is used. Ball valves have excellent shut-off characteristics. Both the butterfly and ball valve require a rotary shaft motion. This is usually obtained from a linear actuator acting on a lever, as in Fig. 16.28e.

The dynamic forces on the restriction act on the actuator shaft. In Fig. 16.29a the flow opposes the opening of the valve, whereas in Fig. 16.29b the flow assists the closing. The latter case is particularly difficult to control at low flows as the plug tends to slam into the seat. This effect can easily be observed by trying to control the flow of water from a bath or basin with the plug. A balanced valve, with little or no reaction on to the actuator shaft, can be constructed with two plugs and seats, as in Fig. 16.29c. With careful design the opposing dynamic

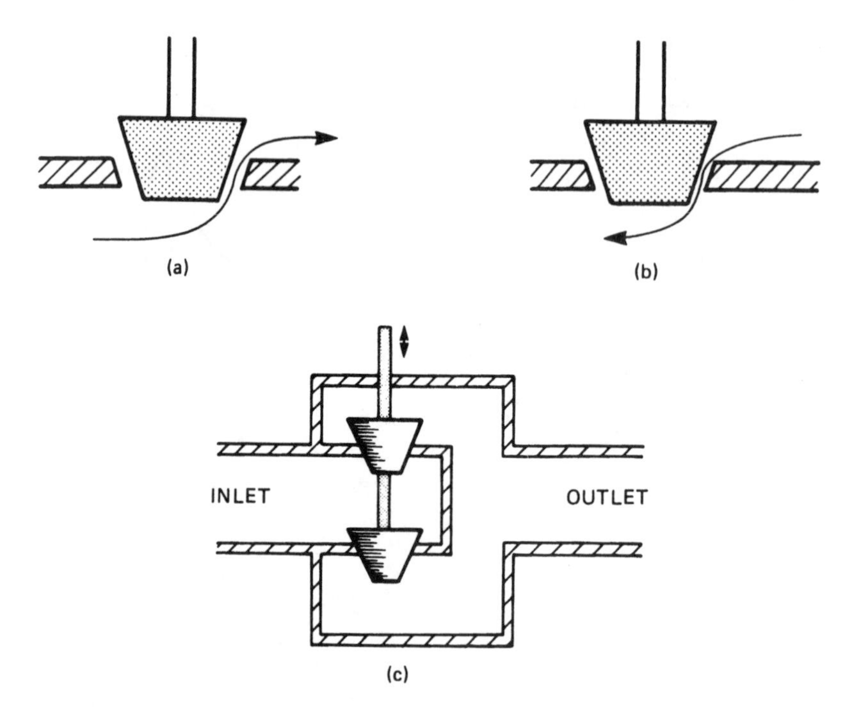

Fig. 16.29 *The bathtub plughole effect. (a) Flow assist opening. (b) Flow assist closing. (c) Balanced valve.*

forces can be made to cancel. Such a valve, however, has a rather high shut-off leakage as manufacturing tolerances will cause one plug to seat before the other.

The valve characteristic relates the flow through the valve to the valve opening. Figure 16.30 shows the three commonly used characteristics. These are specified at constant pressure drop across the valve, a condition which is rare in real-life systems. The choice of valve characteristic is chosen to give a linear flow/position characteristic in the specific application. The differential pressure across a valve in an installation can increase or decrease with increasing flow, dependent on the behaviour of the rest of the system. Quick opening and equal percentage valves compensate, to some extent, for change in differential pressure with flow to give a linear flow/position relationship.

Valve selection can be a complex procedure. Obvious considerations are the pressure, temperature, flow range/turndown and chemical composition of the controlled fluid. The differential pressure across the valve must be calculated from a knowledge of the characteristics of the valve and the rest of the system. With liquids, care must also be taken that cavitation or flashing does not occur in the low pressure region just downstream of the valve. Cavitation and flashing are accompanied by valve damage and excessive noise.

16.6.4. Valve actuators

A pneumatically controlled valve regulates flow by the movement or rotation of the valve shaft, as described in the previous subsection. The controlling signal is a pneumatic pressure, the shaft movement being proportional to the applied pressure. An input signal range of 0.2 to 1 bar, for example, could cause a shaft movement of zero to 50 mm and a flow change from zero to 1000 litres per minute. Such an arrangement could be represented by Fig. 16.31, with an actuator having a gain of 62.5 mm/bar and the valve a gain of 20 litres/min/mm.

A valve actuator operates in a different manner to the linear actuators of Section 16.6.1. A linear actuator produces a *force* which is proportional to the applied pressure. A valve actuator produces a *displacement* which is proportional to applied pressure.

A typical actuator is shown in Fig. 16.32a. The controlling signal is applied to

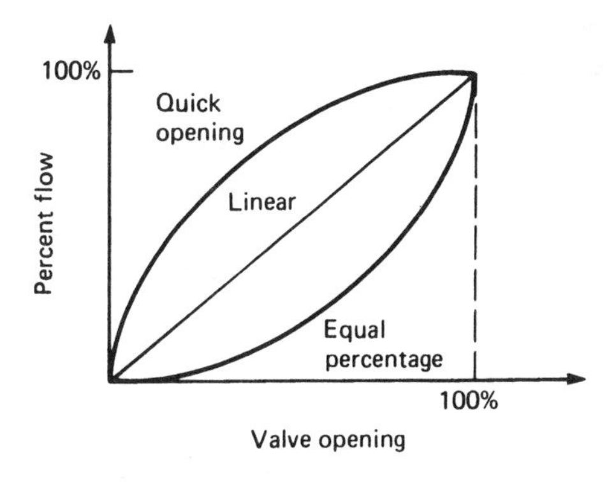

Fig. 16.30 *Valve characteristics.*

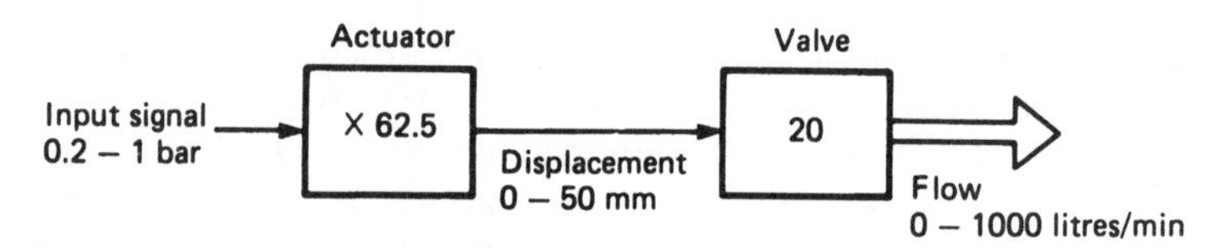

Fig. 16.31 *The gain of pneumatic components.*

the top of a piston which is sealed by a flexible diaphragm. The pressure produces a downward force, given by equation 16.7. This is opposed by a force from the restoring spring. As the control signal increases, the increased pressure causes the piston to move down until the force from the now more compressed spring again balances the force on the piston. Note that there is a fundamental difference between the spring action in Fig. 16.32a and the spring in Fig. 16.26a. The latter is a relatively weak spring to return the piston when the air space is vented.

The actuator gain (movement/applied pressure) is determined by the spring

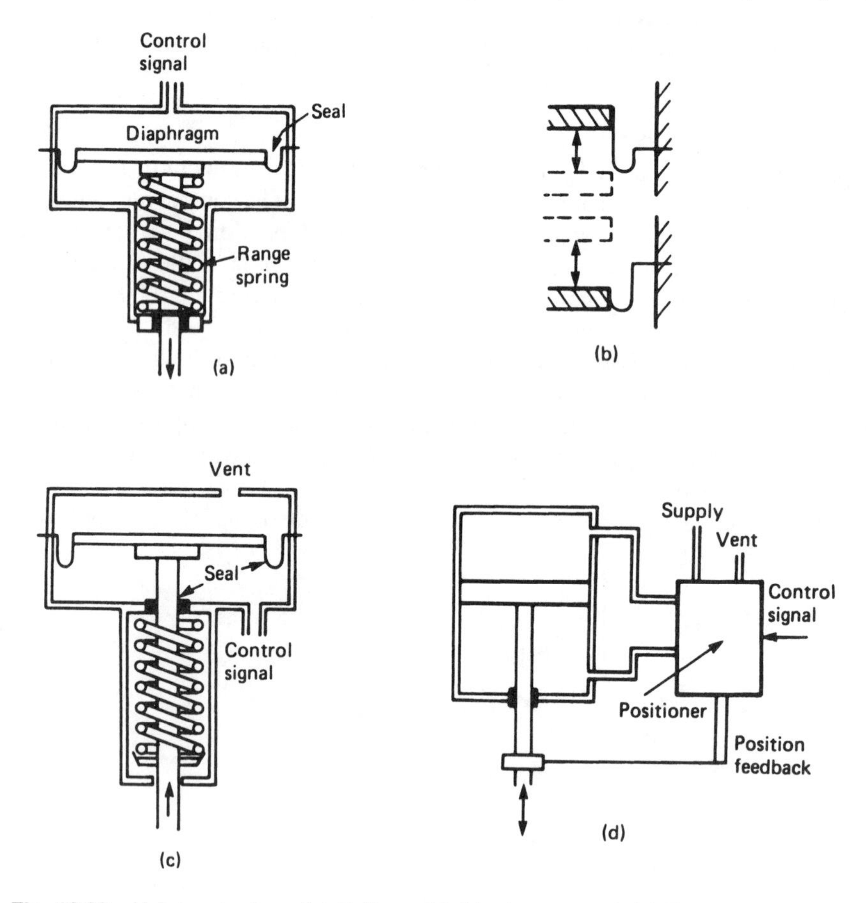

Fig. 16.32 *Valve actuators. (a) Fail up. (b) Diaphragm seal. (c) Fail down. (d) Piston actuator.*

stiffness. The pressure at which the actuator starts to move (0.2 bar in the example above) is set by the pre-tension adjustment.

The action of the rubber diaphragm is shown in Fig. 16.32b. The diaphragm ensures that the effective piston area remains constant over the full range of actuator travel.

The shaft on the actuator of Fig. 16.32a extends for increasing pressure (and fails with the shaft fully in). In Fig. 16.32c the pressure is applied to the underside of the piston and the spring force is reversed. This arrangement gives a shaft motion which extends for decreasing pressure (and fails with the shaft fully out). This is known as a reverse acting actuator. Note that an O ring seal is necessary on the valve shaft.

The net force acting on the actuator shaft is the algebraic sum of the $P \times A$ force from the control signal, the restoring spring force and dynamic forces from the valve being controlled. The effect of valve forces will therefore be the creation of an offset error. This can be reduced by increasing the control signal pressure range or the diaphragm area. There are limits, however: the diaphragm can only withstand relatively low pressures, and the physical size of the actuator must be reasonable. Piston actuators such as in Fig. 16.32d are used, operating at high pressure, where large operating forces are needed. These are similar to the linear actuators of Section 16.6.1, but incorporate a closed-loop positioning measuring device and closed loop position control to make the shaft displacement correspond to the input low pressure control signal. Positioners are described in Section 16.7.5.

16.7. Process control pneumatics

16.7.1. Signals and standards

Process variables (pressure, flow, temperature, level, etc.) are commonly represented by electrical voltages or current. A temperature measurement in the range 0 to 100 °C, for example, could be represented by a current from 4 to 20 mA. In this scaling 50 °C would be represented by a current of 12 mA.

Process variables can also be represented by pneumatic pressure. Liquid level in a tank varying between 0 and 4 metres could, say, be represented by a pressure from 0.2 to 1.0 bar (20 to 100 kPa). With this representation, a pressure of 0.8 bar would correspond to a level of 3 metres.

Pneumatic representation has advantages over electrical methods in certain circumstances. Explosive atmospheres are common in the chemical and petrochemical industries. The use of electrical transducers and actuators is potentially dangerous, and requires the complexity of zener barriers or intrinsically safe equipment which can be installed in dangerous areas without special precautions.

Early process controllers were employed before semiconductors were available, and the only available technology was pneumatics. A great deal of design and application experience has evolved around pneumatic control. Companies with existing pneumatic control understandably will stay with a level of technology that is readily comprehended by their staff.

The usual signal pressure range is 0.2 to 1 bar (20 to 100 kPa) or the imperial equivalent 3 to 15 psig. Other signal ranges tend to be a simple multiple (e.g. 6 to 30 psig). Almost all use an offset zero (e.g. 0.2 bar in a 0.2 to 1 bar system) for speed of response and protection from damage to piping.

Figure 16.33a shows a simple level measurement system where a pneumatic level transducer drives a remote level indicator (which is, of course, a suitably scaled pressure gauge). The mechanism of the transducer need not concern us, but essentially it connects the signal pipe to the supply pressure to increase the reading, or to the vent to reduce the reading. In both cases, the indicator will follow an exponential curve, as in Fig. 16.33b. The speed of response for increasing signal can be increased by raising the supply pressure (a supply of about 2 bar would be used for a 0.2 to 1 bar signal). If a zero pressure corresponded to zero signal, the response of the system to a decreasing signal would be very slow. The use of the offset zero gives a similar response for both increasing and decreasing signals.

If the signal line is fractured it will vent to atmosphere, giving a pressure of 0 bar. This is below instrument zero and will cause the indicator to read off scale negative. An offset zero thus gives protection against damage to the transmission path.

The main disadvantage of pneumatic systems is a slow speed of response. Suppose the transducer of Fig. 16.33a has to follow a step increase of level. This will require a step change of pressure. There are two components to the response of the indicator. There is an inherent transit delay as the pressure step can only travel at the speed of sound (330 m s^{-1}). Piping runs of several hundred metres are common in even medium size plants, so delays of the order of a few seconds are inevitable.

The speed of response can also be adversely affected if the pressure change is associated with a volume change at the indicator or actuator (moving a piston in a cylinder, for example). The change of volume must be supplied by the transducer, and will result in an exponential rise of pressure. A similar effect occurs because the increase in pressure is achieved by a transfer of air into the piping/indicator. The volume of air required is a function of the total system

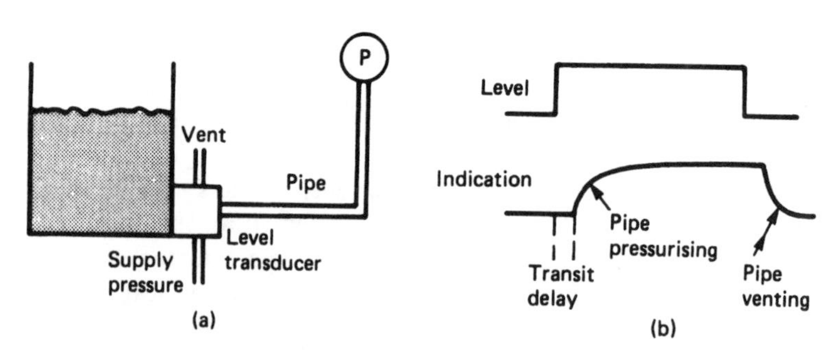

Fig. 16.33 *Response of pneumatic systems. (a) Pneumatic instrumentation system. (b) Response.*

volume. Fast response from a pneumatic system therefore requires short runs, small volume systems and 'loads' which do not change their volume with pressure changes.

16.7.2. The flapper/nozzle

The heart of most pneumatic process control devices is an assembly which converts a small physical displacement to a pressure change. This is invariably a variation on the flapper nozzle shown in Fig. 16.34a.

An air supply, typically at 2 bar, is applied to a nozzle via a restriction. Air flowing from the nozzle will cause a pressure drop across the restriction and the output pressure will be lower than the supply pressure by an amount related to the flow from the nozzle.

The input displacement is applied to a flapper and varies the gap between the flapper and the nozzle. This varies the flow from the nozzle and consequently the output pressure. A typical relationship is shown in Fig. 16.34b; note the small range of the displacement. At the extremes of the curve the relationship is very

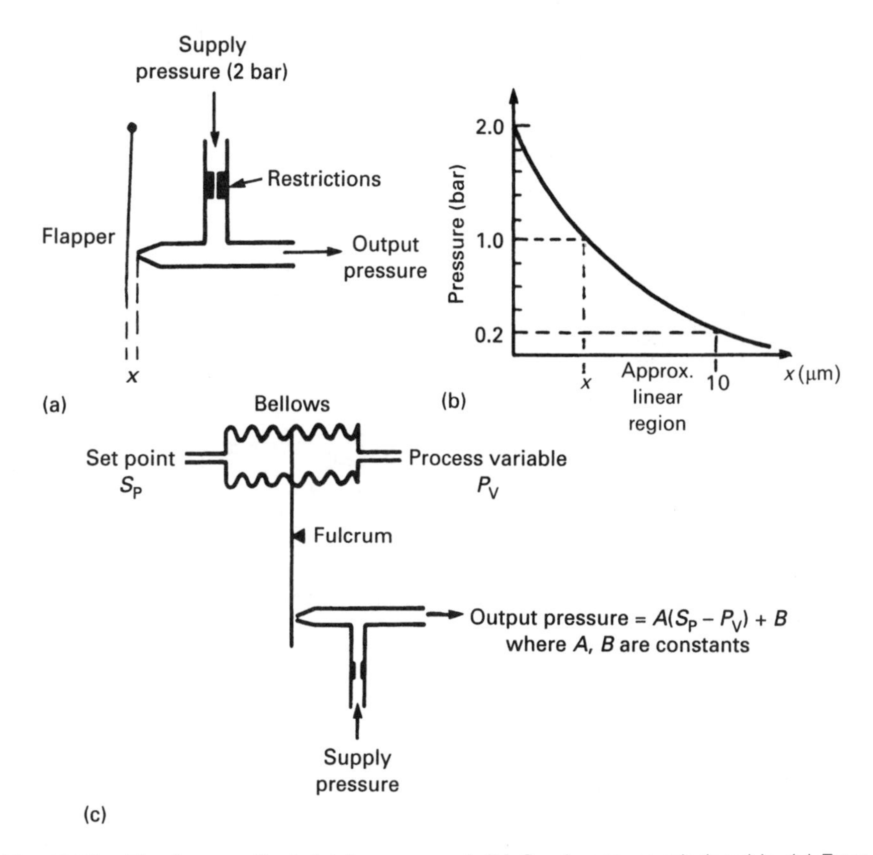

Fig. 16.34 *The flapper effect. (a) Arrangement. (b) Gap/pressure relationship. (c) Error amplifier.*

non-linear, but can be considered to follow a straight line over the normal signal range of 0.2 to 1 bar.

Figure 16.34c shows an application which generates an error signal between pneumatic signals representing a set point and the process variable. These are applied to opposing bellows, and any difference will cause a deflection of the flapper and a corresponding change in output pressure. When the S_P and P_V signals are equal, the output pressure will be mid range (0.6 bar for a 0.2 to 1 bar system). Sections 16.7.4, 16.7.5 and 16.7.6 discuss further applications of flapper/nozzles.

The arrangement of Fig. 16.34a supplies an output pressure, but is incapable of supplying any significant volume of air. For this reason a flapper/nozzle is invariably followed by an air amplifier or volume booster, described in the following section. Force balance techniques, described below, are also used and effectively operate with a fixed gap to overcome the non-linearities of the flapper/nozzle.

16.7.3. Air amplifiers, boosters and relays

Air amplifiers, represented by Fig. 16.35a, are commonly used to increase or decrease pneumatic pressure by a fixed multiple or to supply increased volume of air. A 2 × amplifier, for example, could be used to convert a 0.2 to 1 bar linear signal to a 0.4 to 2 bar signal.

The input signal controls the air flow from the supply to the load in such a way

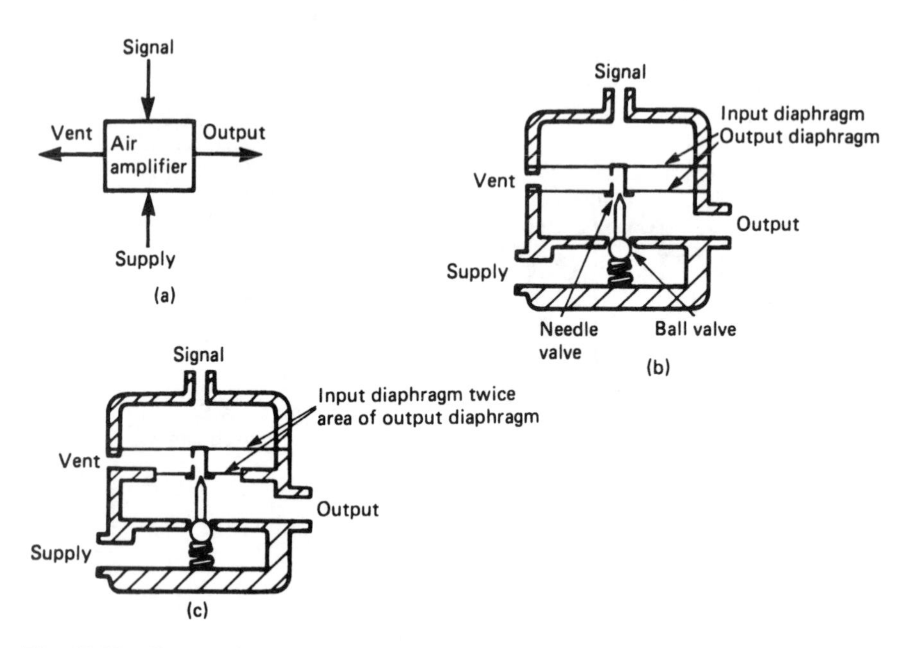

Fig. 16.35 *Pneumatic amplifiers. (a) Block diagram. (b) Unity gain (volume booster). (c) Air amplifier (gain × 2).*

that the correct relationship is maintained. When the input signal falls, the output pressure is reduced by releasing air via the vent port.

Figure 16.35b shows a typical arrangement for a unity gain device. The signal is applied to the input diaphragm, and the supply to the output diaphragm. If the force from the input pressure is larger than the force from the output pressure, both diaphragms will move down, forcing the ball valve off its seat and admitting air to the output until the input and output pressures equalise.

If the input pressure falls both diaphragms move up, closing the ball valve and opening the exhaust valve. Air now vents until the output pressure falls to the correct level. The output pressure thus follows the input pressure.

The input port requires negligible air volume changes, and as such is capable of being driven from a flapper nozzle. The output port can supply a large volume of air. Note, however, that the device will maintain pressure at the outlet port and cannot compensate for flow related pressure drops downstream of the device.

Figure 16.35b has equal area input and output diaphragms, and hence has unity gain. As such it is often called an air booster. If the input and output diaphragm have unequal areas, an air amplifier of non-unity gain is obtained. Figure 16.35c has a larger input diaphragm and acts as an amplifier. The gain is the ratio of the area. A reducing device has a larger output diaphragm.

Figure 16.36 utilises the force balance principle to boost the available air volume from a flapper/nozzle. The air relay of Fig. 16.36a applies the signal to a diaphragm, the extension of which is proportional to the applied pressure. The diaphragm motion is used to admit air to, or bleed air from, the outlet port. In this arrangement there is no feedback, and the inlet pressure controls the flow to, or from, the output.

The air relay of Fig. 16.36a is combined with a flapper/nozzle, as in Fig. 16.37b. The pressure variation from the nozzle is applied to the input of the air relay. The output is applied to the input of the air relay. The output is applied to a force balance diaphragm which opposes the force from the error signal ($P_V - S_P$). Any change in this signal will cause air to flow between the air relay and the force balance to the diaphragm until the flapper/nozzle gap is restored to its operating position.

Because Fig. 16.36 operates at a fixed flapper/nozzle gap, any non-linearities from Fig. 16.34b are overcome, and an output pressure linearly related to ($P_V - S_P$) is obtained. The output signal is taken from the air relay, and can hence deliver a reasonable air volume without introducing any error.

16.7.4. Pneumatic controllers

Figure 16.36b can be represented by Fig. 16.37a. This is identical to a proportional controller (the principles of controllers are discussed in Chapter 18). The gain can be adjusted by moving the position of the pivot. Figure 16.37b shows the applications of a pneumatic proportional-only controller to control a pressure regulating valve.

Figures 16.36 and 16.37 operate in the steady state with a fixed flapper/nozzle gap. To achieve this, a closed loop system will stabilise when the forces from the

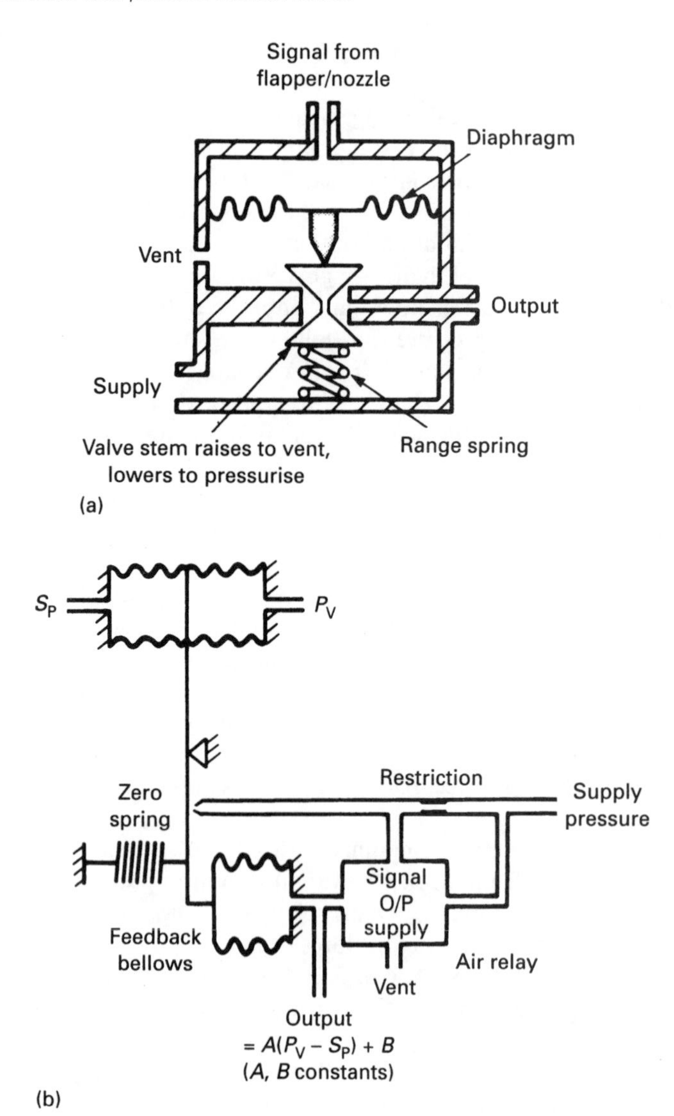

Fig. 16.36 *The force balance principle. (a) Air relay. (b) Proportional controller.*

set point, P_V and feedback bellows balance. This results in a small difference between the P_V and set point pressure called an 'offset error', which is inherent in all proportional only controllers.

Electronic controllers are available to perform three-term control, so called because the output signal is the sum of proportional, integral and derivative terms related to the error. These controllers perform the three-term control function:

$$V_o = K\left(E + \frac{1}{T_i}\int E\,dt + T_d\frac{dE}{dt}\right) \tag{16.11}$$

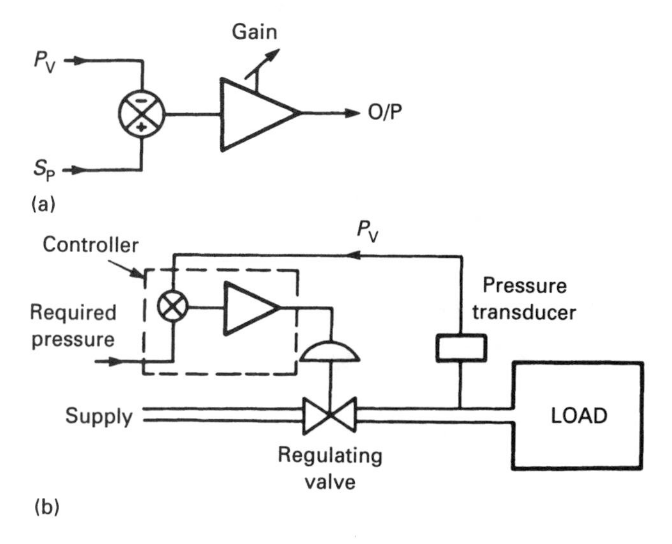

Fig. 16.37 *Use of controller to regulate pressure. (a) Representation of proportional controller. (b) Pressure regulating system.*

where V_o is the output signal, E the error, K the gain, T_i the integral time (sometimes referred to as reset or the inverse repeats per minute) and T_d the derivative time, K, T_i and T_d are adjusted by the process control engineer to give the fastest system response consistent with stability. The integral action of equation 16.11 overcomes the offset error of a proportional controller as the actuator signal will integrate the error. The derivative term adds stability and gives a large correcting signal when the error is changing rapidly.

Figure 16.38a shows how integral action can be achieved in a pneumatic controller. With the integral adjustment valve closed, the controller will act in proportional (P) mode and control with an offset error. Assume the controller is established in P mode, and the integral valve is opened. The integral bellows will oppose the action of the proportional bellows, changing the flapper/nozzle gap and the output signal. The system will stabilise with the flapper/nozzle at the steady state gap, and the set point and P_V equal, and the contributions from the feedback and integral bellows equal. Under these conditions, the output pressure is just correct to make the P_V and the set point equal with no offset.

Derivative action is obtained by restricting the flow to the feedback bellows, giving an output pressure related to the rate of change of error.

Adjustment of K, T_i, T_d in equation 16.11, is achieved by adjusting the beam pivot point (gain K) and the integral and derivative bleed valves. Unlike electronic controllers, pneumatic controllers often exhibit interaction between the three terms and are consequently more difficult to adjust.

16.7.5. Valve positioners

A pneumatic actuator such as those described in Section 16.6.4, gives a displacement which is linearly related to a pneumatic signal. The performance of

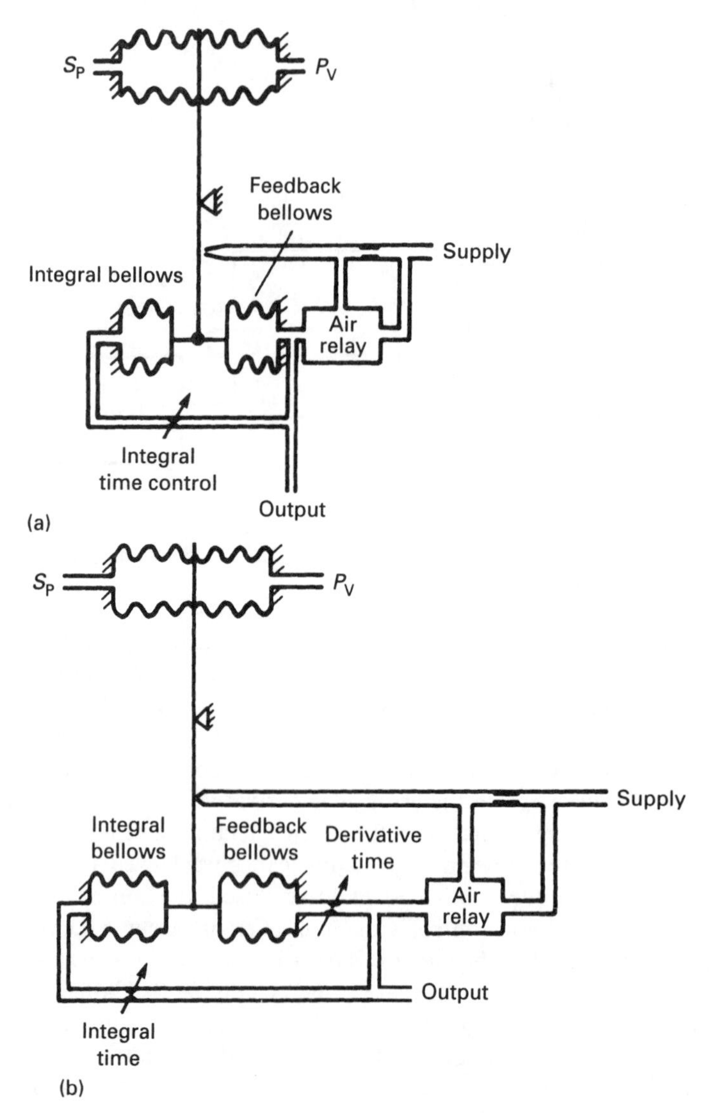

Fig. 16.38 *Pneumatic controllers. (a) Proportional plus integral (P + I) control. (b) Proportional plus integral plus derivative (PID. 3-term) controller.*

actuators can sometimes be improved by the incorporation of a separate actuator position control system, called a valve positioner.

Positioners are mandatory for double acting actuators, such as in Fig. 16.32d, and are useful in the following circumstances:

(a) When accurate valve position is required.
(b) To speed up the response of a valve (particularly in cascade systems where it is essential to have the inner loop faster than the outer loop).

(c) Where volume boosting is needed, e.g. where the device providing the control signal is incapable of driving the valve directly.
(d) Where a pressure boost is required to give the necessary actuator force.

Figure 16.39 shows a positioner using the force balance principle. The position of the actuator is converted to a force by a range spring. This is coupled to an output diaphragm assembly which moves one side of the beam on the flapper/nozzle according to the difference between the spring force and the diaphragm force from the input signal pressure.

The pressure from the flapper/nozzle is fed to the pilot valve diaphragm. If the flapper/nozzle gap is too small, the pilot pressure will increase, moving the spool up. This admits air to the lower connection on the actuator and vents the upper connection, causing the actuator to rise, and reducing the spring force.

Conversely, if the flapper/nozzle gap is too large the pilot pressure will decrease, causing the spool to move down. Air will be admitted to the top actuator connection and vented from the lower, causing the actuator to fall and the spring force to increase.

In both cases the system will balance when the spring force matches the force on the input diaphragm from the input signal, i.e. the actuator position matches the required position.

The positioner zero is adjusted by altering the relative position of the shaft and spring, and the span by altering the effective spring constant.

An alternative positioner, shown in Fig. 16.40, uses the motion balance principle. The valve shaft position is converted to a small displacement by a cam,

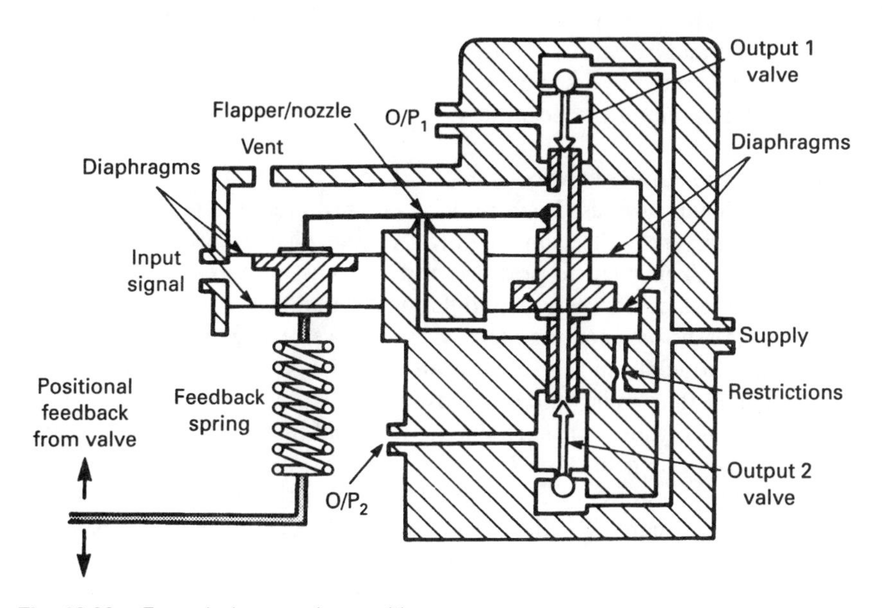

Fig. 16.39 *Force balance valve positioner.*

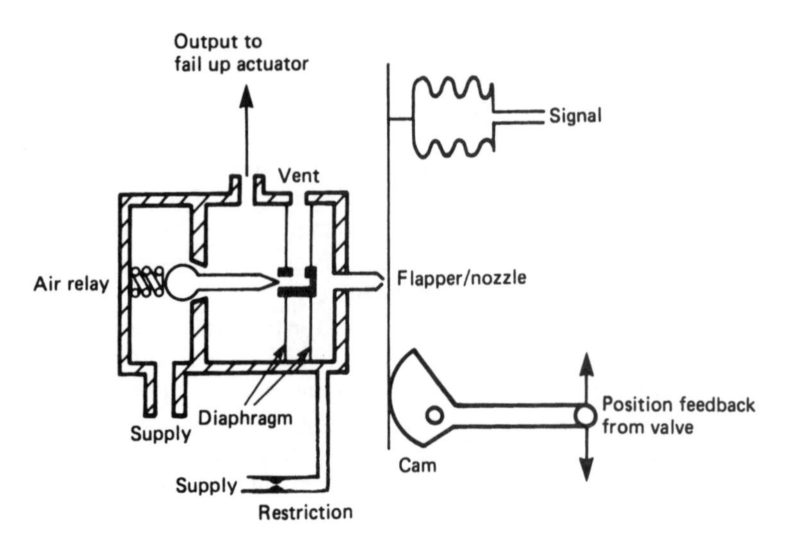

Fig. 16.40 *Motion balance positioner.*

and applied to one end of a flapper/nozzle beam. The input signal pressure is converted to a displacement by bellows at the other end of the beam.

The nozzle pressure is applied to an air relay which admits air to, or vents air from, the actuator until the nozzle gap is correct. At this point the actuator position matches the input signal.

Positioners are usually equipped with gauges indicating supply pressure, signal pressure and actuator pressure(s). Often bypass valves are fitted to allow the signal pressure to be passed direct to the actuator as a temporary measure in the event of a failure of the positioner.

16.7.6. I/P and P/I converters

Pneumatic signals are used where large forces are required or explosive atmospheres are used. Electronic controllers and displays are widely used in control rooms. I/P (current-to-pressure) and P/I (pressure-to-current) converters provide the interface between these differing technologies.

Figure 16.41 shows a common form of I/P converter, again built around the force balance principle and the ubiquitous flapper/nozzle. The signal current is passed through four coils, wired as shown, and causes a rotary torque in the flapper arm. The torque is proportional to the signal current and causes a change in the flapper/nozzle gap.

The resulting pressure controls an air relay, causing the output pressure to rise, or fall until the feedback bellows returns the flapper beam to its balanced position. The output pressure thus follows the input current.

Pressure to current converters are simply pressure transducers (measuring gauge pressure). Pressure transducers are described in Chapter 3.

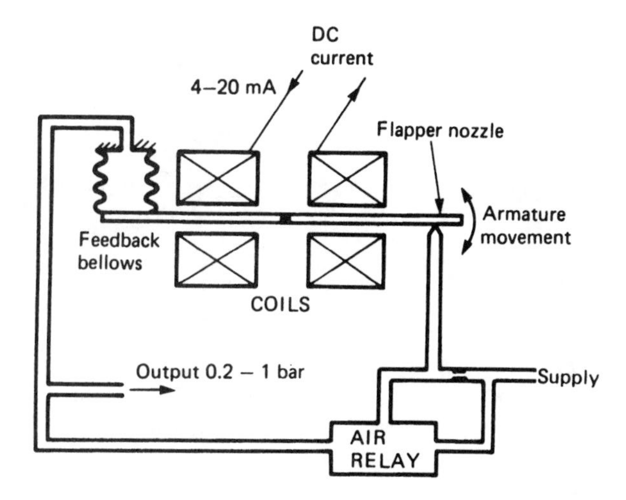

Fig. 16.41 *Current to pressure (I/P) converter.*

16.7.7. Fluidics

Sequencing applications are normally implemented with relays, digital ICs (such as TTL or CMOS) or programmable controllers. It is also possible to devise sequencing systems which are totally pneumatic. Such systems, usually called fluidics, are not particularly cost effective when compared with more conventional devices, but are useful where electronic circuits cannot be used (e.g. in explosive atmospheres or some medical applications).

Figure 16.42 shows a pneumatic equivalent of an electrical limit switch. With the pneumatic sensing outlet exposed, air bleeds to atmosphere and the output port is at atmospheric pressure. With the sensing outlet obstructed, diverted air causes a rise in pressure at the output port.

Logic gates use the 'wall attachment' or 'Coanda' effect shown in Fig. 16.43a. Fluid stream exiting from a jet with a Reynolds number in excess of 1500 (see Section 5.2.2.) tends to attach itself to one wall. If disturbed from the wall for any reason, it will attach itself to the opposite wall.

Figure 16.43b shows a pneumatic SR flip flop. Control ports are used to shift

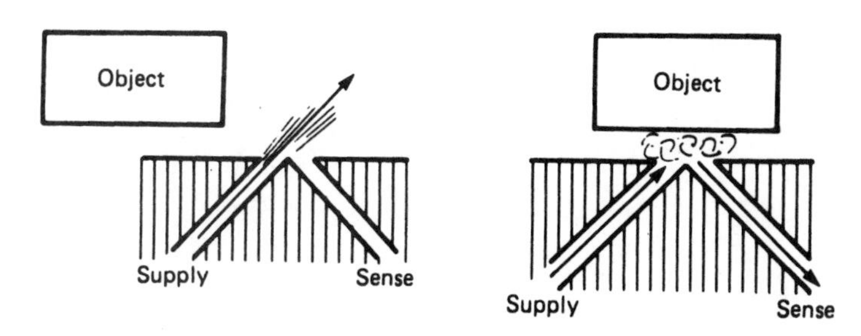

Fig. 16.42 *Pneumatic limit switch.*

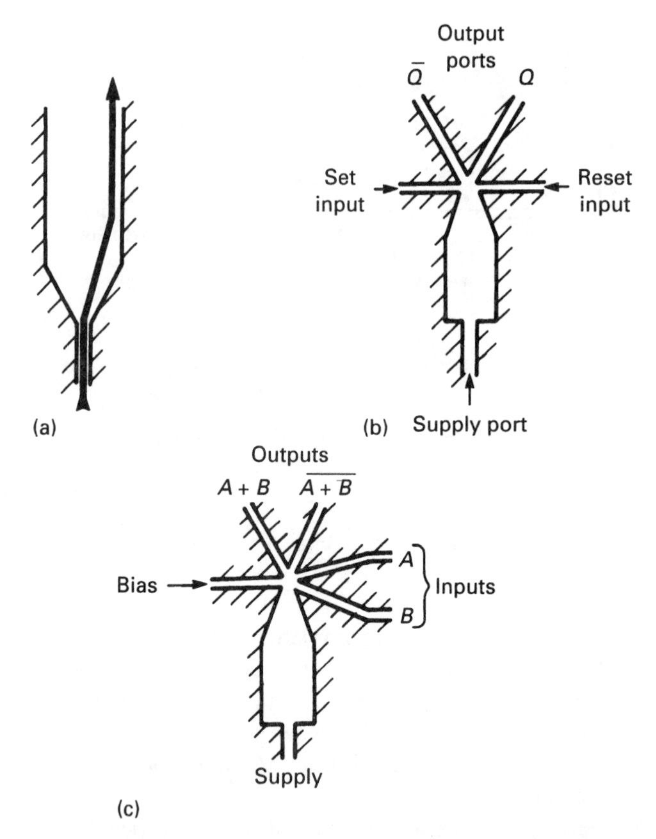

Fig. 16.43 *Fluidic logic. (a) The Coanda effect. (b) Set/reset flip flop. (c) OR/NOR gate.*

the jet stream between output ports. Once switched, the stream will be stable until a control pulse is applied to the other input.

An OR/NOR gate is shown in Fig. 16.43c. The auxiliary inputs bias the jet to the right-hand wall. A signal applied to either input will cause the jet to switch sides. All the common logic elements (AND, NAND, OR, NOR, timers, shift registers, etc.) can be constructed using similar ideas.

Chapter 17

Recording and display devices

17.1. Introduction

All processes require some form of system to allow the operators to observe the plant operation. For a simple motor this can just be lights saying 'Running' and 'Stopped'. At the other extreme, a complete petrochemical plant may have a large mosaic mimic display or a series of computer driven VDUs. Whatever the complexity of the display system, the aim is to give the operator full knowledge of what is going on and to allow fault conditions to be quickly identified.

There is also often a need to provide permanent records of plant performance, sometimes called trend recording. These records can be used for management analysis, maintenance planning and production control. They are also essential

Fig. 17.1 *A typical layout of operator controls, with a mixture of Mimics, VDUs, keyboard, analog and digital instruments and miscellaneous switches and lights. Control layout is always a compromise between control requirements, visibility, available space and ease of use. (Photo courtesy of GEC Electrical Projects.)*

in plants with very long time constants, where instantaneous values of plant variables are of far less importance than indication of direction and rate of change.

This chapter is concerned with devices that are used for the display and recording of plant data. These are often described by the rather grandiose expression 'the man machine interface'.

17.2. Analog instruments

17.2.1. Pointer scale instruments

In many applications the simple moving pointer fixed scale instrument is the cheapest and most effective indicator. Most are constructed around the moving coil fixed magnet principle of Fig. 17.2. The current through the coil is determined by the external voltage and source resistance, and the resistance of the coil itself. This current produces a torque on the armature as described in Section 12.2. The torque is opposed by a fine hair spring, so the armature will balance at an angle where the torque from the spring matches the torque from the coil current. With careful design the angle of deflection can be made proportional to the current.

Moving coil instruments used to measure voltage have high coil resistance and low full scale current. This ensures the coil current is largely independent of the source resistance, and the meter does not affect (load) the voltage being measured. Current measuring instruments (milliameters and microameters) have low coil resistance to give a low voltage across the coil at full scale current. In both cases, the pointer deflection is proportional to coil current.

Any process variable can be represented as an electrical voltage or current by a suitable transducer. Common signal ranges are 4 to 20 mA, 1 to 5 V and 0 to 10 V. The elevated zero signals have protection against open circuit or short circuit cable as the indicator will drive offscale negative.

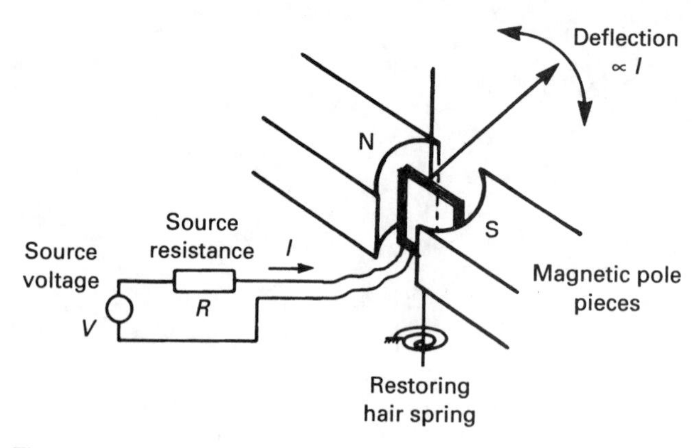

Fig. 17.2 *The moving coil meter.*

The manufacture of, and standards for, electrical pointer scale instruments is defined in the UK by the British Standard BS 89. This describes nine classes of accuracy from ±0.05% to ±5% of full scale deflection (FSD). For industrial applications, accuracies of better than ±1% are generally of little more than academic interest as they are beyond the resolution of the eye of the human observer.

The speed of response of a pointer scale instrument is inherently slow, with time constants of a few tenths of a second. BS 89 specifies that for a step change of 2/3 FSD input, the pointer should settle within 1.5% of FSD in 4 seconds. The construction of the meter makes it behave as a second order system characterised by a natural frequency and damping factor. Normally a damping factor of 0.7 is used, which gives the fastest response for a given natural frequency.

A meter scale should be chosen to be easy to read at the normal viewing distance. A useful rule of thumb is a scale length of 1/15 of the viewing distance (e.g. 20 centimetre scale length for 3 metre viewing distance). Figure 17.3 shows various types of scale; all have the same scale length, but take up different panel areas.

Experiment has shown that observers can quite accurately interpolate to 1/5 of a scale division provided the scale length is chosen as above. The use of 20 scale divisions over the full scale will therefore give a 1% resolution, which is sufficiently accurate for most applications. A greater number of scale divisions does not give increased resolution; the greater number of divisions merely clutters the scale and makes it difficult to read. The design and labelling of instrument scales is covered by BS 3693. Typical scales are shown in Fig. 17.4. Most have twenty minor divisions and four or five major divisions.

Meters should be scaled so that the normal indication is between 40% and 60% of FSD. If this causes an abnormal indication to go offscale, a non-linear scale can be used (see below). In many applications, the operator is required to

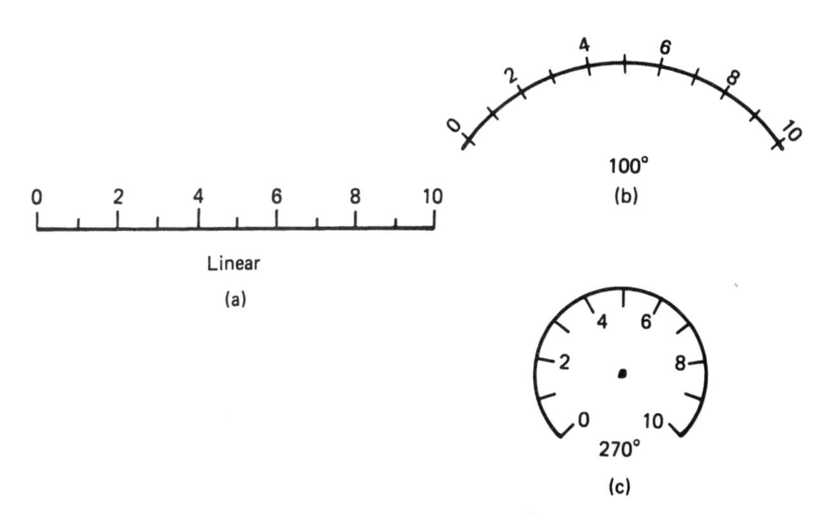

Fig. 17.3 *Different meter scales, all have equal scale length. (a) Linear. (b) 100°. (c) 270°.*

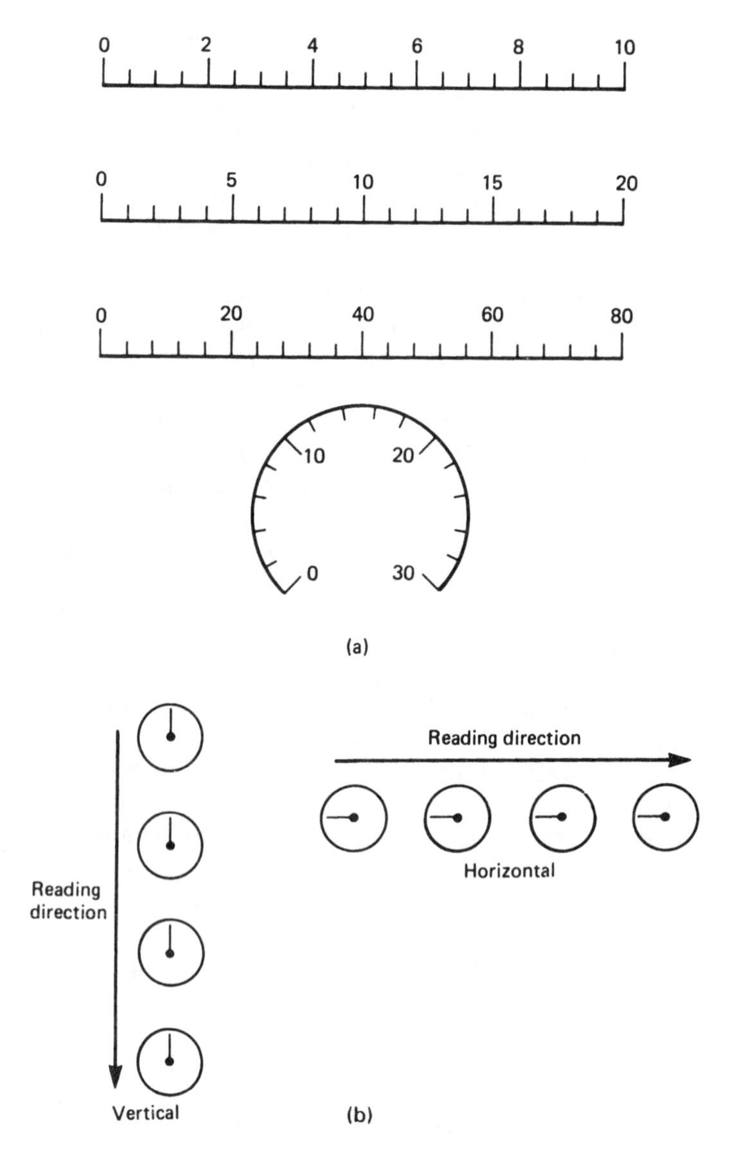

Fig. 17.4 *Moving pointer indicators. (a) Scale markings readable to about 1% resolution. (b) Grouping of meters which are scanned for deviations rather than read precisely. Normal indication arranged as shown.*

detect abnormal conditions rather than accurately read individual indicators. This is simplified if the indicators are arranged in such a way that their pointers are in the same orientation in the normal operating conditions. Ideally the 9 o'clock position should be chosen for horizontal groups of meters and the 12 o'clock position for vertical grouping, as in Fig. 17.4b.

The moving iron meter of Fig. 17.5a is simpler and more robust than the

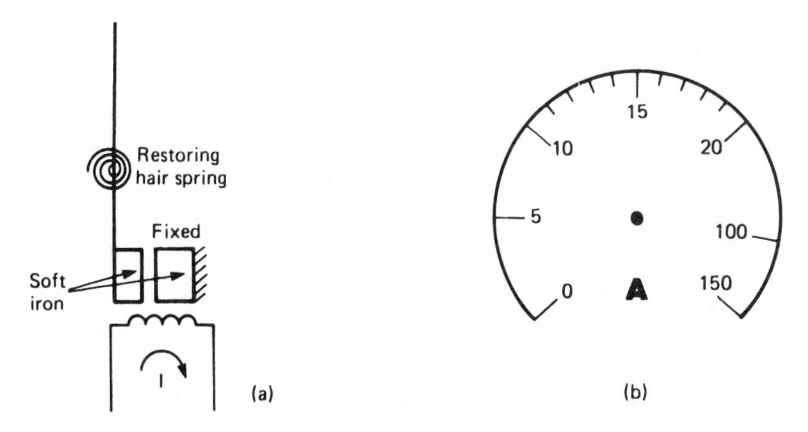

Fig. 17.5 *Moving iron meters. (a) Construction of meter. (b) Non-linear meter scale used with AC Motors.*

moving coil meter. Two pieces of soft iron are under the magnetic influence of the coil. As current flows, the iron pieces become magnetised and repel each other. As before, a restoring force is provided by a hairspring giving a deflection which is a function of the current. It will be seen that the deflection does not depend on the direction of the current as the repulsion occurs because the two pieces of iron have the same sense of magnetisation. Moving iron meters can therefore be used on DC of either polarity or AC.

A moving iron meter is inherently non-linear. This characteristic can be used to give sensible scale deflections for meters where an abnormal condition causes a reading many times higher than a normal condition. Figure 17.5b shows a moving iron ammeter for a fan motor. This gives a normal reading around 50% of pointer deflection but can indicate about ten times overload under starting and fault conditions.

17.2.2. Bar graphs

A bar graph such as Fig. 17.6 is a useful alternative to a pointer scale instrument where a block of similar data is to be displayed. They are particularly useful where precise reading is not required, but a fault condition needs to be visually obvious.

Early bar graphs used a rotating tape scale, as in Fig. 17.7a, or a vertical drum with a spiral red/white division, as in Fig. 17.7b. In both cases the display was moved directly by a meter movement or a position control servo. Most modern devices, however, are electronic and based around the block diagram of Fig. 17.7c, where the input signal is digitised by an ADC and displayed by an array of LEDs or gas discharge tubes. A resolution of 0.5% is common, requiring 200 individual indicator segments. Scaling is achieved by changing the input sensitivity and the plastic indicator overlay.

Electronic bar graphs frequently incorporate alarm detection, giving an indication if the process variable goes outside preset limits. Commonly the bar display is flashed or changes colour for an alarm condition.

Fig. 17.6 *Bar graph indicators showing cooling water temperature. The pattern and a rogue value can be clearly seen.*

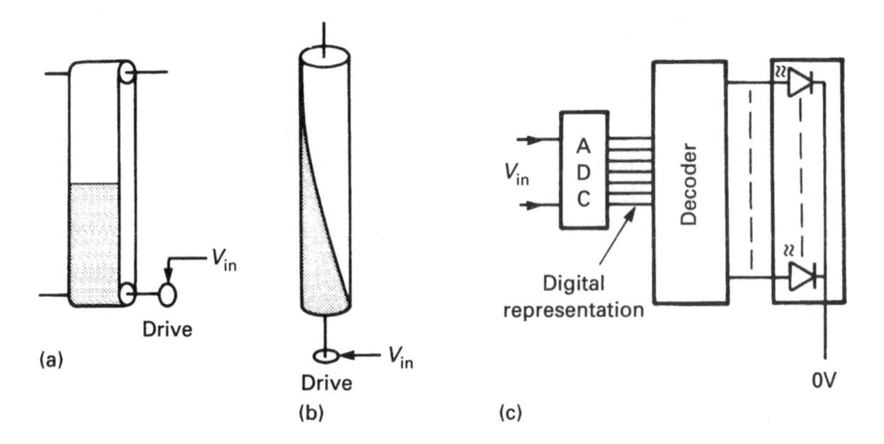

Fig. 17.7 *Bar graph indicators. (a) Belt drive. (b) Drum drive. (c) Electronic.*

Bar graphs are generally more expensive than moving pointer instruments, and require a separate power supply. They are generally easier to read where accurate indication is not required.

17.3. Chart recorders

17.3.1. Introduction

Most plants require records of plant performance to be kept for subsequent analysis. Records of operating conditions may, for example, be needed to show that a plant is being operated within its designed parameters. The operation of unmanned plant such as pump houses must be recorded and checked at regular intervals. Recording can overcome problems with the speed of plant operation. High speed action can be captured and viewed at leisure, and slow speed changes over hours, or days, can be analysed for trends. Multichannel recording can show up relationships, expected and unexpected, between plant variables.

The commonest instrument is probably the chart recorder, which produces a continuous graph of a process variable plotted against time. This can exist in a variety of forms according to the speed and nature of the variables to be recorded.

The first consideration is chart speed which must be chosen to match the likely rate of change of the variable. Too low a speed will blur detail, the minimum resolvable time being limited by the pen width. Too high a speed makes trends difficult to see and is wasteful of paper. Chart speeds are available from about 1 mm per day to tens of millimetres per second for pens and several metres per second for UV recorders, described later. High speeds can obviously only be used intermittently. Chart drives are based around stepper motors with electronic speed control (see Section 12.11) or synchronous motors with speed changes being effected by replaceable gear trains.

The response speed of the pen is also critical. Most pen positioning systems behave as a second order system, characterised by a natural frequency and a damping factor. Figure 17.8 shows the response and phase shift for a recorder driven by a sine wave input at various frequencies and damping factors. Figure 17.8 is normalised, i.e. the frequency axis is expressed in terms of the natural frequency f_n.

Figure 17.8 shows that the widest frequency response is obtained with a damping factor of 0.7, and a recorder can be used up to about 0.5 f_n without significant error.

Measured variables do not follow sine waves, but Fourier analysis shows that any regular waveform can be expressed as a sum of sine and cosine waveforms. With a knowledge of these frequency components, the natural frequency of the recorder can be specified to give the required accuracy. In practice, the specification of natural frequency for a chart recorder is usually more empirical than analytical and a figure is chosen which is related to the expected rate of change of the plant variable. Commonly the time taken for a full scale movement across the chart is specified. Too low a natural frequency will result in an inability to follow fast changes and a loss of detail.

Pen friction can limit the natural frequency and stiction results in a minimum pen movement, i.e. a minimum resolvable error. Early chart recorders used capillary tubes and ink reservoirs. These needed regular maintenance and cleaning (turning white-collar workers into blue-spotted-collar workers in the process). The modern tendency is to combine ink/pen cartridges, usually in the form of felt-tip pens.

At high chart and pen speeds the viscous friction of pens becomes excessive, and other recording methods are used. Heat sensitive paper changes colour when

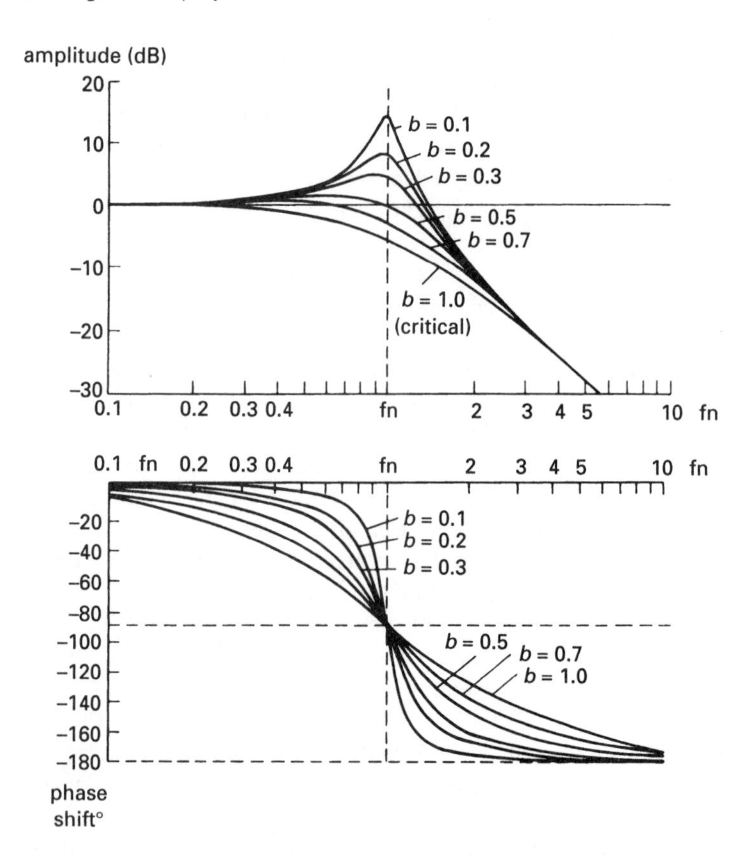

Fig. 17.8 *Response of second order systems.*

heated, and is used in conjunction with a heated stylus. Electrosensitive paper changes colour under the action of an electric current (which burns away the top paper layer), and is used with a stylus connected to a current source. Both methods allow very high chart speeds, but the specialist paper is quite expensive.

Chart recorders usually operate on the standard signal ranges of 4 to 20 mA, 1 to 5 V and 0 to 10 V, the signal standards with offset zeros giving protection against cable faults and transducer failure. Temperature recorders for use with thermocouples and resistance thermometers usually connect direct to the sensor and incorporate the drive circuits (for PTRs), cold junction compensation (for thermocouples) and linearisation. The recorder can usually be arranged to drive offscale high or low in the event of a cable or sensor fault.

The accuracy of a chart recorder is specified in terms of FSD, with a deadband that is the smallest signal which will move the pen. Typical figures are, respectively, 1% and 0.2% of FSD. Speed of response can be specified in terms of natural frequency (for galvanometric instruments) or time to traverse the chart (for servo recorders).

17.3.2. Galvanometric and open loop recorders

In the simplest (and cheapest) chart recorders the process variable signal is used directly to move the pen. In the circular chart recorder of Fig. 17.9a, for example, a pressure signal moves the pen via a set of bellows. The records on a circular chart are difficult to read and interpret because of non-linearities from the circular chart and the arc of the pen. The main advantage is the relationship of chart to a set period (e.g. a shift, a day, a week) and an ease of filing (a simple dowel).

Strip chart recorders are easier to read, but direct driven recorders have

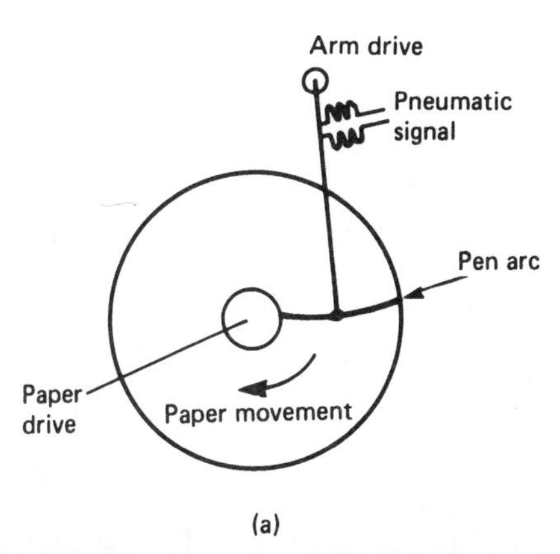

(a)

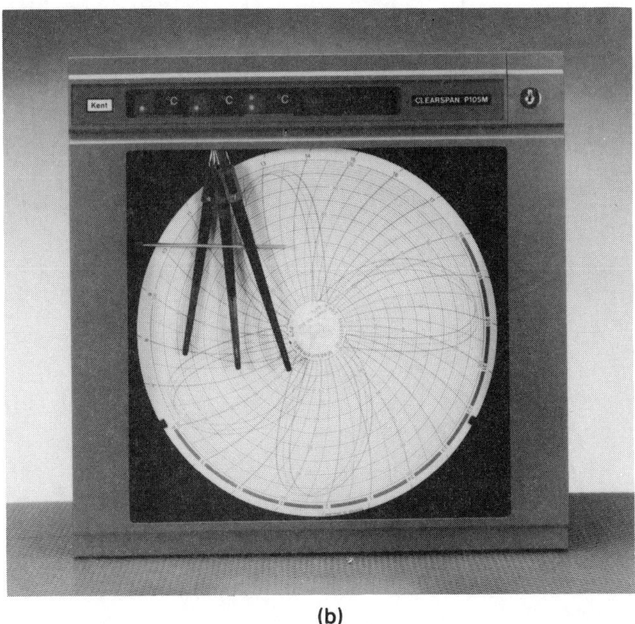

(b)

Fig. 17.9 *Circular chart recorder. (a) Construction. (b) Multipen circular chart recorder. (Photo courtesy of Kent Industrial Measurements.)*

inherently non-linear traces. Most are based on the moving coil mechanism described in Section 17.2.1, with the pen being connected to the end of the pointer. The pen tip traverses an arc, as shown in Fig. 17.10a, with radius equal to the pen arm. This requires chart paper pre-printed with arcs, as in Fig. 17.10b. Although these are easier to read than circular charts, it can still be difficult to infer relationships. The distances along the arcs are directly proportional to the input signal, but the distance from the centre line is given by:

$$x = r \sin \theta \qquad (17.1)$$

To minimise error θ is kept small, usually by employing a long arm.

A curved grid can be avoided by passing the recording paper over a straight knife edge, as in Fig. 17.11. Obviously a pen cannot be used, heat sensitive or electrosensitive paper (see Section 17.3.1) being required. Although this arrangement

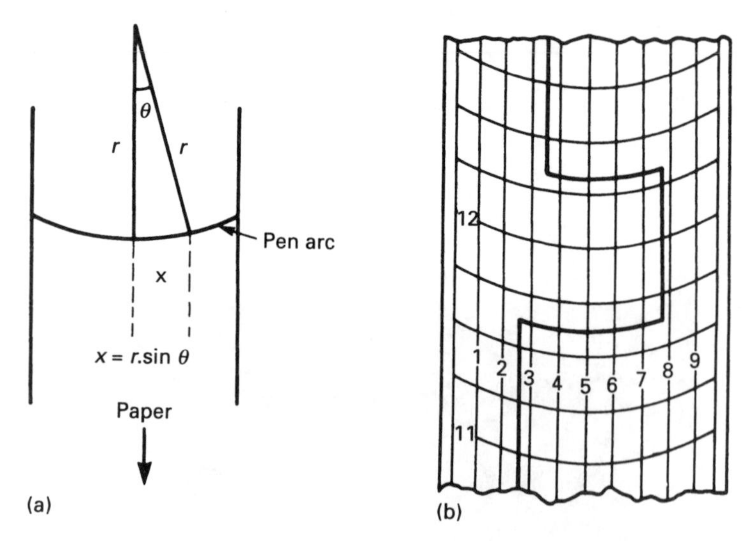

Fig. 17.10 *Direct drive chart recorder. (a) Construction. (b) Sample chart.*

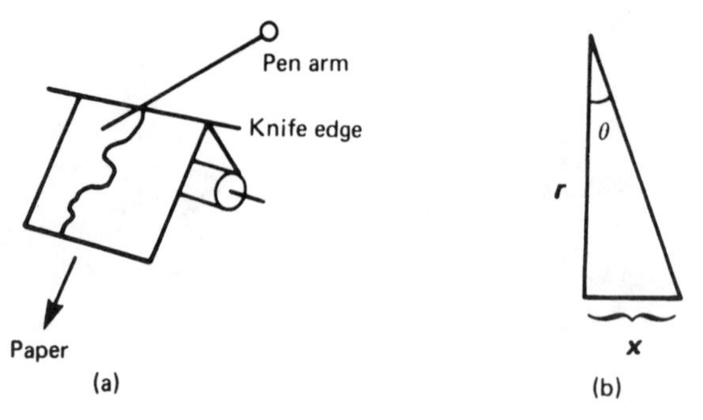

Fig. 17.11 *Knife edge recorder. (a) Construction. (b) Deflection.*

gives a rectilinear chart, it is non-linear with grid spacings getting larger away from the centre line. As shown in Fig. 17.11b:

$$x = r \tan \theta \tag{17.2}$$

As before, errors can be reduced by using a small θ, usually obtained by making the pen arm as long as possible.

The torque available from galvanometric recorders is small, and pen friction is a possible source of error as well as a limiting factor on speed of response. A common technique is to use a 'dotting' pen technique, illustrated in Fig. 17.12. The pen is normally allowed to move freely just above, but not contacting, the paper. A clamping bar above the pen is supported by motor driven cams. At regular intervals the bar is released on to the pen arm, simultaneously clamping it and leaving a dot on the paper.

An alternative approach is the use of a power amplifier to boost the input signal, allowing a more powerful galvanometer to be used.

17.3.3. UV recorders

The natural frequency of a galvanometric device is given by:

$$f_n = \frac{1}{2\pi}\sqrt{\frac{K}{J}}\ \text{Hz} \tag{17.3}$$

where K is the restoring string constant, and J the moment of inertia of the armature assembly. A high natural frequency is required for high speed measurements, and this can be obtained by increasing K or reducing J.

Ultraviolet, or UV, recorders (sometimes called oscillographs) use mirror galvanometers which have very low inertias. A typical arrangement is shown in Fig. 17.13. A mirror galvanometer is encased in a permanent magnet block. Collimated light from a UV lamp is reflected from the mirror on the galvo to

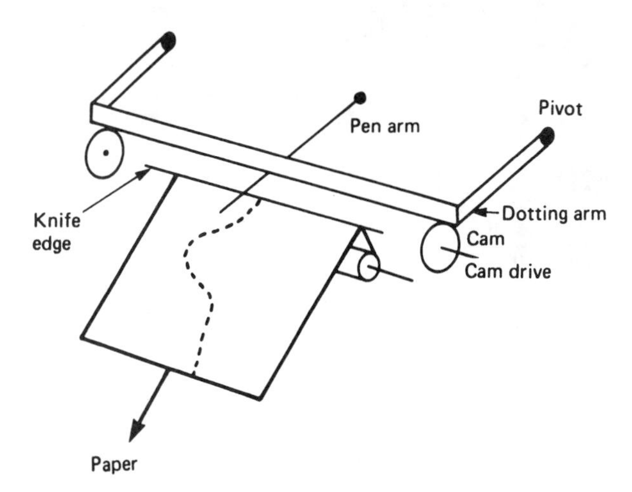

Fig. 17.12 *Dotting recorder.*

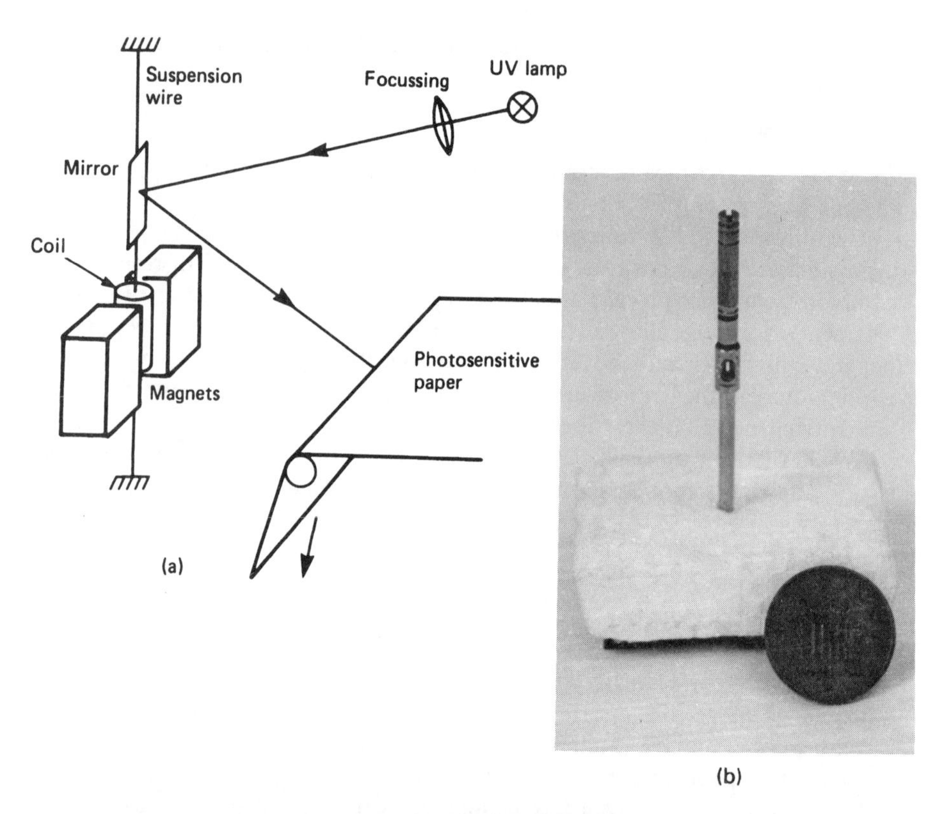

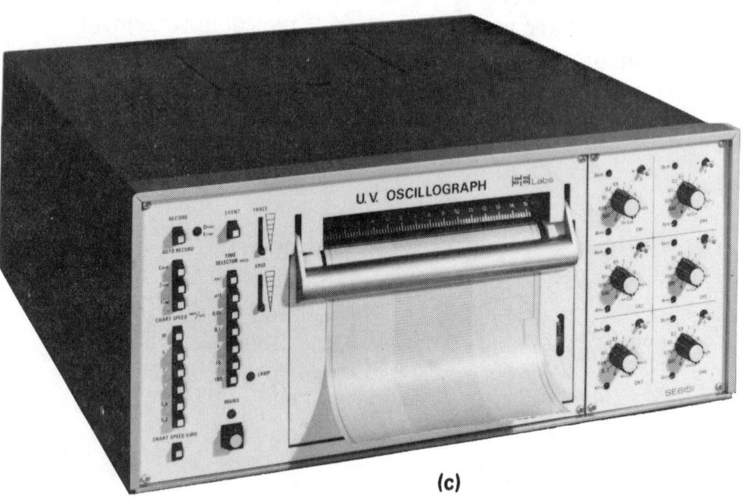

Fig. 17.13 *The ultraviolet recorder. (a) Construction. (b) A UV recorder galvanometer. The small size gives a good frequency response. (c) UV recorder SE6151 manufactured by SE Labs. (Photo courtesy of Thorn EMI Datatech.)*

produce a spot on the UV sensitive paper. Current passing through the galvo deflects the mirror, and hence the trace on the paper. Figure 17.13 has the same inherent non-linearity as Fig. 17.11, but this is reduced to acceptable levels by the use of a long path length.

Galvanometers can be purchased with natural frequencies well in excess of 50 kHz. Equation 17.3 shows that once a practical lower limit of inertia has been reached, the natural frequency can only be raised by increasing the spring constant K, which in turn implies a decrease in sensitivity. High frequency galvos tend to be less sensitive than their lower frequency cousins.

Damping is determined by viscous damping within the body of the galvo, and the impedance of the signal source. As the galvo armature moves in the magnetic field it acts as a generator, and the induced voltage opposes the applied current, consequently providing a damping force. The lower the source impedance, the higher the damping factor. The damping factor can be increased by connecting a damping resistor across the galvanometer leads. A damping factor of 0.7 gives the best response.

Galvanometers are current operated devices, and if connected directly to a voltage source must be used in series with external current determining resistance boxes. More expensive UV recorders incorporate voltage-to-current amplifiers to separate the voltage source and the galvos. The use of integral amplifiers also makes the damping factor independent of the signal source.

Commercial recorders generally have magnet blocks that can take several galvos (six being typical) to give multitrace recording. The user generally purchases the galvanometers separately to cover a range of sensitivities and natural frequencies. Chart speeds of several metres per second are possible, but at these speeds a roll of paper will be consumed in less than a minute.

The photographic paper used is self-developing in a few seconds in natural and incandescent light (but develops rather slowly in fluorescent light). Exposure to sunlight will fog the paper, and the trace slowly disappears in natural light. Fixing sprays are available to make the trace permanent.

17.3.4. Servo recorders

Servo recorders use a closed-loop position control system to drive the pen. This gives an accurate, fast and linear response, and overcomes errors from pen friction. The principle is shown in Fig. 17.14a. The pen position is measured by a linear potentiometer, which produces a wiper voltage proportional to the pen position. This is compared with the input signal by an error amplifier, which in turn drives the servo motor to move the pen until the error voltage is zero. The error amplifier drives the servo motor via a power amplifier to overcome the effect of pen friction.

Figure 17.14a uses a rotating servo motor. Some recorders use the linear motor of Fig. 17.14b. The pen carriage is connected to a coil which can move linearly between the poles of a permanent magnet. If a current is passed through the coil the carriage will move, being attracted to one pole and repelled from the other. The carriage movement can therefore be controlled by the direction and

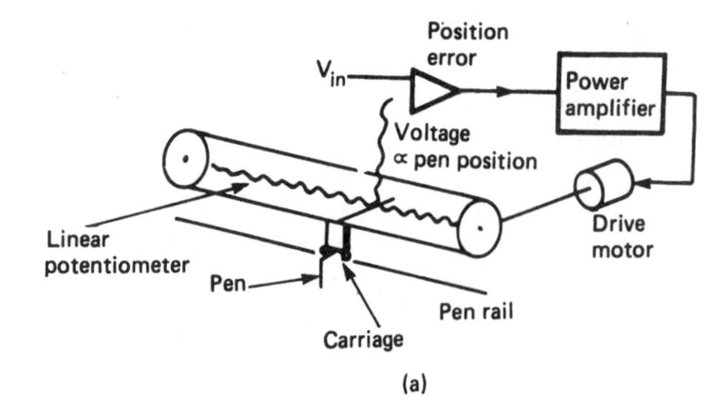

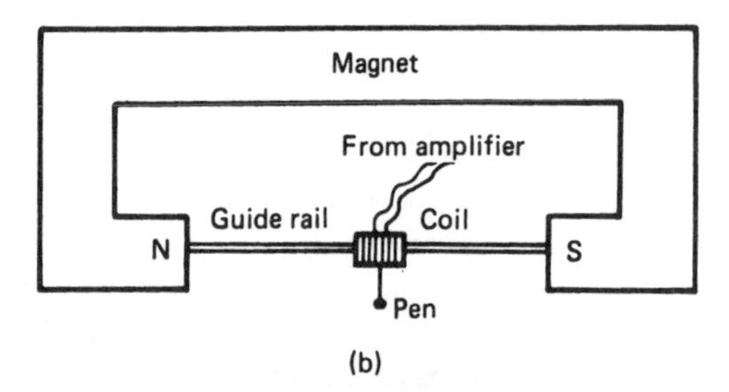

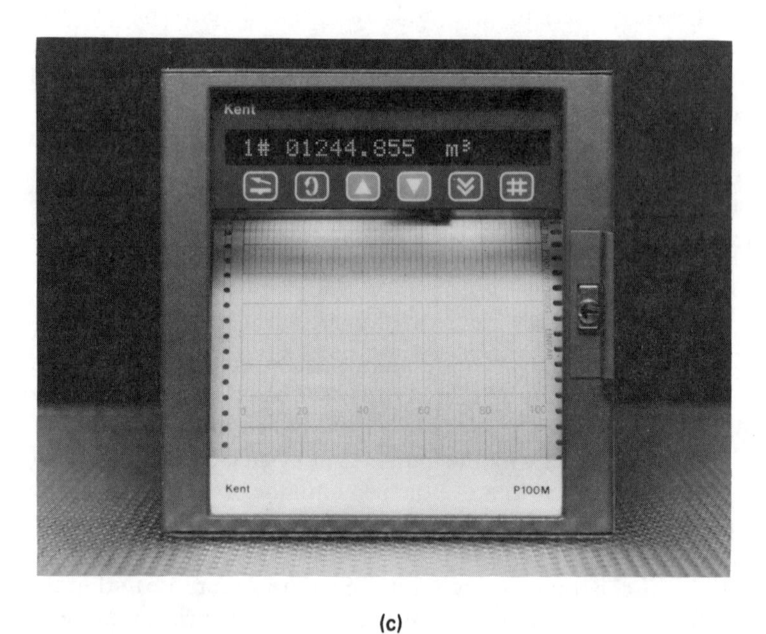

(c)

Fig. 17.14 *Servo chart recorders. (a) Block diagram. (b) Pen motor. (c) Modern servo chart recorder with readout in engineering units. (Photo courtesy of Kent Industrial Measurements.)*

magnitude of the coil current. The arrangement of Fig. 17.14b has the advantage of only one moving part and is consequently more reliable and robust.

Stepper motors, described in Section 12.11, are also used as positioning devices in microprocessor based recorders.

17.3.5. Multichannel recorders

The need often arises for several variables to be recorded on one chart to show interactions between variables or purely to reduce costs and panel space. Multipen recorders giving continuous records can be overlapping (or side by side). Both methods have disadvantages; the overlapping pens need to be offset to allow them to cross giving a time shift between traces, and the side by side arrangement reduces the chart width per record and hence the resolution.

A common approach, useful for slowly changing signals, is the combination of the dotting principle of Fig. 17.12 with a multiplexor as in Fig. 17.15. The pen is replaced by a print head with a different symbol, or ink colour, for each variable. Signals are sequenced in turn to the pen-positioning servo, the print head being stepped to the correct symbol or colour each time. A typical scan rate is 5 seconds per point, giving an overall cycle time of 1 minute for twelve channels. Obviously this approach cannot be used for fast changing variables, but it is sufficiently fast for most temperature monitoring applications.

The UV recorder of Section 17.3.3 can also handle multichannel signals, but is more of a laboratory or maintenance instrument, and is not really suitable for permanent panel mounted installation.

17.3.6. Event recorders

Full proportional representation is not required where signals such as limit switch or valve operation are to be recorded. Recorders which log on/off signals are known as event recorders, and use far simpler pen mechanisms. Most are simply a solenoid which deflects a pen a small distance (typically 5 to 10 mm) giving a trace similar to Fig. 17.16. Event recorders can show relationships between signals and be a useful fault finding tool, particularly when intermittent faults are being pursued. Event recorders are available with 24 channels, and a range of input signals from 5 to 240 V AC and DC.

17.3.7. Flat bed (XY) plotter

The flat bed, or XY, plotter has two separate pen position servos, as in Fig. 17.17a. This allows relationships between two variables to be plotted by connecting one variable to each servo. The plotter is really a laboratory instrument, but can be found in industry where a plant must be operated with certain conditions. Figure 17.17b shows a possible application where a plant must be kept within a certain range of pressure and temperature. An XY plotter with one axis connected to a pressure transducer and the other axis to a temperature transmitter will produce a spider's web trace that can easily be examined for excursions outside the permitted region. Flat bed plotters are also

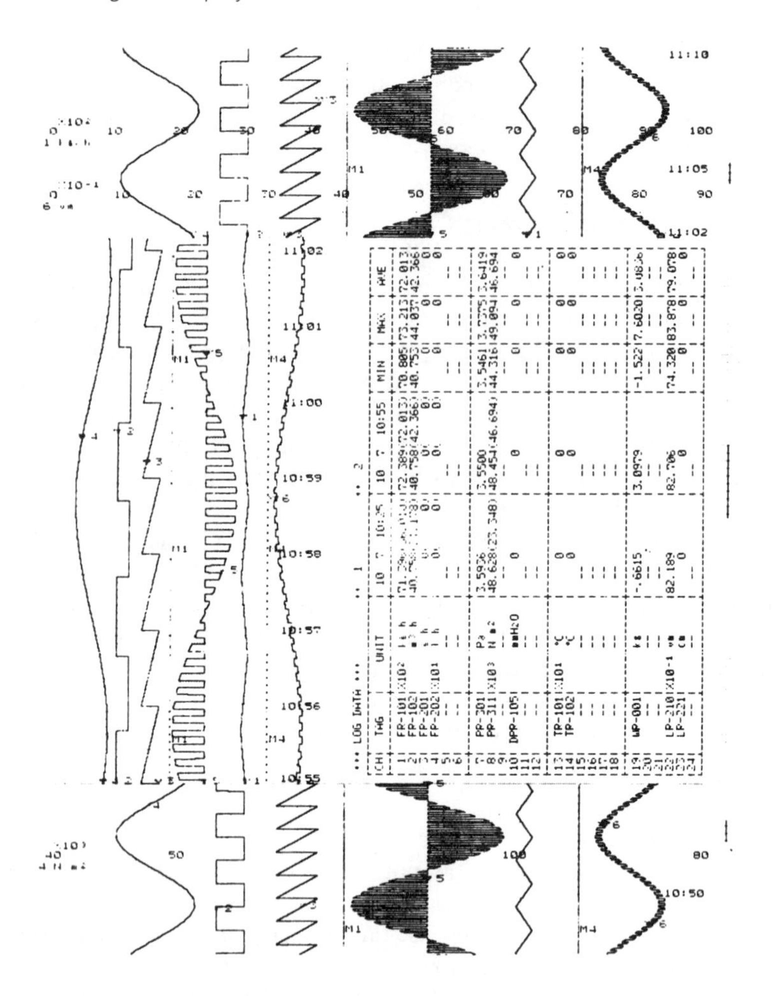

Fig. 17.15 *Multivariable chart recorder. High quality thermal chart with analog trace and report generation facility. (Chart reproduced courtesy of Toshiba and their UK agents Protech Ltd.)*

useful output devices for computers, allowing accurate graphs and sketches to be drawn. Computer aided design (CAD) machines use plotters to produce the final drawings.

17.4. Display devices

17.4.1. Introduction

The multidiscipline subject of opto-electronics was discussed in Chapter 8. The present section covers the application of light emitters in various forms from simple indicator lamps to multiline alphanumeric displays for the purpose of displaying plant status to operators.

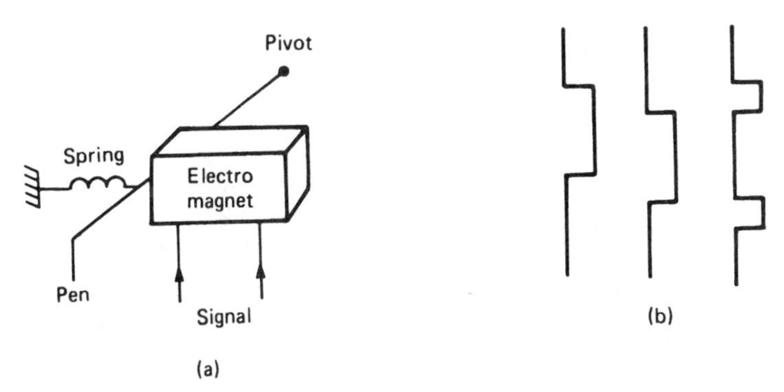

Fig. 17.16 *The event recorder. (a) Construction. (b) Typical trace.*

17.4.2. Indicators

A single indicator lamp can convey an on/off indication to an operator, and as such they are used to display the status of plant items such as limit switches, pumps, motors, etc. Indicators come in a variety of shapes, sizes and intensities.

LEDs and neons are relatively low intensity devices and consequently best suited to small mimic panels. Most indicators are incandescent bulbs which give high intensity and low operating currents when operated on higher voltages (e.g. 110 V AC). The use of transformer indicators, taking 110 V AC drives but utilising 6 V bulbs via a transformer integral with the indicator body, gives increased operator safety.

Indicator colours can be a source of contention. The relevant British Standard, BS 4099, defines (somewhat simplified):

Red	Warning of potential danger or situation requiring action.
Amber	Caution, change, or impending change, of condition.
Green	Indication of safe condition. Authority to proceed.
White	Any meaning when doubt exists about the application of red, amber or green.
Blue	Any meaning not covered by red, amber or green.

but this does not really cope with a simple motor starter using illuminated push buttons, as in Fig. 17.18a. It is really a matter of site convention as to the colouring of the lenses and the light sequence. At the author's plant, the start button would be green and the stop button red, with the indicator showing which button was pressed last. In many plants both the colours and light sequence would be reversed. The most important factor, however, is consistency.

Plants should not be overindicated; the operator should have adequate status information, but not be swamped by a visual overkill. Crucial indication should be given in both pass/fail states to cover for bulb failure. The eye is very good at recognising patterns, so a gap in a row of green lights with a red light below is easily spotted. Needless to say, a lamp test push button should be fitted for all but the most non-critical indications.

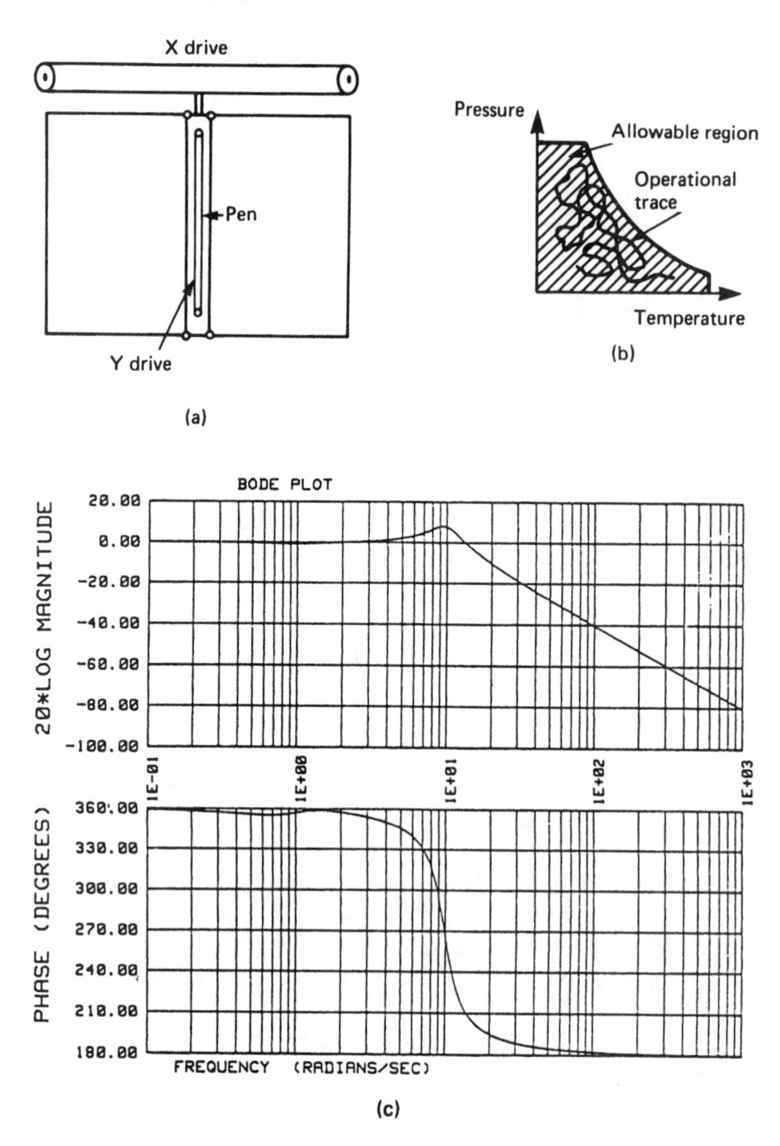

Fig. 17.17 *XY plotter. (a) Construction. (b) Typical application. (c) Output from a high quality Hewlett Packard XY plotter.*

Bulbs have a limited life, and a large inrush current from cold. Bulb life can be improved and the inrush current reduced by the use of lamp warming resistors, as in Fig. 17.18b.

17.4.3. Numerical indicators

Numerical indicators, or digital displays, are used as an alternative to the analog instruments of Section 17.2. Their primary advantage is improved resolution. A good analog instrument with moving pointer can be read to a resolution of 1% of FSD. A four digit display, showing 0 to 9999, can have a resolution, and accuracy, of 0.01%. Digital indicators are also used to display the state of devices which inherently move in finite steps – tap numbers on multitap transformers, for example.

Digital indicators have some disadvantages. They do not convey the 'feel' of a process that can be obtained with well laid out analog instruments or bar graphs. Compare Fig. 17.19a, b and c which shows the same temperatures, of cooling water, say, displayed in bar graph, meter and digital form. The two analog

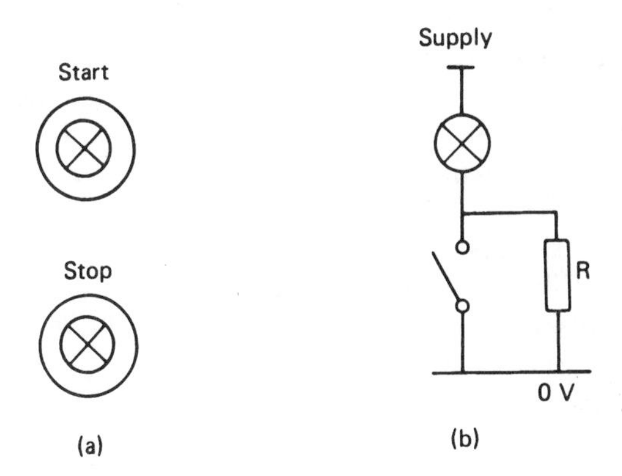

Fig. 17.18 *Using indicators. (a) The illuminated pushbutton problem. What colour are the buttons? (b) Lamp warming resistor. R should be chosen to give one tenth of normal operating currents.*

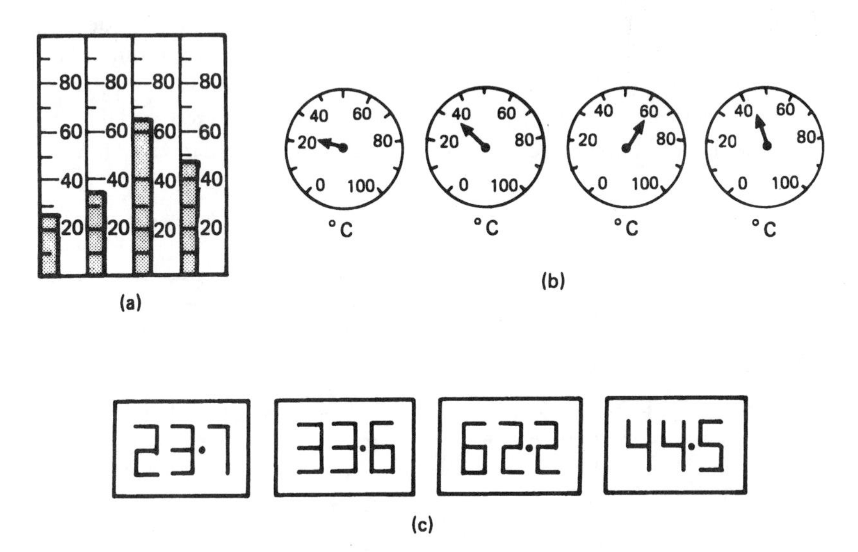

Fig. 17.19 *Comparison of display methods. (a) Bar graph. (b) Analog dials. (c) Digital displays.*

displays can be scanned in one glance; the digital display needs careful reading.

Digital displays are also poor when used with process variables which are subject to rapid change. In these circumstances the display becomes a blur of digits from which rate of change, or under extreme conditions even the direction, cannot be inferred. The update rate of digital displays needs to be carefully chosen; too fast and the display flickers in an annoying manner at points such as 999 to 1000, too slow and the display makes large jumps as the process variable changes.

Analog displays are therefore best suited to applications where the process variable can change quickly, is subject to continuous change, needs relatively low accuracy reading or is part of a block of instruments that need to be scanned for an anomalous reading. Digital displays are required where accuracy is important – say, better than 1 or 2% – the variable is changing slowly (e.g. most temperatures) or the variable is inherently digital in nature. The author views with some suspicion the trend towards digital indication in motorcars, particularly for speedometers.

The basic display devices – incandescent, light-emitting diodes (LEDs), liquid crystal displays (LCDs), and gas discharge – are described in Chapter 8.

Seven-segment displays use seven bars, as in Fig. 17.20, which can be arranged into the digits 0 to 9 plus some letters as shown. ICs are available to decode a 4-bit BCD number to the seven segments. Obviously several displays can be used to give any required resolution.

LED displays are available with character sizes up to 30 mm. These can be viewed at distances up to several metres in normal room lighting, but are difficult to read in daylight. LCD indicators can operate either in the transmissive mode (with internal lighting) or the reflective modes of Fig. 17.21. LCDs are not as clear as LEDs in normal room lighting, but the contrast of reflective mode displays improves with increasing light levels. LCD indicators require minimal current, and units operating on 4 to 20 mA signals can derive their power supply from the loop current without degrading the signal. Life expectancy is reduced if LCDs are operated with a DC bias. The circuit of Fig. 17.22 is often used to give an AC drive.

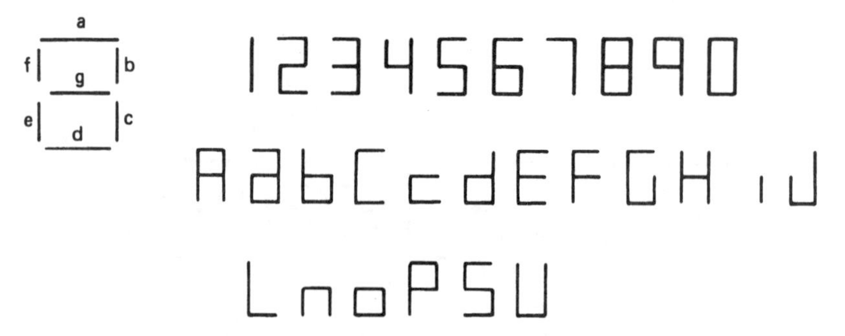

Fig. 17.20 *The seven-segment display and available characters.*

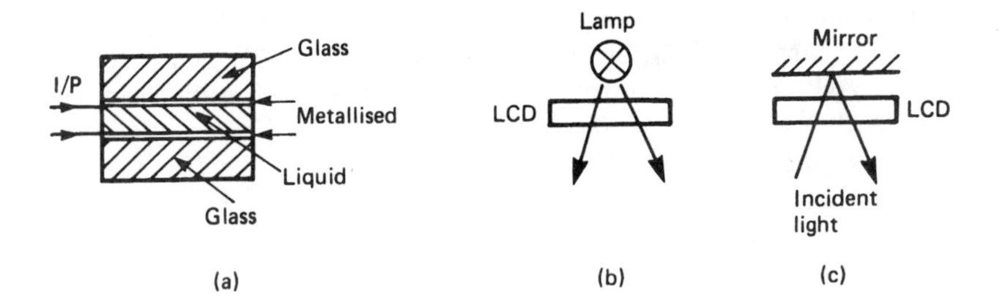

Fig. 17.21 *Liquid crystal displays. (a) Construction. (b) Transmissive mode. (c) Reflective mode.*

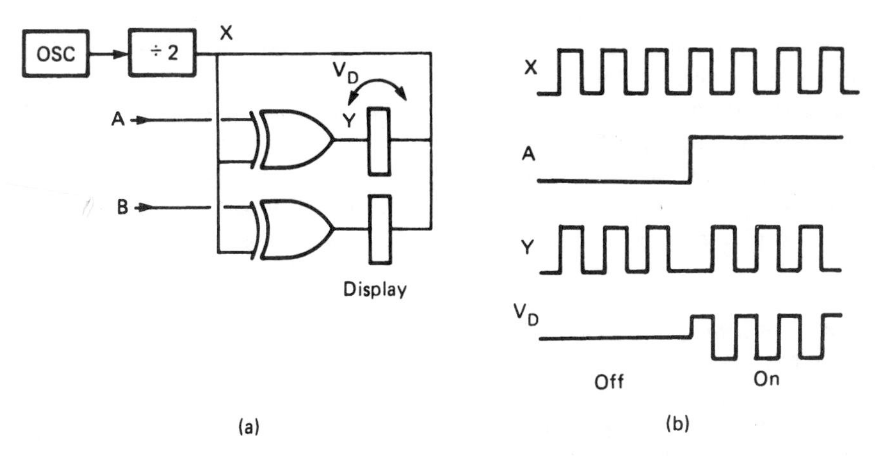

Fig. 17.22 *Driving LCDs without DC bias. (a) Circuit. (b) Operation.*

Gas discharge devices have a bright easy-to-read orange display. In general they are far clearer than LED and LCD indicators and can be read in daylight. A variation is the Nixie tube (a registered trade name of the Burroughs Corporation). This uses shaped cathodes behind a mesh anode, as in Fig. 17.23, and gives natural shaped digits.

Although shaped cathode displays such as the Nixie tube give natural looking

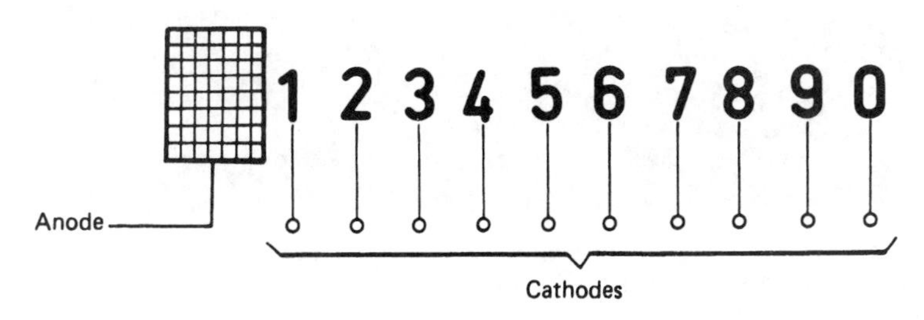

Fig. 17.23 *The Nixie tube.*

digits, they suffer from a varying depth of display, with '1's, say, looking significantly nearer the front of the display than '9's and '0's. The depth also restricts the angle of viewing to around 45° from the normal. The anode voltage in Nixie and other gas discharge displays is around +180 V, posing a possible service hazard for technicians used to working on low voltage display devices. The +180 V anode supply is normally derived inside the display from a DC to DC inverter operating from the 5 V or 12 V logic supply. Ready made encapsulated inverters are also readily available. The high voltage anode supply also increases the cost of the units as high voltage driver transistors are required.

Where large scoreboard displays are needed, incandescent or fluorescent tube based displays are required. Figure 17.24 shows an LED multi-segment display in use on a crane weighing application. Mechanical seven segment displays are also available. These rotate coloured bars with solenoids.

Numerical displays can be made easier to read by the use of leading and trailing zero suppression. Without leading zero suppression, the number 27 on a four-digit display would appear as 0027. Leading zero suppression blanks the unnecessary zeros. Trailing zero suppression works after the decimal point, so 27.1 on a 99.99 display would appear correctly, and not as 27.10. Zero suppression is provided on most display driver ICs; Fig 17.25 shows how the ripple blanking pins can be used to this end on the popular 7447 seven-segment driver IC.

17.4.4. Multiplexing

Digital displays require a large number of signal lines: twelve for a three-digit BCD display plus two power supply connections. Time division multiplexing can considerably reduce the complexity and the number of lines. Figure 17.26 shows a typical scheme for a four decade display.

The four BCD inputs are applied to four data selectors $IC_1 - IC_4$. Note that IC_1 takes the A (LSB) inputs, IC_2 the B inputs, and so on. An oscillator, running typically at 1 kHz, drives a 2 bit counter which selects the inputs to the data selectors in turn.

Fig. 17.24 *A large (1500 mm) alphanumeric display based on a 14-segment display. Each display segment is built up from many individual indicators; in this case LEDs. Displays similar to this are used in crane weighing and similar applications where information is to be read at long range. (Photo courtesy of Displait.)*

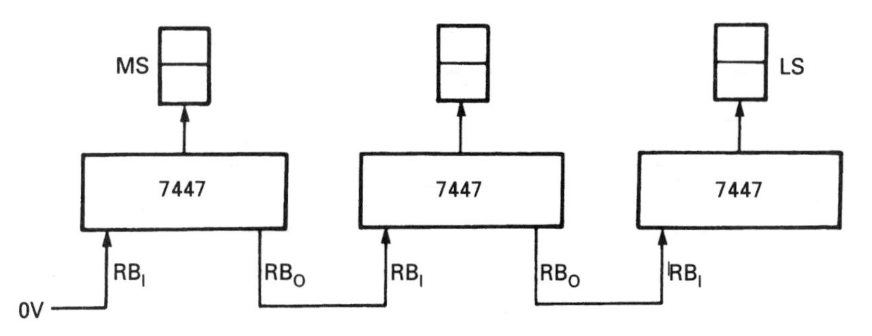

Fig. 17.25 *Ripple blanking (leading zero suppression).*

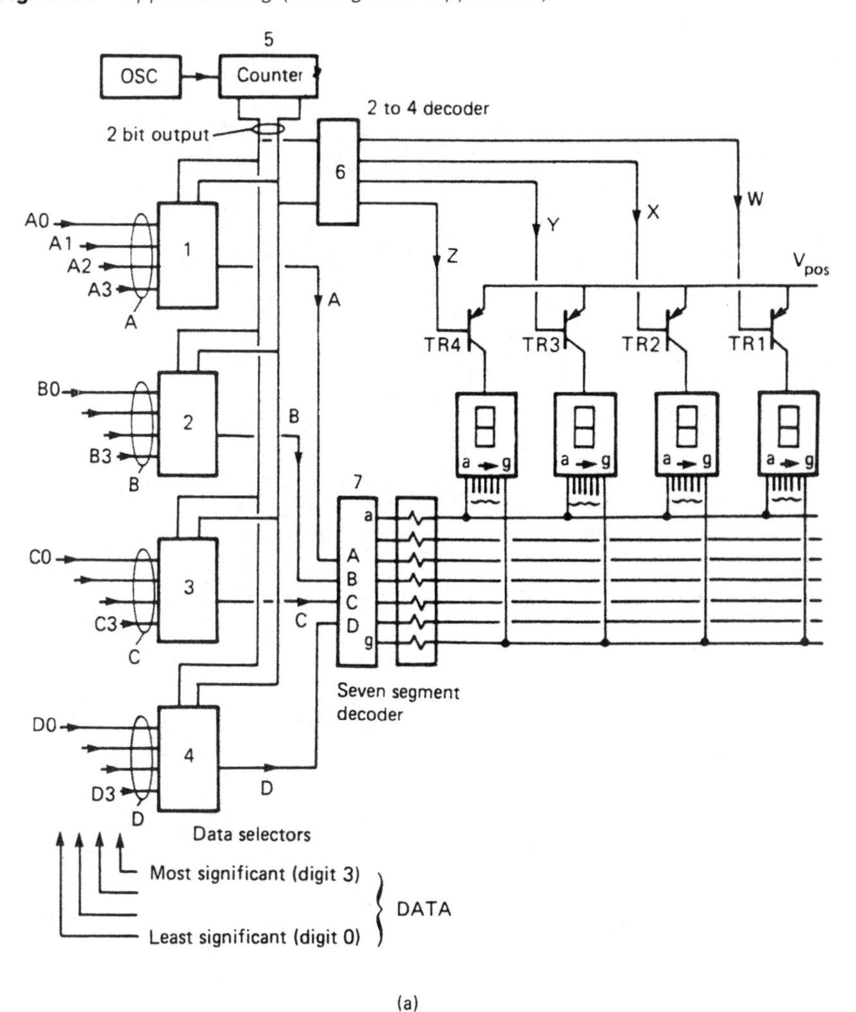

Fig. 17.26 *Multiplexed outputs to reduce cabling. (a) Multiplexed displays. (b) Waveforms for four-digit display. (c) Multiplexed displays on an instrument panel.*

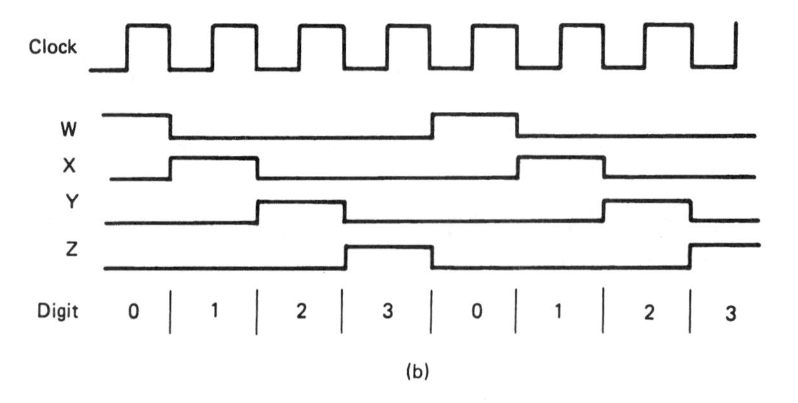

(b)

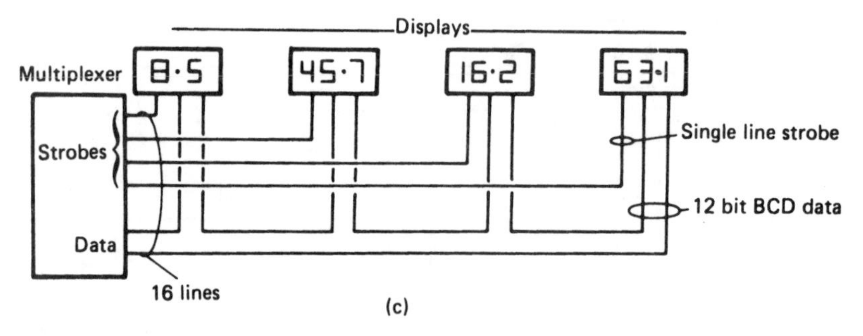

(c)

Fig. 17.26 *contd.*

The counter is decoded to drive the strobe transistors $TR_1 - TR_4$ which are connected to the common anodes of the seven segment displays. The seven-segment decoder provides the seven-segment data for each display in turn, following the strobes from the transistor. Only one digit is illuminated at one time, but the strobing is too fast for the eye to follow and all four digits appear lit.

Multiplexing can reduce cabling significantly. Figure 17.26c shows 4×3 digit display driven with multiplexing. This requires twelve data lines and four strobe lines. Driven directly they would need 48 lines. Multiplexing does reduce the intensity of the displays, and if a large number of devices are being driven, each digit requires a local 4-bit latch to hold the data locally.

The technique of multiplexing can also be used to read push button and decade thumbwheel switches with reduced cabling costs. Figure 17.27 shows a typical scheme. Each decade switch is read sequentially, and the data strobed into storage latches. Note that diodes are necessary to avoid sneak paths through unselected switches.

17.4.5. Alphanumeric displays

A seven-segment display can show a limited range of alphabetic characters. A sixteen segment display, shown in Fig. 17.28, can display all alphanumeric

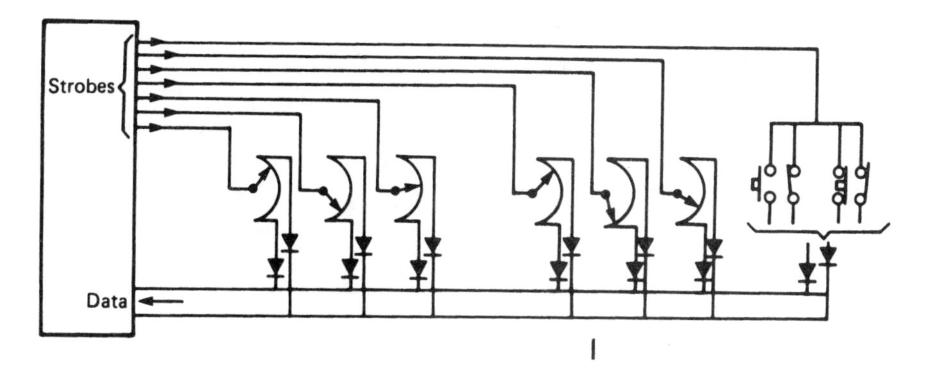

Fig. 17.27 *Multiplexing of inputs.*

characters. These are usually obtained with internal storage and decoding and accept 7-bit ASCII coded data.

Dot matrix LCD displays similar to Fig. 17.28 are a convenient way of displaying text in an easily read form. They have, however, a rather narrow viewing angle. Devices displaying four lines of 64 characters, accepting serial data down an RS232 link, provide a method of displaying large amounts of text with minimal installation cost.

17.5. Alarm annunciators

Faults occur in all process control systems. Fault conditions need to be brought to the operator's attention clearly and concisely so that rectifying action can be taken. This is usually achieved with alarm annunciator panels, such as in Fig. 17.29. Each indicator identifies a fault condition which requires attention.

Figure 17.30 shows a typical response for one indicator. The light flashes on the occurrence of a fault, and an audible alarm sounds. When the alarm is acknowledged, the light goes steady if the alarm condition is still present and the alarm ceases. The steady alarm clears when the fault condition is removed.

Annunciator panels can display a myriad of fault indications which arise from a single cause. An overkill of fault annunciation was apparently a contributory factor of the Three Mile Island nuclear incident. Alarm indication can be simplified by the use of first up groups.

Consider a simple hydraulic system. This could annunciate on pump stopped or low hydraulic pressure or low tank level. These will all interact. A pump stopped alarm will also cause a low hydraulic pressure alarm; a low pressure alarm will necessitate stopping the pump to prevent damage or an oil spill. Similarly low tank level may cause low pressure, and should stop the pump to prevent air being drawn in. If these alarms are linked into a first up group, the initial alarm condition will be displayed and the consequential alarms ignored or queued. The operator can thus determine the source of the problem quickly and without ambiguity.

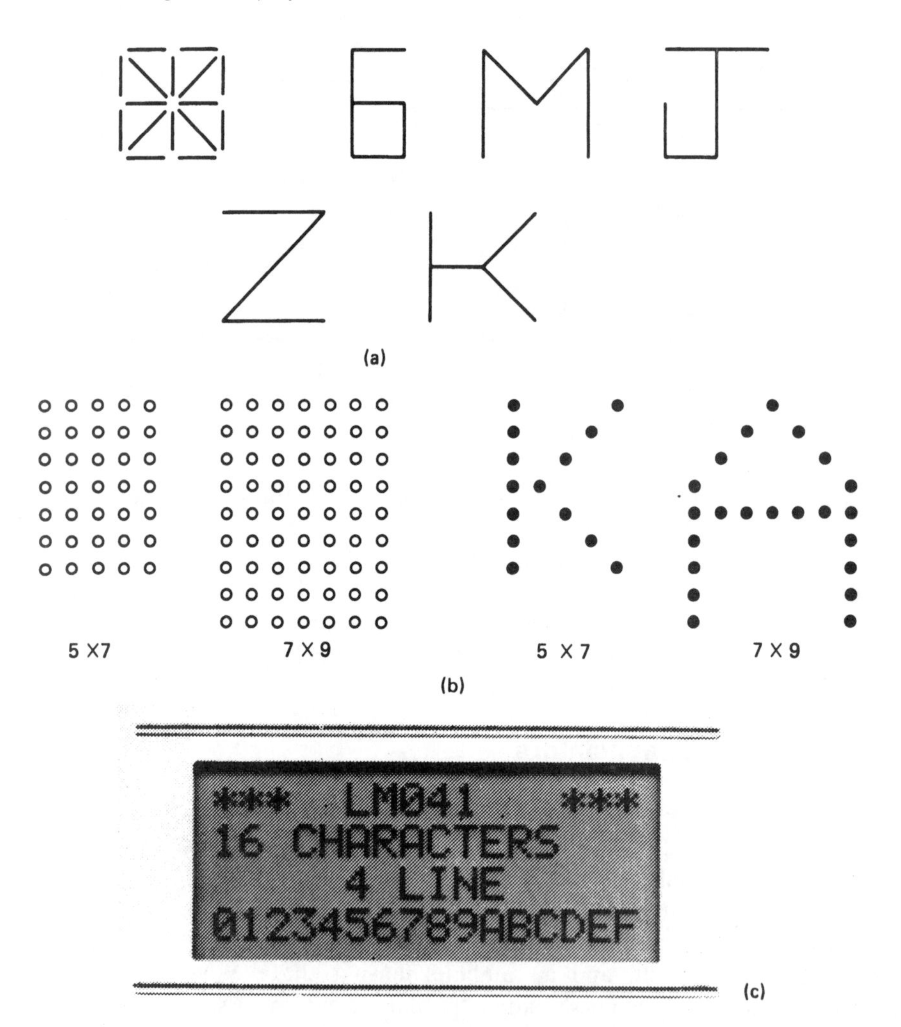

Fig. 17.28 *Alphanumeric displays. (a) 16-segment display. (b) Dot matrix displays. (c) Alphanumeric dot matrix liquid crystal display. The device, with an area of 62 × 25 mm can display 4 lines of 16 characters. (Reproduced from the Hitachi catalogue with permission.)*

VDUs, described in Section 14.6, usually incorporate one or more alarm pages. Alarm status can be shown by changing the colour attributes of alarm text: flashing red for unaccepted alarms, say, steady yellow for accepted, but still present, alarms and blanking the text (attribute the same as the background) for a healthy state.

Alarms can be generated from contact closure or opening, or direct from analog values which go above, or below, a preset value. In general, normally closed contacts which open on a fault are preferred as these will give a fault indication under cable fault, supply failure or transducer failure conditions.

Fig. 17.29 *Conlog alarm annunciator and alarm printer manufactured by Bowthorpe Controls. (Photo courtesy of Bowthorpe Controls and Wm McGeoch & Co.)*

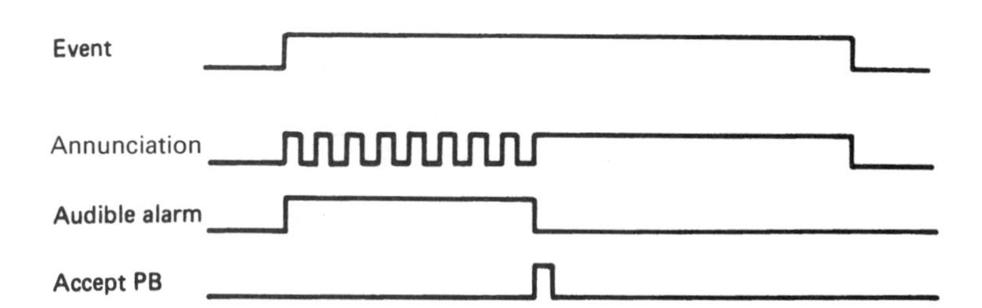

Fig. 17.30 *Typical annunciator action.*

17.6. Data logging and recording

17.6.1. Data loggers

A working plant produces a vast amount of data which can be useful for subsequent analysis of performance, efficiency, running costs, etc. Section 14.2.2 described logging as one of the roles of an industrial computer, where plant data is archived for later examination.

If a computer is not thought necessary, or desirable, for control, similar archiving can be achieved by means of a commercial data logger. These aim to provide a record of plant operation by:

(1) Recording plant analog values at regular timed intervals.

(2) Performing continuous checking for alarm conditions, and recording plant state and time when an alarm occurs.
(3) Recording digital events, with time, as they occur.

The resulting data is usually printed out as it occurs to give a continuous event log, and archived to disc or magnetic tape for later analysis.

Figure 17.31 shows a block diagram of a typical data logger, and it can be seen that this is effectively an industrial computer without an output unit.

Analog values are scanned by a multiplexer, and frozen by a sample-and-hold unit prior to digitisation by an ADC. Signals are normally presented to the data logger in a standard form (e.g. 4 to 20 mA) and converted to engineering units (degrees C, psi, etc.) in the control unit program. This is commonly achieved by

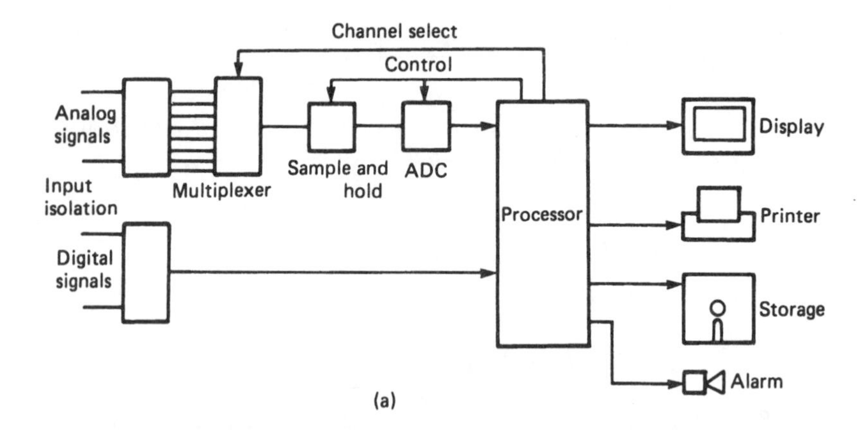

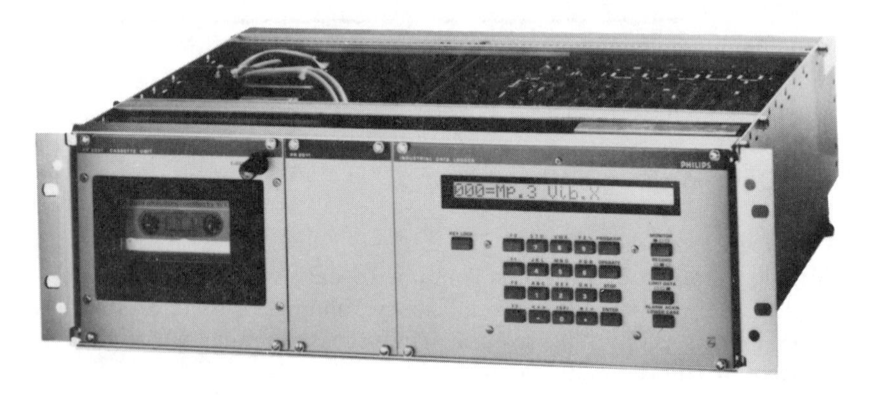

(b)

Fig. 17.31 *Data logger. (a) Block diagram. (b) The Philips PR2011 data logger. This unit can handle up to 256 channels for display, recording and alarm annunciation. (Photo courtesy of Philips Scientific & Industrial Equipment.)*

specifying high and low range engineering values and any linearisation routines (e.g. square root) for each channel as part of the set-up procedures.

Input channels need not be scanned sequentially; flows, for example, may need to be scanned more frequently than temperatures. The set-up procedure will usually allow a logging scan rate and an alarm scan rate (with alarm limits) to be specified for each channel.

Digital inputs will have an event text associated with them. In simple units this can be as terse as 'input 27 on'. More sophisticated units allow the user to define that input 7 corresponds, say, to 'reactor B auto mode selected', and choose if the digital event is to be recorded for contact closure, contact opening or both.

17.6.2. Instrumentation tape recorders

The magnetic tape recorder is an alternative way of archiving plant data for later analysis. The falling costs of computer systems and disc storage (particularly Winchester discs), however, make instrumentation recorders less attractive for industrial users, and tape recorders are more likely to be found in laboratory or experimental environments.

An instrumentation recorder works on the same principle as a domestic tape recorder, although many simultaneous tracks (typically sixteen) are recorded. Plastic tape, coated with ferromagnetic material, is passed over an electromagnetic record head, as shown in Fig. 17.32a. The current passed through the record head coil induces a high magnetic field across the record head gap, and magnetises the ferromagnetic material on the tape. To retrieve the information, the tape is passed over a playback head, similar in construction to the record head, and the variations in magnetisation of the tape induce a voltage in the playback head coil.

Although simple in theory, the process is fraught with many difficulties. The first of these is hysteresis, illustrated in Fig. 17.32b. If a magnetic material is magnetised by an external field, it will follow a curve similar to AB. If the field strength is reduced to zero, the magnetisation of the material does not fall to zero but remains magnetised at point C, called remnant magnetisation. Field reversals cause the material to traverse the curve CDEB.

Hysteresis produces a very non-linear response between applied field and resulting magnetisation, as shown in Fig. 17.32c. This can be overcome by the addition of an AC bias signal, as in Fig. 17.33. The bias signal is typically 30 to 50 kHz (dependent on tape speed) and is around 3 to 5 times the amplitude of the maximum input signal amplitude. It should be noted that the bias signal is *added* to the input signal, i.e. it is not a form of modulation. The bias signal causes the magnetic material on the tape to traverse both sides of the hysteresis curve, and the resulting non-linearities cancel. The bias signal is removed at playback by a simple low pass filter.

The response of the tape is also highly frequency dependent, as both the tape magnetisation and the play back signal are proportional to the rate of change of the magnetic flux across the head gap. Almost all process variables are essentially low frequency; very few can change faster than 50 Hz. It follows that the process signals must be encoded in some way before they can be recorded.

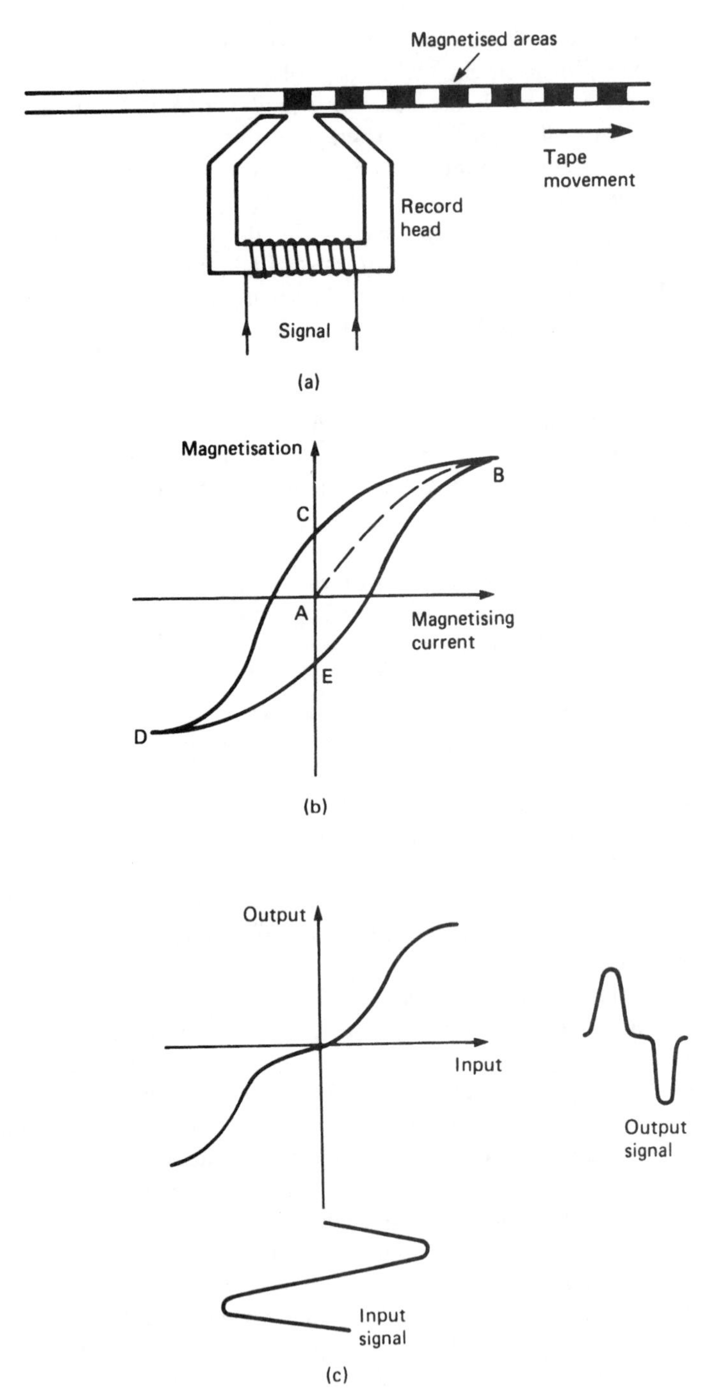

Fig. 17.32 *Magnetic recording. (a) Principle of tape recording. (b) Hysteresis curve. (c) Non-linear response.*

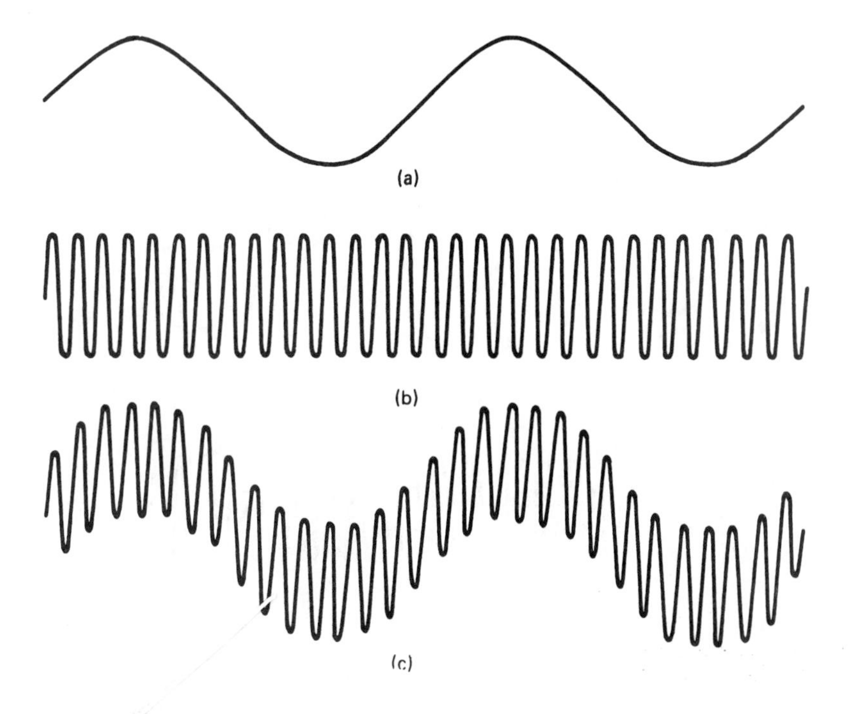

Fig. 17.33 *Overcoming hysteresis non-linearly. (a) Signal to be recorded. (b) Modulating waveform. (c) Resultant.*

Frequency modulation (FM) modulates the frequency of a carrier wave according to the level of input signal, as illustrated in Fig. 17.34. Because the information about the signal is conveyed in the carrier frequency and not its amplitude, errors from hysteresis and the tape/head frequency response are overcome.

Pulse height recording, as in Fig. 17.35, is a way of encoding data and multiplexing signals on to one recorder track. Each signal is sampled in turn and converted to a pulse whose height is proportional to the input signal level. A related technique uses pulses whose height is fixed, but whose width is proportional to signal level.

Increasingly, variables are digitised by a multiplexer and ADC similar to the input section of Fig. 17.31 to give a digital representation before recording. Typically 12-bit representation will give 0.01% resolution. Sample rates must be chosen according to Shannon's sampling theorem to give a faithful representation of the measured variable. Digital representation gives almost total immunity from distortion at the cost of greater tape usage. Typically a recording density of 63 bits per millimetre can be achieved with nine tracks across the 12.7 mm tape.

Digital data can be recorded serially along one track per signal, or in parallel across, typically, nine tracks. With parallel recording, one track is used as a parity check to give error indication.

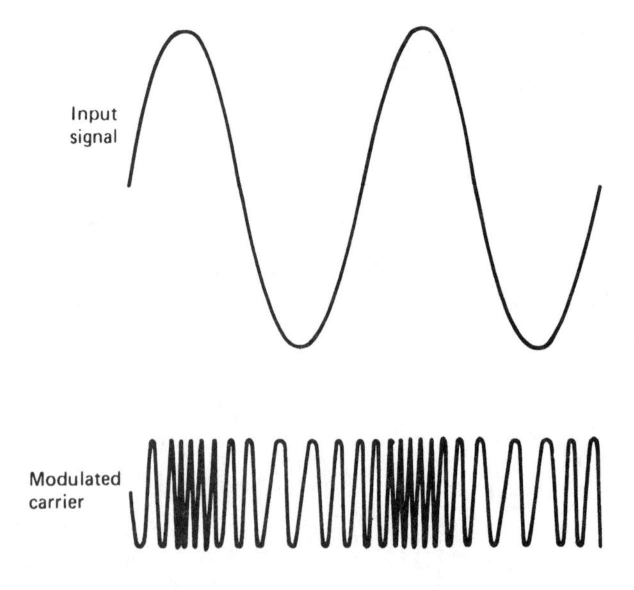

Fig. 17.34 *Frequency modulation (FM).*

Various digital encoding methods are used; Fig. 17.36 shows the more common. Digital signals can be FM modulated, or used directly to drive the tape material into saturation – point B and D on Fig. 17.32b.

Tape speed accuracy and consistency are important particularly for FM recording where speed variations (called wow for low frequency variations and flutter for high frequency variations) will appear directly as signal changes. A reference signal track is often used to control tape speed. This is a very stable high frequency signal recorded on the tape which is compared with a crystal oscillator on playback to correct tape speed. On digital systems a clock track is sometimes used to synchronise the tracks and control tape speed.

17.7. Ergonomics

Ultimately, every plant can be represented by Fig. 17.37 which shows what is often called, rather grandly, the MMI, or man machine interface. Ergonomics is the study and design of this interface such that the operator can perform his duties efficiently, in comfort and with minimum error.

Display of plant data has been the topic of most of this chapter and is crucial to the operator's effectiveness. An overkill of data which has to be painstakingly scanned for crucial information can be as detrimental as too little. Grouping of displays and consistency in the displays are important in allowing the operator to form a mental picture of what is actually happening in the plant under his control. The displays should be arranged so they can be scanned with minimum effort. Figure 17.38 shows boundaries of human perception, and preferred angles for desk tops.

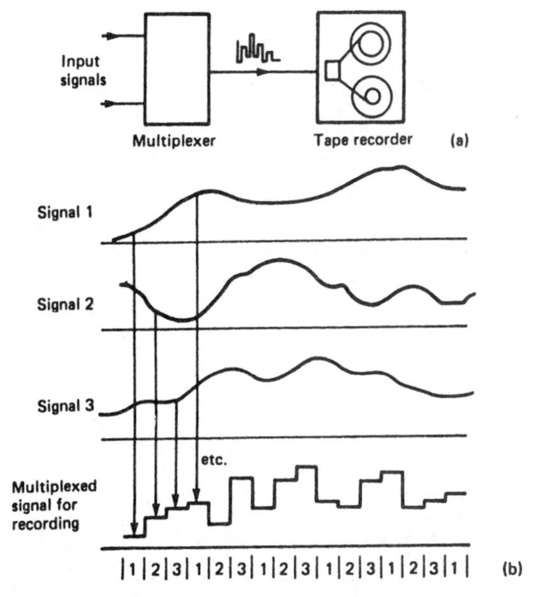

Fig. 17.35 *Pulse height recording. (a) Multiplexed recording system. (b) Typical signals. (c) Instrumentation tape recorder, showing input scaling cards. (Photo courtesy of Thorn EMI Datatech.)*

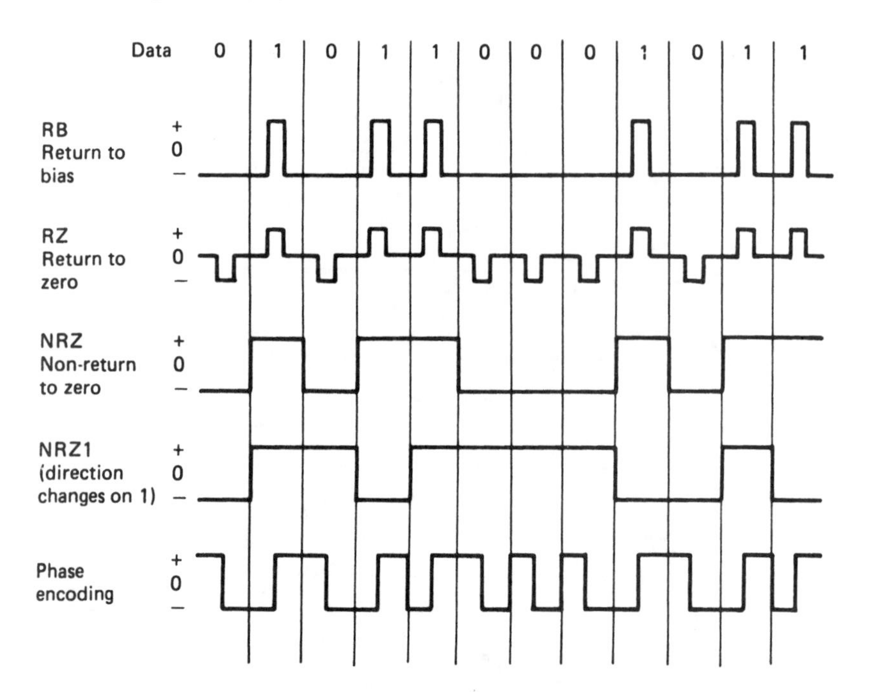

Fig. 17.36 *Digital recording methods.*

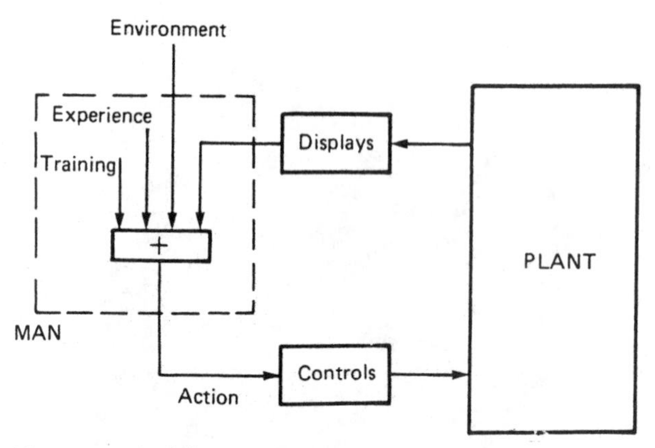

Fig. 17.37 *Man as part of the process.*

Consistency is important in the action of control devices; we all expect to turn something on, to increase some variable, by turning a knob clockwise. Similar expectations are shown in Fig. 17.39. Relative positioning of control devices and displays is important; people expect to see a reaction to an action, even if it is only a light that says the action is being acted upon (the lift call-button effect).

Possibly the most important aspect is the worker's immediate work space and environment. Reliable work cannot reasonably be expected from an operator who has a headache or a sore back within an hour of starting work. Factors such

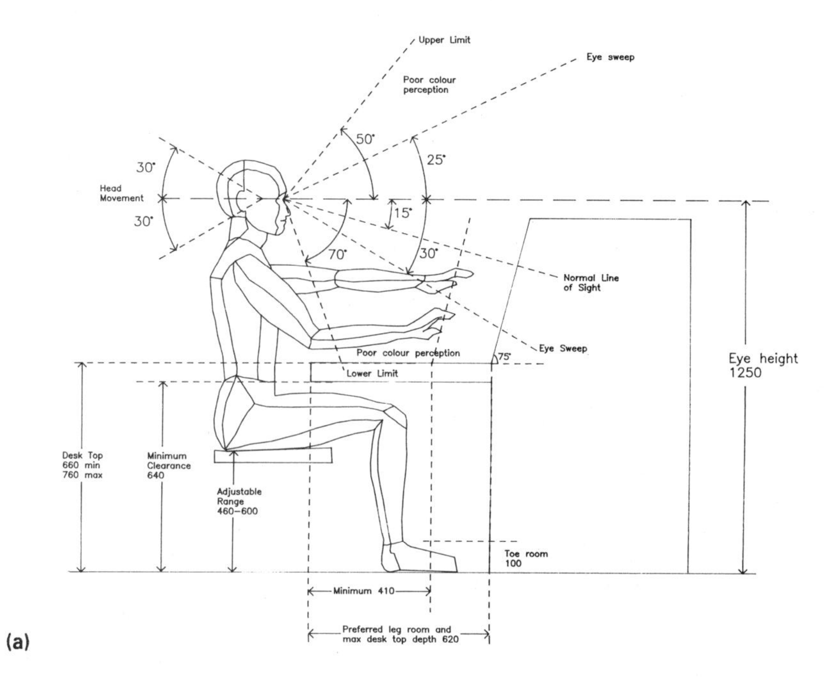

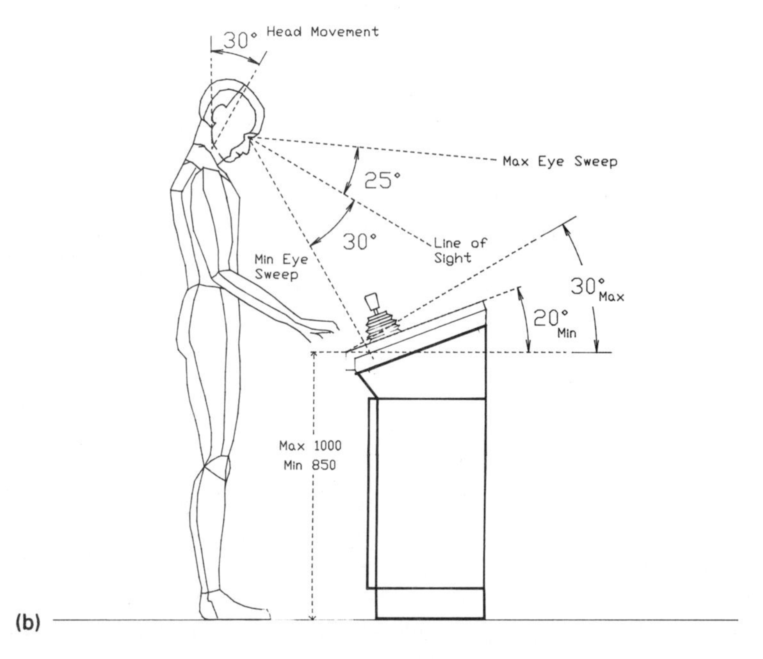

Fig. 17.38 *Control panel design. (a) Desk design. (b) Standing operator. (c) Reach and sweeps for multisection panels and desks. All these dimensions are guidelines as people vary greatly in size.*

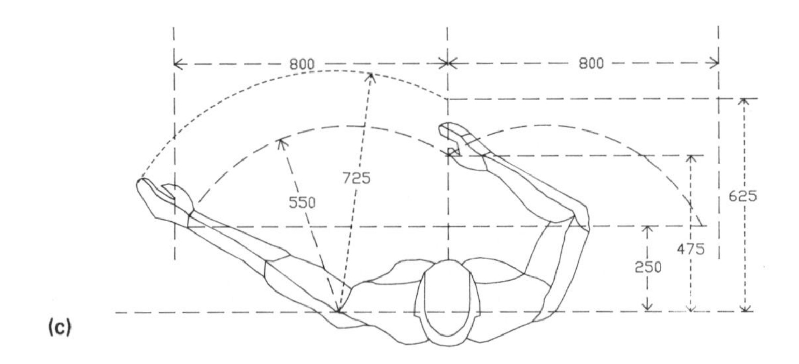

Fig. 17.38 *contd.*

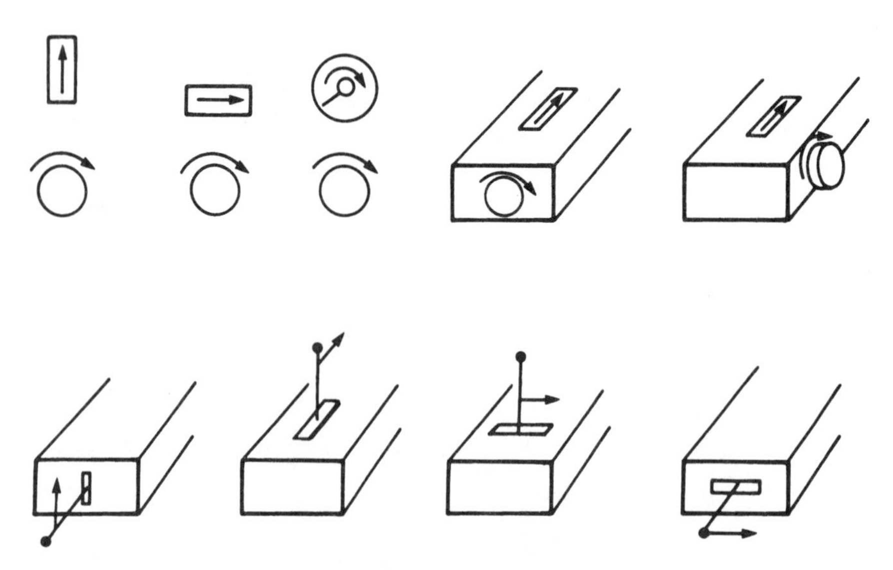

Fig. 17.39 *Human expectations of controls and indications. Arrows show direction for increase (of flow, power, speed, temperature, etc.). It is odd to note that UK light and power outlet switches are reversed from normal expectation. Controls relating to motions (e.g. crane controls) should follow plant movement.*

as noise, dust, smell, vibration, temperature (and temperature changes), lighting levels (and glare) and humidity all contribute to a worker's ability to concentrate. Psychological factors such as the degree of concentration necessary, and the ability to mentally rest and 'coast' for a short period are also important.

The layout of controls, displays and seating for convenience of operation are often overlooked. It is unfortunately true, for example, that most machine tool lathes and drilling machines are designed for operation by workers 4 feet tall with 8-foot-long arms, as can be seen by the postures that have to be adopted in use.

Figure 17.40 shows comfortable working spaces and reaches which suit the majority of the population.

Users of VDUs (visual display units) have long complained about health problems, complaining of sore backs, aching wrists, headaches and sore eyes. There has also been concern about possible side effects arising from the low intensity emissions from the screen.

From the 1st January 1993 the The Health and Safety (Display Screen Equipment) Regulations 1992 became operative in the UK. This covers all workers 'who habitually use VDUs for a significant part of their normal work'. From a process control viewpoint, the whole act may hinge on the word 'habitually', but it is good practice to follow the legislation, all of which is very sensible. The legislation dismisses medical side effects from screen emissions, but draws attention to ergonomic details.

The back and foot support are very important and covered in some detail along with lighting and the hand position during typing. Hands should be unbent and fingers not stretched. The user should be able to adjust the position (angle and height) of both the keyboard and screen. With fixed displays and keyboards usually built into desks, this could cause considerable expense in industrial applications.

New equipment has to conform with the legislation when installed, but companies have until 1996 to modify existing equipment.

The new legislation is linked to an EEC directive 90/270/EEC which is concerned with 'the minimum safety and health requirements for work with display screen equipment'. This goes further than the HSE legislation and covers ease of use of the software and other aspects of Human Computer Interaction (HCI).

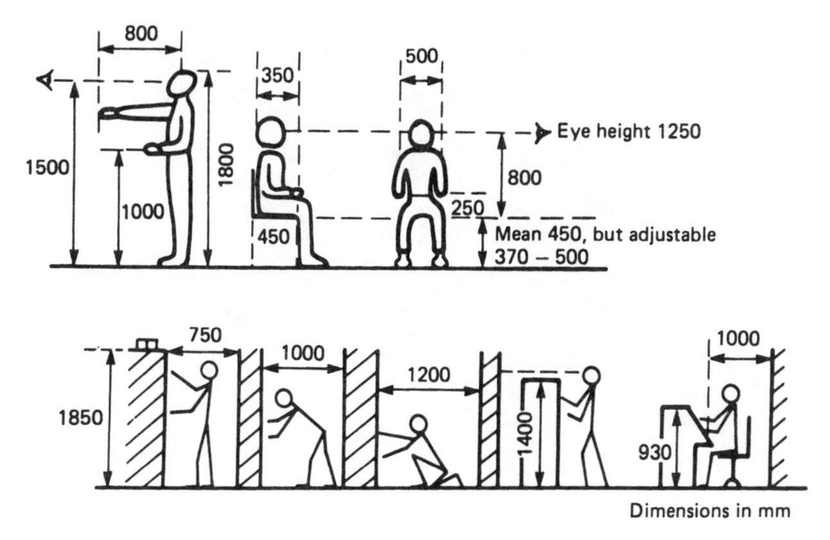

Fig. 17.40 *The human body and the environment.*

17.8. The man machine interface (MMI) and human computer interaction (HCI)

Some time ago I came across the entrance shown in Fig. 17.41. This was sensibly arranged with two doors swinging in different directions and clearly labelled IN and OUT so that people don't collide. What was most odd about these doors is

Fig. 17.41 *An example of a poor man-machine interface. The handle on the out door screams 'Pull me' so everyone goes in via the wrong door.*

that practically everyone (including me) went in through the door marked OUT, and out through the door marked IN.

There are several reasons why the doors didn't work. The major one is the handle which served no purpose at all and positively screams 'Pull me!' It is compounded by the complete absence of any indication (push plate or whatever) on the correct left-hand door. Most people are right handed and the approach to the door is from the right of the picture. In the absence of any other clues, the natural tendency is to go for the right-hand door. Finally there is a slight uphill slope to the door. This slope aids pulling but hinders pushing. Even if you knew all of this, it was still difficult to use the correct door!

A set of doors must be the simplest machine that human beings have to deal with. If most people cannot operate a clearly labelled set of doors correctly, what chance do we have of operating a video recorder, personal computer or a complex process control system?

Some time ago I was using a terminal on a mainframe computer when a message window suddenly appeared announcing:

'Job 014564/M/Bar254 started 24/05/87 @09:05:05 in subsystem QINTER in message queue M allocated to another job
Press Enter to continue'

A very uninformative communication. I duly did as requested, and sure enough the system continued working, but it left me feeling most uneasy. What did it mean? Was it serious? What *was* I supposed to do? Was my work lost? Was the data I was getting still valid? If it wanted me to *do* something why didn't it say so in clear English, and if it *didn't* want me to do anything why on earth did it interrupt me in the first place? Again we will return to this later.

These examples are typical of a breakdown in communication between people and the machines they are using; machines as simple as a pair of doors or as complex as a site-wide mainframe. Both are trivial, even mildly amusing, but they are symptoms of a larger, and more serious, failing in many modern process control systems.

As control systems become more automated, people have become increasingly remote from the things they are controlling, and have to rely on an artificial man-machine link. The design of this link is of major importance if a plant is to work safely and economically. There has been much study of good design practices and the topic has acquired the two jargon terms 'the man machine interface' (MMI) and 'human computer interaction (HCI).

With the arrival of computers, more areas required study. The subject of 'Cognitive Psychology' developed and is concerned with examining how humans handle the vast deluge of information that swamps us every day. It aims to explore human perception, attention, learning, memory and problem solving; all factors of major importance when dealing with the link between operators and a complex automated plant.

If a human being is trying to operate anything from a door to a nuclear reactor, there are essentially three steps:

First we *form a goal*; we describe to ourselves what we wish to achieve.
Next is *execution* of some activity which affects the world, hopefully in a way that moves us towards our goal.
Finally we *evaluate* the outcome of our activity and compare it with our desired goal.

In many respects these three steps are very similar to traditional closed loop control described in Chapter 18. For this simple sequence to work the operator must be able to:

- *Determine the system state*: what it is doing, what its capabilities are.
- *Easily decide* if any changes are needed.
- *Determine what actions are possible and relevant*, and how they are to be performed.
- *Examine the system and interpret its response* to the actions.

The pilots of a Boeing 737 became victims of a poor implementation of this loop and the result was the Kegworth air crash. Shortly after take-off they became aware of vibration and noise from an engine. They were unable to determine from their instruments which engine was faulty, so they reduced power on an engine at random to try to isolate the problem. Their first choice removed the symptoms, so they decided that engine was faulty and left it shut down.

The control system had tricked them. The Boeing 737 is a two-engine plane, and the shutdown of one engine can skew the controls. The designers had tried to reduce this effect by automatically reducing the power in the remaining engine when an engine is shut down. This automatic power reduction, not the engine shutdown, had removed the symptoms. The pilots, for perfectly logical reasons, had shut down a good engine and were flying on a single damaged engine. This engine operated for some time, but failed with tragic consequences when asked for increased power during the landing approach.

The pilots:
could not determine the system state,
were given faulty information (reduction of symptoms during their trial on the wrong engine)
and hence
reached a wrong evaluation of the problem.

The user will always form a mental model of how a system operates. The accuracy of this model is not really that important, the crucial aspect is whether it provides sufficient information for tasks to be performed and unexpected circumstances to be dealt with. Many people, for example, drive a motor car without knowing precisely how it works, or what the true operation of a clutch or choke is. They will, though, have some mental model of what is going on under the bonnet.

Designers naturally think of details, and one of the difficulties of designing a good interface is the need to bridge the gap between the designer's detailed and the user's simplified (and possibly incorrect) view of a system. The designer must try to think how someone else, probably inexperienced, will build the mental model of how it all works. The interface should then be built to match this simplified user's view. If the system is very complex and needs detailed knowledge, the designer must provide the simplified mental model for the user, either by training or by mapping the plant onto the controls.

At the bottom level of a control system there will be simple controls such as switches, push buttons and indicator lamps. The function of these controls should be readily apparent (and linked to the user's mental model.) Techniques like grouping controls in blocks by sectors, or even using controls from different manufacturers to emphasise areas can help the user. Operators at a nuclear plant in the USA broke up a line of identical joysticks by replacing the tops with beer keg controls, the different beers identifying the function. Above all, avoid a regular grid like array of identical controls identifiable only by the label text. It may look good, but be impossible to use.

These controls will offer 'affordances'; suggestions to the user of what they do. Pushbuttons are pressed, switches are turned, handles are pulled and so on. People have natural expectations of how these controls operate. Clockwise turns or away pushes, for example, increase speed, heat, pressure, flow or whatever. Sometimes these affordances are unexpected. I was recently involved in the design of the controls for a remotely controlled lance. We found the operators were having trouble with the vertical movement, controlled with a simple joystick. In the design we had thought about up/down and the controls were laid out accordingly. The operators had a mental model of 'flying' the lance and were thinking of 'climb' and 'dive'. The controls were the opposite way round to the operator's expectations.

Any operator action must have a response from the system. In many cases this is built in; the user asks for something and it can be seen, heard or felt. If there is no obvious feedback the designer must provide one. The simplest is a lamp or screen response acknowledging the request. Speed is important, a user will expect a response within 0.5 s. Above one second, and the request will probably be repeated. A second is a long time when you expect something to happen. The actual system need not respond that quickly, purely the response.

The curse of many systems is flashing lights. A flashing light should only mean one thing; action is needed NOW!! and when the user responds the flashing should stop. If a control room is full of flashing lights, the designer has seen too many episodes of *Star Trek*.

The final rule is be consistent from area to area. Water taps are moved in the opposite direction to our normal clockwise-implies-increase expectation but this anomaly is not apparent because it is consistent.

Computers bring new potential problem areas. Many of these arise because the mental gap between designers and users is wider. Some commonly seen problems are:

- No hints given of what operations are possible or what actions are expected from the operator: the 'Blank Screen' phobia.
- No feedback that commands have been received and are being acted on.

- Arbitrary use of controls, commands and actions. Something is done one way in one place and a totally different way somewhere else. This often arises where different parts of a system were designed by different people with no overall system standard.
- Use of non-obvious abbreviations and meaningless numbers (Undefined descriptor error 227 at 11 : 32 : 05 11/07.93.)
- The inevitable operator errors unhelpfully treated as breaches of contract with snarls and insults from the machine (invalid data entry, redo).
- Dangerous actions with no protection apart from a single line warning on page 619 of an 845-page manual (allowing the designer to say 'But didn't you read the instructions?') Everybody with PCs must have fallen foul of del*.*, but there used to be a popular desk top publisher with a group of file commands that defaulted to *.* and had an entry buffer such that if you quickly selected FILE/DELETE from the pull-down menu and clicked once too often with the mouse you would lose your root directory.

Most interaction between humans and computers take place via screens. These are often cluttered and confusing and, in all honesty, are sometimes designed for visitors or to provide advertising copy for the manufacturer's catalogue. By all means have impressive jazzy screens with lots of continually updated data and flashing lights, but remember the poor operator who has to live with it. Black or white backgrounds strain the eye, grey is most restful. Screens should be gentle with subdued colours and *only* show that data needed to run the plant. They are not video games.

Ideally colour should be used to enhance the display, and not be an integral part of the control. A significant part of the population (about 10%) is colour blind, and many sufferers are not aware of their problem. If colour sensitivity is important, users should be tested for colour blindness.

Designers often overlook the relatively small amount of data that can be clearly displayed on a single screen and try to overcome this by packing everything closely together. This leads to eyestrain. A good design should use a top-down approach, starting with a summary screen with the minimum data necessary, leading to more and more detailed screens. The system should guide the operator to the screen needing attention.

Flashing indicators on screens are as bad as flashing lights; they distract and annoy. If there must be flashing indication, consider bright/dim or a small flashing block alongside fixed text. Colour reversal between text and background should never be used.

Many designers produce screens which SHOUT AT YOU all the time. Only Death in the *Discworld* novels SPEAKS IN UPPERCASE. Lower case text is much easier to read, (if in doubt look at road signs.)

Consistency is, again, of upmost importance. Great care must be taken to lay down system standards when several people are involved in the design, each of whom will have their own peculiar ideas of the way to do things. The result is confusion for the operator.

Consider what the plant will do when a screen fails; if there is only one screen and no standby system it could run blind and uncontrolled for the ten or so minutes it might take to fetch and fit a spare.

Windows is seen as the panacea for screen ills. It does produce systems which look good, and the click and point idea is intuitive. It can, though, be rather uncontrolled leading to an operator getting lost in a display cluttered with overlayed screens and poorly chosen icons. Packages such as Visual Basic and ICOM's WinView for PLC5s do work very well and can be made very easy for an operator to use if the designer ensures that the application, and not the operating system, is the important part of the system.

Too many systems, particularly with Windows, adopt an 'I can do it so I will do it' approach without thinking of its relevance or use.

At the author's home we have a remote control unit for a Teletext TV. This must get an award for the worst ever user interface. The first Teletext controllers had just sixteen buttons. (0–9 plus a few simple options like TV/Text.) They were simple and intuitive to use. Our new controller has acquired a grand total of 56 buttons arranged in a rectangle of fourteen rows of four. This impressive array provides a host of features (which we do not use,) but the basic Teletext has become very confusing. The old three buttons for TV/Text/Mix, for example, have become a single button where you step through a sequence on each press. To go from TV to Text is two presses, from Text to TV is two presses. If you press once too often you have to go right round the loop again.

The 56 very small buttons are identified by peculiar dark grey even smaller icons on a black background making them impossible to read, and even the keypad is non-standard. Two standards in the world are bad enough (telephone 123 and calculator 789) but this controller has 1234, 5678, 90. Even after two years we still have to look hard at the keypad rather than the screen.

The problem here is that the designers packed in features without thinking what the end user really wanted. Much consumer electronics (particularly digital watches and VCRs) suffer from this type of creeping featurism. When the instruction manual for a TV set starts approaching the size of a novel, (as ours does) things have gone too far.

The shortcomings of our TV are minor, but similar problems can be found in most control interfaces. The author once had cause to re-design the controls for an existing process. This involved talking to the operators and finding out what they used, and what information they needed. It was surprising to find how many desk controls were never used (with the purpose of many unknown to some operators) and how much data on the computer screen was really irrelevant clutter. The original designers *could* do it so they *had* done it, and, of course, it looked impressive for visitors.

Based on these findings around 25% of the controls were removed and the displays were vastly simplified to the relief of the operators. With a working plant the people who genuinely *know* what is needed are the operators; listen to them and don't implement things just because the designer can.

The Kegworth aircrash was put down to pilot error; in the author's opinion a slightly unfair judgement. All humans make errors, so good designers should ensure that a system will detect and deal predictably with the inevitable.

Less obviously the design should avoid circumstances which may cause error. Any computer system which uses, say, six-digit numbers for a stores reference *and* six digits for the location *then* compounds this obvious trap by asking for Bin Number/Location on one screen and Location/Bin Number on another deserves all the problems it will get.

Designers should be aware of the seven types of error that are commonly made, and try to ensure the system does not actively encourage them.

- First is *data driven error*, the bin number/location problem above is of this type.
- Next is *capture error* where two or more sequences have the same steps in common at the start, and the commonest sometimes takes over when the mind is on auto pilot.
- Plain forgetfulness is known as a *loss of activation error*. This type of error was common with early hole in the wall bank machines. People were 'driven' by the task of getting money and would walk away leaving their card sticking out of the machine. This was cured by not dispensing the money until the card had been withdrawn.
- *Mode errors* occur when the designer has multiple functions on a set of controls. Common bad examples are multi-function phones with sequences such as **3 to pick up a call from another phone and #9nnn to transfer a call. Digital watches with features selected by how long a button is pressed are another bad example.
- *Associative errors* are Freudian slips, and appear when the user thinks 'I mustn't' then does. These are commoner than might be thought!
- *Descriptive errors* occur when the right action is performed on the wrong (but very similar) device. Laying out a desk with ten identical joysticks in a straight line is asking for descriptive errors. This type of error can be overcome by breaking up the layout in some way.
- Finally we have simple keyboard mistypes: drpped digits, <letetr sawps> and so on.

With careful design, a system can be made to live with all of these common errors.

Chapter 18
Closed loop control

18.1. Introduction

In many plants, process variables (e.g. temperature, pressure, flow, position, chemical analysis) are required to follow, or hold, some desired value. This control is achieved by manipulating plant actuators. Temperature control of a kiln, for example, could be achieved by manipulating the gas/air flow into a burner. It would be possible to control a process manually, but it is usually more efficient to use some form of automatic control.

The basis of an automatic control system is shown in Fig. 18.1. A control algorithm looks at the desired value, the actual value and possible outside influences affecting the plant, and on the basis of these observations adjusts the plant actuators to bring the process variable to the desired value.

The control algorithm has to cope with two circumstances. The desired value may be changing continuously (as in a position controlled telescope or an oven following a heating/cooling curve), or the process variable itself may be affected by disturbances. These disturbances can arise from changes in throughput (called load changes) or from outside influences over which the engineer has no influence. The temperature in a continuous gas-fired oven, for example, will be affected by product throughput, outside ambient temperature, input product temperature, gas and combustion air pressure plus a multitude of other influences.

The simplest control strategy is the bang/bang servo shown controlling the temperature of an oven in Fig. 18.2a. The actual temperature is subtracted from the desired temperature to give an error signal. This error signal is passed to a comparator with hysteresis which controls a power control relay. If the temperature is too low, the relay will be energised and the heater turned on; if the temperature is too high the heater will be turned off. Domestic central heating systems use this approach.

The control system thus maintains the desired temperature by cycling power to the heater. The hysteresis in the comparator is necessary to prevent high speed

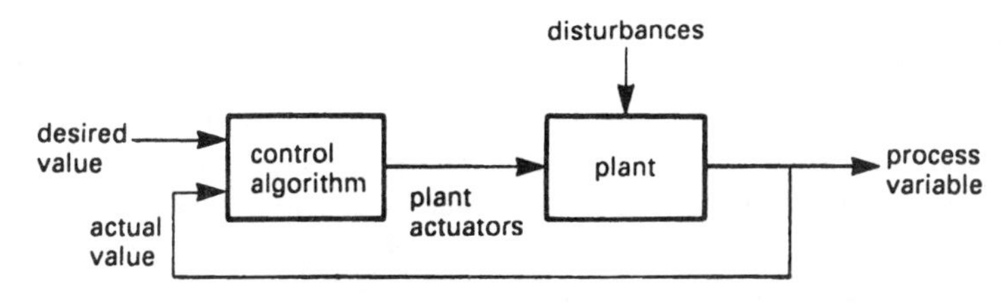

Fig. 18.1 *An automatic control system*

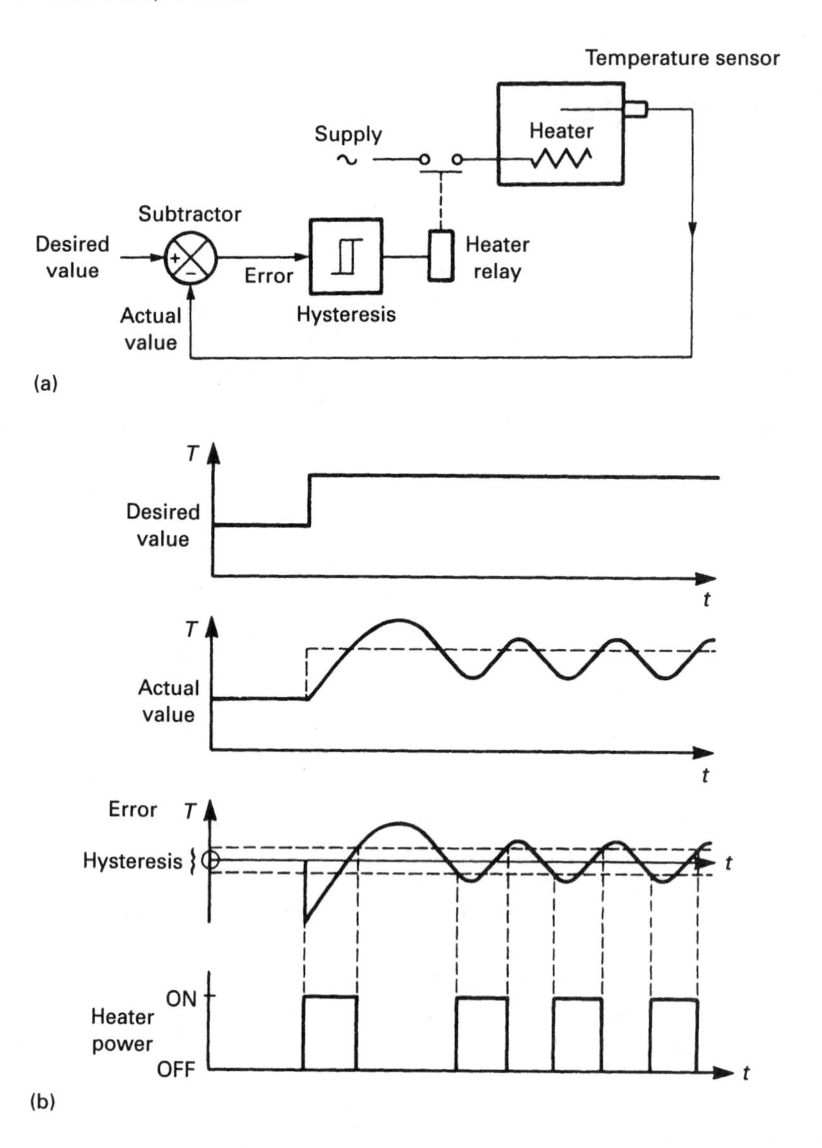

Fig. 18.2 *A bang/bang servo system. (a) System block diagram. (b) System operation.*

chatter in the relay when the desired temperature is achieved.

Figure 18.2b shows what happens after a change in desired value. It can be seen that in the steady state the process variable oscillates about the desired value, with period and amplitude determined by the hysteresis and the characteristics of the process. Decreasing the hysteresis reduces the amplitude of the oscillations and decreases the period (i.e. the oscillations become faster) but in any practical system there is a limit below which the increasingly rapid actuator movement leads to excessive wear and early failure.

In Fig. 18.2 the actuator can only be full on or full off, and in the steady state this leads to continuous oscillations. What is intuitively required is a large corrective action when the error is large and a small corrective action when the error is small. This necessitates proportional plant actuators which can give a controlled response over their full range.

The principle is shown in Fig. 18.3. An error signal is produced by subtracting the actual value (denoted P_v for process variable) from the desired value (denoted S_p for set point). The error is amplified and passed directly to the plant actuator.

The output P_v signal is fed back for comparison with the desired S_p value, so Fig. 18.3 is commonly called feedback control. Because the actuator signal is proportional to the error the term proportional control is also used.

18.2. Proportional control

18.2.1. Proportional only (P) control

Examination of Fig. 18.3 shows that the actuator signal is an amplified version of the error signal, i.e. an error must exist for the actuator to operate. Inherently, therefore, a proportional control system operates with an error between S_p and P_v. This error is called the offset and is analysed in Fig. 18.4. The assumption is made that the plant is operating under steady conditions so we can ignore transient effects. The error, E is given by:

$$E = S_p - P_v \tag{18.1}$$

The actuator signal is given by:

$$A = K(S_p - P_v) \tag{18.2}$$

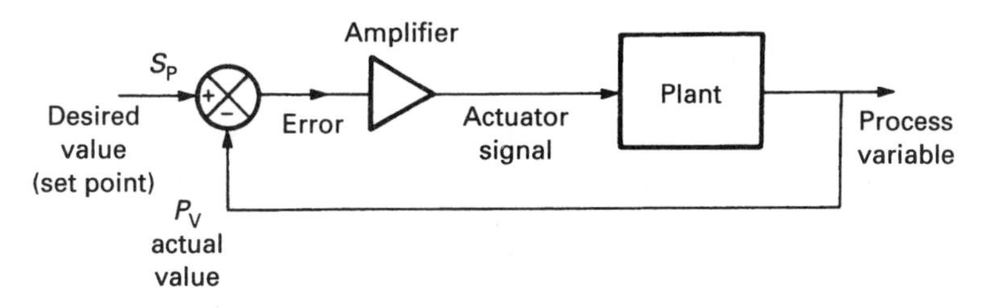

Fig. 18.3 *Proportional (feedback) control system.*

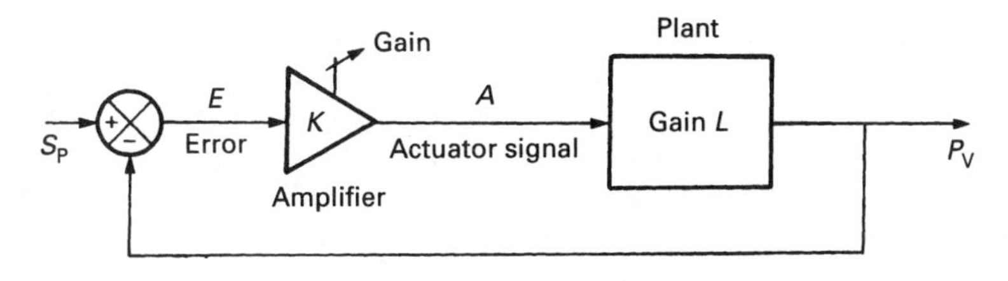

Fig. 18.4 *Steady state representation of a feedback system.*

where K is an adjustable gain.

Assuming the process is linear the plant output signal is:

$$P_v = LA \tag{18.3}$$

$$P_v = LK(S_p - P_v) \tag{18.4}$$

$$P_v(1 + LK) = LKS_p \tag{18.5}$$

or

$$P_v = \frac{LKS_p}{(1 + LK)} \tag{18.6}$$

The product LK is often called the open loop gain, as it is the gain that would be seen if the loop was broken at any point and a signal injected. If we denote the open loop gain by G (i.e. $G = LK$) then:

$$P_v = \frac{GS_p}{1 + G} \tag{18.7}$$

and the error is found by substitution into equation 18.1.

$$E = \frac{S_p}{1 + G} \tag{18.8}$$

It can be seen that for large values of G, the value of P_v approaches S_p, and the offset error, given by equation 18.8, approaches zero. If G is 20, for example, the value of P_v is 0.95 S_p and the error is about 4.7%.

The implication of equations 18.7 and 18.8 is that the open loop gain G should be as large as possible. Usually the plant gain is fixed so, to reduce offset error, K should be made as high as possible. Unfortunately open loop gain cannot be increased indefinitely without the system becoming unstable, a topic discussed further in Section 18.3.

The equations were derived for a change in set point. A similar set of equations could be developed for disturbances from load changes or outside influences. Proportional-only control will again compensate partially for disturbances, but an offset error will occur, the size of which decreases for increasing open loop gain.

The amplifier gain of a proportional controller is often referred to as the 'proportional band' or P_B which is expressed in per cent. It is an indication of the range over which the controller operates, and can be seen by reference to Fig. 18.5. Here, an amplifier of gain K is used to amplify an error signal. Assume both input and output voltage range is the same, ± 15 V say.

The input signal range that causes the output to saturate is therefore $\pm 15/K$ V, signals outside this range cause no further change in output. The amplifier thus has an input proportional band of $\pm 15/K$ V or $100/K$ expressed as a percentage of full input range. If input and output have the same range, the P_B is $100/K$%; a gain of 5 corresponds to a P_B of 20% for example.

If the input and output swings are unequal (or of different quantities, as could

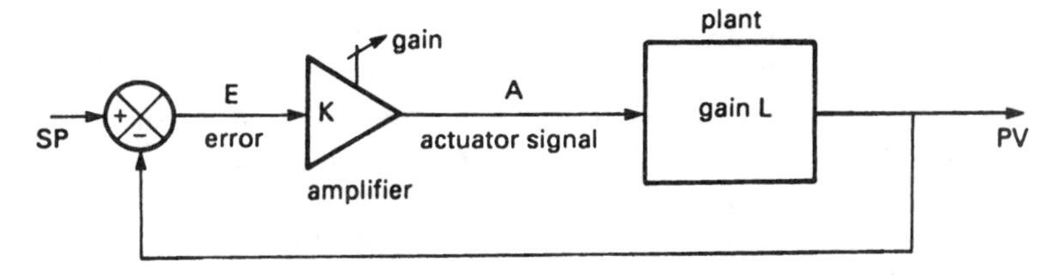

Fig. 18.5 *Relationship between gain and proportional band.*

occur for example, on a voltage to pneumatic pressure amplifier) the P_B is the percentage of input range over which the output does not saturate.

18.2.2. Proportional plus integral (PI) control

Figure 18.6 shows an, albeit manual, way in which the offset caused by set point changes can be overcome. The operator has an additional control which, historically, is labelled 'reset'. After a change of set point, the operator waits for transient effects to die away, then slowly adjusts the reset control until the observed error is zero. The offset error from the change in set point has been removed, but offsets from disturbances will still be present.

The circuit of Fig. 18.6 is called proportional with manual reset control. An obvious improvement would be the addition of automatic reset. This is provided by the circuit of Fig. 18.7. The signal to the control amplifier has two terms, a straight proportional signal as before, and a signal which is a scaled time integral

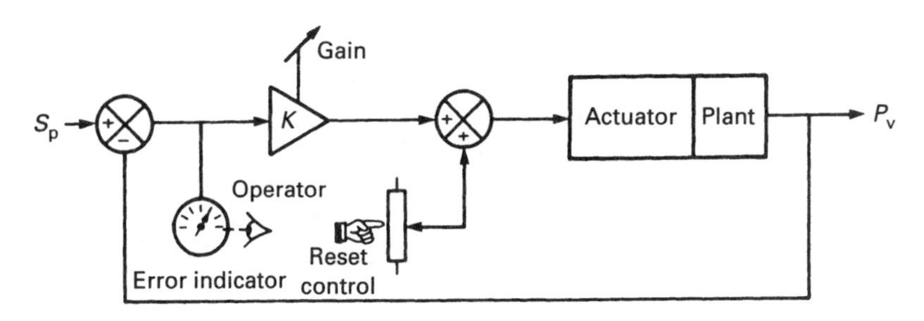

Fig. 18.6 *Removal of offset with operator adjusted reset control.*

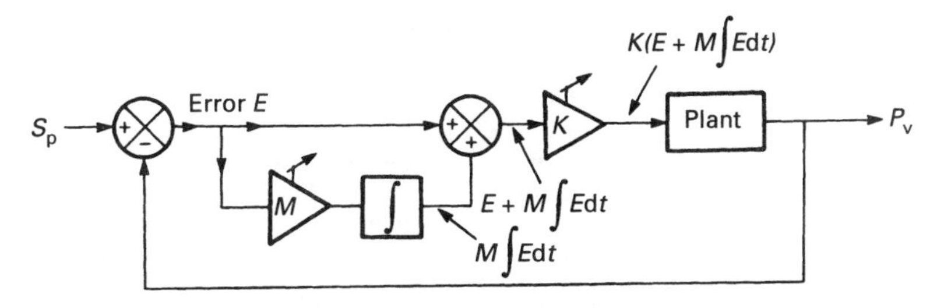

Fig. 18.7 *Removal of offset with integral action.*

of the error. The arrangement is called, for obvious reasons, a PI controller.

Suppose a steady offset error is present. The integral signal will ramp, causing the actuator signal to change and reduce the error. When the error is zero, the integral signal will be steady. Intuitively, therefore, the effect of the integrator will, in the steady state, bring the error to zero.

Similar considerations will show that offsets caused by disturbances will also be reduced over a period of time. Integral action acts to reduce $(S_P - P_V)$ to zero, but the time taken to remove the error depends on the settings of the gain K and M. It may be thought odd that the proportional and integral terms are added before the gain setting amplifier, but as will be seen in later circuits this arrangement simplifies mathematical analysis.

The action of a PI controller is summarised in Fig. 18.8. This shows the result of a step error of E%. In a closed loop system of course, the error would reduce, but a step response is easy to analyse. The initial response of the controller will be a step of height KE% from the proportional term. The integral term will then cause the controller output to ramp up with a slope of KME% per unit time.

The 'gain' of the integral term has the units of inverse time, and can be specified in two ways. The first notes that for a step error input the integral action will repeat the proportional step KE at fixed time intervals. This is termed the

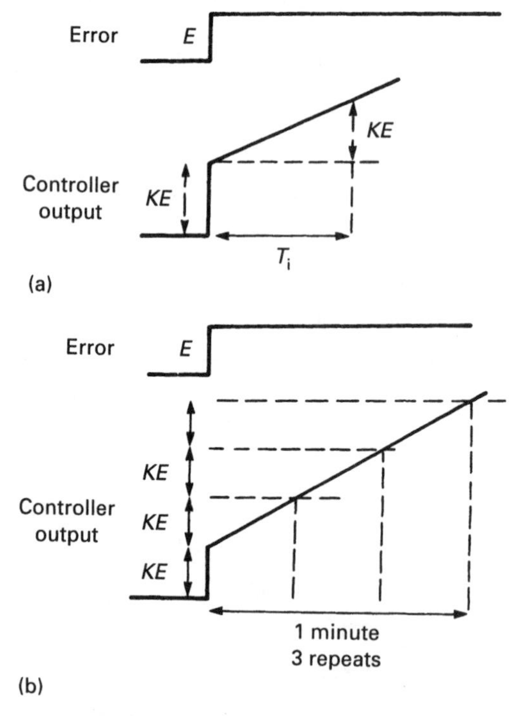

Fig. 18.8 *Definition of controller integral action, (a) Definition in terms of integral time* T_i*; (b) Definition in terms of repeats/minute.*

'integral time', usually denoted by T_i. It can be seen that $M = 1/T_i$, so the controller output is given by:

$$V_o = K\left(E + \frac{1}{T_i}\int E\,dt\right) \tag{18.9}$$

T_i has the units of time, and can be given in seconds or minutes.

The second way of specifying integral action specifies how often the initial proportional step is repeated by the integral action per unit time (usually minutes are used). An integral time of 20 seconds corresponds to 3 repeats per minute as shown in Fig. 18.8b. It can be seen that increasing the repeats per minute or decreasing the integral time will speed up the integral contribution.

For the fastest response to a set point change or a disturbance a short integral time is required. Decreasing the integral time, however, tends to de-stabilise the system to the point of instability.

18.2.3. Derivative (rate) action

If we consider how an operator manually controls a plant, it is possible to identify an additional factor that can be usefully added to a proportional controller. In Fig. 18.9a the voltage from a steam-driven generator is being controlled manually. The operator observes the voltage, and adjusts the steam valve to keep the voltage steady.

Suppose a sudden load is applied to the system. This will cause the generator speed and voltage to fall. The inertia of the system will, however, stop, the voltage falling instantly, and if left uncorrected the voltage would fall over a period of time as Fig. 18.9b. If the operator applied proportional and integral action alone, the voltage would be brought back to the correct level eventually, but the inertia has extended the duration of the droop as both P or I actions are dependent on the magnitude of the error.

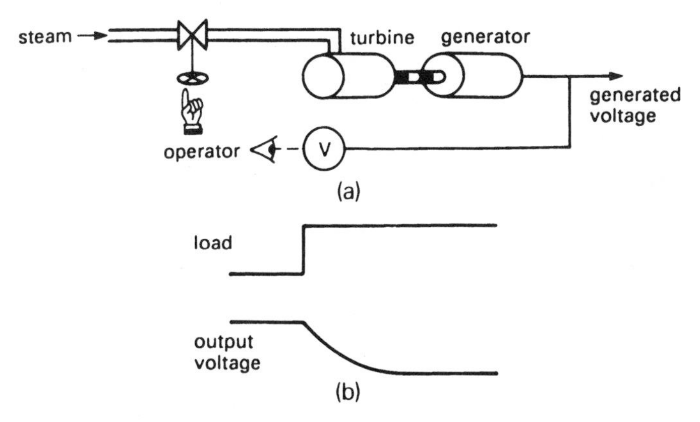

Fig. 18.9 *System requiring operator action based on rate of change of error. (a) Manually-controlled generation system. (b) Effect of load change.*

The operator, however, can anticipate what action is needed by looking at how *fast* the error is changing, and adding more corrective actions when the error is changing rapidly. If the error is increasing, the actuator signal is increased beyond that supplied by the P and I terms. This anticipates, to some extent, the control action required and compensates for the inertia induced time lag.

When the error is reducing, a similar anticipation will reduce the actuator signal such that the error comes to zero as fast as possible without overshoot.

The operator has anticipated the error signal and compensated to some extent for plant inertia, by adding corrective action based on the rate of change of error. This is known as derivative, or rate, action and can be achieved with the block diagram of Fig. 18.10a which combines proportional, integral and reset action. The actuator signal is given by:

$$V_o = K\left(E + \frac{1}{T_i}\int E\,dt + T_d\frac{dE}{dt}\right) \tag{18.10}$$

where K and T_i were defined earlier for equation 18.9 and T_d is a scaling factor for the rate of change signal. This scaling factor has the dimensions of time, and is usually given in minutes or seconds according to the speed of the process. It is known as the derivative time or the rate time.

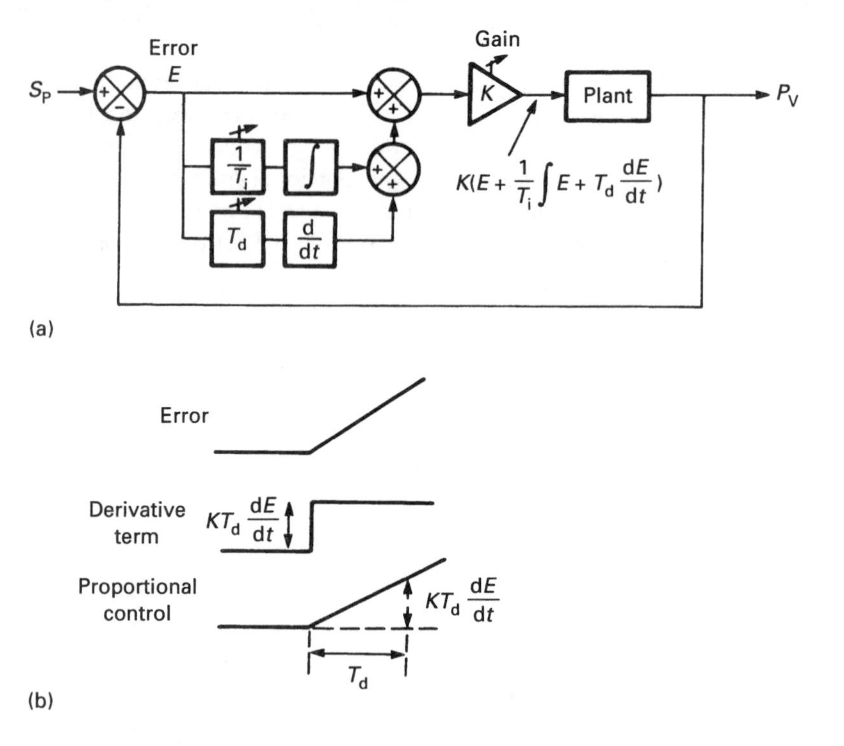

Fig. 18.10 *The addition of derivative action. (a) Block diagram of three-term controller. (b) Definition of derivative (rate) time T_d.*

The derivative time can be visualised as Fig. 18.10b, where a ramp error signal is applied to the controller. The derivative term will contribute a step to the actuator signal, and the proportional term a ramp, as shown. The derivative time T_d is the time taken for the proportional signal to equal the derivative signal.

It can be seen that the longer the value of T_d the more contribution the rate of change of error will make to the output actuator signal.

Derivative action can cause problems in some circumstances. Consider the effect of applying a step change of set point to Fig. 18.10a. This will be seen as an infinitely fast rate of change of error and result in a large, fast, and possibly damaging, actuator signal. For this reason, many controllers implement rate action as Fig. 18.11, where a rate of change of P_v (as opposed to error) signal is used. With a steady set point this behaves exactly as Fig. 18.10a, but does not 'kick' the actuator on a change of set point. If the controller is, however, required to follow a continually changing set point (as in ratio or cascade control described later) the form of Fig. 18.10a is preferred.

Derivative action can also be problematical where the measurement of the process variable is inherently noisy. A typical example is level control where ripples and liquid resonance tend to give a continually fluctuating level signal even in the steady state. If derivative action is used, these rapid fluctuations will be seen as rapid changes of error, causing rapid unnecessary actuator movement and premature wear.

18.2.4. Three-term (PID) control

Equation 18.10 describes an actuator signal which has three components:

(1) proportional to error;
(2) proportional to the time integral of error;
(3) proportional to the rate of change of error.

The strategy of equation 18.10 is therefore called three term or PID control, and is widely used in process control.

There are three adjustable constants in equation 18.10, the gain K, the integral

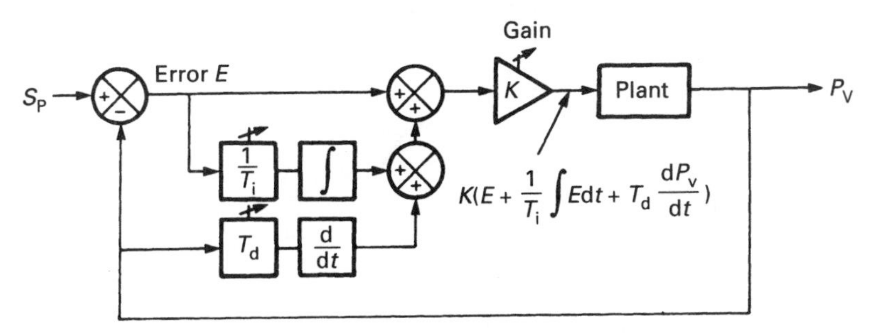

Fig. 18.11 *Derivative action based on rate of change of P_V rather than error.*

time T_i, and the derivative time T_d. The setting of these is crucial to the performance of the controlled plant. The theory and practice of tuning a controller is discussed in later sections.

18.3. Stability

18.3.1. Introduction

The description of proportional control in Section 18.2 implies that perfect control can be obtained by utilising a large proportional gain, short integral time and long derivative time. The system will then respond quickly to disturbances, alterations in load and set point changes.

Unfortunately life is not that simple, and in any real life system there are limits to the settings of gain T_i and T_d beyond which uncontrolled oscillations will occur. Much of this chapter is concerned with analysing a system to give controller settings which allow adequate performance with stability.

Consider the position control system of Fig. 18.12a. The load has high inertia, denoted by a flywheel. This will limit the acceleration after a change in set point. As the load approaches the new set point, the inertia will keep the speed up despite the decreasing motor armature volts, and the load will still be moving as the set point is reached. This will lead to one or more overshoots as shown in Fig. 18.12b. Increasing the gain will speed up the initial acceleration (assuming no element saturates) but will lead to a higher running speed and a larger overshoot. Decreasing the gain will reduce the overshoot, but give a slower initial response.

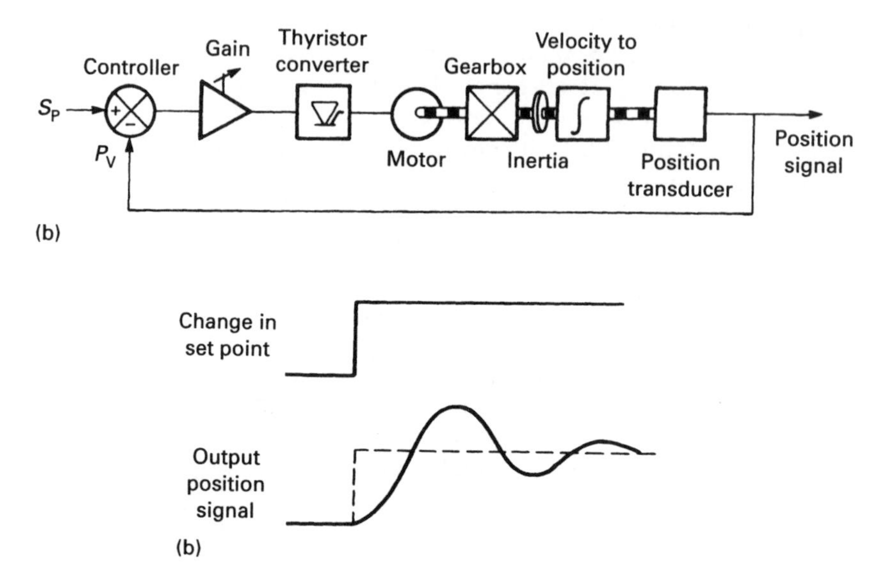

Fig. 18.12 *Response of position control system. (a) Position control system with high inertia. (b) Response to a step change in set point.*

Like many engineering systems, the setting of the controller is a compromise between conflicting requirements.

18.3.2. Definitions and performance criteria

It is often convenient (and not too inaccurate) to consider that a closed loop system behaves as a second order system, with an angular frequency ω_n and a damping factor b as equation 18.11.

$$\frac{d^2x}{dt^2} + 2b\omega_n \frac{dx}{dt} + \omega_n{}^2 x = f(t) \tag{18.11}$$

It is then possible to identify five possible performance conditions, shown for a set point change and a disturbance in Fig. 18.13a and b.

An unstable system exhibits oscillations of increasing amplitude. In practice the oscillations will continue until some element saturates or fails. A marginally stable system will exhibit constant amplitude oscillations; a theoretical condition which is almost impossible to achieve in practice, but is of importance in selecting controller gains.

An underdamped system will be somewhat oscillatory, but the amplitude of the oscillations decreases with time and the system is stable. (It is important to appreciate that oscillatory does not necessarily imply instability.) The rate of decay is determined by the damping factor. An often used performance criteria is the 'quarter amplitude damping' of Fig. 18.13c which is an underdamped response with each cycle peak one quarter of the amplitude of the previous. For many applications this is an adequate, and easily achievable, response.

An overdamped system exhibits no overshoot and a sluggish response. A critical system marks the boundary between underdamping and overdamping and defines the fastest response achievable without overshoot.

For a simple system Fig. 18.13a and b can be related to the gain setting of a P only controller; overdamped corresponding to low gain with increasing gain causing the response to become underdamped and eventually unstable.

It is impossible for any system to respond instantly to disturbances and changes in set point. Before the adequacy of a control scheme can be assessed, a set of performance criteria is usually laid down by production staff. Those defined in Fig. 18.14 are commonly used. These assume the closed loop response is similar to a second order system.

The 'rise time' is the time taken for the output to go from 10% to 90% of its final value, and is a measure of the speed of response of the system. The time to achieve 50% of the final value is called the 'delay time'. This is a function of, but not the same as, any transit delays in the system. The first overshoot is usually defined as a percentage of the corresponding set point change, and is indicative of the damping factor achieved by the controller.

As the time taken for the system to settle completely after a change in set point is theoretically infinite, a 'tolerance limit' or 'maximum error' is usually defined.

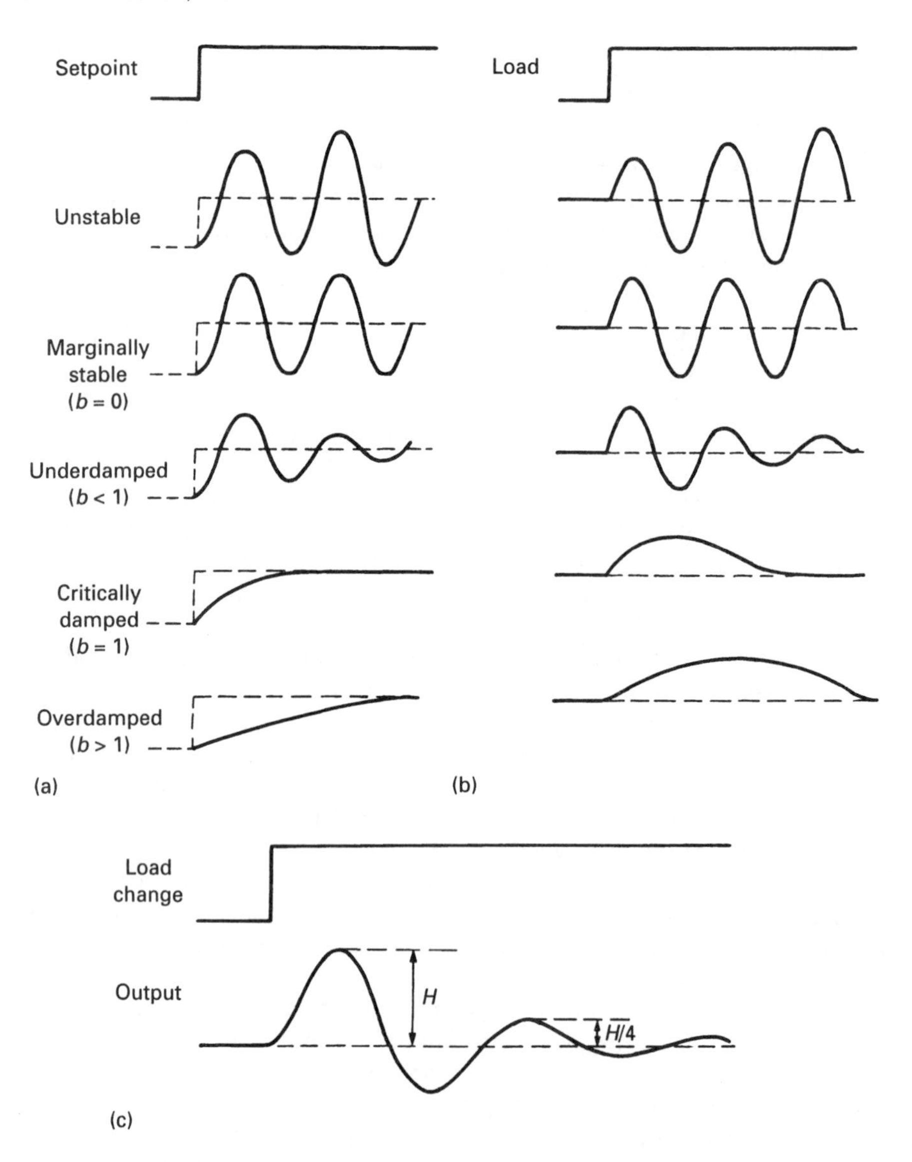

Fig. 18.13 *Various forms of system response. (a) Step change in set point. (b) Step change in load. (c) Quarter amplitude damping.*

This is usually expressed as a percentage of a set point change; once the plant is within the tolerance limit it is considered to have achieved the set point. A typical value for the tolerance limit is 5%. The settling time is the time taken for the system to enter, and remain within, the tolerance limit.

The shaded area is the integral of the error and can be used as an index of performance. Note that for a system with a standing offset (as occurs with a P only controller) the area under the curve will increase with time and not converge

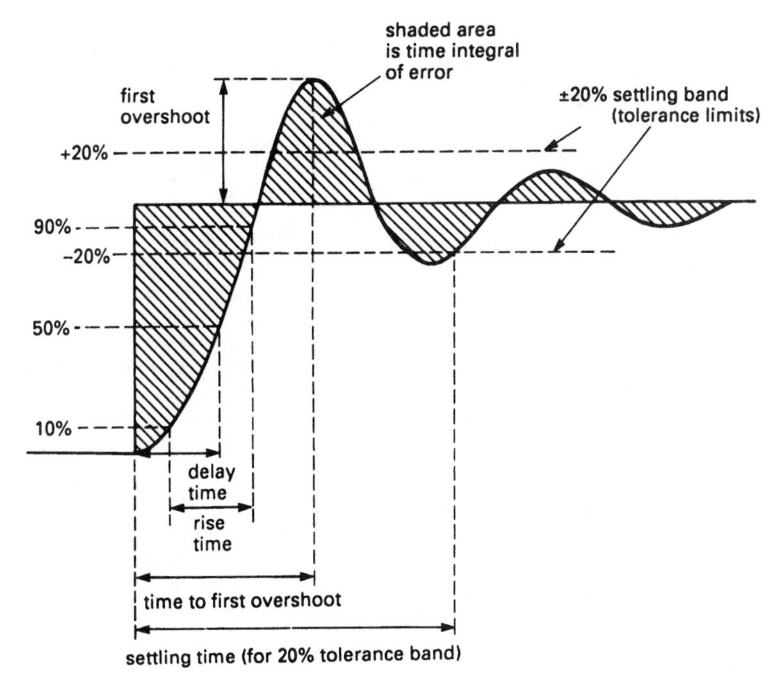

Fig. 18.14 *Definitions of system response.*

to a final value. Stable systems with integral action control have error areas that converge to a finite value. The area between the curve and the set point is called the integrated absolute error (IAE) and is an accepted performance criterion.

An alternative criterion is the integral of the square of the instantaneous error. This weights large errors more than small errors, and is called integrated squared error (ISE). It is used for systems where large errors are detrimental, but small errors can be tolerated.

The performance criteria above were developed for a set point change. Similar criteria can be developed for disturbances and load changes.

18.4. Commercial controllers

18.4.1. Introduction

The commercial three-term controller is the workhorse of process control and has evolved to an instrument of great versatility. This section describes the features of a modern microprocessor based controller. The description is based on the 6360 controller manufactured by Eurotherm Process Automation Ltd of Worthing, Sussex.

18.4.2. Front panel controls

The controller front panel is the 'interface' with the operator who may have little or no knowledge of process control. The front panel controls should therefore be

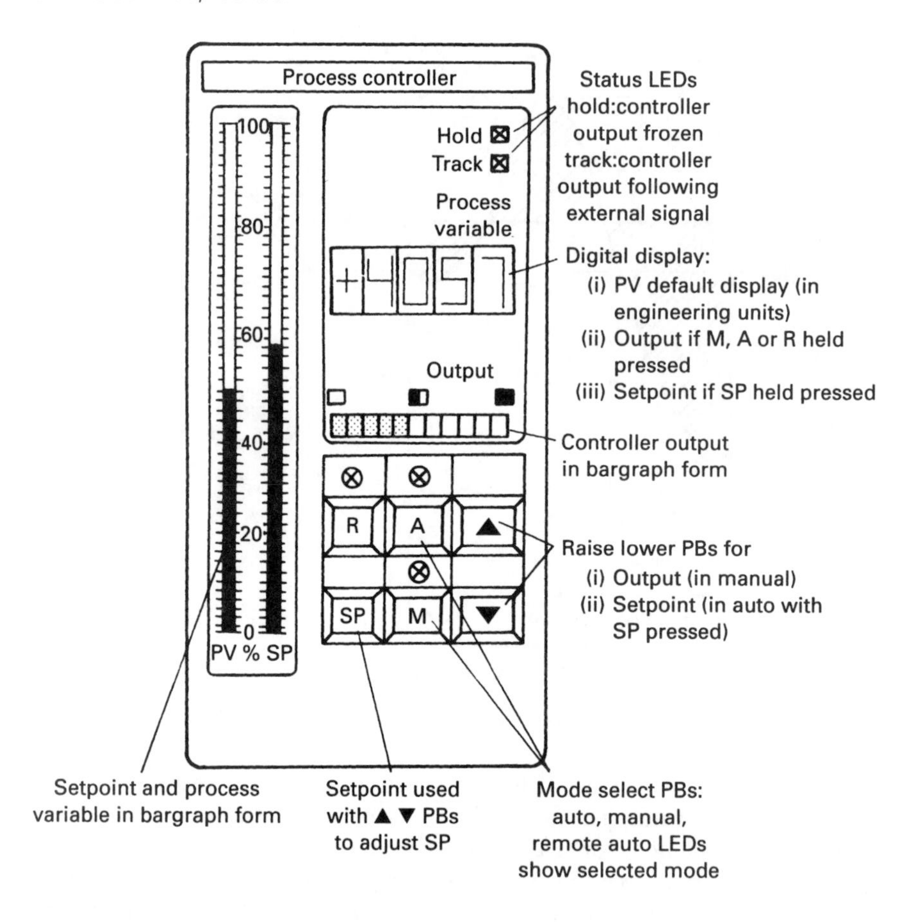

Fig. 18.15 *Controller front panel controls (based on Eurotherm Process Automation 6360 Controller).*

simple to comprehend. Figure 18.15 shows a typical layout.

The operator can select one of the three operating modes – manual, automatic or remote – via the three pushbuttons labelled M, A, R. Indicators in each pushbutton show the current operating mode.

In manual mode, the operator has full control over the driven plant actuator. The actuator drive signal can be ramped up or down by holding in the M button and depressing the ▲ or ▼ buttons. The actuator position is shown digitally on the digital display, whilst the M button is depressed and continuously in analog form on the horizontal bargraph.

In automatic mode the unit behaves as a three term controller with a set point loaded by the operator. The unit is scaled into engineering units (i.e. real units such as °C, psi, litres/min) as part of the set-up procedure so that the operator is dealing in real plant variables. The digital display shows the set point value when the SP button is depressed and the value changed with the ▲ and ▼ buttons. The set point is also displayed in bargraph form on the right-hand side of the dual vertical bargraph.

Remote mode is similar to automatic mode except the set point is derived from an external signal. This mode is used for ratio or cascade loops and batch system where the set point has to follow a predetermined pattern (annealing furnaces are a common application). As before the set point is displayed in bargraph form and the operator can view, but not change, the digital value by depressing the SP button.

The process variable itself is displayed digitally when no push button is depressed, and continually on the left-hand bargraph. In automatic or remote modes the height of the two left-hand bargraphs should be equal; a very useful quick visual check that all is under control.

Alarm limits (defined during the controller set up) can be applied to the process variable or the error signal. If either move outside acceptable limits, the process variable bargraph flashes, and a digital output from the controller is given for use by an external annunciator audible alarm or data logger.

18.4.3. Controller features

Figure 18.10 showed a simple block diagram representation of a controller. In reality, the large number of available options make a commercial controller far more complex.

Common industrial signal standards are 0–10 V, 1–5 V, 0–20 mA and 4–20 mA. These can be accommodated by two switchable ranges 0–10 V and 1–5 V plus suitable burden resistors for the current signals (a 250 ohm resistor, for example, converts 4–20 mA to 1–5 V).

4–20 mA and 0–20 mA signals used on two wire loops (see Section 1.2) require a DC power supply somewhere in the loop. A floating 30 V power supply is provided for this purpose.

Open circuit detection is provided on the main P_V input. This is essentially a pull up to a high voltage via a high value resistor (typically several hundred K ohms.) A comparator signals an open circuit input when the voltage rises. Short circuit detection can also be applied on the 1–5 V input (the input voltage falling below 1 V). Open circuit or short circuit P_V is usually required to bring up an alarm and trip the controller to manual, with the output signal driven high, held at last value, or driven low according to the nature of the plant being controlled. The open circuit trip mode is determined by switches as part of the set up procedure.

The P_v and remote S_P inputs are scaled to engineering units and linearised. Common linearisation routines are thermocouples, platinum resistance thermometers and square root (for flow transducers). A simple adjustable first order filter can also be applied to remove process or signal noise.

The output from the PID algorithm is a function of time and the values of the set point and the process variable. When the controller is operating in manual mode it is highly unlikely that the output of the PID block will naturally be the same as the demanded manual output. In particular the integral term will probably cause the output from the PID block to eventually saturate at 0% or 100% output.

If no precautions are taken, therefore, switching from auto to manual, then back to auto again some time later will result in a large step change in controller output at the transition from manual to automatic operation.

To avoid this 'bump' in the plant operation, the controller output is fed back to the PID block, and used to maintain a PID output equal to the actual manual output. This balance is generally achieved by adjusting the contribution from the integral term.

Mode switching can now take place between automatic and manual modes without a step change in controller output. This is known as manual/auto balancing preload or (more aptly) bumpless transfer.

Large changes in S_P or large disturbances to P_V can lead to saturation of the controller output or a plant actuator. Under these conditions the integral term in the PID algorithm can cause problems.

Figure 18.16 shows the probable response of a system with unrestricted integral action. At time A a step change in set point occurs. The output rises first in a step ($K \times$ set point change) then rises at a rate determined by the integral time. At time B the controller saturates at 100% output, but the integral term keeps on rising. With fixed controller output P_V rises in an approximately linear manner.

At time C P_V reaches, and passes, the required value, and as the error changes sign the integral term starts to decrease, but it takes until time D before the controller desaturates. Between times B and D the plant is uncontrolled, leading to an unnecessary overshoot and possibly even instability.

This effect is called 'integral windup' and is easily avoided by disabling the integral term once the controller saturates either positive or negative. This is naturally a feature of all commercial controllers, but process control engineers should always be suspicious of 'home brew' control algorithms constructed (or written in software) by persons without control experience.

In any commercial controller, the integral term would be disabled at point B on Fig. 18.16 to prevent integral windup. The obvious question now is at what point it is re-enabled again. Point C is obviously far too late (although much better than point D in the unprotected controller).

A common solution is to de-saturate the integral term at the point where the rate of increase of the integral action equals the rate of decrease of the proportional and derivative terms. This occurs when the slope of the PID output is zero, i.e. when

$$e = -T_i\left(\frac{de}{dt} + T_d\frac{d^2e}{dt^2}\right) \tag{18.12}$$

18.5. Controller tuning

18.5.1. introduction

Before a three-term controller can be used, values must be set for the gain (or proportional band), integral time and derivative time. In theory, if a model of the

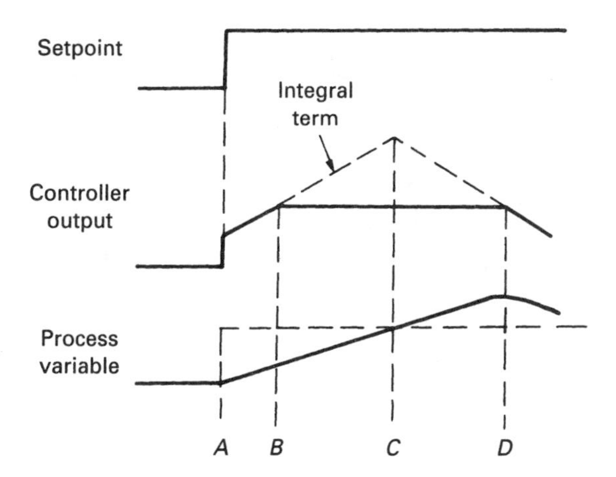

Fig. 18.16 *The effect of integral windup.*

controlled plant is available, suitable values can be found in advance from Nichols charts or Nyquist diagrams using empirical methods based on hill climbing techniques.

In the usual situation, however, the plant model is not known (except in the most general terms) and the controller has to be tuned based on measurements which can be made on the plant with simple equipment.

This section describes practical methods of tuning a control loop. It should be noted that all involve disturbing the plant and some involve driving the plant into instability. The safety considerations of the tests should be clearly understood by all concerned.

Most of the tests aim to give a quarter cycle decay and assume the plant consists of a transit delay in series with a second order block (or two first order lags) plus possible integral action. A useful first step (if only to gain confidence) is to attempt to control the plant manually. This should identify the broad outline of the plant's characteristics (e.g. the rough value of the dominant time constant, if the plant exhibits a transit delay and any obvious non-linearities such as hysteresis).

In conducting the tests, it is useful to have a two pen recorder connected as shown in Fig. 18.17. The ranges of both pens should be the same.

In the tests below, controller gains are referred to in proportional band ($PB = 100\%/\text{gain}$), with time used for integral and derivative action. Conversion to gain or repeats per minute is obviously straightforward.

18.5.2. Ultimate cycle methods

The aim of these methods is to determine the controller gain which will just sustain continuous oscillation. The technique is based on a paper 'Optimum Settings for Automatic Controllers' published by J.G. Ziegler and N.B. Nichols in 1942.

The integral and derivative terms are disabled (to give a P only controller) and the gain slowly increased. Step disturbances are introduced and the system

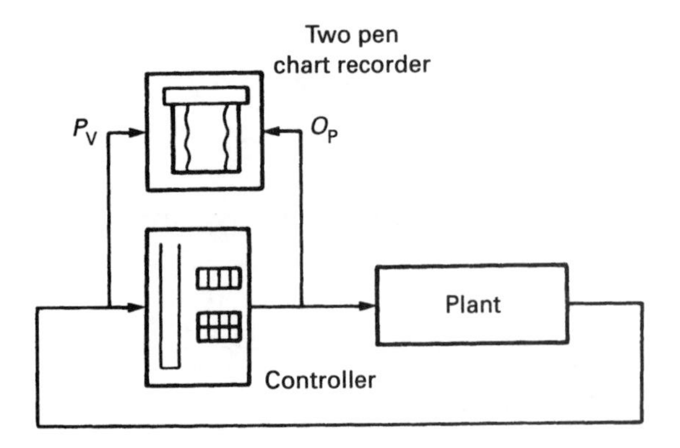

Fig. 18.17 *Equipment setup for controller tuning.*

response observed. (One easy way to do this is by switching the controller to manual, changing the output slightly, say by 5–10%, and switching back to auto). As the gain is increased, the response will become progressively underdamped, and finally continuous oscillation will result. Care must be taken (and patience exercised) to allow transients to die away before each new value of gain is tried. The ultimate PB for stability is denoted by P_u. The period of oscillation T_u should now be measured from the chart recorder.

The required controller settings are now

P only

PB	$2\ P_u\%$

PI

PB	$2{\cdot}2\ P_u\%$
T_i	$0{\cdot}8\ T_u$

PID

PB	$1{\cdot}67\ P_u\%$
T_i	$T_u/2$
T_d	$T_u/8$

This sets $T_i = 4T_d$ which is often a useful starting point.

An alternative setting for a PID controller (attributable to Atkinson) gives

PB	$2P_u\%$
T_i	T_u
T_d	$T_u/5$

This aims to give a damping ratio of $b = 0{\cdot}45$.

The American control engineer F.G. Shinskey gives slightly different values

PI

PB	$2P_u\%$
T_i	$0{\cdot}43T_u$

PID

PB	$2P_u\%$
T_i	$0 \cdot 34T_u$
T_d	$0 \cdot 08T_u$

The slight differences between all these methods are not really significant; all only aim to set the controller in the right area with final adjustment being made by trial and error. In general, it is advisable to leave T_i and T_d alone, and adjust the gain to give the required damping.

Shinskey gives a useful rule of thumb for identifying the plant characteristics. If the manual test (suggested earlier) shows a dead time T_d to be present, then if

$T_u/T_d = 2$	The plant is pure transit delay
$2 < T_u/T_d < 4$	The plant is dominated by transit delay
$T_u/T_d = 4$	There is a single dominant first order lag
$T_u/T_d > 4$	There are several first order delays of similar magnitude.

18.5.3. Decay method

This is a variation on the ultimate cycle method. With the controller in P only mode, trial and error methods similar to those above are used to give a decay ratio of approximately 4:1 (see Fig. 18.13c). If the *PB* which achieves this is denoted P_q and the period of the decaying oscillations is T_q the required controller settings are

P only

PB	$P_q\%$ (obviously)

PI

PB	$1 \cdot 5P_q\%$
T_i	T_q

PID

PB	$P_q\%$
T_i	$0 \cdot 67T_q$
T_d	$0 \cdot 17T_q$

18.5.4. Bang/bang oscillation test

This is the fastest tuning method as it requires only a single test. It is however, the most vicious of the methods, and can be misleading if the plant is non-linear. The controller is set for P only mode, with gain as high as the controller will allow (ideally infinite gain). This turns the controller into a bang/bang servo.

When the controller is switched to auto, oscillation will result as Fig. 18.18. The period of the oscillations T_o and the peak to peak height of the oscillations should be noted H_o as a percentage of the controller input range.

The required controller settings are

P only

$PB = 2H_o\%$

PI

$PB = 3H_o\%$

$T_i = 2T_o$

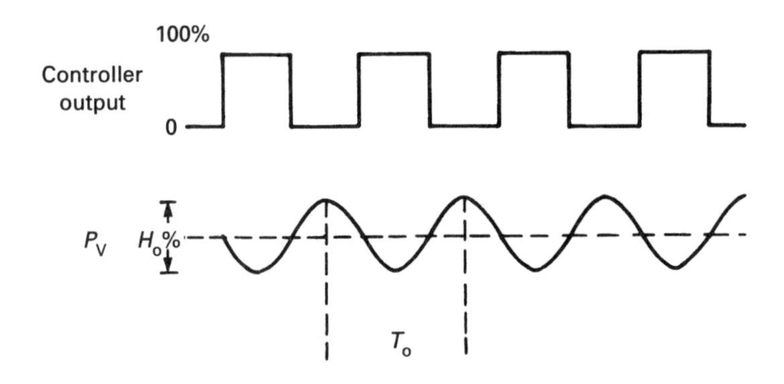

Fig. 18.18 *The ultimate cycle method.*

PID

$PB = 2H_o\%$
$T_i = T_o$
$T_d = T_o/4$

If the P_V oscillations themselves saturate a $PB > 100\%$ is required (i.e. gain less than unity). The range of the controller output swing should be progressively limited until the P_V signal is within the input range. The controller output swing O_P and the P_V swing can then be read from the chart recorder (both as a percentage of full scale). H_o then is simply $100 \times P_V/O_P\%$ allowing the above relationships to be used.

18.5.5. Reaction curve test

This open loop test was proposed by the American control engineers G.H. Cohen and G.A. Coon in 1953 (based on earlier work by Ziegler and Nichols). It assumes a measurable transit delay, and cannot be applied to plants with integral action (e.g. level controls).

The controller output is adjusted manually to bring the plant to near the desired operating point. After transients have died away, a small manual output step ΔO_P is applied, which results in a change ΔP_V in the process variable as Fig. 18.19.

The process gain K_p is then simply $\Delta P_V/\Delta O_P$.

A tangent is drawn to the process variable curve at the steepest point from which an apparent transit delay T_t and apparent time constant T_c can be inferred. Controller settings are then

P only

$PB = 100\ K_pT_t/T_c\%$

PI

$PB = 110\ K_pT_i/T_c\%$
$T_i = 3{\cdot}3\ T_t$

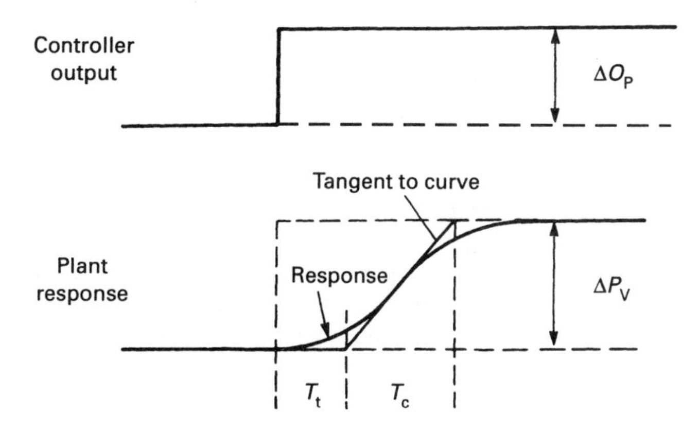

Fig. 18.19 *The reaction curve test.*

PID

$PB = 80\ K_p T_t / T_c\%$
$T_i\ = 2 \cdot 5\ T_t$
$T_d\ = 0 \cdot 4\ T_t$

The reaction curve test is the gentlest of the tuning methods.

18.5.6. General comments

The results of the initial tuning exercises above should not be viewed as rules of law written in stone. At best they put the engineer in the right area, at worst they can actually mislead. They should be viewed as a starting point for further manual adjustment of tuning constants.

It is always useful to have the proportional gain as high as possible to give large initial control action to changes and disturbances (unless the measured variable signal is noisy) so the gain should be adjusted first with integral and derivative setting left at their original settings. The gain should be adjusted to give the desired overshoot and damping.

Integral action is best adjusted next to give best removal of offset error. It may be necessary to decrease the gain again as the integral time is decreased. It is useful to keep a fixed ratio between T_i and T_d ($T_i = 4T_d$ is a useful starting point) whilst adjusting T_i. A useful rule of thumb is that T_i/gain is an 'index' of stability for a given system, i.e. a T_i of 12 sec and a gain of 2 will give similar performance to a T_i of 24 sec and a gain of 4.

The derivative action should be adjusted last as a final stability/overshoot adjustment. Too much derivative action, however, can make the system noise prone and lead to 'twitchy' actuators. Care should be taken with derivative action in systems with significant transit delays. Here, derivative action can be a destabilising influence.

It should be remembered that the loop conditions will change with time and temperature and as items in the loop fail and get replaced with spares. Loop tuning should always be rather conservative, aiming for acceptable rather than perfect control. A finely tuned loop may become unstable under very slight changes in operating conditions.

18.6. Self-tuning controllers

Tuning a controller is more of an art than an exact science and can be unbelievably time consuming. Time constants of tens of minutes are common in temperature loops, and lags of hours occur in some mixing and blending processes. Performing, say, the ultimate cycle test of Section 18.5.2 on such loops can take several days.

Self-tuning controllers aim to take the tedium out of setting up a control loop. They are particularly advantageous if the process is slow (i.e. long time constants) or the loop characteristics are subject to change (e.g. a flow control loop where pressure/temperature changes in the fluid alter the behaviour of the flow control valve).

Self-tuning controllers give results which are generally as good, if not slightly better, than the manual method of Section 18.5 (possibly because self-tuning controllers have more patience than humans!). In the author's experience however, the results from a self tuner should be viewed as recommendations or initial settings in the same way as the results from the manual methods described previously. One early decision to be made when self tuners are used is whether they should be allowed to alter control parameters without human intervention. Many engineers (of whom the author is one) view self tuners as commissioning aids to be removed before a plant goes into production.

There are essentially two groups of self tuners. Modelling self tuners try to build a mathematical model of the plant (usually second order plus transit delay) then determine controller parameters to suit the model. These are sometimes called explicit self tuners.

Model identification is usually based on the principles of Fig. 18.20. The controller applies a control action O_P to the plant and to an internal model. The plant returns a process variable P_v and the model a prediction. These are compared, and the model updated (often via the statistical least squares technique). On the basis of the new model new control parameters are calculated and the sequence repeated.

A model-building self tuner requires actuation changes to update its model, so it follows that self tuners do not perform well in totally static conditions. In a totally stable unchanging loop the model, and hence the control parameters, can easily drift off to ridiculous values. To prevent this, most self tuners are designed to 'kick' the plant from time to time, with the size and repetition rate of the kick being set by the control engineer.

The second group of self tuners (sometimes called implicit tuners) do not form a plant model, but rely on tests such as automated versions of the manual tests of

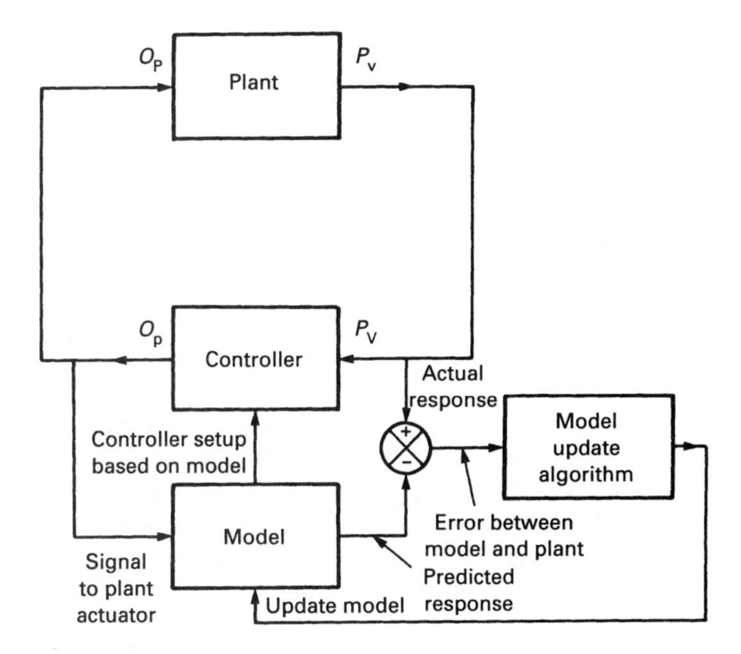

Fig. 18.20 *A modelling self-tuning controller.*

Section 18.5. A typical technique will vary the controller gain until a damped oscillatory response is observed. The control parameters can then be inferred from the controller gain, the oscillation period and the oscillation decay rate.

The useful bang/bang test of Section 18.5.4 can be performed automatically by a controller which forces limit cycling in the steady state via a comparator as Fig. 18.21. The limit block after the comparator restricts the effect on the plant.

Implicit self tuners, like their modelling brothers, do not perform well on a stable unchanging loop, and can be equally confused by outside disturbances.

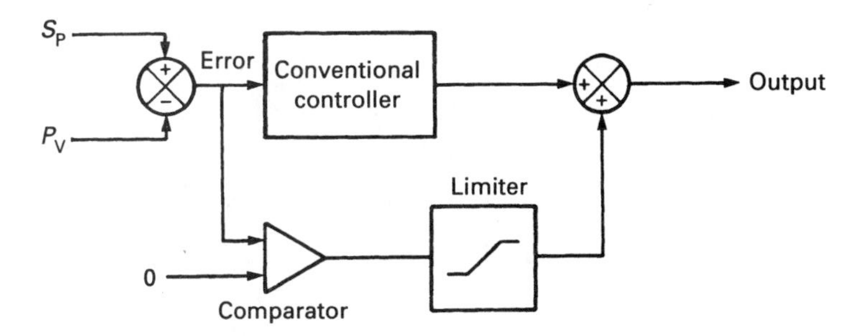

Fig. 18.21 *Implicit self-tuner based on the bang/bang test.*

18.7. Digital systems

18.7.1. Digital controllers

The cheapness and versatility of microprocessors has led to the majority of modern controllers being based on digital techniques rather than the traditional analog circuits of the previous section. Paradoxically most of these devices emulate conventional analog controllers and the digital nature is not apparent to the user (although many of the features of a modern controller such as scaling to engineering units on displays could not easily be provided on an analog device).

A block diagram of a typical device is shown in Fig. 18.22. Input analog signals (two in the example) are scanned sequentially by a multiplexer (usually based on CMOS switches). The signals are digitised by an analog to digital converter (ADC), commonly to a resolution of 12 bits (1 part in 4096) and read by the processor. The control algorithm is performed in software, and the digital actuator signal converted to an analog output by a digital to analog converter (DAC). Digital input and output circuits read the front panel pushbuttons and update the set point, process variable and output front panel displays.

Digital controllers are often made configurable. For example the Eurotherm Process Automation 6366 advanced programmable controller contains analog and digital input/output blocks, filter blocks, set point blocks, PID blocks, delay blocks, lead–lag blocks and totaliser blocks which are linked together by a program to produce a required control strategy.

18.7.2. Digital algorithms

The algorithms used in digital systems are generally based on difference equations rather than differential equations, with the sample time Δt being substituted for dt. In the continuous graph of Fig. 18.23a, for example, the slope is dy/dt.

The digital representation is sampled (a topic discussed below) and the equivalent waveform is shown on Fig. 18.23b. If $y(t)$ is the value of y at time t, and $y(t - \Delta t)$ the value of y at the previous sample Δt seconds earlier, the slope at time t approximates to

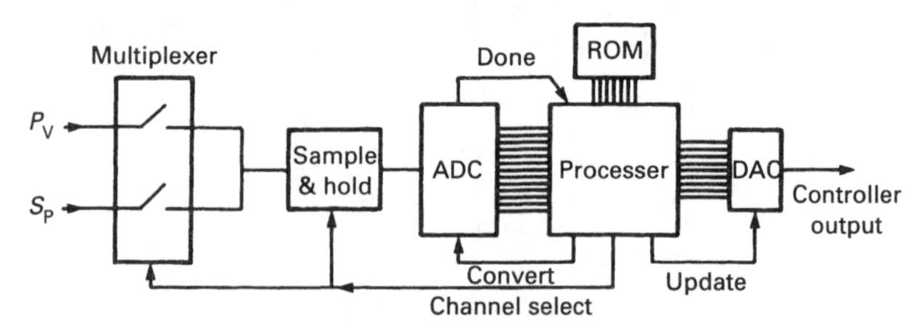

Fig. 18.22 *Block diagram of digital controller*

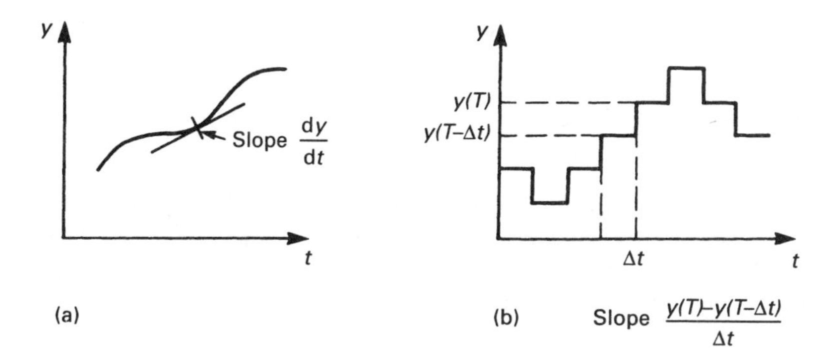

Fig. 18.23 *Difference equations and digital systems. (a) Analog system. (b) Digital system.*

$$\frac{y(t) - y(t - \Delta t)}{\Delta t} \tag{18.13}$$

An alternative representation is to say y_n is the 'nth sample', then the slope becomes

$$\frac{y_n - y_{n-1}}{\Delta t} \tag{18.14}$$

Figure 18.24a shows the integration of a continuous waveform as being equivalent to the area under the curve. The equivalent digitised representation is shown in Fig. 18.24b. The area under the curve from sample y_0 at $t = 0$ to sample y_n at $t = n\Delta t$ is the sum of the individual rectangular areas, i.e.

$$y_1\Delta t + y_2\Delta t + \ldots + y_n\Delta t$$

or more concisely

$$\sum_1^n y_n\Delta t$$

A more accurate integration algorithm is the trapezoid representation of Fig. 18.24c, where each rectangle has area

$$\Delta t(y_n + y_{n-1})/2$$

giving a total area of

$$\sum_1^n \Delta t(y_n + y_{n-1})/2$$

The familiar three-term control algorithm

$$O_P = K\left(e + \frac{1}{T_i}\int e\,dt + T_d\frac{de}{dt}\right) \tag{18.15}$$

becomes, when e_n is the nth sample of the error and Δt the sampling interval

$$O_P = K\left(e_n + \frac{1}{T_i}\Sigma\Delta t(e_n + e_{n-1})/2 + \frac{T_d}{\Delta t}(e_n - e_{n-1})\right) \qquad (18.16)$$

This can easily be converted to a computer program. The example below is written in pseudo code. PV is the value of the process variable, SP the set point. E is the value of the error, and 'olderror' the error value at the last sample. The sample interval is ts, and 'time' is an internal hardware clock which can be read, and reset, by the software. The integral summation is formed in a variable called 'sum'.

```
olderror: = 0    (Initialise variables)
sum: = 0
hellfreezesover: = false
repeat                                        (start of indefinite loop)
  time: = 0                                   (reset hardware clock)
  anin PV                                     (read PV from analog input)
  anin SP                                     (read SP from analog input)
  error: = (SP - PV)
  sum: = sum + (error + olderror)/2           (integral term)
  diff: = (error-olderror)                    (derivative term)
  OP: = K*(error + sum*ts/TI + diff*Td/ts)
  anout OP                                    (write OP to analog output)
  olderror: = error                           (update olderror for next sample)
  repeat                                      (wait until next sample time)
  until time > = ts                           (time for next sample)
until hellfreezesover                         (go back for next sample)
```

The above program, which loops indefinitely, needs some refinement. It works in 'per cent of range' rather than engineering units, and has no protection against integral wind up for example. Additional routines for driving displays and auto/manual changeover would also be needed.

Similar digital techniques can also be applied to produce first order lags, lead-lag compensators, etc. The difference equation for a first order lag with time constant T seconds becomes

$$V_f = oldvf + \frac{\Delta t}{T}(V_n - oldvf) \qquad (18.16)$$

where V_f is the current filtered value based on the raw unfiltered value V_n, and *oldvf* is the last filtered value. As before Δt is the sample interval. The difference equation assumes a rate of change of $(V_n - oldvf)$ for the next sample interval.

A first-order filter can thus be simulated by the listing below. T is the time constant and ts the sample time.

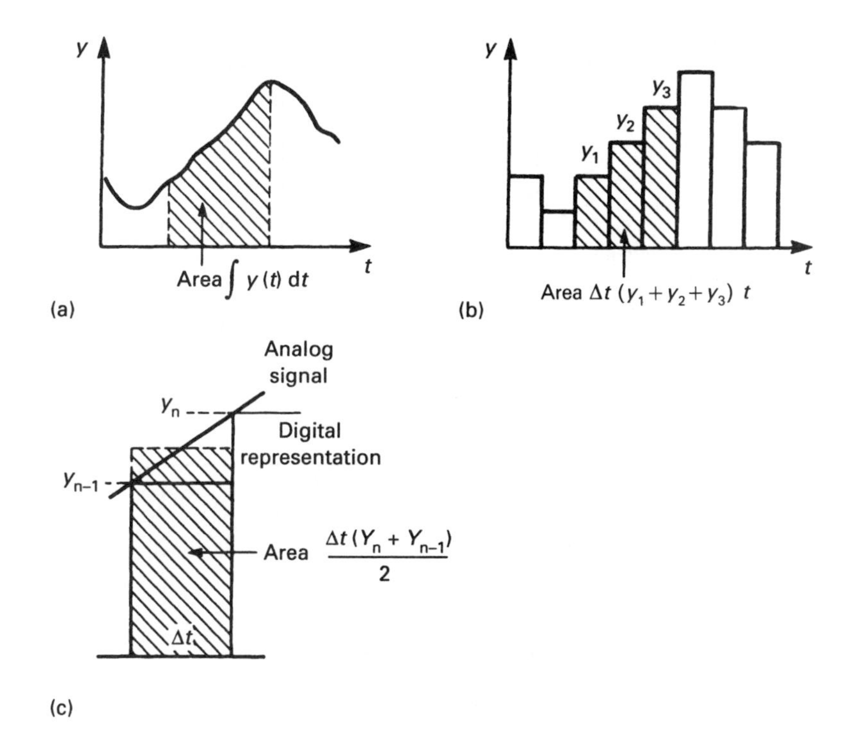

Fig. 18.24 *Integration of digital signals. (a) Area under analog signal. (b) Area under a digital signal. (c) More accurate integration of digital signal.*

```
oldvfilter:=0
repeat
    time:=0
    anin V
    vfilter:=oldvfilter + (V - oldvfilter)*ts/T
    oldvfilter:=vfilter
    repeat:until time>=ts
until hellfreezesover
```

This is represented by the signal diagram of Fig. 18.25a. By similar reasoning, a second-order system can be represented by Fig. 18.25b.

A straight time delay can be represented by an analog bucket chain moving analog values along every sample as Fig. 18.25c. With a sample time of Δt seconds, a pure delay of T seconds is equivalent to taking the value from the INT($T/\Delta t + 0{\cdot}5$) position.

The accuracy of a digital controller obviously depends on the resolution and accuracy of the ADC/DAC circuits. Less obviously, the sample time Δt appears in all the difference equations, so the consistency and accuracy of the sample clock is of equal importance to the analog circuitry. Digital controllers generally employ a hardware crystal controlled clock which uses interrupts to initiate

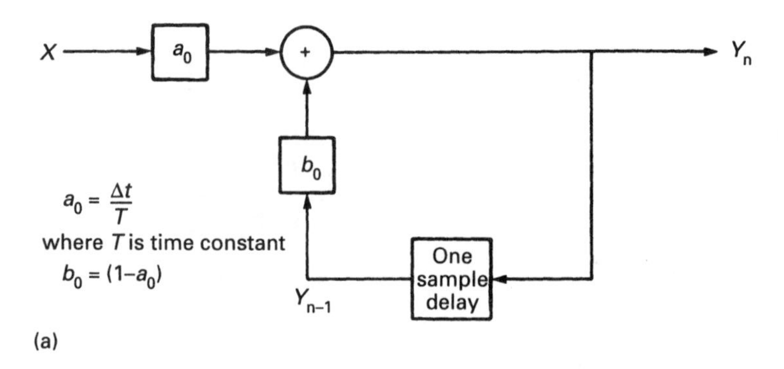

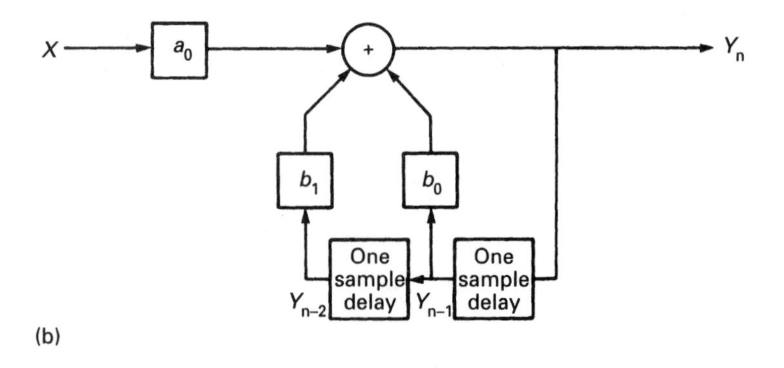

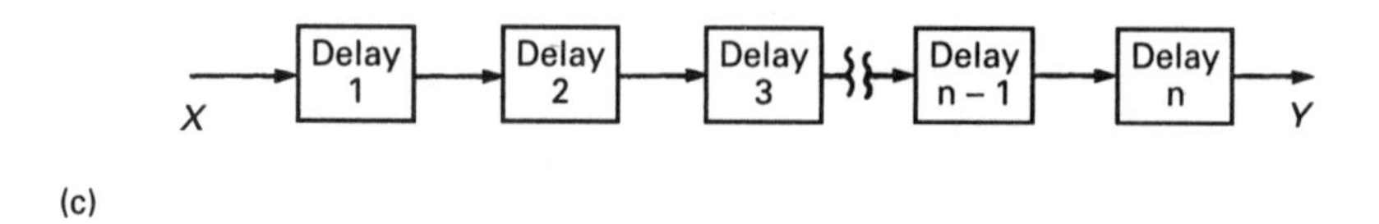

Fig. 18.25 *Digital representation of common analog blocks. (a) First-order lag. (b) Second-order system. (c) A transit delay.*

controller routines at regular intervals. A typical controller will perform a complete input/compute/output cycle every 50 ms (i.e. 20 samples per second).

18.8. Further control strategies

18.8.1. Ratio control

Two process variables are commonly required to be kept in precise ratio to each other. An oxy-gas burner, for example requires a ratio of 2.1 : 1 between oxygen and gas flows to give no wasteful burning and a flame of maximum efficiency. In combustion engineering, this is known as the *stoichiometric ratio*.

The obvious solution is to use ratio itself as the controlled variable as Fig. 18.26. Unfortunately this will not work as the loop gain will increase as the flow decreases making it impossible to obtain a stable loop for all flows.

In Fig. 18.27, one flow A has been declared to be master and its set point comes from an external signal. In a burner control system, for example, the external signal would be provided by the temperature control. The flow of A is then controlled by a straightforward feedback loop. The resultant flow is multiplied by the required ratio and used to provide the set point for a second controller which controls flow B. The two flows are thus kept at the ratio set by R. As the loops are

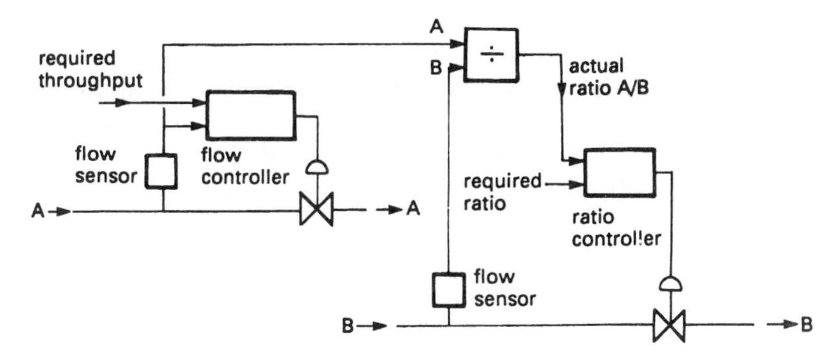

Fig. 18.26 *An intuitive, but incorrect, method of ratio control. The loop gain varies with throughput.*

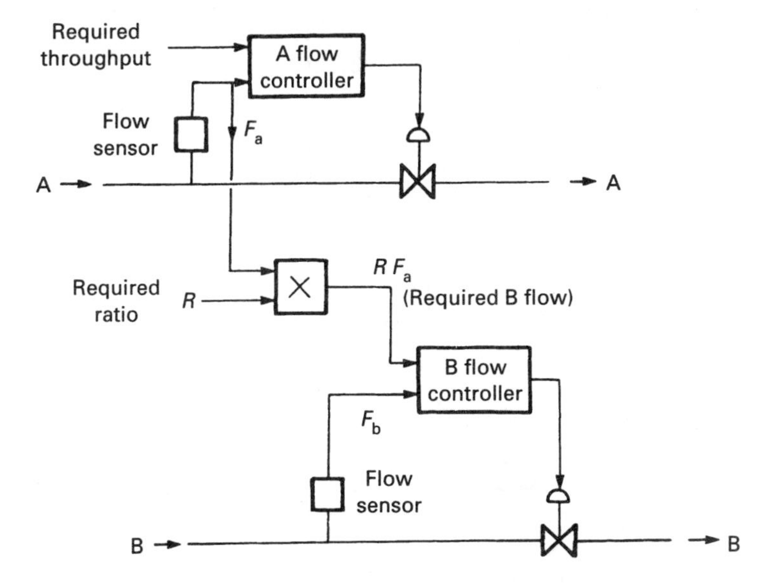

Fig. 18.27 *Acceptable ratio control with stable loop gains.*

simple, the flows do not affect the loop gain and the system can be made stable for all flows.

In burner control, the external signal usually sets the required air or oxygen flow, and the fuel follows. Such schemes are commonly called Gas Follow Air Control.

18.8.2. Lead-lag control

Gas Follow Air is simple, but one side effect is that the flame runs lean for increasing fire rate and rich for decreasing fire rate because the air flow must change first before the fuel can follow. There is also a possible safety implication because a failure of the slave valve or controller could lead to a gross error in the actual ratio such as the fuel valve wide open and the air valve closed.

Large boilers therefore use a system called Lead-lag control shown on Fig. 18.28. This uses cross linking and selectors to provide an air set point which is the *highest* of the external power demand signal or ratioed fuel flow. The fuel set point is the *lowest* of the external power demand or ratioed air flow.

This cross linking provides better ratio during changes; both air and fuel will change together. There is also higher security; a jammed open fuel valve will cause the air valve to open to maintain the correct ratio and prevent an explosive atmosphere of unburnt fuel forming.

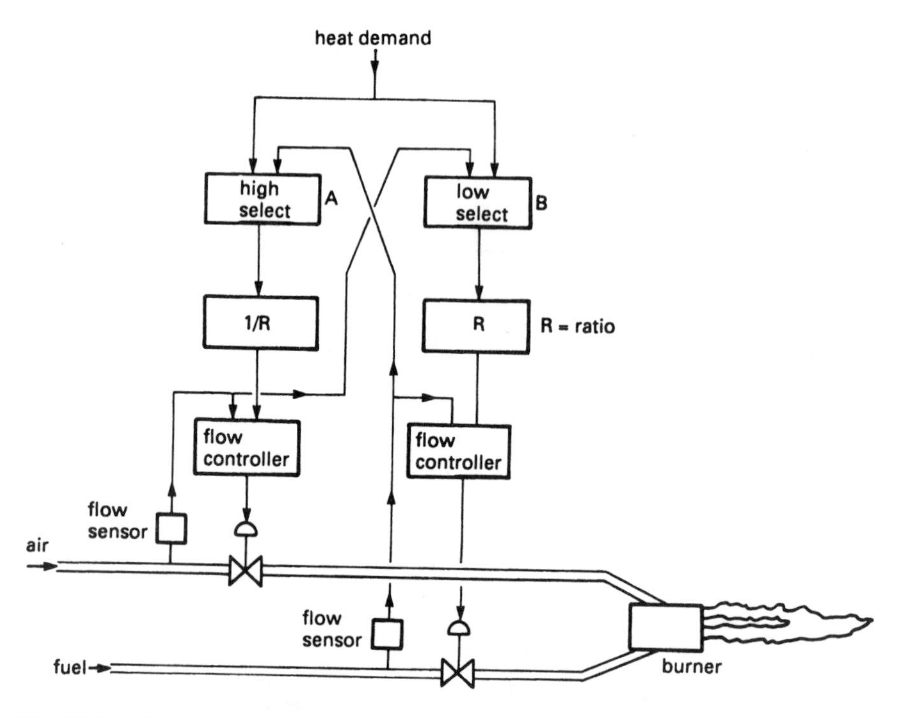

Fig. 18.28 *Lead lag combustion control. The air leads for increasing heat and lags for decreasing heat.*

18.8.3. Cascade control

Closed loop control gives increased performance, so inner control loops are often added around plant items which would otherwise degrade performance. Figure 18.29 shows a typical example where the output of the outer loop becomes the set point of the inner loop. Such schemes are known as cascade control. Any problems in the inner loop (disturbances, non-linearities) are dealt with by the inner controller and do not affect the outer loop. Common examples are valve positioners (see Section 16.7.5) and the speed/current loop controllers in DC drives (see Section 12.9.6).

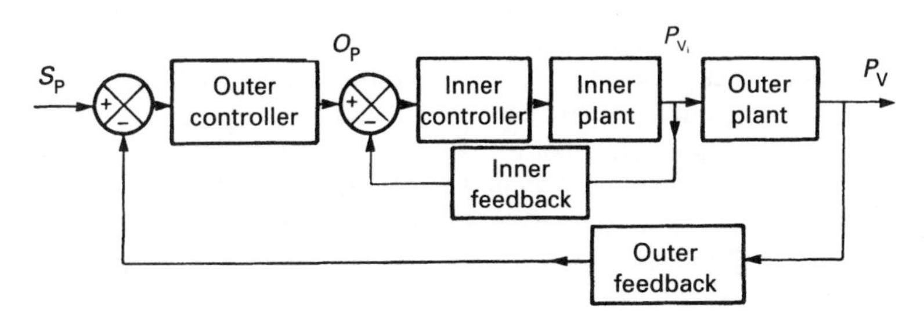

Fig. 18.29 *A system with cascade control.*

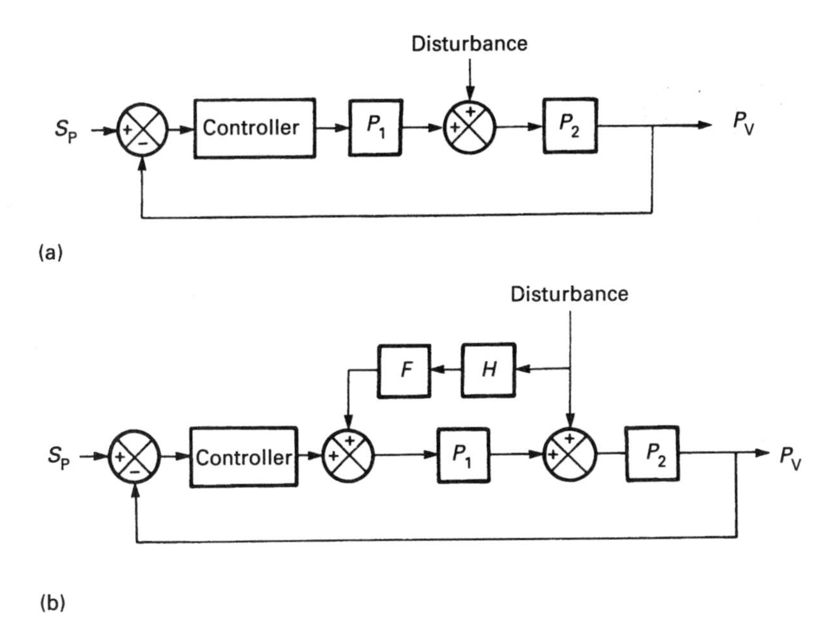

Fig. 18.30 *Disturbance effect reduced by feedforward. (a) System with disturbance to which cascade control cannot be applied. (b) Correcting signal derived by measuring the disturbance.*

For cascade control to work, the inner loop must be faster than the outer loop, and must be tuned first during commissioning.

18.8.4. Feedforward control

Cascade control can reduce the effect of disturbances occurring early in the forward loop, but has little effect on disturbances arising from load or demand changes which directly affect the process variable. These disturbances must produce an error before the control strategy can react.

If a disturbance can be measured (the temperature of the feed liquid to a vessel under temperature control, for example) and an appropriate model of the plant is known, a correcting signal can be added to the controller output as shown in Fig. 18.30. This is known as feedforward.

The characteristics of the plant (P_1, P_2) and the transducer H are fixed. Block F is an inverse model of the plant such that $F = -(1/HP_1)$ and the output will cancel the effects of the disturbance.

Ideally the feedforward block F should model time effects. Perhaps surprisingly, though, the model need not be perfect. Even a crude model will show significant improvements, and any remaining error will be dealt with by the main controller.

Chapter 19

Distributed systems

19.1. Parallel and serial communications

Cabling is one of the most costly parts of any control scheme. There is the cost of the cable itself, the support structure and cable tray plus the labour costs of pulling cable, ferruling and terminating the ends. If, in the course of commissioning, it is discovered that some extra signals are needed and there are insufficient spare cores, another expensive cable will have to be pulled, with all the attendant costs and time delay.

Figure 19.1 shows two computer systems that need to exchange data. As shown there are 8 digital signals one way, 12 signals the other plus two 16-bit numbers. Along with supplies, neutrals and DC-returns this represents 56 cores needing, probably, one 27-core and one 37-core steel wire armoured cable, and ten I/O cards. All the cards require labour to terminate them inside the cubicles at each end. All told, it is not a cheap exercise.

In Section 14.3.4 we described how remote I/O can be used to reduce cabling costs. In this chapter we will see how similar ideas can be developed to provide communication between PLCs, computers and intelligent instruments.

Figure 19.1 is a form of parallel transmission; all the data to be sent is passed simultaneously. This method is widely used (at lower voltages) to connect computers to printers and for bus based computer instrumentation schemes such as the IEEE-488 bus described earlier in Section 14.4.2.2.

In Figure 19.2 a single data line (plus a return) connects the transmitter and the

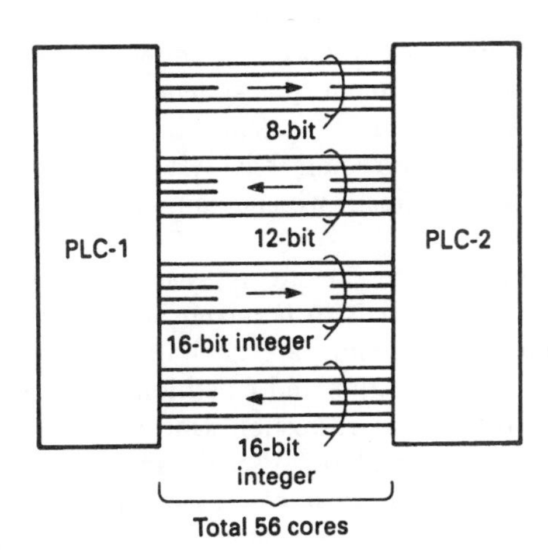

Fig. 19.1 *Parallel data transfer.*

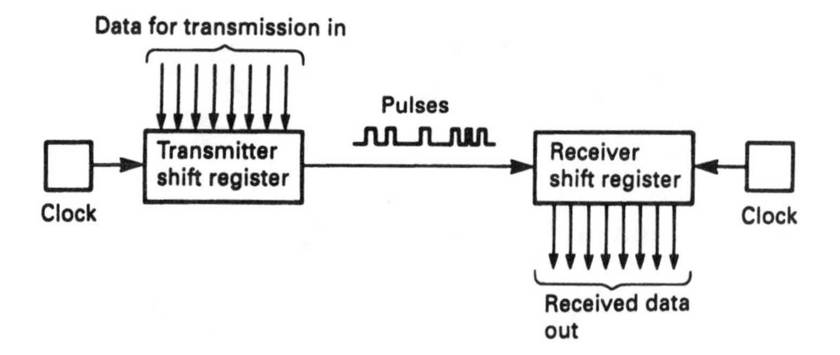

Fig. 19.2 *Serial data transmission.*

receiver, and the data is transmitted as a serial string of bits. Since computers, peripherals, PLCs, etc. all work internally in parallel for speed, parallel to serial conversion is required at the transmitter and serial to parallel conversion at the receiver. The simplest way of achieving this is to use shift registers into which data can be loaded in parallel and shifted out one bit at a time. Specialist integrated circuits called UARTS (universal asynchronous receivers-transmitters) are used to provide this conversion and the control functions. Not surprisingly, this is known as serial transmission.

The advantages of serial transmission arise from cost and flexibility. All that needs to be installed for bidirectional communication is a small, cheap usually four-core (two pair) screened cable, although the signal levels are small and there is usually a cost penalty in that trunking or conduit needs to be used for protection.

Once installed, a serial communication system is not really constrained in the amount of data that can be passed (although there will be a time penalty for large amounts of data). Additional data items can be added with no installation costs.

The disadvantages are speed, noise immunity, safety and program comprehensibility. Serial communication is obviously slower than parallel transmission (by a factor equal to the number of parallel lines). This is generally not a problem, on a dedicated point to point communication system a response time of 0.5 seconds is easily achievable (and remote I/O systems normally achieve around 30 ms). Response times can be longer on commercial systems such as Ethernet, but these are generally not interfacing directly with human beings or a plant in time-critical applications.

The voltages in serial transmission are low, usually of the order of 10 V, and hence prone to noise. Care needs to be taken in the installation (conduit or trunking, separation and screening are advisable) and proprietary systems include methods for error detection and repetition of faulty messages.

Despite these error-detecting correction schemes, a serial communication system should never be considered totally secure, and must not be used for purely safety functions such as emergency stops. These must always be hardwired.

Finally we have program comprehensibility. The idea of serial communications can be difficult to comprehend in the middle of a fault at 3.00 a.m. Essentially what we are achieving is to link two areas of memory in separate computers or PLCs. This added complexity can bring great confusion if it is not supported by good plant documentation.

19.2. Serial standards

19.2.1. Introduction

For a serial communication system to work, there needs to be a consistency between the transmitter and the receiver. There must be definition of:

(a) Signal voltage levels.
(b) The transmission code (what the bit patterns mean, how the message is built up).
(c) Transmission rates (the speed at which the bit pattern is sent).
(d) Synchronisation. In Fig. 19.2 we showed clocks at both ends of the link. If these have a small difference in frequency (as they inevitably will) the receiver will get out of alignment with the transmitter. Some method must be provided to give synchronisation between transmitter and receiver.
(e) Protocols. Apart from the data, there will need to be some method for the transmitter and receiver to interchange control signals such as 'I am unable to receive a message at present'.
(f) Error-checking methods and recovery procedures ('that last message didn't make sense, please send it again').

Getting equipment from different manufacturers to work together over a serial link can sometimes be very difficult. The problems usually arise out of differences in one (or more) of the above points.

19.2.2. Synchronisation

The theoretically simplest way to achieve synchronisation is to have a common clock for both the transmitter and the receiver, as the two can never, in theory, get out of alignment. This is known as synchronous transmission.

Most systems, however, are asynchronous and use separate clocks as Fig. 19.2. The messages are broken down into characters, (typically 5 to 8 bits in length) and the two clocks are synchronised at the start of each character.

The idle state of the line is a '1' signal (called a 'Mark' in telecommunications) and each character in the signal has the form of Fig. 19.3. The character starts

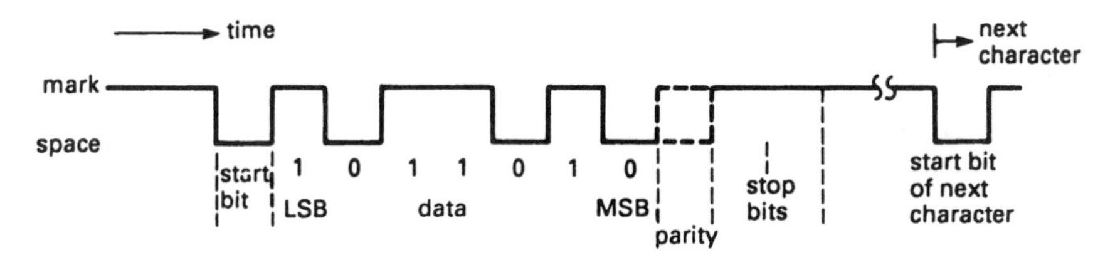

Fig. 19.3 *An asynchronous data character.*

with a '0' signal (called a space) followed by the data bits, usually with least significant bit first. An error-correcting bit (called the parity bit) is sometimes added after the data bits. This is discussed in Section 19.2.7. Finally the signal returns to the idle mark state for a time before the next character can be sent. This is known as the stop bit and can be 1, 1.5 or 2 bits in width dependent on the system. The next character can follow a random time after the stop bit. The transmitter and receiver clocks are synchronised at the start bit, and only have to stay aligned for the 10 or so bits needed to send a character.

It may be thought that, with noise, mark to space transitions in the data could be mistaken for start bits. In practice, the link will pull itself back into synchronisation in a few characters as shown on Fig. 19.4. A framing error is signalled by the UART when it receives a zero where it would expect a stop bit.

19.2.3. Character codes

Many types of character code have evolved over the years, but now the almost universal standard is the ASCII code (American Standard Code for Information Interchange, also known as ISO 646) shown in Table 19.1. Variations on this are the CCITT alphabet No.5, and national options such as the £ symbol in the UK.

ASCII is a 7-bit code giving 128 different combinations covering full upper/lower case alphanumeric characters along with punctuation and 32 control characters that we will return to in Section 19.2.6.

19.2.4. Transmission rates

The transmission signalling rate is expressed in baud which is the number of signal transitions per second. For the majority of serial links we shall consider, with two signalling states (0 and 1), the baud rate and the bits/second are identical (although this is not always true.) For linking a PLC with an instrument, a rate of

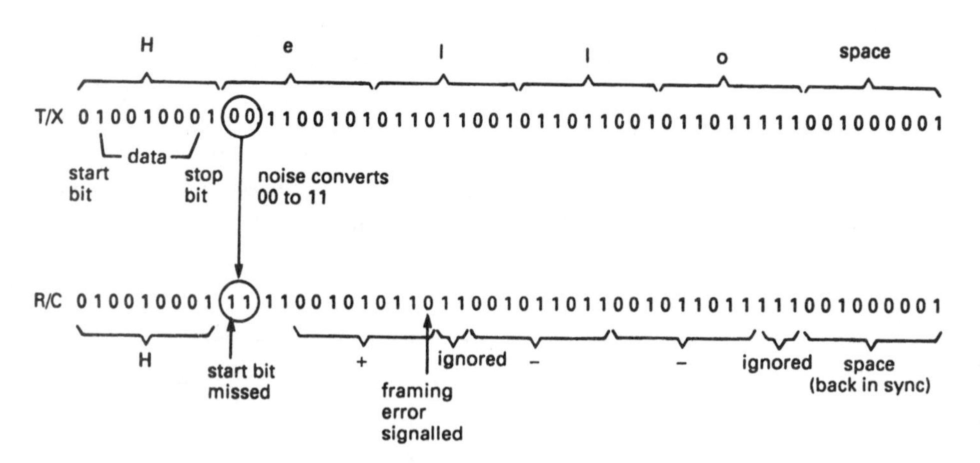

Fig. 19.4 *Framing errors and the ability of an asynchronous transmission to recover from a fault. 7 data bits and a single stop bit are used with ASCII coding. Parity is not used.*

Table 19.1 ASCII codes

Control characters

Decimal	*Hex*	*Char*	*Decimal*	*Hex*	*Char*
0	00	NUL	14	0E	SO
1	01	SOH	15	0F	SI
2	02	STX	16	10	DLE
3	03	ETX	17	11	DC1
4	04	EOT	18	12	DC2
5	05	ENQ	19	13	DC3
6	06	ACK	20	14	DC4
7	07	BEL	21	15	NAK
8	08	BS	22	16	SYN
9	09	HT	23	17	ETB
10	0A	LF	24	18	CAN
11	0B	VT	25	19	EM
12	0C	FF	26	1A	SUB
13	0D	CR	27	1B	ESC

Control characters can be obtained via the use of the CONTROL key. Backspace (BS) for example is ctrl-H.

Printable characters

Decimal	*Hex*	*Char*	*Decimal*	*Hex*	*Char*
28	1C	FS	78	4E	N
29	1D	GS	79	4F	O
30	1E	RS	80	50	P
31	1F	US	81	51	Q
32	20	space	82	52	R
33	21	!	83	53	S
34	22	"	84	54	T
35	23	#	85	55	U
36	24	$	86	56	V
37	25	%	87	57	W
38	26	&	88	58	X
39	27	'	89	59	Y
40	28	(	90	5A	Z
41	29	)	91	5B	[
42	2A	*	92	5C	\
43	2B	+	93	5D	]
44	2C	,	94	5E	^
45	2D	-	95	5F	_
46	2E	.	96	60	`
47	2F	/	97	61	a
48	30	0	98	62	b
49	31	1	99	63	c

Table 19.1 (*cont.*)

Printable characters

Decimal	*Hex*	*Char*	*Decimal*	*Hex*	*Char*
50	32	2	100	64	d
51	33	3	101	65	e
52	34	4	102	66	f
53	35	5	103	67	g
54	36	6	104	68	h
55	37	7	105	69	i
56	38	8	106	6A	j
57	39	9	107	6B	k
58	3A	:	108	6C	l
59	3B	;	109	6D	m
60	3C	<	110	6E	n
61	3D	=	111	6F	o
62	3E	>	112	70	p
63	3F	?	113	71	q
64	40	@	114	72	r
65	41	A	115	73	s
66	42	B	116	74	t
67	43	C	117	75	u
68	44	D	118	76	v
69	45	E	119	77	w
70	46	F	120	78	x
71	47	G	121	79	y
72	48	H	122	7A	z
73	49	I	123	7B	{
74	4A	J	124	7C	|
75	4B	K	125	7D	}
76	4C	L	126	7E	~
77	4D	M	127	7F	DEL

1200 baud might be typical. For proprietary PLC to PLC or remote I/O links, with high-quality communication cable, rates as high as around 115 kilobaud are used. High speed links such as Ethernet operate at 10 Mbaud.

A speed of 115 Kbaud should not be interpreted as an ability to send 115 000 bits of data down the cable in one second. We have already seen in Fig. 19.3 that splitting the data into characters with start/stop bits involves some overheads, which increase with the error checking when full messages are sent.

19.2.5. Modulation of digital signals

So far we have considered a serial link transmitting digital data in its 'raw' form, i.e. as a series of voltage levels directly representing the bit pattern we wish to send. This is known as the baseband transmission.

A digital signal has a bandwidth from 0 Hz (DC corresponding to a string of continuous zeros or ones) to at least half the bit rate. Many transmission media, such as radio telemetry and the telephone network, have inherent low frequency limitations and cannot handle baseband signals.

The data is therefore often modulated onto a carrier wave. There are three different ways of achieving this: amplitude shift keying (ASK), frequency shift keying (FSK) and phase shift keying (PSK) all summarized on Fig. 19.5. One advantage of modulation is that it allows several independent signals to be modulated onto different carrier frequencies and carried on the same cable. A modulated digital signal is said to be using broad band or carrier band transmission. Often the term 'carrier band' is used to imply FSK with one signal on the cable, and 'broad band' used where several signals share the cable.

Broad band and carrier band both require devices to interface the digital signals at the receiver and transmitter to the transmission media. These modulate the signal at the transmitter, and demodulate it again at the receiver. Such devices are known as modulators/demodulators or modems.

Figure 19.6 shows a typical two-way arrangement using the public telephone network and FSK. 'Originator' refers to the station which originally established the link; subsequent communications are bidirectional.

On Fig. 19.6 there are two types of equipment whose names, and more commonly their abbreviations, appear widely in data transmission and are the source of much confusion. The equipment at the transmitting and receiving ends are known as Data Terminal Equipment (DTE). This covers computers, PLCs, printers, terminals, VDUs, graphics displays, etc. The communication equipment (i.e. the modems) is known as Data Communication Equipment (DCE).

The confusion arises because communication standards and protocols are concerned with connecting a DTE and a DCE. When we link a computer and an

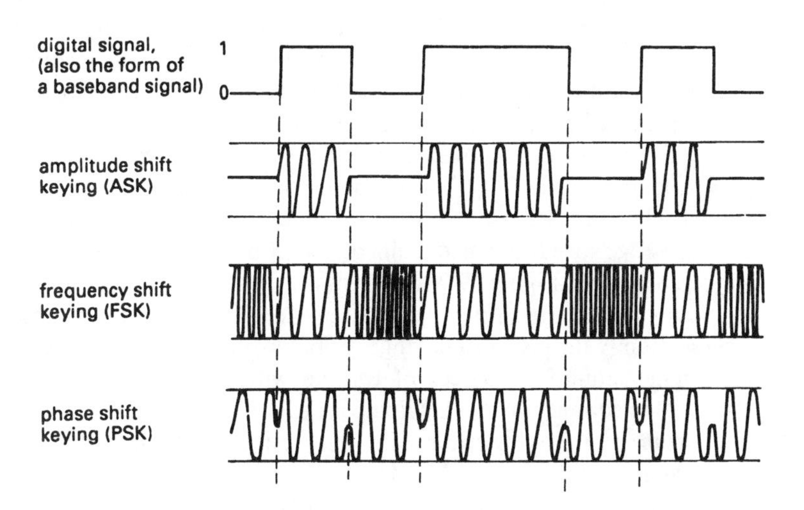

Fig. 19.5 *Various forms of modulation for digital signals.*

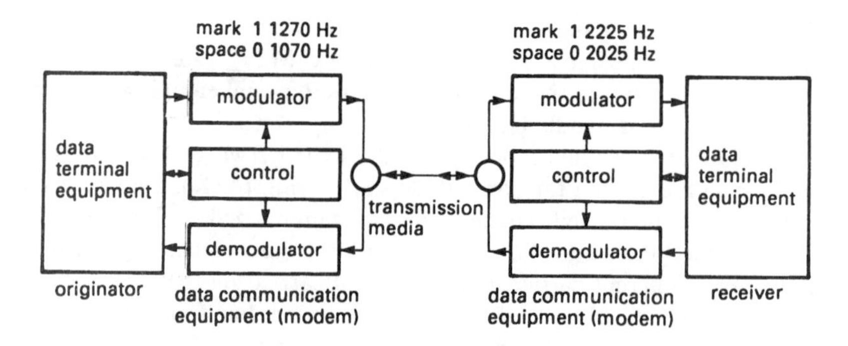

Fig. 19.6 *A typical digital communication system.*

intelligent instrument we are linking two DTEs, and will probably have difficulties. We will return to this problem in the following section.

19.2.6. Standards and protocols

19.2.6.1. RS232C

For successful communications to take place, a set of rules must exist to govern the transmission of data. These rules can be split into standards which govern voltage levels, the connection and control of the DTE-DCE interface and protocols which determine the content and control of the message itself.

Much of the early work on data transmission was done by the Bell Telephone company in the USA, and the result of their work was formalised by the Electrical Industries Association (EIA) into 'A standard for the interface between DTEs and DCEs employing serial binary interchange'. This standard is known as RS232 and is usually used at revision C.

Worldwide standards are set by the Comite Consultatif International Telephonique et Telegraphique (CCITT) which is a part of the United Nations International Telegraph Union. The CCITT publishes standards and recommendations; those for data transmission being prefixed by letters V or X. Standard V24 is, for all practical purposes, identical to RS232C.

Signal levels defined for RS232 and V24 are +6 V to +12 V at the source for a space (zero) and −6 V to −12 V for a mark (one). These are allowed to degenerate to +3 V and −3 V at the receiver. Other characteristics such as line capacitance and edge speeds are also defined. The connections are made with a 25-pin D-type connector. Figure 19.7 summarises the main connections between a DTE and a DCE and the meanings attached to each. These are only a subset of the full specification (which is a rather lengthy document and very heavy going).

There are many common sources of trouble with 'standard RS232'. The standard covers the connection of a DTE and a DCE. Connecting a computer or PLC to an instrument is linking two DTEs. Theoretically, a 'null modem cable' which crosses signals such as pins 2 and 3 (data transmit and receive) should work, but usually doesn't. Manufacturers usually assign their own, often peculiar, ideas to the pin allocation. Many printer manufacturers, for example,

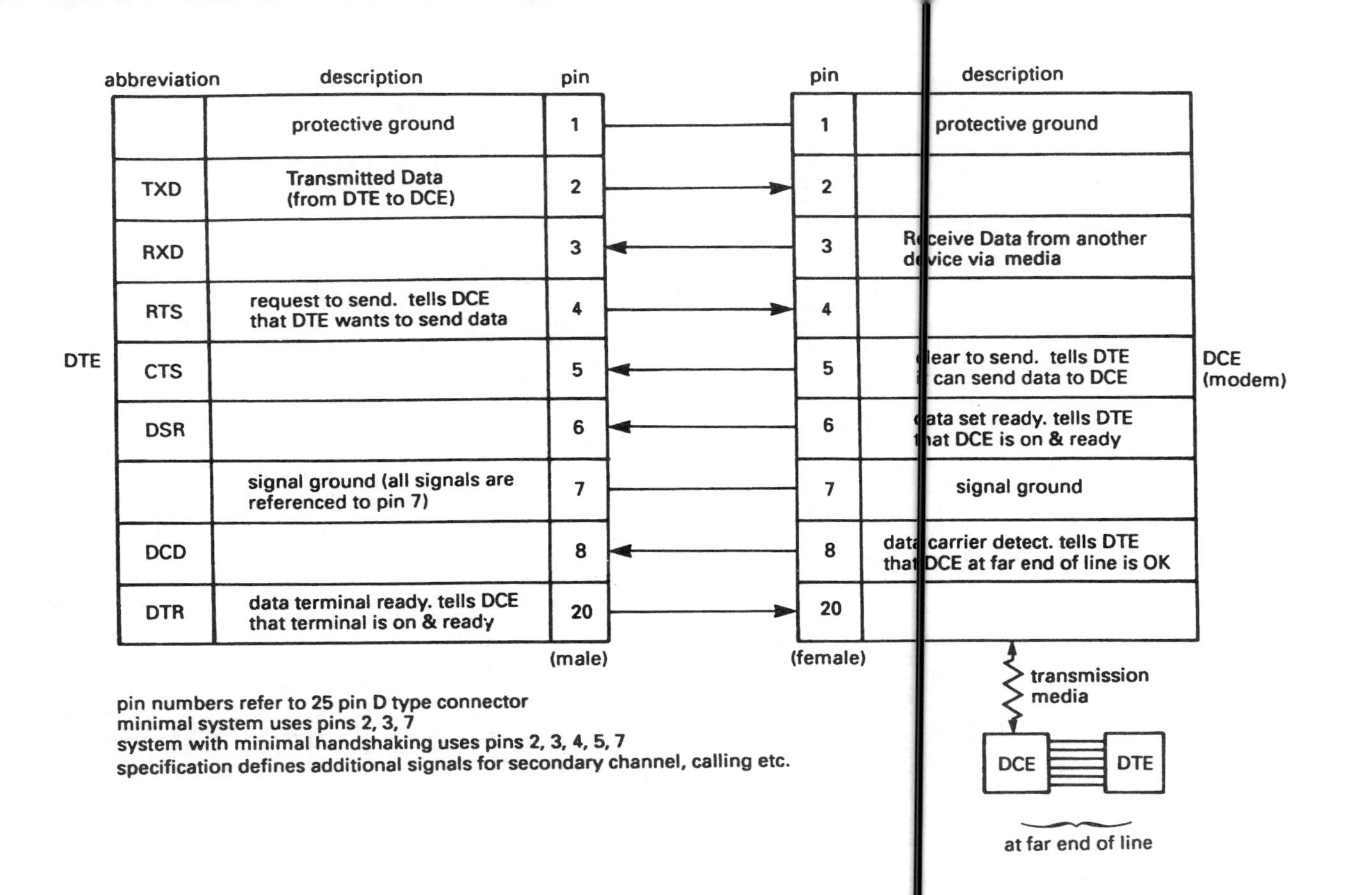

Fig. 19.7 *Connections between DTE and DCE as specified by RS232C.*

use pin 2 to receive data and make the printer a DCE. Even IBM uses a 9-pin D-Type connector (rather than the standard 25-pin) for RS232 connections on their AT range of computers.

Allied with this is an almost random interpretation of the use of the control signals. It is not unknown for an 'RS232 compatible instrument' to have just two connections (corresponding to pins 2 and 7 on the DTE in Fig. 19.7). Such a device can have no data flow control at all.

'RS232 compatible' is thus an expression which strikes dread into the heart of the author. It nearly always means an extended period with a breakout box or line analyser (both essential equipment for use with serial links) and a collection of crimp plugs/sockets and D-type shells.

19.2.6.2. RS422 and RS423

RS232 was designed for a short-haul link between a DTE and a DCE, usually within the same room. If RS232 is used at high speeds over long distances (greater than a few metres), problems will occur.

The EIA have acknowledged the limitations of RS232 for DTE/DTE communications, and have issued two other standards illustrated in Fig. 19.8. One major problem with RS232 is the referencing of signals to a common ground (pin 7) as Fig. 19.8a. RS423 and RS422 (Fig. 19.8b and c) use differential receivers to dispense with the ground connection and overcome common mode noise.

Nominal transmitter voltages are ± 6 V with the signal sense being determined by the relative polarity. Connection A is negative with respect to B for a mark (one) and vice-versa for a space.

RS423 uses a single-ended transmitter and a differential receiver, allowing a standard RS232 transmitter to be used provided the difference in ground potentials does not exceed 4 V. RS422 uses both a differential transmitter and receiver. The mechanical (37-pin connector) details are defined in RS449.

RS422 and RS423 are commonly used on a bus system with a master/slave arrangement as Fig. 19.18.

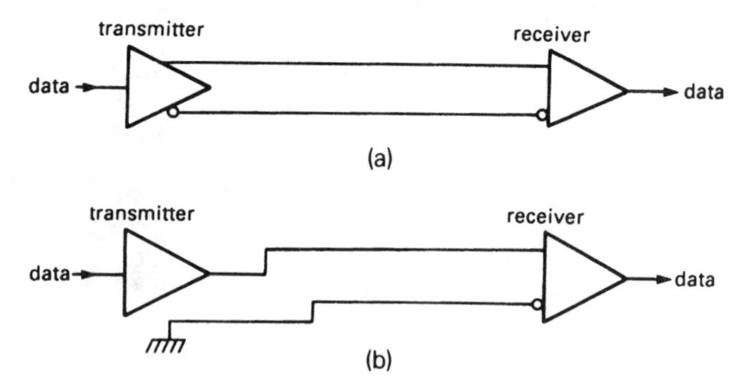

Fig. 19.8 *Variations on RS232. (a) RS422 with differential transmitter and receiver. (b) RS423 with differential receiver and single ended transmitter.*

19.2.6.3. The 20-mA loop

An unofficial early 'standard' is the 20-mA loop. This originated with the early electromechanical teleprinters, but is still found in many applications. It consists of a switch driven by data at the transmitter, a current source, and a current sensor at the receiver. Presence of current is a mark (one) and absence a space (zero).

The current loop, isolated from earth, gives good common mode noise immunity, and overcomes differences between ground potentials at either end of the loop. This is the main reason for its continued use.

High voltages (around 90 V) can exist on a 20-mA loop and unfortunately there are no common standards for control, or even which end of the loop provides the current source. Figure 19.9a is known as active transmit, passive receive and Fig. 19.9b is passive transmit, active receive. Little communication can take place between a passive transmitter and a passive receiver.

19.2.7. Message protocols

The standards described above cover the 'mechanics' of data transmission. The message content is defined by the protocol used. In addition to defining the form of a message (i.e. what group of bits form characters, and what groups of characters form a message), the protocol must define how communication is initiated and terminated, and what actions must be taken if the link is broken during a message. The protocol must also cover how errors are detected, and what action is then to be taken.

There are essentially three types of protocol in use as shown on Fig. 19.10. Character-based protocols (Fig. 19.10a) use control characters from the ASCII set of Table 19.1 to format the message. Most character-based protocols are based, to some extent, on IBM's BISYNC standard.

Bit pattern protocols, such as IBM's SDLC, ISO's HDLC and CCITT X 25 are based on Fig. 19.10b. Flag characters define the start and end of the message, with the end flag being preceded by some form of error control.

The final type of protocol uses a byte count, shown on Fig. 19.10c. The start of the message is signalled by a start flag character followed by a count showing the total number of characters in the message. The receiver counts in the message characters then validates the message with the error-checking data. A common example of this type of protocol is DEC's DDCMP.

Of these, character-based protocols which are variations on BISYNC are probably most commonly used (sometimes called BSC for binary synchronous

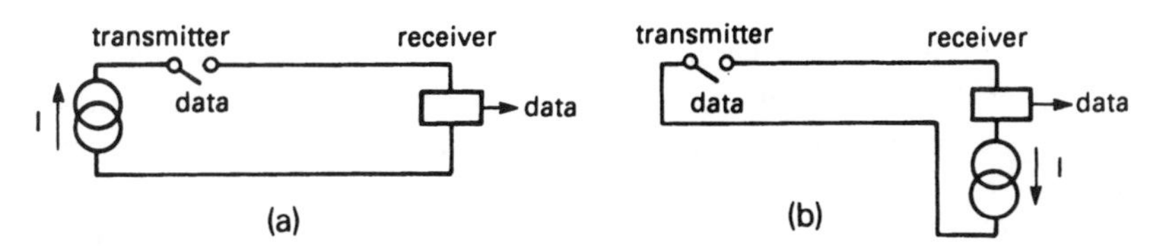

Fig. 19.9 *The two forms of the 20 mA loop, (a) Active transmitter, passive receiver; (b) Passive transmitter, active receiver.*

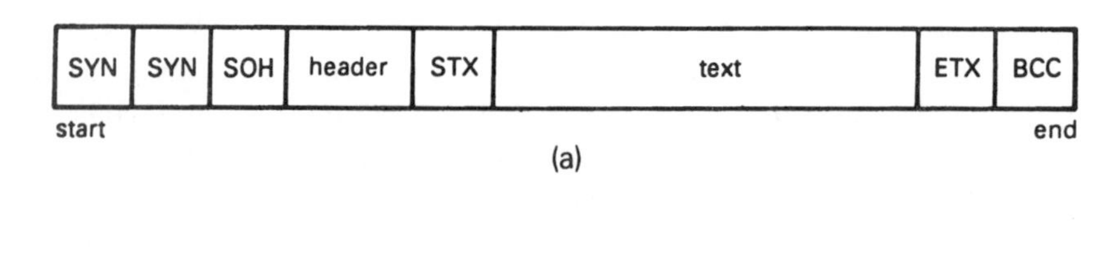

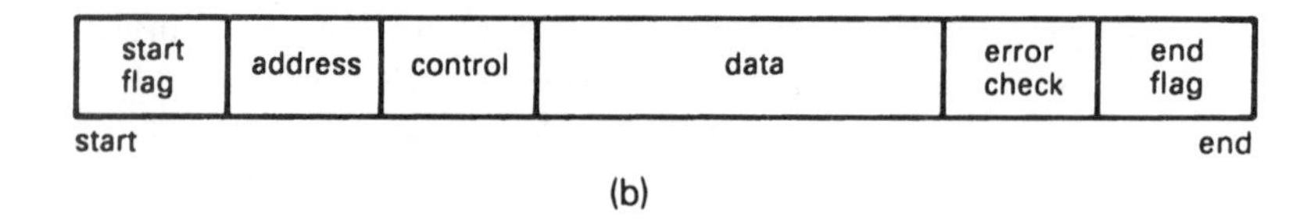

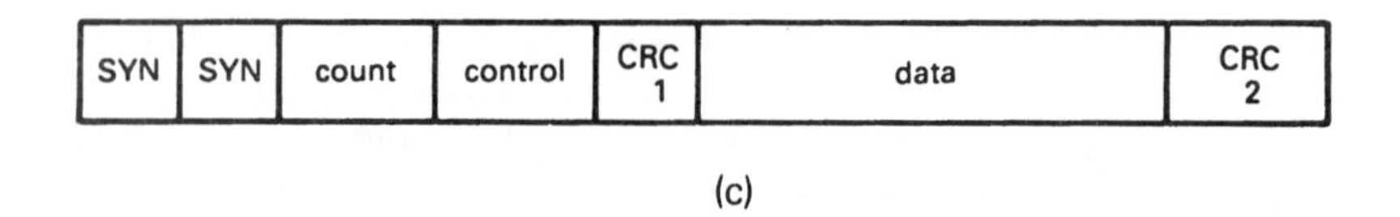

Fig. 19.10 *The three types of protocol commonly used in data communication, (a) The basic form of a character based protocol; (b) The basic form of a bit pattern based protocol; (c) The basic form of a byte count protocol. Note that two error checks are used; one for the header (count and control) and one for the data.*

protocols). They are easy to implement and have the advantage that they can be monitored with a simple terminal across the signal lines.

The control characters from the ASCII set commonly used are:

- Hex 04 EOT End of transmission (often used as a reset to clear the line).
- Hex 16 SYN Synchronising character, establishes synchronisation (i.e. start) and sometimes used as a fill character.
- Hex 05 ENQ Enquiry, used to bid for the line in a multidrop system (see later in Section 19.3.4.)
- Hex 02 STX Start of text. What follows is the message.
- Hex 01 SOH Start of header. What follows is header information, e.g. message type.
- Hex 17 ETB End of transmission block. Data commenced with STX or SOH is complete.
- HEX 03 ETX End of text. Data commenced with STX or SOH is complete and the end of a sequence block. ETX is normally followed by some form of error-checking information, which is validated by the receiver which replies with either:
- Hex 06 ACK Acknowledgement. Message received error free and I am ready for more data. Also used to acknowledge selection on a multidrop system (see later), or:

- Hex 15 NAK Negative acknowledgement. Message received with errors, please retransmit. Also used to say 'not available' when selected on a multidrop system.

19.2.8. Error control

The addition of noise to a digital signal does not necessarily result in corruption. The original signal can be regenerated at the receiving end providing the noise has not been sufficiently severe to turn a '1' into a '0' or vice versa.

Noise generally has a power density distribution similar to Fig. 19.11, with zero mean and tails going off to infinity. If the digital signal has voltage levels $+V$ and 0 V, noise in region A will corrupt a '0' to a '1', and noise in region B will corrupt a '1' to a '0'. The probability of error is thus the sum of areas A and B divided by the total area under the curve.

This probability depends on the ratio between the magnitude of the signal and the noise. The signal to noise ratio, SNR, is defined as:

$$\text{SNR} = \frac{\text{Mean square value of signal}}{\text{Mean square value of noise}} \tag{19.1}$$

An SNR of 20 is normally achievable. There is, however, no 'cut-off' value of noise, and there is a possibility of error whatever the value of SNR. This probability can be calculated (using statistical mathematics) and has the form of Fig. 19.12. From this graph, a link with an SNR of 20 will have an error rate of 10^{-5}. This sounds good, but it represents some 30 corrupt bits in the transmission of a 360K byte floppy disk (which contains 2.88 Mbits).

Rather interestingly, as the signal gets swamped by noise (SNR < 1) the error rate does not tend to one (as might be first thought) but 0.5. What will be received will be a random stream of '1's, and '0's, half of which will, on average, be correct by chance.

With even high SNRs, 100% reception cannot be guaranteed. A single bit in error can have severe results, changing the sign of a number, or turning an 'Open' command to a 'Close' command, so some form of error control is generally needed.

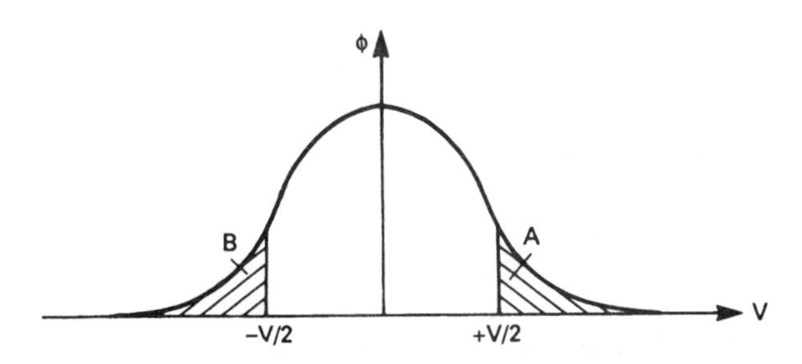

Fig. 19.11 *Power density spectrum of a noise signal. Noise in regions A, B will corrupt a digital signal with V volts between a '1' and a '0'.*

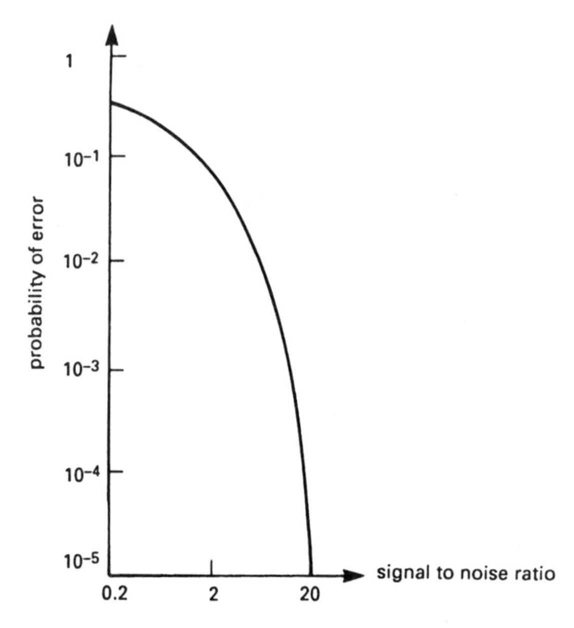

Fig. 19.12 *Relationship between error probability and SNR.*

An error rate of 1 in 10^5 implies a single error bit followed by 99 999 correct bits. This is not a true picture. Anyone who has used a phone will be aware that interference normally has the form of 'clicks' or 'pops' introduced by the switching of inductive loads local to the line. This is similar to the noise found on data transmission lines. A click of 0.05 seconds is ignored in speech, but represents the demise of 60 bits of data at 1200 baud. Noise, therefore, tends to introduce short error bursts separated by periods of error-free transmission, and the error rate represents the average over an extended period of time.

There are generally two ways of handling error control. The simplest, used in almost all industrial systems, detects that an error has occurred, and the receiving station asks for a retransmission. This is known as automatic transmission on request, or ARQ. The ASCII characters ACK (received OK) and NAK (received with errors, please send again) are used for handshaking and control.

The second method attempts to detect and correct any errors by adding redundant characters into the message. This is known as forward error control (FEC). The English language contains a lot of redundancy (allowing communication by speech in difficult circumstances). Given the question:

```
Ar@ y5u co#+ng w%&h ud to{}y
```

which has an error rate of 40%, and the fact that it concerns an outing, it is straightforward to fill in the missing characters to give 'Are you coming with us today'.

FEC is needed for radio links to and from satellites, and is used for the page addressing on Teletext (which uses a technique called the Hamming code). It adds significantly to the message length, and is consequently not widely used on industrial networks.

The simplest form of error detection is the parity bit. This is an extra bit added to ensure that the number of bits in a single character or byte is always odd as shown in Fig. 19.13a. This is known as odd parity; even parity (parity bit added to make number of bits in each character even) is equally feasible, but odd parity is more commonly used. An ASCII character has seven bits, so the addition of a parity bit increases the length to eight bits.

Parity is easily calculated with exclusive OR gates as shown for an 8-bit character on Fig. 19.13b. Parity-calculating ICs are readily available, such as the TTL 74180 and the CMOS 4531.

Parity (or to give it its full title of vertical parity check) can detect single (or 3, 5, 7) error bits, but will be defeated by an even number (2, 4, 6) of error bits.

Additional protection can be provided by breaking the message down into blocks, each character of which is protected by a parity bit, and following the block with a block check character (BCC) which contains a single parity bit for each column position as shown on Fig. 19.14. Normally even parity is used for the

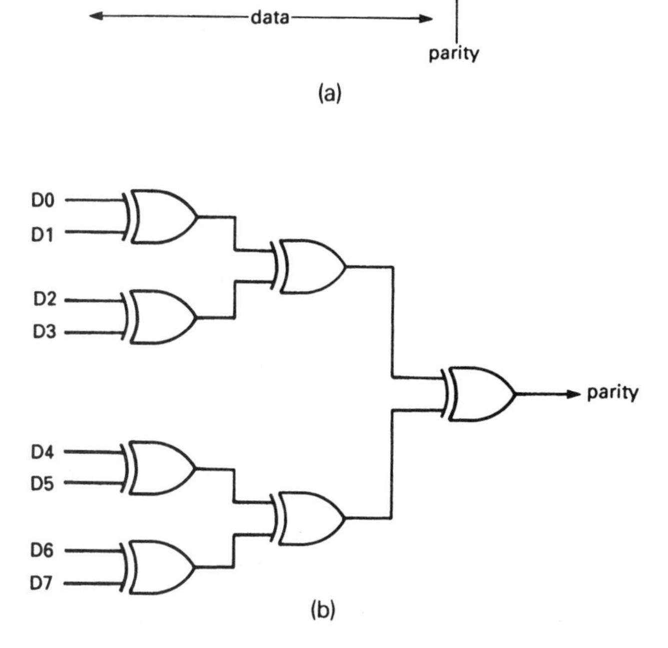

Fig. 19.13 *Error checking with a parity bit. (a) The parity bit makes the number of bits in the word an odd number. (b) Parity circuit for an 8-bit word using exclusive OR gates.*

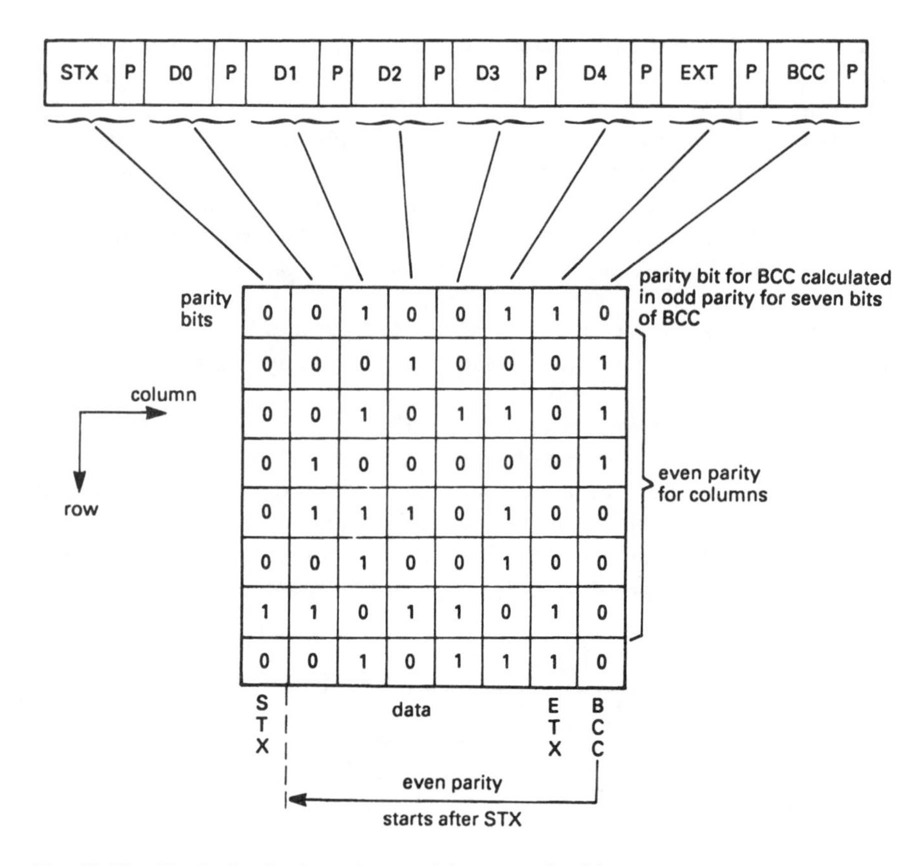

Fig. 19.14 *Block check character used for error checking.*

column parity bits. This is known as longitudinal parity checking. The BCC character has its own odd parity bit which is calculated from the BCC character, not the parity bits in the message. The initial STX or SOH are excluded from the BCC calculations, but the terminating ETB or ETX are included.

BCC can detect all odd numbers of errors, and many multiple-bit combinations. It is defeated by an even number of errors spaced symmetrically around the block.

The most powerful error detection method is known as the cyclic redundancy code (or CRC). Like the BCC method, this splits the message into blocks. Each block is then treated like a (large) binary number which is divided by a pre-determined number. The remainder from this division, called the CRC, is sent as a 16-bit number (two characters of 8 bits) after the message. The same calculation is performed at the receiver, errors being detected by differences in the CRC.

The calculation of the CRC is performed with a shift register and exclusive OR gates, a typical example being the CRC-CCITT circuit of Fig. 19.15. A typical CRC system detects:

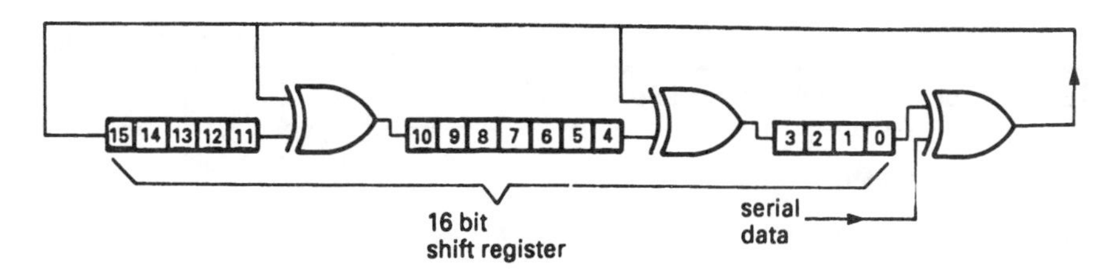

Fig. 19.15 *CRC-CCITT, a common cyclic redundancy code circuit.*

- all single bit errors
- any odd number of errors
- all single and double errors in the message format
- any two burst errors of two bits in the message format
- any single burst of 16 bits or less
- all but 1 in 32768 bursts of exactly 17 bits
- all but 1 in 65536 bursts of greater than 17 bits

The use of CRC greatly improves the error rate. Typical improvements of the order of 10^5 are achieved, giving an undetected error rate of 1 in 10^{10} for a circuit with a basic error rate of 1 in 10^5.

Normally, ARQ systems provide an acknowledgement or an error signal to the initiating device or procedure at the transmitting end, good reception being determined by the reception of the ACK from the receiver. On receipt of a NAK (or no reception of ACK or NAK within a predetermined time) the transmitter will re-send the message. To stop a line being clogged with re-tries, it is usual to set a limit on the number of retries (often three or five) before an error is declared.

19.2.9. Point to point communication

A PLC or computer is often required to establish a simple serial link with a device. Typical applications are reading data from an instrument or a bar code reader, sending data as a setpoint to an instrument or producing a report on a printer. In this section we will look at some of the problems of achieving this simple aim.

Point to point links are usually simple, employing, at most, parity checks for error control. Where data is being read from an instrument, the port on the instrument was probably designed to be connected to a printer, and few, if any, of the control signals on Fig. 19.7 will be used. The first step for reading or writing data is therefore to determine:

(a) the connections on the instrument/device
(b) the baud rate
(c) the data format (ASCII, number of bits, parity used, number of stop bits)
(d) the way the control signals are used
(e) how message transfer is initiated (when reading data)
(f) the form of the message. A temperature transmitter, for example, could send a character string similar to Fig. 19.16 where the three required temperature digits are buried in amongst irrelevant data.

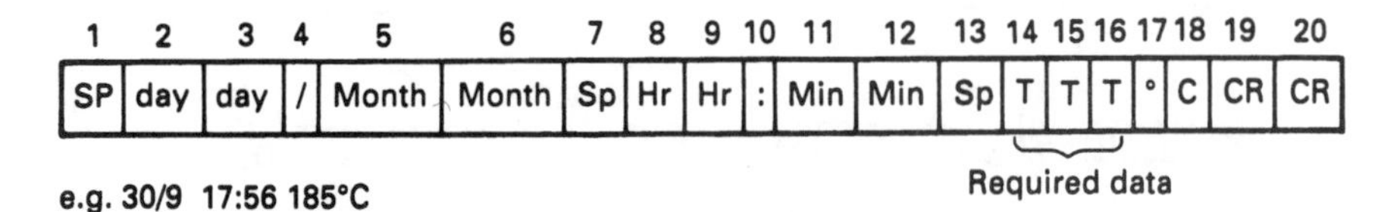

Fig. 19.16 *A typical ASCII string from a transducer.*

The character string must be read into a buffer and the end detected (either by a character count or detection of distinct patterns such as the double CR of Fig. 19.16). The required data can then be found at the correct place in the buffer. In this type of application the link operation can be monitored with a simple dumb terminal connected across the data lines.

19.3. Area networks

19.3.1. Introduction

So far we have considered point to point links. For a true distributed control system we need a method where several PLCs or computers can be linked together to allow communication to freely take place between any member of the system.

To achieve this we need to establish a connection topology, some way of sharing the common network that prevents time-wasting contention and an address system that allows messages to be sent from one member to another. Such systems are known as local area networks (LANS) or wide area networks (WANS) dependent on the size of the area and the number of stations.

19.3.2. Transmission lines

Any network will be based, to some extent, on cable, and at the high speeds used there are aspects of transmission line theory that need to be considered.

Consider the simple circuit of Fig. 19.17. At the instant that the switch closes, the source voltage does not know the value of the load at the far end of the line. The initial current step, i, is therefore determined not by the load, but by the characteristics of the cable (dependent on the inductance and capacitance per unit length). A line has a characteristic impedance, typically 75 ohms or 50 ohms for coax, and 120 to 150 ohms for biaxial or screened twisted pair. The initial current step will therefore be V/Z where Z is the characteristic impedance.

After a finite time, this current step reaches the load R, and produces a voltage step iR. If R is not the same as Z, this voltage step will not be the same as V, and a reflection will result. Typical results are shown on Fig. 19.17b.

This effect occurs on all cables and is normally of no concern as the reflections only persist for a short time. If, however, the propagation delay down the line is similar to the maximum frequency rate of the signal, the reflections can cause problems. It follows that a transmission line should be terminated by a resistance equal to the characteristic impedance of the line. Normally, devices for connecting onto a transmission line have a high input impedance to allow them

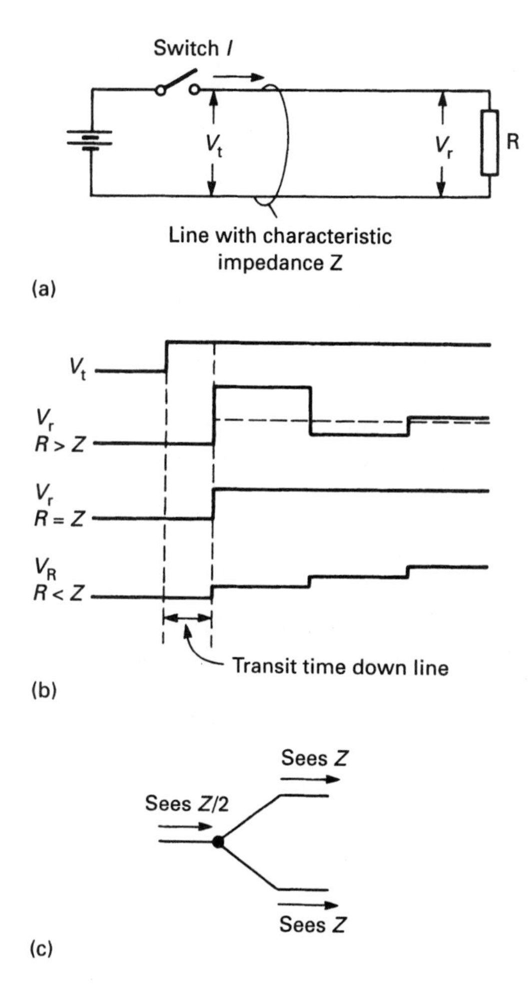

Fig. 19.17 *Transmission lines and characteristic impedance. (a) A transmission line. (b) Effect of terninating resistor. (c) Effect of a branch.*

to tap in anywhere, with terminating resistors being used at the ends of the line.

A side effect of this is that T connections, or spurs, are not allowed (unless the length of the spur is short). In Fig. 19.17c a T has been formed. To the signal, coming from the left, the two legs appear in parallel giving an apparent impedance of $Z/2$ and a reflection.

19.3.3. Network topologies

From the previous section it should be apparent that any network can sensibly only be based on a ring (which needs no terminating resistors) or a line (with a terminating resistor at each end).

Figure 19.18 is a master/slave system where a common master wishes to receive or send data from/to slave devices, but the slaves never wish to talk to each other.

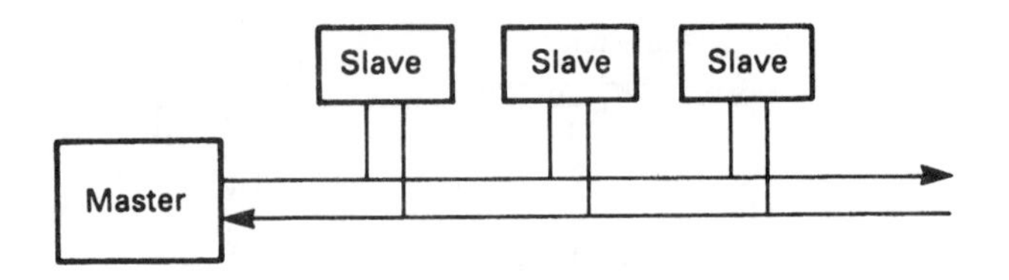

Fig. 19.18 *A master/slave network.*

All the slaves have addresses, which allows the master to issue commands such as 'Station 3; give me the value of analogue input 4' or 'Station 14; your setpoint is 751.2'. Such systems are often based on RS422.

The Star network of Fig. 19.19 is again based on a master with a point to point link to individual stations. This arrangement is commonly used for high level computer systems. Communication control is performed by the master station. Station to station communication is possible via, and with the co-operation of, the master.

In Fig. 19.20 all the stations have been connected in a ring. There is no master,

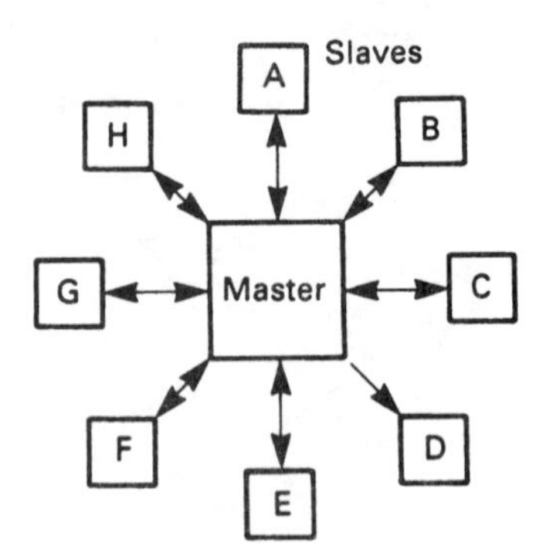

Fig. 19.19 *A star network.*

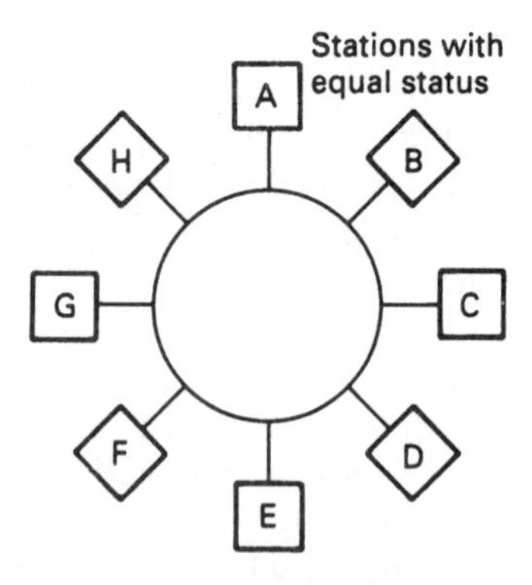

Fig. 19.20 *Masterless peer to peer link or ring.*

and all stations can talk to any other station and all have equal right of access. The term 'peer to peer link' is often used for this arrangement. With Figs. 19.18 and 19.19 control was firmly in the hands of the master. With the ring, some technique is needed to avoid clashes when two stations wish to use the line at the same time. We will discuss this in the following section.

Figure 19.21 is probably the commonest type of network used by computers and PLCs. It is a single line with terminating resistors and, like the ring, is a peer to peer link where all stations have equal standing.

19.3.4. Network sharing

A peer to peer link allows many stations to use the same network. Inevitably two stations will want to communicate at the same time. If no precautions are taken, the result will be chaos. Various methods are used to govern access to the network.

One idea is to allocate time slots into which each station can put its messages. This is known as time division multiplexing, or TDM. Whilst it prevents clashes, it can be inefficient as a station will have to wait for its time slot even if no other station has a message to send. To some extent a mismatch between the frequency of messages from different stations can be overcome by giving more slots to a hardworking station. With a five-station network and stations labelled A to E, if A has a high workload an order ABACADAEAB etc. might be adopted. This is sometimes known as statistical TDM.

The empty time slot of Fig. 19.22 uses a packet which continuously circulates around the ring. When a station wishes a send a message it waits for the empty slot to come round, when it adds its message. Suppose station A wishes to send a message to station D. It waits until the empty packet comes round. Then it puts its message onto the network along with the destination address D. Stations B and C pass the message but ignore it because it is not for their address. Station D matches the address, reads the contents (and appends that it has received the message). Stations E-H ignore it, but pass it on. Station A receives the message back again, sees the acknowledgement and removes its message leaving the empty packet circulating the ring again.

A similar idea is a token passing, where a 'permit to send' token circulates round the network. A station can only transmit when it is in possession of the token, which is released when the acknowledgement that the message arrived is received.

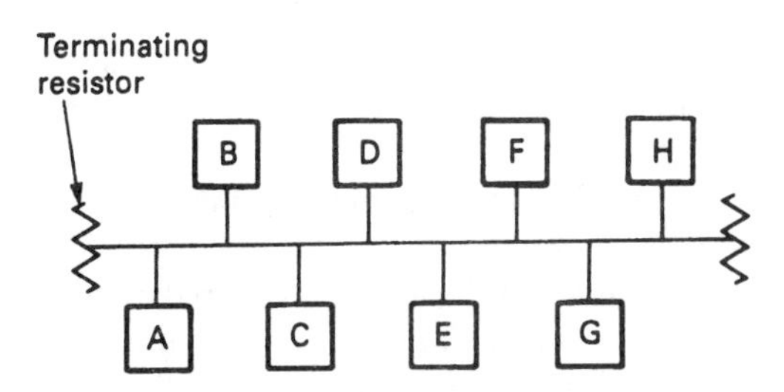

Fig. 19.21 *Peer to peer link arranged as a single highway with terminating resistors.*

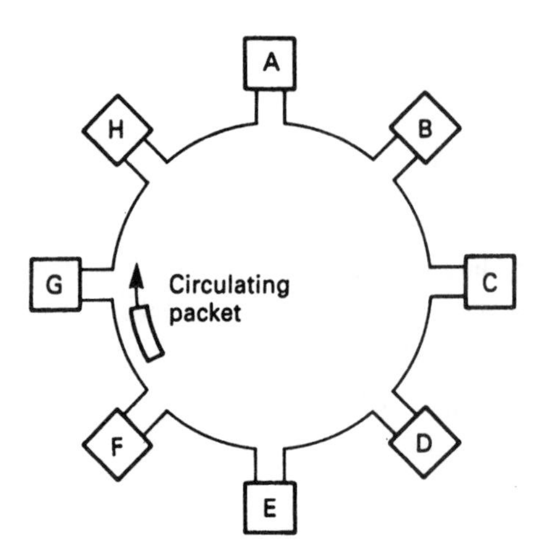

Fig. 19.22 *Empty slot and token passing.*

Both empty slot and token passing require some way of reinstating the packet or token if the network is corrupted by noise or broken. This is usually provided by a master station, or monitor station, but it should be noted this is not fulfilling the same role as the masters on Figs. 19.18 and 19.19.

Empty slot and token passing are usually associated with rings, although they can be used with a bus-based system if the stations are arranged as a logical ring.

Bus systems usually employ a method where a station wishing to send a message listens to the network to see if it is in use. If it is, the station waits. If the network is free, the station sends its message (thereby locking out any other station until the message ends). This is known as carrier sense multiple access (CSMA).

Situations can still arise, however, where two stations simultaneously start to send a message, and a collision (and garbage) results.

This situation can easily be detected, and both stations then stop and wait for a random time before trying again. A random time is used to stop the two stations clashing again. This is known as carrier sense multiple access with collision detection (CSMA/CD).

There is a fundamental difference between TDM, empty slot, and token passing as one group and CSMA. With the former there is a certain amount of time wasting, but every station is guaranteed access within a specified time. With CSMA there is little time wasting, but a station can, in theory, suffer repeated collisions and never get access at all.

A useful analogy is to consider traffic control. TDM/token passing approximates to traffic lights, CSMA to roundabouts. In heavy traffic the best solution is traffic lights; everyone gets through and the waiting is shared evenly. Roundabouts can 'lock out' one road when the traffic flow is heavy and uneven from one direction. In light traffic, however, roundabouts keep the traffic flowing smoothly; and there are few things more annoying than being brought to a halt by a red light, then have nothing go past in the other direction.

19.3.5. A communication hierarchy

Early process control systems tended to be based on a single large computer or PLC. The advent of cheap computers and PLCs along with good communication methods has led to the development of a hierarchy of machines which split the tasks between them. This is generally arranged as Fig. 19.23 with a hierarchy split into four levels.

Level 0 is the actual plant, with devices linking to the next level by direct wiring or simple RS232/422 serial links.

Level 1 consists of PLCs and small computers directly controlling the plant.

Level 2 is microcomputers, such as the DEC VAX, acting as supervisors for large areas of plants.

Level 3 is the large company mainframes, such as IBM's AS400.

Usually the layout is not as clear cut as Fig. 19.23 implies. There are also differences between different companies, some number the layers from top to bottom and some ignore level 0. Normally there will be a split of responsibility in the hierarchy; engineering is usually responsible for levels 0 and 1 and data processing for levels 2 and 3.

There are many advantages to distributed systems. The resulting tree is conceptually simple, and as such is easy to design, commission, maintain and modify. A correctly designed system will be, for short periods, fault tolerant and can cope in a limited mode with the failure of individual stations or the links between them. A distributed system can also bring about an increase in performance as lower level machines take the work off higher level machines.

19.4. The ISO/OSI model

Neat as Fig. 19.23 is, the interconnection between different machines can bring even more problems than linking two 'RS232-compatible devices'. Common problems are different baud rates, flow control, routing and protocols.

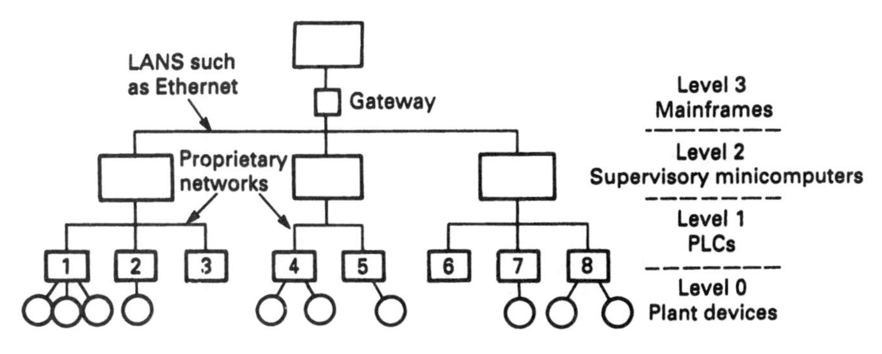

Fig. 19.23 *A typical plant-wide network of linked computers, PLCs and intelligent instruments.*

In 1977 the International Standards Organisation (ISO) started work on standards to try to ensure compatibility between different manufacturer's equipment. This is known as the open systems interconnection (OSI) model, and is primarily concerned with communication between level 2/3 systems on Fig. 19.23.

It consists of definitions for the seven layers of Fig. 19.24. Each layer at the transmission end has a direct relationship with the same layer at the receiving end. The function of each layer is, from the bottom:

(1) The physical link layer – which is concerned with the coding and physical transmission of the message. Requirements such as transmission speed are covered.
(2) Data link layer – controls error detection and correction. It ensures integrity within the network and controls access to it by CSMA/CD or token passing.
(3) Network layer – performs switching and makes connection between modes.
(4) Transport layer – provides error detection and correction for the whole message by ARQ, and controls message flow to prevent overrun at the receiver.
(5) Session layer – provides the function to set up, maintain and disconnect a link, and the methods used to re-establish communication if there are problems with the link.
(6) Presentation layer – provides the data in a standard format (which may require the data to be converted from its original form in the initiating application).
(7) Application layer – links the user program into the communication process and determines what functions it requires.

As a very rough analogy, consider the placing of a verbal order by telephone.

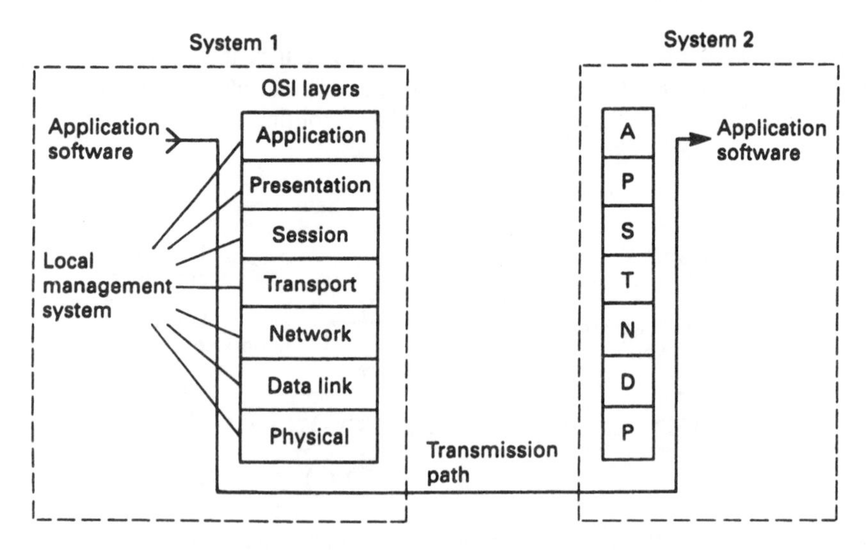

Fig. 19.24 *The OSI model.*

This analogy is based on Siemens material published in their brochure *Communications Setting the Pace in Automation.*

(1) Physical link layer – the phone is lifted and connected to the telephone network. A dialling tone is heard.
(2) Error detection and control – it is a good line with no noise.
(3) Network layer – the number is dialled, 9 for an outside line then the number. The phone rings at the other end.
(4) Transport layer – the telephone is lifted at the receiving end. 'This is ACME products, could you hold on please, I'm handling another call. OK, go ahead now. Sorry, I didn't get that could you repeat please'.
(5) Session layer – 'This is Aphrodite Glue Works, I have a verbal order for you, number CAP4057, my account number is 7322D'. The receiver makes a note of these details in case the call is broken prematurely.
(6) Presentation layer – 'I am using order numbers from the June 1994 catalogue.'
(7) Application layer – 'I require 100 off 302-706 and 50 off 209-417, delivery by datapost.'
'OK, 100 off 302-706 and 50 off 209-417 will be dispatched by datapost this afternoon. Total cost £147.20, invoice to follow.'

At any stage, the lower layers can interact. A burst of noise on the line, for example, will cause the transport layer to ask for a repeat of the last message.

It can be seen that layers 1 to 4 are concerned with processing functions for the particular applications.

19.5. Proprietary systems

19.5.1. Introduction

The ISO/OSI model is mainly concerned with higher level communications such as linking minicomputers. At the PLC level every manufacturer has their own standard (Modicon's MODBUS, Texas Instrument's TIWAY, CEGELEC's ESP) and these link their own equipment in a straightforward manner. If, on Fig. 19.23, PLCs 1, 2 and 3 were Allen Bradley, 4 and 5 were GEM-80s and 6, 7 and 8 were Siemens there would be no real problems linking similar PLCs. Allen Bradley Data Highway would be used for the first three, CEGELEC's CORONET for the GEMs, and Siemens SINEC L1 or L2 for units 6, 7 and 8. Each of these are simple to use and, in the author's experience, very reliable. Linking between the different PLC systems, however, is another story.

In this section we will look at a typical proprietary PLC and computer system and at the tentative steps taken to provide standards that allow linking between different manufacturers. All are similar in principle, and provide useful internal diagnostics for fault finding, and less useful pretty green and red communication LEDs on the cards. (These, in the author's experience, tell you the pretty green and red LEDs are on, off or flickering in a very impressive manner. They do, however, impress visitors!.)

Some of the common proprietary systems are:

Allen Bradley	– Data Highway
ABB	– Masternet and Master Fieldbus
Gould/Modicon	– Modbus
General Electric	– GENET
Mitsubishi	– MelsecNET
Square D	– SYNET
Texas Instruments	– TIWAY

These are all dedicated to their own machines and all use similar ideas which are often tantalisingly close; a topic we shall return to in Section 19.6.

19.5.2. Allen Bradley Data Highway

PLC-5s communicate with each other on a peer to peer (no master) token passing highway based on twin axial cable and operating at 57.6 kbaud. Their trade name is Data Highway Plus (an earlier version called Data Highway linked the predecessor of the PLC-5, the PLC-2 range). The PLC stations addresses are set on DIP switches on each PLC, and up to 64 stations can exist on one line with Octal addresses 0–77.

Communication is established with a single message (MSG) instruction. This can be set up to read or write a block of data, the programmer specifying:

(a) the start address at the local end
(b) the start address at the target end
(c) the length of the block to be transferred (in words)
(d) the station address at the remote end.

In Fig. 19.25a, station 5 is performing an MSG write, sending six words starting from store location N10:40 to a block starting from N7:15 at station 12. In Fig. 19.25b, station 7 is performing a MSG read, taking 8 words starting from N10:0 at station 12 and copying them into a block starting at its own N7:32.

The MSG instruction appears in a ladder program as Fig. 19.26, the transfer being initiated every time the rung goes true. The ENable bit goes true when the transfer is started, and the DoNe bit goes true when it has been successfully completed. The ERRor flag goes true when an error occurs. Common errors are a line fault, a non-existent address at the far end or the PLC at the far end shutdown. The cause of the fault is given in flags set in the message control word. Link statistics (e.g. number of retries) are kept in the processor for diagnostic purposes.

The data highway is also used by the programming terminal, so a programmer can connect anywhere onto the Data Highway and link into any machine on the network.

19.5.3. Ethernet

Ethernet is a very popular bus-based LAN originated by DEC, Xerox and Intel and commonly used to link the computers at level 2 in Fig. 19.23. It uses 50 ohm

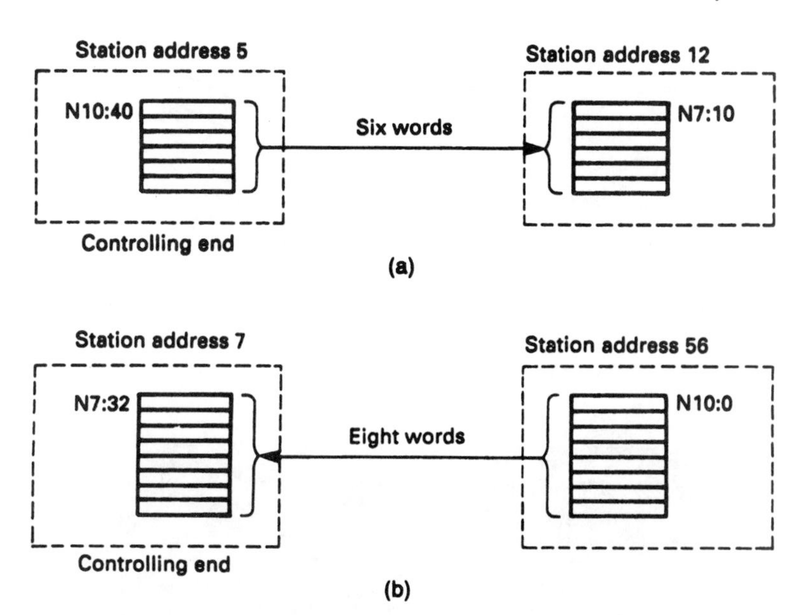

Fig. 19.25 *The Allen Bradley message (MSG) instruction. (a) Write message. (b) Read message.*

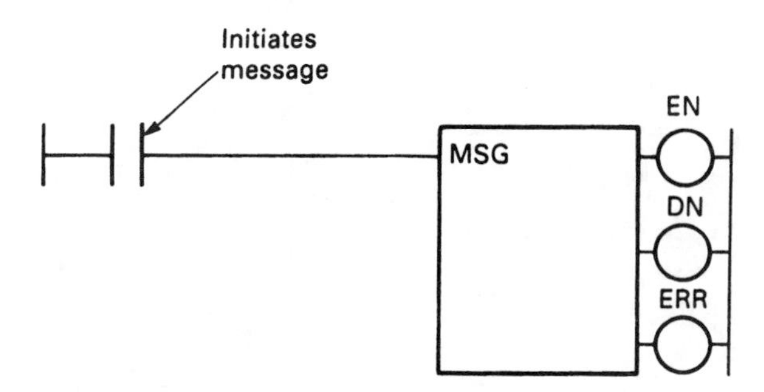

Fig. 19.26 *The PLC-5 message instruction as it appears in the ladder diagram program.*

coaxial cable, with a maximum cable length of 500 m (although this can be extended with repeaters). Up to 1024 stations can be accommodated, although in practical systems the number is far lower. Baseband signalling is used with CSMA/CD access control. The raw data rate is 10 Mbaud, giving very fast response at loading levels up to about 20-30% of the theoretical maximum. Beyond this, collisions start to occur.

Stations are connected onto the cables by transducers known as nodes on the network. Commonly 'vampire technology' is used for these transceivers as shown on Fig. 19.27a. The transceiver clamps onto the cable, with a sharp pin piercing the cable and contacting the centre conductor. The arrangement of the pin shrouding prevents it contacting the screen. This approach allows transceivers to

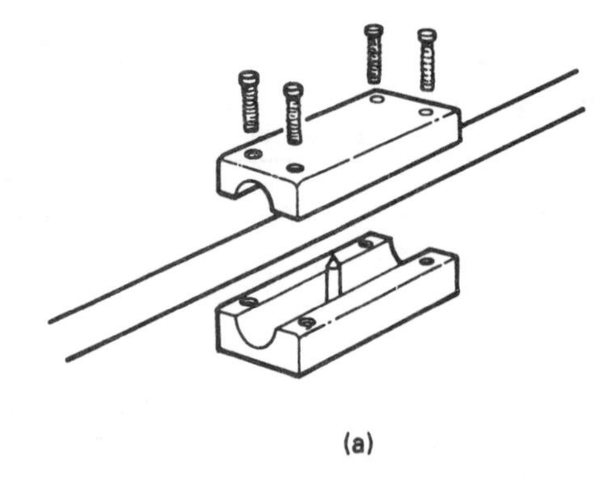

(a)

(b)

Fig. 19.27 *Ethernet connections. (a) Vampire connector. (b) Ethernet cable arranged in loops to provide connection separation. (c) Break the line screw connector.*

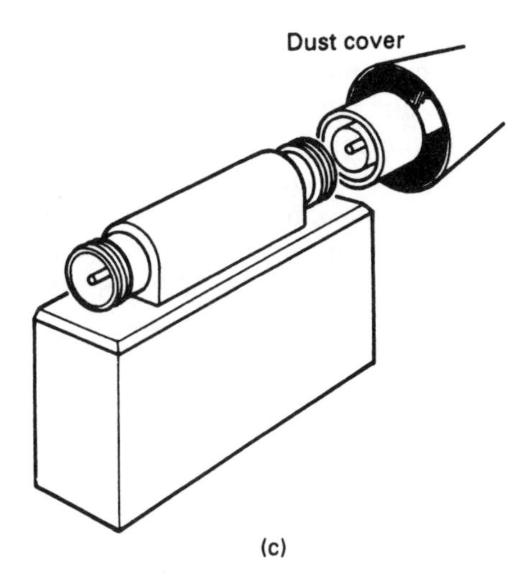

Fig. 19.27 *contd.*

be added, or removed, without disturbing the rest of the network. The author must admit to being more than a little apprehensive about vampire technology, but it does seem to work. To avoid reflections (as discussed earlier in Section 19.3.2.) a minimum spacing of 2.5 m must be maintained between nodes. To assist the user, Ethernet cable has 'tap in' points marked on its sleeving. Figure 19.27b shows an arrangement commonly used to maintain node spacing.

An alternative to the vampire transceivers is the plug-in transceiver using coaxial cable plugs as Fig. 19.27c. These are obviously more secure, but have the disadvantage that the network is disrupted if a node is added or removed.

The transceivers are connected to a local controller which performs the access control. Ethernet has three layers (physical, data-link and client) which approximate to the functions performed by the bottom three layers in the OSI model discussed in Section 19.4.

Ethernet is possibly the most successful and widely used LAN.

19.6. MAP and Fieldbus, towards standardisation

We have already discussed the difficulties of linking different equipment. There is normally little problem linking PLC networks to higher level computers. PLC manufacturers publish their message format and protocols, and interfacing software (called 'drivers') has been written for all common computers and PLCs. The difficulty comes when you want to link two machines at level 1 in Fig. 19.23. In many cases, the only economical solution is to do it through the computers and the higher level link.

General Motors (GM) in the USA were faced with this problem and attempted to specify a LAN for industrial control. This was called MAP (Manufacturing Automation Protocol). A similar office-based LAN called TOP (Technical Office

Protocol) was conceived at the same time. With GM's purchasing muscle, it involved several automation equipment manufacturers. A firm commitment to the OSI model was made, and the network based on broadband token bus as specified in IEEE 802.4 (compare Ethernet; baseband CSMA/CD, to IEEE 802.3). The token bus was chosen as it is deterministic; the response time can be predicted (see discussion on roundabouts and traffic lights in Section 19.3.4).

MAP (currently at version 3.0 in early 1994) seems in the author's opinion to have gone slightly off the rails. In the course of the research for a previous book, major PLC manufacturers were visited. Each could interface MAP, but for each (with the exception of Siemens SINECH2B) it seemed to be an expensive add-on which the customer could 'have if he really wanted'.

There appear to be several reasons for this distinct lack of enthusiasm. The first is a bureaucratic organisation and a changing specification. The term 'moving target' was used independently on several occasions by different manufacturers. The second reason is cost; MAP links often cost more than the computer or PLC to which they are connected. The expression 'Designed by big organisations, for big organisations' was used, and seems apt. The third reason is speed; by using token passing MAP is slow by comparison with Ethernet and the OSI model is not really designed for time-critical applications. The non-deterministic nature of CSMA/CD does not seem to cause any problems up to about 30% of the theoretical maximum loading, and real systems normally operate below 10% loading. The final, and perhaps most crucial, fact is that MAP seems to have settled at a level where it is in direct competition with established LANs such as Ethernet rather than the proprietary systems at level 1 of Fig. 19.23.

In the mid-1980s MAP was going to be the common standard of industrial control. MAP systems have been installed, both in Europe and the USA, but it has not yet achieved anything like acceptance.

There are other attempts at standardisation. In conjunction with the Instrument Society of America, specifications for a low cost (twisted pair) low level network called Fieldbus has emerged. Its full specification was due to be completed by 1992, but (inevitably) has been delayed by commercial and political infighting. Demonstrations linking different manufacturers equipment can be seen at most automation and control exhibitions and it could, perhaps, fulfil the role that MAP was publicised for.

The factors that have limited the adoption of Fieldbus are the vast number of contending vendor-led bus standards, all with odd names, (FIP, HART, LonWorks, Batibus, Profibus and ISP, to name just a few) coupled with a wish not to be 'the man who bought Betamax' from the user. At the time of writing (in late 1994) there appear to be two main groups.

The first is WorldFIP which is probably the closest to the IEC/ISA Fieldbus standard, and is backed by major vendors such as Honeywell, Allen Bradley, Square D, Télémécanique, Cegelec and Electricité de France plus many others. It uses a fast time-service approach and permits field device to field device communication without using the master host.

The second major contender is ISP (for InterOperable System Project) which is backed by Siemens, Fisher, Rosemount and, again, many other major companies. Unlike WorldFIP it is driven by a master host which controls all communication.

The attractions of the Fieldbus concept are obvious: simple and cheap serial communication cables linking controllers, sensors and actuators. There are additional benefits in ease of expansion, remote diagnostics and commonality of spares. The 4-20 mA loop has become a de facto standard for analog instrumentation, and there is no problem linking, say, a Rosemount transmitter to a Eurotherm controller to a Vickers valve. All use the same standard, and the user knows that the resulting control loop will work. Unfortunately Fieldbus has not reached the stage where the end user can buy with similar confidence. When there is an agreed standard and interchangeability between different vendors things will rapidly change. Watch this space.

19.7. Safety and practical considerations

Figure 19.28 shows a fairly common situation where a switch connected to one PLC is, via a serial link, causing a motor to run in another. Supposing the motor is started and the link is severed. The bit corresponding to 'motor run' which is set inside PLCB will not be cleared by the link failure, and PLCA will be unable to stop the motor.

When the switch is turned off, the serial link control in PLCA will signal an error, but this is of no use to PLCB which does not know that PLCA is trying to communicate with it. This may, or may not, be a problem depending on the application, but it is obvious there are implications that need to be considered.

One approach is to define how long an output driven by the link is allowed to be uncontrolled, say 2 seconds. The originating PLC then sends a toggling signal via the link at a slightly shorter period, say 1.5 seconds, as Fig. 19.29. Inside PLCB, the true and complement forms of this signal trigger two TOF's (delay off) set for 2 seconds. With the link healthy coil energised, the link-driven outputs can be energised. If the link fails, one TOF will de-energise (and one stay energised), causing the 'link healthy' signal to de-energise and all link-controlled outputs to go to a safe state.

A network introduces extra delays into the system. These delays obviously depend on the loading on the network, but are typically of the order of 0.2 to 0.5 seconds on proprietary networks, and a bit slower on Ethernet and MAP.

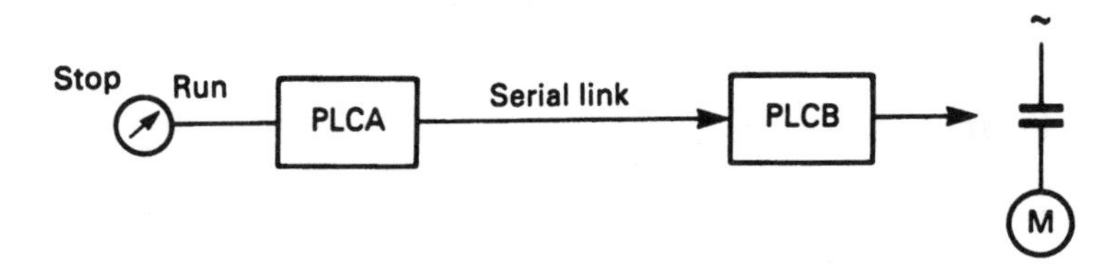

Fig. 19.28 *Safety considerations with a serial link.*

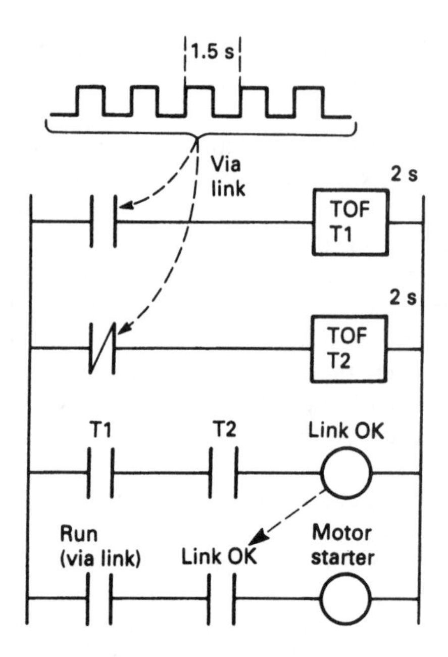

Fig. 19.29 *A way of checking a serial link.*

Noise is a major source of problem, and normally manifests itself as an increase in the delay time introduced by the network (caused by a large number of re-tries). Because of checking and CRC discussed earlier, noise rarely causes operational problems, and when it does (in really severe cases) the effect is almost always something not working when requested (rather than something starting unexpectedly). Noise prevents signals getting through, it does not usually cause faulty signals to be accepted.

Obvious precautions against noise are separation from power cables, and the use of conduit or trunking (mainly to identify low-signal-level cables). Cable screens should be continuous and earthed in one, and only one, place. Great care should be taken to prevent screens accidentally grounding inside junction boxes.

Most proprietary networks have monitoring facilities. Some errors are inevitable on all systems (see Fig. 19.12 earlier) and it is worth logging the rate when a network is first commissioned. This allows checks to be made at a later date, and any deterioration noticed early before problems start to arise.

Fibre optics, discussed in the next section, give almost total freedom from interference.

19.8. Fibre optics

When light passes from one medium to another, the beam is bent as Fig. 19.30a. This is known as refraction, and is the cause of water appearing shallower than it really is. If the angle of incidence, x, is greater than a certain critical angle θ_c, the

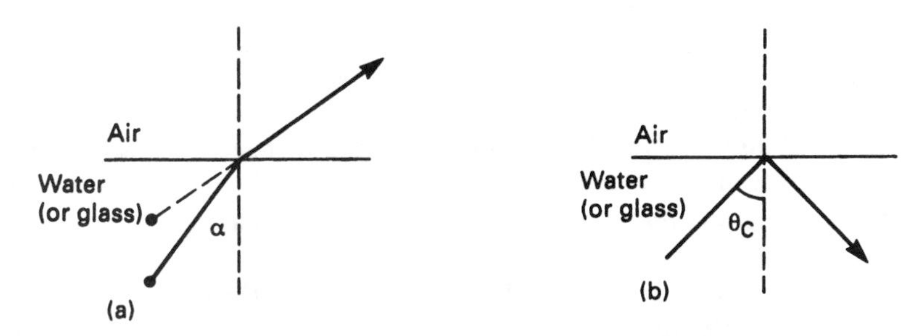

Fig. 19.30 *Light refraction as a light beam passes into a medium of lower density (e.g. water to air). (a) Refraction. (c) Total internal reflection.*

light beam does not emerge from the surface, but is reflected internally as Fig. 19.30b. This is known as total internal reflection, and it can be shown that

$$\sin \theta_c = 1/\mu$$

where μ is the refractive index for the two materials.

In Figure 19.31a, a small diameter tube of glass has been constructed. Light entering at a shallow angle will be conveyed down the tube with little loss by repetitive total internal reflection. This principle, known as fibre optics, is the basis of an interference-free form of data communication.

The principle is very simple. Data at the transmitter is converted into light pulses which are conveyed down the fibre optic cable and detected by a photo sensor at the receiver. Fibre optic cable has a very large bandwidth, so modulation or time domain signal multiplexing allows several high-speed serial channels to be carried down one cable. Fibre optic cable is, however, unidirectional in all practical applications, so to achieve full duplex operation two fibre optic cores must be run.

There are many advantages to fibre optic cables. The transmission is totally free from problems caused by noise, crosstalk, and ground loops and gives total isolation between transmitter and receiver. It can also pass through explosive atmospheres with total safety, as a cable breakage cannot result in sparks.

There are two basic types of fibre. Step index fibre operates as Fig. 19.31a, with reflections occurring at the fibre wall. Graded index fibre has a non-uniform refractive index, causing the light beam to follow a gentler curve as shown on Fig. 19.31b. Graded index fibres have lower losses.

The optical signal is attenuated as it passes down the cable, these losses are usually given in dB/km (typically 5 to 20 dB/km). Further losses occur at curves (the minimum bending radius is usually related to losses rather than mechanical

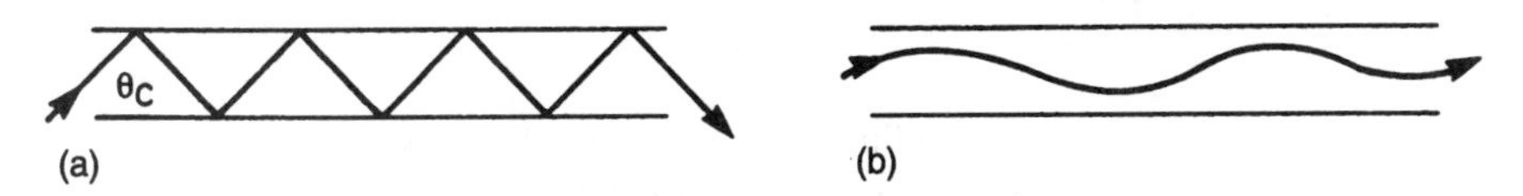

Fig. 19.31 *Fibre optic cable. (a) Step index. (b) Graded index.*

damage) and at the couplings at each end of the cable. A typical link can operate for 1 to 2km without repeaters.

A data transmission cable will usually consist of two fibres (one for each direction) of 200 μm diameter loose inside a protective sheath. Loose sheathing reduces the chance of impact and stretch damage. At each end, the sheathing has to be removed, and protective sleeves put onto the individual fibre optic cores as shown on Fig. 19.32.

The commonest connector is the SMA connector used on Fig. 19.32. This allows fibre optic cables to be disconnected and re-connected like a normal signal cable. Fitting these connectors to the cable is, however, a skilled job. The method with least subsequent signal attenuation uses epoxy resin to hold the core which is then cleaved and made square with a laborious polishing routine. A simpler (and commoner) method uses a crimping tool and specialist cutter.

Cables can be jointed by fusing the cores together with little, or no, resulting transmission loss. This technique, though, requires specialist knowledge and equipment. All jointing methods (resin, crimp or fusing) require a fairly high level of skill. After the joint has been made, the resulting termination must be carefully checked with a microscope viewing a light source sent down the fibre from the other end. Checking the connections on a jointed multicore fibre optic cable can be very time-consuming, as there is usually no core identification.

There are several disadvantages to fibre-optic-based links. The first is the link is strictly point to point. Topologies such as Ts, multidrops or buses can only be achieved by the use of (expensive) repeaters at each node.

Fibre optic cables are also vulnerable to damage. The cable is not only less robust than conventional cable, but it cannot be easily (or quickly) jointed. With normal coaxial cable, a damaged length can be quite readily cut out and a new length spliced in with through connectors and little, or no, ill-effects.

Through connectors are possible with fibre optic cables, but they introduce high losses into the link and may even prevent it from working. It is not unknown

Fig. 19.32 *Fibre optic cable showing the core split block, the sheathing and the commonly-used SMA connector.*

for a new run of cable to be needed, or a repeater introduced, as a result of a single cable break. One break may, of course, need two joints if a length of cable has been damaged and there is no slack. Cables can be jointed using any of the three methods outlined above, but these are all time-consuming and require staff training to achieve an acceptable level of skill.

Fibre optic cable should therefore always be well protected with conduit or robust trunking to minimise damage. Although there is no technical reason why fibre optic links should not share cable tray with 33-kV cables, it is not good practice as they will probably be damaged if any more power cables are added.

A final important point is safety. Most fibre optic links use high-power optical sources, sometimes lasers. Never look down a cable 'to see if the transmitter is working'. If there is any doubt about cable continuity, disconnect the cable at both ends (taking care to ensure the right cable has been disconnected in multicable applications) and use low power incandescent sources for testing. In many cases anyway, the source used for data transmission is outside the visible range, and cannot be seen. It can still, however, cause damage to the eye.

Chapter 20

New ideas: fuzzy logic and neural networks

20.1. Introduction

This chapter examines some new ideas which are appearing in control systems. They do not fit neatly elsewhere; fuzzy logic has no relationship with the digital logic of Chapter 13, and neural networks are not a form of computing, so they have been placed in a chapter of their own.

20.2. Fuzzy logic

20.2.1. Introduction

Suppose we are interested in measuring the temperature in a room. If we were to ask several people at the same time, we would probably get a range of answers such as ‘cool’, ‘warm’, ‘a little on the cold side’, ‘a bit hot’ and so on, with each person having their own view.

A thermostat for the heating system, however, would only have a single crisp view, the room is either ‘hot’ (turn the heating off) or ‘cold’ (turn the heating on). Figure 20.1 compares the difference between a human approach of using vague terms and a crisp binary representation of the same facts.

Fuzzy logic attempts to build a control system which works by classifying and dealing with data in a similar way to human beings.

Fuzzy logic categorises data by probabilities. Suppose we did a study of people’s view of temperature. We might find that below 10 °C everybody said it was cold. At 20 °C there might be a 50/50 split between cold and warm and above 30 °C everyone might say it was warm. We could plot this as probability graphs as Fig. 20.2. With this arrangement a temperature of 25 °C could be described as being both ‘slightly cold’ (probability 0.25) and ‘quite warm’ (probability 0.75).

20.2.2. Data classification

In Fig. 20.2 we classified data into two groups, warm and cold. If we were to study people’s ages we would probably find five defined groups: children, youths, young-adult, middle-aged and old as one possible set of terms. If we attempted a crisp definition we would probably arrive at something similar to Fig. 20.3. This leads to anomalies such as at midnight just before your fortieth birthday you suddenly switch from young-adult to middle-aged. This does not really correspond with reality, we all know jet-skiing seventy-year olds and people in their thirties who behave like elderly retired pensioners.

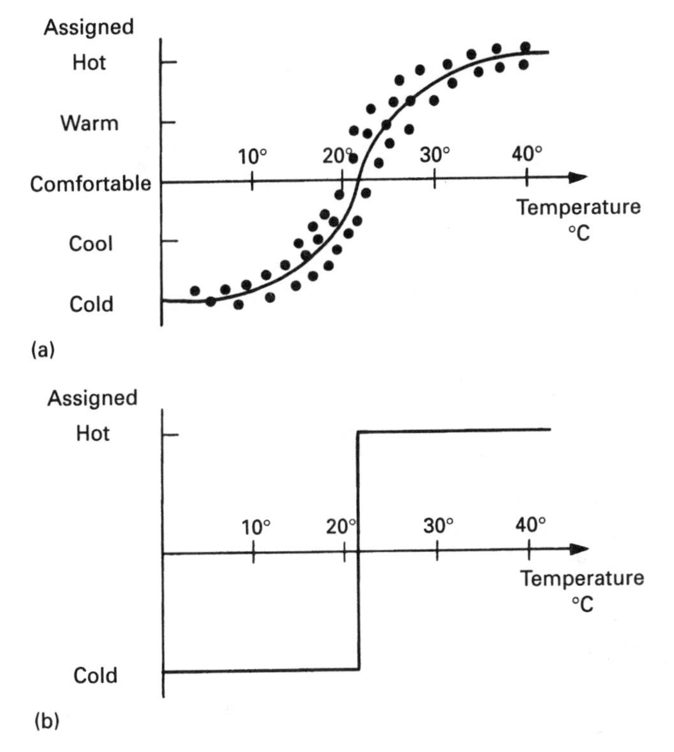

Fig. 20.1 *The difference between fuzzy and crisp assignments. (a) People's view, each '•' is a person. (b) Thermostat's view.*

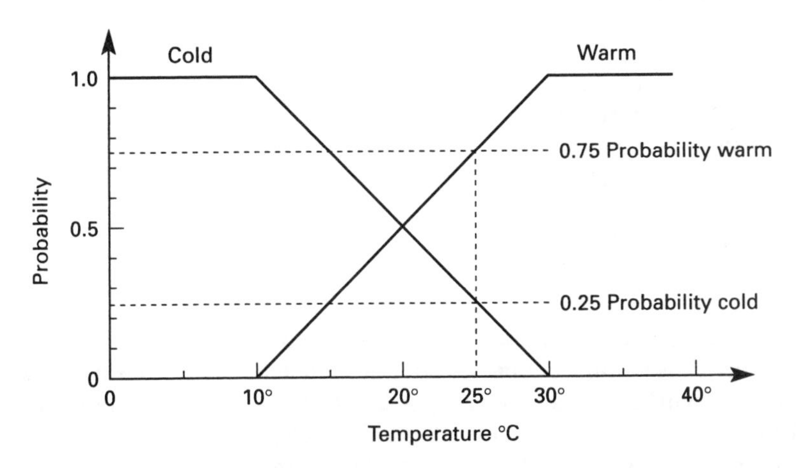

Fig. 20.2 *Temperature assigned as probabilities.*

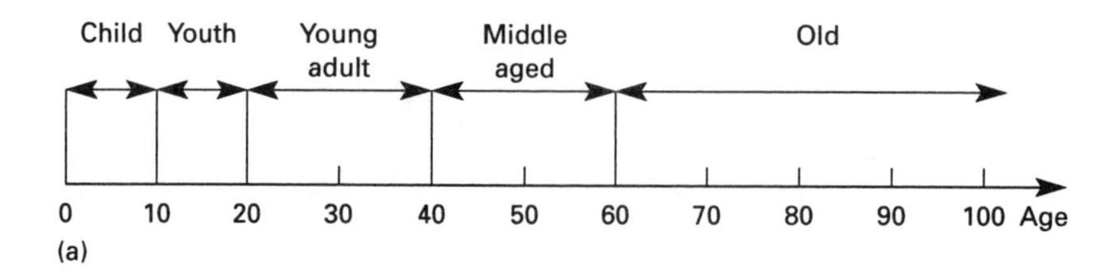

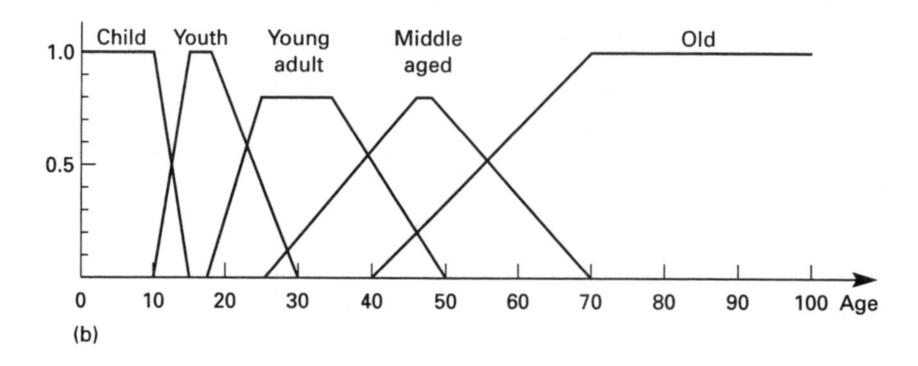

Fig. 20.3 *Assignment of age into five groups. (a) Crisp assignment of age. (b) Probability assignment of age.*

A more realistic arrangement would be to again classify sets as Fig. 20.3b where probabilities are defined for each age group. Someone at age twenty-seven could be described 'a rather old youth' (probability 0.05), 'young-adult' (probability 0.85) or 'starting to approach middle age' (probability 0.1).

The first stage in fuzzy logic is to consider how data is classified. One of the leaders in fuzzy logic is Omron, who use seven classification groups:

PL Positive Large
PM Positive Medium
PS Positive Small
ZR Approximately Zero
NS Negative Small
NM Negative Medium
NL Negative Large

Not all of these have to be used in a control scheme.

We must classify our data into groups. The first step is to consider how the probabilities will be spread. So far in Figs. 20.1 to 20.3 we have used straight lines. Statistics tend to use the three distribution curves of Fig. 20.4: triangular, bell (Gaussian) and trapezoid. Although the Gaussian is a more realistic curve, triangular and trapezoid have tended to be more commonly used.

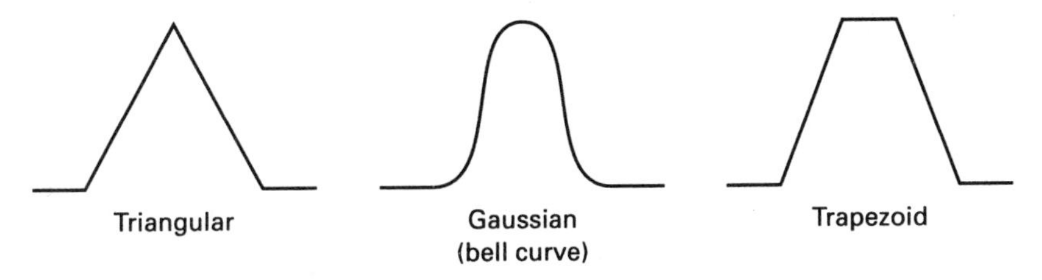

Fig. 20.4 *Distribution curves.*

We are going to design an automatic braking system for motor cars which would reduce multiple vehicle pile-ups caused by 'tail-gating'. This example is based on material from Omron's *Fuzzy Logic Guide Book.*

We need two analog signals: vehicle speed and a measure of the distance from the vehicle in front. A fuzzy logic control system will process these giving a single output which controls the brakes. Figure 20.5 is a block diagram of our control system.

We will first classify our two inputs using four of the seven possible groupings: approximately zero (ZR) and the three positive classes (PS–PL).

We could start by saying that a vehicle has nearly zero speed from stationary to zero to 20 km/h (about 12 mph), could be driving slowly between zero and 40 km/h and so on. The four classifications are shown on Fig. 20.6a–d and combined (which is the usual representation) on Fig. 20.6e.

We next classify the distance from the car in front. This is shown in combined form in Fig. 20.7. Comment should, perhaps, be made at this stage that a real automatic braking system would have to be more complex as the definition of distance grouping will be speed dependent (or, more accurately, time dependent between you and the car in front passing the same point. Fuzzy logic can achieve this, but it would complicate the description at present.)

Our control system will have a single output: the braking force. This also needs to be placed into fuzzy sets. Figure 20.8 uses three groups, PS, PM and PL.

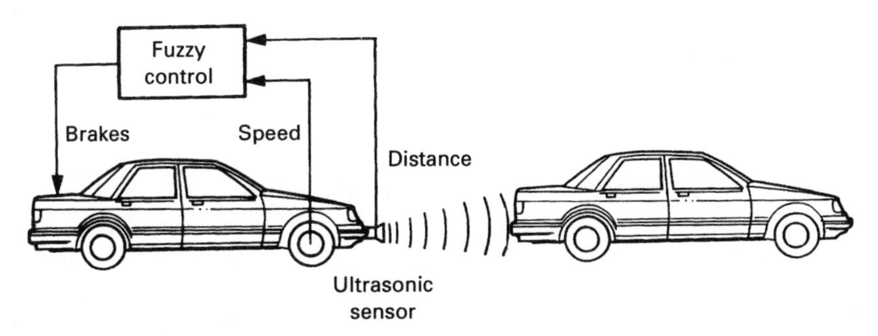

Fig. 20.5 *Braking control system.*

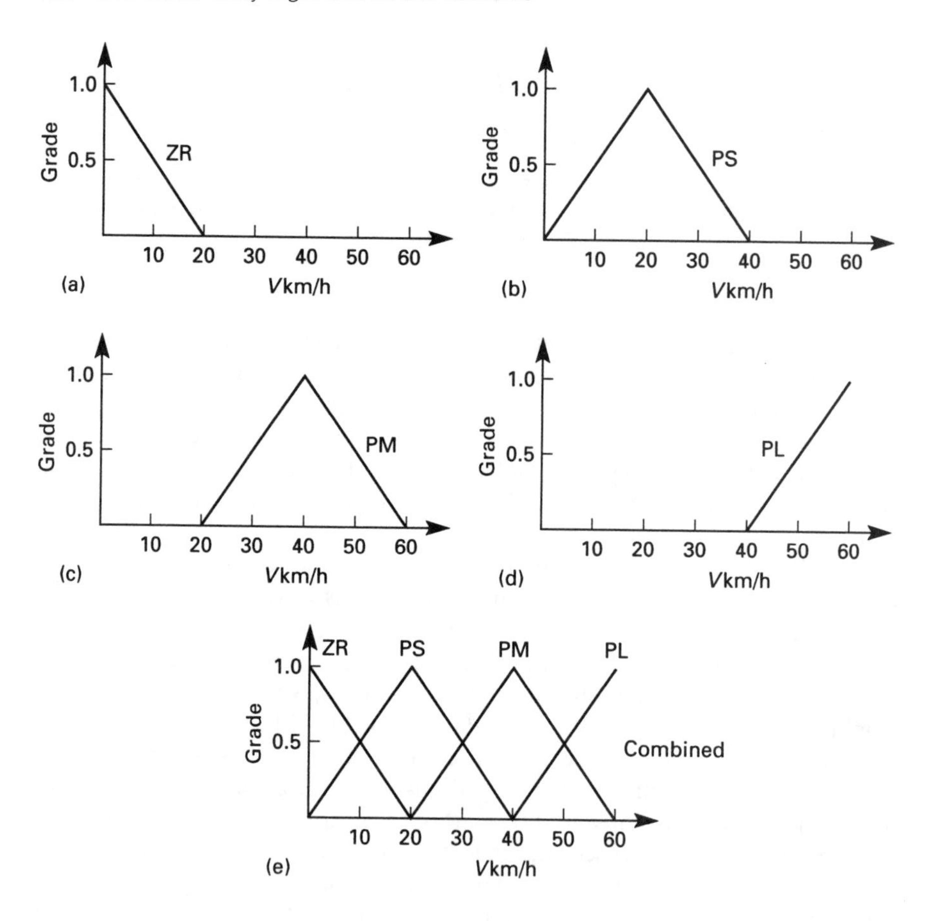

Fig. 20.6 *Representations of vehicle speed.*

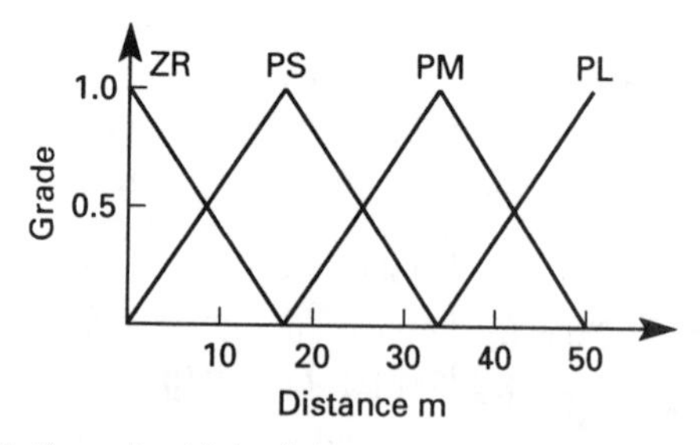

Fig. 20.7 *Representation of vehicle distance.*

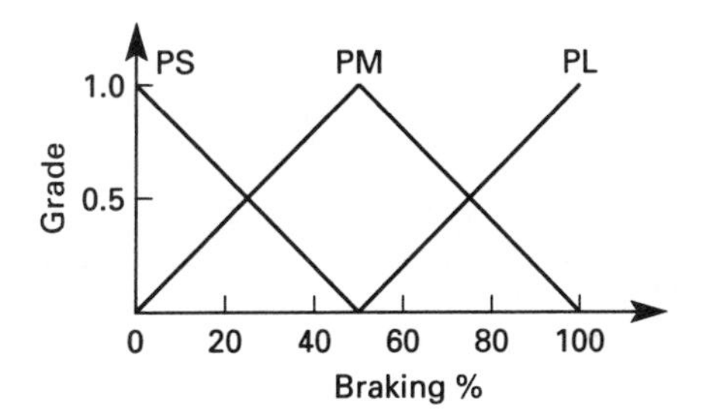

Fig. 20.8 *Representation of braking force.*

20.2.3. Rule definition

If we had to describe a braking strategy in English, we would probably come up with something like:

Rule 1

If the distance between the cars is small and the car speed is rather high then brake hard.

Rule 2

If the distance between the cars is moderately long and the car speed is high then brake moderately hard.

If we define our signals

Speed: V
Distance: D
Braking force: B

we can write these rules using an IF/THEN construct:

Rule 1: IF D = PS AND V = PM THEN B = PL
Rule 2: IF D = PM AND V = PL THEN B = PM

These two rules can be shown diagrammatically on Fig. 20.9 (for Rule 1) and Fig. 20.10 (for Rule 2).

20.2.4. Rule evaluation

To demonstrate how these rules are applied, let us assume we are travelling in a queue at 55 km/h and we are 27 m from the car in front.

First we look at the values of each term in the rules. These are known in the jargon as antecedent blocks. We talked about probabilities earlier; fuzzy logic uses the term 'grade' but the meaning is the same. From Rule 1 the grade of V = PM is 0.25, and the grade of D = PS is 0.38. This is shown on Fig. 20.9 along with the mapping onto B = PL. The results from a rule are known as the consequent blocks.

From Rule 2, shown on Fig. 20.10, the grade of V = PL is 0.75 and the grade of D = PM is 0.62.

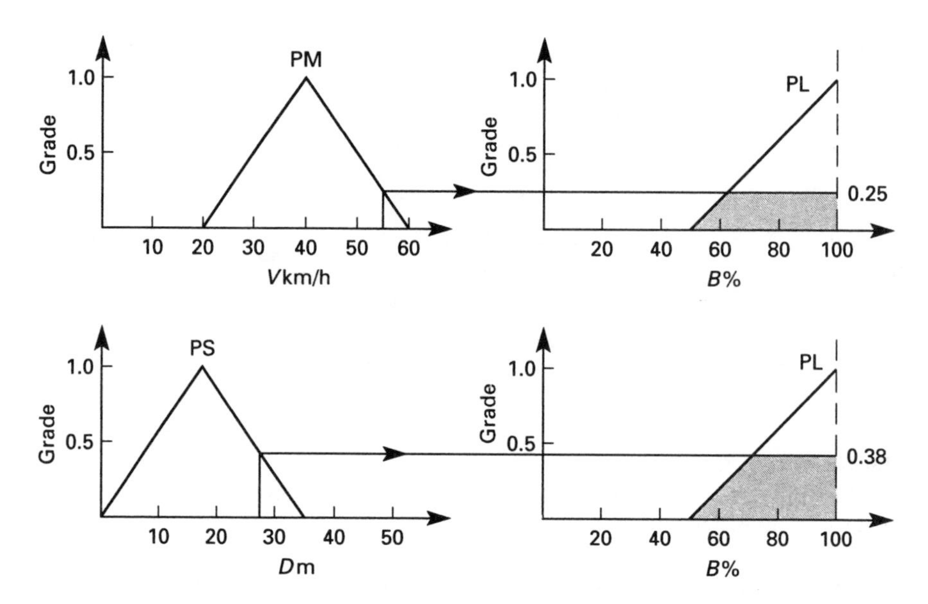

Fig. 20.9 *Rule 1: If D = PS and V = PM then B = PL for V = 55, D = 27.*

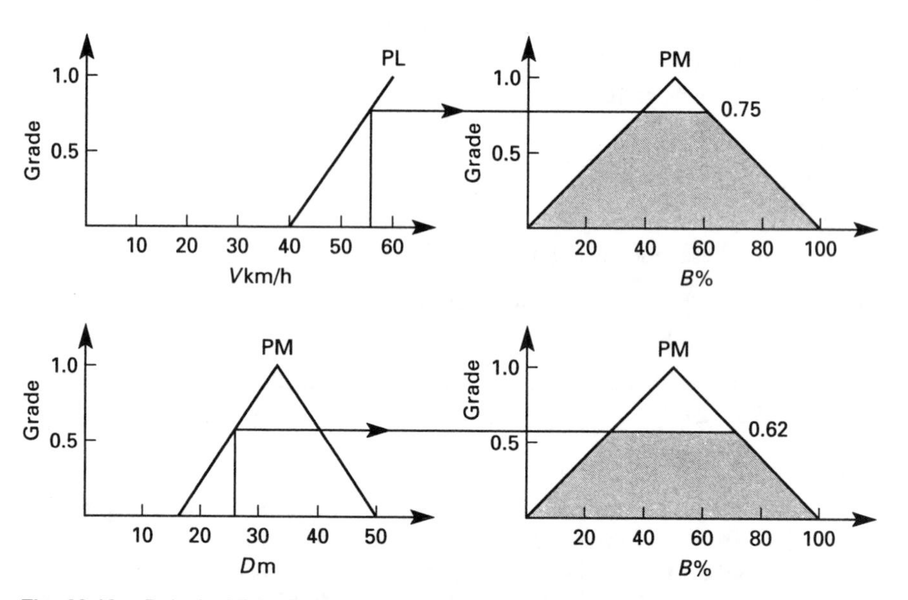

Fig. 20.10 *Rule 2: If D = PM and V = PL then B = PM for V = 55, D = 27.*

We now have four values for B; 0.25 and 0.38 from Rule 1 plus 0.75 and 0.62 from Rule 2. The next stage is to go from these fuzzy set values to a single value which can be sent to the brakes. This is called *de-fuzzification*.

The first stage is to take a single value from each rule. If the rule uses AND (as our two rules do) the *lowest* value is taken. If the rule uses OR the *highest* value is taken. Precedence defines how combinations of AND/OR/() are interpreted.

The output of Rule 1 is thus 0.25 and the output of Rule 2 is 0.62. as shown on Fig. 20.11. The output areas A_1, A_2 below these minimum values are then combined. The output to the real world is then the centre of gravity of the combined areas.

The evaluation process, summarized on Fig. 20.12, operates continuously, working on the input speed and distance data with fuzzy logic to produce a safe braking force.

20.2.5. A real application

Fuzzy logic works well in applications where it is difficult to mimic human intuition and experience with conventional analog or digital control. Successful applications are as diverse as wheel slip minimization for railway locomotives and air conditioning control.

The application briefly described below is designed to reduce the swinging of a load suspended from an overhead bridge crane. It is again based on material from Omron's *Fuzzy Logic Guide Book*. As the crane accelerates along the shop, the suspended load will swing, and the swing will be reinforced and magnified when the crane stops. Part of the skill of a crane driver is controlling the acceleration and deceleration of the crane such that the swing does not become dangerous. Talk to a crane driver and you will be given rules like, 'well if the swing gets a bit large you steer the crane into the swing a bit' – typical fuzzy logic descriptions.

Figure 20.13 shows the crane we are controlling and a block diagram of the control system. We will deal only with long travel, with a real system two almost identical systems will be required, one for long travel and one for cross travel.

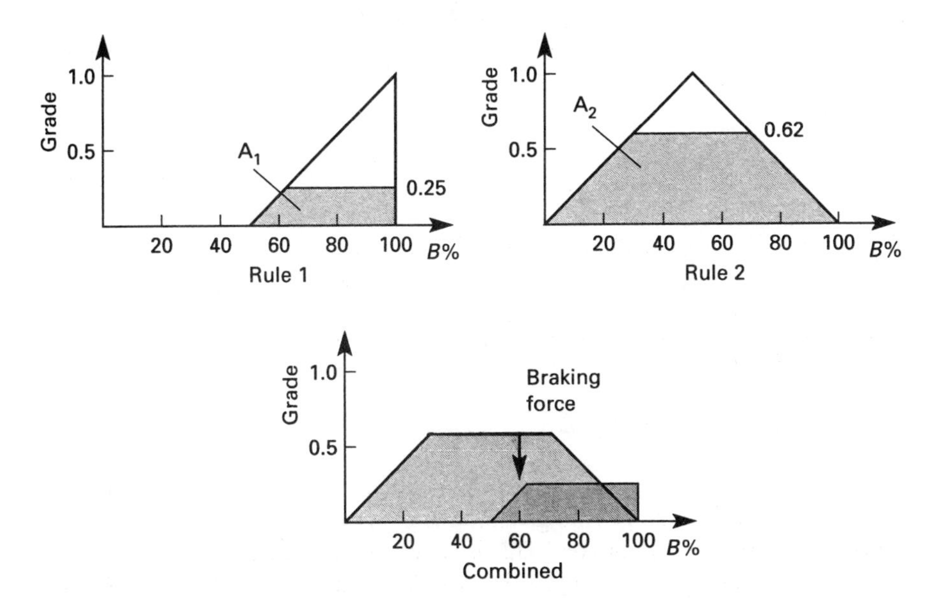

Fig. 20.11 *The de-fuzzification process.*

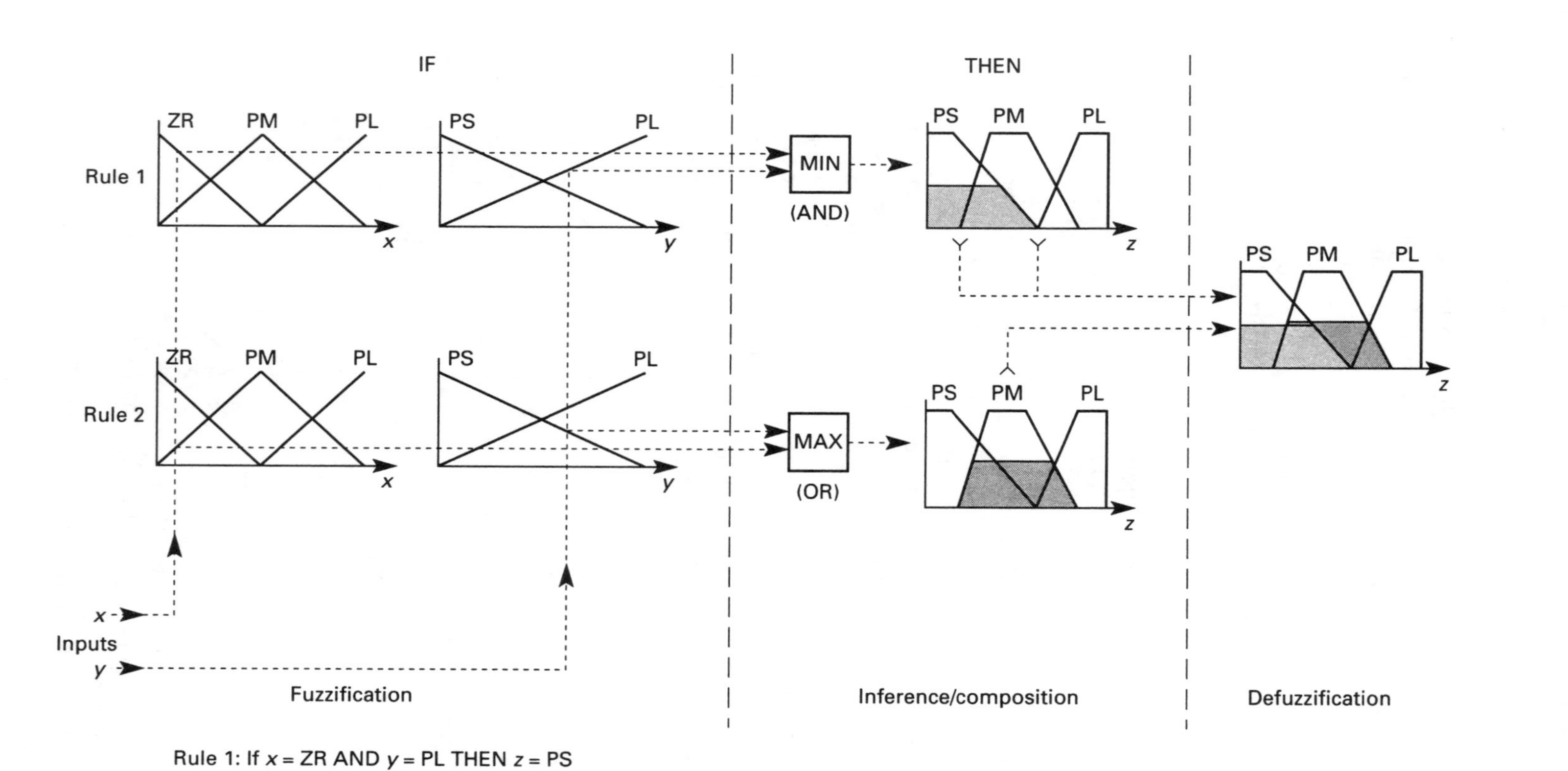

Fig. 20.12 *The fuzzy logic process.*

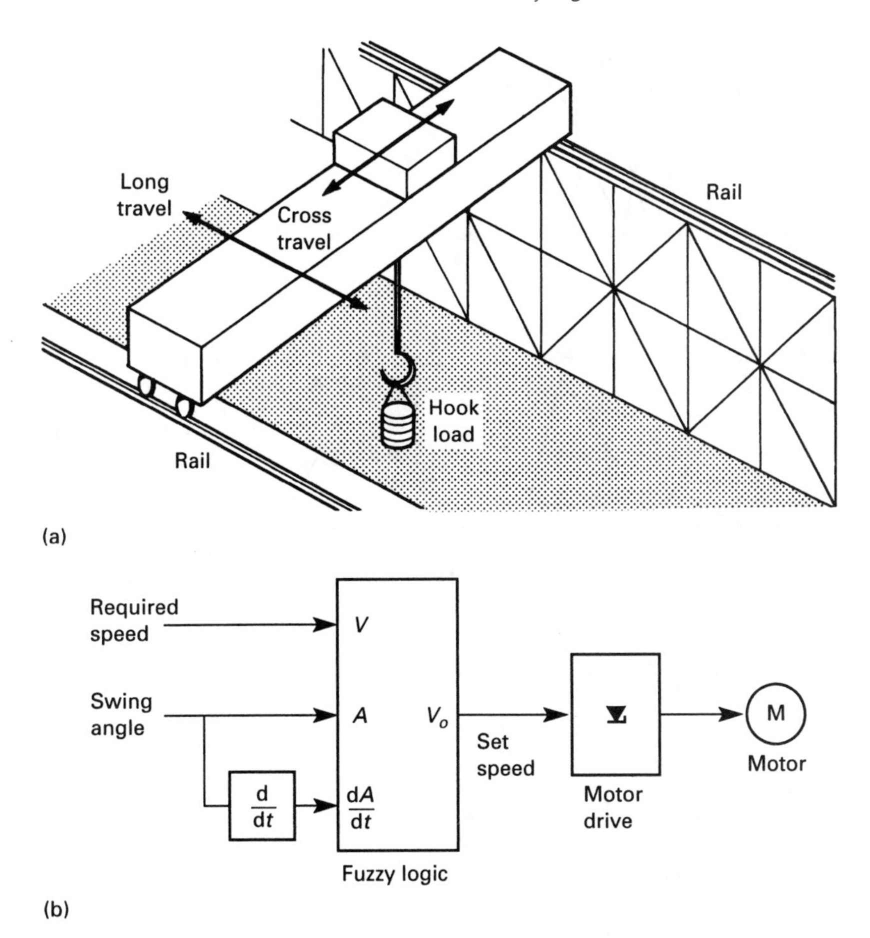

Fig. 20.13 *Fuzzy logic control of a bridge crane. (a) An overhead bridge crane. (b) Control schematic for one motion.*

For each axis we need two analog input signals: one for the desired speed V, and one for the swing angle A of the load (sensed by a rotational potentiometer moved by the load rope). We derive a third analog signal, the rate of change of swing, $\mathrm{d}A/\mathrm{d}t$, by performing a digital differentiation of A. If A is sampled at time interval T, then

$$\frac{\mathrm{d}A}{\mathrm{d}t} = \frac{A_n - A_{n-1}}{T} \qquad (20.1)$$

where A_n is the current angle and A_{n-1} is the last sample T seconds ago. The subject of digital algorithms is discussed in Section 18.7.2.

We have one output signal; the actual motor speed V_o sent to the long travel VF drive control.

Our three input signals and our output are put into groups as Fig. 20.14. We

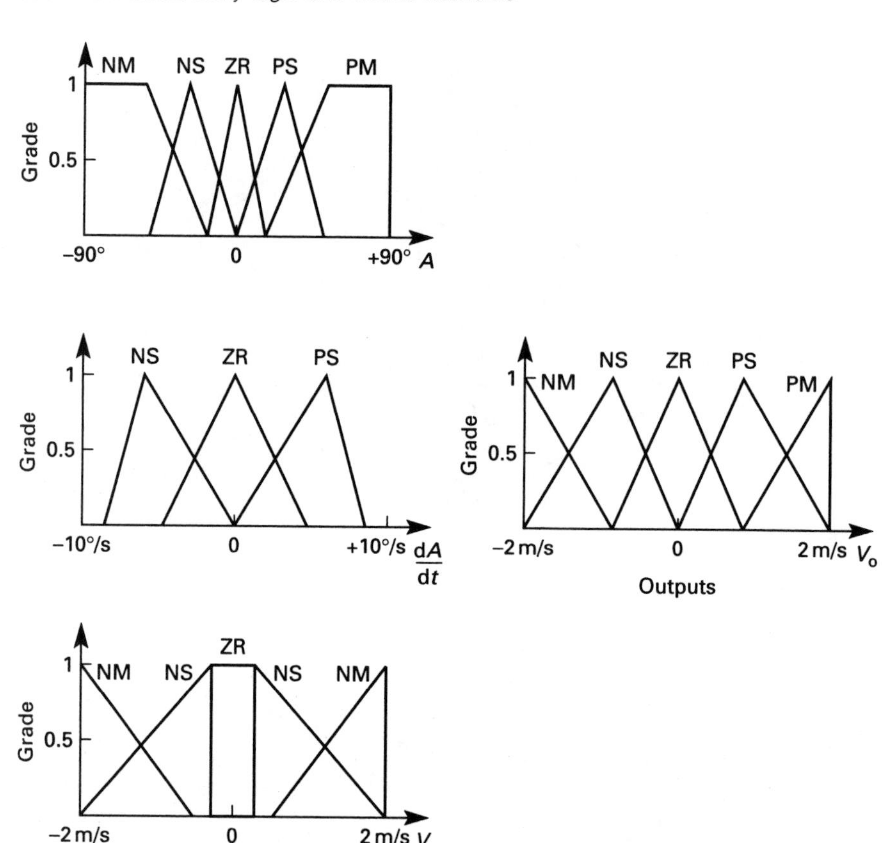

Fig. 20.14 *Grade allocation for bridge crane control.*

can now write the rules. Based on experience from human crane drivers we write rules such as:

'IF I want to stop (V = nearly zero) AND the swing angle is a little negative AND the swing velocity is positive (i.e. the swing angle is decreasing) THEN move the crane into the swing a little.'

This rule becomes

$$\text{IF } V = \text{ZR AND } A = \text{NS AND } \frac{dA}{dt} = \text{PS THEN } V_o = \text{NS}$$

which is rule (4) below.

The full set of deceleration rules (V = ZR) is:

(1) IF V = ZR AND A = NM THEN V_o = NM

(2) IF V = ZR AND A = NS AND $\frac{dA}{dt}$ = NS THEN V_o = ZR

(3) IF $V = \mathrm{ZR}$ AND $A = \mathrm{NS}$ AND $\frac{\mathrm{d}A}{\mathrm{d}t} = \mathrm{ZR}$ THEN $V_o = \mathrm{NS}$

(4) IF $V = \mathrm{ZR}$ AND $A = \mathrm{NS}$ AND $\frac{\mathrm{d}A}{\mathrm{d}t} = \mathrm{PS}$ THEN $V_o = \mathrm{NS}$

(5) IF $V = \mathrm{ZR}$ AND $A = \mathrm{ZR}$ AND $\frac{\mathrm{d}A}{\mathrm{d}t} = \mathrm{NS}$ THEN $V_o = \mathrm{PS}$

(6) IF $V = \mathrm{ZR}$ AND $A = \mathrm{ZR}$ AND $\frac{\mathrm{d}A}{\mathrm{d}t} = \mathrm{ZR}$ THEN $V_o = \mathrm{ZR}$

(7) IF $V = \mathrm{ZR}$ AND $A = \mathrm{ZR}$ AND $\frac{\mathrm{d}A}{\mathrm{d}t} = \mathrm{PS}$ THEN $V_o = \mathrm{NS}$

(8) IF $V = \mathrm{ZR}$ AND $A = \mathrm{PS}$ AND $\frac{\mathrm{d}A}{\mathrm{d}t} = \mathrm{NS}$ THEN $V_o = \mathrm{PM}$

A similar set of rules is derived for acceleration.

20.3. Neural networks

20.3.1. Introduction

Fuzzy logic and computer expert systems (which also use the

```
IF <conditions> THEN <result or deduction>
```

construct) aim to go from the crisp binary decisions of conventional control towards the woolly way in which humans think. Neural networks go further by aiming to build models (albeit much simplified) of the way a human being thinks and reaches conclusions. One of the very intriguing consequences of this approach is that you can end up with a control system which works, but the designer doesn't know precisely how it does it!

20.3.2. The biological neuron

The computers described in Chapter 14 are all based on the Von Neumann architecture with a store, input, output and processing units. The operation is essentially serial; instructions are brought from the store and obeyed one at a time. Vast increases in speed have been made, and techniques such as parallel processing can improve performance further, but there is still little difference, in operating principle at least, between the latest PC and the first computers from the early 1950s.

The human brain operates on a totally different principle. It is based on small individual analog processing units called 'neurons'. There are an amazing 10^{11} of these in the average brain, and each connects to 10^4 others. It is hard to think of numbers like 10^{11}, but it approximates to the number of drops of water in several Olympic-sized swimming pools. All of these neurons operate relatively slowly (at about 100 Hz) but there is vast computing power because they are all operating in parallel.

Figure 20.15 shows the essential features of a biological neuron. It consists of a

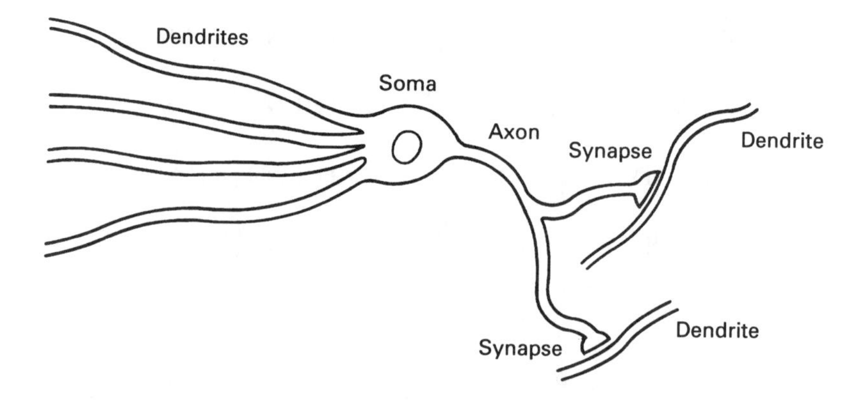

Fig. 20.15 *A biological neuron.*

body, called the soma, and receives signals from other neurons via the dendrites. The signals received can be permissive or inhibitory. The signals arriving at the neuron are summed, and if the result is greater than a threshold value the neuron fires.

The output from the neuron appears on axons. These terminate as synapses which link the output and a dendrite input to another neuron. There is no actual connection between the synapse and the dendrite, the data transfer takes place by the release of chemicals called neurotransmitters.

Figure 20.16 is a simplified model of this operation. This is a gross simplification, it should be remembered that each neuron links to 10^4 others! Input signals x_1, x_2, x_3 are multiplied by weights w_1, w_2, w_3 and summed. The weights can be positive (permissive) or negative (inhibitory). If the sum is greater than the threshold, the neuron will fire and give a signal on its output. It is thought that learning occurs by varying the weights and thresholds.

20.3.3. Artificial neural networks (ANNs)

The first attempts to build an operational model of the neuron used the simple binary comparator (on/off output) of Fig. 20.16. This is known as a binary decision neuron. The standard neuron model at present, known as the perceptron, uses a threshold of zero, and a sigmoid transfer function as Fig. 20.17. The threshold can be changed by applying the required threshold as a fixed negative signal with a weight of one to an input. Figure 20.18 shows several ways that a neural network can be built to simulate an exclusive OR (XOR) gate operation. It is interesting to observe that none of these are 'right' or 'definitive': one of the characteristics of ANNs.

An ANN is built up in layers, the most common being the three-layer structure of which Fig. 20.19 is typical. These consist of an input layer, a hidden layer and an output layer. Information in these simple networks flows in one direction only from input to output.

An alternative, but less common, arrangement is the fully connected Hopfield

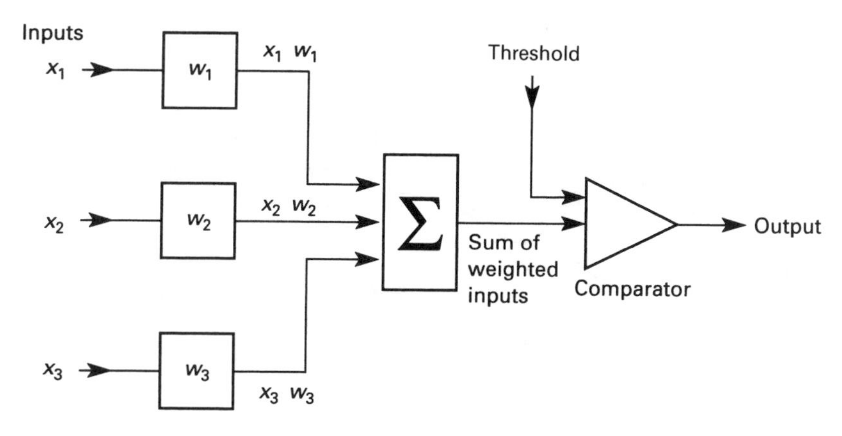

Fig. 20.16 *Simplified neuron model.*

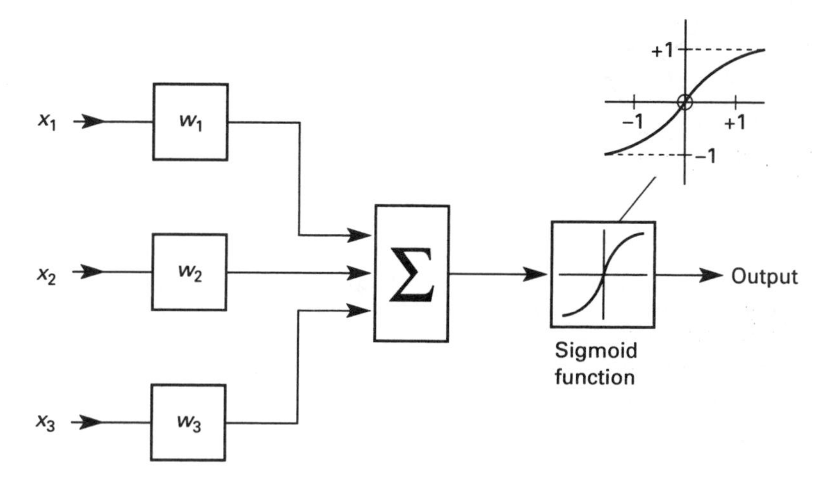

Fig. 20.17 *The perceptron neuron model.*

net of Fig. 20.20. This has no real input (or output for that matter!). Data is set into the net as an input pattern. This causes transitory unstable states to run round the net. It can be shown that the net will always settle into a final stable state which is the output corresponding to the input pattern.

Human intelligence is acquired through education and learning. ANNs also have to be trained. This is achieved by adjusting the weights and thresholds of the individual neurons in the network.

The operation can be summarised by Fig. 20.21. The network is built and a (possibly large) training set of trial data (typical inputs and correct outputs) collected. The data is presented to the network and the weights and thresholds adjusted so the correct answers are given. A training set for a credit rating system, for example, could be the personal data (salary, marital status, age, etc.) for past

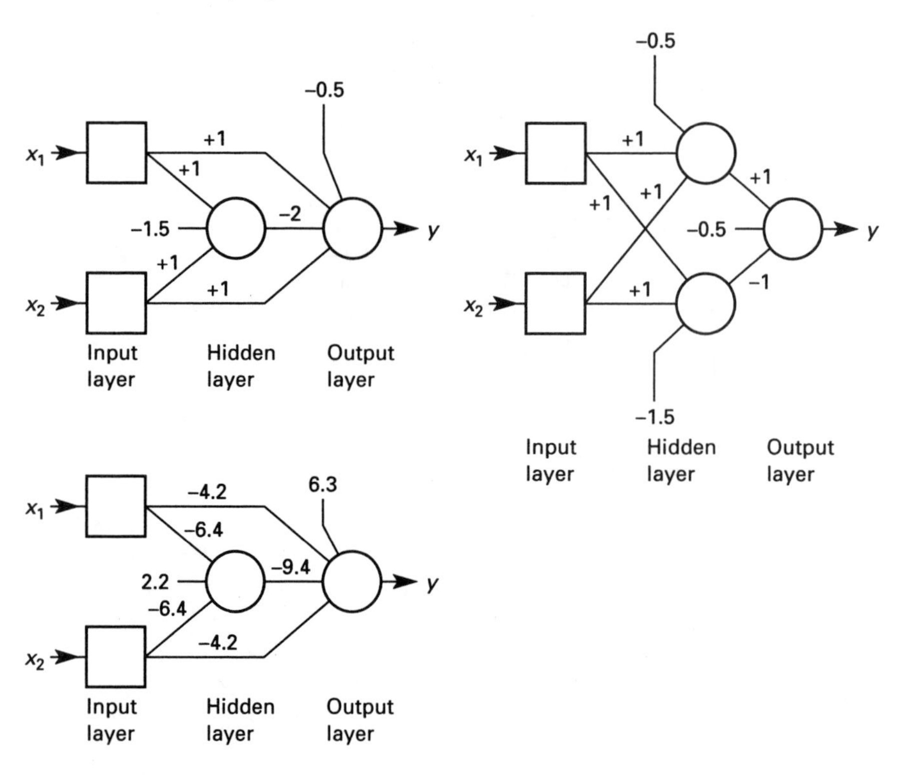

Fig. 20.18 *Three neural network XOR gates. The weights are shown on the lines and the threshold as an unconnected input. None of these are definitive or 'optimum' solutions.*

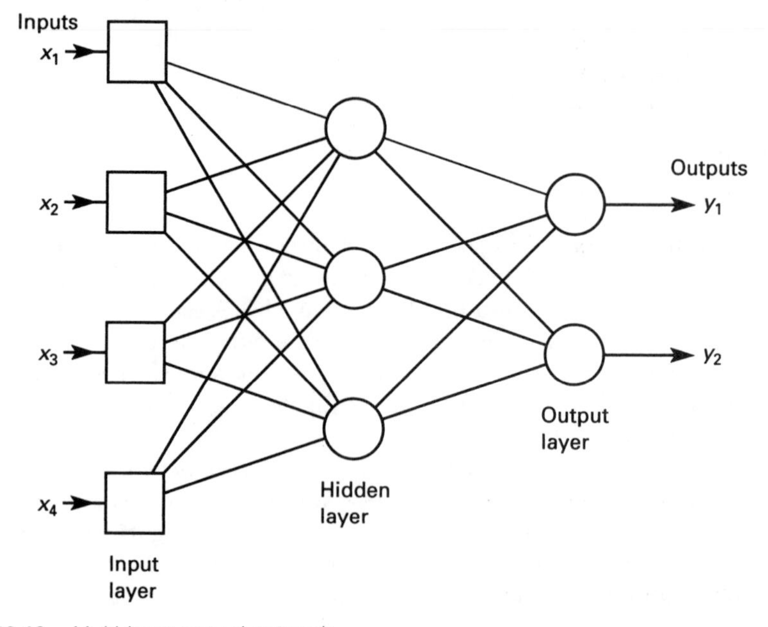

Fig. 20.19 *Multi-layer neural network.*

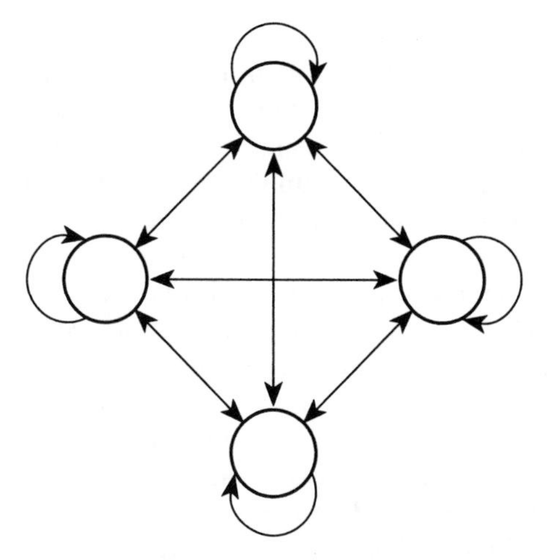

Fig. 20.20 *Fully connected neural network–the Hopfield net.*

loan applications and the result of those applications (accept, reject, refer to higher authority).

Figure 20.21 identifies what is commonly overlooked in ANNs; building the network is the easy part, the training is much more complex and many training algorithms have been developed. We will return to these in Section 20.3.5.

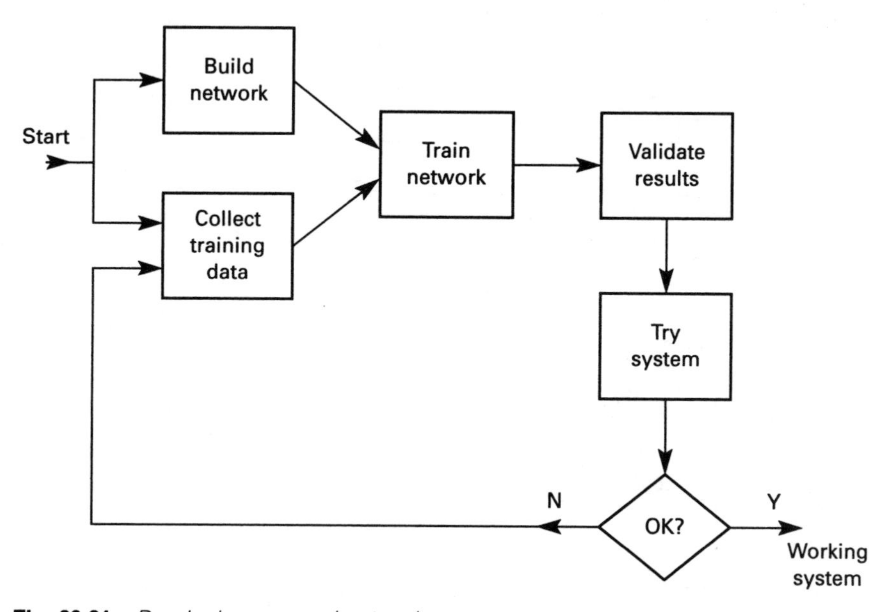

Fig. 20.21 *Developing a neural network.*

20.3.4. Data presentation

The figures have so far implied that a neuron operates rather like a logic gate and works with digital On/Off signals. Many signals will be of this type, but often other types of data will be needed as an input or output.

Neural networks are commonly found in classification systems (they are rather good at identifying groups.) These need input data in the form of '1 of N' as Fig. 20.22a.

Numeric data causes problems. The speed with which an ANN can be trained depends largely on the way in which the data is presented. Suppose, for example, you wanted to train an ANN to evaluate the area of a circle by presenting it with data for the radius and the resultant area. To do this it would have to deduce that the radius depends on r^2 and pi. If the data was presented in the form radius squared and resultant area, the network, and training, would be a lot simpler.

The obvious way to present data is pure binary, thirteen being presented as four signals 1101. Whilst efficient in input usage this can present problems as there is no real correlation between adjacent numbers such as 10111 and 11000. The Gray code (see Section 13.6.4) is better. If the range of a number is small, the thermometer representation of Fig. 20.22b has benefits as the data is in a form

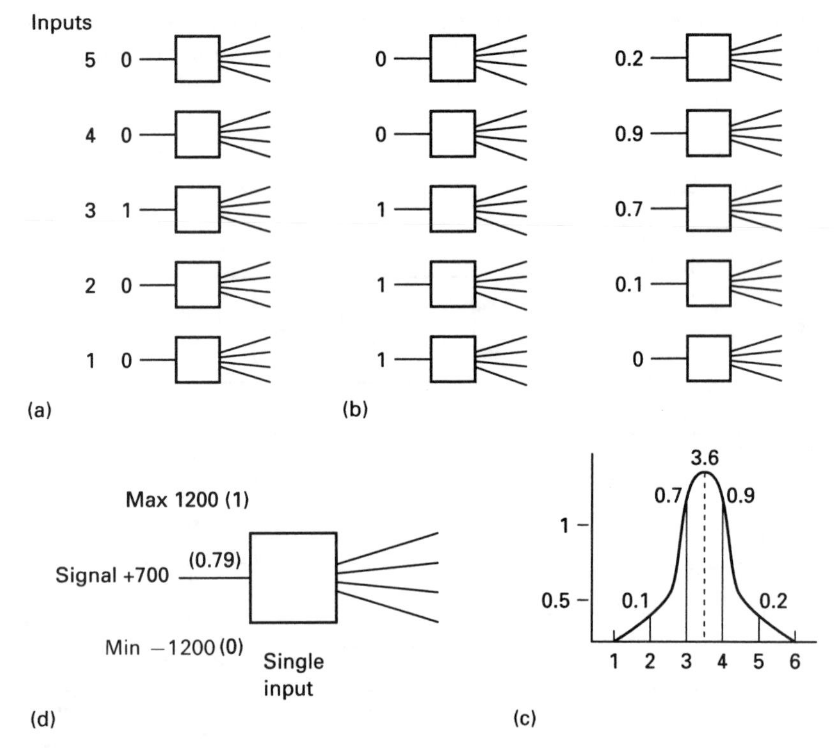

Fig. 20.22 *Dealing with numbers in a neural network. (a) One of N. (b) Thermometer. (c) Moving hump (shown for 3.6). (d) Normalised input.*

that an ANN can easily deal with. It is, however, very prodigal in its use on inputs.

Inputs to an ANN can have any value between 0 and 1 (they are not just binary logic on/off signals, 0.3726 is a perfectly valid ANN input), and the sigmoid transfer function permits numeric values to propagate through the network. A more efficient, and widely used, method presents data between 0 and 1 to several inputs by using a moving Gaussian hump centred on the data value as Fig. 20.22c. Input signals are assigned at regular intervals over the range of interest, and their values (between 0 and 1) read off the Gaussian distribution as shown.

Numeric data can also be scaled to lie within the range 0 to 1 as Fig. 20.22d, although this is sometimes less successful than 1 of N, thermometer or moving hump method.

20.3.5. Training algorithms

The whole 'knack' of using ANNs is the training, and much research is being done in this area. The commonest method in real (as opposed to research) applications is the Error Back Propagation algorithm (familiarly known as BackProp). This uses a steepest descent search to adjust weights and thresholds in order to minimise the error between real ANN output and desired output for each training set. For each set, the algorithm looks at the error and works back from the output (hence the name) finding the direction and size of the steepest downward slope at each layer. It then moves the weights by a small amount in that direction. Part of the trick is deciding how 'a small amount' is determined. The operation is summarised by Fig. 20.23. One often overlooked aspect of training is that there is often no totally correct solution; terms such as 'best' or 'acceptable' should be applied.

One of the problems can be false minima, where the algorithm lands in a 'bowl'

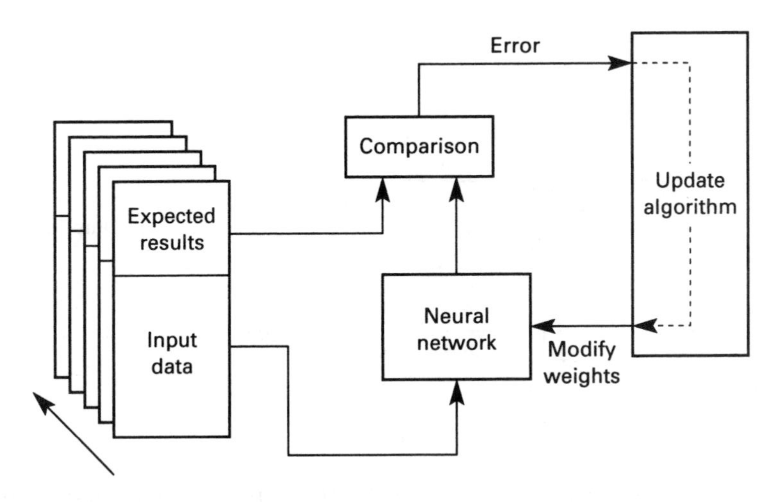

Fig. 20.23 *Training a neural network.*

where all the slopes are up, but the bowl is not at the minimum error value. Simple examples such as the incorrect XOR gate of Fig. 20.24 are common. Oddly enough noisy training sets help to kick an ANN out of false minima.

The training works on training sets (inputs and corresponding outputs) provided by the designer. These data sets must be typical of the real data to be used, and cover the range expected. The data must also be random and have no in-built bias. An ANN cannot be expected to extrapolate outside the range of the training data. There is a (possibly apocryphal) story of a military ANN which was designed to identify the presence of camouflaged vehicles hidden under trees. ANNs are very good at dealing with visual identification, so the ANN was presented with many photographs taken from high flying aircraft along with the required result (present/absent). The ANN was quickly trained, and presented with trial photographs from the same series as the training set (but not used for training). The ANN correctly identified the presence or absence of vehicles. When, however it was presented with data from elsewhere it failed miserably.

What had happened was that all the training set photographs with vehicles had been taken in the morning, and all the photographs without vehicles had been taken in the afternoon. The ANN had latched onto shadow lengths and direction, and was evaluating time of day, not the presence of hidden vehicles. If the data is not random, ANNs can deliver surprises like this.

BackProp is an example of supervised learning, and roughly approximates to how we learn with a teacher at school. Reinforced learning uses a penalty/reward approach (and is often described as training with a critic rather than a teacher). Totally unsupervised networks can also exist, these look for group patterns and find a common thread. Kohonen networks are an example of unsupervised learning and more closely mimic the way the brain is organised. Most of the research effort in ANNs is investigating better ways of training.

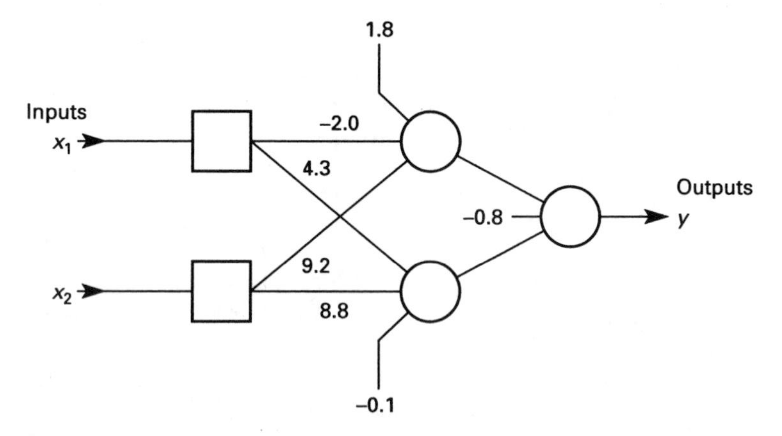

Fig. 20.24 *A non-working XOR neural network which lies at a false minima. A simple downhill slope algorithm could not correct this and would remain trapped.*

20.3.6. Hardware or software

Until quite recently ANNs were usually implemented in computer software. This rather reneges on the statements made about the Von Neumann architecture in Section 20.3.2! Hardware implementations in the form of RAM-based neural networks and content addressable memories have become available which will allow purely hardware-based ANNs to be built. These will still be trained by computer programs.

20.3.7. Applications

Neural networks are useful at solving problems where the algorithm used to achieve a solution, or the solution itself, is difficult to define. An ANN would not be a good approach where a crisp clear mathematical algorithm can be written down on paper. Here a conventional control scheme is probably easier and cheaper to implement.

Neural networks are more robust than conventional systems. Like the brain they can continue to work with failures and tolerate bad, incomplete or noisy data. This ability to plod on regardless can be advantageous in some applications.

One major disadvantage is that training is required, and the amount of training data can be large. The training can also be time-consuming. The result should be a working system, but many people feel uneasy about not knowing how it works or reaches its conclusions. In the (totally distinct) area of expert systems much research is being done into methods of providing 'explanation' facilities. (I am doing < .. > because of < .. > type of output.) Such ideas may be developed for ANNs.

A common application area is pattern recognition from photographs and video cameras. At the steel plant where the author works an ANN-based video system reads alphanumeric characters for material tracking. The characters are sprayed onto a steel surface and can be read at any angle, even if the spraying is poor and incomplete. Similar ANN schemes are used for fault detection in fabric and paper manufacturing ANNs have also appeared in commerce, forecasting share and currency movements and, more mundanely, dealing with loan and insurance applications.

The ability of ANNs to deal with poor, noisy and incomplete data also has led to them being used in seismic analysis, military radar recognition and medical diagnosis.

20.4 General comments

Both fuzzy logic and neural networks are not replacements for conventional control systems. They are additional tools which can be added to, and used alongside, more traditional methods.

As both aim to mimic the operation of the human brain to some extent, an obvious question is 'are we building artificial intelligence?' There is a long way to go is the simple answer. The human brain has, it will be remembered, some 10^{11} neurons, each with 10^4 connections. ANNs built today have of the order of 10^2

perceptrons, each with at most 10 connections. The perceptron operation is also much simplified compared with a neuron. At present we are adrift by a factor of about 10^8, but the author would not like to bet on where we will be in a hundred years' time.

At the time of writing (late 1994) the British Department of Trade and Industry have recognised the importance of neural networks and are running an awareness programme.

Chapter 21

Maintenance, fault finding and safety

21.1. Introduction

It is the duty of the design engineer to ensure that in any new system:

(a) At least one item is obsolete.
(b) At least one item is on twelve months delivery.
(c) At least one item is experimental.
(d) The drawings arrive six months after the equipment and do not include commissioning modifications.
(e) The instruction manuals are written in a confusing mixture of the banal and the impossibly complex which assumes that the user is a moron with an engineering doctorate.

21.2. System failures

21.2.1. Introduction

Equipment inevitably fails, and there is no such thing as a totally reliable system. It is not possible to predict when an item will fail; it is not even possible to say, with certainty, that an item will not fail in the next 30 seconds. Discussions of reliability are consequently based on statistical analysis, rather than predictions for one specific piece of equipment. In a similar way, one can talk with great accuracy of the life expectancy of a human being, but this cannot be used to predict the lifespan of one person. It is similarly possible to say, in general terms, what are the effects of smoking, alcohol, lack of exercise, overeating, etc., but everyone knows someone who indulged in all the vices and lived to a ripe old age. In industrial terms, it is not feasible to say that this system, operated under these conditions, will run for 1000 hours, plus or minus 5 hours, without failure. It is, however, realistic to say that if a very large number of systems are studied, the average lifetime will be 1000 hours.

Production management often expresses the wish for fault free systems. This is unachievable, but by using techniques such as redundancy (described later) any desired level of reliability can be achieved. High reliability may not, surprisingly, be what is really required.

Low reliability is achieved at low cost, but brings additional costs in terms of repair effort and lost production. As reliability increases, production losses decrease, and maintenance costs increase to a point where reliability improvements require more staff to 'stand guard' and continuously check the plant. High reliability techniques such as redundancy are also expensive to install. The overall reliability cost curve thus has a shape similar to Fig. 21.1, with the optimum operating point being the minimum overall cost. Identifying this point is the art of plant maintenance.

Most plants are, to some degree, failure tolerant and tend to operate in some form of failure mode for most of the time. Good plant design considers the effect of failures and the ability of the plant to continue operating safely and reasonably economically for a period of time whilst a fault is identified and rectified.

21.2.2. Reliability

Because it is not realistic to predict when one item will fail, statistical methods are used to discuss reliability. The reliability of an item or system is the probability that it will perform correctly under the defined operating conditions for a specified time period. For example, a transducer may have a 95% probability of operating without failure for two years when used in accordance with the manufacturer's data sheets.

Reliability measurements are based on tests done on a large number of items. If N items are run for a time t, and N_s are still working at the end of the test and N_f have failed, the reliability for time t is:

$$R_t = \frac{N_s}{N} = \frac{N - N_f}{N} \tag{21.1}$$

The unreliability, Q_t, is defined as:

$$Q_t = \frac{N_f}{N} = \frac{N - N_s}{N} \tag{21.2}$$

Note that $0 < R_t < 1$, $0 < Q_t < 1$ and $R_t + Q_t = 1$.

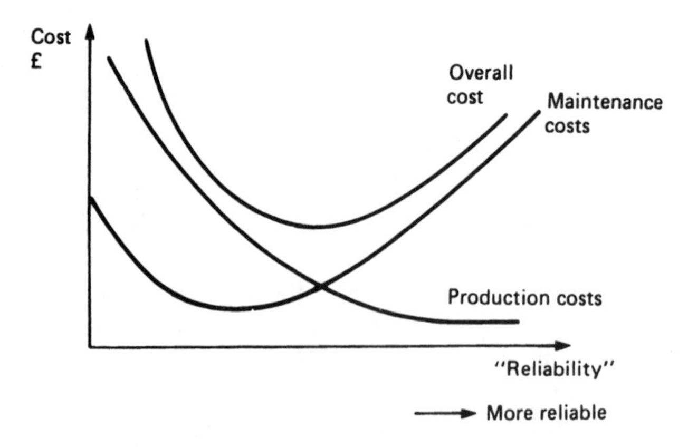

Fig. 21.1 *The financial implications of reliability.*

21.2.3. MTTF and MTBF

Reliability is related to a specified time period (e.g. 1000 hours, one year), but an estimate of life expectancy is more useful in most circumstances. This is given by mean time to failure (MTTF) for non-repairable items (e.g. lamp bulb and disposable equipment) and mean time between failure (MTBF) for repairable items. Both MTTF and MTBF are obtained from tests based on a large number of items.

Table 21.1 shows results of a typical test to determine MTTF for, say, light bulbs. A 1000 bulbs are lit and observed daily; 25 fail during the first day, 17 during the second, and so on. The mean number of bulbs alight each day multiplied by 24 gives the total operational time, in hours, each day. At the end of 10 days, 145 bulbs have failed, and 220 596 operational hours have accumulated. The MTTF is therefore 220 596/145 or 1521 hours.

MTBF is a measure of the time between failures for repairable items. A MTBF test for a lighting system would set up 1000 bulbs as before, but bulbs would be replaced as they failed and a record kept of the number of failures over a given time. Strictly speaking, the MTBF is the operational time/number of failures, with the operational time being the length of the test multiplied by the number of

Table 21.1 *MTTF calculation*

Day	*Hours*	*Failures*	*Cumulative failures*	*Survivors*	*Mean survivors*	*Total time operational*
0	0			1000		
		25			987.5	23700
1	24		25	975		
		17			966.5	23196
2	48		42	958		
		15			950.5	22812
3	72		57	943		
		11			937.5	22500
4	96		68	932		
		13			925.5	22212
5	120		81	919		
		16			911.0	21864
6	144		97	903		
		14			896.0	21504
7	168		111	889		
		12			883.0	21192
8	192		123	877		
		9			872.5	20940
9	216		132	868		
		13			861.5	20676
10	240		145	855		
					Total time	220596

$$\text{MTTF} = \frac{\text{total time}}{\text{cum. failures}} = \frac{220596}{145} = 1521 \text{ hours}$$

items under test less the repair time; but as all realistic systems will have MTBFs far larger than the repair time, it is usual to define MTBF as the length of the test multiplied by the number of items divided by the number of failures.

Suppose 1000 bulbs are kept alight for 10 days, with bulbs being replaced as they fail. At the end of 10 days 136 bulbs have failed, so the MTBF is $240 \times 1000/136$ or 1765 hours. In general, for most items MTTF and MTBF will be similar and the two terms are (incorrectly) used indiscriminately.

Note that both MTTF and MTBF are average values, not a prediction for one bulb. Table 21.1 shows that some bulbs had a life of less than 24 hours, and some would, of course, have a life far greater than the MTTF or MTBF.

21.2.4. Maintainability

When equipment fails, and all equipment can, it is important that it is returned to an operational state as soon as possible. The term maintainability describes the ease with which an item can be repaired, and is defined as the probability that a piece of faulty equipment can be returned to an operational state within a specified time. A DC drive with a thyristor bridge fault, for example, could have a 0.85 probability of being operational within 30 minutes of the fault occurring.

Mean time to repair (MTTR) is another measure of maintainability, and is defined as the mean time taken to return a failed piece of equipment to an operational state. Like MTTF and MTBF, it is a statistical figure derived from a large number of observations. The DC drive mentioned above, for example, might have a MTTR of 20 minutes.

Maintainability is determined partly by the designer and partly by the user. Important factors are:

(1) The design and use of the equipment should be such that the failure is immediately apparent and can be quickly localised to a changeable item. This requires good documentation, readily identifiable test points and indication, and modular construction. Simple and cheap techniques, such as running the neutral out to the limit switch junction box in Fig. 21.2 to allow input signals to be checked, can make significant impact on plant maintainability.

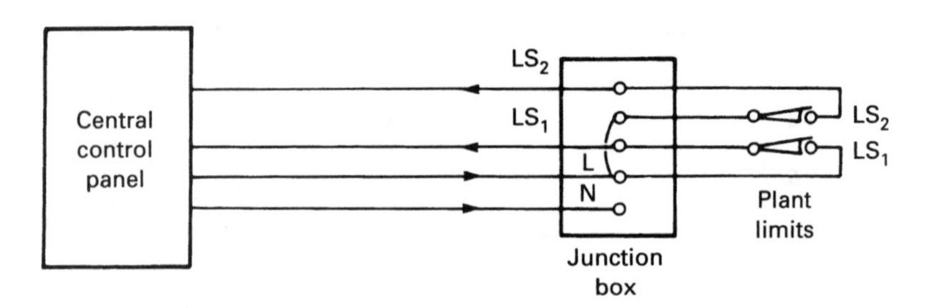

Fig. 21.2 *Designed in maintainability. The neutral in the junction box serves no functional purpose, but allows plant limit states to be checked with a multimeter at the box.*

(2) Vulnerable components should be easily accessible. On a DC drive, for example, a high proportion of faults may lead to the rupturing of the HRC protection fuses, so these should be provided with indicator flags and the mounting designed so they can be replaced quickly.
(3) The maintenance personnel should be competent, well trained and have suitable tools and test equipment. MTTR will obviously depend on how long it takes for maintenance personnel to respond to a fault call.
(4) Adequate spares should be carried and be accessible. MTTR will usually be reduced if a policy of unit replacement rather than unit repair on site is adopted. In the case of the DC thyristor drive referred to earlier, it may be advantageous to treat the whole unit as a replaceable item to be repaired later in the workshop.

Items (1) and (2) are the responsibility of the designer, items (3) and (4) the responsibility of the user.

Plant availability is the percentage of the time that equipment is functional, i.e.

$$\text{availability} = \frac{\text{uptime}}{\text{uptime} + \text{downtime}} \tag{21.3}$$

$$= \frac{\text{MTBF}}{\text{MTBF} + \text{MTTR} + \text{PMT}} \tag{21.4}$$

where PMT is the planned maintenance time, during which the plant is off-line for essential scheduled servicing. Normal availability of a plant will be well over 95%.

21.2.5. Failure rate

If N_s components are in operation, and ΔN_f components fail over time Δt, the failure rate $\lambda(t)$ (also called the hazard rate) is defined as:

$$\lambda(t) = \frac{1}{N_s}\frac{\Delta N_f}{\Delta t} \tag{21.5.}$$

As Δt tends towards zero we can write:

$$\lambda(t) = \frac{1}{N_s}\frac{dN_f}{dt} \tag{21.6}$$

This is, of course, a theoretical model, as in reality we are dealing with a finite number of components and finite time intervals.

The failure rate for most systems follows the bathtub curve of Fig. 21.3. This falls into three distinct regions. The first, called 'burn in' or 'infant mortality', lasts at most a few weeks and exhibits a high failure rate as faulty components, bad soldering, etc., become apparent. Manufacturers generally employ heat cycling tests to provoke infant mortality before equipment is shipped to the user. In complete instrumentation and process control systems an initial 'problem period' is not uncommon as the designer's mistakes and computer software bugs become apparent.

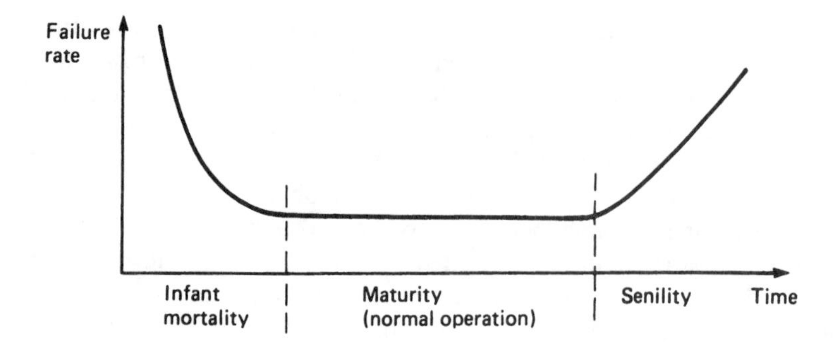

Fig. 21.3 *The bathtub curve*

The centre portion, called 'maturity', exhibits a low constant failure rate. During this period failures are random. The final period, often called 'senility', is characterised by a rising failure rate. Generally this increased unreliability is caused by structural old age: connectors oxidising and losing their spring, electrolytic capacitors drying out, open circuits resulting from temperature cycling induced strain, and so on.

In the centre portion, which is where plant operates continuously, it is possible to predict the probability that a piece of equipment will operate without failure for a given time.

If the failure rate is constant, we can write:

$$\lambda = \frac{1}{N_s}\frac{dN_f}{dt} \tag{21.7}$$

or

$$\frac{dN_f}{dt} = \lambda N_s \tag{21.8}$$

From equation 21.1, the reliability is

$$R = \frac{N_s}{N} = \frac{N - N_f}{N} = 1 - \frac{N_f}{N} \tag{21.9}$$

Differentiating gives:

$$\frac{dR}{dt} = -\frac{1}{N}\frac{dN_f}{dt} \tag{21.10}$$

Substituting from equation 21.8

$$\frac{dR}{dt} = -\lambda\frac{N_s}{N} \tag{21.11}$$

or from equation 21.9:

$$\frac{dR}{dt} = -\lambda R \tag{21.12}$$

Integrating from $t = 0$ to t:

$$R = Ae^{-\lambda t} \tag{21.13}$$

where A is an integration constant. However, at $t = 0$, R must be 1 because the equipment must work for zero time, hence:

$$R = e^{-\lambda t} \tag{21.14}$$

For the centre 'maturity' period, the failure rate is simply:

$$\lambda = \frac{1}{\text{MTBF}} \quad \text{or} \quad \lambda = \frac{1}{\text{MFFT}} \tag{21.15}$$

dependent on whether the item under consideration is repairable or not.

For example, a transducer has an MTBF of 17 500 hours (approximately two years). The probability that it will run at least 8750 hours (approximately one year) without failure is:

$$\begin{aligned} R = e^{-\lambda t} &= \exp(-t/\text{MTBF}) \\ &= \exp(-8750/17\,500) \\ &\simeq 0.6 \end{aligned}$$

The probability that an item will run for at least its MTTF or MTBF is simply (but rather surprisingly):

$$\begin{aligned} R &= e^{-1} \\ &= 0.37 \end{aligned}$$

The 0.5 reliability time (i.e. the time for which the probability of no failure is 0.5) is approximately 0.7 MTBF (or 0.7 MTTF).

21.2.6. Series and parallel reliability models

Figure 21.4 shows a typical instrumentation system, where a power supply provides 24 V DC for a transducer whose output is shown on a bar graph display. The failure of any component will result in the loss of the display and a 'system failure'. Given the reliability, or MTBF, of each unit, it is possible to predict the reliability, and MTBF, of the whole system.

Suppose the MTBFs, and the corresponding λs, are:

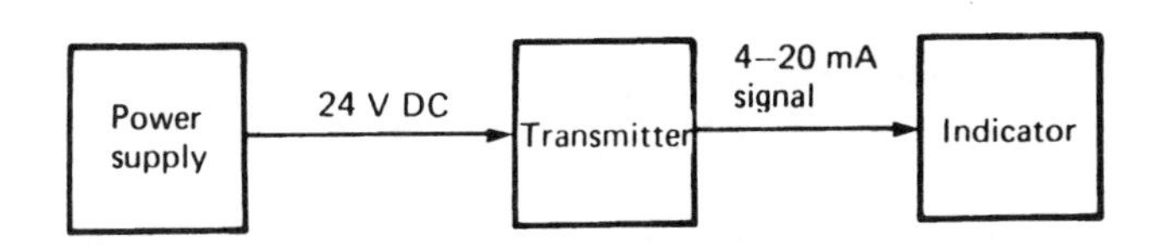

Fig. 21.4 *Series reliability model.*

	MTBF	λ
PSU	10000	0.0001
Transmitter	15000	0.000067
Display	20000	0.00005

The reliability of a component for time T is the probability that it will run for time T without failure. The probability of three components running for time T is the product of their three reliabilities (assuming there is no interaction and no common influence such as external damage):

$$R = R_{1T} . R_{2T} . R_{3T} \tag{21.16}$$

$$= e^{-\lambda_1 T} . e^{-\lambda_2 T} . e^{-\lambda_3 T} \tag{21.17}$$

$$= e^{-(\lambda_1 + \lambda_2 + \lambda_3)T} \tag{21.18}$$

i.e. the resulting λ is the sum of the individual component λs. For the example above:

$$\lambda = 0.0001 + 0.000067 + 0.00005$$
$$= 0.000217$$

so the system MTBF:

$$= \frac{1}{0.000217}$$
$$= 4600 \text{ hours}$$

i.e. much lower than the MTBF of any individual unit.

The arrangement of Fig. 21.4 is called a series reliability model, because the failure of any unit results in the failure of the whole system.

In Fig. 21.5, the temperature of a vat is measured by two independent temperature measuring systems. Such an arrangement is called a parallel reliability model, because it requires a failure of both sensors to lead to a loss of temperature display. It is assumed that there is no interaction, and there is no common failure route. (The loss of mains supply is an obvious potential common failure; the analysis below assumes a totally reliable supply.)

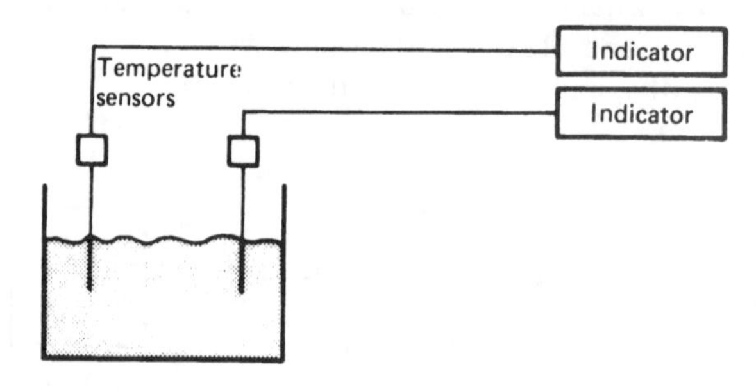

Fig. 21.5 *Parallel reliability model.*

For a failure to occur within a specified time, both systems must fail. The probability of this is the product of the unreliability of each unit, i.e. the system unreliability is given by:

$$Q = Q_{1T} \cdot Q_{2T} \tag{21.19}$$

$$(1 - R) = (1 - R_{1T}) \cdot (1 - R_{2T}) \tag{21.20}$$

$$R = R_{1T} + R_{2T} - R_{1T} \cdot R_{2T} \tag{21.21}$$

Suppose that the two systems of Fig. 21.5 have an MTBF of 15 000 hours. The reliability for 8000 hours for one sensor is:

$$\begin{aligned} R_{1T} &= e^{-\lambda T} \\ &= \exp(-8000/15\,000) \\ &= 0.59 \end{aligned}$$

The reliability for the two in parallel from equation 21.21 is:

$$\begin{aligned} R &= 0.59 + 0.59 - 0.59 \cdot 0.59 \\ &= 0.83 \end{aligned}$$

i.e. the parallel arrangement exhibits a higher reliability than the individual units. The above reliability corresponds to an MTBF of over 40 000 hours.

The parallel arrangement is often called 'redundancy' and is used in critical applications (e.g. nuclear reactors, petrochemical plants) where very high reliability is required. If three parallel paths are used, with each path having a reliability of 0.85, for example, a system reliability of 0.997 is achieved.

It is important for systems employing redundancy that protection against common failure routes (called common mode failures) is included. Separate and independent power supplies, for example, are essential. Less obvious precautions are different sensor positions and cable routes to avoid common mechanical damage, different sensor types (e.g. thermocouples and RTDs) to avoid problems affecting one type of sensor, and precautions against outside events affecting all channels simultaneously. In Fig. 21.6, for example, duplicate temperature sensors are used to indicate the exit water temperature from a cooling jacket. If

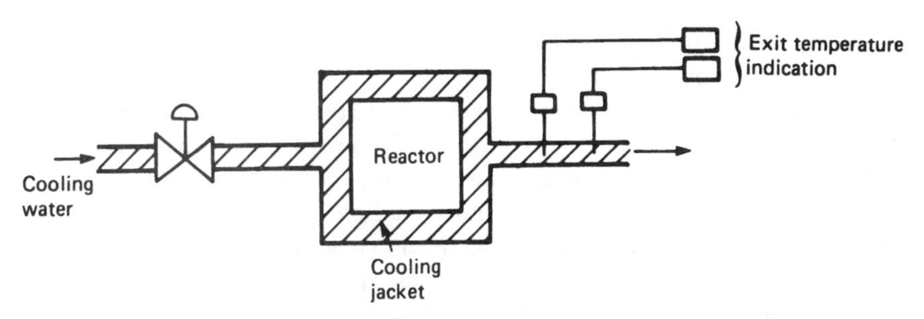

Fig. 21.6 *Potential common mode failure caused by incorrect sensor positioning.*

the water flow fails – caused by a pipe blockage, say – both indicators will show the same, probably incorrect, temperature. A common mode failure similar to Fig. 21.6 was a contributing factor to the operator's confusion during the Three Mile Island nuclear incident.

Variations on redundancy are the majority vote and highest (or lowest) voting systems. With two parallel systems, it can be problematical establishing which unit is working, and which has failed, in a fault condition. Majority voting uses 3, 5 or 7 systems, and takes the majority signal (2 of 3, 3 of 5, etc.) as being correct. Highest (or lowest) voting is normally employed where safety is the prime consideration (e.g. shutdown systems). The most unsafe signal (e.g. the highest temperature reading) is assumed to be correct from a control viewpoint. Majority voting gives the high reliability of redundancy; high voting gives very high safety, but the operational reliability is worse than a single system as any failure can lead to a potential shutdown.

Other approaches employing redundancy are arithmetic averaging of several signals, or discounting readings from the highest and lowest of three sensors (called median redundancy). All redundancy techniques are expensive in initial installation costs.

21.3. Maintenance philosophies

Even with the best planned maintenance and preventative maintenance procedures, faults will occur in all equipment at some time. Maintenance departments, responsible for the servicing, repair and operational improvement of plant, are often seen as a necessary evil, but are in reality a crucial part of a production team. Plant downtime is invariably expensive, and a maintenance department should not just 'fix it when it breaks' but consider the whole economic cost of maintaining the equipment in their care. Admittedly such a broad view is difficult to take at 3 a.m. with a shift manager asking the three inevitable questions 'What's wrong?', 'Do you think you can fix it?' and 'How long will it take?'

There is a fundamental difference between most process control/electronics problems and, say, mechanical faults. In the latter case the fault is usually obvious, often to non-technical persons, but a repair is lengthy. Typical examples are seized bearings, broken couplings and so on. Process control problems tend to be more subtle. Symptoms are noticed by production staff, and a logical fault finding procedure is needed to locate the fault. Once diagnosed correctly, the repair is usually straightforward and quick.

The reliability of modern equipment can create problems for the maintenance staff. With MTBFs measured in years, it is possible that the maintenance technician really sees a piece of equipment for the first time when it fails (and the instruction manuals are collecting dust in a cupboard). More reliable equipment also means that a maintenance technician can cover, and hence needs to know about, much more plant. It is therefore essential that maintenance staff are involved in the installation and commissioning of all new plant, as this is the best time to observe test procedures and learn how it works. Refresher training at regular intervals is also helpful. Design staff can assist by standardising on types of equipment in all areas, so the maintenance staff only have to learn the details of, say, one type of programmable controller. Small price differences between competitive equipment can be totally wiped out by production losses at the first fault.

A fault manifests itself as a symptom; a gas-fired burner is stuck on low fire, say. Investigation shows that the air valve is almost closed and the gas valve, correctly following the air flow, is also on a low setting. The non-operable air valve is also a symptom. Further investigation reveals the diaphragm in the pneumatic actuator has ruptured; this is the fault which is causing the symptoms. A repair is effected by replacing the diaphragm or, more probably, by changing the actuator. The fault-finding procedure splits naturally into realization that there is a fault, location of the fault and rectification of the fault. Symptoms and faults should not be confused; a blown fuse, for example, is a symptom, not a fault.

There is, however, an often overlooked fourth stage which is the analysis of the fault to see if it was a random occurrence or was due to some underlying cause. A continuing failure of the diaphragm in the example above would suggest a possible environmental or application problem. Faults are repaired usually by shift staff who inevitably only see one quarter of the faults and are naturally mainly concerned with restoring production. A managerial fault-recording and analysis system will allow a broader view of problems and possible solutions.

Apart from revealing reasons for higher than expected fault rates, such studies may suggest a need for better diagnostic aids and instruments or lack of knowledge at the first line maintenance level, or may show that the level of repair being carried out on site is too deep (and it would be more cost effective to use replaceable units with the repairs being effected off-line in the workshop). The fault-finding procedure does not stop when the plant goes back into production.

Fault finding can be considered as first line maintenance (repairs carried out on plant) and second line maintenance (repairs carried out in a workshop). In both cases it is a logical procedure which homes in on the fault along the lines of Fig. 21.7. Symptoms are studied, and from the available information possible causes are postulated. Tests are conducted to confirm, or refute, these possibilities. Further information gained from these tests allows the probable causes to be narrowed down and further tests made. The procedure is repeated until the fault is found.

One of the arts of fault finding is the balancing of the probabilities of the various possible causes of a fault against time, effort and equipment necessary to perform the tests required to confirm or refute them. (This manifests itself in the sensible old rule 'check the power supply first'. The probability of a supply problem may be less than 10%, but it usually is a possible cause that can be checked out in a few seconds.) Good equipment design should provide diagnostic aids such that tests for the most probable faults can be performed quickly without specialised test instruments. It does not help the MTTR if the most probable faults can only be located with a 50 kg test box which has to be signed out of the stores and carried 20 metres up a vertical ladder. Such problems should be revealed by the managerial fault analysis system.

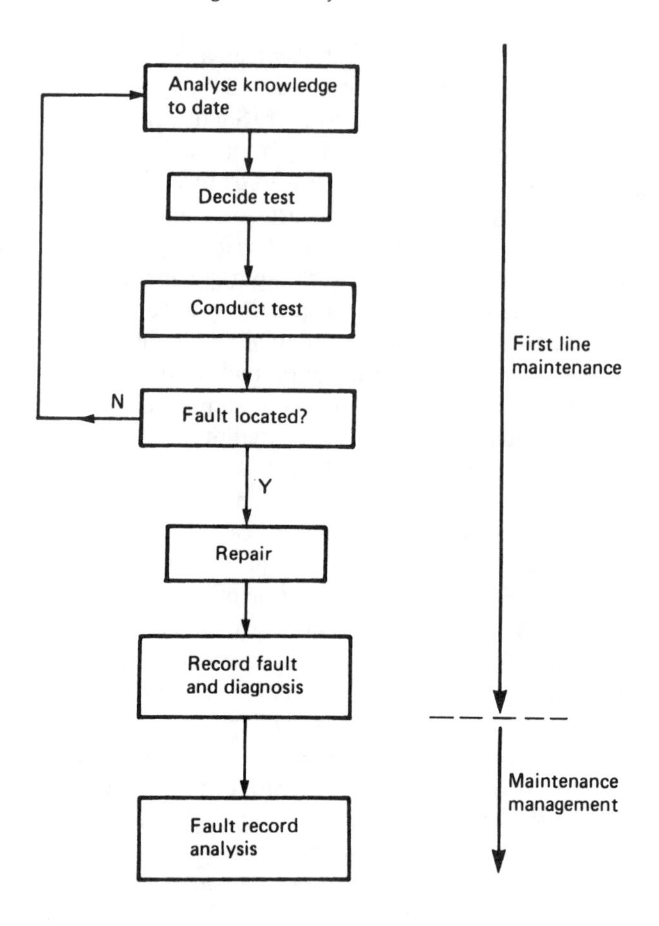

Fig. 21.7 *The fault finding procedure.*

Fault finding is simplified if equipment is conceptually modularised, a topic discussed further in Section 21.6. Figure 21.8 shows a flow control loop which can be broken down to six modules, each of which can be tested independently. With care, most systems can be considered as a series chain of modules, as in Fig. 21.8b. The loop of Fig. 21.8a could be broken, for example, by putting the controller in manual and considering the controller output as the input to the chain and the feedback signal as the output.

Fault finding is based on signal injection and signal tracing. Signal injection, shown in Fig. 21.9a, introduces test signals whose effects are observed. Signal tracing, shown in Fig. 21.9b, follows normal signals through the chain. In both cases, the so-called half split method of Fig. 21.9c, where the possible fault area is reduced by half with each test, is the best approach.

Plant reliability can be improved by a sensible planned maintenance programme. It is possible to identify the life of essentially mechanical plant items (e.g. filters,

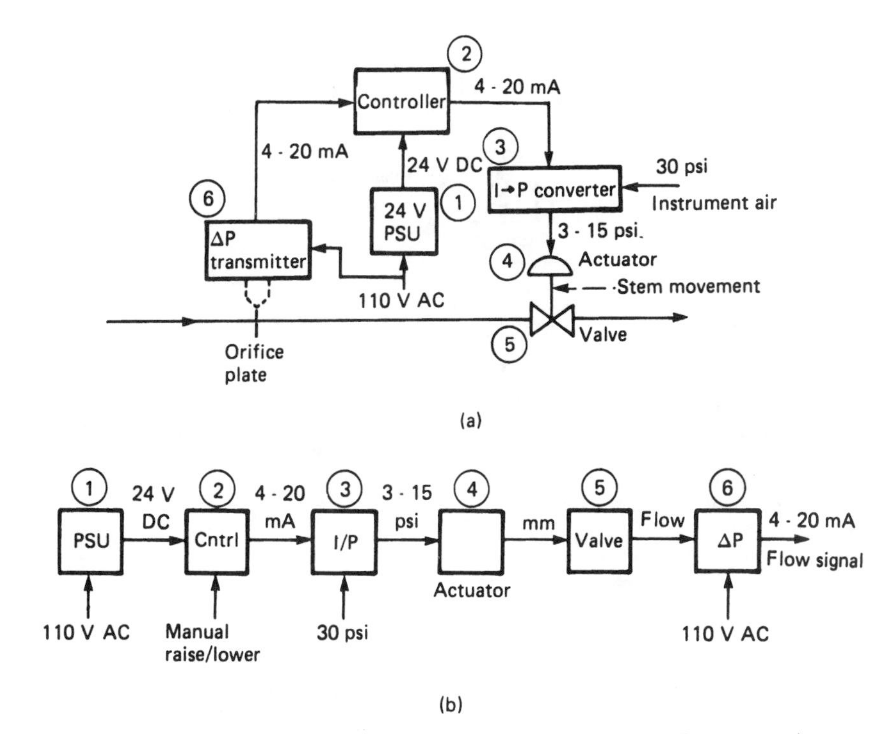

Fig. 21.8 *Fault finding methodology. (a) Flow control system. (b) Flow control system modularised for fault finding.*

oils, seals, etc.) with a fair degree of accuracy from manufacturers' data sheets or from plant records accumulated from the fault recording process described earlier. For such items a routine schedule of replacement can easily be devised.

Most process control equipment, however, does not 'wear out' but exhibits the bathtub reliability curve of Fig. 21.3. Once the maturity period is reached, routine replacement will not improve reliability and may even take the equipment back into the infant mortality region. Planned maintenance therefore usually takes the form of regular calibration checks (such as the setting of span and zero on transducers).

The most important factor in determining reliability is probably the competency and experience of the maintenance personnel. Training is often overlooked, but a well-trained staff is the best investment a company can make. Often, however, the training is of the wrong type. If a technician, say, goes on a manufacturer's course on a specific PLC there is a natural tendency to assume thereafter that plant faults can only exist inside the PLC. There is also a defensive tendency for first line personnel, when in doubt, to do what they can do well. When the ideas run out, a technician freshly trained in stripping down and servicing the instrument air compressor will invariably strip down and service the instrument air compressor. Training should therefore deal with a broad view of a plant; what the various items do and how they interact. On large plants it is useful to have a course on the

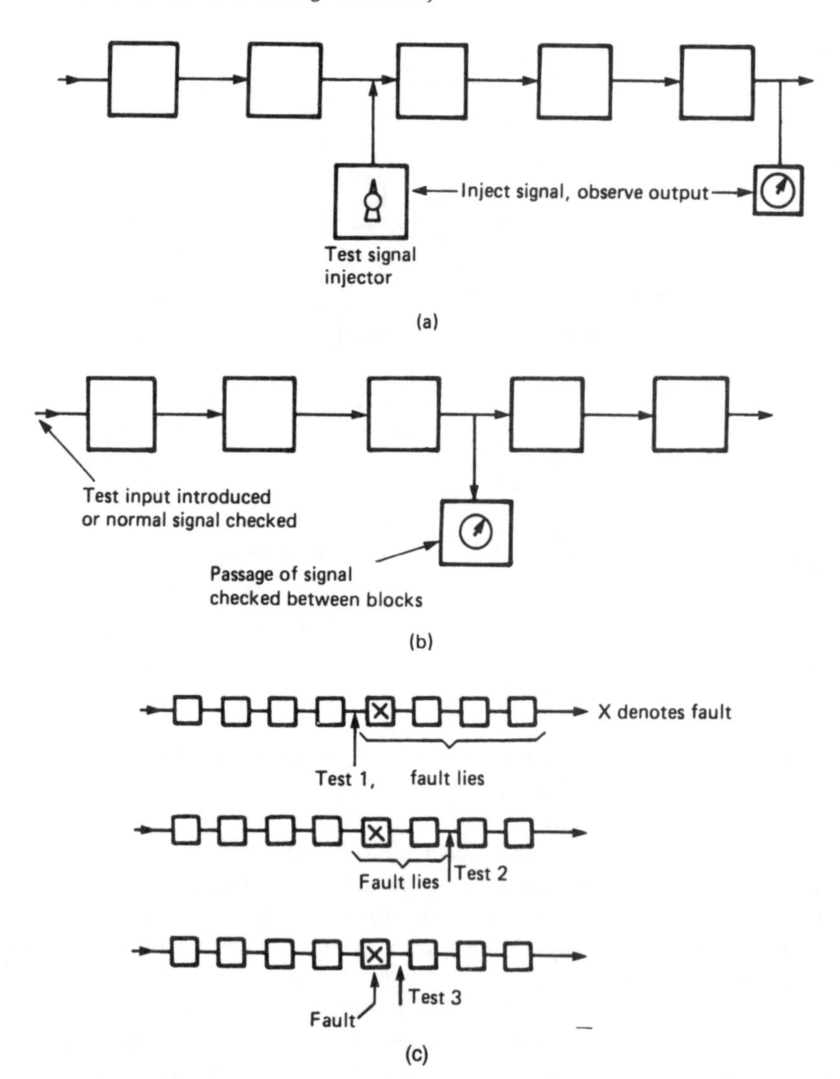

Fig. 21.9 *Fault-finding techniques. (a) Signal injection. (b) Signal monitoring. (c) The half split method.*

physical locations of plant equipment. Many wrongly identified plant mounted limit switches and transducers have been changed in error!

21.4. Fault-finding instruments

The previous section showed that testing is essentially of a signal injection or signal tracing nature. Instruments for assisting in the tracing of faults therefore tend to be devices for injecting test signals or displaying the value of signals at strategic points.

The classical (and in many organisations the only) fault-finding aid is the multimeter, used to measure voltages and currents. Digital multimeters are useful where accurate readings are needed but can be confusing where values change quickly, for reasons outlined in Section 17.4.3. Auto-ranging meters can be confusing if the signal goes across several range changes. Ideally auto-ranging meters should have optional manual range selection or a hold high range option.

In the author's experience, the high impedance of digital meters (normally a desirable characteristic) can cause misleading indications in some quite common circumstances. A typical problem is shown in Fig. 21.10 where a triac output from a programmable controller is used to operate a pneumatic solenoid. The coil of the solenoid has gone open circuit, and a digital meter has been connected across the coil. A triac has a small leakage current in the off state; this is insufficient to pick up the coil or displace a moving coil meter, but *will* be enough for the digital multimeter which will register full line volts regardless of whether the PLC output is on or off. The technicians could thus be misled into suspecting a PLC or output card fault.

The other classical fault-finding device is the oscilloscope, but its bulk and relative delicacy tends to limit its use to control rooms and other centralised points. Storage scopes are useful for capturing transients, a role also filled by the UV recorder described in Section 17.3.3.

A large proportion of instrumentation signals use 4 to 20 mA and similar signals, so a test device for the injection and monitoring of current signals is essential for the maintenance and calibration of process control equipment. A typical device is shown in Fig. 21.11.

Monitoring and injection of millivolt signals is required where thermocouples are used. Devices such as Fig. 21.11 usually allow the monitoring and injection of both voltage and current signals. The use of millivolt sources/display with thermocouples requires thermocouple tables and a knowledge of the cold junction temperature. Thermocouple principles are described in Section 2.5, but the procedures are:

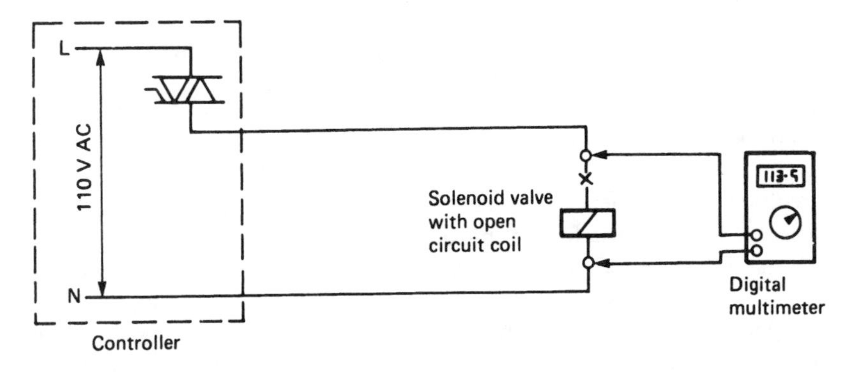

Fig. 21.10 *The need for care with digital multimeters.*

Fig. 21.11 *An instrumentation injection/monitor/calibration unit; the SUPERCAL manufactured by Rochester Instruments. This unit provides and monitors mA, mV, V for thermocouple and transducer testing, and can simulate and monitor resistance temperature transducers. Additional facilities include a calculator function, frequency monitoring and injection plus displays in engineering units. (Photo courtesy of Rochester Instruments.)*

Injection

(1) Measure the cold junction temperature (usually ambient) and note the voltage V_a corresponding to this from the tables.
(2) Note the voltage V_t from the tables corresponding to the required temperature.
(3) Set the injection voltage to $V_t - V_a$. The system under test should indicate the required temperature.

Monitoring

(1) Find the ambient temperature voltage V_a from the tables, as above.
(2) Measure the thermocouple voltage V_t at the place where the ambient temperature measurement was made.
(3) Find the temperature corresponding to $(V_a + V_t)$ from the tables; this is the thermocouple temperature.

Direct-reading temperature indicators are, of course, readily available but these are not as versatile for test purposes as a millivolt source/monitor.

Pneumatic test devices are also available for the injection and monitoring of pneumatic signals. A typical device is shown in Fig. 21.12. Similar devices are available for hydraulic systems, but hydraulic monitoring (i.e. pressure gauges) should be built into the system at the design stage as most hydraulic test sets are very bulky and difficult to transport.

Fig. 21.12 *A pneumatic injection/monitoring unit being used to calibrate a differential pressure transmitter.*

Faults on sequencing systems can be difficult to trace particularly if (as is usual) many events happen in a short period of time. So-called 'signature analysers' are a useful, if somewhat expensive, approach to this problem. Essentially signals from the plant are brought back to the analyser which has been pre-programmed with acceptable patterns of plant behaviour. The analyser compares the plant operation with its internal model and flags any deviation (e.g. input 27 late coming on). Like expert systems, the use of signature analysers is best 'built into' a plant at the design stage rather than grafted on to an existing plant.

21.5. Noise problems

21.5.1. Introduction

Electrical interference, usually called 'noise', is often the cause of poor or erratic performance of a process control system. Many so-called intermittent faults are found to originate with noise problems, particularly in digital sequencing systems. Chapter 19 discussed the theoretical aspects of data transmission through a noisy environment; the present section is primarily concerned with the practical aspects of noise elimination.

The tracing and removal of noise problems can be more of an art than a science, and requires an almost detective-like instinct. Is the noise always present, or occurring randomly? Is the noise related to events occurring on a plant (e.g. a compressor starting)? Has any new plant been added and tied into the grounding system? Does the noise have a recognisable waveform? These and similar questions should be asked when noise problems occur. Good design with screening, isolation and sensible grounding should, however, prevent most noise problems.

21.5.2. Types of noise

It is useful to categorise types of noise. There are, in general, three types of noise that can be encountered, as summarised by Fig. 21.13. Random, or white, noise is a wideband random fluctuation in a signal. In audio circuits this appear as a 'hiss'. All electrical components generate white noise with resistors, thermionic valves and zener diodes being particularly effective noise generators. Internally-generated white noise should not be a problem in process control except with high gain instrumentation amplifiers. Poor connections and badly soldered joints can, however, produce random noise.

Impulsive noise appears as random spikes on an otherwise steady signal. In process control these spikes are invariably associated with the switching of adjacent heavy or inductive loads. Poor or dirty connections can again generate impulsive noise.

Cross-coupled noise appears as capacitive or inductive, 50 Hz (UK)/60 Hz (USA) interference from power cables to adjacent signal cables.

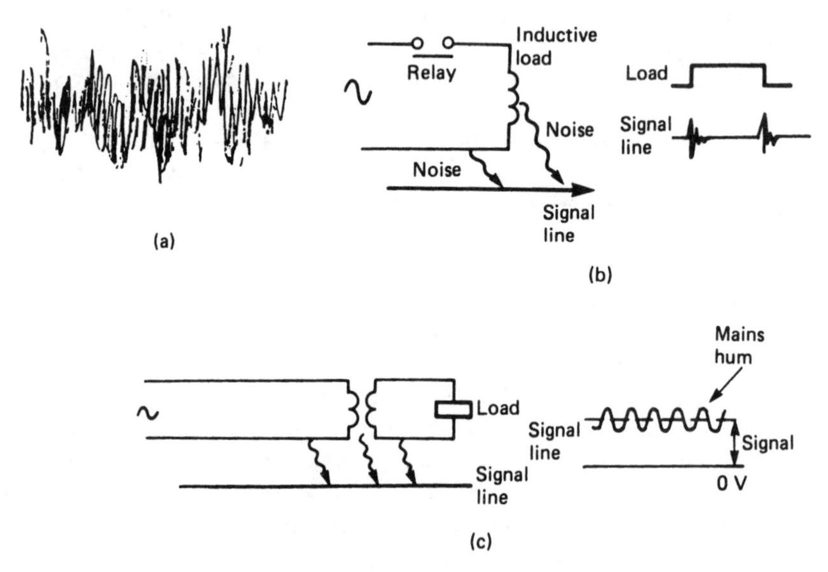

Fig. 21.13 *Types of noise. (a) Random noise. (b) Impulsive noise. (c) Cross-coupled noise.*

21.5.3. Noise sources

Figure 21.14 shows the common routes by which noise can be introduced into a system. For simplicity no noise sources are shown affecting the primary sensor, but these are, of course, subject to the same noise sources as the rest of the system. A noise problem will usually arise as a combination of the effects of Fig. 21.14.

Many process variables are inherently noisy; level measurement is particularly difficult, for example, as the surface is affected by ripple and turbulence. Other problems arise from pressure spikes in compressor systems, flame flicker in temperature measurement and suspended solids or entrapped gases in flow measurement. All these, and similar effects, contribute to a noisy primary signal.

Noise coupled on to signal lines is a common problem. This noise usually affects all signal lines in a particular cable equally, and is called common mode noise. The effect is shown in Fig. 21.14b. Series mode noise occurs when circulating currents are induced around signal pairs, and is far less common (but more difficult to deal with). Signal line induced noise, often called pickup noise, can be reduced by utilising sensible spacing between signal and power cables (typically 1 to 2 metres minimum), the avoidance of long parallel runs, and the use of screening on signal cables or running signal cables in conduit. Electrical currents are far less affected by common mode noise than are voltages and this is

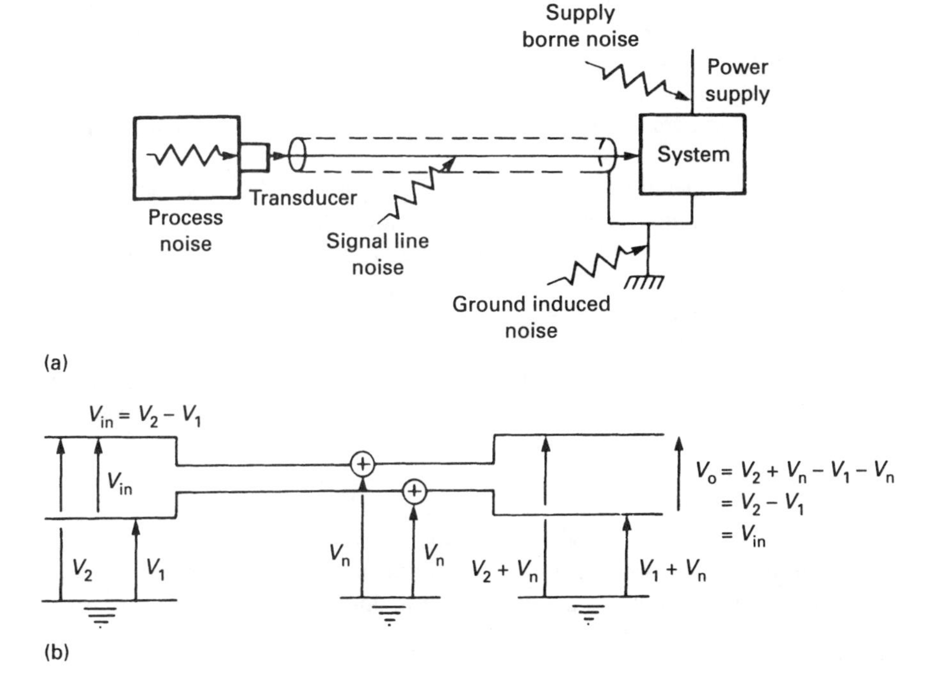

Fig. 21.14 *Noise and real-life systems. (a) Noise routes into a system. (b) Common mode noise.*

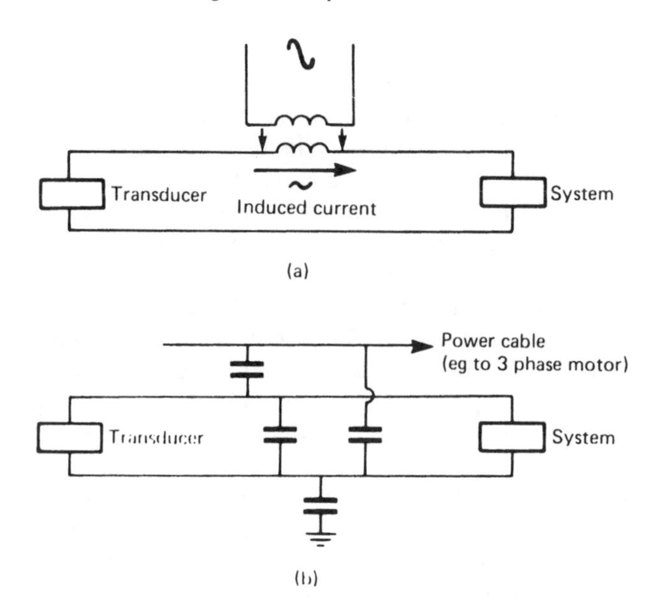

Fig. 21.15 *Methods by which noise enters system. (a) Electromagnetic coupling. (b) Capacitive coupling.*

one reason why standards such as the 4 to 20 mA current loop are widely used in instrumentation. The best solution, though, is to keep a reasonable separation between power and signal cables.

Ground induced noise is often overlooked. The earthing arrangements in instrumentation systems require special care, and the topic is treated further in Section 21.5.4.

Some systems generate internal noise, TTL-based logic systems being particularly vulnerable. Internal noise prevention is largely a matter of sensible PCB layout and power supply rail decoupling. This subject is treated further in Section 13.10.

Noise can also be introduced via the supply lines. The switching of large loads, the use of thyristor drives and random lighting strikes on the supply authority's overhead power lines all cause erratic system behaviour. The use of line LC filters or constant voltage transformers (CVTs) is highly recommended for line sensitive equipment.

Noise problems can be overcome by removing the cause(s) of the noise at source, or by making the system less sensitive to the noise. The former method should be used wherever possible. The suppression of an inductive load or the construction of a stillwell around a level transducer are examples of noise removal. The use of filters is an example of making a system noise tolerant.

21.5.4. Grounding

Ill-conceived grounding arrangements can be the cause of particularly elusive noise problems and intermittent faults. Problems arise when different equipment or parts of the same equipment, share a common route to ground, as in Fig. 21.16a. The common impedance Z_3 causes interaction between the two items. Generally the resistance of the connections is small, and it is the inductive impedance that causes the problem on fast changing currents.

One common example of poor grounding called an earth loop occurs when screened cable is used to connect equipment on different earthing circuits, as in Fig. 21.16b. There will inevitably be a significant AC potential (as high as a few volts in extreme cases) between the two earths, causing a large current to flow down the screen and any other 0 V link. Screened cable must have the screen

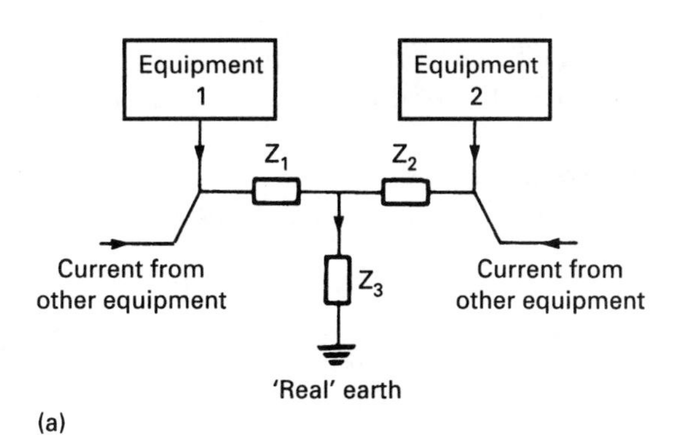

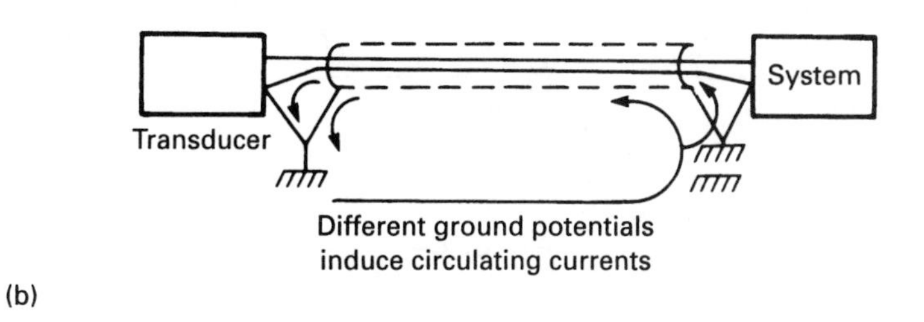

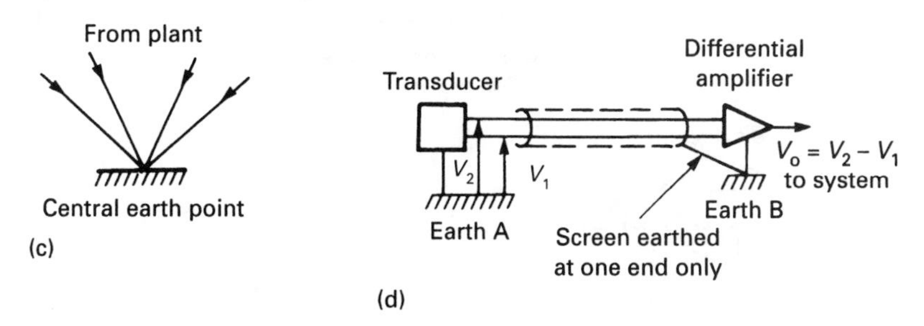

Fig. 21.16 *Noise induced via ground lines. (a) Noise from shared ground routes. (b) Problems with multiple earths. (c) The CEP. (d) Earthing problems minimised by differential amplifier.*

connected to ground at one place only. Particular care should be taken to ensure screen continuity through junction boxes whilst avoiding multiple earths.

Ideally, it is possible to identify at least three different 'grounds' in a system, namely a safety ground for earthing cases, cubicles, instruments, etc., for the prevention of electrical shock, a power ground for current returns from power supplies and high current loads such as relays and solenoids, and signal grounds from low current instrumentation. These should all meet at one, and only one, central earthing bus which has a direct low impedance route to ground, as in Fig. 21.16c.

This ideal state of affairs cannot exist on a widely separated plant, and in these circumstances differential or isolation amplifiers can be used to remove the common mode noise arising from different earth potentials, as in Fig. 21.16d. Particular care should again be taken to avoid the creation of earth loops.

21.5.5. Suppression of inductive loads

In Fig. 21.17a a current is flowing through an inductor L. The voltage across the indicator is given by:

$$V = L\frac{dI}{dt} \tag{21.22}$$

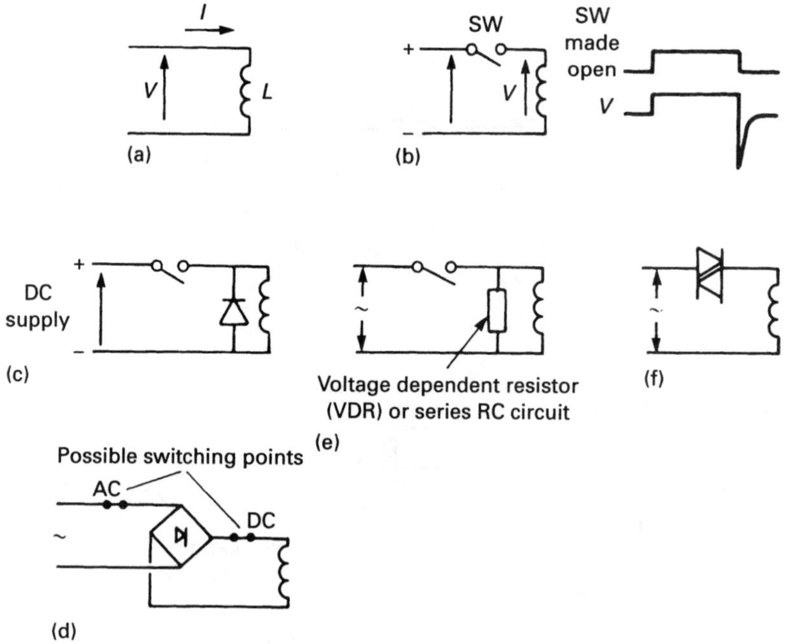

Fig. 21.17 *Noise from inductive loads. (a) Inductive circuit. (b) DC circuit. (c) Spike suppression diode. (d) AC/DC circuit. (e) AC circuit. (f) Triac circuit.*

This predicts that the voltage induced is proportional to the rate of change of current. The switching off of inductive loads results in large transient voltages, arcing across contacts and a very real source of interference, as shown in Fig. 21.17b.

DC switching can be made noise-free by the use of a spike suppression diode, as in Fig. 21.17c. This provides a route for the current in the inductor to decay gracefully and without an excessive transient. If a DC load is being fed from a rectified AC supply, as in Fig. 21.17d, AC side switching is preferred as the bridge rectifier provides a route for circulating currents and acts as a suppression diode.

The switching of AC loads presents more problems as diodes cannot be used. For small loads, series RC snubbers can be connected across the load or switching contact. For larger loads, voltage dependent resistors (VDRs, also known as metrosils) can be used. The ideal solution is to use triacs (see Section 12.5.3), with zero voltage crossing control circuits. These will turn on at the zero voltage point of the AC supply and inherently turn off when the AC load current is zero. The induced voltage, and hence the interference with other equipment, is minimal.

21.5.6. Differential amplifiers

The differential amplifier of Fig. 21.18a (analysed previously in Section 11.3.3) amplifies the difference between its two input voltages, i.e.:

$$V_o = V_1 - V_2 \qquad (21.33)$$

assuming all the resistors have equal values. This feature makes it particularly useful for separating out a signal from large amounts of common mode noise, as in Fig. 21.18b.

Differential amplifiers are commonly employed where signals are to be passed between equipment with different earth potential. Figure 21.18c shows a particularly useful technique where the signal is converted to a current for transmission (as explained earlier, current signals are less affected by noise). The current signal is converted back to a voltage at the receiver by a burden resistor and a differential amplifier. The voltage/current circuit is described in Section 11.6.1.

The performance of a differential amplifier is defined by the common mode rejection ratio (CMRR). This is tested as shown in Fig. 21.18d. The inputs are linked and driven from an oscillator simulating common mode noise. The output should, theoretically, be zero, but in practice will follow V_{in} to some extent. If the amplifier gain is A, the CMRR is defined as:

$$\text{CMRR} = A\frac{V_i}{V_o} \qquad (21.24)$$

The CMRR is very large, and is usually expressed in decibels (dB). A typical value for an instrumentation amplifier is around 140 dB.

21.5.7. Isolation techniques

Differential amplifiers can handle common mode voltages of a few tens of volts. Beyond this, or where faults can induce large voltages on to inputs, or equipment

R

R

V_i

$V_o = V_1 - V_2$

R

V_2

V_1

R

(a)

V_{in}

Noise

V_1 0 V

V_2 0 V

V_o
$= V_1 - V_2$
$= V_{in}$

(b)

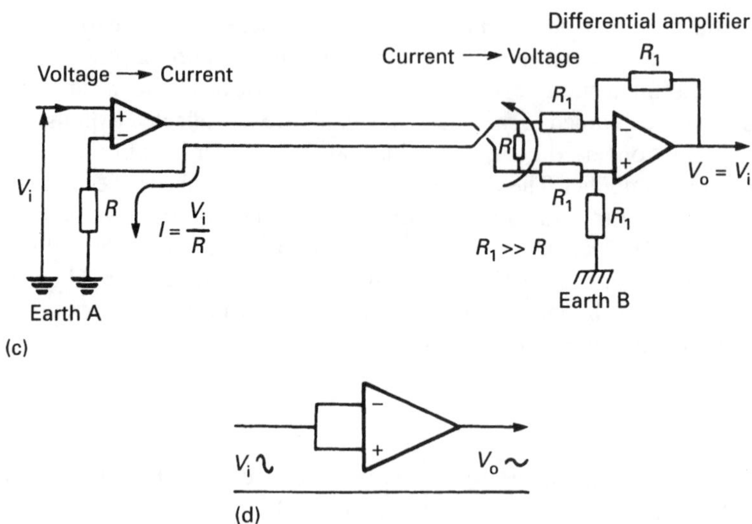

Fig. 21.18 *Removal of common mode noise. (a) The differential amplifier. (b) Typical signals. (c) Current transmission of signal. (d) Definition of common mode rejection ratio.*

is to be used in a particularly electrically noisy environment, devices called isolation amplifiers should be used. These have no direct electrical connections between the input and output terminals (called four-port isolation). A typical device can withstand 1 kV between input and output and reject common mode noise of a similar value.

An early technique, but one still used because of its simplicity, is the flying capacitor of Fig. 21.19b. Reed relay pairs A and B close alternately; with A closed, the capacitor charges to V_{in}. With B closed, the capacitor voltage is presented to the differential amplifier input.

A more modern technique, shown in Fig. 21.19c, uses modulation of an AC waveform with transformer isolation. A similar technique uses pulse width modulation of a square wave and opto isolators, as in Fig. 21.19d.

Figure 21.19 shows isolation techniques for analog signals. Isolation of digital signals is, if anything, more important. Usually this is provided by opto isolation, as described in Section 13.10.

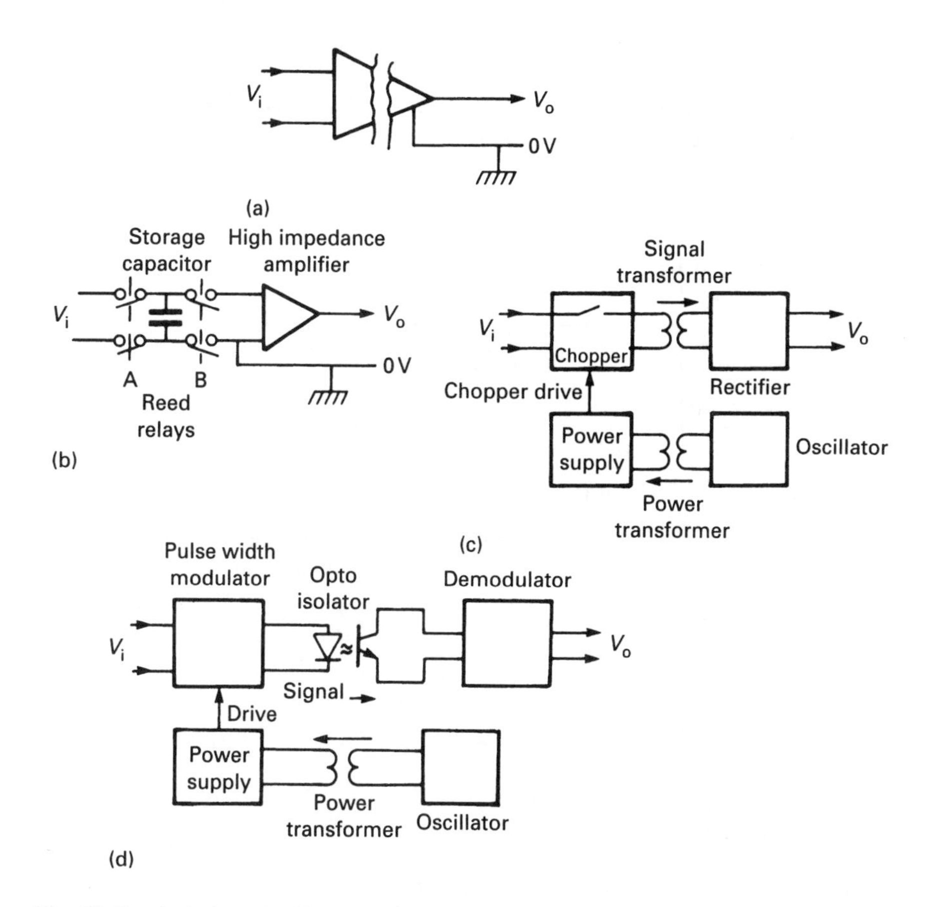

Fig. 21.19 *Isolation amplifiers. (a) Symbol for isolation amplifier. (b) Flying capacitor. (c) Transformer coupled isolation. (d) Opto isolation.*

21.5.8. Filtering

If noise cannot be removed at source, usually the best solution is to apply filtering to the signals before it is used by the control system. Before suitable filtering can be chosen, it is first necessary to determine the bandwidth of the measured signal so that the filtering will remove the noise but not degrade the signal information content. Loop stability can be affected by filtering; an often overlooked fact is the phase shift introduced by the simple low pass filter at frequencies well below the corner frequency, as in Fig. 21.20.

Filtering normally takes the form of low pass filters (to remove random high frequency noise and impulsive noise) or notch filters to remove noise of a specific frequency (e.g. mains induced hum or level resonance – see Section 7.1). Low pass filters are described in Section 11.5.1 and notch filters in Section 11.5.3.

Digital filtering is essentially an averaging procedure, and consists of taking a rolling average of the last 'N' samples. The effective time constant is determined by the number of samples and the sample time:

$$V_n = V_{n-1} + \frac{\Delta t}{T}(V - V_{n-1}) \qquad (21.25)$$

Where V_n is the current filtered value, V_{n-1} the last filtered value, Δt the sample time, T the time constant and V the raw unfiltered sample.

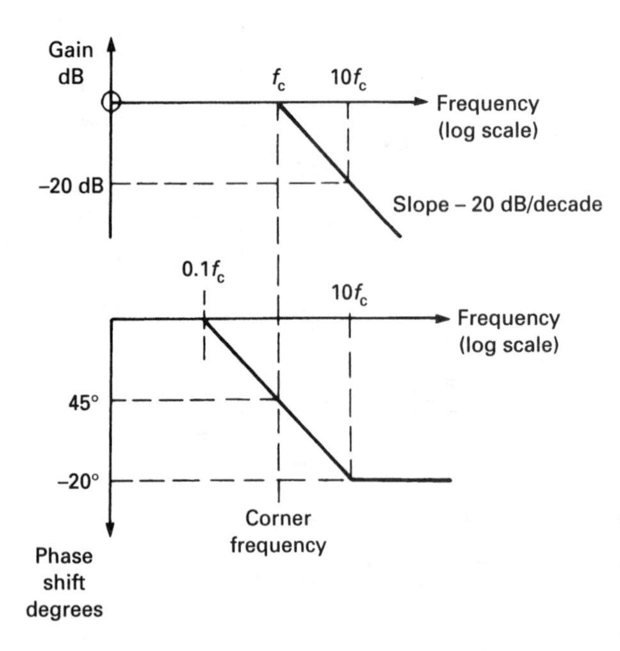

Fig. 21.20 *Straight line approximation Bode diagram of first-order filter.*

21.6. Documentation

Modern plants are both complex and reliable. Together, these characteristics create problems for maintenance staff, as it is not usually possible to build up experience of faults on specific items. (If it is, the fault recording and analysis procedures should have flagged a possible design, application or environmental problem.) The plant documentation is therefore crucial in reducing repair time. It also follows that the documentation should be available, and not sitting in the bookcase in the chief engineer's office.

Figure 21.21 is a common example, familiar to most people, of a car wiring diagram. This, like a politician's answer, is completely factual and truthful but of little use. The drawing has been produced for constructional purposes and not for fault finding. The circuit has been redrawn in a style suitable for fault finding in Fig. 21.22, with a logical flow of signals from left to right.

Figures 21.21 and 21.22 illustrate a common failing of documentation. There are two types of plant drawing. The first is produced by the manufacturer to construct and interconnect the plant. Such drawings are essential, but of little use for subsequent fault finding unless there is a major fire or similar disaster. These construction diagrams tend to be of a locational nature (such as Fig. 21.21) or panel orientated (such as Fig. 21.23, which is an extract from the drawings for a PLC system). The latter drawing would not be much use in finding out why a particular motor will not start.

Day-to-day maintenance requires drawings which are functionally orientated whilst retaining locational information. Figure 21.24 shows part of a functional drawing of the same PLC system as Fig. 21.23, showing the plant signals and where they can be found and traced for fault finding purposes.

Unfortunately, many manufacturers and designers only provide constructional and locational orientated drawings, making the task of maintenance personnel more difficult than it need be.

In the 1960s, the Royal Navy were concerned about the increasing complexity of ship-borne equipment and the problems of maintenance. A team at HMS *Collingwood* devised an approach called FIMs, for functionally identified maintenance system. FIMs is diagnostic documentation which is supplementary to the main constructional or functional drawings. It is based on functional blocks whose inputs and outputs can be identified and tested. These blocks are arranged in a hierarchy, as in Fig. 21.25. By identifying blocks in which a fault lies, a technician will naturally follow the half split method of Section 21.3. FIMs works well with equipment that is modular in nature and has been designed with FIMs in mind, but can be applied to any plant.

Figure 21.26 shows a complete FIMs documentation for a simple modular power supply, and Fig. 21.27 part of the documentation for a thyristor drive. In each case the FIMs charts will lead the technician to a replaceable unit (denoted by a triangle in the bottom right-hand corner of the block) or a simple circuit diagram.

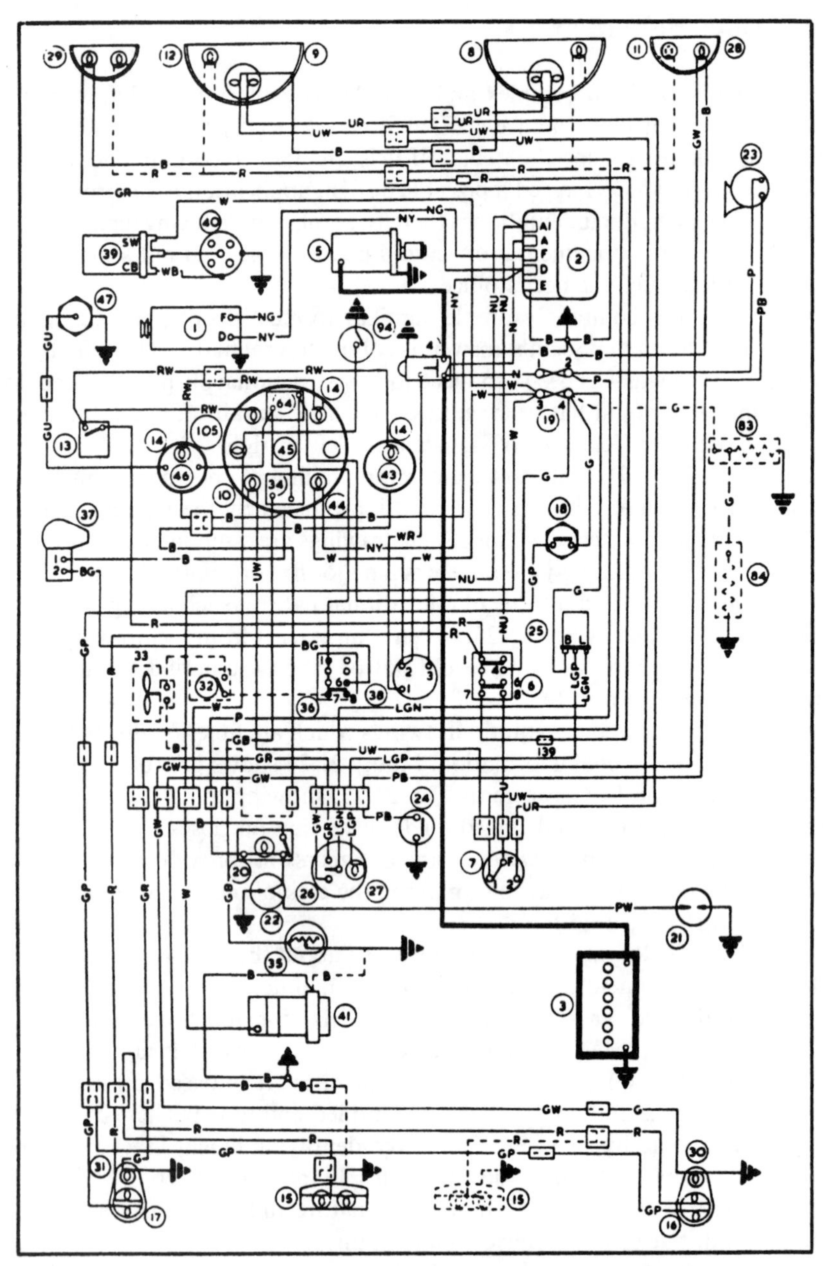

Fig. 21.21 *A typical car wiring diagram with which most people are familiar. It emphasises spatial relationships in that the layout of components follows, to some extent, the physical arrangement in the car. This results in the diagram having a large number of wiring crossovers and parallel runs, and a lack of any 'direction' or functional flow. In consequence it appears cluttered and is difficult to follow. (Reproduced from IBA Technical Review with permission. The concept of Fig. 21.21 and Fig. 21.22 is based on material published by the IBA.)*

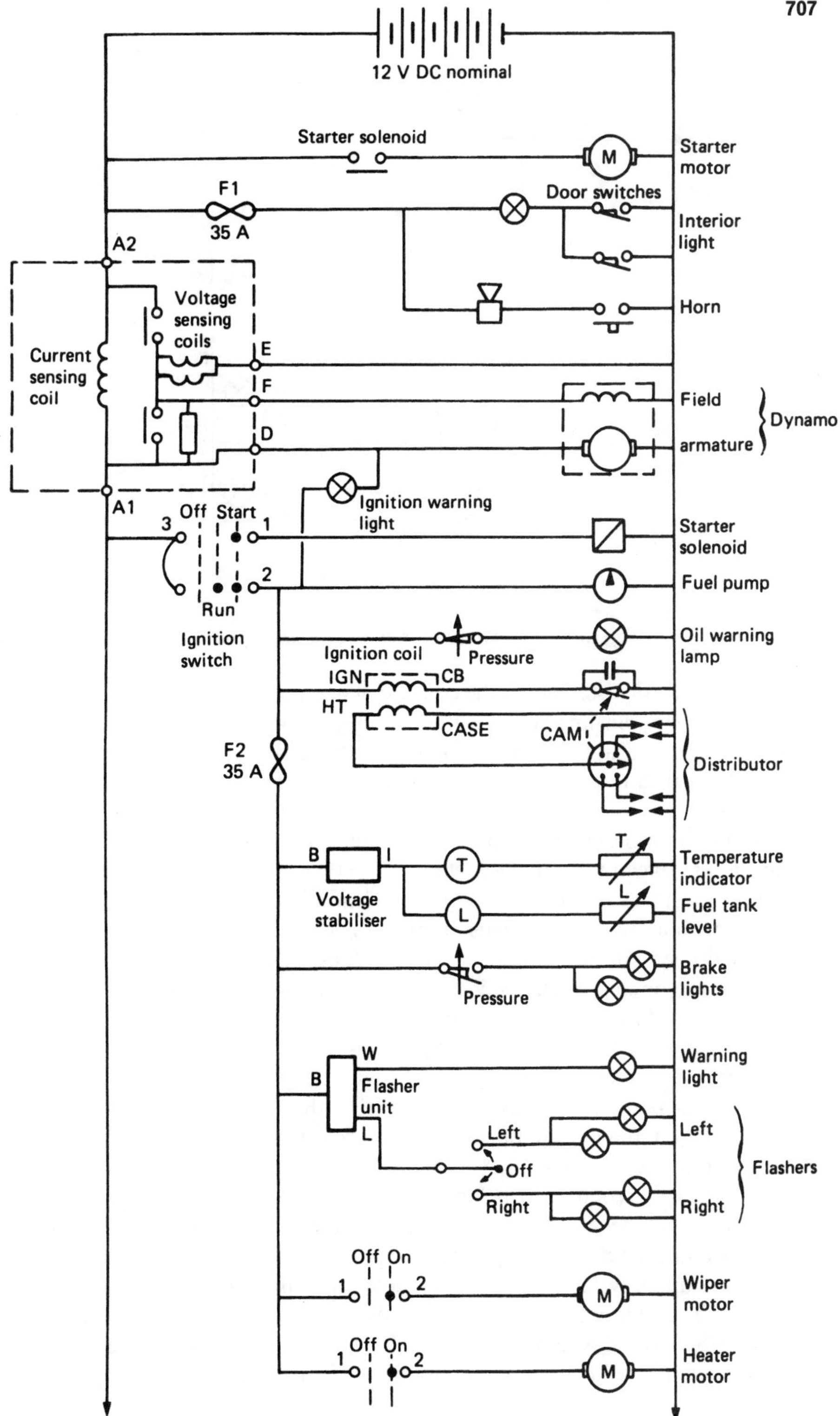

Fig. 21.22 *continued overleaf.*

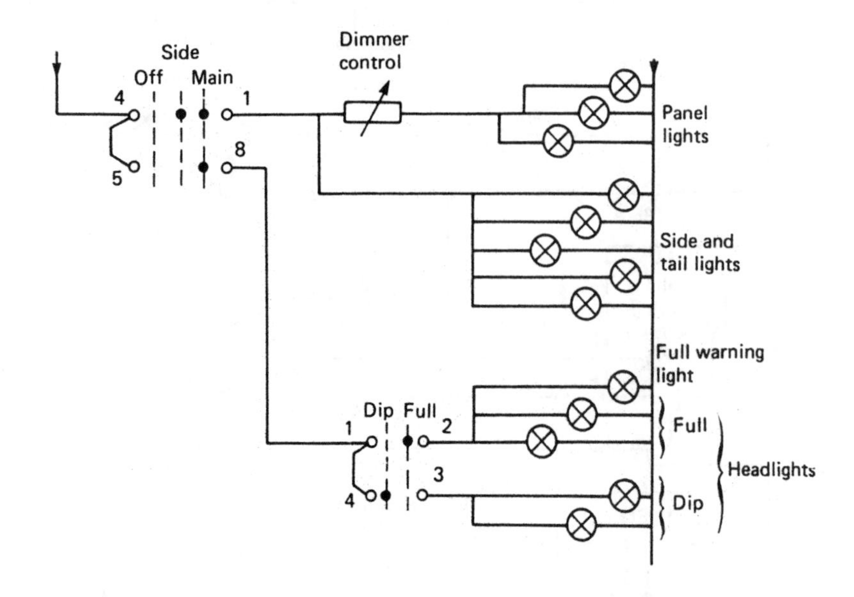

Fig. 21.22 *Car wiring diagram redrawn for ease of understanding.*

The FIMs team also devised the concept of a maintenance dependency chart. This shows relationships between items, using the symbols of Fig. 21.28. The charts are used for fault finding by following the signals back to their origin. On Fig. 21.28, for example, output d requires unit Z, which in turn requires signals e, f, g, etc.

A common form of fault-finding documentation is the flow chart (also known variously as a symptom analysis chart or, rather grandly, an algorithmically based diagnostic chart). Flow charts aim to lead the technician to the fault via a series of predetermined tests. This approach is well suited to a computer-based maintenance system. It is important that all faults are covered (even the subtle ones). Far too often flow charts deal only with the simple, obvious faults such as 'check the fuses' and leave the technician just when he needs most help. Flow charts linked with FIMs can be a powerful tool.

Documentation should include plant and item descriptions and operating parameters at plant commissioning. (Knowing that a feed line is operating at 6 bar is no use if the normal operating pressure is not known.) These should be given by the manufacturers, but it is good practice for maintenance staff to record normal operating conditions before the first fault occurs. The documentation should also include a full spares equipment list with manufacturer's names, addresses and manufacturers' part numbers.

Process control system can be very complex, and are best described by process drawings. The symbols of Fig. 21.29 have evolved (Instruments Society of America (ISA), BS 1646 and IS 03511) to show the operation of complex systems. These are often called P and IDs, for piping and instrumentation drawings (not to

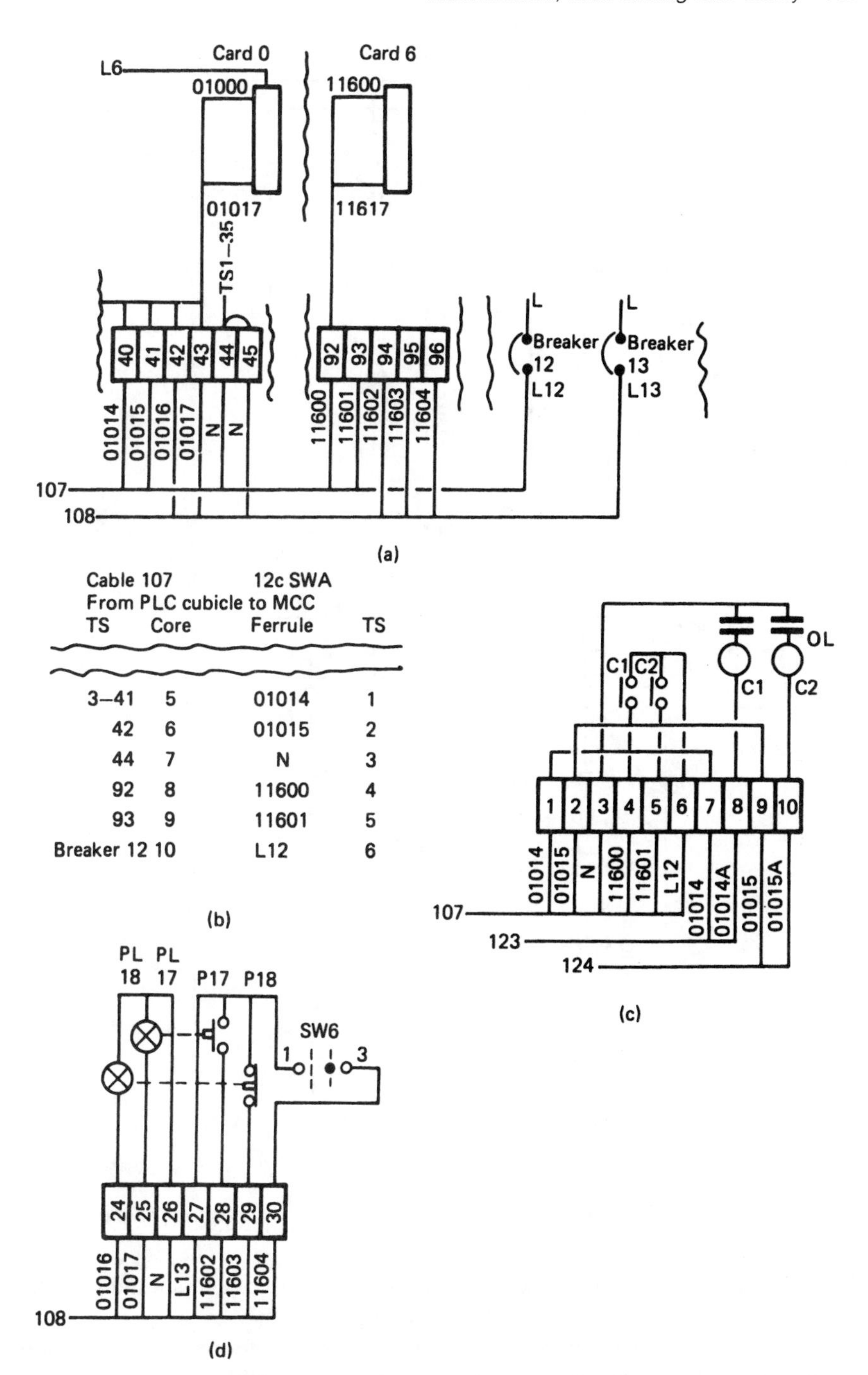

Fig. 21.23 *Plant documentation as usually supplied for first line maintenance. (a) Part of PLC cubicle constructional drawing. (b) Part of cable schedule. (c) Part of MCC constructional drawing. (d) Part of panel constructional drawing.*

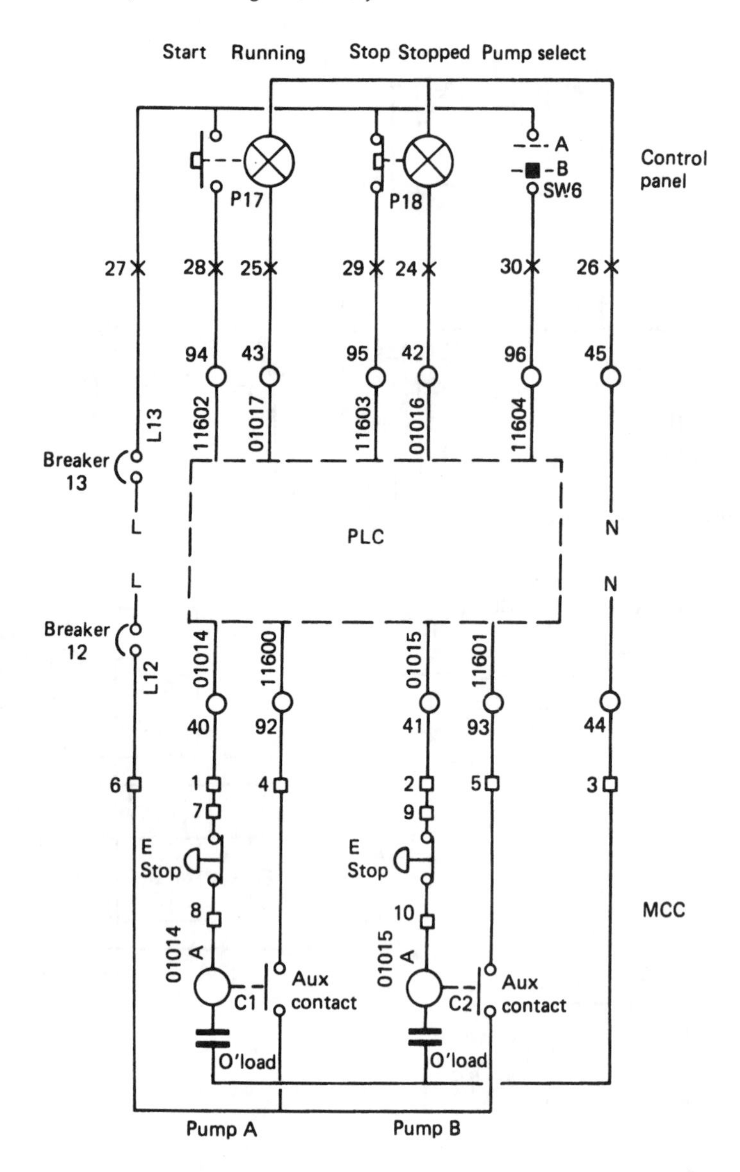

Fig. 21.24 *The information in Fig. 21.23 redrawn to show functional relationships.*

be confused with PID controllers). Plant mounted equipment should be tagged with the drawing reference identifications (e.g. FE107).

Standard graphical symbols are covered by the following international standards: DIN 40700–40717 (Germany), BS 3939, ANSI 32.2 (USA), NEMA ICS (USA), CEMA ICS (Canada) and International Electrotechnical Commission (IEC) publication 117. Unfortunately most manufacturers seem to derive their

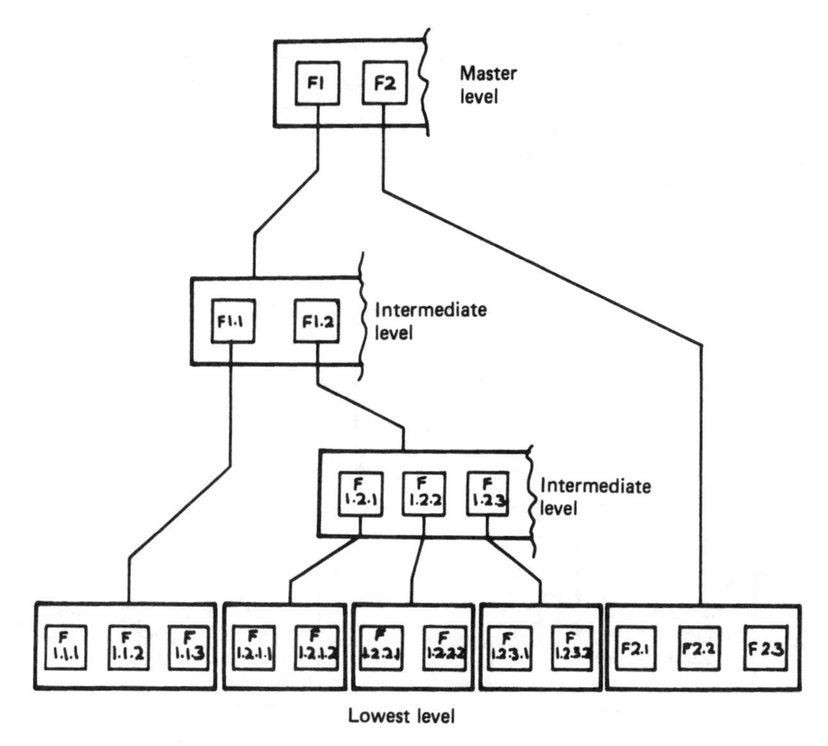

Fig. 21.25 *The FIMs hierarchy (based on an IBA technical review drawing with permission).*

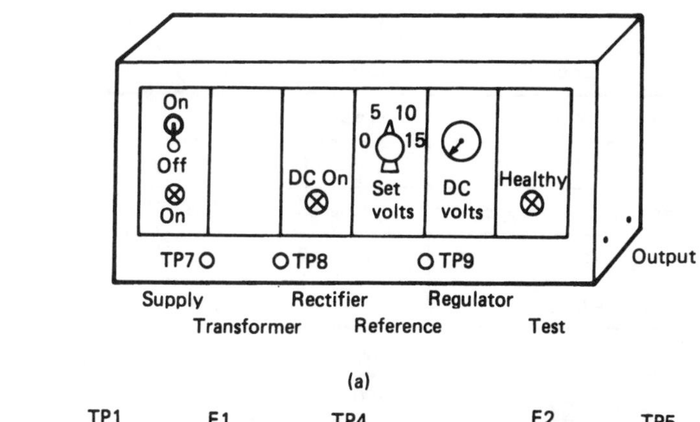

(a)

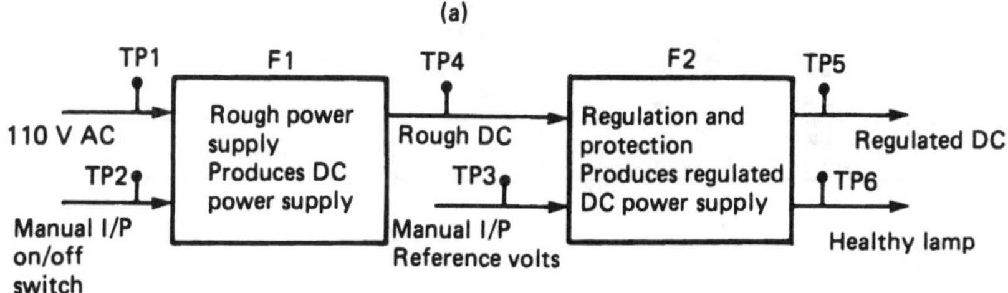

(b)

Fig. 21.26 *continued overleaf.*

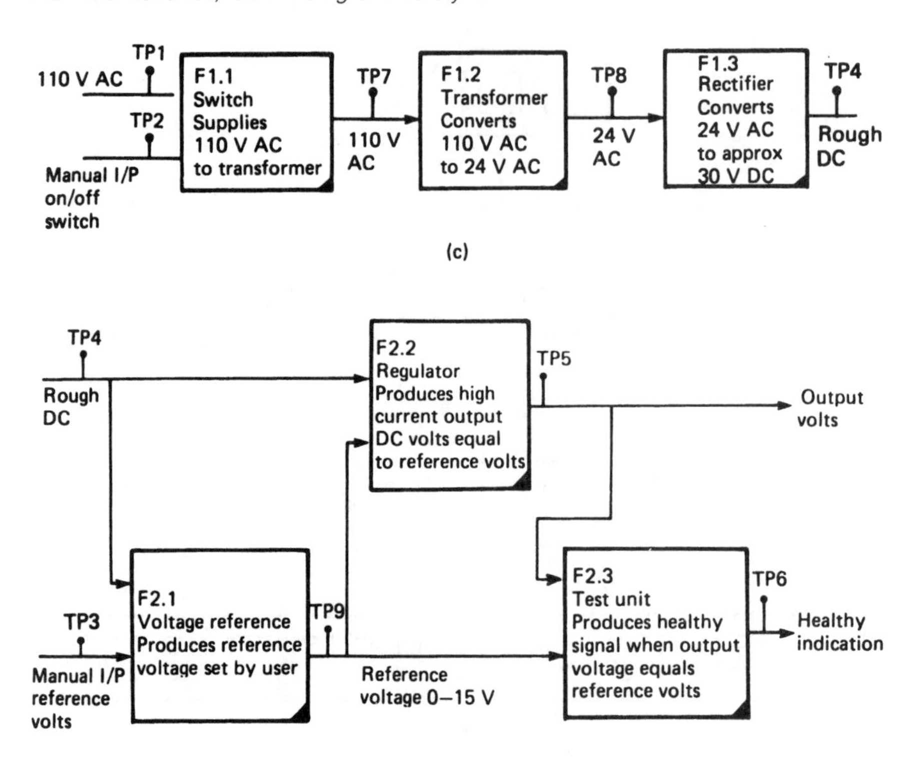

Test point	Description	Normal indication
1	Neon lamp LP1	ON
2	Supply switch	ON position
3	Reference voltage knob	Pointer shows required voltage
4	Rough DC lamp LP2	ON
5	Output volts	0 – 15 V according to TP3
6	Healthy lamp LP3	ON
7	TP7 multimeter on 250 V AC range	100 – 130 V AC
8	TP8 multimeter on 50 V AC range	27 – 36 V AC
9	TP9 multimeter on 30 V DC range	0 – 15 V according to TP3

(e)

Fig. 21.26 *Complete FIMs chart for a power supply. (a) Physical construction. (b) Master level. (c) F1 lowest level. (d) F2 lowest level. (e) Test point table.*

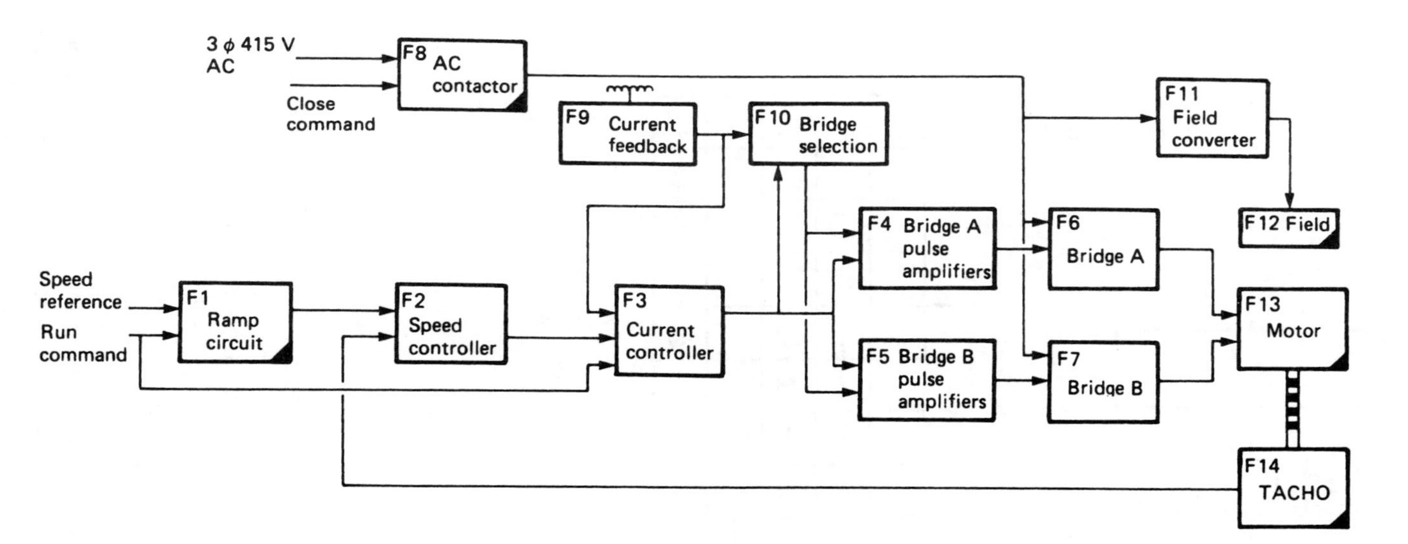

Fig. 21.27 *Top level FIMs chart for a thyristor drive.*

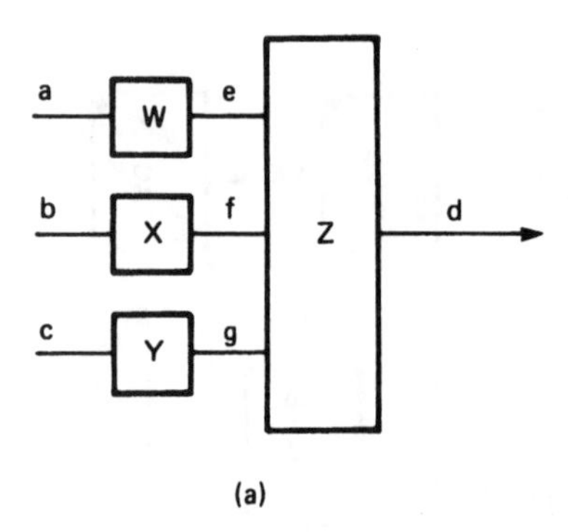

(a)

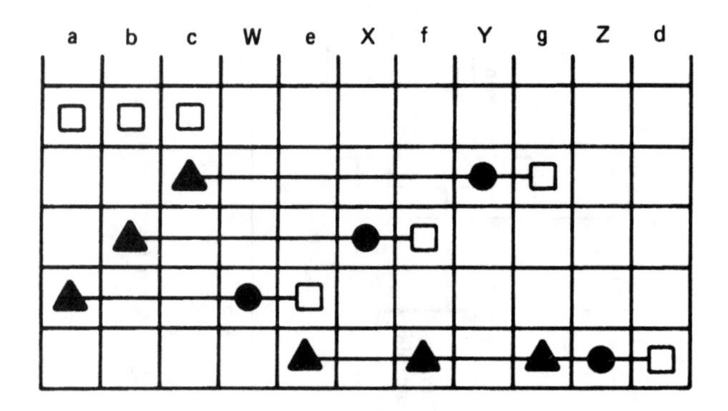

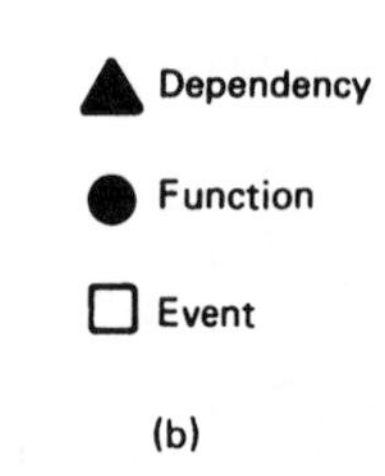

(b)

Fig. 21.28 *The dependency chart. (a) A simple system. (b) Chart for simple system.*

own. Large organisations can probably impose standards on suppliers, but the small purchaser usually does not have sufficient influence.

21.7. Environmental effects

The environment in which equipment operates has a large impact on its reliability. Pressure, humidity, temperature (particularly temperature cycling), corrosive atmospheres and vibration all have an adverse effect on equipment.

Equipment protection is defined in BS 5420 (IEC 144) by the letters IP (for

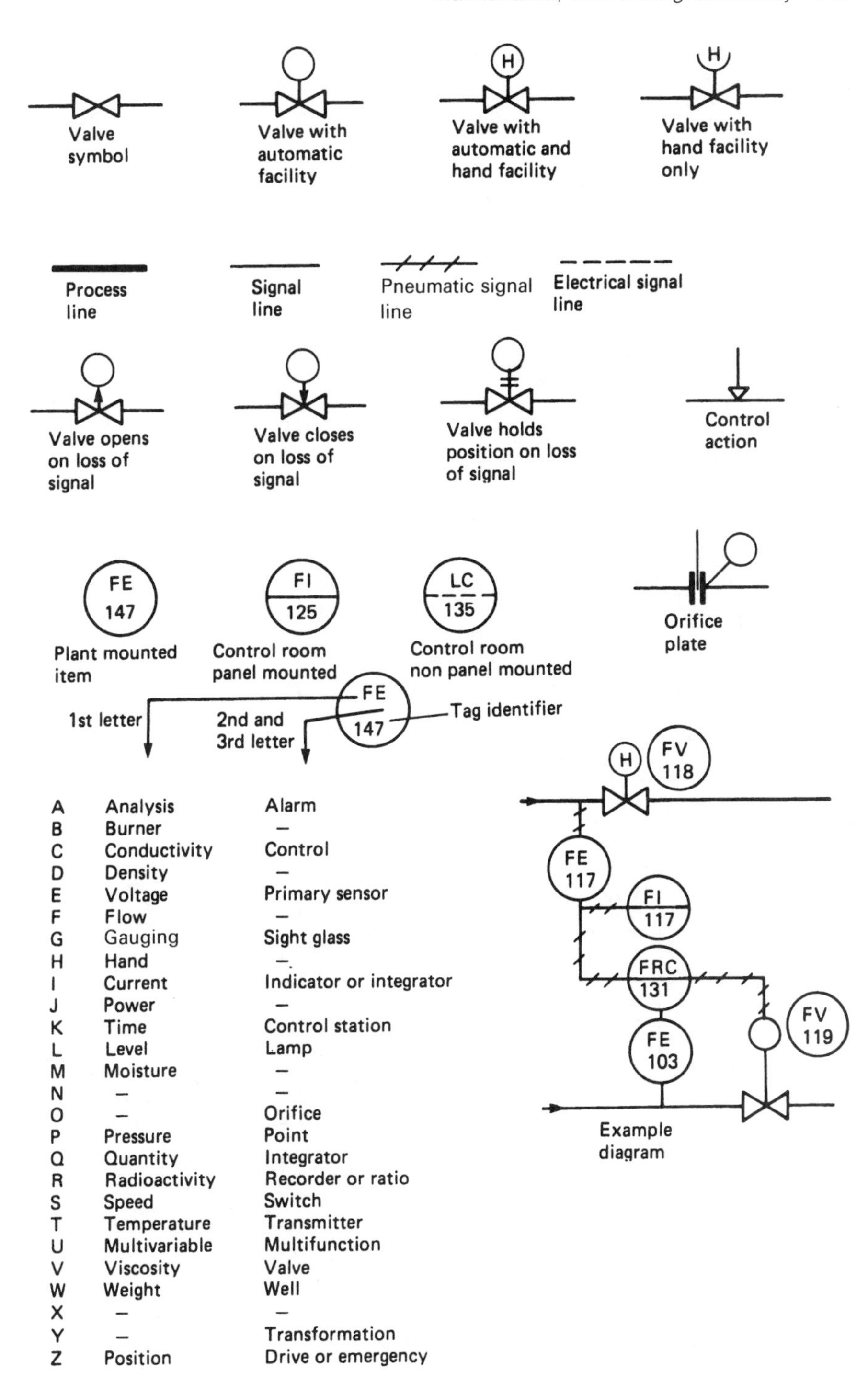

Fig. 21.29 *P and ID symbols.*

ingress protection) followed by two digits (e.g. IP54). The first digit refers to the degree of protection against solid objects, the second to the protection against liquids, as in Table 21.2. Some IP numbers have commonly used names, but these have no official standing:

Table 21.2

Solid Bodies			Liquids		
First number			Second number		
0	/	No protection	0	/	No protection
1	50 mm	Protection against large solid bodies Hand cannot come into contact with live parts	1		Drops of condensed water falling on enclosure shall cause no harm
2	12 mm	Protection against medium solid bodies Fingers cannot come into contact with live parts	2	leak/roof	Falling liquid shall have no harmful effect up to 15° from vertical
3	2.5 mm	Protection against objects > 2.5 mm diam. Tools (e.g. screwdrivers) cannot contact live parts	3		Falling liquid shall have no harmful effect up to 60° from vertical
4	WIRE 1 mm	Protection against objects > 1 mm	4		Protection against splashing from any direction
5	Talc	Totally enclosed Dust may enter but not in harmful quantities	5		Protection against hose pipe water from any direction. Water may not enter in harmful quantities
6		Dust may not enter Total protection	6		Protection against conditions on ships decks. Occasional immersion. Water must not enter
—	—	—	7	1 m	Permanent immersion up to 1 metre. Water must not enter
—	—	—	8	X	Permanent immersion to specified depth and/or pressure

Common ratings are IP11, IP21, IP22, IP23, IP44, IP54, IP55

Drip proof	IP22
Rain proof	IP23
Splash proof	IP34
Dustproof	IP54
Weatherproof	IP55
Watertight	IP57
Dust tight	IP65

Note the subtle difference between 'shall have no harmful effect' and 'shall not enter' between digits 5 and 6 in Table 21.2. Contrary to some beliefs, water and dust can enter IP55 enclosures in small amounts. It should also be noted that the IP rating is only achieved with the seals in good condition and the door closed.

The temperature range over which equipment will operate needs to be investigated. Most equipment will have ranges specified for storage, operation (with reduced accuracy) outside of which damage will occur. Equipment in the open air in the UK can expect to experience a range of −10 °C to +40 °C, but this can obviously vary considerably according to local conditions and the proximity of heat sources. Operating conditions for process control equipment is defined in BS 5967, 1980. This defines, for example, temperature and humidity ranges for control rooms (18 °C to 27 °C) and sheltered outside locations (−25 °C to 55 °C or −40 °C to +70 °C), for example, along with humidity and rate of change of temperature limits.

IEC publication 364 gives a more extensive classification. This utilises a three character code (e.g. AD6) to define external influences. The first character defines:

- A environment
- B utilisation
- C construction of buildings

The second character (A...Z) defines the nature of the influence. For environmental effects, A is temperature, B is humidity, C is altitude, D is water, E is dust, F is corrosive substances, G and H are mechanical stress and vibration, and so on to R (wind effects). The final character is a number which defines the degree of the effect. AA4, for example, defines a temperature range of −5 °C to +40 °C. Classifications can be combined to give an environment definition.

Utilisation covers the capability of people coming into contact with the equipment (e.g. BA5 skilled technician), electrical resistance of the body, and the proximity of earthed equipment (e.g. BC4m earthed metallic surroundings) and similar considerations.

21.8. Safety considerations

21.8.1. General aspects

The well-known author Isaac Asimov postulated three laws of robotics which are, slightly modified:

(1) No robot shall, through action or inaction, allow harm to come to a human being.
(2) No robot shall, through action or inaction, allow harm to come to itself except where this conflicts with the first law.
(3) No robot shall disobey the legitimate orders of a human being except where these conflict with the first two laws.

These three simple laws can be considered to be the basic requirements for process control design if a word 'plant' is substituted for 'robot'. In essence, the priorities must be human safety first, plant protection second, and production a poor third.

A plethora of legislative might ensures that most employees are in a safer environment at work than at home. For people who travel to and from work by car or bike, the most dangerous part of the working day is that journey.

The majority of industrial accidents are not electrocutions or burns or spectacular petrochemical explosions but a series of relatively trivial cuts, bruises, sprains, etc., from slipping on oily floors, incorrect use of tools, poor housekeeping or short cuts around safe working procedures. This is not grounds for complacency, however. Most industrial plants have the capacity to maim or kill, and great care must be taken to ensure that a safe working environment is maintained.

Safety legislature mainly falls under the Health and Safety at Work Act (1974) which puts the responsibility for the safe use of equipment on the manufacturer (who must provide sufficient information for the equipment to be used safely), the user (who must ensure that a piece of equipment is safe by virtue of its application and location) and employees (who must be competent to use the equipment and follow safe working procedures). In the USA, the Occupational Safety and Health Act (OSHA) affords similar protection.

Electrical installations generally fall under the Institute of Electrical Engineers (IEE) wiring regulations, currently the sixteenth edition, and the Electricity (Factory Act) Special Regulations (1908 and 1944). There is also a wide range of legislation for special circumstances such as mines, quarries and petrochemical industries.

Hazards can be considered to fall under hazards during normal operation, hazards during fault conditions, and hazards whilst plant is under maintenance or repair.

Hazards during normal operation cover normal design precautions such as using correctly specified materials, correctly stressed pressure vessels, etc., and ensuring that people cannot come into contact with hazardous material, moving machinery or exposed electrically live equipment. Safe working procedures need to be laid down (and followed!) for potentially dangerous operations.

Fault conditions can introduce additional hazards. The failure of a temperature sensor, for example, could lead to overheating of a chemical reactor and a subsequent fire or explosion. Risks can be reduced by making plants fault tolerant; if a temperature sensor failed, for example, the pressure vessel could be designed so that it could contain the maximum conceivable temperature and pressure. Alternatively, techniques such as redundancy or majority voting can be used. Care must be taken to ensure that the failure of one element of a redundancy based system is detected, or a supposedly two-out-of-three system could be operated unbeknown as a single route system.

Redundancy based systems are also vulnerable to common mode failures, which affect all parallel paths. Typical examples are services such as water, instrument air and electrical supply. Often the operators should be considered as a potential common mode failure.

Maintenance activities are possibly the most hazardous times. Maintenance work, particularly fault rectification, is usually carried out in an atmosphere of haste and stress, both of which are conducive to dangerous short cuts and a potential overlooking of hazards. (The author speaks from personal experience of fault finding on a hydraulic system without having carried out the fundamental step of blowing down the accumulator.) All plants must have a formal written procedure for isolating plant and making it safe – both electrically dead and immobile. Similarly, safe working practices need to be defined for all potentially hazardous jobs. Needless to say, staff training is essential.

Usually, it is the ad hoc repair jobs that result in accidents rather than routine maintenance work. These jobs are less controlled, the work ill-defined and the plant usually operational and electrically live. There is also a tendency to bypass safety interlocks to 'get the plant away'. Once out, interlocks tend to stay out. Three Mile Island, Flixborough and Chernobyl all originated, to some extent, from ill-advised maintenance work. A cool, logical atmosphere is required for fault finding. Undue pressure from production management can all too easily lead to an accident, maybe weeks after the fault has been 'repaired'.

21.8.2. Explosive atmospheres

An explosive mixture is formed when combustible materials are mixed with air. Combustible vapours occur in many chemical and petrochemical processes and, less obviously, powders from coal dust and even such apparently harmless materials as flour and custard powder, can ignite explosively.

Precautions must obviously be taken to prevent ignition of potentially explosive mixtures. Such ignition can occur from electrical sparks, hot surfaces, mechanical sparks (e.g. formed by rubbing surfaces) and electrostatic discharges. Equipment used in hazardous areas must be designed to prevent the above sources of ignition. The legislation governing such installations is profuse and complex, and the descriptions below should only be taken as a guide to the techniques used.

There are three factors which determine the degree of hazard in any particular location. The first of these determines the probability of an explosive gas being present. In the UK, three zones are defined:

Division 0, where an explosive mixture is present for long periods or continuously under normal operation.

Division 1, where an explosive mixture is likely to occur in normal operation.
Division 2, where an explosive mixture is not likely to occur in normal operation but may occur in abnormal or fault conditions. If an explosive mixture does occur, it will only persist for a short period. (This implies adequate ventilation.)

In the USA divisions 0 and 1 are combined, and called division 1. By default areas not classified as divisions 0 to 2 are deemed non-hazardous.

The second consideration is the ease of ignition of the mixture. There are, unfortunately, several ways of grouping gases, with notable differences between Europe and the USA. Table 21.3 gives an *approximate* relationship between the different standards.

The final consideration is the ignition temperature of the explosive mixture. This is the temperature of a surface which will ignite the gas, and should not be confused with flash point. The latter is the temperature at which sufficient vapour is produced for the vapour to ignite when in contact with a naked flame. Flash point temperature is lower than ignition temperature. Six classes are defined from T1 (450 °C) to T7 (85 °C). These six classes are further subdivided in the USA.

All the above classifications are applied to the equipment which is being considered for use in a hazardous environment (i.e. it is the equipment which is really being classified, not the environment, but the subtle difference is more pedantic hairsplitting than a practical consideration).

Equipment intended for use in hazardous areas is generally termed 'explosion proof'. There are several techniques for achieving this (see BS 5501, parts 1 to 9, and BS 5345). The commonest are flameproof enclosures (permitted in divisions 1 and 2), pressurisation (again permitted in divisions 1 and 2) and intrinsic safety (permitted in practically all locations).

A flameproof enclosure is one designed in such a way that it can withstand an internal explosion, and potential flamepaths (e.g. joints) are of such a length and cross-section that the flame from the explosion cannot propagate to the outside atmosphere. The surface temperature of the enclosure must, at all times and under all conditions, not exceed the specified temperature classification. Flameproof equipment is, of necessity, bulky and heavy, and particular care

Table 21.3 *Approximate relationships for gas grouping*

Test gas	*IEC*	*BS* 1259 *Intrinsic safety*	*BS* 229 *Flameproof*	*American NEC* 50°	*German VDE* 0171
Ammonia		2a			
Propane	IIA	2c	II	D	1
Ethylene	IIB	2d	IIIA Ethylene IIIb coal gas	C	2
Hydrogen	IIC	2e	IV	B	3a
Acetylene	IIC	2f	IV	A	3b, c and n

IEC group 1 is intended for use, in mines subject to fire damp (methane).

needs to be taken to ensure that the integrity is not affected by maintenance work. Flameproof enclosures are the only practical solution for electrical motors.

With pressurisation, the equipment is kept separate from the explosive atmosphere by a positive pressure differential maintained by a purging gas (e.g. nitrogen). In division 2 applications, loss of pressurisation should raise an alarm, whilst in division 1 an automatic shut down interlock on pressure loss is required. The interlocks, which will operate in a *non-safe* atmosphere, will probably use the next technique: intrinsic safety.

Intrinsically-safe equipment is designed such that no possible normal or fault condition can result in the ignition of an explosive mixture. Somewhat simplified, this limits operating voltages to 30 V *and* operating currents to 100 mA. Intrinsically safe equipment is subdivided into 'ia' where safety is maintained with two simultaneous faults, and 'ib' where safety is maintained with one fault.

The principle advantages of intrinsically safe equipment are ease of use,

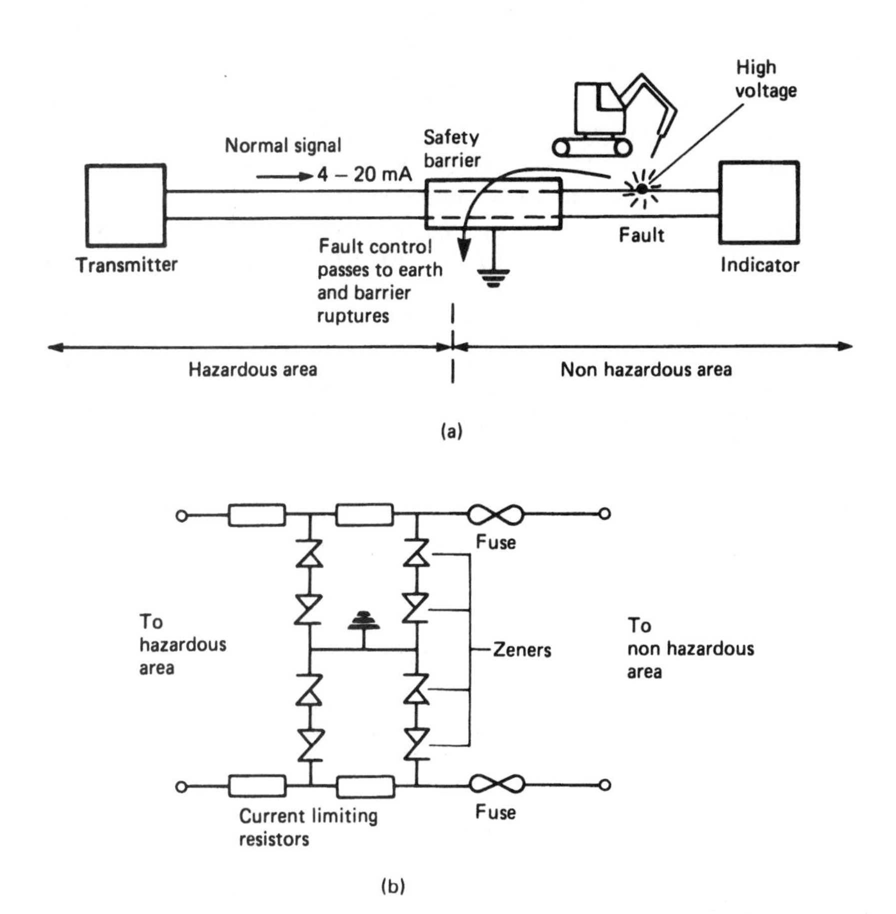

Fig. 21.30 *Zener safety barriers. (a) Principle of barrier protection. (b) Construction; note this is a certified device and cannot be 'home built'.*

simplicity and the ability (in some circumstances) to be maintained and adjusted whilst live.

Some portable equipment can be made inherently safe (e.g. meters and radios) but problems exist with, say, transducers. These operate at low voltages and currents, but their cabling could, under fault conditions and intercable shorts, introduce high voltages and currents into the hazardous area. Safety is maintained by the use of safety barriers at the transition between safe and hazardous areas. These have no effect on normal signals, but prevent fault voltages or currents entering the hazardous area, as in Fig. 21.30a.

Figure 21.30b shows a typical zener barrier. As long as the operating voltages are less than the zener voltage, the barrier has no effect. If the input voltage rises due to a fault, one zener will conduct, causing the fuse to blow. Barriers are certified, and consequently the fuses are not user replaceable.

The use of zener barriers allows great flexibility in design, and almost total freedom in the choice of 'safe area' equipment. Care should be taken to ensure that the 'hazardous area' equipment cannot store electrical energy, either capacitively or inductively, which could result in a spark in a fault condition. Cable parameters (self-capacitance and inductance) are theoretically important, but unlikely to be significant in practical installations.

Legislation governing the use of equipment in explosive atmospheres is complex, and places responsibility on both manufacturer and user. Compliance with the legislation is best demonstrated by certification by an independent recognised authority (such as BASEEFA in the UK). Electrical equipment for use in mines is covered by separate, and very specialised, legislation.

Pneumatic, hydraulic and fibre optic signals, of course, present no hazard in an explosive atmosphere, and are often chosen as an alternative to electrical instrumentation in difficult applications.

Index